HYDROTHERMAL REACTIONS FOR MATERIALS SCIENCE AND ENGINEERING

An Overview of Research in Japan

HYDROTHERMAL REACTIONS FOR MATERIALS SCIENCE AND ENGINEERING

An Overview of Research in Japan

Edited by

Shigeyuki Sōmiya

The Nishi-Tokyo University, Japan

ELSEVIER APPLIED SCIENCE
LONDON and NEW YORK

ELSEVIER SCIENCE PUBLISHERS LTD
Crown House, Linton Road, Barking, Essex IG11 8JU, England

Sole Distributor in the USA and Canada
ELSEVIER SCIENCE PUBLISHING CO., INC.
655 Avenue of the Americas, New York, NY 10010, USA

WITH 64 TABLES AND 505 ILLUSTRATIONS

Softcover reprint of the hardcover 1st edition 1989

British Library Cataloguing in Publication Data

Hydrothermal reactions for materials science and
engineering.
1. Powder metallurgy
I. Sōmiya, Shigeyuki
671.3'7

ISBN-13: 978-94-010-6819-2 e-ISBN-13: 978-94-009-0743-0
DOI: 10.1007/978-94-009-0743-0

Library of Congress Cataloging-in-Publication Data

Hydrothermal reactions for materials science and engineering: an
overview of research in Japan / edited by Shigeyuki Sōmiya.
p. cm.
Bibliography: p.
Includes index.
1. Powder metallurgy. I. Sōmiya, Shigeyuki.
TN695.H93 1989
671.3'7—dc20 89-35565

To Professor Emeritus E. F. Osborn,
The Pennsylvania State University,
University Park, Pennsylvania, USA

Without the benefit of study at the Department of Geochemistry,
College of Mineral Industries,
The Pennsylvania State University,
under a Fulbright Travel Grant,
I would not have been able to commence my studies on
hydrothermal reactions

Preface

According to the late Professor Emeritus Seitaro Tsuboi,[1] the word 'hydrothermal' was used as early as 1849 by a British geologist, Sir Roderick Murchison (1792–1871), in relation to the action of heated water in bringing about change in the earth's crust. The term abounds in later geological literature, and is most frequently met in connection with the processes that take place at a stage near the closing in the course of consolidation of magma. When a cooling magma reaches that stage, the residual liquid contains a large proportion of volatile components, chiefly water, and further cooling results in the formation of minerals of special interest or ore-deposits.

A great concern of Tsuboi's as a petrologist was to elucidate the details of the nature of various actions involved in these 'hydrothermal processes', of which little was known. It is remarkable that, in the last few decades, extensive high-temperature and high-pressure experiments, in which water plays an important role, have become practicable in laboratories, owing to the development of new apparatus and new methods. As a result, the knowledge essential to the elucidation of 'hydrothermal processes' has been improved, but is still far from complete.

On the other hand, we are informed that, apart from the geological interest, research workers in the synthesis of engineering materials are greatly concerned with 'hydrothermal reactions', and that their studies along this line are being carried on efficiently. Armed with this information, Tsuboi considered it highly desirable to provide arrangements for a sort of interdisciplinary conference to discuss subjects of common interest, which would be of benefit to both earth scientists and materials engineers. Over the years, the idea of such a joint meeting occurred repeatedly to him.

In the field of materials science and engineering, ordinary use of the term 'hydrothermal synthesis' usually refers to heterogeneous reactions in aqueous media above 100°C and 1 bar.[2]

'Hydrothermal crystal growth' is

> a process in which an aqueous solution is held at high temperature and pressure in order to dissolve a material (the nutrient) in one part of the system, transport it to another part of the system and there deposit it, either on a seed, or on small crystals attached to the containing vessel by epitaxy. This process is usually carried out in sealed vertical autoclaves with a temperature difference (ΔT) established between the top and bottom of the vessel. Usually the nutrient is placed in the lower, hotter part of the autoclave and the seeds are mounted in the cooler upper part. The seeds are single crystal plates properly oriented and mounted on a suitable wire frame.
>
> Transport of the hot solution from the region containing the nutrient to the region containing the seeds is by convection. Once the solution reaches the growth region it becomes supersaturated with respect to the seed material and deposits on the seeds. The cooler depleted solution then returns to the hotter zone by convection and dissolves more nutrient.[3]

There are many types of hydrothermal reaction:

1. Hydrothermal treatment.
2. Hydrothermal metamorphism (hydrothermal alternation).
3. Hydrothermal growth.
4. Hydrothermal synthesis (hydrothermal reactions).
5. Hydrothermal dehydration.
6. Hydrothermal decomposition.

7. Hydrothermal extraction.
8. Hydrothermal sintering.
9. Hydrothermal reaction sintering.
10. Hydrothermal hot isostatic processing.
11. Hydrothermal crystallization.
12. Hydrothermal deposition for thin films.
13. Hydrothermal corrosion.
14. Hydrothermal etching.
15. Testing under hydrothermal conditions.

After 1950, Professors E. F. Osborn, O. F. Tuttle and Rustum Roy of the Pennsylvania State University used hydrothermal experiments for mineral synthesis and for phase equilibrium studies. Preparation of homogeneous starting materials is very important in phase equilibrium studies, which is one of the reasons why hydrothermal conditions are used for their preparation.

In Japan, hydrothermal work first appeared in 1926, by Dr Tominosuke Katsurai. His report, entitled 'Soda treatment of the aluminum ore', dealt with hydrothermal extraction. On hydrothermal synthesis, Professor Shoichiro Nagai reported on calcium silicate in 1931. The Morey bomb used after the Second World War was reported by Professor Raisaku Kiyoura in 1952 for hydrothermal alternation and hydrothermal synthesis.

After the mid-1970s there were several reports related to hydrothermal reactions, since such reactions can be considered for use in ceramic processing, such as for powder preparation, reaction sintering, etc. Powder prepared by hydrothermal preparation is not agglomerated, is fine-grained, has a narrow grain size distribution and consists of single crystals. Such processing therefore received wide attention. Many papers have been published in Japan and several companies produce powders such as ZrO_2 under hydrothermal conditions. Reports relating to hydrothermal reactions in such fields as powder preparation, sintering, corrosion, crystal growth, etc., also appeared.

Following the First International Symposium on Hydrothermal Reactions in March 1982, it was decided to maintain communications between workers in different fields, and on 25 September 1982 the Hydrothermal Reactions Study Group was established. Since then papers related to hydrothermal reactions have increased, both in Japanese and in English.

This book is a compilation of English language reports by Japanese workers on hydrothermal reactions related to materials science and engineering. The book shows the development and the present status of R & D in Japan. The papers are scattered in many journals, which is one reason for collecting them together. I am convinced that this book will promote the R & D of hydrothermal reactions.

Shigeyuki Sōmiya

References

1. S. Tsuboi, in S. Sōmiya (ed.), *Proceedings of the First International Symposium on Hydrothermal Reactions*, 22–26 March 1982 (Gakujutsu Bunken Fukyu-kai, 1983).
2. A. Rabenau, *Angew. Chem. Int. Ed. Engl.*, **24** (1985) 1026–1040.
3. R. A. Laudise, *Solid State Phys.*, **12** (1961) 150.

Acknowledgements

I wish to express my gratitude to the following authors for agreeing to the reproduction of their papers:

Dr Shin-ichi Hirano
Dr M. G. M. U. Ismail
Dr Kiyoshi Matsuoka
Dr Taiko Mitsusio
Dr Kazuo Nakamura
Dr Sadao Taki
Dr Eiji Tani
Dr Hideo Toraya
Dr Nakamichi Yamasaki
Dr Kazumichi Yanagisawa
Dr Masahiro Yoshimura

I would also like to thank the following publishers for granting permission to reproduce copyright material:

Academic Press, Inc.
American Ceramic Society
American Chemical Society
Atomic Energy Society of Japan
Center for Academic Publications, Japan
Ceramic Society of Japan
Chapman & Hall Ltd
Chemical Society of Japan
Elsevier Science Publishers BV
Elsevier Science Publishing Co., Inc.
Japanese Association of Mineralogists, Petrologists and Economic Geologists
Kodansha, Japan
Materials Research Society
Nagaoka Technical College
National Bureau of Standards
Pergamon Press Ltd
Plenum Press
Research Laboratory of Engineering Materials, Tokyo Institute of Technology
Verlag J. A. Barth, Leipzig
Verlag Schmid GmbH, Freiburg

and also the organizers of the following conferences:

Expermat 87 (Bordeaux)
8th AIRAPT Conference/19th EHPRG Conference (1981, Sweden)
4th International Conference on Crystal Growth (1974, Tokyo)
4th International Conference on High Pressure (1974, Kyoto)
2nd International Symposium on Ceramic Materials and Components for Engines (1986, Germany)
International Symposium on Factors in Densification and Sintering of Oxide and Non-oxide Ceramics (1978, Japan)
1st International Symposium on Hydrothermal Reactions (1982, Japan)
Powder Metallurgy Conference (1986, Germany)

Contents

Part 8: Hydrothermal Hot Pressing

Part 9: Hydrothermal Hot Isostatic Pressing

Part 10: Nitridation

Part 11: Phase Diagrams

Part 1

Hydrothermal Reaction Sintering

Reprinted from *Journal of the American Ceramic Society*, **59** (11–12) (1976) 534.

Hydrothermal Reaction Sintering of Pure Cr_2O_3

SHIN-ICHI HIRANO and SHIGEYUKI SŌMIYA*

MUCH work has been done in an effort to fabricate a pure high-density Cr_2O_3. High volatilization of chromium oxide compounds, however, made its densification difficult. The studies reported to date on the sintering of Cr_2O_3 have been concerned mainly with controlling the sintering atmosphere and the additives. Hagel *et al.*[1] studied the initial sintering of Cr_2O_3 in an Ar atmosphere with a p_{O_2} of $\approx 10^{-2}$ mm Hg but were unable to obtain a high density. Ownby and Jungquist[2] studied the final sintering of Cr_2O_3, giving attention to the effect of oxygen activity and the addition of MgO on the sintering of Cr_2O_3. At 1600°C, their product approached theoretical density at $p_{O_2} = 2\times10^{-12}$ atm, whereas the high weight loss due to volatilization of chromium oxides was observed at $p_{O_2} < 10^{-12}$ atm where the maximum density was achieved.

In the present study, pure Cr_2O_3 was sintered at temperatures up to 1000°C, without any additive, by the newly developed hydrothermal reaction sintering method. Reagent-grade metallic chromium powder (purity >99.9%) was used as a starting material. Scanning electron microscopy revealed that the particle size was $< \approx 50$ μm. Water as a reaction reagent was prepared with a quartz distiller from deionized water which had previously been distilled with a copper distiller. The chromium powder and water were mixed at the molar ratio of 2 to 3 in a Pt capsule 3 or 5 mm ID, 0.15 mm thick, and 35 mm long. The capsule was hermetically sealed in air with an electric arc and tested for leakage. The sealed capsule was placed in a test-tube-type pressure vessel or a high gas pressure apparatus,[3] and then subjected to the hydrothermal reaction sintering process under the desired pressure and temperature. The pressure was first applied at 1000 kg/cm^2 at room temperature through the capsule wall with water or Ar gas; subsequently the temperature was raised to the desired value under the constant pressure. Pressure and temperature during a run were held within ± 20 kg/cm^2 and ± 3°C, respectively. The density of a sintered specimen was measured by the liquid displacement method with redistilled water. Fracture surfaces of sintered specimens were examined by scanning electron microscopy.

The metallic chromium powder reacted with supercritical water under pressure and temperature to form fine-grained, active chromium oxide (Cr_2O_3) and hydrogen at temperatures above 500°C under 1000 kg/cm^2 pressure through the simple reaction $2Cr + 3H_2O \rightarrow Cr_2O_3 + 3H_2$. Hydrogen thus formed was gradually diffused out through the capsule wall, which caused the reaction to proceed and the specimen to be compressed more efficiently. The particle size of the starting metallic chromium powder did not influence the final density of the sintered specimen, because each metallic particle had transformed to many oxide grains during the reaction with water. Well-sintered specimens having a density above 95% of theoretical value were prepared by this hydrothermal reaction sintering at temperatures above 850°C under 1000 kg/cm^2 pressure. Very dense sintered Cr_2O_3 specimens (99.2% of theoretical density) were obtained at 1000°C under 1000 kg/cm^2 pressure for 3 h. The average grain size of the sintered specimen was ≈ 10 μm and the range of the grain size was ± 3 μm. A typical scanning electron micrograph of the fracture surface of the specimen is shown in Fig. 1. Fracture occurred mostly through grains, and fracture surfaces contained typical brittle fracture cleavage steps and river patterns. The detailed mechanism of this kind of sintering is not yet understood. It is concluded that this newly developed hydrothermal reaction sintering method can be used to prepare a very dense oxide specimen at relatively low temperature without additives by controlling the sintering atmosphere spontaneously with the formed hydrogen; it is also concluded that the sintering is accelerated by use of a hydrogen-permeable capsule, such as a Pt capsule, instead of a gold one so that the formed hydrogen is gradually diffused out of the sintering system.

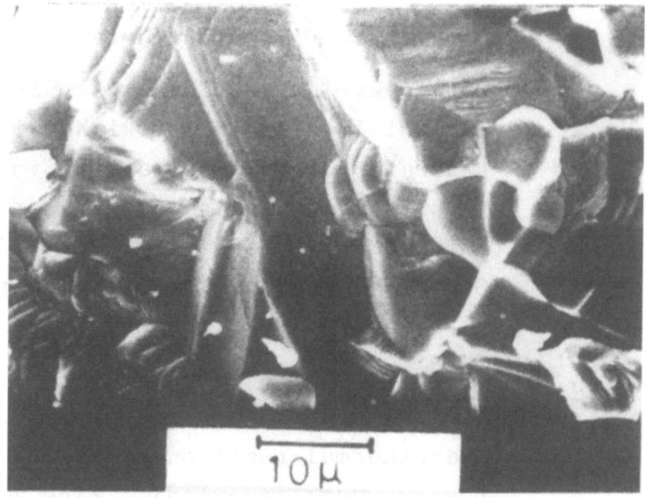

Fig. 1. Scanning electron micrograph of fracture surface of pure Cr_2O_3 specimen sintered under hydrothermal conditions, 1000°C, 1000 kg/cm^2, 3 h.

Acknowledgments: The writers are indebted to J. A. Pask and R. M. Fulrath, University of California, and to R. L. Coble, Massachusetts Institute of Technology, for their interest and helpful discussions. They also acknowledge F. Dachille of The Pennsylvania State University for his continuous encouragement.

[1] W. C. Hagel, P. J. Jorgensen, and D. S. Tomalin, "Initial Sintering of α-Cr_2O_3," *J. Am. Ceram. Soc.*, **49** [1] 23–26 (1966).
[2] P. D. Ownby and G. E. Jungquist, "Final Sintering of Cr_2O_3," *ibid.*, **55** [9] 433–36 (1972).
[3] Shin-ichi Hirano, Kazuo Nakamura, and Shigeyuki Sōmiya, pp. 418–23 in Proceedings of the International Conference on High Pressure, 4th, 1974. Edited by Jiro Osugi. Published 1975.

Received June 18, 1976.
The writers are with the Research Laboratory of Engineering Materials and the Department of Materials Science and Technology, Tokyo Institute of Technology, Ookayama, Meguro-ku, Tokyo 152, Japan.
*Member, the American Ceramic Society.

Reprinted from *Proceedings of International Symposium on Factors in Densification and Sintering of Oxide and Non-oxide Ceramics*, Japan, 1978, pp. 267–277.

Hydrothermal Reaction Sintering of Cr_2O_3 and Iron Oxides

Shigeyuki Sōmiya, Shin-ichi Hirano*, Masahiro Yoshimura,
Shuichi Itoh and Hideyuki Kanai.
Laboratory for Hydrothermal Syntheses attached to the Research
Laboratory of Engineering Materials and Department of Materials
Science and Engineering, Tokyo Institute of Technology.
Ookayama, Meguro, Tokyo 152, Japan
* Present Address: Nagoya University

Abstract

Hydrothermal reaction sintering of chromium and iron metal powders were described. Fine metal powders were reacted with high temperature and high temperature solution to form oxides under certain hydrothermal conditions.
Reactions were

$$2Cr + 3H_2O \longrightarrow Cr_2O_3 + 3H_2$$
$$3Fe + 4H_2O \longrightarrow Fe_3O_4 + 4H_2$$

Well sintered Cr_2O_3 body was obtained at 1000°C for 15h. under high pressure.

1. Introduction

Chromium oxide has difficulty to get high density by sintering. Stone and Lockengton[1)2)] reported that addition of carbon powder and heated in vacuum at high temperature of 1430°C and 1600°C and 0.25% carbon addition showed good results for getting.
Hagel, Jorgensen and Tomalin[3)] studied initial sintering of α-Cr_2O_3 between 1050°C and 1300°C in Argon atmosphere containing oxygen at Po_2 of about 10^{-2} mmHg. They showed volume-diffusion process.
Ownby and Jungquist[4)] showed control of Po_2 was essential to the final sintering of Cr_2O_3. The maximum denstity obtained from sintering of Cr_2O_3 for 1 hr. at 1600°C occurred at $Po_2 = 2 \times 10^{-12}$ atm. and also they showed addition of 0.1wt%MgO to Cr_2O_3 has increased density at $Po_2 = 2 \times 10^{-12}$ at 1600°C for 1hr.
$Po_2 = 2 \times 10^{-12}$atm. is equilibrium pressure for the reaction $4Cr + 3O_2 = 2Cr_2O_3$ at 1600°C.

Ownby[5)] was investigated sintering of Cr_2O_3 at 1600°C under various pressure of Po_2. 10^{-11} —— 10^{-12} atm. of Po_2 was effective for sintering at 1600°C.
Claussen and Exner[6)] were used nitrogen, hydrogen, and air atmosphere at 1400°C for 1h. for study of sintering. Hydrogen was effective for the lowest specific surface, and nitrogen was effective for the highest specific surface.
Halloron and Anderson[7)] and Neve and Coble[8)] mentioned equilibrium partial pressure for the reaction $4Cr + 3O_2 = 2Cr_2O_3$ at experimental temperatures for initial sintering of Cr_2O_3.
High density Cr_2O_3 crucibles were reported by Toker and Darken[9)] in 1975 by addition of 1 ——2% of metallic chromium at high temperature of 1500°C or 1600°C under very low oxygen pressure.
Hirano and Sōmiya[10)] published their report in 1976 using Cr metal powder under hydrothermal conditions. As for iron oxide, Hirano and Sōmiya[11)] reported in detail in 1976. Aims of this paper are presentation results of oxidation of Cr and/or Fe metal powder under hydrothermal conditions and of a making solid body like a sintered body or a single crystal.

2. Experimental

Regent-grade metalic chromium powder (purity > 99%, 200mesh pass) of Wako Junyaku Kogyo Co. Ltd., Tokyo, Japan was used as a starting material in this investigation and metallic chromium powder (purity > 99.9%, particle size $<\approx$ 50m, Merck, Germany) was used in 1976. Water as a reaction regent was prepared with a quartz distiler from deionized water which had previously been distiled with copper distiler.
The chromium powder and water were mixed at the molar ratio of 2 to 3 in a Pt capsule 3mm ID 0.15mm thick and 35mm long. The capsule was hermetically sealed in air with an elctric arc and tested for leakage. The sealed capsule was placed in a gas high temperature and pressure apparatus of internal heating type and then subjected to the hydrothermal reaction sintering process under the desired pressure and temperature. Heating element was a Mo wire.
Pressure and temperature during a run were held within ± 20kg/cm² by strain gauge of Shinko type and ± 3°C by 96%Pt·6%Rh ——

70%Pt·30%Rh thermocouple using EC-61 controler of Ohkura Electric Co. Ltd., Tokyo, Japan, respectively.
Sōmiya, Hirano, Fukuda, and Sawada[12,13], and Hirano, Nakamura, and Sōmiya[14] reported this apparatus already.
Runs for the experiment, the pressure was first applied at 1000kg/cm^2 at room temperature through the capsule wall with Ar gas, subsequently the temperature was rised to desired value of 1000°C within 3hrs. Pressure gages were calibrated once in a year by standard gages at authorized office in Japan.
Starting materials for iron, analytical grade, purity > 99% Merck, Germany, consisted of reduced iron particle of spherical shape with an average size of about 7μm. Analysis of the iron powder reported by supplier was as follows;
Soluble part in water < 0.01%, insoluble part in sulfuric acid < 0.1%, S < 0.05%, N < 0.01%, Cu < 0.002%, Zn < 0.005%, Mn < 0.01% and As < 0.0002%. Sodium hydroxite, potassium hydroxide and ammonium chloride were extra regent grade of Wako Junyaku Kogyo Co. Ltd., and they were used for the case of oxidation of iron metal powder. Weighed amount of iron powder, pure water and solution were sealed hermetically without any seed crystal into a gold capsule, 5mm in outer diameter, 0.15mm thick and 50mm in length. Typical amounts of these substances for an experiment at the crystallization or growing zone of 500°C under 1000kg/cm^2 were as follows;

1) 111.7mg iron powder
   48.0mg pure water
   330 μl 10mNaOH solution
2) 223.4mg iron
   96.0mg pure water
   330 μl 10mNaOH solution
3) 335.1mg iron
   144.0mg pure water
   330 μl 10mNaOH solution

The sealed capsule was placed into a test tube of Tuttle-Roy type, made of Stellite 25. Thermocouples of platinel for the test tube runs were calibrated against the melting points of gold (1064.4°C) and zinc (419.5°C).
The crystals formed in the capsule were observed with an optical

microscope (Vanox, Olympus Optical Co. Ltd., Tokyo, Japan) and a scanning electron microscope (HSM-2, Hitachi Ltd., Tokyo, Japan) and also identified with a X-ray diffractometer (JDX-7F, JEOL Co. Ltd., Tokyo, Japan) using Fe-filtered $CoK\alpha$ or $CuK\alpha$ radiation through a graphite monochromator. Some of the crystals thus grown were pulverized into powder in order to measure the precise unit cell parameter. The calculated lattice parameters from higher angle peaks above 60° in θ corrected with an internal standard of hydrothermally grown calcium fluoride beforehand defined with 99.999% silicon, were plotted against $\cos^2\theta$ and were extrapolated to $\cos^2\theta = 0$.
Content of chromium and chromium oxide were determined by height of peak intensity of (104) of Cr_2O_3 and (110) of Cr of X-ray powder diffraction patterns as comparison with standard lines of patterns by the mixtures of pure Cr_2O_3 and Cr. The density of the sintered specimens was measured by the liquid displacement method with redistilled water.

3. Results and Discussion

3.1. Chromium Oxide

The Reaction from the metalic chromium powder to form chromium oxide (Cr_2O_3) started above $500^\circ C$ under $1000 kg/cm^2$ [10]. The capsul before and after experiments is shown in Fig. 1. Hydrogen gas goes out of the thin wall of capsule. Reaction and following sintering proceeded more easily in the Pt capsule than in the Au capsule. Hydrogen is permeable through Pt wall. In Fig. 2, we can see many bright Cr grains and dark pores. It means insufficient reaction. In fact, we observed an residual water in the capsule. A phase analysis indicated 67.5wt% Cr_2O_3 + 32.5wt% Cr for this specimen by an X-ray powder diffraction method. The observed density was 78.0% of the calculated one for this composition. It seems to indicate that outer gas pressure could not compress inside powders due to the residual H_2O vapour. As increased outer pressure of $3000 kg/cm^2$ did not significant effect either the production of Cr_2O_3 or sintering of it. Cr_2O_3/ (Cr_2O_3 + Cr) was 69.5wt%. Relative density: ρ_{obs}/ρ_{calc} was calculated to be 91.3%, this high value, however, might be wrong, because so many pores could be seen in

Fig. 3, and in the SEM photographs (Fig. 9).
Photomicrograph of the polished surface of hydrothermal reaction sintered Cr_2O_3 specimen at 1000°C, 5000kg/cm^2 for 3h. is shown in Fig. 4. Even the pressure was increased up to 5000kg/cm^2, the microstructure did not be changed after 3 hours treatment. Cr_2O_3/ (Cr_2O_3 + Cr) was 65.0wt% ρobs/ ρcalc was 75.6%.
Photomicrograph of the polished surface of hydrothermal reaction sintered Cr_2O_3 specimen at 1000°C,.5000kg/cm^2 for 15h. is shown in Fig. 5. When the treatment time was elongated to 15 hours, the un-reacted Cr content and pores could be decreased. Cr_2O_3/ (Cr_2O_3 + Cr) was 74.5wt%, ρobs/ ρcalc was 90.2%. Photomicrograph of the polished surface of hydrothermal reaction sintered Cr_2O_3 specimen at 1000°C, 1000kg/cm^2 for 15h. is shown in Fig. 6. Even at low pressure as 1000kg/cm^2, an elongated treatment of 15 hours resulted in relatively high Cr_2O_3 content of Cr_2O_3/ (Cr_2O_3 + Cr) was 90wt% and density of ρobs/ ρcalc was 93.5.
SEM photograth of the fracture surface of hydrothermal reaction sintered Cr_2O_3 specimen at 1000°C, 2000kg/cm^2 for 3h. is shown in Fig. 7. This SEM photograph indicated more clearly the microstructure of the specimen. We can see a porous structure, and also the original Cr grains with about 10 —50μ in diameter were converted into many small grains of Cr_2O_3.
Fig. 8 was the enlarged photograph of Fig. 7. Right-below is a fractured surface, and the remainder was free surface showing aggregates of small grained Cr_2O_3. Bright mass at left-below was rather rare particles in their shape, may be some impurity. SEM photograph of the fracture surface of hydrothermal reaction sintered Cr_2O_3 specimen at 1000°C, 3000kg/cm^2 for 3h. is shown in Fig. 9. Pressure of 3000kg/cm^2 did not effect in the microstructure of the specimen. We can see that original Cr-grains almost converted into the aggregates of the small Cr_2O_3 grains, but were still preserved in their shapes. SEM photograph of the surface of hydrothermal reaction sintered Cr_2O_3 specimen at 1000°C, 3000kg/cm^2 for 3h. is shown in Fig. 10. This picture is almost the same as the Fig. 9. SEM photograph of the fracture surface of hydrothermal reaction sintered Cr_2O_3 specimen at 1000°C, 5000kg/cm^2 for 3h. is shown in Fig. 11. From Left to right-below was showing a fractured surface, and the top to Right-top

was showing the free surface in backward. They were still porous, and essentially the same as the previous one shown in Fig. 7 —10. $Cr_2O_3/(Cr_2O_3 + Cr)$ was 65.0wt% and ρobs/ρcalc was 75.6%. Residual water could be detected after this treatment. SEM photograph of the fracture surface of hydrothermal reaction sintered Cr_2O_3 specimen at 1000°C, 5000kg/cm^2 for 15h. is shown in Fig. 12. An elongated treatment of 15 hours instead of 3 hours resulted in a completely different microstructure from Fig. 7 —10. This was much dense, ρobs/ρcalc= 90.2%. Original grains could no longer be seen, then a homogeneous compact of small Cr_2O_3 grains was seen in this photograph of Fig. 11. SEM photograph of the fracture surface of hydrothermal reaction sintered Cr_2O_3 specimen at 1000°C, 5000kg/cm^2 for 15h. is shown in Fig. 13. This is enlarged photograph of the Fig. 12, which shows 1 —2μ grains with well developed faces, where typical revers or steps indicating brittle fracture in intergrains can be seen. Small hole in the center of each grain was presumably showing the diffusion of Cr into Cr_2O_3 layer produced on the outside of the original Cr grains. SEM photograph of the surface of hydrothermal reaction sintered Cr_2O_3 specimen at 1000°C, 5000kg/cm^2 for 15h. is shown in Fig. 14. This shows a free surface of the same specimen as before. You can see well-shaped small Cr_2O_3 grains with apparently uniform size. There were no small pores on the free surfaces. It also was a proof of intergrain-fracture of this materials. SEM photograph of the fracture surface of hydrothermal reaction sintered Cr_2O_3 specimen (Exceptionally Large Grain) at 1000°C, 5000kg/cm^2 for 15h. is shown in Fig. 15. This shows an exceptionally large grain found in the fractive surface. It also contained small pores produced by Cr-diffusion as mentioned before. Pressure dependence of relative density and Cr_2O_3 content in the hydrothermal reaction sintered Cr_2O_3 specimens at 1000°C for 3h. and for 15h. are shown in Fig. 16.

This figure summarize the results obtained in this work. The effects of the external pressure are small both on the production of Cr_2O_3 and on the densification. For example, after 3 hours Cr_2O_3 contents were about 65 to 70wt% in every specimens, and relative densities were 75 —91%. But prolonged treatment of 15 hours yielded more Cr_2O_3 and denser bodies up to 93.5. We could prepare denser one with 95.6% density in a just recent experiment. The hydrothermal reaction sintering seems to proceed in the following two stages.

(1) Oxidation of Cr to produce Cr_2O_3 fine grains and hydrogen gas.

(2) As the hydrogen gas diffuse out of the capsule, the external pressure beyond the internal one effect to compress the internal Cr_2O_3 powder like as a hot isostatic pressing.

Thus the reaction duration can affect more strongly than the external pressure.

3.2. Iron Oxide[11)]

The reaction for $3Fe + 4H_2O$ ——— $Fe_3O_4 + 4H_2$ occures above 250°C under 1000kg/cm^2 of under hydrothermal conditions. Experiments in detail were described already in J. Cryst. Growth[11)]. Therefore this part is mentioned only temperature effect on crystal growth.

Solution was 10m NaOH and duration was 1 day under 1000kg/cm^2 of hydrothermal conditions. The size of Au capsule was 5mm in outer diameter and 4.7mm in inner diameter, and 50mm in length. Temperature of the capsule was 550°C for bottom of the capsule and 500°C for top of the capsule. At the bottom well sintered mass of magnetite appeared and single crystals appeared at top of the capsule as shown in Fig. 17. The lattice constant of the magnetite crystals grown under hydrothermal conditions was 8.3962 ± 0.0003 Å.

4. Summary

Hydrothermal reaction sintering of chromium and iron metal powders were described. Well sintered oxide bodies were obtained by hydrothermal reaction sintering.

Acknowledgements

The authors desire to express their appreciation to Dr. Shinroku Saito of the President, Tokyo Institute of Technology for his continuous encouragements. Equipments for the hydrothermal reaction sintering were supported by the Ministry of Education, Japan.

References

1) H.E.N. Stone and N.A. Lockington, Powder Metallurgy 7 (14) 113 —124 (1964)

2) Ibid, 8 (15) 81 —91 (1965)
3) W.C. Hagel, P.J.Jorgensen, and D.S. Tomalin, J. Am. Ceram. Soc., 49 (1) 23 —26 (1966)
4) P.D. Ownby and G.E. Jungquist, J. Am. Ceram. Soc., 55 (9) 433 —436 (1972)
5) P.D. Ownby, Sintering and Related Phenomena, ed. by C.C. Kuczynski, Plenum Press, (1973) P.431 —437
6) N. Claussen and H.E. Exner, Powder Metallurgy, 15 (30) 202 —215 (1972)
7) J.W. Halloran and H.U. Anderson, J. Am. Ceram. Soc., 57 (3) 150 (1974)
8)J.M. Neve and R.L. Coble, J. Am. Ceram. Soc., 57 (6) 274 —275 (1974)
9) N. Toker and L.S. Darken, Geochemica et Cosmochemika, Acta. 39 847—852 (1975)
10) S. Hirano and S. Sōmiya, J. Am. Ceram. Soc., 59 (11—12) 534 (1976)
11) S. Hirano and S. Sōmiya, J. Cryst. Growth, 35 No. 3, Oct. 273 —278 (1976)
12) S. Sōmiya, S. Hirano, T. Fukuda, M. Sawada, Koatsu Gasu, 10 (6) 16 —27 (1973)
13) Shigeyuki Sōmiya, Shin-ichi Hirano, Tamotsu Fukuda, and Masahisa Sawada, Koatsu Gijutsu, 12 (1) 15 —22 (1974)
14) S. Hirano, K. Nakamura, and S. Sōmiya, Proceedings the 4th International Conference on High Pressure 418 —423 (1975)

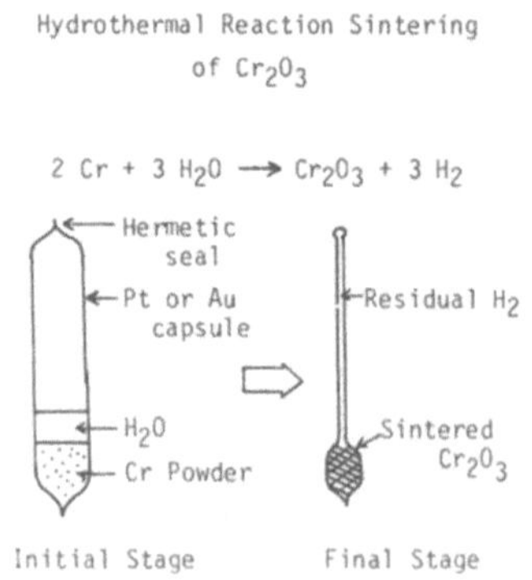

Fig. 1

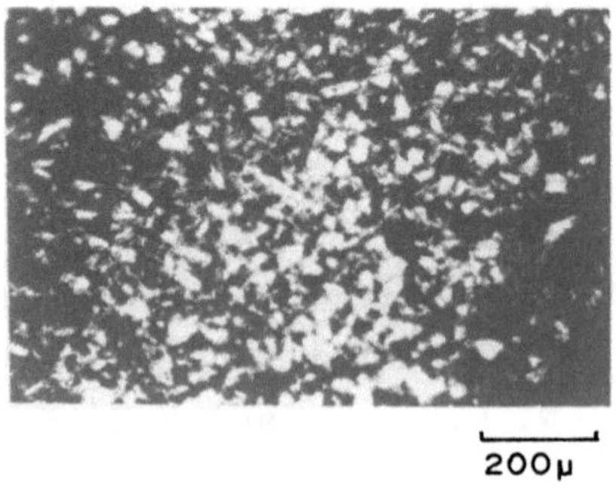

Fig. 2

Photomicrograph of the Polished Surface of Hydrothermal Reaction Sintered Cr_2O_3 Specimen at 1000 °C, 2000 Kg/cm^2 for 3 h.

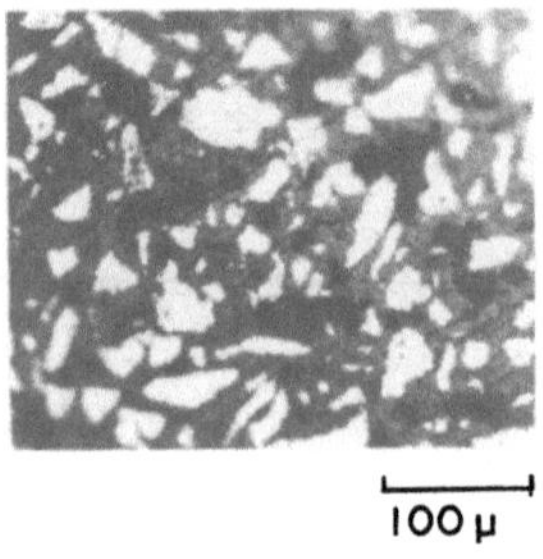

Fig. 3

Photomicrograph of the Polished Surface of Hydrothermal Reaction Sintered Cr_2O_3 Specimen at 1000 °C, 3000 Kg/cm^2 for 3 h.

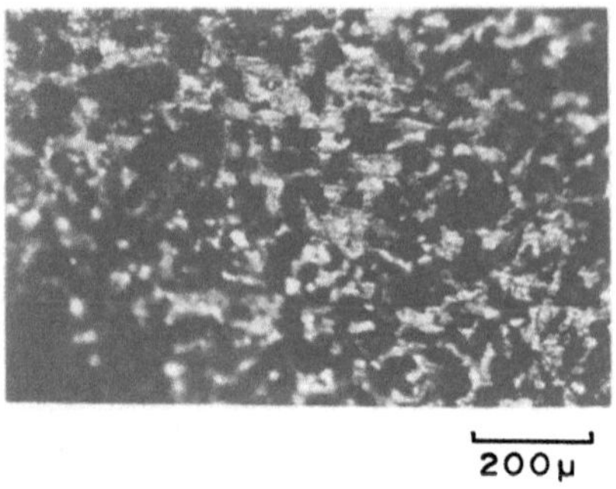

Fig. 4

Photomicrograph of the Polished Surface of Hydrothermal Reaction Sintered Cr_2O_3 Specimen at 1000 °C, 5000 Kg/cm^2 for 3 h.

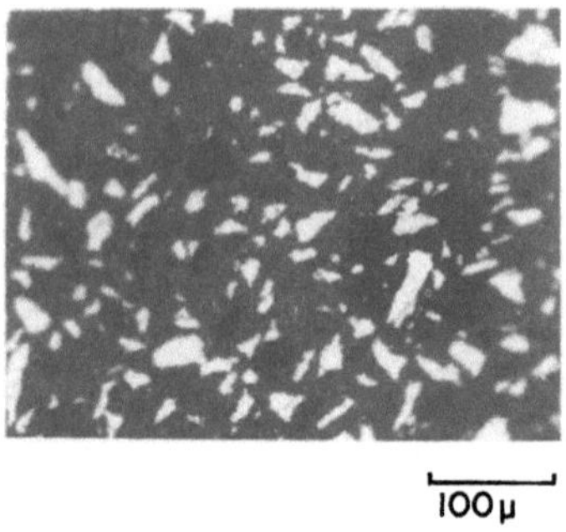

Fig. 5

Photomicrograph of the Polished Surface of Hydrothermal Reaction Sintered Cr_2O_3 Specimen at 1000 °C, 5000 Kg/cm^2 for 15 h.

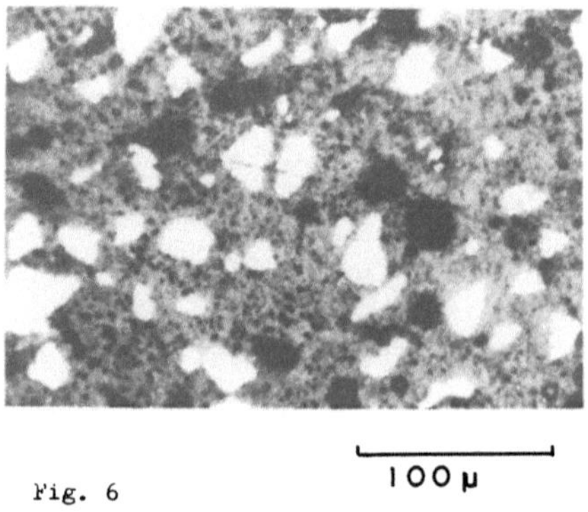

Fig. 6

Photomicrograph of the Polished Surface of Hydrothermal Reaction Sintered Cr_2O_3 Specimen at 1000 °C, 1000 Kg/cm^2 for 15 h.

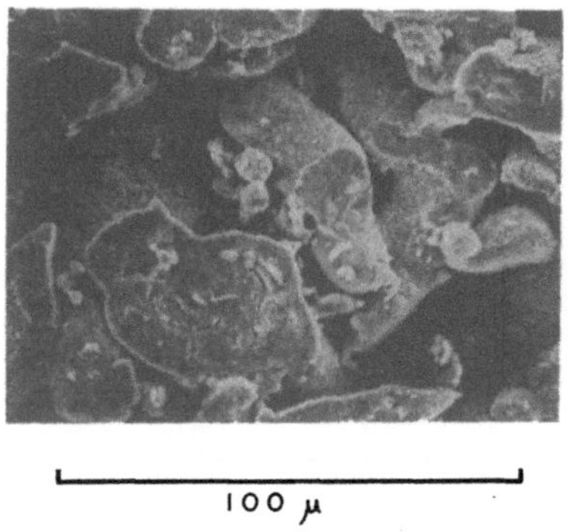

Fig.7

SEM Photograph of the Fracture Surface of Hydrothermal Reaction Sintered Cr_2O_3 Specimen at 1000°C,2000 Kg/cm^2 for 3 h

Fig. 8

SEM Photograph of the Surface of Hydrothermal Reaction Sintered Cr_2O_3 Specimen at 1000 °C, 2000 Kg/cm^2 for 3 h.

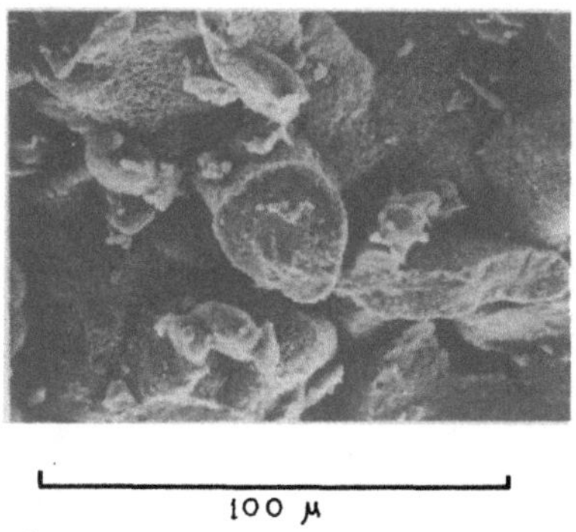

Fig. 9

SEM Photograph of the Fracture Surface of Hydrothermal Reaction Sintered Cr_2O_3 Specimen at 1000 °C, 3000 Kg/cm^2 for 3 h.

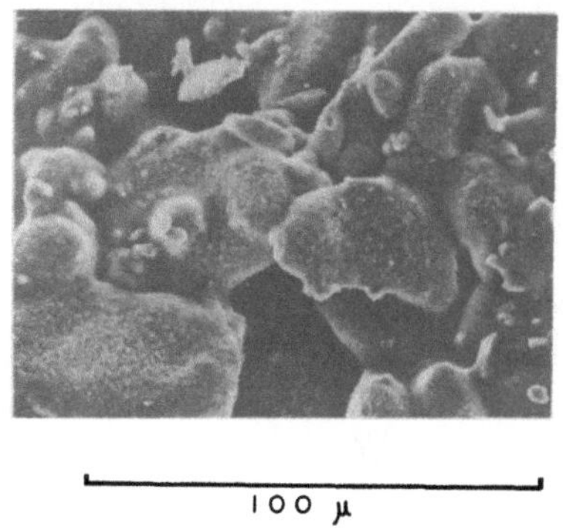

Fig.10

SEM Photograph of the Surface of Hydrothermal Reaction Sintered Cr_2O_3 Specimen at 1000 °C, 3,000 Kg/cm^2 for 3 h.

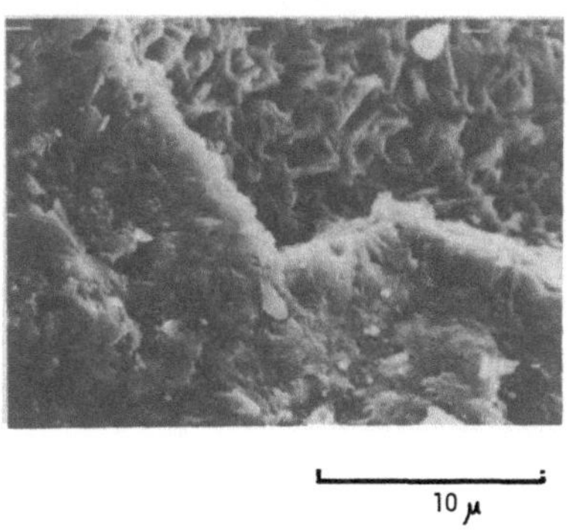

Fig. 11

SEM Photograph of the Fracture Surface of Hydrothermal Reaction Sintered Cr_2O_3 Specimen at 1000 °C,5000 Kg/cm^2 for 3 h.

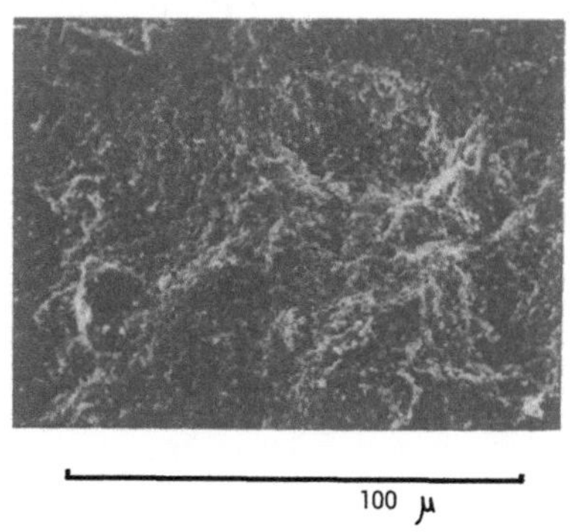

Fig. 12

S E M Photograph of the Fracture Surface of Hydrothermal Reaction Sintered Cr_2O_3 Specimen at 1000 °C, 5000 Kg/cm^2 for 15 h.

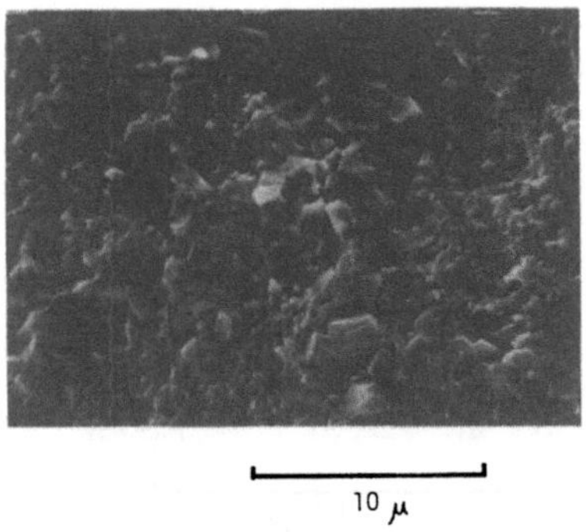

Fig. 13

SEM Photograph of the Fracture Surface of Hydrothermal Reaction Sintered Cr_2O_3 Specimen at 1000 °C,5000 Kg/cm^2 for 15 h.

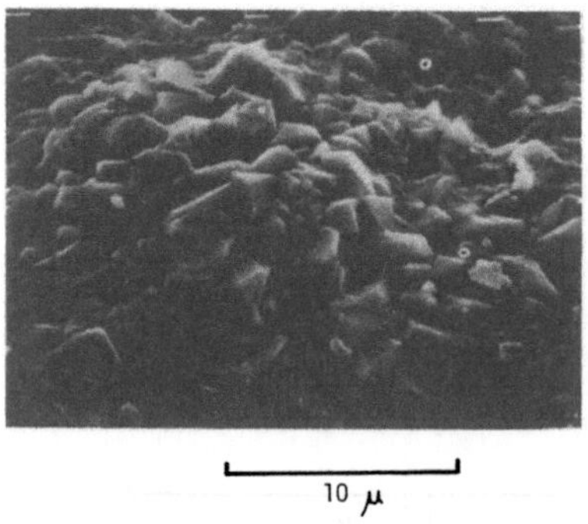

Fig. 14

SEM Photograph of the Surface of Hydrothermal Reaction Sintered Cr_2O_3 Specimen at 1000 °C, 5000 Kg/cm^2 for 15 h.

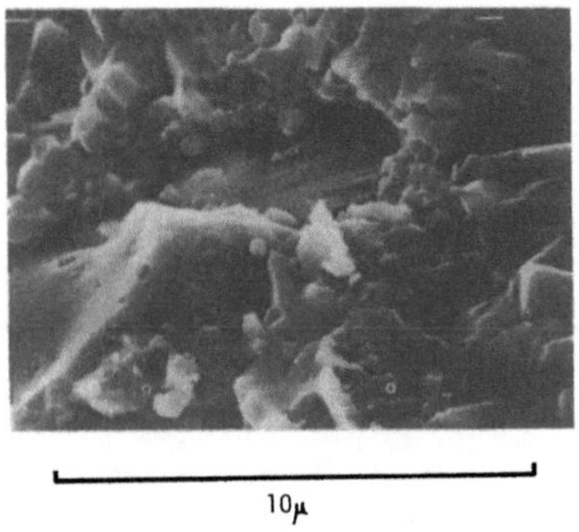

Fig. 15

SEM Photograph of the Fracture Surface of Hydrothermal Reaction Sintered Cr_2O_3 Specimen (Exceptionally Large Grain) at 1000 °C, 5 000 Kg/cm^2 for 15 h.

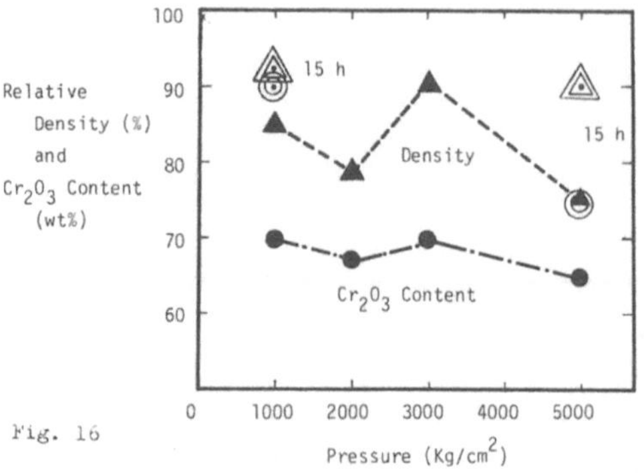

Fig. 16

Pressure Dependence of Relative Density and Cr_2O_3 Content in the Hydrothermal Reaction Sintered Cr_2O_3 Specimens at 1000 °C for 3 h.

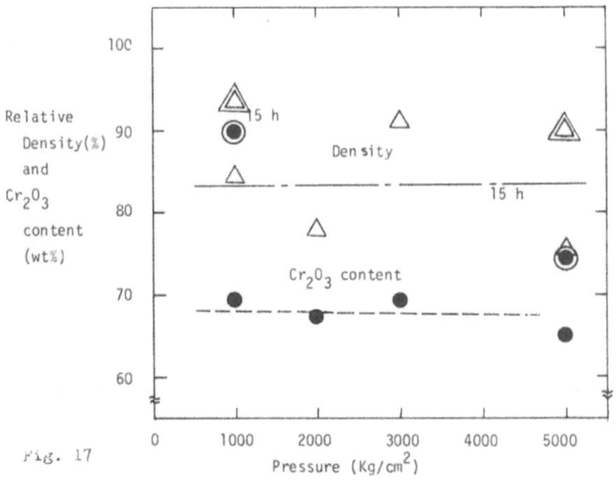

Fig. 17

Pressure Dependence of The Relative Density and Cr_2O_3 Content in the Hydrothermal Reaction Sintered Cr_2O_3 Specimen at 1000 °C for 3 h.

Reprinted from *American Ceramic Society Bulletin*, **59**(2) (1980) 246.

Fabrication of Dense, Nonstabilized ZrO_2 Ceramics by Hydrothermal Reaction Sintering

MASAHIRO YOSHIMURA* and SHIGEYUKI SŌMIYA*

Tokyo Institute of Technology, Nagatsuta, Yokohama, Japan

Hirano and Sōmiya[1] prepared high density Cr_2O_3 ceramics by a method known as hydrothermal reaction sintering, in which chromium metal and water are reacted and then sintered under hydrothermal conditions. The reaction is written simply as $2Cr+3H_2O=Cr_2O_3+3H_2$. The mechanism of this reaction, however, appears to be complicated because it should involve 1) reaction between Cr and H_2O, 2) sintering of formed Cr_2O_3, and 3) diffusion of H_2 out of a capsule, at high temperatures under high pressures. A subsequent study[2] that studied the effects of pressure, temperature, and treating period on hydrothermal reaction sintering of Cr_2O_3 showed less reproducibility of densities and microstructures in the final product. In spite of such difficulties in the mechanism of the reaction, this technique has great merit in preparing a dense, isostatically sintered oxide with fine grains at relatively low temperatures without any additives.

In the present note we applied this method to fabricate dense, nonstabilized ZrO_2 ceramics, which are difficult to prepare by normal sintering techniques because of cracking due to the phase transition[3-4] monoclinic⇌tetragonal.

Zirconium metal powder* (99.9% pure; 300 mesh) and redistilled water with a Zr/H_2O mole ratio of 1:2 were sealed with an electric arc in a 3-mm-ID Pt capsule 0.2 mm thick and 35 mm long. The sealed capsule was heated in a test-tube-type pressure vessel or a high-gas-pressure apparatus[5] under the desired pressure of water or argon gas. Temperature and pressure were held within ±20°C and ±1.96 MPa, respectively, during a run, which was typically 3 h long. Heated specimens were decapsuled and examined by X-ray powder diffraction and scanning electron microscopy. The density of sintered specimens was measured by Archimedes' method using pure toluene and fine tungsten wires 0.05 mm in diameter.

The reaction between Zr metal and supercritical water under 98 MPa pressure was observed above 300°C to yield micrograins of monoclinic ZrO_2 and ZrH_2. At higher temperatures active, fine-grained crystalline m-ZrO_2 particles and hydrogen gas were produced. As the hydrogen gas permeated the capsule wall during heating, these m-ZrO_2 particles were compressed more effectively and then allowed to sinter.

Well-sintered m-ZrO_2 ceramics having a density of 5.17–5.85 g/cm³ were prepared by this method at 1000°C for 3 h under 98 MPa pressure. This density range is one of the highest obtained among the dense, nonstabilized ZrO_2 ceramics previously fabricated by reactive hot-working methods,[6-7] reactive hot pressing,[3] or reoxidizing of oxygen-deficient ceramics sintered in vacuo.[4] Furthermore, the ZrO_2 ceramics prepared in the present study are expected to be essentially crack-free because the fabrication temperature did not exceed the phase transition temperature of ZrO_2.

The observed maximum density of 5.85 g/cm³ (relative density 99.0%) corresponds to the X-ray density of 5.91 g/cm³ for 98% ZrO_2-2% HfO_2 solid solution. A typical scanning electron photomicrograph of the fractured surface of this specimen is shown in Fig. 1(A). Many zirconia polyhedra, with a grain size of 1–5 μm, indicating intergranular fracture, and fractured grains are evident in this figure.

Fig. 1. Fracture surfaces of m-ZrO_2 specimens sintered hydrothermally at 1000°C for 3 h under 98 MPa. (A) $H_2O/Zr=1.74$, density=5.85 g/cm³; (B) $H_2O/Zr=2.056$, density=5.38 g/cm³.

The presence of more water than is required for stoichiometry of the reaction $Zr+2H_2O=ZrO_2+2H_2$ seemed to prevent effective sintering and densification. Figure 1(B) shows a scanning electron photomicrograph of the fracture surface of the specimen with the initial mole ratio of $H_2O/Zr=2.056$ sintered at 1000°C for 3 h under 98 MPa pressure. The microstructure of this specimen, with a density of 5.38 g/cm³ and composed of 0.5–1.0 μm fine grains, aggregated inhomogeneously. With more excess water, no well-sintered bodies could be prepared even at temperatures as high as 1200°C or under 490 MPa pressure.

The mechanism of hydrothermal reaction sintering has not yet been completely explored, but it appears that this technique is applicable to the preparation of pure, dense, fine-grained oxides or oxide-metal composites at lower temperatures. Further investigation of the effects of pressure, temperature, heating time, and the H_2O/Zr ratio is in progress.

Acknowledgments

The authors are indebted to S. Itoh and H. Kanai for their assistance with the experiments.

References

[1] S. Hirano and S. Sōmiya, "Hydrothermal Reaction Sintering of Pure Cr_2O_3," *J. Am. Ceram. Soc.*, **59** [11–12] 534 (1976).
[2] S. Sōmiya, S. Hirano, M. Yoshimura, S. Itoh and H. Kanai, "Hydrothermal Reaction Sintering of Cr_2O_3 and Iron Oxides"; pp. 267–77 in Proceedings of the International Symposium on Factors in Densification and Sintering of Oxide and Non-oxide Ceramics. Edited by S. Sōmiya and S. Saito. Gakujutsu Bunken Fukyukai, Tokyo Institute of Technology, Tokyo, Japan, 1979.
[3] A. C. D. Chaklader and V. T. Baker, "Reactive Hot-Pressing: Fabrication and Densification of Non-Stabilized ZrO_2," *Am. Ceram. Soc. Bull.*, **44** [3] 258–59 (1965).
[4] H. J. Garrett and R. Ruh, "Fabrication of Specimens from Pure Dense Oxidized Zirconia," *ibid.*, **47** [6] 578–79 (1968).
[5] S. Hirano, K. Nakamura and S. Sōmiya, "Graphitization of Carbon in Presence of Calcium Compounds under Hydrothermal Condition by Use of High Gas Pressure Apparatus"; pp. 418–23 in Proceedings of the 4th International Conference on High Pressure. Special Issue of the Review of Physical Chemistry of Japan. Edited by J. Osugi. Nippon Butsurikagaku Kai, Kyoto, Japan, 1975.
[6] G. K. Bansal and A. H. Heuer, "Transformational Hot-Working of ZrO_2 Polycrystals," *J. Am. Ceram. Soc.*, **58** [1–2] 76–77 (1975).
[7] C. F. Smith and W. B. Crandall, "Calculated High-Temperature Elastic Constants for Zero Porosity Monoclinic Zirconia," *ibid.*, **47** [12] 624–27 (1964). □

*Member, the American Ceramic Society.
*Zr with ≈2% Hf, Kōjundo Chemicals Lab., Inc., Wako, Japan.

Received September 11, 1979; approved October 19, 1979.

Reprinted from *Advances in Ceramics*, Vol. 3, *Science and Technology of Zirconia* (American Ceramic Society, 1981) pp. 455–463.

Hydrothermal reaction sintering of monoclinic zirconia

MASAHIRO YOSHIMURA AND SHIGEYUKI SŌMIYA

Laboratory for Hydrothermal Syntheses
Research Laboratory of Engineering Materials
and Dept. of Materials Science and Engineering
Tokyo Institute of Technology
Nagatsuta, Yokohama 227, Japan

A dense monoclinic zirconia ceramic having a maximum density of 5.85 g/cm³ and grains of 1–5 μm has been prepared from Zr metal powder and water in a Pt capsule by hydrothermal reaction sintering at 1000 °C for 3 h under 98 MPa. The effects of temperatures, pressures, and initial compositions on the densification and sintering were studied. The reaction between Zr metal powder and H_2O started above ≈300 °C under 98 MPa, then produced fine grains of monoclinic ZrO_2 and H_2. Sintering of these grains started above 700 °C, where H_2 had permeated the capsule wall. Well-sintered bodies could be prepared at 1000 °C under 98 MPa. Higher pressures or higher temperatures did not produce further densification. The highest densifications were observed in the samples with the ratio of $H_2O/Zr = 2$ in the initial compositions. The samples with the ratio $H_2O/Zr > 2.1$ did not give well-sintered bodies.

Dense monoclinic nonstabilized zirconia ceramics are difficult to fabricate by normal sintering techniques because of cracking due to the phase transformation from tetragonal to monoclinic form during cooling.[1–4] They have been prepared previously by reactive hot-working,[1,2] reactive hot-pressing,[3] or reoxidizing of oxygen-deficient ceramics sintered in vacuo[4] as summarized in Table I. Comparison among them appeared to be difficult since the compositions of the starting materials were different in these studies, and some authors reported relative densities while the others reported bulk densities. Dense monoclinic zirconia ceramics are generally prepared at high temperatures (1600° to 2300 °C), then heat-treated at lower temperatures near the transition temperature.

Recently, zirconia ceramics with among the highest densities have been fabricated by the hydrothermal reaction sintering method at 1000 °C.[5] They were prepared at temperatures below the transition temperature of ZrO_2, so that essentially no cracks were expected to have occurred in the specimens. In this hydrothermal reaction-sintering method, Zr metal powder and water were used as starting materials encapsulated in a Pt capsule, as illustrated in Fig. 1. They reacted, then yielded fine grains of monoclinic ZrO_2 and hydrogen gas at high temperatures under pressure. As the hydrogen gas permeated the capsule wall during heating, these monoclinic ZrO_2 grains were compressed more effectively and then allowed to sinter.

In the present study, the effects of the starting compositions, water contents, temperatures, and pressures on the sintering and densification were investigated. Similar hydrothermal reaction sintering of Cr_2O_3 from Cr metal powder and water has already been reported.[6-8]

Experimental Procedure

Zirconium metal powder,* 99.9% pure, containing ≈2% Hf, was the starting material. It consisted of particles 1 to 10 μm in diameter (Fig. 2). This zirconium and redistilled water were hermetically sealed with an electric arc in a 2.6–2.7 mm inside diameter platinum capsule 0.15–0.2 mm thick and 35 mm long. The sealed capsule was heated in a test-tube-type pressure vessel or a high-gas-pressure apparatus under the desired pressure of water or argon gas. Water was used for runs up to 700 °C, where temperature was measured by two platinel thermocouples calibrated against the melting point of Zn (419.5 °C) and controlled within ±3 °C during a run. The pressure was measured by a calibrated Heise gage and controlled within ±0.2 MPa.

In the experiments using the high-gas-pressure apparatus,[9] the temperature and pressure were held within ±20 °C and 1.96 MPa, respectively, during a run, which was typically 3 h long. The capsules were weighed to check the leakage and/or the decrease resulting from the loss by permeation of hydrogen gas. Heated specimens were removed from the capsules, then examined by X-ray powder diffraction using a diffractometer† with a rotating Cu anode and by scanning electron microscopy.‡ The density of sintered specimens was measured by a liquid replacement method, where toluene and tungsten wires 0.05 mm in diameter were used to reduce the influence of surface tension.

Results and Discussion

The reaction between Zr metal powder and supercritical water under 98 MPa was observed above 300 °C to yield micrograins of monoclinic ZrO_2. They were much smaller than initial Zr metal grains and formed a crust-like structure, as seen in Fig. 3. The average crystallite size was calculated to be ≈24 nm from the measurement of the half-width of the X-ray diffraction peak, using the Warren equation[10] for these monoclinic ZrO_2 grains prepared at 400 °C. This formation of active fine grains is one of the important features in the hydrothermal reaction sintering. At 400° to 700 °C, monoclinic ZrO_2 grains of ≈24 nm and hydrogen gas existed in the capsules. Since hydrogen gas started to permeate the capsule wall above 700 °C, the monoclinic ZrO_2 grains thus formed were compressed and sintered.

Well-sintered bodies having bulk densities of 5.17–5.85 g/cm³ were obtained at 1000 °C after 3 h under 98 MPa (Table II). Figure 4 shows a typical scanning electron photomicrograph of the fractured surface of the specimen with the maximum density of 5.85 g/cm³, or relative density of 99.0% compared to the X-ray density of 5.91 g/cm³ for a 98% ZrO_2-2%HfO_2 solid solution. Polyhedral zirconia grains, with a grain size of 1–5 μm, are evident in this figure.

The presence of more water than is required for stoichiometry in the reaction $Zr + 2H_2O = ZrO_2 + 2H_2$ seemed to prevent effective sintering and densification. Figure 5 shows a scanning electron photomicrograph of the fractured surface of the specimen with the initial mole ratio of

$H_2O/Zr = 2.095$ sintered at 1000 °C for 3 h under 98 MPa pressure. The microstructure of this specimen, with a density of 5.38 g/cm^3 and composed of 0.5–1.0 μm fine grains, aggregated inhomogeneously. With more excess water, no well-sintered bodies could be prepared, even at temperatures as high as 1200 °C or under 686 MPa pressure (Table II). Pressures above 98 MPa did not always promote densification, nor did temperatures of 1100 ° or 1200 °C.

The ratio of water to Zr apparently has the greatest effect on the densification. Figure 6 indicates that the highest densities might be attained in the samples with the stoichiometric amount (or slightly less) of water vs Zr and that excess water, even if the excess is small, prevents densification. These results can be explained by the presence of the residual gas(es) in the capsule during sintering; the compression of the monoclinic ZrO_2 particles by outer gas pressure would stop when the inner pressure became equal to the outer ambient gas pressure. The most abundant residual gas appears to be water, which cannot permeate platinum. For example, the initial mixture of 1 mol Zr and 2.10 mol H_2O ($H_2O/Zr = 2.10$, a 5% excess of water over stoichiometry), would yield 1 mol ZrO_2 and 0.10 mol H_2O as the final products in the capsule. This water would occupy a volume of 10.62 cm^3 at 1000 °C under 100 MPa, which was calculated from the specific volume of water 5.900 cm^3/g observed under the same conditions by Kennedy.[11] That volume corresponds to 50.2% of the molar volume of m-ZrO_2, 21.14 cm^3.[12] That is, the monoclinic ZrO_2 compacts with a porosity of 33.4%, 50.2/(100 + 50.2) and could no longer be compressed by the ambient gas pressure of 100 MPa at 1000 °C. This is qualitatively in agreement with the experimental results, in which no well-sintered bodies could be prepared when the initial compositions had H_2O/Zr ratios larger than 2.10.

Quantitative assessment of the density seems to be difficult, particularly near theoretical density, because other gases such as N_2 in the ambient atmosphere might affect the final sintering stage, and the mechanism of this sintering is not completely understood. It seems likely, however, that sintering and grain growth occurred simultaneously, because the samples with higher densities are generally composed of grains with larger sizes, as seen in Table II with the exception of sample No. 28.

Note that this hydrothermal reaction method enables us to prepare dense fine-grained ceramics at lower temperatures than the other methods, even though the sintering is accompanied by grain growth.

Acknowledgments

The authors are grateful to S. Itoh and H. Kanai for assistance during the study.

References

[1]G. K. Bansal and A. H. Heuer, "Transformational Hot-Working of ZrO_2 Polycrystals," *J. Am. Ceram. Soc.*, **58** [1-2] 76-77 (1975).

[2]C. F. Smith and W. B. Crandall, "Calculated High-Temperature Elastic Constants of Zero Porosity Monoclinic Zirconia," *ibid.*, **47** [12] 624-27 (1964).

[3]A. C. D. Chaklader and V. T. Baker, "Reactive Hot-Pressing: Fabrication and Densification of Non-Stabilized ZrO_2," *Am. Ceram. Soc. Bull.*, **44** [3] 258-59 (1965).

[4]H. J. Garrett and R. Ruh, "Fabrication of Specimens from Pure Dense Oxidized Zirconia," *ibid.*, **47** [6] 578-79 (1968).

[5]M. Yoshimura and S. Sōmiya, "Fabrication of Dense, Nonstabilized ZrO_2 Ceramics by Hydrothermal Reaction Sintering," *ibid.*, **59** [2] 246 (1980).

[6]S. Hirano and S. Sōmiya, "Hydrothermal Reaction Sintering of Pure Cr_2O_3," *J. Am. Ceram. Soc.*, **59** [11-12] 534 (1976).

[7]S. Sōmiya, S. Hirano, M. Yoshimura, S. Itoh, and H. Kanai; pp. 267-77 in Proceedings of the International Symposium on Factors in Densification and Sintering of Oxide and Nonoxide Ceramics. Edited by S. Sōmiya and S. Saito. Gakujutsu Bunken Fukyukai. Tokyo Institute of Technology, Tokyo, Japan, 1979.

[8]S. Sōmiya, M. Yoshimura, and H. Kanai, "Effects of Temperature, Pressure, Reaction Period, and Solutions on the Hydrothermal Reaction Sintering of Chromic Oxide"; to be published in *Zairyo Kagaku (J. Mater. Sci. Soc. Japan).*

[9]S. Sōmiya, S. Hirano, T. Fukuda, and M. Sawada, "Development and Application of a High Gas Pressure and High Temperature Apparatus up to 1500°C and 10 Kb," *Koatsu Gasu,* **10** [6] 368-79 (1973).

[10]L. V. Azároff; pp. 562-71 in Elements of X-ray Crystallography. McGraw-Hill, New York, 1968. Japanese edition by Maruzen Co. Ltd., 1973.

[11]G. C. Kennedy, "Pressure-Volume-Temperature Relations in Water at Elevated Temperatures and Pressures," *Econ. Geol.*, **45**, 629 (1950).

[12]Handbook of Chemistry and Physics. Edited by R. C. Weast. Chemical Rubber Co. Press, Cleveland, Ohio, 1975-76; p. B217.

---

*Kōjundo Chemical Lab Inc., Wako, Saitama, Japan.
†RU-200; Rigaku Electric Co. Ltd., Tokyo, Japan.
‡JSM T-200; JEOL Co. Ltd., Tokyo, Japan.

Table I. Fabrication of dense nonstabilized monoclinic ZrO_2 ceramics

Fabrication method	Bulk density (g/cm^3)	Relative density (%)	Composition (%)	Ref.
Reactive hot-pressing	5.55	99.8	98 $ZrO_2 + HfO_2$	3
55.8 MPa, 800–1200 °C	5.45	98.0	99.7 ZrO_2	3
Reactive hot-pressing 27.5 MPa, 800–1200 °C		99	99.99 ($ZrO_2 + HfO_2$)	1
Reactive hot-pressing, 165 MPa, 800–1200 °C		98	”	1
Reactive hot-working, 97.6 MPa, 1000–700 °C, after hot-pressing		95	93.77 ($ZrO_2 + HfO_2$) 3.90 SiO_2	2
Vacuum sintering at 2300 °C, then oxidized at 1000 °C	5.43–5.52	94–96	High purity ZrO_2	4
Hydrothermal reaction sintering, 1000 °C, 98 MPa or above	5.79–5.85	98–99	99.9 ($ZrO_2 + HfO_2$) 2 Hf	5

Table II. Selected results of hydrothermal reaction sintering of monoclinic ZrO_2 for 3 h

Sample No.	H_2O/Zr	Deg. of fill (%)	Temp. (°C)	Pressure (MPa)	Bulk density (g/cm^3)	Relative density (%)	Avg grain size (μm)
3	1.773	19.3	1000	98	5.85	99.0*	3
5	2.020	15.5	1000	98	5.51	93.2	1
29	2.063	14.8	1000	98	5.17	87.5	0.5
30	2.095	15.0	1000	98	5.38	89.2	0.5
10	2.415	18.4	1000	98	Not sintered		
25	2.038	15.1	1000	196	5.51	93.2	0.3
28	1.985	15.1	1000	294	5.96	100.9†	0.1
8	2.012	22.7	1000	490	5.83	98.7	3
7	2.010	20.4	1000	686	5.77	97.7	3
23	2.038	41.2	1000	686	5.70	96.4	
32	2.016	14.8	1000	98	5.68	96.1	2
40	2.142	14.3	1100	98	Not sintered		
41	2.089	14.2	1200	98	Not sintered		
39	2.091	14.1	1000	490‡	Not sintered		0.5

*With gray skin. †Gray fine body, rare case. ‡Capsule leaked slightly.

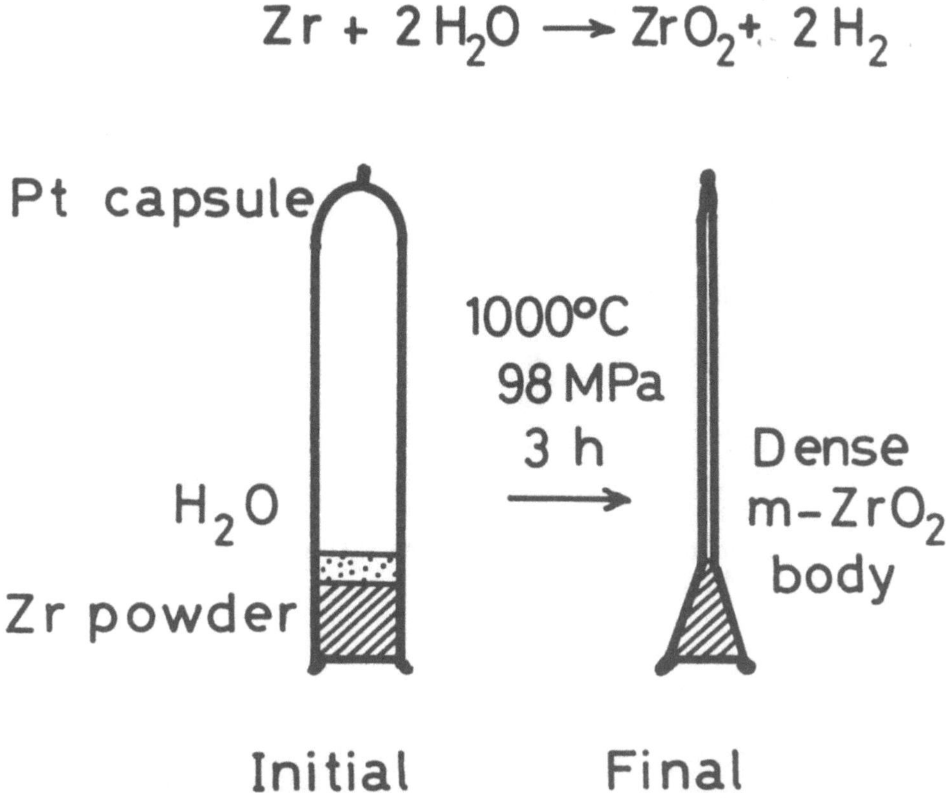

Fig. 1. Schematic illustration of the hydrothermal reaction sintering of monoclinic ZrO_2.

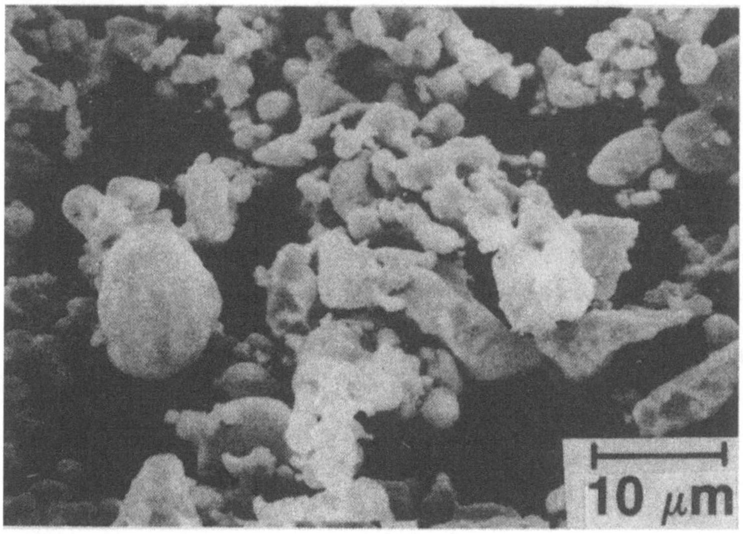

Fig. 2. Starting Zr metal powder.

Fig. 3. Zr metal powder reacted with H_2O at 300 °C for 3 h under 98 MPa.

Fig. 4. Fracture surface of m-ZrO_2 specimen sintered hydrothermally; 1000 °C, 3 h, 98 MPa, $H_2O/Zr = 1.773$, fill 19.3%.

Fig. 5. Fracture surface of m-ZrO_2 specimen sintered hydrothermally; 1000 °C, 3 h, 98 MPa, $H_2O/Zr = 2.095$, fill 15.0%.

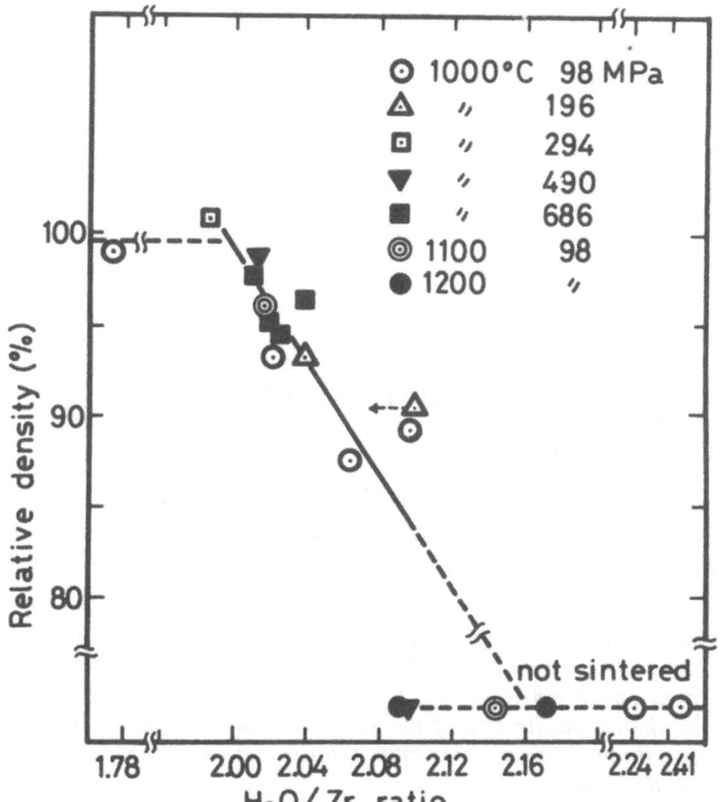

Fig. 6. Relative density of m-ZrO_2 ceramics sintered hydrothermally as a function of the H_2O/Zr ratio.

Reprinted from *Journal of the American Ceramic Society*, **65**(9) (1982) C159–C160.

Hydrothermal Reaction-Sintering of Monoclinic HfO_2

HIDEO TORAYA, MASAHIRO YOSHIMURA,* AND SHIGEYUKI SŌMIYA*

Laboratory for Hydrothermal Syntheses, Research Laboratory of Engineering Materials, and Department of Materials Science and Engineering, Tokyo Institute of Technology, Nagatsuta, Midori, Yokohama 227, Japan

Sintered monoclinic HfO_2 bodies were fabricated below the transformation temperature from Hf metal and water by hydrothermal reaction-sintering. Sintering was observed above 900°C under 100 MPa for 3 h. Generally, both the bulk density and the crystallite size of the sintered bodies increased with increasing temperature. Bodies with the maximum relative density (0.98) were obtained by treatment above 1000°C.

HAFNIA (HfO_2) has a higher monoclinic⇌tetragonal transformation temperature and a lower thermal expansion coefficient than isomorphous zirconia (ZrO_2).[1] Hafnia ceramic bodies are therefore expected to be thermally tough. Hydrothermal reaction-sintering is a combined process consisting of (1) hydrothermal oxidation of a metal, (2) diffusion of H_2 from the capsule, and (3) sintering of the oxide powder formed.[2,3] This technique, which has been used to fabricate dense Cr_2O_3,[2] $LaCrO_3$,[4] and nonstabilized ZrO_2[3,5] ceramics at relatively low temperatures, is well suited for forming an HfO_2 ceramic below the transformation temperature. Thus, a sintered HfO_2 (or ZrO_2) body will be essentially free from the cracks associated with the phase transformation which are usually inevitable in reactive hot-pressing[6] or oxidation of oxygen-deficient

CONTRIBUTING EDITOR–H. K. BOWEN

Received March 24, 1982; revised copy received June 3, 1982. Approved June 3, 1982.

Supported by Grant-in-Aid for Developmental Scientific Research, No. 585207.

*Member, the American Ceramic Society

ceramics[7] at high temperatures. The preparation of fine monoclinic HfO_2 powders from Hf metal and high-temperature, high-pressure water by hydrothermal oxidation has been reported.[8,9] This communication describes the fabrication of a sintered monoclinic HfO_2 body by hydrothermal reaction-sintering.

Hafnium metal chips* (1 by 0.2 by 0.02 mm) and redistilled water (2:1 H_2O/Hf mole ratio†) were sealed, using an electric arc, in a Pt capsule 2.7 mm in ID, 0.15 mm thick, and 35 mm long. The water was prepared with a quartz still from deionized water which had been previously distilled with a copper still. Three capsules in each run were heated at 800°, 900°, 1000°, and 1200°C under 100 MPa for 3 h in a high-gas-pressure apparatus[10] using argon gas as the medium. The run at 900°C was repeated to confirm experimental reproducibility. The desired temperatures and pressures were held within ±20°C and ±2 MPa, respectively, during a run. Their bulk densities were measured by Archimedes' method. The specimens were examined by X-ray powder diffractometry‡ and scanning electron microscopy.§ The HfO_2 crystallite sizes were calculated from the Scherrer formula, using the integral breadth of the deconvoluted diffraction line profile[11] of the $\bar{1}11$ reflection. A spherical shape was assumed for the crystallite, and the effect of lattice strain was disregarded. The 101 reflection of low quartz was used to determine the instrumental line profile.

Sintered monoclinic HfO_2 bodies were obtained by treatment above 900°C. Two capsules among three heated in the first run at 900°C were, however, partly inflated after the release of external pressure on cooling, indicating the presence of residual H_2O and/or H_2 gas. The starting materials in these capsules were converted to HfO_2 powders by hydrothermal oxidation,[8,9] but they were not sintered; in the one remaining capsule, sintered HfO_2 was obtained. In the next run at the same temperature, all three specimens were sintered. At 800°C, no sintering was observed. Therefore, the low-temperature limit for hydrothermal reaction-sintering of monoclinic HfO_2 is near 900°C under the present conditions of pressure and time. In hydrothermal-reaction sintering of ZrO_2, excess H_2O above the stoichiometric amount for oxidation (>5% excess, i.e. $H_2O/Zr=2.1$) prevented densification of ZrO_2. The residual H_2 gas produced by the overall reaction $Hf+2H_2O \rightarrow HfO_2+2H_2$ might also prevent the sintering of HfO_2 in the capsule at 900°C.

The bulk density and crystallite size generally increased with increasing temperature. The bulk densities of sintered bodies ranged from 9.15 g/cm^3 (relative density 0.92) at 900°C to 9.36 g/cm^3 (0.94) at 1000°C and 9.77 g/cm^3 (0.98) at 1200°C; relative density is defined as the ratio of bulk density to the X-ray density (9.94 g/cm^3) for HfO_2 containing 3.9% ZrO_2. Crystallite sizes at these respective temperatures were 73, 89, and 110 nm. One specimen among those obtained at 1200°C, however, had a relative density >1 and contained much smaller crystallites than the other two. This specimen was grayish, similar in color to a sintered ZrO_2 body with a relative density of 1.01.[5] Hafnium hydride, which is an intermediate in the hydrothermal oxidation,[8,9] is the most probable source of this color and increased relative density, although its presence was not observed by X-ray powder diffraction.

Scanning electron micrographs of fracture surfaces of sintered HfO_2 bodies are shown in Fig. 1. The individual grain sizes were ≈0.2 to 0.5 μm for samples sintered at 1000°C and 0.5 to 1.0 μm for those sintered at 1200°C. The surface of the body sintered 900°C appeared to be rough; the grain size would be much less than those in bodies sintered at 1000° and 1200°C, as indicated by the average crystallite sizes. Thus, increases in temperature above 1000°C induce grain growth and densification, which are needed to obtain well-sintered HfO_2 bodies by hydrothermal reaction-sintering.

The hydrothermal reaction-sintering process of HfO_2 is currently under study in relation to hydrothermal oxidation.[8,9]

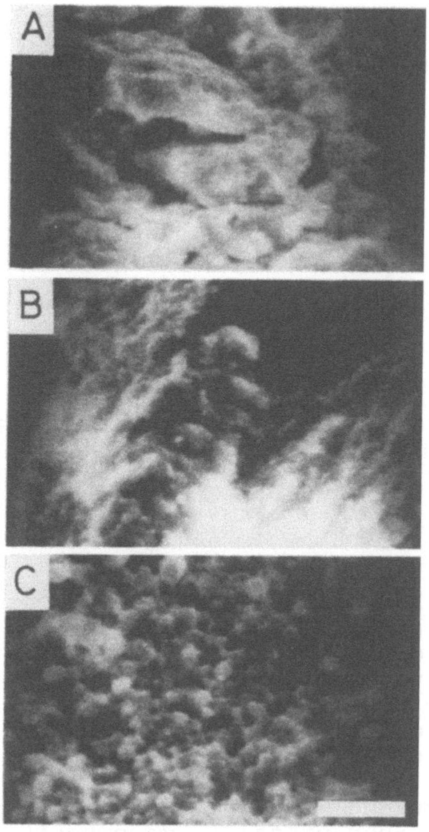

Fig. 1. Scanning electron micrographs of sintered HfO_2 bodies obtained by treatment at (*A*) 900°C, (*B*) 1000°C, and (*C*) 1200°C under 100 MPa for 3 h (bar=5 μm).

*Hf in ppm, 0:545, H:7, N:27, Al<20, C:110, Cl<50, Cu<10, Fe<50, Mg<10, Ti<20, W<15, U:1, Zr:3.9%.
†Mole ratio calculated by accounting for 3.9% Zr in Hf.
‡RU-200, Rigaku Denki Co., Ltd., Tokyo, Japan.
§JSM T-200, JEOL Co., Ltd., Tokyo, Japan.

ACKNOWLEDGMENT

The authors thank Nippon Mining Co., Ltd. for supplying Hf metal.

REFERENCES

[1] C. T. Lynch; pp. 193–216 in Refractory Materials: High Temperature Oxides, Pt. II, Vol. 5. Edited by A. M. Alper. Academic, New York, 1970.
[2] S. Hirano and S. Sōmiya, "Hydrothermal Reaction Sintering of Pure Cr_2O_2," *J. Am. Ceram. Soc.*, **59** [11–12] 534 (1976).
[3] M. Yoshimura and S. Sōmiya, "Fabrication of Dense Non-Stabilized ZrO_2 Ceramics by Hydrothermal Reaction Sintering," *Am. Ceram. Soc. Bull.*, **59** [2] 246 (1980).
[4] M. Yoshimura, S. Song, and S. Sōmiya; pp. 429–32 in Ferrites. Edited by H. Watanabe, S. Iida, and M. Sugimoto. Gakkaishi Kanko Center, Tokyo, 1981.
[5] M. Yoshimura and S. Sōmiya; pp. 455–63 in Advances in Ceramics, Vol. 3. Edited by A. H. Heuer and L. W. Hobbs. The American Ceramic Society, Columbus, OH, 1981.
[6] A. C. D. Chaklader and V. T. Baker, "Reactive Hot-Pressing: Fabrication and Densification of Non-Stabilized ZrO_2," *Am. Ceram. Soc. Bull.*, **44** [3] 258–59 (1965).
[7] H. J. Garrett and R. Ruh, "Fabrication of Specimens from Pure Dense Oxidized Zirconia," *Am. Ceram. Soc. Bull.*, **47** [6] 578–79 (1968).
[8] H. Toraya, M. Yoshimura, and S. Sōmiya, "Preparation of Fine Monoclinic Hafnia Powders by Hydrothermal Oxidation," *J. Am. Ceram. Soc.*, **65** [5] C-72 (1982).
[9] H. Toraya, M. Yoshimura, and S. Sōmiya; unpublished work.
[10] S. Hirano, K. Nakamura, and S. Sōmiya; pp. 418–23 in Proceedings of the 4th Conference on High Pressure, 1974. Edited by J. Osugi. Nippon Butsurikagaku Kai, Kyoto, Japan, 1975.
[11] H. P. Klug and L. E. Alexander, X-Ray Diffraction Procedures, 2d ed., Wiley Interscience, New York, 1974. □

Reprinted from *Proceedings of the First International Symposium on Hydrothermal Reactions*, Japan, 22–26 March 1982, pp. 887–903.

Hydrothermal Reaction Sintering of High Density Sintered Oxides

Shigeyuki Sōmiya

Laboratory for Hydrothermal Syntheses
Research Laboratory of Engineering Materials,
Tokyo Institute of Technology
4259, Nagatsuta, Midori, Yokohama 227, Japan

Abstracts

Hydrothermal Reaction Sintering is one of methods to produce high density oxides ceramics. Oxide formation occurs between metal and high temperature and high pressure solutions at certain temperature, pressure and time.

$$Me + H_2O \longrightarrow MeO_x + H_2$$

Grain size of sintered oxides is smaller than that of starting metal powders. Characteristics of hydrothermal reaction sintering oxides are as follows:

1) Low temperature sintering - able to make sintered body even materials have high vapor pressure, decomposition, and/or transitions.
2) Able to make a very fine grain size body.
3) Able to make high purity body.
4) Able to make high density body.
5) Able to make uniform microstructure.
6) Able to save energy due to low temperature sintering.
7) Able to control valency.
8)

Introduction

There are several methods to make high density sintered body. Hydrothermal reaction sintering is also one of methods to produce high density materials with 1 to 5 m fine grains. The reaction occures between metal and high temperature - high pressure solution to form oxide. Oxide powders were obtained by reactions between solutions or mon-aqueous liquid and metals. Hydrothermal reaction sintering is a kind of reaction hot isostatic processing.
Fine powders and/or new compounds formed by chemical reactions or new compounds formed by chemical reactions between solution or non-aqueous liquid and solids such as metals are pressed by fluid or gas pressure isostatically.
For example, metals put into Pt capsule and reacted with high pressure-high temperature solutions. Then metal oxides and hydrogen gas are formed in the capsule. Hydrogen diffuses out from the capsule, the capsule is pressed by pressure outside the capsule. More exactly speaking, it may be called hydrothermal oxidation reaction sintering.

Characteristics by hydrothermal reaction sintering are as follows:
1) It is possible to sinter oxides which have trasformation, sublimation and high vapor pressure due to low temperature sintering.
2) The sintered bodies are very fine and uniform grains.
3) It is possible to make a high purity body.
4) It is able to make a high density body.
5) It is possible to make a uniform microstructure body.
6) It is able to save energy due to low temperature sintering.
7) It is possible to control valency such as Fe^{+2}, Fe^{+3}.

Processing of hydrothermal reaction sintering is complicated and it is 4 or 6 processing.
1) Crack formation due to reaction between metal and solution. Grain size reduction.
2) Oxide fine powder formation due to reaction between metal and solution.
3) To diffuse out of hydrogen through the capsule.
4) Hot isostatic processing of oxide fine powder.

In some case,
1) Crack formation due to reaction between metal and solution. Grain size reduction.
2) Formation of oxide film by reaction between metal and solution.
3) Formation of hydride by reaction between metal and solution.
4) Decomposition of hydride and formation of fine oxide powder.
5) Diffusion of hydrogen through thin wall of the capsule and escape hydrogen from inside of the capsule.
6) Sintering of fine oxide powder under hot isostatic pressure condition.

Effect of factors such as purities grain size, shape, etc. of starting materials, reaction temperature, pressure, duration time, ratio of metal and water, kinds of salt in solution, and so on hydrothermal reaction sintering are not clear at present time.
Sintered bodied by hydrothermal reaction sintering have characteristic properties, especially grain size is very fine.

2. Experimental method

Metals powders, and/or chips, and redistilled water or solution were hermetically sealed with an electric arc in a 2.6 ∼ 2.7 mm inside diameter platinum capsule 0.15 ∼ 0.2 mm thick and 35 mm long. Schematic illustration of the hydrothermal reaction sintering is shown in Fig. 1.

3. Magnetite

$$3Fe + 4H_2O \rightarrow Fe_3O_4 + 4H_2\uparrow$$

Magnetite formation occured above 250°C under 98 MPa with NaOH, NH_4Cl, KOH solution. Good sintered body was obtained at 550°C, 98 MPa for 1 day. Effect of reaction time on sintering is shown in Fig. 2. Even 1 h oxidation occured.

A good sintered body was obtained after 24 h. Fig. 3 shows SEM photograph of fracture surface of hydrothermal reaction sintering of magnetite. Fig. 4 shows relations between relative density and sintering time with different ratios of $Fe:H_2O$ and capsule. Stoichiometric ratio of $Fe:H_2O$ and using Pt capsule was better than gold as you have seen in Fig. 4.

If there is excess water in capsule, it was very difficult to obtain theoretical density even for 500 hrs. In the case of stoichiometric mixture in the Pt capsule, it is easy to get theoretical density as low as at temperature of 550°C due to high reaction rate.

Fig. 5 shows relation between grain size and sintering time for magnetite. Ratio of Fe:H_2O was (a) 3:4 and (b) 3:6. This picture also shows effect of water in the capsule on grain size. Grain size was increased by excess water. Fig. 6 shows relation between pressure and relative density and also grain size. Grain size and relative density increased by increasing pressure, but not so much effect on grain size and relative density over 98 MPa at 550°C for 72 hrs.

Lattice constant of hydrothermally magnetite which was grown at 500 550°C under 98 MPa, in 10 M NaOH solution of 55% of fill was 0.83964 ± 0.00003 nm.

4. Wüstite[6)]

$$Fe + H_2O \longrightarrow FeO + H_2 \qquad (2)$$

Wüstite was able to produce under 98 MPa above 650°C. Lattice constant was a_o= 0.4305 nm. Grain size received a large influence by water content.

5. Monoclinic ZrO_2[8-10)]

There are many crystal forms for zirconia. It is very difficult to make sintered monoclinic zirconia by high temperature sintering due to transformation temperature of tetragonal ⇄ monoclinic forms.

$$Zr + H_2O \longrightarrow ZrO_2 + H_2 \qquad (3)$$

Reaction between zirconium metal and high temperature-high-pressure water occured at 250°C under 98 MPa to form monoclinic ZrO_2 and also ZrH_x.

Platinum or gold was used as a capsule material. More high temperature above 400°C under 98 MPa, ZrH_x disappeared to form oxide. Starting metal of zirconium devided to small particle by through foarmed bread state reactions then it completed to form oxide. This is one of reasons to produce very fine oxide powders. Between 400°C and 700°C, under 98 MPa, in the capsule, oxide powders and hydrogen coexisted. Above 700°C under 98 MPa, hydrogen passed through thin wall of capsule. Then the capsule was the same condition of hot isostatic processing.

Schematic ilustration of effect of temperature on hydrothermal reaction products is shown in Fig. 7. This figure shows powder formation and sintered bodies.

Sintered materials of high density such as 5.17~5.82 g/cm^3 were obtained at 1000°C under 98 MPa for 3 h. Table 1 shows some results of hydrothermal reaction sintering of monoclinic ZrO_2 for 3 h.

Fig. 8,9,10 shows SEM photograph of fracture surface, starting materials and reactioned powders for high density hydrothermal reaction sintering bodies respectively. Density of this sample was 5.85 g/cm^3. It was 99.0% of relative density, if density of ZrO_2 98% and HfO_2 2% sample was 3.91 which have been obtained by X-ray method. Grain size of this sample was 1~5 μm. If the ratio of H_2O/Zr was more than 2, it is difficult to get high density body due to excess water.

Relation between relative density and ratio of H_2O/Zr and also effect of pressure and temperature on relative density of the hydrothermal reaction sintered bodies are shown in Fig. 11. This figure shows not so much effect of pressure and temperature on relative density and a large influence of ratio of H_2O/Zr upon relative density. Relative density of smple No.28 in Table 1 shows more than 100%. This sample showed gray skin. It may be existence of Zr metal which is very fine and uniform.

Evne size of starting zirconium metal were different, grain sizes of hydrothermal reaction sintered body were 1～5 μm. It is one of the most characteristics of the hydrothermal reaction sintered body.

6. Chromia[11-13)]

Fine powder of Cr_2O_3 was obtained above 500°C under 98 MPa. After hydrogen was passed away from above 800°C, the capsule was received hot isostatic pressure. Then sintered body was obtained. Addition of HCl and/or HNO_3 were effective on oxide formation. SEM photograph of fracture surface of high density hydrothermal reaction sintered body was shown in Fig. 12. Oxidation process for Cr metal was complicated and it is not clear until today.

It, however, is as follows tentatively.

$$Cr \longrightarrow Cr^3 + 3e- \quad (a)$$
$$H_2O + 2e- \longrightarrow H_2 + O^{2-} \quad (b)$$
$$2Cr^3 + 3O^{2-} \longrightarrow Cr_2O_3 \quad (c)$$
$$Cr + H_2O \longrightarrow Cr_2O_3 + H_2O \quad (4)$$

Table 2 and 3 show effect of size of starting metals and solution on yield of Cr_2O_3. Pure water and higher purity of Cr metal did not show good yields. Acidic solution showed good yield.

Fig. 13,14 and 15 show effect of sintering temperature, pressure and reaction time on yield and density of Cr_2O_3.

Density was increased up to 1000°C, but above 1000°C did not show change of apparent density of hydrothermal reaction sintered bodies. Average grain size of Cr_2O_3 was 2～3 μm at 1000°C and 1100°C, and 30～40 μm at 1400°C in the case of 0.01 M HNO_3, 100 MPa, 3 h and 16% fill. Therefore grain growth occured above 1200°C.

Rate of reaction decreased increasing pressure from 100 MPa to 700 MPa and apparent density increased up to 300 MPa, it, however decreased above 300 MPa in the case of 0.01 M HNO_3 at 1000°C, for 3 h experiments. This specimen showed high density of 5.28 g/cm^3 and 2～3 μm average grain size.

One of important factors for hydrothermal reaction sintering is evacuation of water or gas in the capsule. If there is liquid or gas in the capsule, it is very difficult to get high density sintered bodies.

7. Monoclinic HfO_2[14-16)]

$$Hf + H_2O \longrightarrow HfO_2 + H_2 \quad (a)$$
$$Hf + H_2 \longrightarrow H_fH_x \quad (b)$$
$$H_fH_x + H_2O \longrightarrow HfO_2 + H_2 \quad (c)$$

$$H_f + H_2O \longrightarrow H_fO_2 + H_2 \quad (5)$$

Reaction (a) occured only surface of Zr metal, then reaction (b) occured on the surface. Cracks appeared on the surface of the grain. Reaction (a) and (b) occured more and more and also cracks developed more and more. It means H_fH_x is produced more and more.

Cycling of (a), (b), cracks, oxide formation, hydride formation appeared. At 300~400°C under 100 MPa for 3 h, HfO_2 was identified. Small amount of H_fH_x appeared as low as 250°C. At 500°C, H_fH_x was about 60% and H_f metal existed with H_fH_x.
At 700°C, H_fH_x decomposed to oxide and the hydrogen diffused out through wall of Pt wall. Grain size was 0.2~0.5 μm at 900°C, 0.5~1.0 μm at 1000°C and 0.5~1.0 μm at 1200°C respectively.

8. $LaCrO_3$ [10-16]

$LaCrO_3$ was able to synthesis hydrothermally.

$La_2O_3 + Cr + H_2O \longrightarrow LaCrO_3 + H_2$ (6)

Rate of this reaction was very slow even at 1000°C.

$La_2O_3 + Cr(OH)_3 \cdot nH_2O \longrightarrow LaCrO_3 + H_2O$ (7)

Then starting materials were changed from Cr metal and La_2O_3 to La_2O_3 and $Cr(OH)_3 \cdot nH_2O$. Reaction (7) started above 300°C under 100 MPa. Above 700°C, reaction product was only $LaCrO_3$. Grain size was 0.7 ± 2 μm at 700°C under 100 MPa for 3 h.

It is necessary to evacuate gas from the capsule. Then

$La_2O_3 + Cr(OH)_3 \cdot nH_2O + Cr \longrightarrow LaCrO_3 + H_2$ (8)

was made. This reaction sintered material at 1000°C under 100 MPa for 3 h was porous and its grain size was less than 1 μm.

Fig. 16 and 17 show fine grains $LaCrO_3$ by reaction (7) and SEM fracture surface reaction (7) sintered body respectively.

The third experiment was adopted 2 steps.
In the first place, $LaCrO_3$ fine powder was synthesized at 1000°C under 100 MPa.

Make a hole in the capsule for degassing and then seal the capsule again. Then the capsule was put into the high pressure-high temperature apparatus at 1400°C, under 100 MPa for 3 h. Average grain size of this sample was 5 μm. SEM picture was shown in Fig. 18.

9. Summary

Hydrothermal reaction sintering has many characteristics. This process and products have a future's potential.

Acknowledgements

I wish to express Professor Shin-ichi Hirano (Nagoya University) Professor Masahiro Yoshimura, Dr. Hideo Toraya, Mr. Shen-tai Song (Central Iron and Steel Institute, Academy of Science, Peoples Republic of China) and Mr. Hideyuki Kanai for their experimental works and discussion.

References

1) Shin-ichi Hirano and Shigeyuki Sōmiya
"Synthesis and Sintering of Magnetite under Hydrothermal Conditions"
The 12 Symposium on Basic Ceramics, p.33, January 30, 1974

2) Shigeyuki Sōmiya and Shin-ichi Hirano
"Sintering of Magnetite"
Showa 50 nen (1975) Sogokenkyu; Mechanism of Mass Transport in Oxide System p.33 (1976); Chairman Wazo Komatsu

3) Shin-ichi Hirano and Shigeyuki Sōmiya
"Hydrothermal Crystal Growth of Magnetite in the Presence of Hydrogen"
J. Cryst. Growth; 35 [3] [Oct.] 273-278 (1976)

4) Shin-ichi Harano and Shigeyuki Sōmiya
"Crystal Growth of Magnetite under Hydrothermal Condition with Hydrogen"
Forth International Conference on Crystal Growth,
March , 1974 Tokyo; Cellected Abstracts 683-4 (1974)

5) Shin-ichi Hirano and Shigeyuki Sōmiya
"Sintering and Crystal Growth of Magnetite under Hydrothermal Conditions"
Sogo Kenkyu; Diffusion and Sintering Mechanism in Inorganic Materials p.22 (1975)
Chairman Yasumichi Oishi

6) Shin-ichi Hirano and Shigeyuki Sōmiya
"Synthesis and Sintering of Wüstite under Hydrothermal Conditions"
Abstracts for 1974 Annual Meeting of the Ceramic Society
p.116 May 29 (1974)

7) Masahiro Yoshimura and Shigeyuki Sōmiya
"Fabrication of Dense, Non-stabilized ZrO_2 Ceramics by Hydrothermal Reaction Sintering"
J. Am. Ceram. Soc., 59 [2] (Feb.) 256 (1980)

8) Masahiro Yoshimura and Shigeyuki Sōmiya
"Hydrothermal Reaction Sintering of Monoclinic Zirconia"
Advances in Ceramics, Vol.3, Science and Technology of Zirconia
455-463 (1981) Am. Ceram. Soc., Ed. by A. H. Heuer and L. W. Hobbs

9) Masahiro Yoshimura and Shigeyuki Sōmiya
"Synthesis and Sintering of Zirconia Fine Powders by Hydrothermal Reactions From Zirconium Metal and High-Temperature High Pressure Solutions"
Proceedings of the Round Table Conference on Sintering,
September, 1981 417-42, Ed. by Kolar, et al.

10) Shigeyuki Sōmiya and Masahiro Yoshimura
"Hydrothermal Reaction Sintering of Oxides"
Sogo Kenkyu (A) "Formation Processing of Inorganic Materials and its Evaluation" - No. 435044; Chairman Toshiyuki Sata
p. 69-75 (March, 1977)

11) Shin-ichi Harano and Shigeyuki Sōmiya
"Hydrothermal Reaction Sintering of Pure Cr_2O_3"
J. Am. Ceram. Soc., 59 [11.12] 534 (1976)

12) Shigeyuki Sōmiya, Shin-ichi Hirano, Masahiro Yoshimura
Shuichi Itoh and Hideyuki Kanai
"Hydrothermal Reaction Sintering of Cr_2O_3 and Iron Oxides"
Proceedings of the International Symposium on Factors in Densification and Sintering of Oxide and Non-oxide Ceramics,
1978 Japan, 267-277 (1979) Ed. by S. Sōmiya and S. Saito
(Gakujutsu Bunken Fukyu-kai, Tokyo Institute of Technology, Meguro, Tokyo Japan)

13) Shigeyuki Sōmiya, Masahiro Yoshimura, and Hideyuki Kanai
"Hydrothermal Reaction Sintering of Chromic Oxide"
J. Materials Sci. Soc., 19 [3] 176-182 (1982) (Sept.)

14) Hideo Toraya, Masahiro Yoshimura and Shigeyuki Sōmiya
"Hydrothermal Reaction Sintering of Monoclinic HfO_2"
J. Am. Ceram. Soc., 65 [9] 1-159-C-160 (1982)

15) Ibid., Preparation of Monoclinic Hafnia Fine Powders by Hydrothermal Oxidation Method,
J. Am. Ceram. Soc., 65 [5] C - 72, 1982

16) Masahiro Yoshimura, Shen-tan Song and Shigeyuki Sōmiya "Hydrothermal Synthesis and Sintering of $LaCrO_3$" Ferrites, Proceeding of the International Conference September-October 1980 Japan 429-432

17) Shigeyuki Sōmiya, Shin-ichi Hirano, Tamotsu Fukuda and Masahisa Sawada
"Problems in Design and Applications of Internal Heating High Pressure Gas Apparatus up to 1000 kg/cm^2, up to 1500°C" Koatsu Gasu 10 [6] Dec. 16-27 (1973)

Table 1. Selected results of hydrothermal reaction sintering of monoclinic ZrO_2 for 3 h

Sample No.	H_2O/Zr	Deg. of fill (%)	Temp (°C)	Pressure (MPa)	Bulk density (g/cm^3)	Relative density (%)	Avg grain size (μm)
3	1.773	19.3	1000	98	5.85	99.0*	3
5	2.020	15.5	1000	98	5.51	93.2	1
29	2.063	14.8	1000	98	5.17	87.5	0.5
30	2.095	15.0	1000	98	5.38	89.2	0.5
10	2.415	18.4	1000	98	Not sintered		
25	2.038	15.1	1000	196	5.51	93.2	0.3
28	1.985	15.1	1000	294	5.96	100.9†	0.1
8	2.012	22.7	1000	490	5.83	98.7	3
7	2.010	20.4	1000	686	5.77	97.7	3
23	2.038	41.2	1000	686	5.70	96.4	
32	2.016	14.8	1000	98	5.68	96.1	2
40	2.142	14.3	1100	98	Not sintered		
41	2.089	14.2	1200	98	Not sintered		
39	2.091	14.1	1000	490‡	Not sintered		0.5

*With gray skin. †Gray fine body, rare case. ‡Capsule leaked slightly.

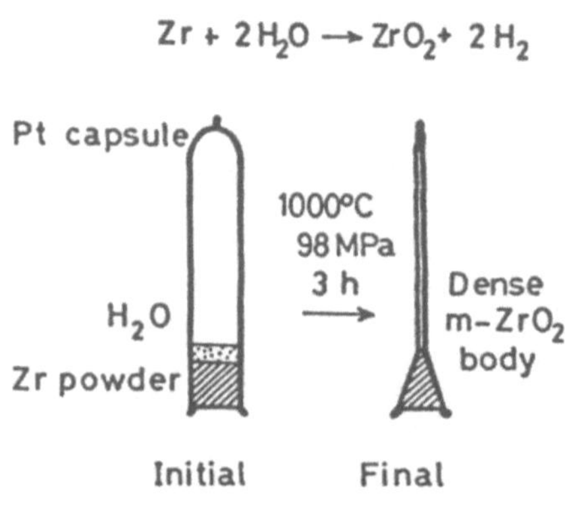

Scematic illustration of the hydrothermal reaction sintering of monoclinic ZrO_2.

Fig. 1

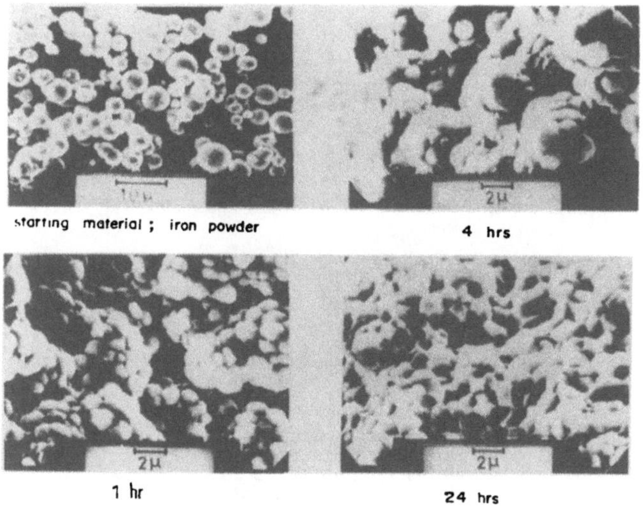

SEM on fracture surface of magnetite sintered under hydrothermal condition 1000 kg/cm², 550°C

Fig. 2

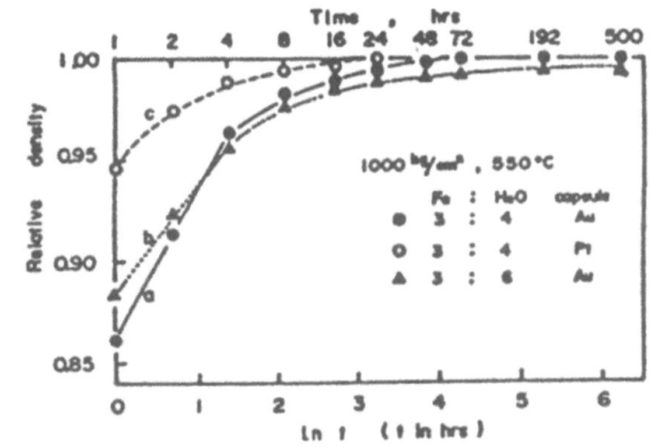

Fig. 4 Relations between relative density and sintering time for hydrothermal reaction sintered magnetite

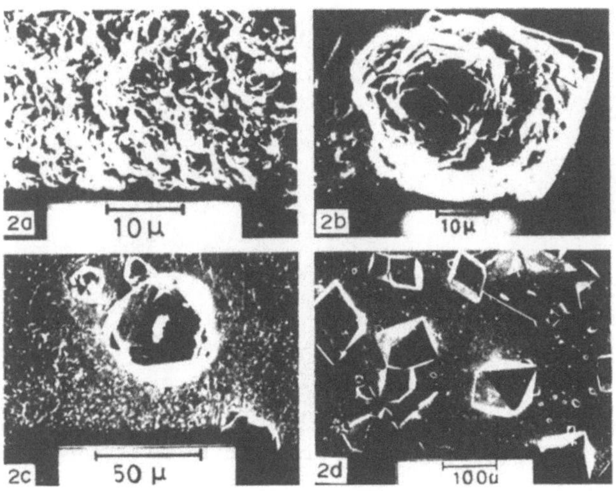

Fig. 3 Scanning electron micrographs of magnetite crystals on inner wall of cut-opened capsule at different position (at temperature as below); 10m NaOH, 1 day, 1000 kg/cm²: (a) 550°C bottom of capsule); (b) 540°C; (c) 525°C; (d) 500°C (top of capsule).

Fig. 3

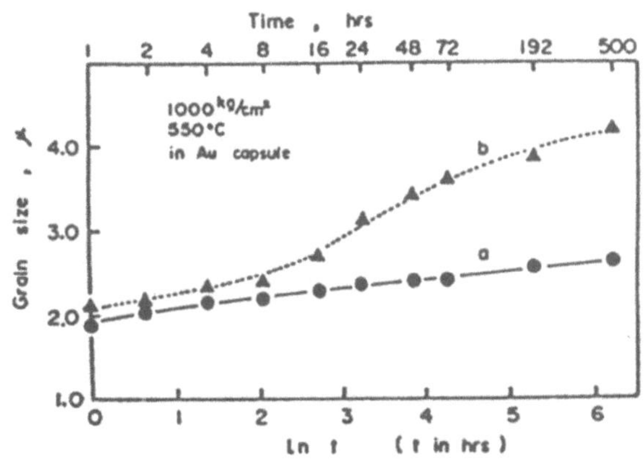

Fig. 5 Relation between grain size of magnetite and sintering time (a) $Fe:H_2O=3:4$ (b) $Fe:H_2O=3:6$

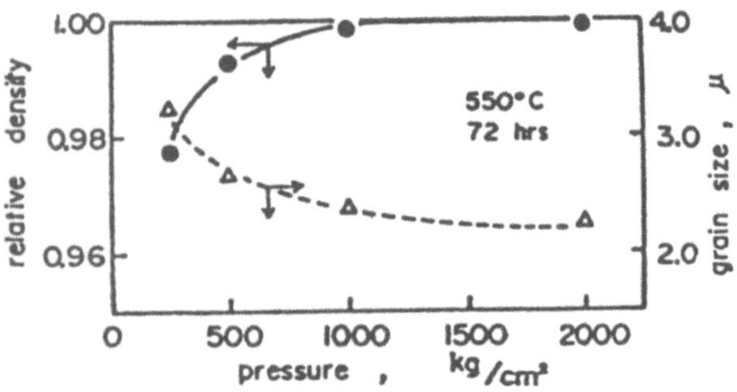

Fig. 6 Relation between pressure and grain size or relative density hydrothermal reaction sintering of magnetite

Fig. 8 SEM Fracture surface of high density monoclinic zirconia
98 MPa, 1000°C 3 hrs. 1-3 μm
d=5.85 g/cm^2, Relative density 99.1%
H_2O/Zr = 1.74

SCHEMATIC ILLUSTRATION OF THE PROCESSES IN THE HYDROTHERMAL REACTION SINTERING OF MONOCLINIC ZrO_2 UNDER 100 MPa IN PT CAPSULE

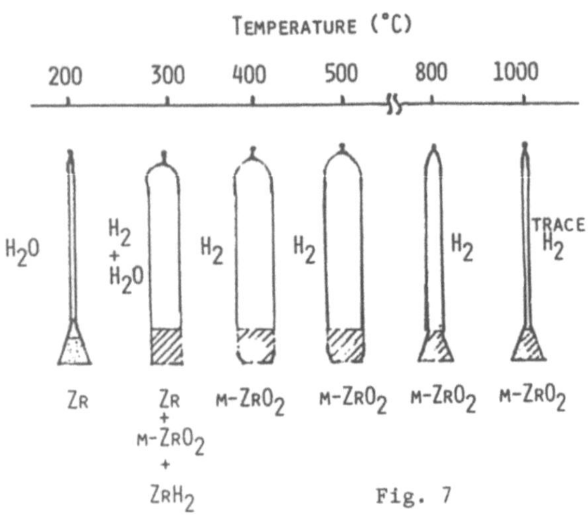

Fig. 7

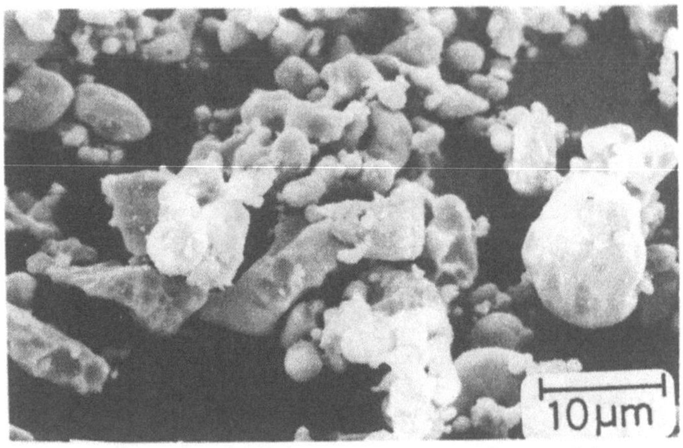

Fig. 9 Starting Zr metal powder

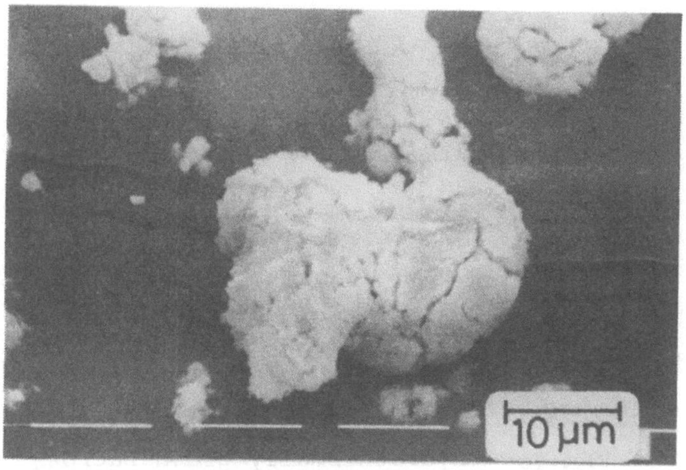

Fig. 10 Zr metal powder reacted with H_2O at 300°C for 3 hrs under 98 MPa.

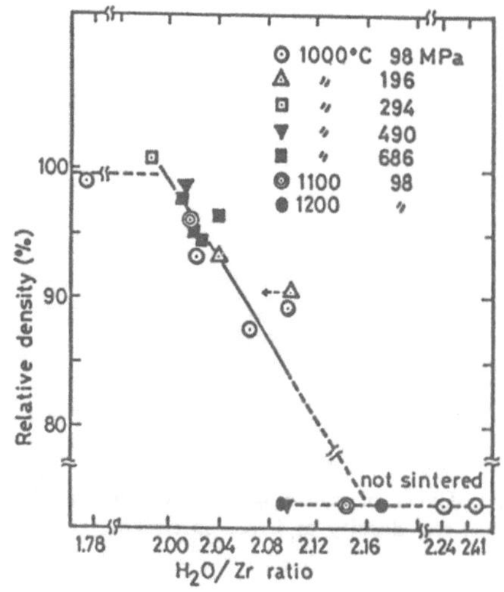

Relative density of m-ZrO_2 ceramics sintered hydrothermally as a function of the H_2O/Zr ratio.

Fig. 11

Fig. 12 SEM Fracture surface of high density hydrothermal reaction sintered Cr_2O_3. 100 MPa, 1000°C for 3 hrs. 0.01 m HNO_3 solution was used. Relative density was 99.1%

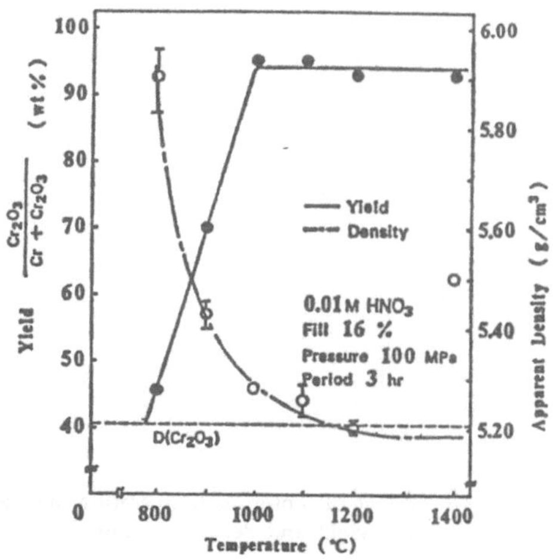

Fig. 13 Effect of sintering temperature on the yield and density of Cr_2O_3 specimen.

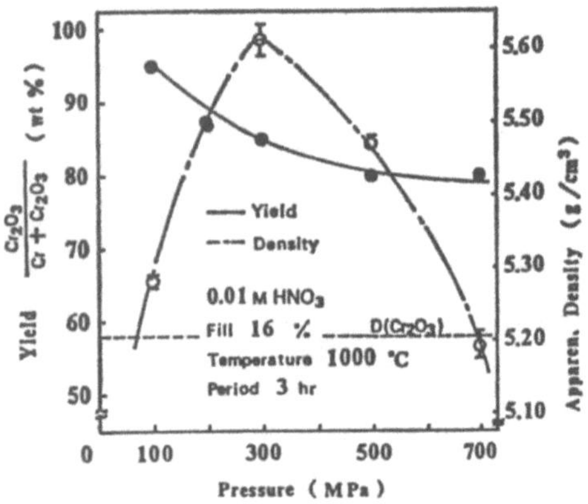

Fig. 14 Effect of sintering pressure on the yield and density of Cr_2O_3 specimen.

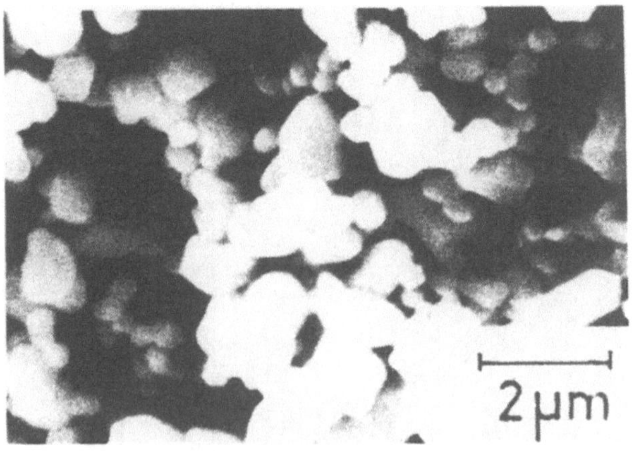

Fig. 16 Hydrothermally grown $LaCrO_3$ fine crystals.
100 MPa, 700°C 3 hrs
Grain size was 0.7 ± 0.2 μm.

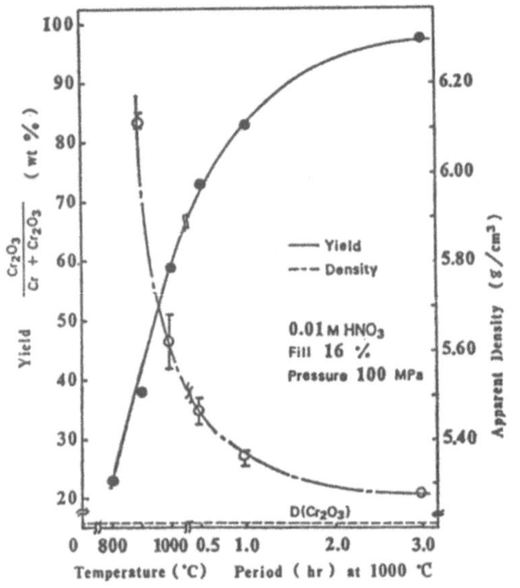

Fig. 15 Effect of sintering temperature and period on the yield and density of Cr_2O_3 specimen.

Fig. 17 SEM fracture surface of $LaCrO_3$ by hydrothermal reaction sintering
100 MPa, 1000°C 3 hrs.

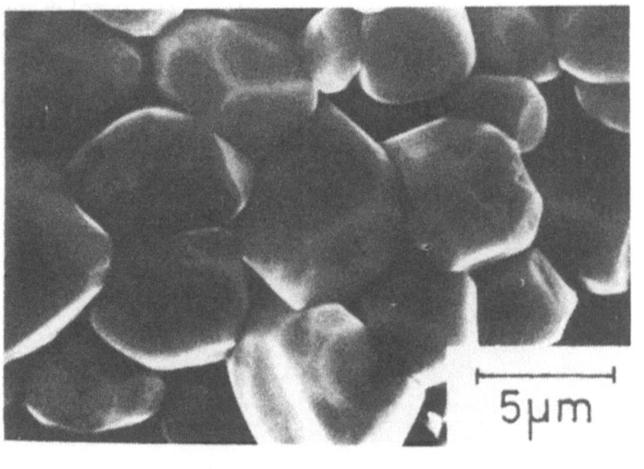

Fig. 18 SEM fracture surface of hydrothermal reaction sintered $LaCrO_3$.
The first at 1000°C and the second at 1400°C 100 MPa 3 hrs.
Grain size was about 5 μm.

Reprinted from G. S. Upadhyaya (ed.), *Sintered Metal–Ceramic Composites* (Elsevier Science Publishers BV, Amsterdam, 1984) pp. 97–105.

Hydrothermal Reaction Sintering of Oxides

Shigeyuki Sōmiya
Laboratory for Hydrothermal Syntheses, Research Laboratory of Engineering Materials, Tokyo Institute of Technology,
4259 Nagatsuta, Midori, Yokohama 227 Japan

ABSTRACT

Metal powders or chips were reacted with high temperature and high pressure solution to form oxides under certain hydrothermal conditions.

$$Me + XH_2O \longrightarrow MeO_X + XH_2$$

Well sintered oxide body was obtained at 1000°C for 3 to 15 h under high pressure of 100 MPa. One of the most characteristics of the hydrothermal reaction sintered body was fine grains as low as 5 μm even starting materials were over 10 to 50 μm.

INTRODUCTION

There are several materials for new high technology ceramics. Hydrothermal reaction sintering is one of the new processing for making high density sintered bodies with fine grains.

Metals or alloys were reacted with high temperature and high pressure water or solutions to form oxides. At high temperatures hydrogen diffuse out from a capsule. The capsule is located at similar condition of HIPing processing. There is reaction between metal and/or alloy and high temperature-high pressure solution.

$$Me + H_2O \longrightarrow MeO_x + H_2 \qquad (1)$$

Then the author obtained high density sintered materials. The author called this processing is the hydrothermal reaction sintering.[1)]

EXPERIMENTAL PROCEDURE

Apparatus used in this experiments was a test-tube-type hydrothermal pressure vessel.[2)] It is able to use 200 MPa at 700°C or 100 MPa at 900°C for a long time by Stellite 25. In the case of more high pressure and temperature, a high gas pressure apparatus was used. This is able to use as high as 1000 MPa and up to 1500°C.[3)] Gold and platinum were used as capsule materials.

RESULTS AND DISCUSSION

Monoclinic ZrO_2

There are several crystal modifications such as monoclinic, tetragonal, cubic, and rhombohedral forms.[4] In many cases, it is difficult to get a sintered body of monoclinic ZrO_2 due to trnasformation. There are reactive hot pressing[5,6,7] for monoclinic ZrO_2 ceramics. The hydrothermal reaction sintering is one of the methods of reaction hot isostatic processing.

Zirconium metal powder, 99.9% pure, containing ≈ 2% Hf, was the starting material. It consisted of particles 1 to 10 μm in diameter (Fig. 1). This zirconium and redistilled water were hermetically sealed with an electric arc in a 2.6-2.7 mm inside diameter platinum capsule 0.15-0.2 mm thick and 35 mm long. The sealed capsule was heated in the pressure vessel under the desire pressure of water or argon gas. Water was used for runs up to 700°C. The density of sintered specimens was measured by a liquid replacement method, where toluene and tungsten wires 0.05 mm in diameter were used to reduce the influence of surface tension.

The reaction between Zr metal powder and supercritical water under 98 MPa was observed above 300°C to yield micrograins of monoclinic ZrO_2. They were much smaller than initial Zr metal grains and formed a crust-like structure, as seen in Fig. 2. The average crystallite size was calculated to be ≈ 24 nm from the measurement of the half-width of the X-ray diffraction peak, using the Warren equation for these monoclinic ZrO_2 grains prepared at 400°C. This formation of active fine grains is one of the important features in the hydrothermal reaction sintering. At 400° to 700°C, the monoclinic ZrO_2 grains of ≈ 24 nm and hydrogen gas existed in the capsules. Since hydrogen gas started to permeate the capsule wall above 700°C, the monoclinic ZrO_2 grains thus formed were compressed and sintered.

Well-sintered bodies having bulk densities of 5.17-5.85 g/cm^3 were obtained at 1000°C after 3 h under 98 MPa (Table 1). Figure 3 shows a typical scanning electron photomicrograph of the fractured surface of the specimen with the maximum density of 5.85 g/cm^3, or relative density of 99.0% compared to the X-ray density of 5.91 g/cm^3 for a 98% ZrO_2-2% HfO_2 solid solution polyhedral zirconia grains, with a grain size of 1-5 μm, are evident in this figure

The presence of more water than is required for stoichiometry in the reaction $Zr + 2H_2O = ZrO_2 + 2H_2$ seemd to prevent effective sintering and densification. Figure 4 shows a scanning electron photomicrograph of the fractured surface of the specimen with the initial mole ratio of H_2O/Zr = 2.095 sintered at 1000°C for 3 h under 98 MPa pressure. The microstructure of this specimen, with a density of 5.38 g/cm^3 and composed of 0.5-1.0 μm fine grains, aggregated inhomogeneously. With more excess water, no well-sintered bodies could be prepared, even at temperatures as high as 1200°C or under 686 MPa pressure. Pressures above 98 MPa did not always promote densification, nor did temperatures of 1100° or 1200°C.

The ratio of water to Zr apparently has the greatest effect on the densification. Figure 5 indicates that the highest densities might be attained in the samples with the stoichiometric amount (or slightly less) of water vs Zr and that excess water, even if the excess is small, prevents densification. These results can be explained by the presence of the residual gas(es) in the capsule during sintering; the compression of the monoclinic ZrO_2 particles by outer gas pressure would stop when the inner pressure became equal to the outer ambient gas pressure. The most abundant residual gas appears to be water, which cannot permeate platinum. For example, the initial mixture of 1 mol Zr and 2.10 mol H_2O (H_2O/Zr = 2.10, a 5% excess of water over stoichiometry), would yield 1 mol ZrO_2 and 0.10 mol H_2O as the final products in the capsule. This water would occupy a volume of 10.62 cm^3 at 1000°C under 100 MPa, which was calculated from the specific volume of water 5.900 cm^3/g observed under the same conditions by Kennedy.[8] That volume corresponds to 50.2% of the molar volume of m-ZrO_2, 21.14 cm^3. That is, the monoclinic ZrO_2 compacts with a porosity of 33.4%, 50.2/(100 + 50.2) and could no longer be compressed by the ambient gas pressure of 100 MPa at 1000°C. This is qualitatively in agreement with the experimental result, in which no well-sintered bodies could be prepared when the initial compositions had H_2O/Zr ratios larger than 2.10.

Quantitative assessment of the density seems to be difficult, particularly near theoretical density, because other gases such as N_2 in the ambient atmosphere might affect the final sintering stage, and the mechanism of this sintering is not completely understood. It seems likely, however, that sintering and grain growth occured simultaneously, because the samples with higher densities are generally composed of grains with larger sizes, as seen in Table 1 with the exception of sample No. 28.

Magnetite

In the case of formation of magnetite, pure iron powder was used as a starting materials. Iron powder and water mixed each other and put into the Pt-or Au capsule. Both ends were sealded by electric arc welding. Then the capsule was put into the bomb to pressurized at the certain temperature for a long time. Ratio of iron powder and water was 3:4 and 3:6. After attained at certain pressure, the capsule was heated by rate of 10?min. Then the capsule was quenched into the water. Sample was taken from the capsule to measure density, grain size, etc. Relations among ratio of iron powder and water, relative density and the capsule materials are shown in Fig. 6. This is the results under 98 MPa and at 550°C. If there is an excess of water, relative density did

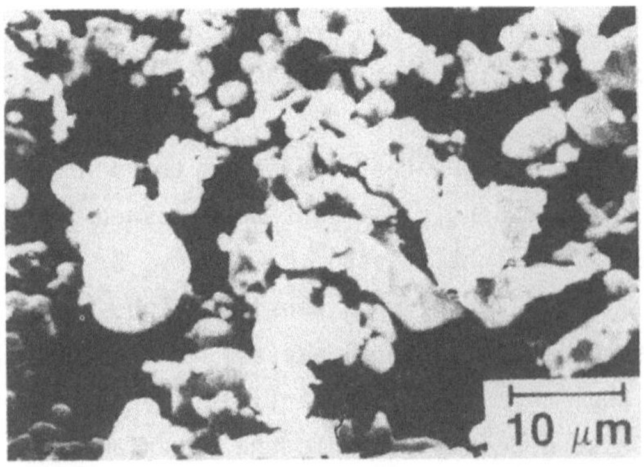

Fig. 1

Starting Zr metal powder.

Fig. 2

Zr metal powder reacted with H_2O at 300°C for 3 h under 98 MPa.

Fig. 3

Fracture surface of m-ZrO_2, specimen sintered hydrothermally; 1000°C, 3 h, 98 MPa, H_2O/Zr = 1.773, fill 19.3%.

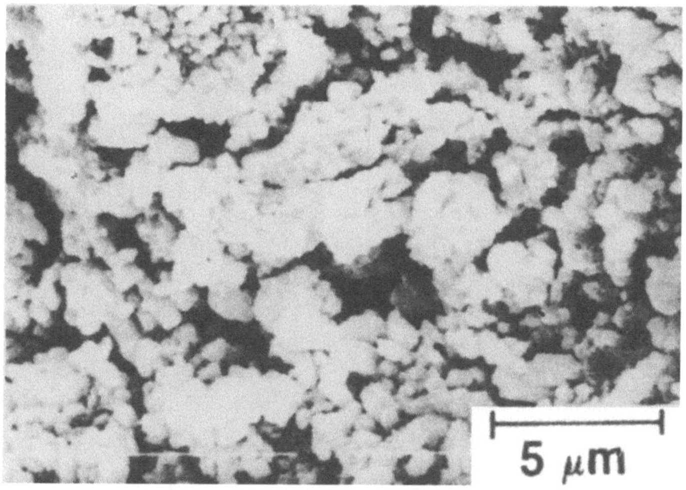

Fig. 4 Fracture surface of m-ZrO_2 specimen sintered hydrothermally; 1000°C, 3 h, 98 MPa, $H_2O/Zr = 2.095$ fill 15.0%.

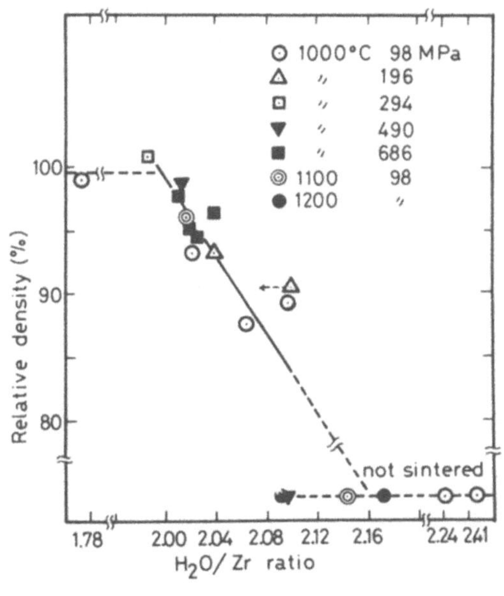

Fig. 5 Relative density of m-ZrO_2 ceramics sintered hydrothermally as a function of the H_2O/Zr ratio.

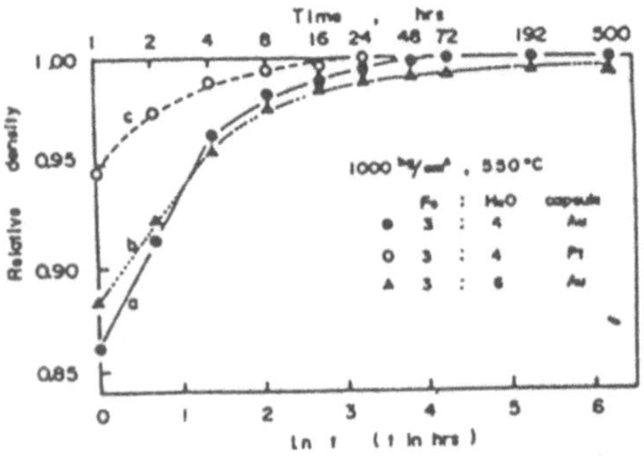

Fig. 6

Relations between relative density, deuration time, the capsulre materials of Pt or Au, and ratio of $Fe:H_2O$ = 3:4 or 3:6.

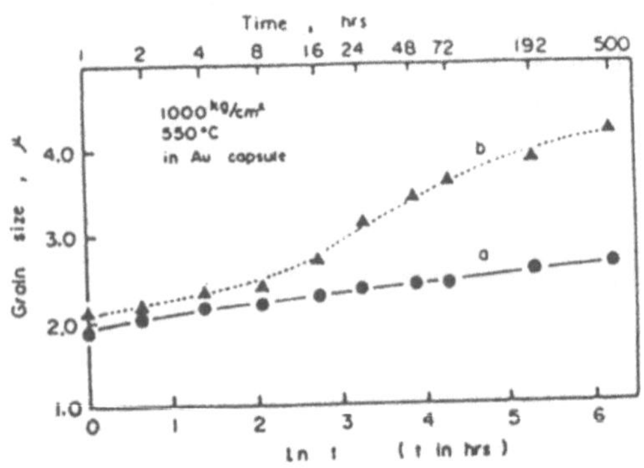

Fig. 7

Relation between grain size and duration time for hydrothermal reaction sintering of magnetite.
a) $Fe:H_2O$ = 3:4
b) $Fe:H_2O$ = 3:6

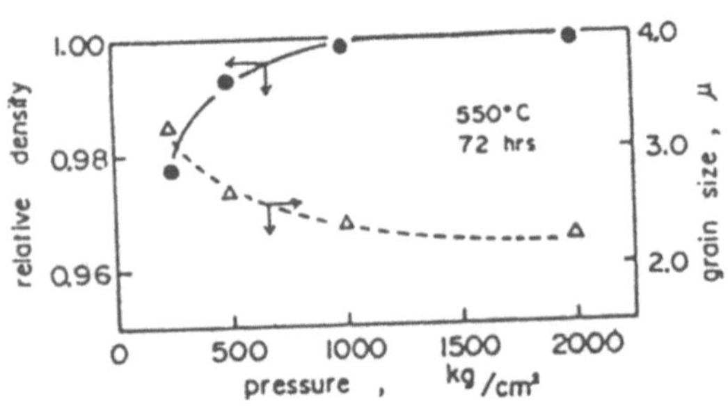

Fig. 8

Relation among relative density, grain size, pressure for hydrothermal reaction sintering of magnetite.

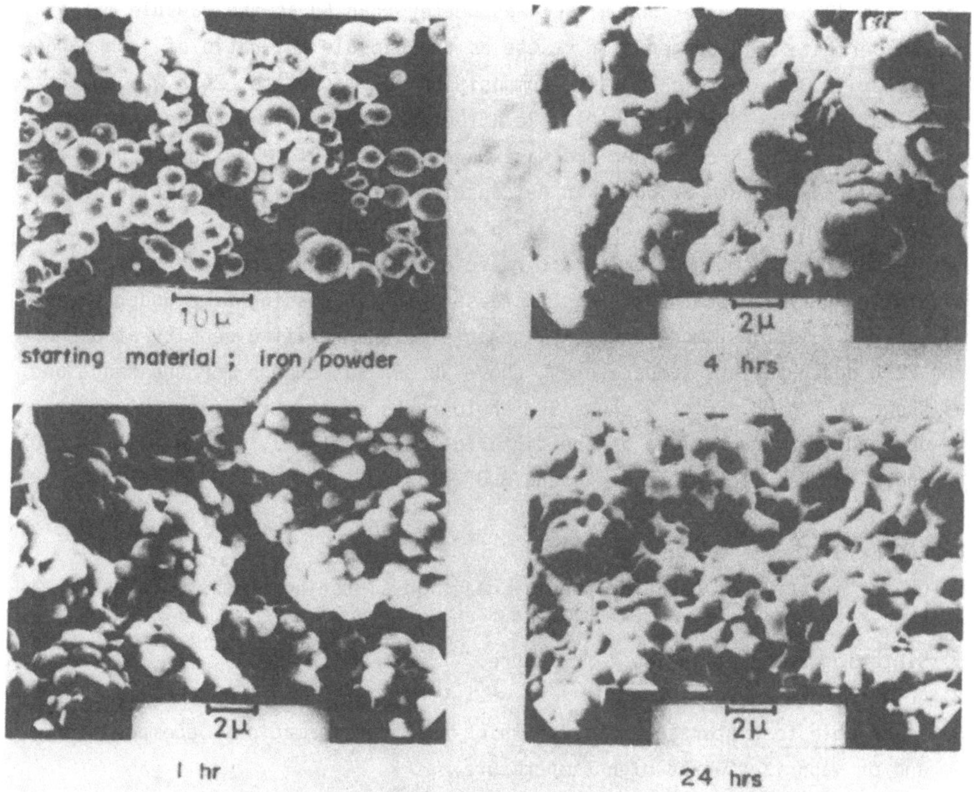

SEM on fracture surface of magnetite sintered under hydrothermal condition 1000 kg/cm^2, 550°C

Fig. 9

not reach 100% even after 500 h. Pt was better than Au as the capsule material due to permeate of hydrogen. If Pt use as the capsule and ratio of iron powder and water is 3:4 of stoichiometric composition for magnetite, 24 h was enough to make 100% relative density bodies. Relations between grain size and sintering time with ratio of iron powder and water are shown in Fig. 7. The results was taken under 98 MPa and at 550°C in Au capsule. Grain size of magnetite in the sintered body increase from 2 μm to 2.5 μm in 3:4 ratio of iron powder and water and from 2 μm to 4 μm in 3:6 ratio of iron powder and water after 6 h of sintering time. Grain size was increased with increasing time under excess of water. Relations among pressure, grain size and relative density at 550°C for 72 h under various temperatures, above 98 MPa was enough for the reaction as shown in Fig. 8. Fig. 9 shows SEM photographs of starting material of iron powder and fracture surface of the sintered bodies after 1, 4, and 24 h at 550 °C under 98 MPa. The ratio of Fe and H_2O was 3:4.

CHARACTERISTICS

Characteristics of the hydrothermal reaction sintering process are as follows:

1) Sintering at relative low temperature.
2) Able to save energy.
3) To be able to sinter the material which has transformation, decomposition, and/or vaporization at high temperature.

As for the sintered body, characteristics are as follows:

1) Fine grain.
2) Isostatically formed.
3) High purity.
4) High density.

ACKNOWLEDGEMENTS

Experimental works done by Professor S. Hirano, Professor M. Yoshimura, Dr. H. Toraya, Mr. S. Itoh, Dr. E. Tani, Mr. K. Yanagisawa, Mr. H. Kanai, Mr. S. Song, Mr. K. Yamazaki, Mr. T. Nakazawa, Mr. S. Kikugawa, Mr. K. Yamazawa, Mr. M. Suzuki, Mr. K. Kim, Mr. H. Ohira, and so on.
Some of figures and tables are reproduced a paper from Advances in Ceramics Vol. 3, 455-463 (1981), American Ceramic Society.

REFERENCES

1 For example
a) S. Hirano and S. Sōmiya, J. Am. Ceram. Soc., 59 [11,12] (1976) 534 pp.
b) S. Sōmiya, S. Hirano, M. Yoshimura, S. Itoh, and H. Kanai, in S. Sōmiya, S. Saito (Eds.), Proceedings of the International Symposium on Factors in Densification and Sintering of Oxide and Nonoxide Ceramics, (1979)

pp. 267-277, Gakujutsu Bunken Fukyukai, Tokyo 1.
c) M. Yoshimura and S. Sōmiya, Am. Ceram. Soc. Bull., 59 [2] (1980) 246 pp.
d) M. Yoshimura and S. Sōmiya, Advances in Ceramics, Vol. 3 (1981) 455-463.
e) H. Toraya, M. Yoshimura and S. Sōmiya, J. Am. Ceram. Soc., 65 [9] (1982) C-159-C-160.
2 R. Roy and O.F. Tuttle, Physics and Chemistry of Earth, Vol. 1 (1956) 138 Pergamann Press.
3
a) S. Hirano, K. Nakamura and S. Sōmiya, in J. Osugi (Ed.), Proceedings of the Fourth International Conference on High Pressure, (1975) pp. 418-423.
b) S. Sōmiya, S. Hirano, T. Fukuda and M. Sawada, Koatsu Gasu, 10 [6] (1973) 368-79.
4 Advances in Ceramics Vol. 3 and 11, Science and Technology of Zirconia, American Ceramic Society.
5 A.C.D. Chaklader and V.T. Baker, J. Am. Ceram. Soc., 47 [12] (1964) pp. 624-627.
6 H.J. Garrett and R. Ruh, J. Am. Ceram. Soc., 47 [6] (1968) pp. 578-579.
7 G.K. Bansal and A.H. Heuer, J. Am. Ceram. Soc., 58 [1,2] (1975) pp. 76-77.
8 G.C. Kennedy, Econ. Geol., 45 (1950) 629 pp.
9 S. Sōmiya and S. Hirano, in W. Komatsu (Ed.), Mass Transport of Mechanism on Oxides, (1976) p.33.

Table 1.

Selected results of hydrothermal reaction sintering of monoclinic ZrO_2 for 3 h

Sample No.	H_2O/Zr	Deg. of fill (°c)	Temp. (°C)	Pressure (MPa)	Bulk density (g/cm³)	Relative density (%)	Avg grain size (μm)
3	1.773	19.3	1000	98	5.85	99.0*	3
5	2.020	15.5	1000	98	5.51	93.2	1
29	2.063	14.8	1000	98	5.17	87.5	0.5
30	2.095	15.0	1000	98	5.38	89.2	0.5
10	2.415	18.4	1000	98	Not sintered		
25	2.038	15.1	1000	196	5.51	93.2	0.3
28	1.985	15.1	1000	294	5.96	100.9†	0.1
8	2.012	22.7	1000	490	5.83	98.7	3
7	2.010	20.4	1000	686	5.77	97.7	3
23	2.038	41.2	1000	686	5.70	96.4	
32	2.016	14.8	1000	98	5.68	96.1	2
40	2.142	14.3	1100	98	Not sintered		
41	2.089	14.2	1200	98	Not sintered		
39	2.091	14.1	1000	490‡	Not sintered		0.5

*With gray skin. †Gray fine body, rare case. ‡Capsule leaked slightly.

Reprinted from *Materials Research Society Symposium Proceedings*, **24** (1984) 255–271.

Hydrothermal Preparation and Sintering of Fine Ceramic Powders

Shigeyuki SŌMIYA
Laboratory for Hydrothermal Syntheses, Research Laboratory of Engineering Materials, Tokyo Institute of Technology, 4259 Nagatsuta, Midori, Yokohama 227, JAPAN

ABSTRACT

Hydrothermal preparation is one of the methods to make fine ceramic powders. There are: (1) hydrothermal oxidation, (2) hydrothermal precipitation, (3) hydrothermal synthesis, (4) hydrothermal decomposition, (5) hydrothermal crystallization, etc. Starting materials for hydrothermal oxidation is mainly metal. Reaction between metal and high temperature-high pressure solution occurs to form oxide. Therefore, starting materials for very fine powders are: (1) metal powders or chips, (2) gel, solutions, etc., (3) oxides and solutions, (4) solids, and (5) hydrous material. This paper describes hydrothermal crystallization and oxidation. Fine powders of ZrO_2, Al_2O_3, HfO_2, and Al_2O_3-ZrO_2 are described.

Sintered bodies are also able to get under hydrothermal conditions with reactions. Hydrothermal reaction sintered bodies of iron oxides, Cr_2O_3, ZrO_2 and HfO_2 were obtained under 100 MPa above 1000°C with 1-5 μm grains.

PART I. HYDROTHERMAL PREPARATION OF FINE POWDERS

There are several methods to prepare fine powder for ceramics. These are grain size reduction and grain size build up. As for grain size reduction, we used to use ball milling, attrition milling, vibrating milling, etc. Relating to grain size build up processing, we had many chemical methods like thermal decomposition, by spray, flame, plasma, etc. freeze drying, precipitation and hydrolysis. These are shown in Table 1.

Hydrothermal preparation is one of the reaction products between solution-solution or solution-gas reaction which occur under hydrothermal condition. Hydrothermal reactions for preparation of fine powders are shown in Table 2. The author's laboratory studies on hydrothermal preparation, especially hydrothermal crystallization and hydrothermal oxidation, are discussed.

APPARATUS

Apparatus used in this experiment were a test tube type of hydrothermal equipment described by Roy and Tuttle[1] and a high temperature-high pressure vessel up to 1500°C, 10 Kbar.[2–4]

1.1. Hydrothermal Oxidation

1) ZrO_2 [5–9]

The starting material of zirconium metal powder is shown in Fig. 1. Grain size was 1 to 30 μm. As for solutions, pure water, a 59 wt% calcium nitrate solution and a 41 wt% yttrium nitrate solution were made from 99.0% $Ca(NO_3)_2 \cdot 4h_2O$, 99.9% $Y(NO_3)_3 \cdot 6H_2O$, $Ca(OH)_2$ were used for stabilizing agents.

Appropriate amounts of Zr and redistilled water or the above solution, or $Ca(OH)_2$ powder and H_2O, were hermetically sealed by arc welding in a platinum capsule 2.7 mm ID, 0.15 mm thick and 35 mm long. (See Fig. 2) The sealed capsule was heated in a test-tube-type pressure vessel under the water pressure of 100 MPa. The pressure was measured by a calibrated Heise bourdon gauge and controlled within ± 0.2 MPa. The temperature was measured by Platinel thermocouples calibrated against the m.p. of Zn (419.5°C) and controlled within ± 3°C.

After the pressure vessel was quenched into a cold water, the capsule was weighed to check the leakage of the content. The solid products were examined by X-ray powder diffraction using a diffractometer with a rotating Cu anode, by scanning electron microscopy and by transmission electron microscopy.

The reaction between Zr metal and H_2O did not occur up to 200°C under 100 MPa, but did above 250°C. The solid products in the capsule at 300°C contained monoclinic zirconia, zirconium hydride and unreacted zirconium metal (Fig. 3). The zirconium hydride was formed by the reaction between Zr and high pressure hydrogen produced by the reduction or water. As the hydride decomposed above 400°C, sole monoclinic zirconia was produced, which indicated the complete reaction of the eq. (1).

$$Zr + 2H_2O \longrightarrow ZrO_2 + H_2 \qquad (1)$$

The average crystallite size was calculated to be 24 nm for the monoclinic zirconia produced at 400°C based upon the line broadening of the X-ray diffraction. This was confirmed by TEM observation. Scanning electron micrograph, Fig. 4, revealed that the monoclinic zirconia micro-grains formed a crust-like structure on an original metal grain.

The growth of crystalline was oberved above 700°C, then well-crystallized monoclinic zirconia ($>$100 nm) was formed at 1000°C as suggested in Fig. 3.

It is noteworthy that no tetragonal zirconia could be produced in the reaction between Zr and H_2O in the present study, even though the average crystallite size was less than 30 nm. Topotactic nucleation on the zirconium metal surface and/or strain-free growth might not allow to crystallize metastable tetragonal zirconia in the hydrothermal oxidation processes.

The heating of the mixture 0.15 $Ca(OH)_2$ + 0.85 Zr + 1.55 H_2O (=$Zr_{0.85}Ca_{0.15}O_{1.85}$ + 1.70 H_2) at 500°C or 600°C under 100 MPa did not produce cubic zirconia but the mixture of $Ca(OH)_2$ and m-ZrO_2, whereas the heating at 700°C under 100 MPa yielded c-ZrO_2 and $CaZrO_3$ in addition to m-ZrO_2.
The products contained also m-ZrO_2, $Ca(OH)_2$, ZrH_x and ZrN by reaction between Zr metal and $Ca(OH)_2$.
The reaction between Zr and H_2O has never resulted in the formation of either cubic zirconia or tetragonal zirconia even though Zr was excess against H_2O; i.c. Zr/O $>$ 0.5. The formation of the cubic zirconia by the reaction between Zr and $Ca(NO_3)_2$ solution should be resulted from the influences of calcium and/or nitrogen. The lattice parameter of the cubic zirconia phase ranged a=0.507~0.510 nm. At 700°C under 100 MPa for 3 h, of the mixture of Zr and NH_4NO_3 solutions could produce cubic zirconia phase though the yield was less than the case of $Ca(NO_3)_2$ solutions. The cubic zirconia containing nitrogen had a lattice parameter around a=0.508 nm. The cubic zirconia phase produced by the reaction between Zr and $Ca(NO_3)_2$ solutions seems to be stabilized both by calcium and nitrogen. The cubic phase had an average crystallite size of 18 nm, but sometimes contained large cubic particles with about 250 nm as seen in Fig. 5. Similar reaction between Zr and $Y(NO_3)_3$ solutions with excess Zr contents could produce small amounts of cubic zirconia phase at 700°C under 100 MPa.

This study indicates that the hydrothermal oxidation method can produce fine grained oxides, pure and doped.

2) Al_2O_3 [10]

Al metal powder was oxidized to α-Al_2O_3 under 100 MPa for 3 h above

500°C. α-Al_2O_3 was formed above 500°C, under 100 MPa for 3 h. The reactions are shown in Fig. 6. Fig. 7 shows SEM photographs of reaction products between Al metal powder and high-temperature high-pressure solution. Grain size by X-ray diffraction patterns is shown in Table 3.

3) HfO_2 [11,12,13,14]

Hafnium metal chips (1 by 0.2 by 0.02 nm) and redistilled water (2:1 H_2O/Hf mole ratio) were Pt capsule. The water used as a reaction reagent was prepared with a quartz distiller from deionized water which had been previously distilled with a copper distiller. The sealed capsules were heated in a test-tube-type pressure vessel under 100 MPa for 3 h. In addition to the closed system runs, open system experiments were also performed without sealing the capsule.

In the closed system, Hf metal chips scarcely reacted with high-temperature high-pressure water at 300° and 400°C. At 500°C half of the Hf metal chips were changed into Hf hydride and a small amount into monoclinic HfO_2. The reaction proceeded rapidly above 600°C, where most of the product consisted of monoclinic HfO_2 powder with a small amount of Hf hydride. At 700°C HfO_2 was produced with only a trace amount of Hf hydride remaining. Fig. 8 shows Hf metal chips used as starting materials and monoclinic HfO_2 powder obtained by hydrothermal oxidation at 700°C in the closed system. The crystallite sizes of the formed HfO_2 were determined to be ≈33 nm by measuring the integral breadth of the X-ray 111 reflection. The particle sizes ranged from 1 to 10 μm as determined by SEM where the large particles are aggregates of small ones.

The reaction mechanism of Hf metal chips with high-temperature high-pressure water is expected to be similar to that of Zr metal powder.

$$Hf + 2H_2O \longrightarrow HfO_2 + 2H_2 \quad (2)$$

$$Hf + \frac{x}{2}H_2 \longrightarrow HfH_x \quad (3)$$

$$HfH_x + H_2O \longrightarrow HfO_2 + 2\frac{x}{2}H_2 \quad (4)$$

The transformation from Hf to HfO_2, however, proceeded at a much higher temperature than that from Zr to ZrO_2 which started above 250°C and was complete above 400°C under 100 MPa for 3 h. The Hf reaction proceeded slowly so that it was possible to observe the dependence of the hydrothermal reaction on temperature-pressure-time more distinctly than that for the Zr reaction which was completed almost immediately. (Fig. 9)

In contrast to the closed system, in the open system Hf metal chips scarcely reacted at 500°C. However, the reaction proceeded very rapidly at 600°C, faster than that of the corresponding run in the closed system. These differences are thought to result from the differences in fugacities in the capsule's atmosphere. Since the amount of H_2O would decrease during reaction in the closed system while pressurized excess H_2O would continue to surround the sample in the open system, the fugacity relation, fH_2O, would be much smaller in the former and larger in the latter system. The open system appears to be more effective than the closed one for the preparation of oxide fine powders by hydrothermal oxidation. The hydriding of Hf at lower temperatures seems to induce the pulverization of the Hf metal chips during the hydrothermal processing.

4) Al_2O_3-ZrO_2 [14,15,16]

Alloys of $AlZr_3$ and Zr_5Al_3 were produced by an xenon arc image furnace (UF-10001). This furnace is able to reach as high as 3000°C in a second and to cool down to room temperature very rapidly. Therefore Al and Zr are mixed in atomic scale in the melting sphere.

Al in alloy was oxidized under hydrothermal conditions then expansion occured. This caused making cracks in the alloy surface. Reactions for oxide formation occured above 500°C not below 500°C under 100 MPa for $ZrAl_3$ and above at 500°C not below 400°C for Zr_5Al_3. Tet. ZrO_2 appeared at 600°C and 700°C under 100 MPa. Reaction products of $ZrAl_3$ and Zr_5Al_3 were α-Al_2O_3 mon. ZrO_2 and tet. ZrO_2. Amount of tet. ZrO_2 decreased increasing heating time and also temperature. These are shown in Fig. 10.

Crystallite sizes of the reaction products under 100 MPa for 3 h are shown in Table 4. Both addition of Zr to Al and Al to Zr showed decreasing crystallite sizes of α-Al_2O_3 and ZrO_2, especially α-Al_2O_3. Tet. ZrO_2 and mon. ZrO_2 in $ZrAl_3$ and Zr_5Al_3 were bellow 24 nm. These are very small value.

1-2 Hydrothermal Crystallization

ZrO_2 [17,18]

The starting material was amorphous hydrous zirconia prepared from Hf-free $ZrCl_4$ solution, precipitated with 3N NH_4OH, filtered, dispersed in distilled water, and refiltered. After four dispersion and filtration processes, the precipitate was dried at 120°C for 48 h. The water content of this material was determined by TG-DTA. Precipitates calcined at 240°, 420°, 620°, or 820°C in air for 60 min at a heating rate of 10°C/min were also used as starting materials. Distilled water and solutions of 8 wt% KF; 30 wt% NaOH; 7, 15 and 30 wt% LiCl; and 10 wt% KBr were used as mineralizers. The Pt or Au tube was filled with the starting material and solution and treated under 100 MPa at 200° to 600°C for 24 h using the test-tube-type pressure vessel; the vessel was then quenched into cold water. The products were washed in distilled water and then fine powders were obtained.

The phases and crystallite sizes produced by hydrothermal crystallization at 300°C for 24 h with several mineralizers are shown in Table 5. Both tet. and mon. ZrO_2 formed when H_2O and LiCl and KBr solutions were used, whereas only the mon. ZrO_2 phase was produced with KF and NaOH solutions. The crystallization process clearly depends on the mineralizer solution. The crystallite size, however, was almost the same ($\approx$ 17 nm) for the monoclinic and tetragonal phases in all products except those obtained with NaOH solution.

Fig. 11 shows fine mon. and tet. ZrO_2 particles formed by hydrothermal treatment at 100 MPa and 300°C for 24 h with 15 wt% LiCl solution. The particles are well shaped and isolated from each other. The pure mon. ZrO_2 particles produced by hydrothermal treatment of the same starting materials with 8 wt% KF solution at 200° to 500°C were similarly shaped and isolated. Thus, hydrothermal crystallization yields well-shaped, isolated fine particles, in contrast to the aggregated particles resulting from calcination of amorphous hydrous zirconia in air.

Fig. 12 shows the variation with temperature of crystallite size and tet.ZrO_2 content for the products of hydrothermal crystallization for 24 h with H_2O and 15 wt% LiCl solution. The crystallite sizes of mon. and tet. ZrO_2 were $\approx$13 and $\approx$15 nm, respectively, below 300°C and increased with treatment temperature above 300°C. Mon. ZrO_2 crystallites were $\approx$ 25 nm in diameter at 500°C, tet. ZrO_2 was not produced when H_2O was used, but 30 nm tet. crystallites resulted with LiCl solution at 500°C. At 600°C, mon. ZrO_2 Crystallites $\approx$32 nm in size were found; no tet. ZrO_2 produced with either solution. Therefore, the maximum crystallite size of tet. ZrO_2 produced by hydrothermal crystallization was $\approx$30 nm.

The mechanism of tet. ZrO_2 formation is thought to be topotactic crystallization on the unclei in the amorphous zirconia.

PART II HYDROTHERMAL REACTION SINTERING

1) Mon. ZrO_2 [9,19,20]

According to the equation (1), hydrogen formed with ZrO_2. Schematic illustration showed in Fig. 13. Hydrogen gas started to permeate the capsule wall above 700°C, the mon. ZrO_2 grains thus formed were compressed and sintered.

Well-sintered bodies having bulk densities of 5.17-5.85 g/cm^3 were obtained at 1000°C after 3 h under 98 MPa. Fig. 14 shows a typical scanning electron photomicrograph of the fractured surface of the specimen with the maximum density of 5.85 g/cm^3, or relative density of 99.0% compared to the X-ray density of 5.91 g/cm^3 for a 98% ZrO_2-2% HfO_2 solid solution. Polyhedral zirconia grains, with a grain size of 1-5 μm, are evident in this fugure.

The presence of more water than is required for stoichiometry in the reaction $Zr + 2H_2O = ZrO_2 + 2H_2$ seemed to prevent effective sintering and densification. Fig. 15 shows a scanning electron photomicrograph of the fractured surface of the specimen with the initial mole ratio of H_2O/Zr = 2.095 sintered at 1000°C for 3 h under 98 MPa pressure. The microstructure of this specimen, with a density of 5.38 g/cm^3 and composed of 0.5-1.0 μm fine grains, aggregated inhomogeneously. With more excess water, no well-sintered bodies could be prepared, even at temperatures as high as 1200°C or under 686 MPa pressure. Pressures above 98 MPa did not always promote dinsification, nor did temperatures of 1100° or 1200°C.

The ratio of water to Zr apparently has the greatest effect on the densification. Fig. 16 indicates that the highest densities might be attained in the samples with the stoichiometric amount (or slightly less) or water vs Zr and that excess water, even if the excess is small, prevents densification.

2) Cr_2O_3 [21,22]

Hydrothermal reaction to form chromium oxide (Cr_2O_3) occured above 800°C under 100 MPa for 3 h. Purity of Cr metal was 90 to 99.3% of 3 kinds and grain sizes were 20-40, or less than 4 μm. Yields of Cr_2O_3 was 2 to 91% under 100 MPa at 1000°C for 3 h in the case of pure water and Cr metal. Yield of Cr_2O_3 with different acidic solution is shown in Table 6. Using dilute HCl and HNO_3 solutions were effective on reproducible results and high yields of Cr_2O_3 under 100 MPa at 1000°C for 3 h. Even addition of H_2CO_3, yield of Cr_2O_3 were 93% and 90%. Rate of reaction increased above 800°C to 1000°C, but it was not increased above 1000°C. This is shown in Fig. 17. Effect of sintering temperature and period on the yield and density of Cr_2O_3 specimen is shown in Fig. 18. From 800°C to 1000°C the yield was increased rapidly. At 1000°C, increasing period was from 1 h to 3 h density was getting constant and its value was 95% of relative density. SEM photograph of fracture surface of Cr_2O_3 by hydrothermal reaction sintering is shown in Fig. 19. Grain size of this body was 2-3 μm and was uniform. Density was 5.28.

3) Mon. HfO_2 [23,24]

Hafnia metal chips (1 by 0.2 by 0.02 mm) and redistilled water (2:1 H_2O/Hf mole ratio) were sealed, using an electric arc, in a Pt capsule. The water was prepared with a quartz still from deionized water which had been previously distilled with a copper still. Three capsules in each run were heated at 800°, 900°, 1000°C and 1200°C under 100 MPa for 3 h in a high-gas pressure apparatus using argon gas as the medium.

Sintered monoclinic HfO_2 bodies were obtained by treatment above 900°C. Two capsules among three heated in the first run at 900°C were, however, partly inflated after the release of external pressure on cooling, indicating the presence of residual H_2O and/or H_2 gas. The starting materials in these capsules were converted to HfO_2 powders by hydrothermal oxidation, but they were not sintered; in the one remaining capsule, sintered HfO_2 was obtained. In the next run at the same temperature, all three specimens were sintered. At 800°C, no sintering was observed. Therefore, the low-temperature limit for hydrothermal reaction-sintering of monoclinic HfO_2 is near 900°C under the present conditions of pressure and time. In hydrothermal reaction sintering of ZrO_2, excess H_2O above the stoichiometric amount for oxidation (> 5% excess, i.e. $H_2O/Z \geq 2.1$) prevented densification of ZrO_2. The residual H_2 gas produced by the overall reaction $Hf + 2H_2O \longrightarrow HfO_2 + 2H_2$ might also prevent the sintering of HfO_2 in the capsule at 900°C.

The bulk density and crystallite size generally increased with increasing temperature. The bulk densities of sintered bodies ranged from 9.15 g/cm^3 (relative density 0.92) at 900°C to 9.36 g/cm^3 (0.94) at 1000°C and 9.77 g/cm^3 (0.98) at 1200°C; relative density is defined as the ratio of bulk density to the X-ray density (9.94 g/cm^3) for HfO_2 containing 3.9% ZrO_2. Crystallite sizes at these respective temperatures were 73, 89 and 110 nm. One specimen among those obtained at 1200°C, however, had a relative density > 1 and contained much smaller crystallites than the other two. This specimen was grayish, similar in color to a sintered ZrO_2 body with a relative dinsity of 1.01. Hafnium hydride, which is an intermediate in the hydrothermal oxidation, is the most probable source of this color and increased relative density, although its presence was not observed by X-ray powder diffraction.

Scanning electron micrographs of fracture surfaces of sintered HfO_2 bodies are shown in Fig. 20. The individual grain sizes were ≈0.2 to 0.5 μm for samples sintered at 1000°C and 0.5 to 1.0 μm for those sintered at 1200°C. The surface of the body sintered 900°C appeared to be rough; the grain size would be much less than those in the bodies sintered at 1000° and 1200°C, as indicated by the average crystallite sizes. Thus, increases in temperature above 1000°C induce grain growth and densification, which are needed to obtain well-sintered HfO_2 bodies by hydrothermal reaction-sintering. Table 7 shows summary of the results for hydrothermal reaction sintering at my laboratory.

PART III CHARACTERISTICS

Characteristics of fine powders prepared hydrothermally are as follows:
1) 20 30 nm grain size in many cases
2) Well crystallized growth powder
3) Not agglomerate in many case
4) Narrow distribution of grain size
5) No stress
6) Preparation at relative low temperature
7) High purity
8) Able to make doped powder

Characteristics for hydrothermal reaction sintering are as follows:
1) Able to sinter at relative low temperature
2) Able to save energy
3) Able to sinter the materials which have high vapor pressure, phase transformation and decomposition at high temperature
4) Able to get fine and uniform sintered bodies
5) Able to get high density bodies
6) Able to get isostatic sintered bodies
7) Able to get high purity bodies

Advantages of hydrothermal processing are as follows:
1) Use of a wide variety of feedstocks
2) Higher quality products
   Fine, narrow grain size distribution,
   Not agglomerate
3) New products

4) No precipitants
5) Pollution minimized

ACKNOWLEDGEMENT

Experimental works were done by Professor S. Hirano, Professor M. Yoshimura, Dr. H. Toraya, Mr. S. Itoh, Dr. E. Tani, Mr. K. Yanagisawa, Mr. H. Kanai, Mr. S. Song, Mr. K. Yamazaki, Mr. T. Nakazawa, Mr. S. Kikugawa, Mr. K. Yamazawa, Mr. H. Suzuki, Mr. K. Kim, Mr. H. Ohira and Mr. Y. Fushii and so on. Pictures, Figures and Tables are reproduced from Publications of American Ceramic Society, Elsevier Scientific Publishing Co., Yogyo Kyokai, Zairyo Kagakukai, etc.

REFERENCES

1. Rustum Roy and O.F. Tuttle, in:Physics and Chemistyr of Earth 1 L.H. Ahrens, K. Pankaina and S.K. Runcorn, eds. (Pergamon Press, Oxford 1955) pp. 138 - 180
2. Shigeyuki Sōmiya, Shin-ichi Hirano, Tamotsu Fukuda and Masahisa Sawada, Koatsu Gasu 10 16 - 27 (1973)
3. Ibid. Koatsu Gijutsu 12 15 - 22 (1974)
4. Shin-ichi Hirano, Kazuo Nakamura and Shigeyuki Sōmiya in: Proceedings Fourth Int. Conference on High Pressure, Jiro Osugi, ed. (Nippon Butsuri Kagakukai, Kyoto 1975) pp. 418 - 421
5. Shigeyuki Sōmiya and Masahiro Yoshimura, Kinozairyo 3 [4] 51 - 58, [5] 58 - 62 (1983)
6. Shigeyuki Sōmiya: in Zirconia Ceramics I, Shigeyuki Sōmiya, ed. (Uchida Rokakuho Publishing Co. Tokyo 1983) pp. 1 - 19
7. Masahiro Yoshimura and Shigeyuki Sōmiya: in Zirconia Ceramics I, Shigeyuki Sōmiya, ed. (Uchida Rokakuho Publishing Co. Tokyo 1983) pp. 45 - 60
8. Masahiro Yoshimura, Shin-ya Kikugawa and Shigeyuki Sōmiya: in High Pressure Research and Industry, Vol. 2, C.M. Backman, T. Johannisson and L. Tegner eds. (1982) pp. 793 - 796
9. Masahiro Yoshimura and Shigeyuki Sōmiya in: Sintering-Theory and Practice, Proceedings of the 5th International Round Table Conference, D. Kolar, S. Pejovnik and M.M. Ristic, eds. (Elsevier Pub. Co., Amsterdam 1982) pp. 417 - 422
10. Masahiro Yoshimura, Shin-ya Kikugawa and Shigeyuki Sōmiya, Funtai Oyobi Funmatsu Yakin 30 207 - 210 (1983)
11. Hideo Toraya, Masahiro Yoshimura and Shigeyuki Sōmiya, J. Am. Ceram. Soc. 85 May, C-72 (1982)
12. Hideo Toraya, Masahiro Yoshimura and Shigeyuki Sōmiya, J. Am. Ceram. Soc. 66 148 - 150 (1983)
13. Masahiro Yoshimura and Shigeyuki Sōmiya, J. Am. Ceram. Soc., in Press. November, 1983
14. Shigeyuki Sōmiya, Masahiro Yoshimura and Hideo Toraya: in Zirconia Ceramics I, Shigeyuki Sōmiya, ed. (Uchida Rokakuho Publishing Co. Tokyo 1983) pp. 61 - 67
15. Masahiro Yoshimura, Shin-ya Kikugawa and Shigeyuki Sōmiya, Yogyo Kyokai Shi 91 182 - 188 (1983)
16. Shigeyuki Sōmiya, Masahiro Yoshimura and Shin-ya Kikugawa: in The 19th North Carolina State University Conference (1982) (In Press), Hayne Palmour III, R.F. Davis, eds. (Plenum Publishing Co. 1983)
17. Eiji Tani, Masahiro Yoshimura and Shigeyuki Sōmiya, J. Am. Ceram. Soc. 64 181 (1981)
18. Eiji Tani, Masahiro Yoshimura and Shigeyuki Sōmiya, J. Am. Ceram. Soc. 66 11 - 14 (1983)
19. Masahiro Yoshimura and Shigeyuki Sōmiya, Am. Ceram. Soc. Bull. 63 246 (1980)

20. Masahiro Yoshimura and Shigeyuki Sōmiya: in Advances in Ceramics 3, A.H. Heuer and L.W. Hobbs, eds. (Am. Ceram. Soc. 1981) pp. 455 - 463
21. Shin-ichi Hirano and Shigeyuki Sōmiya, J. Am. Ceram. Soc. 59 534 (1976)
22. Shigeyuki Sōmiya, Masahiro Yoshimura and Hideyuki Kanai, Zairyo Kagaku 18 176 - 182 (1982)
23. Hideo Toraya, Masahiro Yoshimura and Shigeyuki Sōmiya, J. Am. Ceram. Soc. 65 C-159-C-160 (1982)
24. Hideo Toraya, Masahiro Yoshimura and Shigeyuki Sōmiya, Yogyo Kyokai Shi 91 235 - 240 (1983)

Table 1. Fine Powder Preparation

A; Size Reduction Process
from Solid { Mechanical : Grinding and Attrition Process
{ Physical : Sublimation

B; Particle Growth Process (Size Built up Process)
1) from Solid (a) Chemical : Decomposition
: Solid-Solid Reaction
: Solid-Liquid Reaction { Hydrothermal Oxidation }
: Solid-Gas Reaction { Hydrothermal Decomposition }
2) from Liquid (a) melt : Evaporation, Dispersion
: Flux
(b) solution : Solvent Removal and Decomposition : Drying (Spray, Freeze Flame, Hot Kerosen, Hot Petroleum, Plasma
: Precipitation : Hydrolysis (Salts, Sol-Gel, Alkoxide)
: Coprecipitation { Hydrothermal Precipitation }
{ Hydrothermal Synthesis }
: Crystallization { Hydrothermal Crystallization }
: Electric Discharge in Water
3) from Gas { Physical : Condensation, Deposition
{ Chemical : Vaper Reaction
: Vaper Decomposition

Table 2. Hydrothermal Reactions for Preparation of Fine Powders

Hydrothermal precipitation
Hydrothermal Crystallization
Hydrothermal Synthesis
Hydrothermal Decomposition
Hydrothermal Oxidation

Table 3 Grain size of hydrothermally α-Al_2O_3 100 MPa for 3 h at temperature between 500°C and 800°C.

	Open system	Closed system
500°C	~1 μm	~1 μm
600°C	1~20 μm	~1 μm
700°C	1~28 μm	1~10μm
800°C	—	1~10μm

Table 4. Crystallite sizes of the products by hydrothermal oxidation under 100 MPa for 3 h.

Starting material	Temp. (°C)	Products	Crystallite size (nm)
Zr	250—700	$m\text{-}ZrO_2$	23 - 27
Al	100—500	$\gamma\text{-}AlOOH$	—
	500—700	$\alpha\text{-}Al_2O_3$	110 - 200
$ZrAl_3$	600—700	$m\text{-}ZrO_2$	13 - 20
		$t\text{-}ZrO_2$	9 - 12
		$\alpha\text{-}Al_2O_3$	27 - 35
Zr_5Al_3	500—700	$m\text{-}ZrO_2$	15 - 24
		$t\text{-}ZrO_2$	10 - 12
		$\alpha\text{-}Al_2O_3$	28 - 35

Table 5. Phases Present and Crystallite Size of Hydrothermally Treated* Products

	Avg crystallite size of products (nm)	
Mineralizer	Tetragonal ZrO_2	Monoclinic ZrO_2
KF (8 wt%)	ND	16 nm
NaOH (30 wt%)	ND	40 nm
H_2O	15 nm	17 nm
LiCl (15 wt%)	15 nm	19 nm
KBr (10 wt%	13 nm	15 nm

*100 MPa, 300°C, 24 h.

Table 6. Effects of solutions on the reaction yield of Cr_2O_3 under the following conditions.
Used Cr: Cr(W), Fill: 16%, Temperature: 1 000°C, Pressure: 100 MPa, Period: 3 hr.

Sample No.	Solution	Yield of Cr_2O_3 (wt%)
PTT 50W	1.0 M HCl	96
PTT 51W	1.0 M HCl	97
PT 48W	0.1 M HCl	93
PTT 53W	0.1 M HCl	83
PTT 54W	0.01 M HCl	90
PTT 55W	0.01 M HCl	90
PTT 56W	0.01 M HNO_3	95
PT 146W	0.001 M HNO_3	81
PT 148W	0.0001 M HNO_3	80
PTT 58W	0.01 M H_2O_2	90
PT 169W	H_2CO_3 (pH 5)	93
PT 170W	H_2CO_3 (pH 5)	90
PT 95K*	0.01 M HNO_3	73
PT 96K*	0.01 M HNO_3	64

*: Cr(K) was used.

Table 7. Examples of Hydrothermal Reaction Sintering

Products	Press.	Starting Temp. & Press. for Formation of Oxide		Solution	Temp.	Press.	Time	Average Grain Size		Relative Density	Starting Temp. of Sintering
								Start. Mat.	Sintered Body		
	Mpa	°C	MPa		°C	MPa	h	m	m	(%)	°C
Wüstite "FeO"	49 ~ 490	650	98	Pure water	900		24				
Magnetite "Fe_3O_4"	~ 196	250	98	10 m NaOH, KOH, NH_4Cl	550	98	1 h				
Chromia "Cr_2O_3"	~ 98	500	98	Pure water	1000	98	3	50	10±3	>99.2	
					850	98	3	50		>99	
	~ 693			Pure water	1000	100	3	20 ~ 40		>90	
				0.01 m HCl							
				0.01 m HNO_3	1000	100	3	20 ~ 40	2 ~ 3	~95	
					1100	100	3	20 ~ 40	2 ~ 3	~95	
					1400	100	3	20 ~ 40	30 ~ 40	~95	
Monoclinic Zirconia Mon.-ZrO_2	98 ~ 686	300	98	Pure water	1000	98	3	1 ~ 20	0.1 ~ 3	> 95	700
Monoclinic Hafnia Mon.-HfO_2				Pure water	900	100		0.2x0.02 (mm)	average 73 nm	92	900
					1000	100			average 89 nm 0.2 0.5	94	
					1200	100			average 110 nm 0.5 ~ 1.0	94	
$LaCrO_3$		700	100	Pure water	1000~1400	100	3				

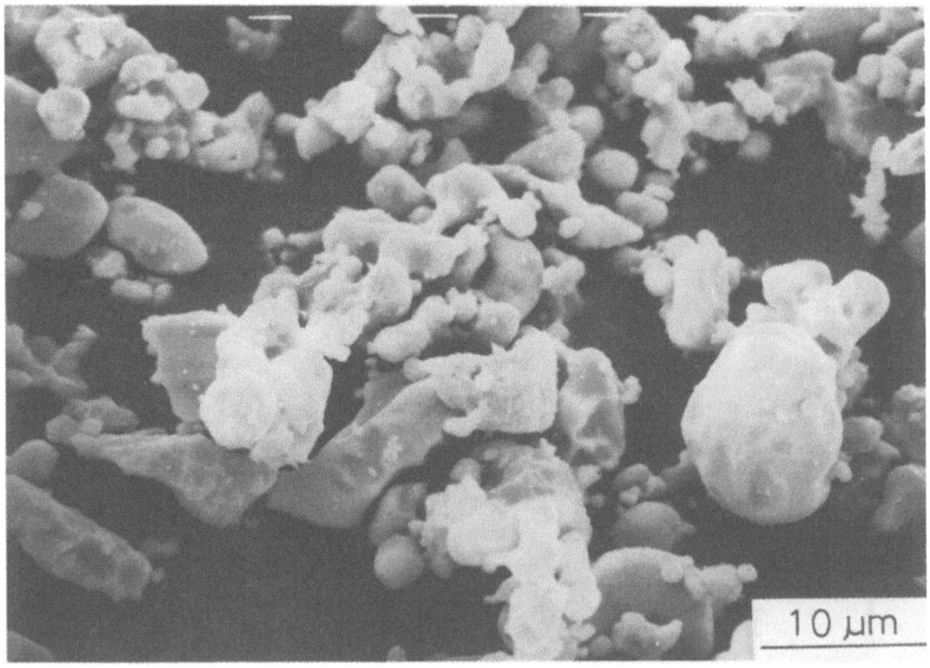

Fig. 1. Starting Zr metal powder.

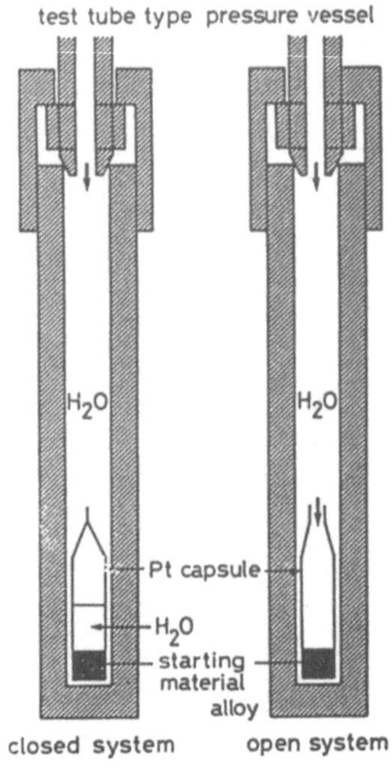

Fig. 2. Schematic illustration for closed and open system of a capsule.

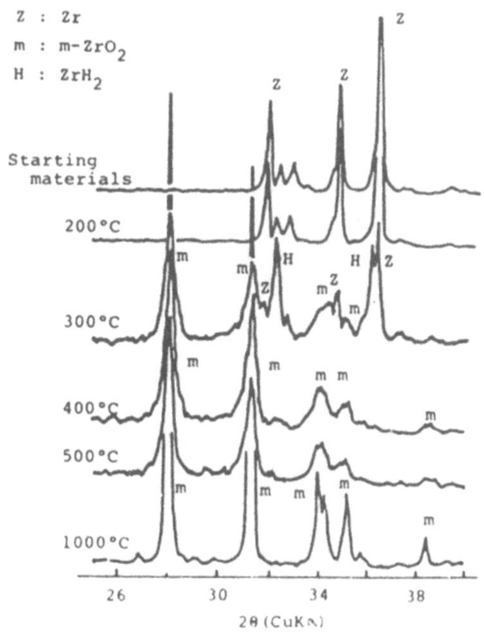

Fig. 3. Reaction products between Zr and H_2O under 98 MPa for 3 h.

FIG. 4. Zr metal powder reacted with H_2O at 300°C for 3 h under 98 MPa.

FIG. 5. Cubic zirconia grains produced from Zr and $Ca(NO_3)_2$ solution at 700°C under 100 MPa for 3 h. bar = 500 nm

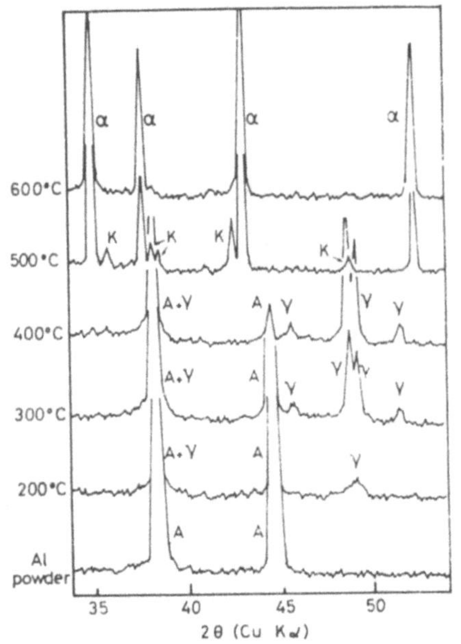

FIG. 6. X-ray powder patterns of the products by hydrothermal oxidation of Al metal powder at various temperatures under 100 MPa for 3 h using open capsule. α: α-Al_2O_3, K: Kl-Al_2O_3, γ: γ-AlOOH A: Al

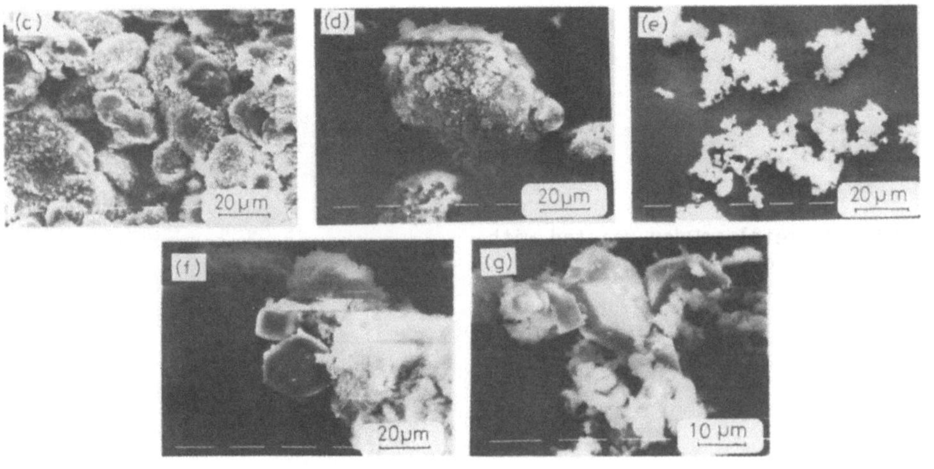

FIG. 7. Scanning electron micrographs of typical products by hydrothermal oxidation in Al metal powder at various temperature under 100 MPa for 3 hrs using open capsule. (a) starting Al metal powder; (b) 200°C; (c) 300°C; (d) 400°C; (e) 500°C; (f) 600°C; (g) 700°C.

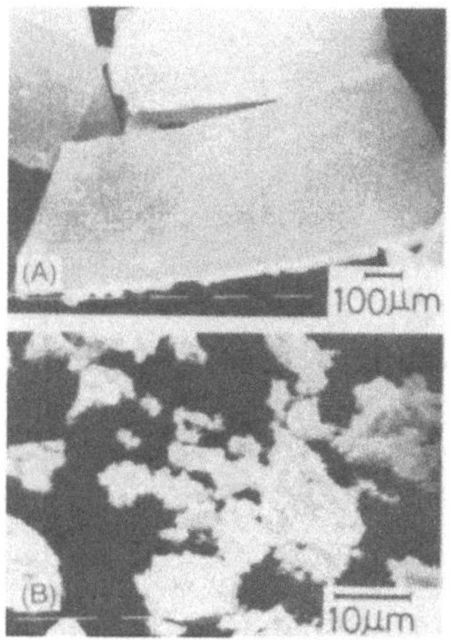

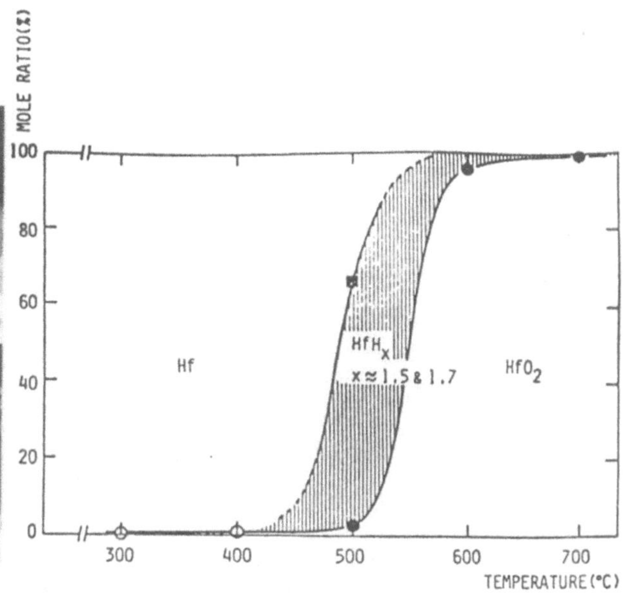

FIG. 8. Scanning electron micrographs of (A) Hf metal chips used as starting materials and (b) monoclinic HfO_2 powder obtained after hydrothermal oxidation at 700°C in a closed system.

FIG. 9. Variation in relative amounts of products with temperature for treatment in closed system under 100 MPa for 3 h.

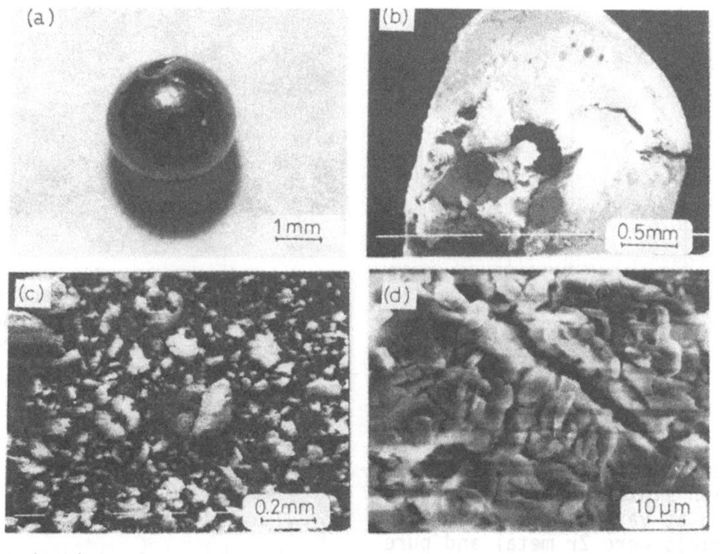

FIG. 10. Hydrothermal oxidation of spherical Zr_5Al_3 alloy. (a) Fused spherical sample; (b) partially reacted (400°C, 100 MPa, 3 hrs); (c) and (d) fully oxidized: α-Al_2O_3 + ZrO_2 powder (700°C, 100 MPa, 3 hrs).

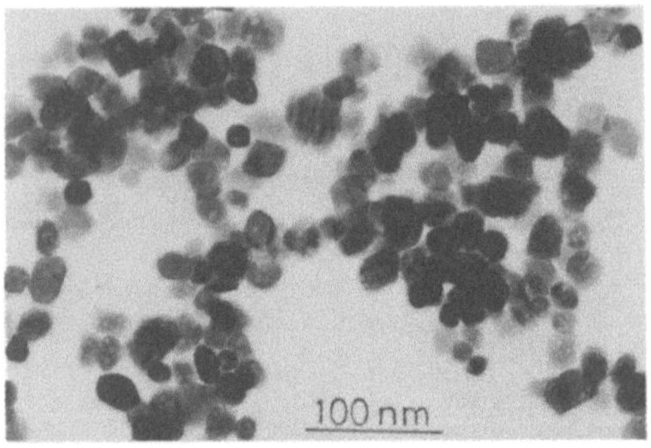

FIG. 11. Transmission electron micrograph of tetragonal and monoclinic ZrO_2 powders produced by hydrothermal treatment at 100 MPa at 300°C for 24 h using 15 wt.% LiCl solution.

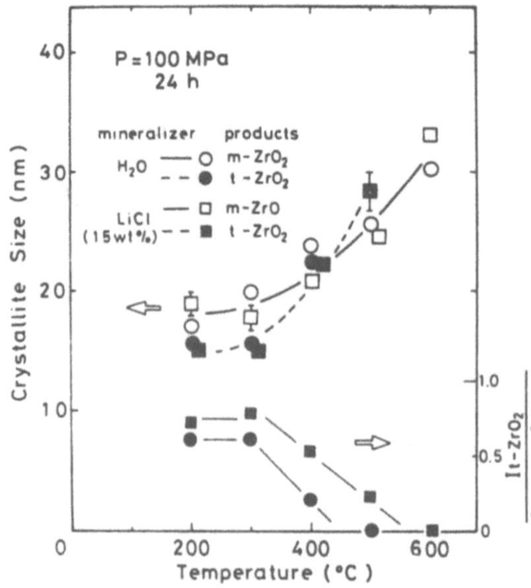

FIG. 12. Variation with temperature of crystallite site and tetragonal ZrO_2 contents in products of hydrothermal treatment at 100 MPa for 24 hrs.

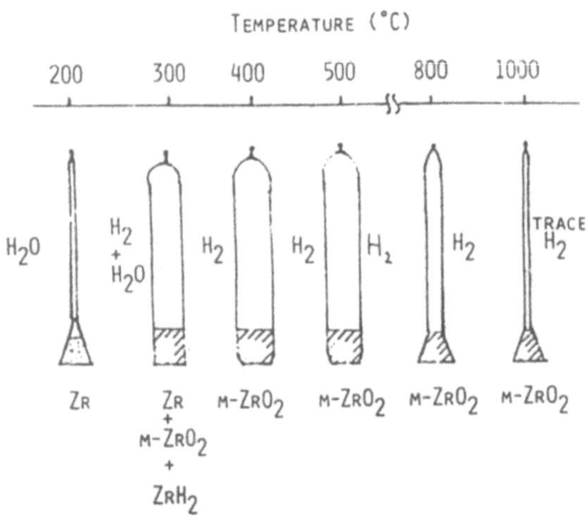

FIG. 13. Schematic illustration for hydrothermal reaction sintering of m-ZrO_2 under 100 MPa for 3 h at various temperatures. Starting materials were Zr metal and pure water. Reaction products were Zr + m-ZrO_2 and ZrH_x at 300°C. At 400°C ZrH_x tended to decompose m-ZrO_2. At 800°C, hydrogen defused out through thin wall of platinum. At 1000°C the capsule was similar condition of HIPing.

FIG. 14. Fracture surface of m-ZrO_2 specimensintered hydrothermally; 1000°C, 3 h, 98 MPa, H_2O/Zr=1.773, fill 19.3%.

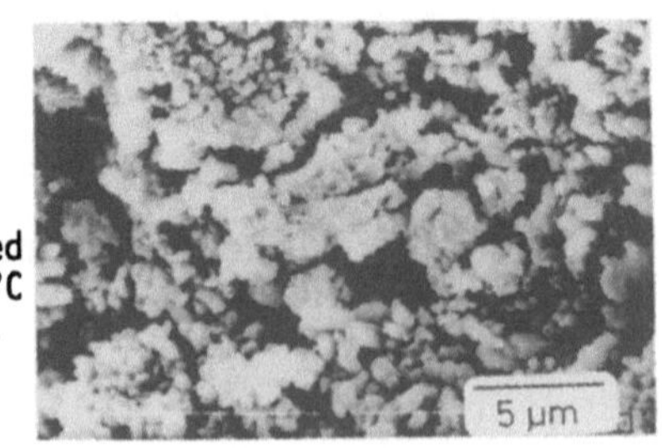

FIG. 15. Fracture surface of m-ZrO_2 specimen sintered hydrothermally; 1000°C, 3 h, 98 MPa, H_2O/Zr=2.095, fill 15.0%.

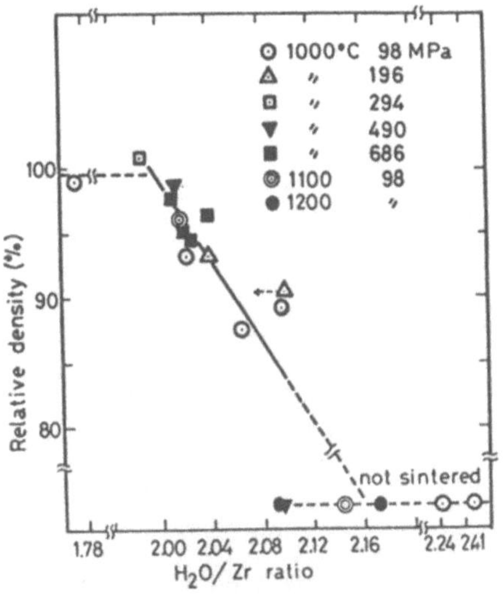

FIG. 16. Relative density of m-ZrO_2 ceramics sintered hydrothermally as a function of the H_2O/Zr ratio.

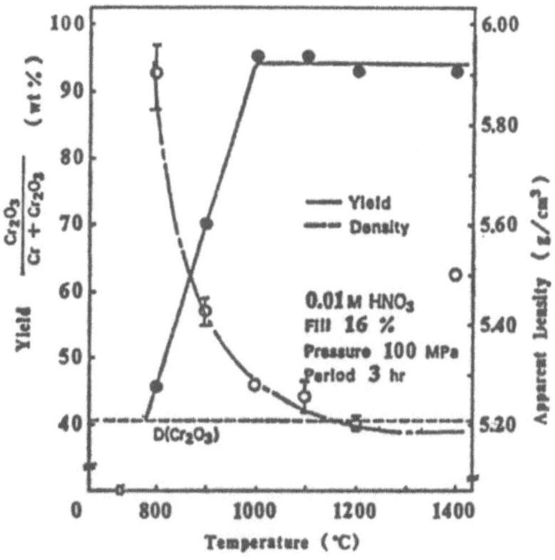

Fig.17. **Effect of sintering temperature on the yield and density of Cr_2O_3 specimen.**

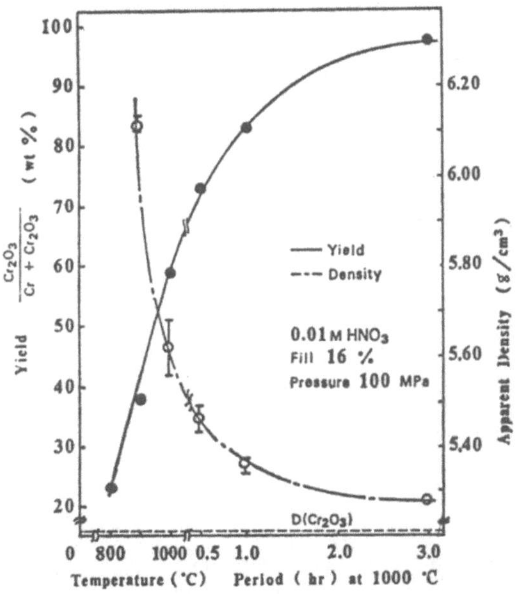

Fig. 18.**Effect of sintering temperature and period on the yield and density of Cr_2O_3 specimen.**

FIG. 19. SEM photograph for hydrothermal reaction sintering of Cr_2O_3 under 100 MPa at 1000°C for 3 h, bar = 1 μm.

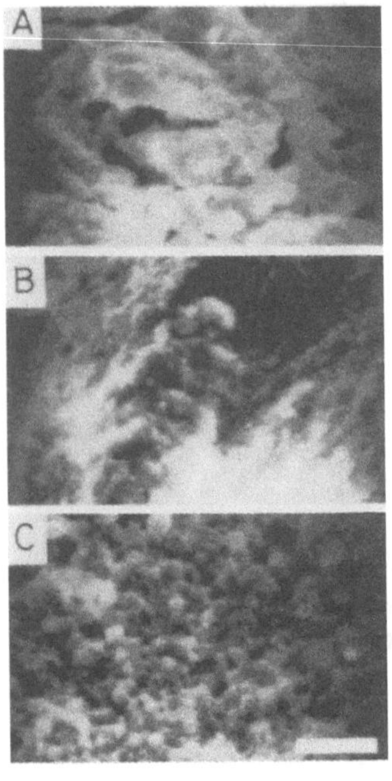

FIG. 20. Scanning electron micrographs of sintered HfO_2 bodies obtained by treatment at (A) 900°C, (b) 1000°C, and (C) 1200°C under 100 MPa for 3 hrs. (bar = 5 μm).

Reprinted from G. C. Kuczynski, D. P. Uskoković, Hayne Palmour III and M.M. Ristić (eds), *Sintering '85* (Plenum Press, New York, 1987) pp. 199–204.

The Processing of a Permalloy–Magnetite Composite by Hydrothermal Reaction Sintering

Masahiro Yoshimura, Shigeyuki Sōmiya, Kimiyuki Kamino*, and Takahiro Nakagawa*

Research Laboratory of Engineering Materials and Department of Materials Science, Tokyo Institute of Technology
4259 Nagatsuta-cho, Midori-ku, Yokohama-shi, 227 Japan
*Research and Development Center, Mitsubishi Seiko Inc.
1-9-31 Shinonome, Koto-ku, Tokyo-to, 135 Japan

ABSTRACT

The water-atomized permalloy powder and redistilled water were hermetically sealed in a platinum capsule and treated in a high pressure, high temperature apparatus using Ar gas as the pressuring medium. The sintered composites thus produced were examined by X-ray diffraction, optical and scanning electron microscopy, and energy dispersive spectroscopy. The sintered composite of Fe-Ni alloy and iron oxide can be made by hydrothermal reaction sintering, where the composition of iron oxide is controlled by the water/alloy ratios and temperatures.

1. INTRODUCTION

Hydrothermal Reaction Sintering is one of the methods to prepare oxide ceramics.[1-3] Metals, intermetallic compounds or alloys are sealed in a Pt capsule with water. On heating they react with high pressure, high temperature water, and produce metal oxides and hydrogen gas in the capsule. At high temperature, hydrogen gas diffuses out of the capsule, then the capsule containing oxide powders is compressed by outside gas pressure. The oxide powders, therefore, can be sintered in the capsule in a manner similar to HIP (hot isostatic pressing). By this method, when we control the ratio of water to metal or alloys in the capsule, we can obtain a metal-oxide composite. Here, we describe the formation of a permalloy-magnetite composite by this technique.

2. EXPERIMENTAL PROCEDURE

The starting material was water-atomized permalloy powder (Fe-47 wt.%Ni<#100). The known quantities of alloy powder and redistilled water were hermetically sealed with an electric arc in a platinum capsule 2.6 - 2.7 mm inside diameter, 0.15 - 0.2 mm thick and 35 mm long. This capsule was treated at 100 MPa, 800 - 1000 °C, for 1 h in a high pressure, high temperature apparatus using Ar gas as the pressurizing medium. The sintered ceramics thus produced were examined by X-ray diffraction, optical and scanning electron microscopy, and energy dispersive spectroscopy (EDS).

3. RESULTS

Figure 1 shows photomicrographs of the cross-section of specimens sintered at 1000 °C under 100 MPa for 1 h with various water/alloy ratios. The gray matrix is the oxide phase, the white dispersed phase is the alloy phase, and the black parts are pores. The density was 6.48 g/cm^3 for the specimen of 10 wt.% H_2O subjected to 100 MPa at 1000 °C for 1 h. The porosity is calculated to be 9.75%, because the calculated density is 7.18 g/cm^3 based on the density for Fe-Ni alloy (d = 8.45 g/cm^3 for Fe - 47 Ni),[4] and that for iron oxide (d = 5.7 g/cm^3 for Wüstite).[5] According to the EDS, the oxide phase contained only iron as cations. Selective oxidation of iron, therefore, had occurred under the hydrothermal conditions, as well as in air.[6] The iron-nickel distributions in the alloy phase were found to be homogeneous by EDS analysis. In addition, X-ray diffraction peaks were found to be neither broadened nor split. The selective oxidation of iron might bring about the depletion of iron in the alloy near the interface of the surface oxide layer. However, when the temperatures are sufficiently high as in this case, 1000 °C, the diffusion rate of nickel or iron in the alloy is so high that the alloy phase becomes homogeneous.

Table I shows the results of X-ray diffraction studies. In the specimen treated at 1000 °C, 100 MPa for 1 h, the oxide phase was Wüstite only, when the wt(H_2O)/wt(Fe-Ni) ratio was in the range 0.05 - 0.10. When wt(H_2O)/wt(Fe-Ni) ratio was 0.15 and 0.20, respectively, Wüstite and Magnetite phases coexisted. Magnetite alone was observed when the wt(H_2O)/wt(Fe-Ni) ratio was 0.30. The treatment at 800 °C yielded Magnetite even for wt(H_2O)/wt(Fe-Ni) = 0.10 or 0.15, i.e., further reduction seemed to occur at 1000 °C. In the hydrothermal reaction sintering of iron powder[7] (purity >99%) with redistilled water, the oxide phase was Magnetite below 650 °C under 98 MPa, but Wüstite and Magnetite coexisted above 650 °C under 98 MPa.

4. DISCUSSION

The formation of iron oxide on the surface of the Fe-Ni alloys can be written by the reaction

$$Fe_{(in\ Fe\text{-}Ni)} + H_2O \rightarrow Fe\text{-}oxide + H_2 \tag{1}$$

Table I. Oxide phases present in the products of a hydrothermal reaction sintering under 100 MPa for 1 h.

Temperature (°C)	wt(H_2O)/wt(Fe-Ni)	Oxide Phases
1000	0.05	W
1000	0.10	W
1000	0.15	W + M
1000	0.20	W + M *
1000	0.30	M *
800	0.10	M *
800	0.15	M *

W : FeO, M : Fe_3O_4

* : porous.

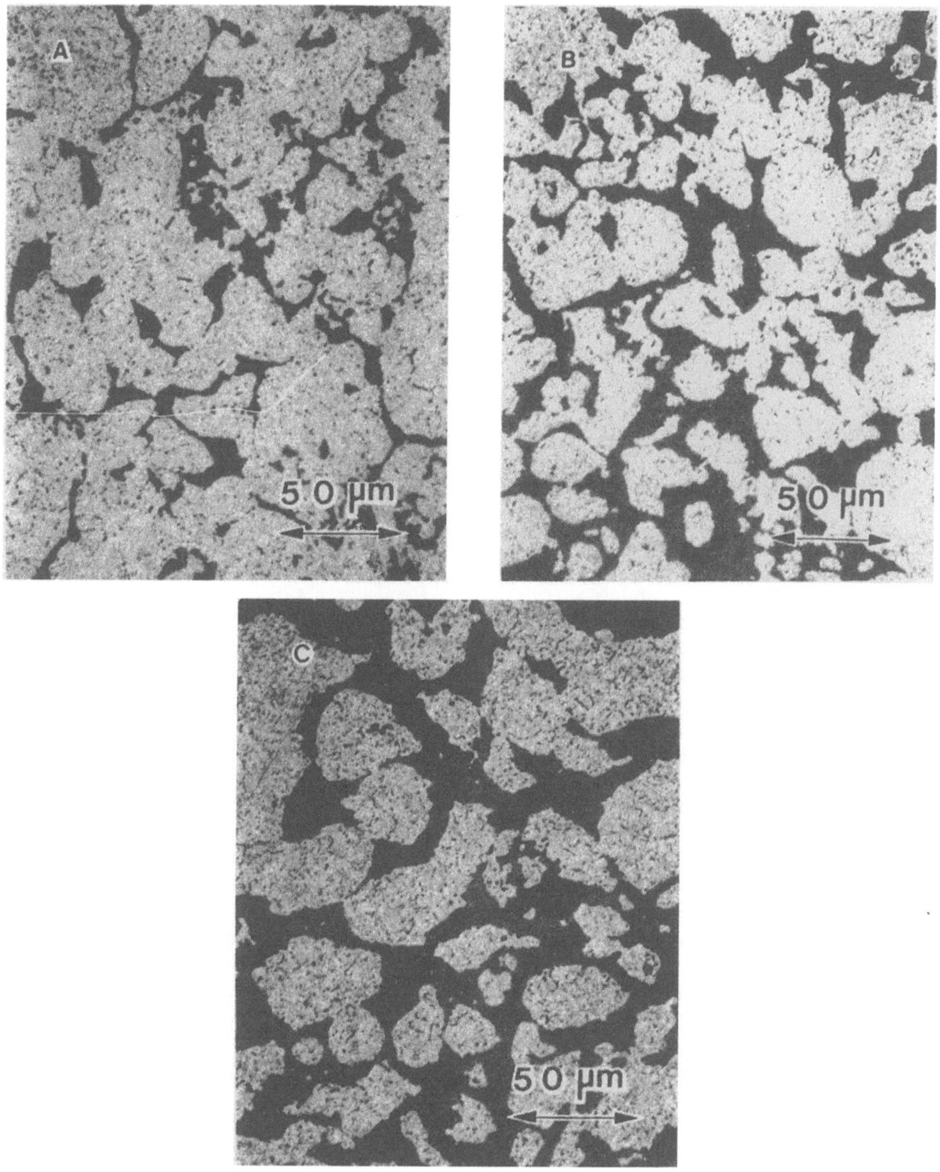

Fig. 1. Photomicrographs of the cross section of the (Permalloy)-(Fe-oxide) composite by hydrothermal reaction sintering. |wt(H_2O)/wt(Fe-Ni) = 0.05 (A), 0.10 (B), 0.15 (C)|.

The composition of Fe-oxide is controlled both by thermodynamics (oxygen and iron activities) and by kinetics (oxygen and iron diffusivities). Here we discuss the mechanism of the iron-oxide formation. Figure 2 shows the

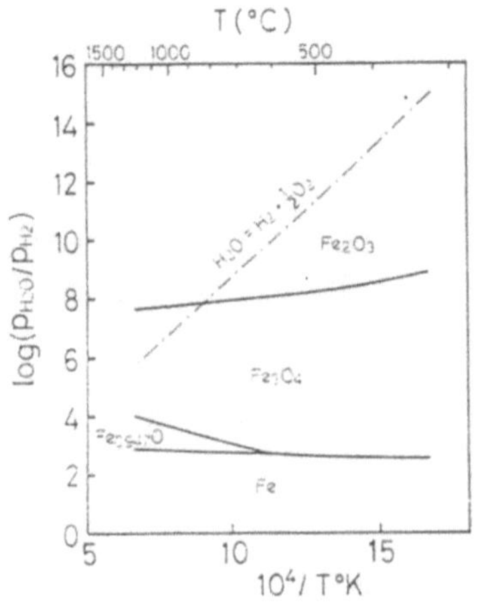

uilibrium diagram of Fe-O-H system under an atmosphere of 0 MPa H_2O.

iagram of the Fe-O-H system under the condition of 100 MPa ermodynamic data of iron oxides were taken from JANAF,[8] and MPa H_2O from the work by Keenan et al.[9] Hydrogen and oxygen as ideal gases. This diagram indicates that at 1000 °C the is Fe_3O_4, but at 800 °C, it would be Fe_2O_3 under the condi- Pa H_2O. However, the actual products were much reduced, i.e., at 1000 °C and Fe_3O_4 at 800 °C, respectively (See Table I). nces indicate that the composition(s) of iron-oxides are so by iron-activity in the alloy.

shows the triple point of (Fe-Ni alloy) - ($Fe_{0.947}O$)-(Fe_3O_4) of the activity of iron (a_{Fe}) in the Fe-Ni alloy. The acti- in the Fe-Ni alloy is a function of the composition. If we eal solution, the activity of the iron becomes a_{Fe} = 0.54 for 53 wt% Fe used in this experiment. As the activity of iron , the critical temperature of Wüstite formation shifts to ature as shown in Fig. 4. This diagram shows the critical es to form Wüstite are a_{Fe} = 0.32 at 800 °C, and a_{Fe} = 0.14 espectively.

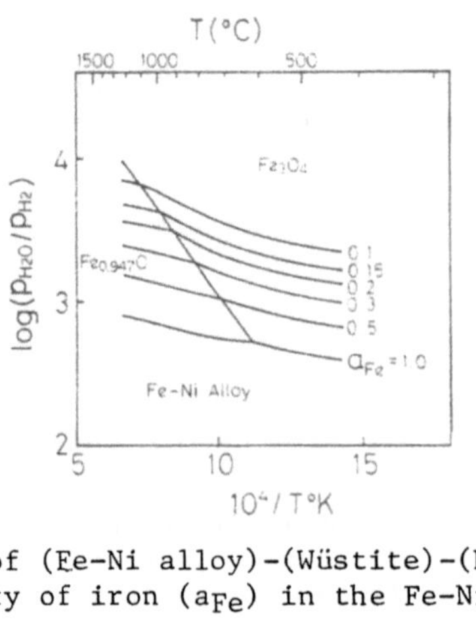

.e point of (Fe-Ni alloy)-(Wüstite)-(Magnetite) as a function ıe activity of iron (a_{Fe}) in the Fe-Ni alloy.

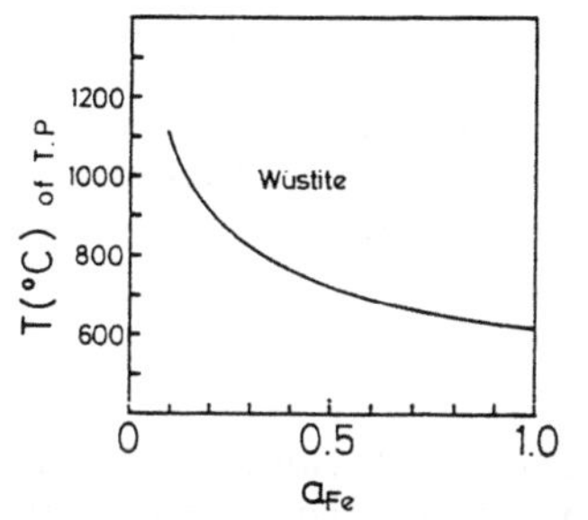

Fig. 4. Critical temperature of Wüstite formation as a function of the activity of iron (a_{Fe}) in the Fe-Ni alloy.

Once an oxide layer has been formed on the surface of alloy particles, the iron activities are controlled also by the diffusivity, because the iron ion is supplied from the inner alloy to the oxide layer. On the other hand, the oxygen ion (which has a slow diffusion coefficient) cannot diffuse in the oxide layer as fast as the iron. Therefore, the oxide should be Wüstite in the contact layer of the alloy coated by oxide layers, as indicated in Fig. 5. In the initial oxidation, the oxide potential in the capsule is to produce Magnetite at 1000 °C as indicated in Fig. 2. So the oxide phases are Magnetite in the outer layer, but Wüstite in the inner layer at the particle surface. When water in the capsule reacts completely, the oxygen potential in the oxide phase goes down, because the supply of Fe from the alloy phase to the oxide phase still continues. In this way the Wüstite phase becomes thickened until finally the Magnetite phase disappears. With an increase in the ratio of H_2O to Fe-Ni, Magnetite can remain longer. In the case where excess water remained in the capsule, only Magnetite was the stable oxide, as indicated in Fig. 2. The excess water cannot diffuse out of the capsule, so that this water pressure prevents the pressure-assisted sintering of the sample in the capsule. This is the case of the samples having wt(H_2O)/wt(Fe-Ni) ratios = 0.20 and 0.30, respectively, as indicated in Table I. In the case of the experiments done at 800 °C, iron depletion (= nickel concentration) has occured at the interface between the alloy and the oxide layer. It causes the decrease of a_{Fe} at the interface so that Magnetite becomes stable as indicated in Fig. 4. At a temperature as low as 800 °C, the reaction rate between Fe and H_2O is slow. In addition, the diffusion rate of H_2 in Pt of the capsule wall is small. These facts resulted in retention of gas

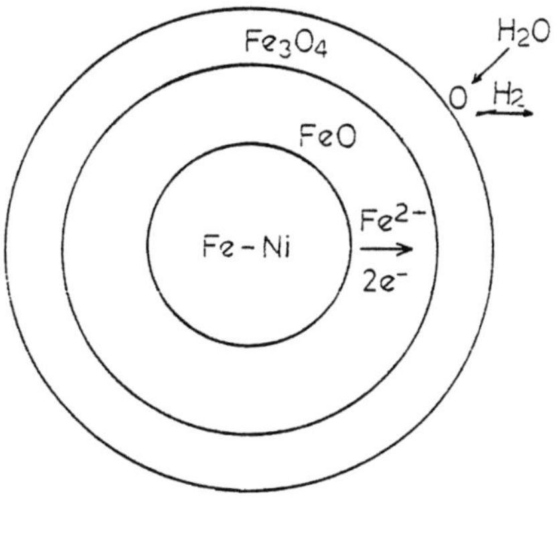

Fig. 5. Schematic illustration of the oxidation of Fe-Ni alloy by hydrothermal reactions.

(H_2O + H_2) in the capsule, which prevents densification.

In conclusion, the sintered composite of Fe-Ni alloy and iron oxide can be made by hydrothermal reaction sintering, where the composition of iron oxide is controlled by the water/alloy ratios and temperatures.

REFERENCES

1. S. Somiya, M. Yoshimura, H. Toraya, E. Tani, M. Suzuki, S. Kikugawa, T. Hattori, "Hydrothermal Processing for Ceramics-Powder Preparation and Sintering", Proc. First Int. Symp. Ceramic Components for Engine, Ed. by S. Somiya, E. Kanai, K. Ando, KTK Sci. Pub. Co., Tokyo and D. Reidel Pub. Co. Dordrecht, 1984, 739.
2. M. Yoshimura and S. Somiya, "Fabrication of Dense, Nonstabilized ZrO_2 Ceramics by Hydrothermal Reaction Sintering", Am. Ceram. Soc. Bull. 63 (1980) 246.
3. S. Hirano and S. Somiya, "Hydrothermal Reaction Sintering of Pure Cr_2O_3", J. Am. Ceram. Soc., 59 (1976) 534.
4. R. M. Bozorth, "Ferromagnetism", D. Van Nostrand Co. Inc., Princeton, New Jersey, 1964, p.105.
5. "LANGE's Handbook of Chemistry", Edited by J. A. Dean, McGraw-Hill Book Company, New York, 1973, p.4.
6. M. Seo and N. Sato, "Selective Oxidation of Fe-30Ni Alloy in a Low-Temperature Range", Oxid. Met., 19 (1983) 151.
7. S. Hirano and S. Somiya, "Synthesis and Sintering of Magnetite under Hydrothermal Conditions", The 12 Symposium on Basic Ceramics, January 30, 1974, p.33.
8. D. R. Stull and H. Prophet, "JANAF Thermochemical Tables Second edition", U. S. Government Printing Office, Washington, D.C., 1971.
9. J. H. Keenan, F. G. Keyes, P. G. Hill and J. G. Moore, "Steam Tables - Thermodynamic Properties of Water Including Vapor, Liquid, and Solid Phases", John Wiley & Sons Inc., New York, 1978.

Part 2

Hydrothermal Oxidation

Reprinted from *High Pressure in Research and Industry*, 8th AIRAPT Conf./19th EHPRG Conf., Sweden, 17–22 August 1981, pp. 793–796.

Preparation of Zirconia Fine Powders by the Reactions Between Zirconium Metal and High Temperature–High Pressure Solutions

Masahiro YOSHIMURA, Shinya KIKUGAWA and Shigeyuki SŌMIYA
Laboratory for Hydrothermal Syntheses, Research Laboratory of Engineering Materials, Tokyo Institute of Technology, Nagatsuta, Midori, Yokohama, 227 JAPAN

A new method, Hydrothermal Oxidation Method, has been developed to prepare fine grained zirconia powders. Under 100 MPa, Zr metal and H_2O sealed in the capsule reacted above 250°C and yielded fine grained ZrO_2 powders. The powders were sole monoclinic modification having the average grain size of ≈24 nm. The use of Ca solutions instead of H_2O led to a partiale formation of cubic zirconia fine grains, particularly in the case of $Ca(NO_3)_2$ with excess an amount of Zr and at high temperatures, 700°C. The cubic zirconia seemed to be stabilized both by calcium and by nitrogen.

INTRODUCTION

Zirconia powder have been prepared by the thermal decomposition of amorphous hydrated zirconia,[1)] zirconium salts[2)] or alkoxide,[3)] or vapor phase reactions.[4)] We have developed two methods to prepare fine zirconia powders using hydrothermal reactions, [1] hydrothermal treatments of hydrated zirconia[5)] and [2] hydrothermal oxidation method, where Zr metal is oxidized by H_2O[6)] or aqueous solutions at high temperatures under pressures. This paper describes the latter in some detail.

EXPERIMENTAL

Zr metal powder, 99.9 % including 2 % Hf, under 300 mesh, and redistilled water were sealed in a capsule of Pt or Au 2.7 mmID and 35 mm long. The capsule was heated in a test-tube type pressure vessel under the pressure of 100 MPa water up to 200° to 800°C.

After quenching, the products in the capsule were examined by X-ray powder diffraction, scanning electron micrography, transmission electron micrography and electron diffraction techniques.

RESULTS AND DISCUSSION

Under the pressure of 100 MPa, the reaction between Zr and H_2O has been observed above 250°C and yielded ZrO_2 and H_2,

$$Zr + 2\,H_2O \longrightarrow ZrO_2 + 2\,H_2 \quad \text{-----(1).}$$

The complete formation of ZrO_2 was achieved above 400°C under 100 MPa. The zirconia prepared by this method has always the monoclinic modification and the average grain size of $\approx$24 nm. This value calculated from the half width of the x-ray diffraction lines was confirmed by TEM observation as shown in Fig. 1, which shows fine ZrO_2 grains formed by the oxidation of original Zr particles having relatively large grain size. It is noteworthy that neither tetragonal nor cubic zirconia could be detected by x-ray diffraction in the products even though the grain size of zirconia was less than a critical size, $\approx$30 nm,[7] to produce "metastable tetragonal zirconia". It seemed to be probable that a topotactic nucleation[8] and/or strain-free growth[9] of zirconia in this method did not allow to crystallize "metastable tetragonal zirconia". No impurities can be expected to enter in the products by this method, so that this Hydrothermal Oxidation Method appears to be an interesting method to prepare pure monoclinic zirconia fine grains, and furthermore fine grains of other oxides.

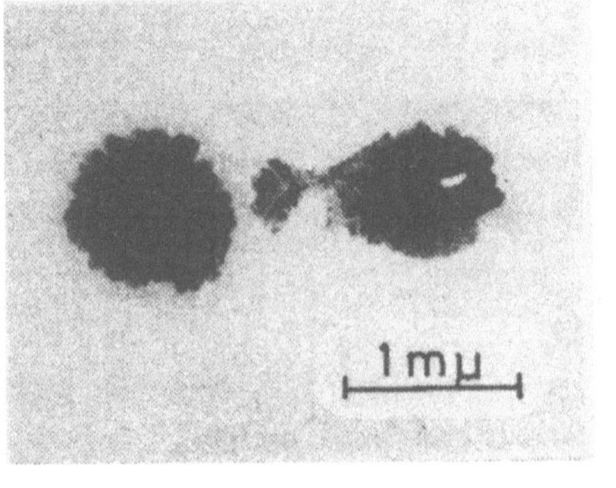

Fig. 1 m-ZrO_2 powders produced from Zr + 2 H_2O at 400°C, 100 MPa, 3 h.

Expecting the direct formation of stabilized zirconia fine grains, we have applied this method to the reaction between Zr and Ca solutions. The use of $Ca(OH)_2$ powder and H_2O instead of $Ca(OH)_2$ solution because of the low solubility of $Ca(OH)_2$ in H_2O for a expected reaction;

$$0.85\,Zr + 0.15\,Ca(OH)_2 + 1.55\,H_2O \rightarrow Zr_{0.85}Ca_{0.15}O_{1.85} + 1.70\,H_2 \quad \text{--(2)}$$

yielded monoclinic ZrO_2 and $Ca(OH)_2$ instead of stabilized zirconia below 600°C. At 700°C, however, a small amount of cubic ZrO_2 phase could be detected in the products. The yields of cubic zirconia did not exceed 0.15 in the case of $Ca(OH)_2$ probably because of its low solubility in water. The yield of cubic zirconia was increased when 59 wt% $Ca(NO_3)_2$ solution, an almost saturated solution, and excess amounts of Zr metal were reacted above 500°C. The excess Zr decomposed water and nitrate completely then produced ZrH_2 and ZrN as seen in Fig. 2. The apparent fraction of cubic zirconia showed a maximum around Zr/O = 0.63 at 700°C.

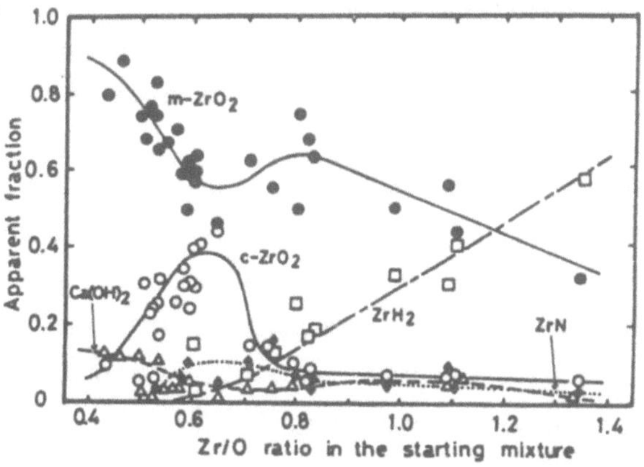

Fig. 2 Products of Zr + $Ca(NO_3)_2$ solution at 700°C

The average grain size was 25 nm for monoclinic zirconia and 18 nm for cubic zirconia in these products, but sometimes large cubic grains (Fig. 3) could be observed. The cubic zirconia had the lattice parameter of a = 0.507 - 0.510 nm, which is smaller than a= 0.5135 nm for $Zr_{0.85}Ca_{0.15}O_{1.85}$[10] but rather close to that a = 0.509 nm for nitrogen stabilized zirconia[11]. This result and the formation of free $Ca(OH)_2$ in the products suggested that the cubic zirconia might be stabilized both by calcium and

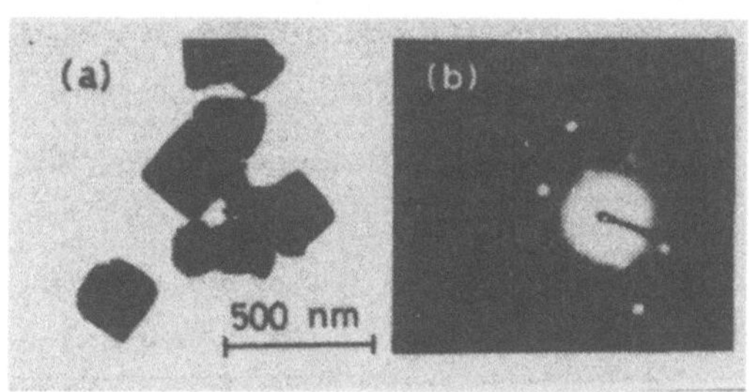

Fig. 3 Cubic ZrO_2 grains formed from Zr + $Ca(NO_3)_2$ soln. at 700°C, 100 MPa, 3 h

by nitrogen. This suggestion was confirmed by the fact that stabilized zirconia could be formed by the reaction between Zr and NH_4NO_3 solutions under the similar conditions: 700°C, 100 MPa, 3 h, excess Zr as seen in Fig. 4. The products consisted of also monoclinic zirconia, cubic zirconia, ZrH_2, and ZrN.

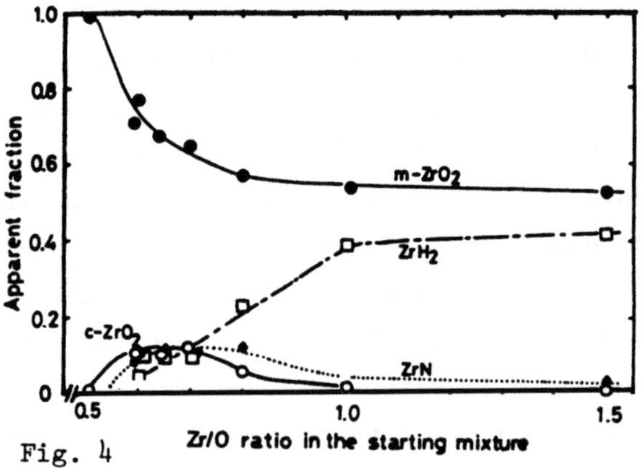

Fig. 4

Products of Zr + NH_4NO_3 soln. at 700°C

Similarly, cubic zirconia could be observed in the products between Zr and $Y(NO_3)_3$ solution, saturated 41 wt%, under the similar conditions by an expected reaction;

$$0.94\,Zr + 0.06\,Y(NO_3)_3 + 1.43\,H_2O \rightarrow Zr_{0.94}Y_{0.06}O_{1.97} + 0.09\,N_2 + 1.43\,H_2 \quad (3)$$

The yield of the cubic zirconia in either the case of NH_4NO_3 or $Y(NO_3)_3$, however, was less than 0.15 even at maxima contrasting to 0.4 in the case of $Ca(NO_3)_2$. This result also supports the suggestion that the cubic zirconia is stabilized both by calcium and by nitrogen in the case of $Ca(NO_3)_2$.

References

1) Cypres R. et al. Ber. Deut. Keram. Ges., 1963, 40, 527
2) Murase Y. et al. Nippon Kagaku Kaishi, 1972 2329
3) Mazdiyasni, K.S. et al. J. Am. Ceram. Soc., 1966, 49, 286
4) Suyama Y. et al. Ceramurgia 1977, 3, 141
5) Tani E. et al. J. Amer. Ceram. Soc., Accepted for publication
6) Yoshimura M. et al. Am. Ceram. Soc. Bull., 1980, 59, 246
7) Garvie R.C. et al. J. Phys. Chem., 1965, 69, 1238
8) Livage J. et al. J. Am. Ceram. Soc., 1968, 51, 349
9) Mitsuhashi T. et al. ibid., 1974, 57, 97
10) Duwez P. et al. ibid., 1952, 35, 107
11) Claussen N. et al. ibid., 1978, 61, 369

Reprinted from *Journal of the American Ceramic Society*, **65**(5) (1982) C-172.

Preparation of Fine Monoclinic Hafnia Powders by Hydrothermal Oxidation

HIDEO TORAYA, MASAHIRO YOSHIMURA,* AND SHIGEYUKI SŌMIYA*

Laboratory for Hydrothermal Syntheses, Research Laboratory of Engineering Materials and the Department of Materials Science and Engineering, Tokyo Institute of Technology, Nagatsuta, Midori, Yokohama 227, Japan

Fine single-phase monoclinic HfO_2 powders were prepared from Hf metal chips and high-temperature high-pressure water by hydrothermal oxidation in closed and open systems. In the closed system, Hf metal was converted to HfO_2 by treatment at ≥600°C under 100 MPa for 3 h. At 500°C half the Hf reacted to produce Hf hydride and a small amount of HfO_2. In the open system, Hf metal scarcely reacted at 500°C, but at 600°C the reaction was more rapid than the corresponding run in the closed system.

HAFNIA (HfO_2) is isomorphous with ZrO_2 and is similarly characterized by a high melting point and chemical stability.[1-2] It has, however, a lower thermal expansion coefficient and higher transition temperature than zirconia,[1-2] and is considered to be potentially important in ceramics despite its lesser abundance than zirconia. Zirconium normally contains from 0.5 to 2.0% hafnium.[2] This communication describes the preparation of hafnia fine powders from hafnium metal chips by the recently developed hydrothermal oxidation technique.[3-5]

Hafnium metal chips* (1 by 0.2 by 0.02 mm) and redistilled water (2:1 H_2O/Hf mole ratio) were sealed with an electric arc in a 2.7 mm ID Pt capsule 0.15 mm thick and 35 mm long. The water used as a reaction reagent was prepared with a quartz distiller from deionized water which had been previously distilled with a copper distiller. The sealed capsules were heated at 300°, 400°, 500°, 600°, and 700°C in a test-tube-type pressure vessel under 100 MPa for 3 h. The desired temperatures and pressures were held within ±20°C and ±2 MPa, respectively, during a run. In addition to the closed system runs, open system experiments were also performed without sealing the capsule at 500° and 600°C. After quenching, heated specimens were examined by X-ray powder diffractometry† and scanning electron microscopy.‡

In the closed system, Hf metal chips scarcely reacted with high-temperature high-pressure water at 300° and 400°C. At 500°C half of the Hf metal chips were changed into Hf hydride and a small amount into monoclinic HfO_2. The reaction proceeded rapidly above 600°C, where most of the product consisted of monoclinic HfO_2 powder with a small amount of Hf hydride. At 700°C, HfO_2 was produced with only a trace amount of Hf hydride remaining. Figure 1 shows Hf metal chips used as starting materials and monoclinic HfO_2 powder obtained by hydrothermal oxidation at 700°C in the closed system. The crystallite sizes of the formed HfO_2 were determined to be ≈33 nm by measuring the integral breadth of the X-ray $\bar{1}11$ reflection. The particle sizes ranged from 1 to 10 μm as determined by SEM where the large particles are aggregates of small ones.

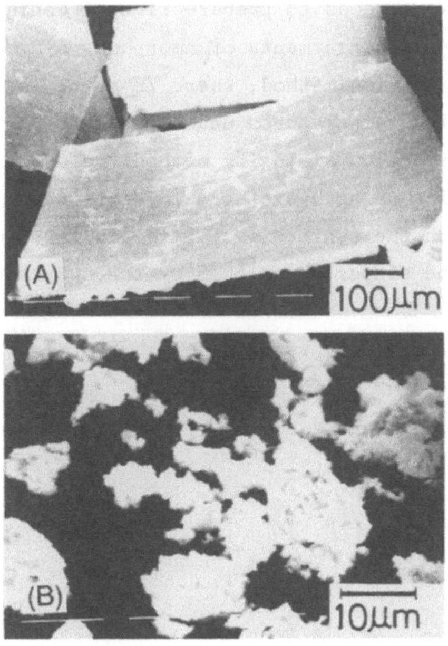

Fig. 1. Scanning electron micrographs of (*A*) Hf metal chips used as starting materials and (*B*) monoclinic HfO_2 powder obtained after hydrothermal oxidation at 700°C in a closed system.

The reaction mechanism of Hf metal chips with high-temperature high-pressure water is expected to be similar to that of Zr metal powder.[3-5] The transformation from Hf to HfO_2, however, proceeded at a much higher temperature than that from Zr to ZrO_2, which started above 250°C and was complete above 400°C under 100 MPa for 3 h.[3-5] The Hf reaction proceeded slowly so that it was possible to observe the dependence of the hydrothermal reaction on temperature-pressure-time more distinctly than that for the Zr reaction which was completed almost immediately.[3-5]

In contrast to the closed system, in the open system Hf metal chips scarcely reacted at 500°C. However, the reaction proceeded very rapidly at 600°C, faster than that of the corresponding run in the closed system. These differences are thought to result from the differences in fugacities in the capsule's atmosphere. Since the amount of H_2O would decrease during reaction in the closed system while pressurized excess H_2O would continue to surround the sample in the open system, the fugacity relation, f_{H_2O}, would be much smaller in the former and larger in the latter system. The open system appears to be more effective than the closed one for the preparation of oxide fine powders by hydrothermal oxidation. The method is also more interesting from the viewpoint of science and technology.

The hydriding of Hf at lower temperatures seems to induce the pulverization of the Hf metal chips during the hydrothermal processing. A study on the mechanism of the hydrothermal oxidation is now in progress.

Acknowledgments

The authors thank Nippon Mining Co., Ltd. for supplying Hf metal.

References

[1]R. Ruh, H. J. Garrett, R. F. Domagala, and N. M. Tallan, "The System Zirconia-Hafnia," *J. Am. Ceram. Soc.*, **51** [1] 23–27 (1968).

[2]C. T. Lynch; pp. 193–216 in Refractory Materials: High Temperature Oxides, Pt. II, Vol. 5. Edited by A. M. Alper. Academic, New York, 1970.

[3]M. Yoshimura and S. Sōmiya, "Fabrication of Dense Nonstabilized ZrO_2 Ceramics by Hydrothermal Reaction Sintering," *Am. Ceram. Soc. Bull.*, **59** [2] 246 (1980).

[4]M. Yoshimura and S. Sōmiya, "Hydrothermal Reaction Sintering of Monoclinic Zirconia," pp. 455–63 Advances in Ceramics, Vol. 3. Edited by A. H. Heuer and L. W. Hobbs. The American Ceramic Society, Columbus, Ohio, 1981.

[5]M. Yoshimura, S. Kikugawa, and S. Sōmiya, "Preparation of Zirconia Fine Powders by the Reaction between Zirconium Metal and High Temperature-High Pressure Solutions," Proceedings of the 8th AIRAPT & 19th EHPHG Conference on High Pressure in Research and Industry, Aug. 17–22, 1981, Uppsala, Sweden. □

CONTRIBUTING EDITOR — J. A. PASK

Received January 15, 1982.

Supported by Grant-in-Aid for Scientific Research under Contract No. 443015.

*Member, the American Ceramic Society.

*In ppm, Hf 0 545, H 7, N 27, Al <20, C 110, Cl <50, Cu <10, Fe <50, Mg <10, Ti <20, W <15, U 1, Zr 3.9%.

†RU-200, Rigaku Denki Co., Ltd., Tokyo, Japan.

‡JSM T-200, JEOL Co., Ltd., Tokyo, Japan.

Reprinted from D. Kolar, S. Pejovnik and M. M. Ristić (eds), *Sintering—Theory and Practice*, Proceedings of the 5th International Round Table Conference on Sintering, Portorož, Yugoslavia, 7–10 September 1981, Material Science Monographs, Volume 14 (Elsevier Scientific Publishing Company, Amsterdam, 1982) pp. 417–422.

Synthesis and Sintering of Zirconia Fine Powders by Hydrothermal Reactions from Zirconium Metal and High-Temperature High-Pressure Solutions

Masahiro YOSHIMURA and Shigeyuki SŌMIYA
Laboratory for Hydrothermal Syntheses, Research Laboratory of Engineering Materials
Tokyo Institute of Technology, Nagatsuta, Midori, Yokohama, 227 Japan.

ABSTRACT

A new method, hydrothermal oxidation method, has been developed to prepare zirconia fine powders by reactions between Zr metal and high-temperature high-pressure water or solutions. Zr metal and water sealed in a Pt capsule reacted under 98 MPa above 250°C and produced monoclinic ZrO_2 fine grains having the average crystallite size of 24 nm. Further heating of this product up to 1000°C yielded a dense sintered body of m-ZrO_2 with the maximum relative density of 99.0%. The hydrothermal oxidation method has been applied to the preparation of stabilized zirconia powders using Ca or Y solutions instead of water. A partial stabilization has been achieved when an excess Zr was existed in the capsule to decompose the solutes and the solvents at higher temperatures.

INTRODUCTION

Zirconia is one of the potential materials for uses in ceramics, electronics, mechanics and other engineerings. The zirconia powder has been made in general by the thermal decomposition of zirconium salts; $ZrOCl_2$, $ZrCl_4$ etc., hydroxide $Zr(OH)_4$ precipitated from aqueous solutions of the salts or from alkoxides, or by vapor phase oxidation of $ZrCl_4$. The powders by these methods are liable to have the contamination of impurities such as Cl^-, and to agglomerate resulting in the difficult dispersion.

We have developed two methods to prepare fine zirconia powders using hydrothermal reactions; (1) hydrothermal treatments of amorphous hydrated zirconia (ref. 1), and (2) hydrothermal oxidation method, where Zr metal is oxidized by H_2O or aqueous solutions at high temperatures under pressures. The latter can be combined with the hydrothermal reaction sintering method (ref. 2~5), which is a technique to prepare dense ceramics using hydrothermal reactions.

This paper deals with the synthesis of zirconia fine powders by the hydrothermal oxidation method and the sintering of them by the hydrothermal reaction sintering method.

EXPERIMENTAL PROCEDURE

Hydrothermal oxidation method

The starting material was zirconium metal powder*, 99.9% pure containing ≈ 2% Hf. It consisted of particles 1 to 10 μm in diameter. A 59 wt% calcium nitrate solution and a 41 wt% yttrium nitrate solution were made from reagent grade 99.0% $Ca(NO_3)_2 \cdot 4H_2O$** and 99.9% $Y(NO_3)_3 \cdot 6H_2O$***. $Ca(OH)_2$** and NH_4NO_3 were also used as stabilized agents.

Appropriate amounts of Zr and redistilled water or the above solution, or $Ca(OH)_2$ powder and H_2O, were hermetically sealed by arc welding in a platinum capsule 2.7 mm ID, 0.15 mm thick and 35 mm long. The sealed capsule was heated in a test-tube-type pressure vessel under the water pressure of 100 MPa. The pressure was measured by a calibrated Heise bourdon gauge and controlled within ± 0.2 MPa. The temperature was measured by Platinel thermocouples calibrated against the m.p. of Zn (419.5°C) and controlled within ± 3°C.

After the pressure vessel was quenched into a cold water, the capsule was weighed to check the leakage of the content. The solid products were examined by X-ray powder diffraction using a diffractometer# with a rotating Cu anode, by scanning electron microscopy## and by transmission electron microscopy###.

Hydrothermal reaction sintering method

This method was originally developed for sintering of metal oxides prepared from metal and water, so that it can be directly combined with hydrothermal oxidation method,

$$Zr + 2H_2O \longrightarrow ZrO_2 + 2H_2 \qquad \text{---------} \ (1).$$

As the hydrogen gas can permeate the capsule wall in further heating, the zirconia fine grains in the capsule are compressed isostatically by high-pressure fluids (gas or liquid). The heating and the compression of the fine grains provide an effective sintering in the capsule.

The capsule containing Zr and H_2O was heated in a high-gas-pressure apparatus using argon as a pressure medium up to 1000°C, 1200°C in particular cases, by the rate of 6°C/min. The temperature and pressure were held within ± 20°C and ± 2 MPa, respectively. After cooling down to a room temperature in 2 h, the capsule was weighed to check the permeation of hydrogen gas according to eq.(1). The sintered specimen was removed from the capsule, then examined by X-ray diffraction and scanning electron microscopy. The density of sintered specimens was measured by an Archimedes' method where toluene and tungsten wire 0.05mm in diameter were used to reduce the influence of surface tension.

* : Kōjundo Chemical Lab. Inc., Wako, Saitama, Japan.
** : Kantō Chemical Co., Ltd., Tokyo, Japan.
*** : Yoneyama Chemical Co., Ltd., Tokyo, Japan.
\# : RU-200, Rigaku Denki Co., Ltd., Tokyo, Japan.
\## : JSM T-200, JEOL Co., Ltd., Tokyo, Japan.
\### : JEM-100CX, JEOL Co., Ltd., Tokyo, Japan.

RESULTS AND DISCUSSION

The reaction between Zr metal and H_2O did not occur up to 200°C under 100 MPa, but did above 250°C. The solid products in the capsule at 300°C contained monoclinic zirconia, zirconium hydride and unreacted zirconium metal (Fig. 1). The zirconium hydride was formed by the reaction between Zr and high pressure hydrogen produced by the reduction of water. As the hydride decomposed above 400°C, sole monoclinic zirconia was produced, which indicated the complete reaction of the eq.(1). The average crystallite size was calculated to be 24 nm for the monoclinic zirconia produced at 400°C based upon the line broadening of the X-ray diffraction. This was confirmed by TEM observation. Scanning electron micrograph, Fig. 2, revealed that the monoclinic zirconia micro-grains formed a crust-like structure on an original metal grain.

The growth of crystalline was observed above 700°C, then well-crystallized monoclinic zirconia (>100 nm) was formed at 1000°C as suggested in Fig. 1.

It is noteworthy that no tetragonal zirconia could be produced in the reaction between Zr and H_2O in the present study, even though the average crystallite size was less than 30nm. Topotactic nucleation (ref. 6) on the zirconium metal surface and/or

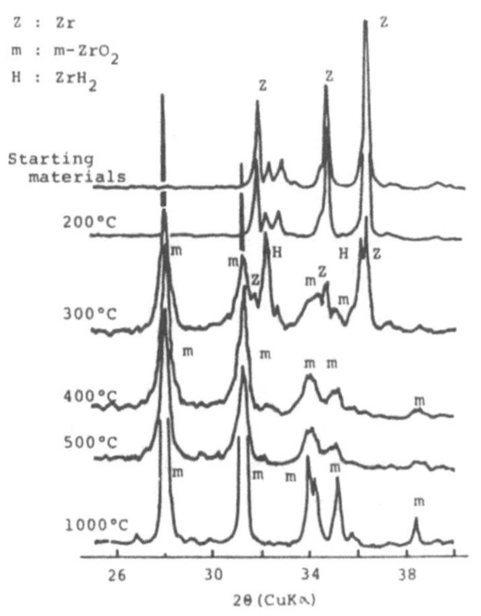

Fig. 1. Reaction products between Zr and H_2O under 98 MPa for 3 h.

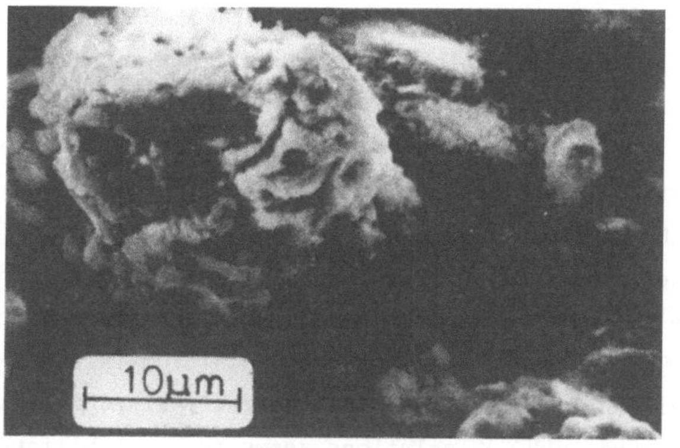

Fig. 2. Monoclinic ZrO_2 grains produced at 400°C.

strain-free growth (ref. 7) might not allow to crystallize metastable tetragonal zirconia in the hydrothermal oxidation processes.

The heating of the mixture 0.15 $Ca(OH)_2$ + 0.85 Zr + 1.55 H_2O (= $Zr_{0.85}Ca_{0.15}O_{1.85}$+ 1.70 H_2) at 500°C or 600°C did not produce cubic zirconia but the mixture of $Ca(OH)_2$ and m-ZrO_2, whereas the heating at 700°C yielded c-ZrO_2 and $CaZrO_3$ in addition to m-ZrO_2. c-ZrO_2 and $CaZrO_3$, therefore, seemed to be formed by solid state reactions between m-ZrO_2 and $Ca(OH)_2$. The contents of c-ZrO_2 did not exceed 0.15 in the products from the starting materials containing $Ca(OH)_2$. On the other hand, relatively high yield ($\approx$ 0.4) of c-ZrO_2 was observed in the products by the reaction between Zr and $Ca(NO_3)_2$ solution at 700°C when excess amounts of Zr existed against total oxygen; Zr/O > 0.5, as seen in Fig. 3. The products contained also m-ZrO_2, $Ca(OH)_2$, ZrH_2 and ZrN. The last would be produced by the reaction between excess Zr and nitrogen gas reduced from nitrate ions. The reaction between Zr and H_2O has never resulted in the formation of either cubic zirconia or tetragonal zirconia even though Zr was excess against H_2O; i.e. Zr/O > 0.5, as described before. The formation of the cubic zirconia by the reaction between Zr and $Ca(NO_3)_2$ solution should be resulted from the influences

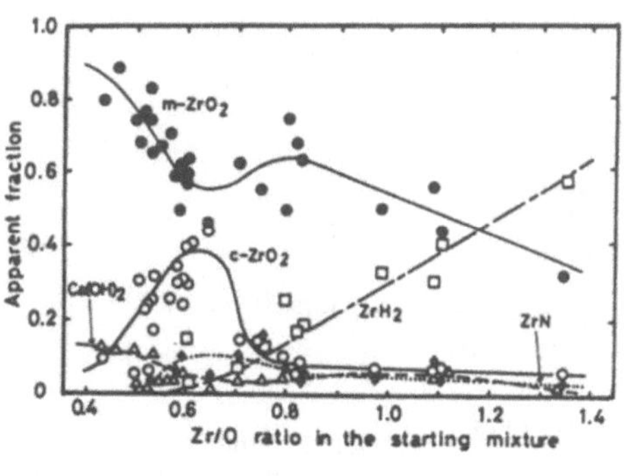

Fig. 3. Reaction products between Zr and $Ca(NO_3)_2$ solution at 700°C.

of calcium and/or nitrogen. However, the lattice parameter of the cubic zirconia phase ranged a = 0.507 ~ 0.510 nm, which is much smaller than that a = 0.515 nm for a calcium stabilized zirconia but rather close to a = 0.509 nm for the cubic phase of pure zirconia (0% Ca). The contribution of nitrogen to stabilize the cubic zirconia has been expected. (Ref. 8). The same treatments, at 700°C under 100 MPa for 3 h, of the mixture of Zr and NH_4NO_3 solutions could produce cubic zirconia phase though the yield was less than the case of $Ca(NO_3)_2$ solutions. The cubic zirconia containing nitrogen had a lattice parameter around a = 0.508 nm. An energy dispersive spectrometry associated with TEM revealed the existence of calcium in the aggregation of such crystallites of zirconia. In conclusion, the cubic zirconia phase produced by the reaction between Zr and $Ca(NO_3)_2$ solutions seems to be stabilized both by calcium and nitrogen. The cubic phase had an average crystallite size of 18 nm, but sometimes contained large cubic particles with about 250 nm as seen in Fig. 4.

Similar reaction between Zr and $Y(NO_3)_3$ solutions with excess Zr contents could produce small amounts of cubic zirconia phase at 700°C under 100 MPa.

The present study indicates that the hydrothermal oxidation method can produce fine grained oxides, pure and doped. However its reaction mechanism seems to be complex.

Hydrothermal reaction sintering

After the preparation of m-ZrO_2 fine grains by hydrothermal oxidation, further heating above 700°C has brought about sintering due to the compression by the ambient gas. Well-sintered bodies having bulk densities of 5.17 ~ 5.85 g/cm^3 were obtained at temperatures as low as 1000°C after 3 h under 100 MPa. Fig. 5 shows a typical scanning electron micrograph of the fracture surface of the specimen with the maximum density of 5.85 g/cm^3, or relative density of 99.0% compared the X-ray density of 5.91 g/cm^3 for 98% ZrO_2 - 2%HfO_2 solid solution. The fracture occured at the grain boundaries and the intergrains of zirconia grains with the size of 1 ~ 5 μm.

As the residual gas (or liquid) in the capsule would compensate the compression by

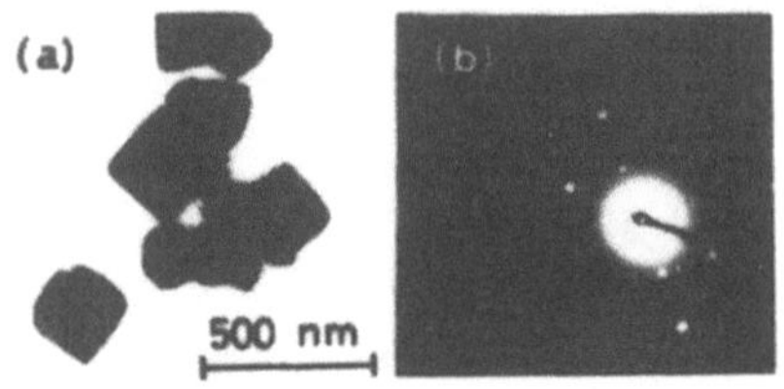

Fig.4 a) Cubic zirconia grains produced from Zr and $Ca(NO_3)_2$ solution at 700°C.
b) Electron diffraction pattern

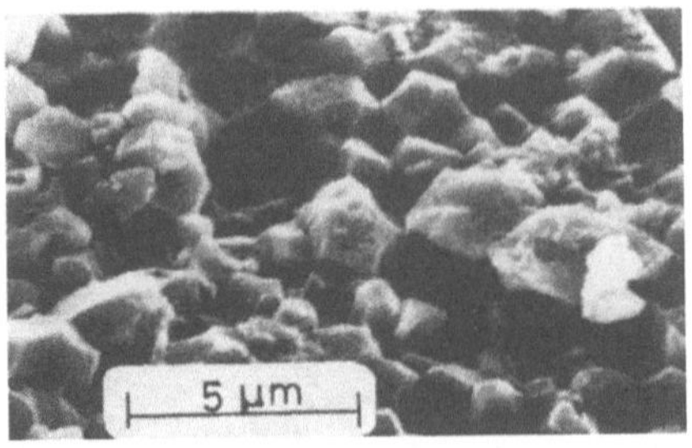

Fig.5. Dense m-ZrO_2 ceramics by hydrothermal reaction sintering at 1000°C, 98 MPa.

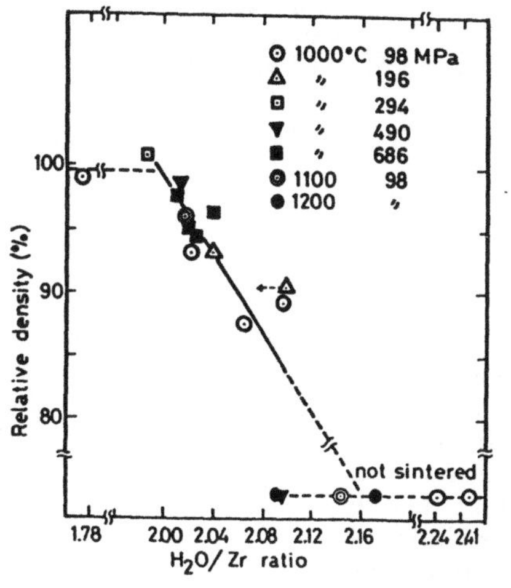

Fig. 6. Relative density of m-ZrO_2 ceramics sintered hydrothermally.

the ambient outer pressure in the hydrothermal reaction sintering processes, the ratio of water to zirconium metal apparently has the greatest effect on the final densification as seen in Fig. 6. (ref. 5).

This sintering method provides to prepare dense fine grained ceramics without any sintering additives at lower temperatures than the other sintering methods, because active fine oxide grains are sintered under the isostatic compression. Cr_2O_3 (ref. 3) and $LaCrO_3$ (ref. 9) also could be effectively sintered by this method.

REFERENCES

1. E. Tani, M. Yoshimura and S. Sōmiya, J. Am. Ceram. Soc., Accepted for publication.
2. M. Yoshimura and S. Sōmiya, Am. Ceram. Soc. Bull., 59 (1980) 246.
3. S. Hirano and S. Sōmiya, J. Am. Ceram. Soc., 59 (1976) 534.
4. S. Sōmiya, S. Hirano, M. Yoshimura, S. Itoh and H. Kanai, in S. Sōmiya and S. Saito (Eds.), Proc. Int. Symp. Factors in Densification and Sintering of Oxide and Non-oxide Ceramics, Hakone, Oct. 3-5, 1978, Gakujutsu Bunken Fukyukai, Tokyo Inst. Tech. Tokyo, 1979, pp. 267-77.
5. M. Yoshimura and S. Sōmiya, Advances in Ceramics, Proc. 1st Int. Conf. Science and Technology of Zirconia, Cleveland, June 16-18, 1980, Am. Ceram. Soc., Ohio, in Press
6. J. Livage, K. Doi and C. Mazières, J. Am. Ceram. Soc., 51 (1968) 349-53.
7. T. Mitsuhashi, M. Ichihara and U. Tatsuke, J. Am. Ceram. Soc., 57 (1975) 97-101.
8. N. Claussen, R. Wagner, L. J. Gaukler and G. Petzow, J. Am. Ceram. Soc., 61 (1978) 369-70.
9. M. Yoshimura, S-t. Song and S. Sōmiya, in H. Watanabe (Ed.) Proc. 3rd Int. Conf. Ferrites, Kyoto, Sept. 30-Oct. 4, 1980, Center for Acad. Publications Japan Tokyo, in Press.

Reprinted from *Journal of the American Ceramic Society*, **66**(2) (1983) 148–150.

Hydrothermal Oxidation of Hf Metal Chips in the Preparation of Monoclinic HfO_2 Powders

HIDEO TORAYA, MASAHIRO YOSHIMURA,* and SHIGEYUKI SŌMIYA*

Laboratory for Hydrothermal Syntheses, Research Laboratory of Engineering Materials and Department of Materials Science and Engineering, Tokyo Institute of Technology, Nagatsuta, Midori, Yokohama 227, Japan

Fine monoclinic HfO_2 powders were prepared from Hf metal chips by reaction with high-temperature high-pressure water. Dependence of the reaction on temperature-pressure-time was examined in both closed and open systems. Hydrothermal oxidation of Hf proceeded through three stages: the surface oxidation of Hf metal chips with H_2O, the reaction of the Hf with released H_2 to form hydrides, and the succeeding oxidation of the hydrides with H_2O to form HfO_2. Most of the fine HfO_2 powders were, however, considered to be formed by the latter two reactions after the cracking and pulverizing of metal chips associated with hydriding. The difference in the reaction rates between the closed and open systems was explained by taking into account the fugacity relations of H_2 and H_2O in the respective systems.

I. Introduction

Hydrothermal oxidation is an efficient technique for preparing fine oxide powders from metals by reaction with high-temperature high-pressure water and solutions. Its advantages are that very fine powders (particle size 20 to 30 nm) can be obtained at relatively lower temperatures (400° to 700°C) and no impurities can enter into the products.[1–6] Cr_2O_3, Fe_3O_4, ZrO_2, and stabilized ZrO_2 have been prepared using this method.[1–6] The reactions were simply written as $m\mathrm{M}+n\mathrm{H_2O}\rightarrow\mathrm{M}_m\mathrm{O}_n+n\mathrm{H_2}$ (M=Cr, Fe, or Zr).[1–6]

Hafnia is a thermally excellent material compared to the isomorphous ZrO_2.[7] In a previous study,[8] the present authors reported the preparation of fine monoclinic HfO_2 powders from Hf metal chips by this method. The formation of metal hydrides was, however, observed as an intermediate product as in the case of ZrO_2,[5,6] indicating that the reactions are more complicated than expected. It was also suggested[8] that this hydriding brings about the pulverization of the original metal chips. Hydriding, oxidizing, and pulverizing processes are, however, not yet completely understood. To elucidate the mechanism of hydrothermal oxidation is important for controlling the particle sizes of formed oxides.

In this study, the reaction rates and particle sizes of the products were examined as a function of temperature, pressure, time, and difference in atmosphere.

II. Experimental Procedure

Closed- and open-system runs were prepared to change the atmospheres. In the closed-system runs, the Hf metal chips[8] (1 by 0.2 by 0.02 mm) and ≈24 mg of redistilled water (2:1 H_2O/Hf mole ratio) were sealed by electric arc welding in a 2.7-mm ID Pt capsule (0.15 mm thick and 35 mm long). In the open-system runs, a capsule with the same dimensions and contents as the closed ones was prepared without sealing the top, leaving an opening of 4 by 0.7 mm. These capsules were heated in a test-tube-type pressure vessel. In the closed-system runs, the temperature was varied from 300° to 700°C under 100 MPa for 3 h, the pressure from 50 to 150 MPa at 600°C for 3 h, and time from 3 to 120 h at 500°C under 100 MPa. In the open-system runs, time was varied from 3 to 120 h at 500°C under 100 MPa and 3 to 15 h at 600°C. The temperature was measured by a platinel thermocouple and the pressure by a Bourdon gage.* They were held within ±20°C and ±2 MPa during a run, respectively.

After the heat treatment, the vessel was quenched in water, then the products in the capsule were examined by X-ray powder diffractometry† and scanning electron microscopy.‡ Their relative amounts were determined by the X-ray powder pattern-fitting technique with a revised version of the program PFLS.[9] The crystallite sizes were determined from the integral breadths of the deconvoluted line profile[10] of the 101 reflection of Hf, 101 of Hf hydrides, and $\bar{1}11$ of HfO_2. The spherical shape was assumed for the crystallite and the lattice strain effect was ignored.

Received February 3, 1982; revised copy received June 28, 1982; approved July 27, 1982.
Supported in part by a Grant-in-Aid for Scientific Research (No. 443015 (1979-1980)).
*Member, the American Ceramic Society.

*Heise Plant, Dresser Industrial Valve and Instrument Div., Dresser Industries, Newtown, CT.
†RU-200, Rigaku Denki Co., Ltd., Tokyo, Japan.
‡JSM T-200, JEOL Co., Ltd., Tokyo, Japan.

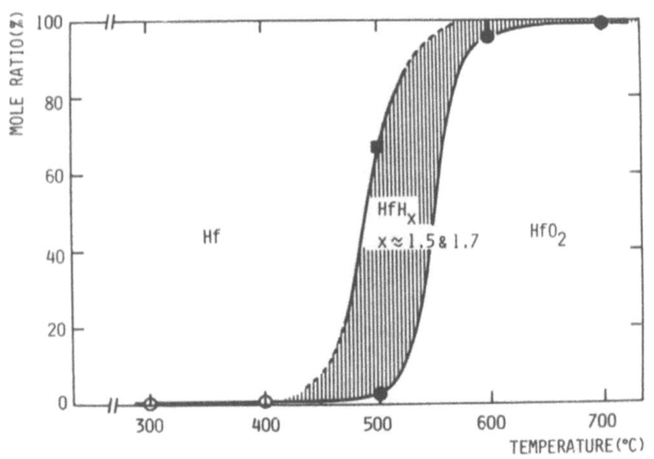

Fig. 1. Variation in relative amounts of products with temperature for treatment in closed system under 100 MPa for 3 h.

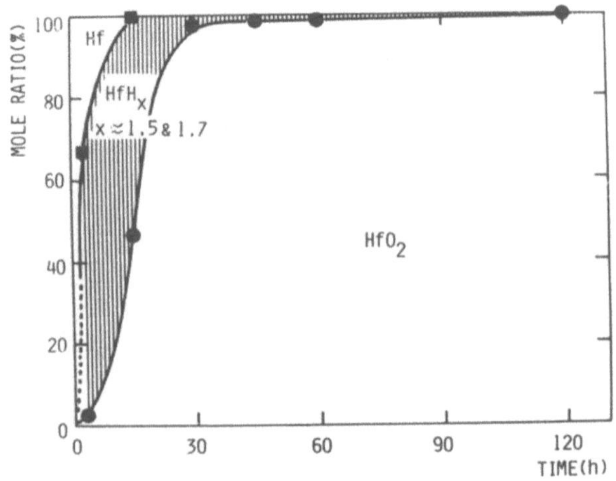

Fig. 2. Variation in amounts of products with time in closed-system runs at 500°C under 100 MPa.

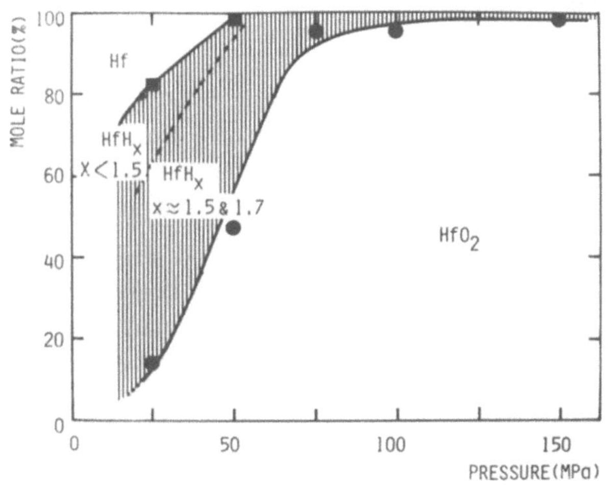

Fig. 3. Variation in amounts of products with pressure in closed-system runs at 600°C for 3 h.

III. Results and Discussion

(1) *Dependence of the Reaction on Temperature-Pressure-Time*

Figure 1 shows the variation of products with temperature in the closed system under 100 MPa for 3 h. At 300° and 400°C, Hf metal chips scarcely reacted with high-temperature, high-pressure water. The X-ray powder diffraction patterns of these products, however, showed a weak reflection at 28.3° in 2θ, indicating the formation of a very small amount of monoclinic HfO_2. The products at 500°C were primarily gray-black Hf hydrides, which still retained the original shape of chips but with little ductility. The broadened peak of the hydride observed at $\approx$38.3° in 2θ was thought to be due to the overlapping of reflections from two phases of hydrides, i. e. $HfH_{1.53}$ with tetragonal (δ' phase) and $HfH_{1.70}$ with cubic symmetry (δ phase).[11] At a high temperature of 600°C, these hydrides were changed into fine monoclinic HfO_2 powder with 5% Hf hydrides remaining. At 700°C, the products were monoclinic HfO_2 with only a trace amount of hydrides.

Figure 2 shows how the products at 500°C in Fig. 1 change with prolonged reaction time. Hafnium metal and, then, Hf hydrides decreased with time, and could be almost completely changed into HfO_2 even at 500°C, with a long enough reaction time. At higher temperatures, the same reactions were faster than those in Fig. 2.

At 300° and 400°C, HfO_2 is considered to be formed by the surface oxidation of Hf metal chips as

$$Hf+2H_2O \rightarrow HfO_2+2H_2 \quad (1)$$

Coating of the metal surface by oxide layers obstructs further oxidation in the interior of metal chips at these low temperatures. Hydrogen released by this reaction, however, starts to react with Hf. Hydrogen has a great affinity for Hf, as well as Zr or Ti, and solubility rapidly increases with increasing temperature.[12] Above 500°C, H atoms permeate actively into Hf metal chips through the coating oxide layers and react with Hf to form hydrides. As can be seen in Fig. 2, the formation of a large amount of HfO_2 follows that of Hf hydrides. Therefore, most of the HfO_2 is considered to be formed from Hf hydrides through two stages of reactions as

$$Hf+\tfrac{x}{2}H_2 \rightarrow HfH_x \quad (2)$$

$$HfH_x+2H_2O \rightarrow HfO_2+2\tfrac{x}{2}H_2 \quad (3)$$

In the hydrothermal oxidation of Zr, it was reported that Zr hydride was formed at 300°C and Zr was almost changed into ZrO_2 at 400°C[5,6], indicating that the reaction of Zr proceeds similarly to Hf at lower temperatures. However, metal powders were used instead of the metal chips in the case of ZrO_2. Therefore, a larger surface area would be effective for the reaction at lower temperatures as well as the difference in elements.

Figure 3 shows the pressure dependence of products in the closed-system runs at 600°C for 3 h. The relative amounts of HfO_2 did not vary above 100 MPa, but decreased abruptly with decreasing pressure below 75 MPa. Under 25 and 50 MPa, Hf metal still remained, together with the hydrides, and an extra phase was observed, which was thought to be an Hf hydride with an H/Hf ratio of less than 1.5.[13] The decrease in pressure lowers the fugacities of H_2 (f_{H_2}) and of H_2O (f_{H_2O}), and is considered to be responsible for the decrease in the reaction rates of Eqs. (2) and (3).

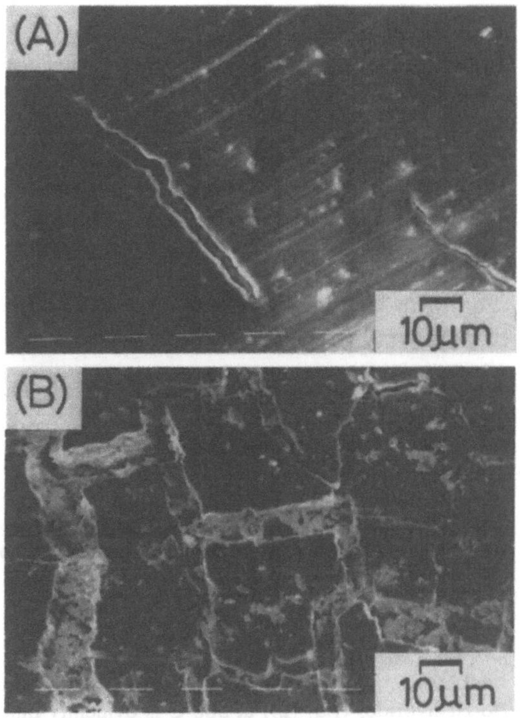

Fig. 4. Scanning electron micrographs of cracks observed on surface of Hf metal chip after runs at (*A*) 500°C under 100 MPa for 3 h and (*B*) 600°C.

(2) *Pulverization Process*

Figure 4 shows (*A*) cracks running on the surface of Hf metal chips after the run at 500°C under 100 MPa for 3 h and (*B*) many cracks over original metal chips at 600°C. The high-temperature, high-pressure water would be introduced into these cracks, and would accelerate the reaction by Eqs. (2) and (3). The diffraction peaks of Hf metal were gradually broadened with increasing temperature, indicating the permeation of H and/or O atoms into metal chips. After the run at 500°C under 100 MPa for 3 h, the crystallite size is 24 nm for Hf metal and 17 nm for Hf hydrides, a substantial reduction from the original sizes ($\gg$100 nm). The formation of $HfH_{1.53}$ and $HfH_{1.70}$ by hydriding Hf accompanies the change in the stacking sequence of the closest packing plane of Hf atoms and the volume increase by about 16%.[11,14,15] The drastic change in both coordination and volume induces microcracks in Hf metal chips and is the driving force of pulverization.

(3) *The Closed and Open Systems*

Figure 5 shows the time dependence of the reaction in the open system under the same conditions of temperature (500°C) and pressure (100 MPa) as those in Fig. 2. The open-system reactions were similar to those in the closed system, but were much more sluggish. On the contrary, at 600°C under 100 MPa, the reactions proceeded more rapidly in the open system than in the closed one; for 3 h, 99% HfO_2 was formed and only 1% Hf hydrides remained, and for 6 and 15 h, only HfO_2 was formed. However, for 3 h, 5% Hf hydrides still remained in the closed system (Fig. 1).

The difference in the reaction rates between the closed and open systems can be seen by distinguishing the reaction by Eq. (2) from

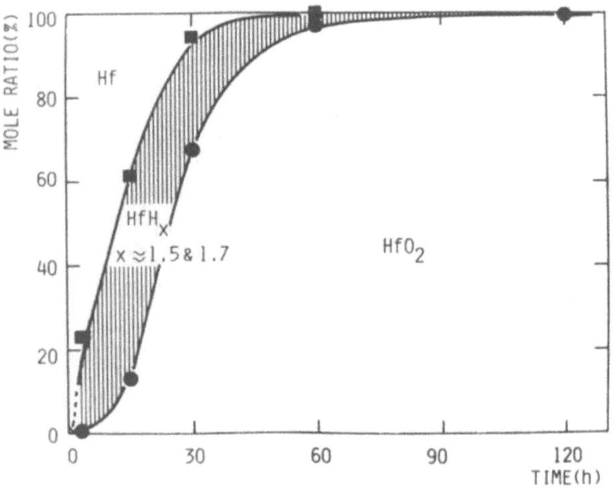

Fig. 5. Variation in amounts of products with time in open system under temperature and pressure conditions the same as those represented in Fig. 2.

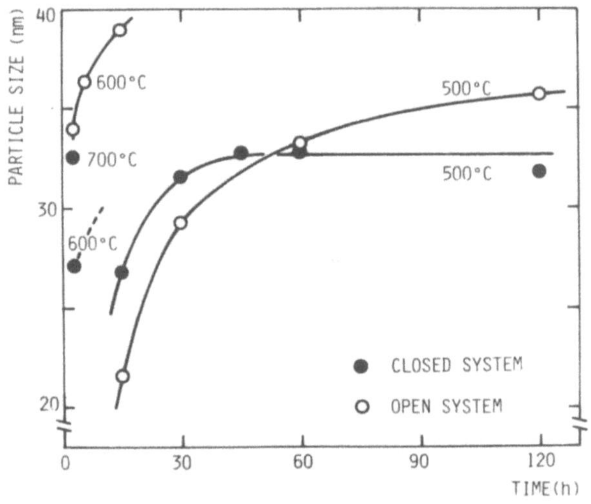

Fig. 6. Variation of crystallite size for monoclinic HfO_2 with time at temperatures indicated in both closed and open systems under 100 MPa.

that by Eq. (3) in each system. At first, the reaction rate from Hf to Hf hydrides (Eq. (2)) at 500°C is evidently much lower in the open system than in the closed one. In the latter system, f_{H_2} increases with the formation of HfO_2, whereas H_2 freely escapes from the former, giving a lower f_{H_2}. As has been demonstrated in the pressure-dependence diagram of Fig. 3, the decrease of f_{H_2} lowers the reaction rate of Eq. (2), resulting in a delayed overall reaction in the open system at 500°C. On the other hand, the reaction rates from Hf hydrides to HfO_2 are generally higher in the open system than the closed one (Figs. 2 and 5). The higher reaction rate of Eq. (3) in the former system is due to the inexhaustible supply of H_2O which keeps a high f_{H_2O}, giving the apparently rapid overall reaction at 600°C compared to the latter.

Figure 6 shows the variation of crystallite size of HfO_2 with time at various temperatures in the respective systems. In each system, crystallite size increases with increasing temperature. Furthermore, the crystallite sizes in the open system are generally larger than those in the closed one. The smaller crystallite size at 500°C in the open system for 0≈50 h is due to the delay of the overall reaction. In the closed system, the crystallite size at 500°C does not increase beyond 45 h, being constant at ≈32 nm. This variation is due to the decrease of H_2O which transports the materials in the system. On the other hand, in the open system, the free circulation of H_2O continually promotes the growth of the HfO_2 crystallite. Since the crystal growth is suppressed, the closed system is better for preparing fine oxide powders.

IV. Summary

The hydrothermal oxidation of Hf metal chips proceeded through three stages: (1) the surface oxidation of Hf metal chips with H_2O, (2) the reaction of Hf with released H_2 to form hydrides above 500°C, and (3) the succeeding oxidation of the hydrides with H_2O. Most of the HfO_2 was, however, formed by the latter two reactions after the cracking of metal chips associated with hydriding. The fugacities of H_2 (f_{H_2}) and of H_2O (f_{H_2O}) are rate determinants in reactions (2) and (3), respectively. The difference in the reaction rates between the closed and open systems could be explained by taking into account the fugacity relations f_{H_2} (closed)> f_{H_2} (open) and f_{H_2O} (closed)<f_{H_2O} (open). These relations are also very important for crystallite growth and, therefore, the fineness of formed oxide powders.

Acknowledgments: The authors thank Nippon Mining Co., Ltd. for supplying the Hf metal.

References

[1]S. Hirano and S. Sōmiya, "Hydrothermal Crystal Growth of Magnetite in the Presence of Hydrogen," *J. Cryst. Growth*, **35**, 273–78 (1976).
[2]S. Hirano and S. Sōmiya, "Hydrothermal Reaction Sintering of Pure Cr_2O_3," *J. Am. Ceram. Soc.*, **59** [11–12] 534 (1976).
[3]M. Yoshimura and S. Sōmiya, "Fabrication of Dense, Nonstabilized ZrO_2 Ceramics by Hydrothermal Reaction Sintering," *Am. Ceram. Soc. Bull.*, **59** [2] 246 (1980).
[4]M. Yoshimura and S. Sōmiya, "Hydrothermal Reaction Sintering of Monoclinic Zirconia"; pp. 455–63 in Science and Technology of Zirconia, Advances in Ceramics, Vol. 3. Edited by A. H. Heuer and L. W. Hobbs. American Ceramic Society, Columbus, OH, 1981.
[5]M. Yoshimura, S. Kikugawa, and S. Sōmiya, "Preparation of Zirconia Fine Powders by the Reactions between Zirconium Metal and High Temperature-High Pressure Solutions"; Proceedings of the 8th AIRAPT & 19th EHPHG Conf., High Pressure in Research and Industry, Aug. 17-22 (1981), Uppsala, Sweden (in press).
[6]M. Yoshimura and S. Sōmiya, "Synthesis and Sintering of Zirconia Fine Powders by Hydrothermal Reactions from Zirconium Metal and High-Temperature High-Pressure Solutions"; Proceedings of the 5th Int. Round Table Conf. on Sintering, Sept. 7-10 (1981), Portoroz, Yugoslavia (in press).
[7]C. T. Lynch; pp. 193–216 in Refractory Materials: High Temperature Oxides, Pt. II, Vol. 5. Edited by A. M. Alper. Academic Press, New York, 1970.
[8]H. Toraya, M. Yoshimura, and S. Sōmiya, "Preparation of Fine Monoclinic Hafnia Powders by Hydrothermal Oxidation," *J. Am. Ceram. Soc.*, **65** [5] C-72 (1981).
[9]H. Toraya and F. Marumo, "Application of Total Pattern-Fitting to X-ray Powder Diffraction Data," *Rept. Res. Lab. Eng. Mater., Tokyo Inst. Technol.*, **1980** [5] 55–64.
[10]H. P. Klug and L. E. Alexander; pp. 618–708 in X-Ray Diffraction Procedures, 2d ed. Wiley & Sons, New York, 1974.
[11]S. S. Sidhu and J. C. McGuire, "An X-Ray Diffraction Study of the Hafnium-Hydrogen System," *J. Appl. Phys.*, **23** [11] 1257–61 (1952).
[12]J. D. Fast; pp. 175–233 in Interaction of Metals and Gases, Vol. 1. Academic Press, New York, 1965.
[13]M. Caillet, J. Vigneron, and J. Besson, "Reaction of Hydrogen with Hafnium" (in Fr.), *Bull. Soc. Chim. Fr.*, **1972** [8] 3041–45.
[14]R. W. G. Wyckoff; pp. 11 in Crystal Structures, Vol. 1. Wiley & Sons, New York, 1963.
[15]R. Ruh and P. W. R. Corfield, "Crystal Structure of Monoclinic Hafnia and Comparison with Monoclinic Ziconia," *J. Am. Ceram. Soc.*, **53** [3] 126–29 (1970).

Reprinted from *Journal of the American Ceramic Society*, **66**(11) (1983) 818–822.

Reaction Kinetics in the Hydrothermal Oxidation of Hf

HIDEO TORAYA, MASAHIRO YOSHIMURA, and SHIGEYUKI SŌMIYA*

Laboratory for Hydrothermal Syntheses, Research Laboratory of Engineering Materials, Tokyo Institute of Technology, Nagatsuta, Midori, Yokohama 227, Japan

The hydrothermal oxidation of Hf was studied using metal powders and chips to change the surface-area effect. Fine monoclinic HfO_2 powders (32 to 34 nm) were formed through two parallel reaction paths, one the oxidation of Hf with H_2O and the other the consecutive reaction via formation of Hf hydride. Based on the experimental results at 500°C under 100 MPa, three reaction rate constants were determined for two kinds of starting materials. The relative widths of two paths were proved to be independent of the size of starting materials, and the pulverization and oxidation of both metal powders and chips could be explained by the same mechanism, where the formation of hydride was a driving force in pulverization.

Received March 11, 1983; revised copy received July 25, 1983; approved August 9, 1983.

*Member, the American Ceramic Society.

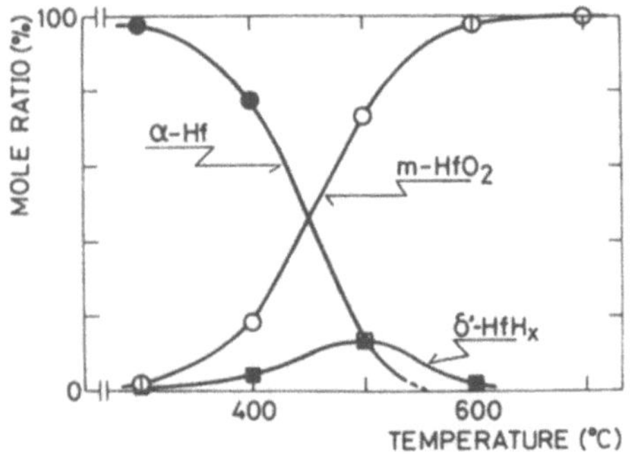

Fig. 1. Variation in relative amounts of products from Hf metal powders with temperature for heat treatment under 100 MPa for 3 h.

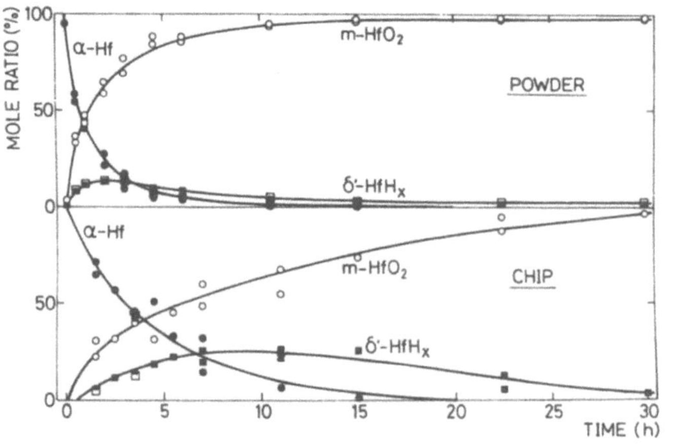

Fig. 2. Variation in amounts of products from metal powders (upper) and chips (lower) with time for heat treatment at 500°C under 100 MPa.

I. Introduction

IN THE hydrothermal oxidation of metals, pulverization and oxidation proceed simultaneously during reaction with high-temperature, high-pressure water. These phenomena have been applied to prepare fine oxide powders (20 to 30 nm) of Fe_3O_4,[1] Cr_2O_3,[2] ZrO_2,[3,4] HfO_2,[5] Al_2O_3,[6] and mixed Al_2O_3-ZrO_2[7] at relatively lower temperatures (400° to 600°C). In a previous report on the hydrothermal oxidation of Hf metal chips,[8] the present authors examined the dependence of the reaction on temperature-pressure-time in both closed and open systems. The reaction was shown to proceed through three stages: the surface oxidation of Hf metal chips with H_2O, the reaction of the Hf with released H_2 to form a hydride, and the succeeding oxidation of the hydride with H_2O to form HfO_2.[8] Most of the fine HfO_2 powders were considered to be formed by the latter two reactions, and formation of the hafnium hydride played an important role in cracking and pulverizing metal chips and accelerating the oxidation. The previous study, however, raised the question of whether the reactions of the Hf with H_2O follow the same reaction sequence when the metal chips become much smaller. If the surface oxidation proceeds at a much faster rate, with an increasing surface area effect due to the decrease in chip size, the role of hydride formation in the hydrothermal oxidation of Hf will become less important. In the hydrothermal oxidation of other metals, the formation of hydride was observed in the case of Zr,[4] with chemical properties similar to those of Hf.[9] Answering the question is, therefore, important for understanding the mechanism of hydrothermal oxidation, not only of Hf but also of Zr.

In the present study, the kinetics of the hydrothermal oxidation of Hf in the closed system was examined based on the experimental results for both metal powders and chips. The mechanism of the hydrothermal oxidation of Hf is discussed.

II. Experimental Procedure

The Hf metal powders* (particles averaging 15 μm in diameter by SEM observation) and ≃24 mg of redistilled water (2:1 H_2O/Hf mole ratio) were sealed by electric arc welding in a 2.7 mm ID Pt capsule (0.15 mm thick and 35 mm long). The same capsule, but with Hf metal chips† (1 by 0.3 by 0.04 mm avg) was prepared to obtain additional data for metal chips. The oxygen in air encapsulated during sealing was estimated to be <0.16% in total oxygen content. These capsules were heated in a test tube-type pressure vessel. In the runs with metal powders, the temperature was varied from 300° to 700°C under 100 MPa for 3 h. Reactions at 500°C are particularly interesting, because they proceed at moderate reaction rates. Thus the time was varied up to 60 h at 500°C under 100 MPa for both metal powders and chips. The heating rate was 16°C/min in all runs. In the kinetic study at 500°C, the time corrections during temperature increases were not applied to the reaction time, because the reaction was sluggish below 500°C. The temperature was measured by a platinel thermocouple (type K, IEC) and the pressure by a Bourdon gauge.‡ They were held within ±10°C and ±1 MPa, respectively, during a run. After the heat treatment, the vessel was quenched in water.

The products in the capsule were identified by X-ray powder diffractometry.§ X-ray powder diffraction intensity data were collected with Cu$K\alpha$ radiation monochromatized with a graphite monochromator using the step scan technique (step width=0.02° in 2θ and fixed time=20 s). The relative amounts of products were determined by comparing the observed integrated intensities of reflections (101 of Hf, 101 of Hf hydride, and $\bar{1}11$ of HfO_2) with the corresponding calculated intensities, where the latter were calculated from the powder diffraction intensity formula[10] and the structure factors based on the atomic coordinates of Hf,[11] Hf hydride,[12] and monoclinic HfO_2.[13] The crystallite sizes were determined from the integral breadths of the deconvoluted line profiles[14] of the same reflections with those used for the relative amount measurements. Line broadening was assumed to be due only to the size effect of the spherical crystallite.[15] The d_{101} values of α-Hf were measured with an internal standard technique.

III. Results and Discussion

(1) Variation of Products with Temperature and Time

Products formed from Hf metal powders were α-Hf, HfH_x (x=1.53: δ′ phase) with deformed cubic lattice[12] and monoclinic HfO_2 as in the case of metal chips.[8] An extra reflection with d of 0.267 nm was observed in the runs at 500°C. This extra phase always coexisted with the δ′ phase (about 5 to 15% of δ′), and it was thought to be HfH_x (x<1.5).[16] Figure 1 shows the variation in amounts of these products with temperature for heat treatment under 100 MPa for 3 h. The variation showed similar features to that obtained for metal chips,[8] but reactions of metal powders proceeded much faster at the respective temperatures. At 300°C, a very small amount of Hf hydride appeared, which did not appear in the corresponding runs using metal chips.[8] At 700°C, the Hf was completely changed into HfO_2 for 3 h, while a trace amount of Hf hydride still remained in the case of metal chips.[8]

In Fig. 2, the variation of products with reaction time at 500°C under 100 MPa is compared between the metal powders and chips. Hafnium in powder decreased more rapidly than that in chips, but

*Maruyasu Sangyo Co., Ltd., Tokyo, Japan.
†Nippon Mining Co., Ltd., Tokyo, Japan.

‡Heise Plant, Dresser Industrial Valve and Instrument Division, Dresser Industries, Newton, CT.
§RU-200, Rigaku Denki Co., Ltd., Tokyo, Japan.

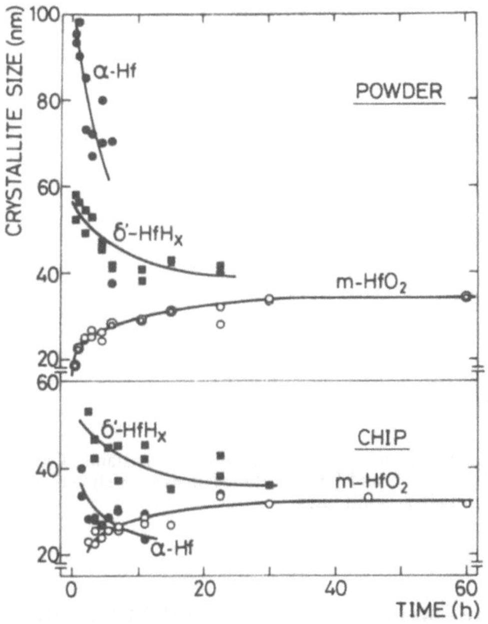

Fig. 3. Variation of crystallite sizes of products represented in Fig. 2.

both disappeared at the same time. On the other hand, HfH_x formed from metal powders increased more rapidly and passed through the maximum earlier than chips by about 7 h. The variation of HfO_2 for metal powders showed a much faster overall reaction rate.

Figure 3 shows the variation of crystallite sizes with reaction time for the same products in Fig. 2. The plastic deformation and the microcracks are associated with the increasing permeation of hydrogen into metals and the phase transition of α phase to δ' phase.[8,17] The rapid decrease in crystallite size of Hf metals is ascribed to both size and strain effects from these phase transitions and, therefore, a large part of the Hf metal should be considered to be a metal hydride of α phase immediately after the start of reaction. The smaller crystallite sizes of α-Hf for metal chips seemed to be affected by the larger strain of chips than powdered grains. On the other hand, the crystallite sizes of HfH_x and HfO_2 at each time generally coincide between the metal powders and chips. The crystallite sizes of HfH_x decreased with decreasing amounts of HfH_x due to the oxidation with H_2O, whereas the crystallite sizes of HfO_2 increased with time, and then gradually approached a constant value of 32 to 34 nm. In the open system runs with $H_2O/Hf \gg 2$, the crystallite size of HfO_2 increases with time monotonically beyond the constant value in the corresponding closed system.[8] In the closed system with $H_2O/Hf=2$, the H_2O decreases with the formation of HfO_2. Therefore, the decrease in the growth rate of HfO_2 crystallite is considered to be due to the decreasing mass transportation effect of H_2O.[8] In the system Hf-H_2O, the maximum crystallite size of HfO_2 is primarily controlled by the ratio H_2O/Hf, and is independent of the size of the starting materials.

The volume of α-Hf generally increases with increasing hydrogen permeated into metals.[17] In the present experiments, the observed d_{101} values of α-Hf did not change with time within the experimental errors, and the volume expansion could not be detected. This is partly due to the small solubility of hydrogen into α-Hf (<2%)[17] at room temperature.

(2) *Determination of Reaction Rate Constants*

Reactions in the hydrothermal oxidation of Hf proceeded through two paths, A and B. Path A is a reaction expressed by

$$Hf+2H_2O \longrightarrow HfO_2+2H_2 \tag{1}$$

and the parallel path B is a consecutive reaction by

$$Hf+\frac{x}{2}H_2 \longrightarrow HfH_x \tag{2}$$

$$HfH_x+2H_2O \longrightarrow HfO_2+2\frac{x}{2}H_2{}^{8} \tag{3}$$

One alternative to the reaction by Eq. (2) is

$$Hf+\frac{x}{2}H_2O \longrightarrow HfH_x+\frac{x}{4}O_2$$

A reaction rate of forming HfH_x was, however, much lower in the open system runs than in the closed system runs, where the fugacity relations are $f_{closed}>f_{open}$ for H_2 and $f_{closed}<f_{open}$ for H_2O.[8] Equation (2) is, therefore, reasonable. The other problem is the presence of two hydride phases at 500°C. Equations (2) and (3) can be written more exactly as

$$Hf+\frac{x_1}{2}H_2 \longrightarrow HfH_{x_1}$$

$$Hf+\frac{x_2}{2}H_2 \longrightarrow HfH_{x_2}$$

$$HfH_{x_1}+2H_2O \longrightarrow HfO_2+2\frac{x_1}{2}H_2$$

$$HfH_{x_2}+2H_2O \longrightarrow HfO_2+2\frac{x_2}{2}H_2$$

($x_1=1.53$ and $x_2<1.5$). However, the extra phase with $x<1.5$ amounts to only 5 to 15% of the phase with $x=1.53$ in intensity ratio. Therefore, the assumption that $[HfH_x]=[HfH_{x_1}]+[HfH_{x_2}]$ was considered to introduce no significant error. In the following calculation, the reactions by Eqs. (1) to (3) were assumed for the hydrothermal oxidation of Hf metal powders and chips. Then, the next step is to determine the rate constants of respective reactions for two kinds of starting materials.

The following five differential equations can be written for the reactions by Eqs. (1) to (3)[18]:

$$-\frac{d[Hf]_A}{dt}=K_1[Hf][H_2O]^2 \tag{4}$$

$$-\frac{d[Hf]_B}{dt}=K_2[Hf][H_2]^{x/2} \tag{5}$$

$$\frac{d[HfH_x]}{dt}=K_2[Hf][H_2]^{x/2}-K_3[HfH_x][H_2O]^2 \tag{6}$$

$$\frac{d[HfO_2]_A}{dt}=K_1[Hf][H_2O]^2 \tag{7}$$

$$\frac{d[HfO_2]_B}{dt}=K_3[HfH_x][H_2O]^2 \tag{8}$$

where K_1, K_2, and K_3 are rate constants for the reactions by Eqs. (1) to (3), and subscripts A and B represent paths A and B. The integrations of Eqs. (4) and (5) give

$$[Hf]=[Hf]_0 \exp\left[-\int_0^t (K_1[H_2O]^2+K_2[H_2]^{x/2})dt\right] \tag{9}$$

where $[Hf]_0$ is an initial concentration of Hf. Equation (6) is a linear differential equation of the first order, and its solution is given by[19]

$$[HfH_x]=\exp\left[-\int_0^t K_3[H_2O]^2dt\right]\left\{\int_0^t K_2[Hf][H_2]^{x/2} \exp\left[\int_0^t K_3[H_2O]^2dt\right]dt\right\} \tag{10}$$

Integration of Eqs. (7) and (8) gives

$$[HfO_2]=K_1\int_0^t [Hf][H_2O]^2dt+K_3\int_0^t [HfH_x][H_2O]^2dt \qquad (11)$$

Under high temperature and high pressure, only H_2 permeates out of the Pt capsule and H_2O remains in it.[1] In Eqs. (9) and (10), however, $[H_2]$ is concerned only when Hf persists in the capsule. At a relatively lower temperature of 500°C, the amount of H_2 diffused out of the Pt capsule can be ignored up to 15 h; beyond that range the Hf is a negligible amount. Further ignoring the amount of hydrogen in α-Hf, $[H_2]$ and $[H_2O]$ with an initial concentration of $[H_2O]_0$ can be calculated by

$$[H_2]=2[HfO_2]-\frac{x}{2}[HfH_x] \qquad (12)$$

$$[H_2O]=[H_2O]_0-2[HfO_2] \qquad (13)$$

From Eqs. (9) to (13), relative amounts of Hf, HfH_x, and HfO_2 can be calculated as a function of time by using the numerical integration, provided the rate constants K_1 to K_3 are known. On the contrary, the three rate constants can be determined by fitting the calculated to the observed values (Fig. 2). The curve fitting was performed by minimizing the following function with the simplex algorithm.[20]

$$\Delta=\sum_i([Hf]_{obs_i}-[Hf]_{calc_i})^2$$
$$+\sum_i([HfH_x]_{obs_i}-[HfH_x]_{calc_i})^2$$
$$+\sum_i([HfO_2]_{obs_i}-[HfO_2]_{calc_i})^2$$

where a subscript obs_i represents the observed value at time t_i and $calc_i$ the corresponding calculated value. Figure 4 shows the results of the minimization in the definition range of 0 to 30 h (time step of 0.01 h) with the value of 1.53 for x in HfH_x, supporting the general scheme described above. The rate constants thus determined are given in Table I.

(3) The Relative Widths of Reaction Paths A and B

In Fig. 5, calculated values of $[HfO_2]_A$ and $[HfO_2]_B$, the first and second terms on the right of Eq. (11), are plotted against time, together with $[HfO_2]$ and $[H_2O]/2$, for both metal powders and chips. Except for the time scale, the variation showed, surprisingly, the same pattern for two kinds of starting materials with different sizes: $[HfO_2]_A$ increased rapidly with time, approaching a constant value of about ⅓ and giving a ratio

$$\frac{[HfO_2]_B}{[HfO_2]_A}\simeq 2 \qquad (14)$$

at a sufficiently long time. This result indicates that path B is about twice as large as path A, irrespective of the size difference of starting materials.

(4) Mechanism of Hydrothermal Oxidation of Hf

A reaction sequence starts from the surface oxidation by Eq. (1). The large quantity of $[HfO_2]_A$, amounting to one-third of the total HfO_2, however, cannot be caused only by the oxidation on the original surfaces of metals. Associated with hydride formation of Hf, new surfaces along the cracks would be continuously exposed to H_2O in the interior of the metals, where the oxidation of α-Hf (path A) and HfH_x (path B) by H_2O proceeded simultaneously. The concentration of released hydrogen $[H_2]$ and, then, $[HfH_x]$ increase with increasing HfO_2, counterbalancing the relative widths of $[HfO_2]_A$ and $[HfO_2]_B$. The reaction through path A persists as long as α-Hf lasts, fading away with decreasing α-Hf in the system. A quantitative result of Eq. (14) demonstrated that the hydrothermal oxidation of Hf proceeds with the same mechanism for both metal powders and chips. The same variation of crystallite sizes of HfH_x and HfO_2 for two kinds of starting materials also supports this inference.

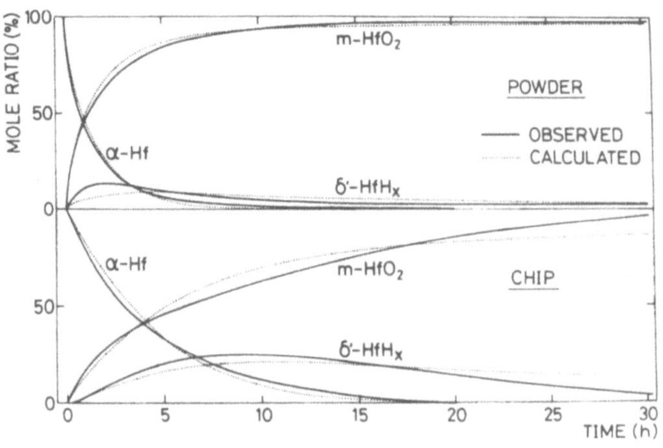

Fig. 4. Results of optimization of three reaction rate constants. Calculated values of [Hf], $[HfH_x]$, and $[HfO_2]$ (dotted lines) are shown with observed values from Fig. 2 (solid lines).

Table I. Rate Constants for Reactions of Hf Metal with High-Temperature, High-Pressure Water

Form of Hf	K_1	K_2	K_3
Powder	0.166	0.395	4.261
Chips	0.037	0.232	0.296

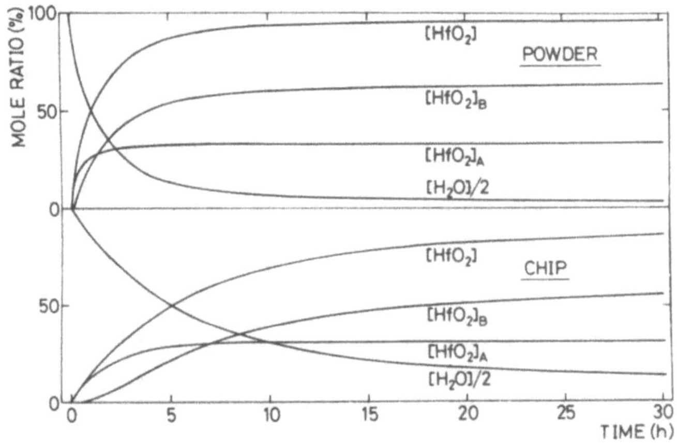

Fig. 5. Variation of $[HfO_2]_A$ and $[HfO_2]_B$ for metal powders and chips, together with $[HfO_2]$ and $[H_2O]/2$.

The time needed for 50% total gain of HfO_2 is 1.3 h for metal powders and 6 h for chips (Fig. 2), giving a ratio of 1.3/6 (=0.22), exactly the same as the ratio K_1(chip)/K_1(powder) of 0.22. The difference in overall reaction rates between the metal powders and chips can be attributed to the difference in rate constants K_1, being a rate determinant of overall reaction in hydrothermal oxidation of Hf. Among many factors for the rate constant, such as surface area effect, grain-boundary diffusion, or diffusion through grains, the difference in surface area per mole of starting materials would be a primary cause of the difference in K_1 between the metal powders and chips. A ratio K_1(chip)/K_1(powder) is in the same order with the ratio of surface area per mole calculated from the geometrical sizes of chip to that of powder (≃0.15). On the other hand, the rate constants K_2 for metal powders and chips do not differ greatly. The diffusion constant of hydrogen through metal grains would be a primary factor for K_2, giving the small difference in K_2 between the two materials.

IV. Summary

The hydrothermal oxidation of Hf proceeds through two parallel reaction paths, one the oxidation of Hf with H_2O and the other the consecutive reaction via formation of Hf hydride as an intermediate product. A kinetic study based on the experimental results for two kinds of starting materials with different surface area effects demonstrated that the relative widths of two paths are independent of the size of starting materials. The surface area of starting materials is considered to be a primary factor determining an overall reaction rate in the hydrothermal oxidation of Hf. Yet the same mechanism of pulverization and oxidation proceeds, irrespective of the size difference. The formation of Hf hydride is very important for the pulverization of original metals, cracking the metals, introducing H_2O into the interior of metals, and accelerating the oxidation in all metal grains.

References

[1]S. Hirano and S. Sōmiya, "Hydrothermal Crystal Growth of Magnetite in the Presence of Hydrogen," *J. Cryst. Growth*, **35**, 273–78 (1976).

[2]S. Hirano and S. Sōmiya, "Hydrothermal Reaction Sintering of Pure Cr_2O_3," *J. Am. Ceram. Soc.*, **59** [11–12] 534 (1976).

[3]M. Yoshimura and S. Sōmiya, "Fabrication of Dense, Nonstabilized ZrO_2 Ceramics by Hydrothermal Reaction Sintering," *Am. Ceram. Soc. Bull.*, **59** [2] 246 (1980).

[4]M. Yoshimura, S. Kikugawa, and S. Sōmiya, "Preparation of Zirconia Fine Powders by the Reactions between Zirconium Metal and High Temperature-High Pressure Solutions"; Proceedings of the 8th AIRAPT & 19th EHPHG Conf., High Pressure in Research and Industry, Aug. 17–22 (1981), Uppsala, Sweden.

[5]H. Toraya, M. Yoshimura, and S. Sōmiya, "Preparation of Fine Monoclinic Hafnia Powders by Hydrothermal Oxidation," *J. Am. Ceram. Soc.*, **65** [5] C-72 (1982).

[6]M. Yoshimura, S. Kikugawa, and S. Sōmiya, "Preparation of Alpha Alumina Fine Powders by Hydrothermal Oxidation Method," *J. Jpn. Soc. Powder Powder Metall.*, **30** [5] 207–10 (1983).

[7]S. Sōmiya, M. Yoshimura, and S. Kikugawa, "Preparation of Zirconia-Alumina Fine Powders by Hydrothermal Oxidation of Zr-Al Alloys"; Proceedings of the 19th University Conference on Ceramic Science, Nov. 8–10 (1982), North Carolina University.

[8]H. Toraya, M. Yoshimura, and S. Sōmiya, "Hydrothermal Oxidation of Hf Metal Chips in the Preparation of Monoclinic HfO_2 Powders," *J. Am. Ceram. Soc.*, **66** [2] 148–50 (1983).

[9]R. J. H. Clark, D. C. Bradley, and P. Thornton; pp. 419–90 in The Chemistry of Titanium, Zirconium and Hafnium. Pergamon, New York, 1973.

[10]R. W. James; pp. 34–52 in The Optical Principles of the Diffraction of X-rays: The Crystalline State, Vol. 2. G. Bell & Sons, London, 1967.

[11]R. W. G. Wyckoff; p. 11 in Crystal Structures, Vol. 1. Wiley & Sons, New York, 1963.

[12]S. S. Sidhu and J. C. McGuire, "An X-ray Diffraction Study of the Hafnium-Hydrogen System," *J. Appl. Phys.*, **23** [11] 1257–61 (1952).

[13]R. Ruh and P. W. R. Corfield, "Crystal Structure of Monoclinic Hafnia and Comparison with Monoclinic Zirconia," *J. Am. Ceram. Soc.*, **53** [3] 126–29 (1970).

[14]H. Toraya, M. Yoshimura, and S. Sōmiya, "A Computer Program for the Deconvolution of X-ray Diffraction Profile with the Composite of Pearson Type VII Functions"; to be published in *Journal of Applied Crystallography*.

[15]A. J. C. Wilson; pp. 37–54 in X-ray Optics, 2d ed. Methuen, London, 1962.

[16]M. Caillet, J. Vigneron, and J. Besson, "Reaction of Hydrogen with Hafnium" (in Fr.), *Bull. Soc. Chim. Fr.*, **1972**, [8] 3041–45.

[17]R. L. Beck and W. M. Mueller; pp. 241–335 in Metal Hydrides. Edited by W. M. Mueller, J. P. Blackledge, and G. G. Libowitz. Academic Press, New York, 1968.

[18]W. J. Moore; pp. 253–322 in Physical Chemistry, 3d ed. Prentice-Hall, Englewood Cliffs, NJ, 1962.

[19]W. Kaplan and D. J. Lewis; pp. 1111–17 in Calculus and Linear Algebra, Vol. 2. Wiley & Sons, New York, 1971.

[20]J. A. Nelder and R. Mead, "Simplex Method for Function Minimization," *Comput. J.*, **7**, 308–13 (1965).

Reprinted from N. Claussen, M. Rühle and A. Heuer (eds), *Advances in Ceramics*, Vol. 12, *Science and Technology of Zirconia II* (American Ceramic Society, 1984) pp. 806–815.

Preparation of Mixed Fine Al_2O_3-HfO_2 Powders by Hydrothermal Oxidation

HIDEO TORAYA,* MASAHIRO YOSHIMURA, AND SHIGEYUKI SŌMIYA

Tokyo Institute of Technology
Laboratory for Hydrothermal Syntheses
Research Laboratory of Engineering Materials
Yokohama 227, Japan

Mixed Al_2O_3-HfO_2 powders were prepared by hydrothermal oxidation of the intermetallic compound $HfAl_3$ and the mixture $HfAl_3$-4Al at 400° to 700°C under 100 MPa. Monoclinic HfO_2 with particle sizes of 10–20 nm was finely dispersed among the α-Al_2O_3 with sizes ranging from nanometers to micrometers. The temperature-time dependence of the reactions of Al, $HfAl_3$, and $HfAl_3$-4Al with H_2O was examined. The hydrothermal oxidation mechanism of the compound $HfAl_3$ was different from those of Hf and Al metals. The mixture with fine precipitation of $HfAl_3$ among the Al matrix is supposed to be effective for the preparation of mixed Al_2O_3-HfO_2 powders.

In hydrothermal oxidation, metals are pulverized and oxidized by reaction with high-temperature, high-pressure water. The oxidation proceeds at relatively lower temperatures (400°–700°C), and the formed oxide powders, such as those of ZrO_2[1,2] or HfO_2,[3,4] have very small particle sizes (25–35 nm) and narrow ranges of size distribution. Recently, $ZrAl_3$ and Zr_5Al_3 alloys were used as starting materials to prepare mixed Al_2O_3-ZrO_2 powders by hydrothermal oxidation.[5,6] Since Zr and Al atoms are distributed on their respective crystallographic positions of these intermetallic compounds, the experiment was expected to result in the fine dispersion of formed oxides, Al_2O_3 and ZrO_2. In that study, well-mixed powders of α-Al_2O_3 and monoclinic and tetragonal ZrO_2 were obtained; the particle sizes of the respective oxides were less than those obtained separately by the hydrothermal oxidation of Al and Zr metals.[5,6]

Mixed Al_2O_3-HfO_2 powders are potentially useful as ceramic materials in a higher temperature range, compared to Al_2O_3-ZrO_2 powders, due to a higher transition temperature of monoclinic $HfO_2 \rightleftharpoons$ tetragonal HfO_2.[7] The present study reports the hydrothermal oxidation of Al-Hf alloys. The Al metal, an intermetallic compound on the Al-rich side, $HfAl_3$, and a mixture, $HfAl_3$-4Al, with an intermediate composition between the Al and $HfAl_3$ were selected as starting materials. The hydrothermal oxidation of Al metal powders has already been reported,[8] whereas metal chips were used in the present study. In the hydrothermal oxidation of Hf metals, the structural change associated with the formation of hafnium hydride plays an important role, inducing microcracks and bringing about the succeeding pulverization and oxidation.[4,9] The formation of zirconium hydride was

*Now with the Nagoya Institute of Technology, Ceramic Engineering Research Laboratory, 10-6-29, Asahigaoka, Tajimi, 507, Japan.

observed in the hydrothermal oxidation of Zr metal[2] as in the case of Hf, whereas no formation of hydride was reported in the case of $ZrAl_3$ and Zr_5Al_3.[5,6] The hydrothermal oxidation mechanism of the compounds $HfAl_3$ and $ZrAl_3$ were, therefore, expected to be different from those of Hf and Zr metals. The pulverization-oxidation mechanism of the compound $HfAl_3$ and the effect of precipitated $HfAl_3$ on the hydrothermal oxidation of the mixture $HfAl_3$-4Al are the points of interest in the present study.

Experimental Procedure

Preparation of Starting Materials

The Al metal ingot (99.999%)[†] was rolled into ribbons 0.7-mm thick, and then cut into chips. The Al and Hf (containing 3.9% Zr)[‡] metals were weighed in a 1:3 Hf-Al atomic ratio for $HfAl_3$ and 1:7 for $HfAl_3$-4Al. These metals were fused by the radiation of a tungsten arc in a vacuum arc furnace.[§] The fused alloys were solidified quickly by cutting off the arc, and they were remelted more than five times to improve the homogeneity. These alloys were crushed into fragments, washed by acetone and alcohol in an ultrasonic washer, and dried in air. Microscopic observation of the polished surface of the mixture $HfAl_3$-4Al by reflected light showed that $HfAl_3$ crystals with acicular form (about 50 μm in length) were precipitated among the Al matrix. The $HfAl_3$ was identified to be the $ZrAl_3$ type by X-ray powder diffraction. The $HfAl_3$ sample included a small amount of $HfAl_2$.

Hydrothermal Oxidation

Closed- and open-system runs were performed to change the atmospheres for each of three kinds of starting materials. For the closed-system runs, Al metal chips (or crushed fragments of $HfAl_3$ or $HfAl_3$-4Al) (about 0.7 mm in length) and ≈35 mg of redistilled water (3:2 H_2O/Al molar ratio for Al, 13:2 $H_2O/HfAl_3$ for $HfAl_3$, and 25:2 $H_2O/HfAl_3$-4Al for $HfAl_3$-4Al) were sealed by electric arc welding in a 2.7-mm ID Pt capsule (0.15-mm thick and 35-mm long). For the open-system runs, a capsule with the same dimensions and contents as the closed ones was prepared without sealing the top, leaving an opening of 4 by 0.7 mm^2. These capsules were heated in a test-tube-type pressure vessel.

For the six systems, the temperature dependence of the reactions was first examined by varying the temperature for heat treatment from 400° to 700°C for 3 h. Next, the time dependence was examined by varying the time for heat treatment from 0 to 120 h at 500°C. The runs at 400° for 120 h and at 600°C for 30 h were also examined. The heating rate was 16°C/min, and the pressure was kept at 100 MPa in all runs. The temperature was measured by a platinel thermocouple and the pressure by a Bourdon gauge. They were held within ±10°C and ±1 MPa during the run. After the heat treatment, the vessel was quenched in water.

X-ray Diffraction and Microscopic Analyses

Products in the capsule were identified by X-ray powder diffractometry. For quantitative analyses, X-ray powder diffraction intensity data were collected with Cu$K\alpha$ radiation monochromated with a graphite monochromator by using the step-scan technique (step width = 0.02° in 2θ and fixed time of 10 or 20 s).[¶] The

[†]Osaka Asahi Metal Kōjō Co., Ltd., Tokyo, Japan.
[‡]Nippon Mining Co., Ltd., Tokyo, Japan.
[§]Daia Sinku Giken Co., Ltd., Tokyo, Japan.
[¶]RU-200, Rigaku Denki Co., Ltd., Tokyo, Japan.

Table I. Products Obtained by Heat Treatment at Various Temperatures under 100 MPa for 3 h.

	Closed system				Open system			
Product	400°C	500°C	600°C	700°C	400°C	500°C	600°C	700°C
				$Al\text{-}H_2O$				
Al	m*	m	w	vw	vw			
γ-AlOOH	m				s			
χ-Al_2O_3	w	vw			vw			
κ_1-Al_2O_3		w	w			w	w	
α-Al_2O_3	vw	m	s	vs	vw	s	vs	vs
				$HfAl_3\text{-}H_2O$				
$HfAl_3$	s	m	w	vw	s	w	vw	
γ-AlOOH	w				w			
Unidentified	w	vw			w			
α-Al_2O_3		m	s	vs		s	vs	vs
HfO_2	w	m	s	vs	w	s	vs	vs
				$HfAl_3\text{-}4Al\text{-}H_2O$				
$HfAl_3$	s	s	w	vw	m	w	vw	
γ-AlOOH	s	vw			s			
κ_1-Al_2O_3		m	m					
α-Al_2O_3		m	m	vs	vw	vs	vs	vs
HfO_2	w	m	s	vs	s	s	vs	vs

*vw, very weak; w, weak; m, medium; s, strong; vs, very strong.

relative amounts of products (Al, $HfAl_3$, α-Al_2O_3, and HfO_2) were determined from observed integrated intensity ratios of relevant reflections of each component compared with the calibration curves previously obtained for the mixtures of known amounts. Crystallite sizes were determined from the integral breadths of true diffraction profiles obtained by the deconvolution[10] of observed line profiles with standard ones. Line broadening was assumed to be due only to the size effect of the spherical crystallite.[11]

The powders obtained by hydrothermal treatment were examined by scanning** and transmission†† electron microscopy.

Results and Discussion

Reactions between the Hf-Al and H_2O

Table I gives the products obtained by the heat treatment at various temperatures under 100 MPa for 3 h. Except for the formation of χ-Al_2O_3 at 400° and 500°C, the products in the system $Al\text{-}H_2O$ generally coincide with those obtained from the corresponding runs using Al metal powders as starting material.[8] Boehmite (γ-AlOOH) was formed in all systems at 400°–500°C, and an unidentified phase, which seemed to be δ-Al_2O_3, was found in the system $HfAl_3\text{-}H_2O$ at 400°–500°C. κ_1-Al_2O_3[12] was formed in both closed- and open-system runs in $Al\text{-}H_2O$ at 500°–600°C but only in the closed-system runs in $HfAl_3\text{-}4Al\text{-}H_2O$.

**JSM T-200, JEOL Co., Ltd., Tokyo, Japan.
††JEM-200CX, JEOL Co., Ltd., Tokyo, Japan.

These intermediate products were changed into α-Al_2O_3 with increasing temperature. In the hydrothermal oxidation of $ZrAl_3$ and Zr_5Al_3,[5,6] both tetragonal and monoclinic ZrO_2 were formed, whereas only monoclinic HfO_2 was formed in the present experiment. The formation of hafnium hydride was not observed in all runs.

In the system Al-H_2O, the amounts of products with time at 500°C under 100 MPa varied in the same manner in both closed and open systems but at a much faster rate in the latter. The Al metal vanished within 1 h in the open system, while its trace amount remained even after 120 h in the closed system. γ-AlOOH and χ-Al_2O_3, which were believed to be formed while raising the temperature to 500°C, disappeared in the early stage of reaction (0–1 h). Instead of these products, κ_1-Al_2O_3 and α-Al_2O_3 appeared after the 1-h run. κ_1-Al_2O_3 first increased and then decreased gradually, while α-Al_2O_3 increased monotonically. A small amount of κ_1-Al_2O_3 persisted even after 120 h in the closed system, while it disappeared after 30 h in the open one. κ_1-Al_2O_3 is transformed into stable α-Al_2O_3 by prolonging the reaction time or raising the temperature or pressure.[12] The high concentration of H_2O will also promote the transformation of κ_1 to α. The cause of the large difference in the reaction rates between the closed and open systems in Al-H_2O is attributed primarily to the difference in the concentration of H_2O in the respective systems.

In contrast to the system Al-H_2O, the concentration of H_2O had little effect on the reaction rates in the system $HfAl_3$-H_2O, where a trace amount of $HfAl_3$ remained after 120-h runs in both closed and open systems (Fig. 1(*A*)). Since the present $HfAl_3$ sample contained no Al metals, when examined by X-ray powder diffraction, γ-AlOOH would be formed by the reaction of Al in $HfAl_3$ with H_2O when the temperature is increased. Small amounts of γ-AlOOH and an unidentified phase appeared only in the initial stage of reaction. The remaining $HfAl_3$, therefore, would be changed into HfO_2 and α-Al_2O_3 without formation of γ-AlOOH.

The reactions in the system $HfAl_3$-4Al-H_2O were the most complex of all the reactions in two systems Al-H_2O and $HfAl_3$-H_2O (Fig. 1(*B*)). As can be seen from the rapid decrease of Al in both closed and open systems, the reaction of the Al with H_2O proceeded at a much faster rate than the reaction of $HfAl_3$. The reaction was completed very rapidly in the open system, while it proceeded more slowly in the closed systems. Furthermore, the ratio of κ_1-Al_2O_3 to α-Al_2O_3 is very large in the closed system. The κ_1-Al_2O_3 was formed in the systems Al-H_2O and $HfAl_3$-4Al-H_2O but could not be observed in the system $HfAl_3$-H_2O with no Al metals.

Pulverization Mechanism

In the system Al-H_2O, the formed aluminum oxides were well-dispersed, white, fine powders. In the system $HfAl_3$-H_2O, the products retained the original bulk shape almost completely, even after the run for 120 h at 500°C; the bulks were hard to grind, in spite of the fact that most of the $HfAl_3$ was changed into HfO_2 and α-Al_2O_3 (Fig. 1(*A*)). Well-developed crystals of α-Al_2O_3 were observed on the surface exposed to water in the open system (Fig. 2(*A*)). In the interior, individual crystals were tightly agglomerated, as observed on the fractured surface (Fig. 2(*B*)). In the composite system $HfAl_3$-4Al-H_2O, the products were loosely gathered.

Figure 3 shows some transmission electron micrographs and energy dispersive spectra (EDS) of mixed Al_2O_3-HfO_2 powders formed in the system $HfAl_3$-H_2O. Very small particles (10–20 nm (Fig. 3(*B*)) adhered to the surface

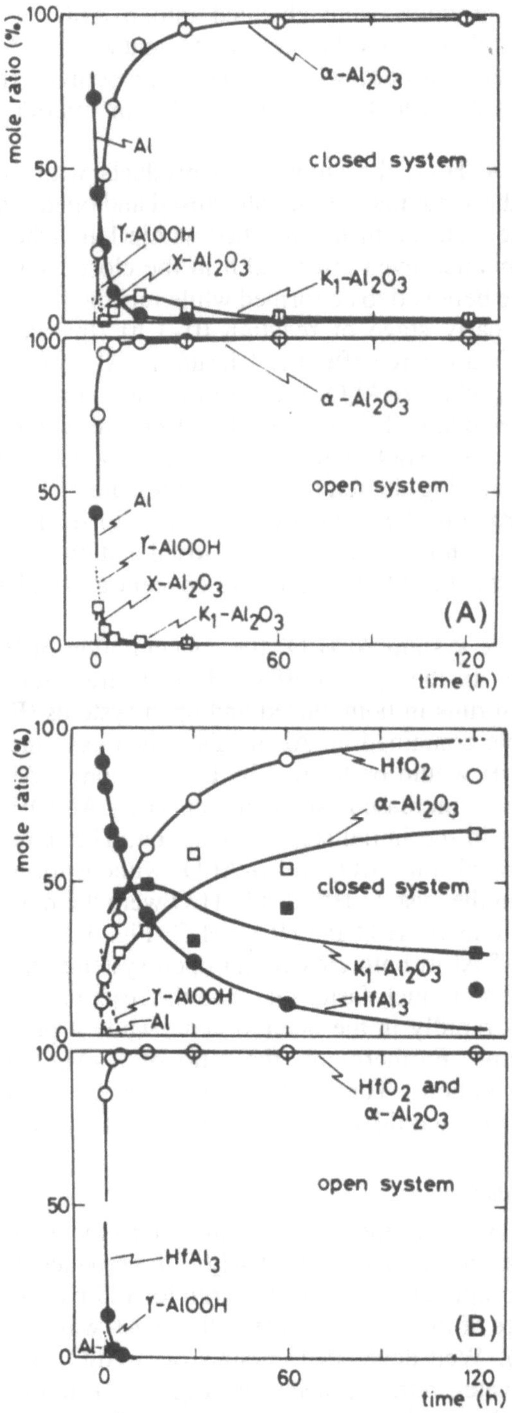

Fig. 1. Variation in relative amounts of products with time at 500°C under 100 MPa in the systems (*A*) $HfAl_3$-H_2O and (*B*) $HfAl_3$-4Al-H_2O.

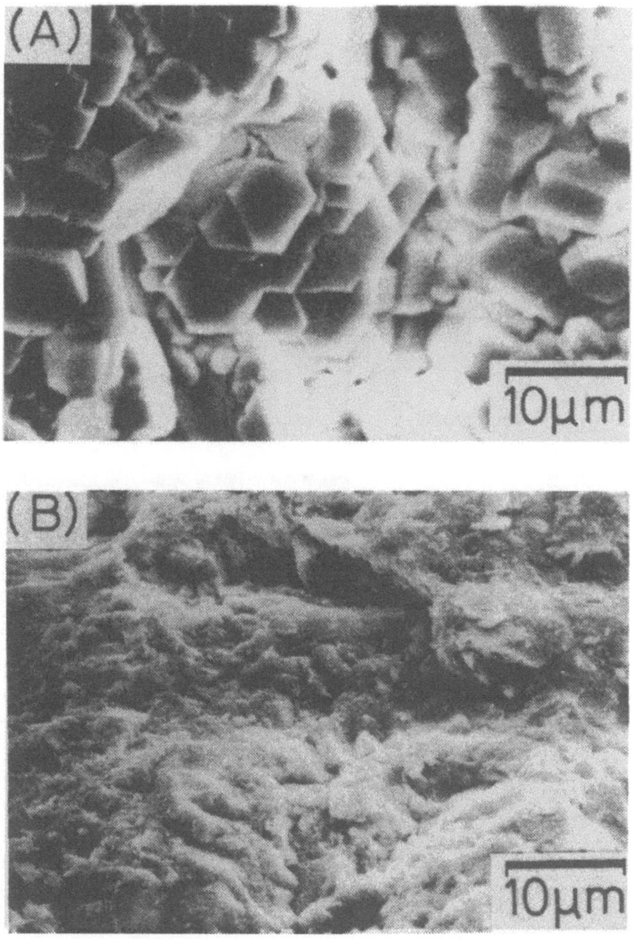

Fig. 2. Scanning electron micrographs of products obtained in the open system run at 500°C for 30 h in $HfAl_3$-H_2O: (*A*) The surface exposed to water and (*B*) the fractured surface.

of large crystallites (Fig. 3(*A*)); isolated particles were revealed to be HfO_2 by EDS (Fig. 3(*C*)). Large platelike crystals (1–10 μm long), which appeared in both $HfAl_3$-H_2O and $HfAl_3$-4Al-H_2O, were confirmed to be α-Al_2O_3 (Fig. 3(*D*)). The EDS spectra obtained by scanning the region, as shown in Fig. 3(*E*), gave both Al*K*α and Hf*K*α lines. Therefore, some α-Al_2O_3 crystals would have a size compatible with that of HfO_2. The average crystallite size of α-Al_2O_3 was estimated to be $\geqslant$150 nm in the systems Al-H_2O and $HfAl_3$-4Al-H_2O by XRD but in the range of 60–90 nm in $HfAl_3$-H_2O. The range of size distribution is, however, very wide as has been observed by TEM.

Figure 4 shows the variation of crystallite size of HfO_2 with time at 500°C, together with values for the runs at higher temperatures. The crystallite sizes of HfO_2 determined by the X-ray technique were in good agreement with those measured by TEM, having a narrow range of size distribution. In the system $HfAl_3$-H_2O, the presence of excess water has no apparent effect on the crystallite size, showing the same variation for both closed and open systems. On the other

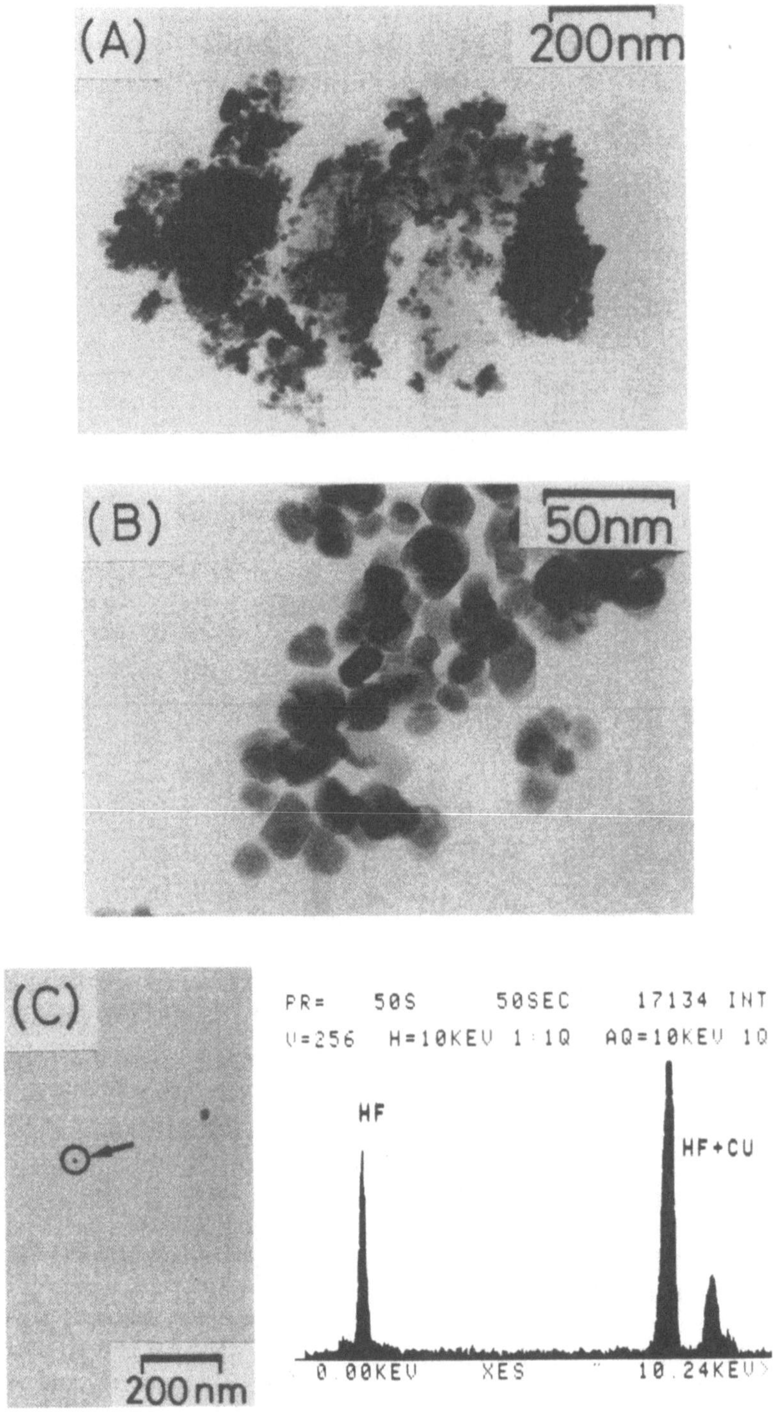
(A)
200nm
(B)
50nm
(C)
200nm
PR= 50S 50SEC 17134 INT
V=256 H=10KEV 1:1Q AQ=10KEV 10
HF
HF+CU
0.00KEV XES 10.24KEV

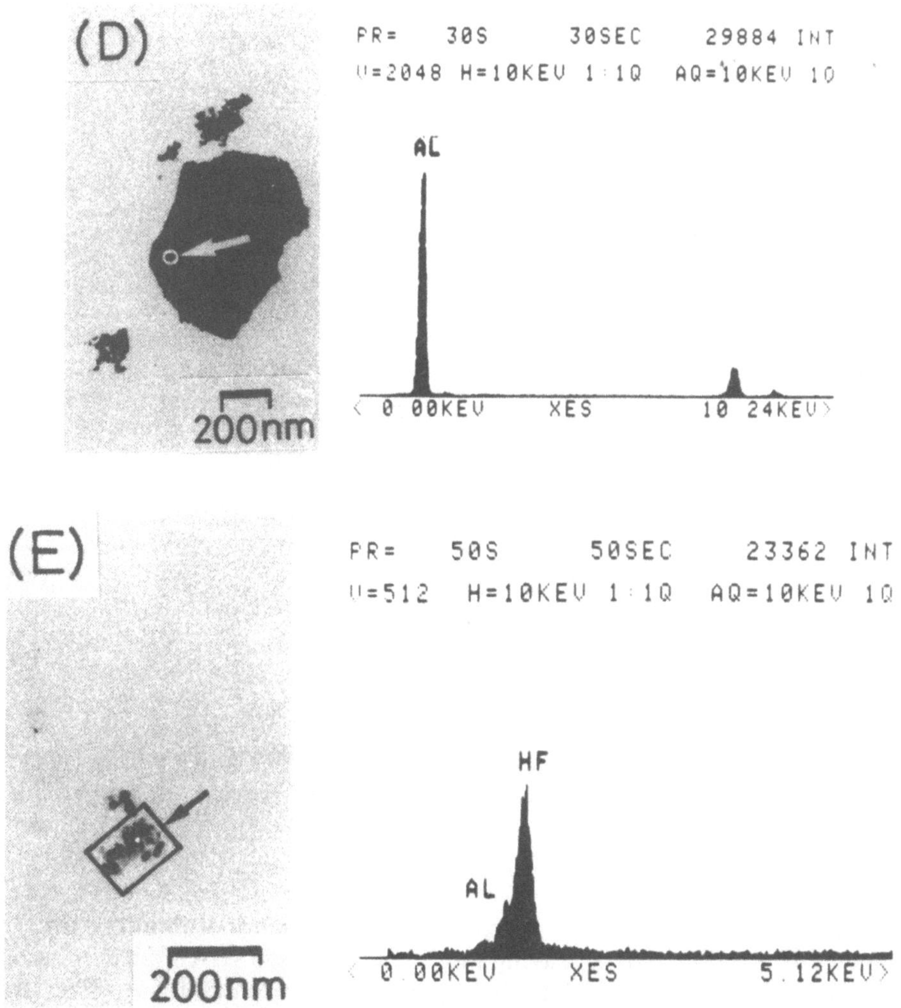

Fig. 3. (*A*) and (*B*) Transmission electron micrographs of the products at 500°C for 120 h in the system $HfAl_3$-H_2O. (*C*), (*D*), and (*E*) EDS spectra obtained by scanning the regions indicated for the products at 600°C for 3 h in the same system as above.

hand, its effect is significant in the system $HfAl_3$-4Al-H_2O, where the variation is similar to that in the system Hf-H_2O.[4,9]

In the system $HfAl_3$-H_2O, hafnium hydride was not formed, so the pulverization mechanism differs from that in Hf-H_2O.[4,9] The Hf and Al atoms react with H_2O mainly on the bulk surface. Thus the diffusion of Hf and Al atoms through $HfAl_3$ grains will primarily control the rate of the reaction on the limited area of interface between the alloys and H_2O. The very small crystallite size of HfO_2 will be due to the interference between the Hf and Al atoms.[5,6] Furthermore, the agglomeration of formed oxides (Fig. 2) will prevent the penetration of H_2O into the interior of the bulk sample, so that the presence of excess water in the open system has little effect. The same variation of crystallite size and reaction rate, therefore, results in both closed and open systems. On the other hand, in the

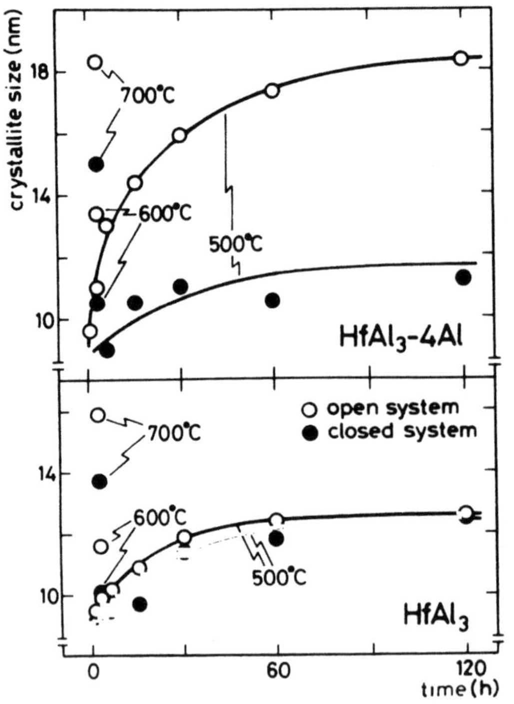

Fig. 4. Variation of crystallite size for monoclinic HfO_2 with time at temperatures indicated in the systems $HfAl_3$-4Al-H_2O (upper diagram) and $HfAl_3$-H_2O (lower diagram).

mixture $HfAl_3$-4Al, the Al in the matrix is oxidized much more rapidly than the precipitated $HfAl_3$ (Fig. 1(*B*)), leaving the finely dispersed $HfAl_3$ crystals among the aluminum oxide powders. The bulk, therefore, becomes loose. The large differences in the crystallite size variation and the reaction rate between the closed and open systems arises from the easy penetration of H_2O into the bulk.

Summary

Al, $HfAl_3$, and $HfAl_3$-4Al were hydrothermally oxidized at 400°–700°C under 100 MPa. Mixed Al_2O_3-HfO_2 powders were formed from the $HfAl_3$ and $HfAl_3$-4Al, where monoclinic HfO_2 with particle sizes of 10–20 nm was finely dispersed among the α-Al_2O_3, with sizes ranging from several nanometers to micrometers. The hydrothermal oxidation mechanism of $HfAl_3$ is different from those of Hf and Al metals. Furthermore, the Al metal has a much faster reaction rate than $HfAl_3$. The formed oxide powders were not perfectly homogeneous because of the wide size distribution of α-Al_2O_3. To suppress the growth of α-Al_2O_3 crystallites, it will be necessary to increase the reaction rate of $HfAl_3$ and change the Al into α-Al_2O_3, without the formation of intermediates, such as γ-AlOOH and κ_1-Al_2O_3. The starting materials of the mixture, with finely precipitated $HfAl_3$ among the Al matrix, are considered to be effective for the preparation of mixed Al_2O_3-HfO_2 powders by hydrothermal oxidation.

Acknowledgments

The authors express their thanks to Professor T. Suzuki and Mr. Y. Ōya of Tokyo Institute of Technology for the preparation of alloy samples. They also thank JEOL Co., Ltd., for use of the transmission electron microscope.

References

[1]M. Yoshimura and S. Sōmiya, "Fabrication of Dense, Non Stabilized ZrO_2 Ceramics by Hydrothermal Reaction Sintering," *Am. Ceram. Soc. Bull.,* **59** [2] 246 (1980).
[2]M. Yoshimura, S. Kikugawa, and S. Sōmiya, "Preparation of Zirconia Fine Powders by the Reactions between Zirconium Metal and High Temperature-High Pressure Solutions"; in High Pressure in Research and Industry. Edited by C. -M. Backman et al. Arkitektkopia ISBN, Uppsala, Sweden, 1982.
[3]H. Toraya, M. Yoshimura, and S. Sōmiya, "Preparation of Fine Monoclinic Hafnia Powders by Hydrothermal Oxidation," *J. Am. Ceram. Soc.,* **65** [5] C-72 (1982).
[4]H. Toraya, M. Yoshimura, and S. Sōmiya, "Hydrothermal Oxidation of Hf Metal Chips in the Preparation of Monoclinic HfO_2 Powders," *J. Am. Ceram. Soc.,* **66** [2] 148–50 (1983).
[5]S. Sōmiya, M. Yoshimura, and S. Kikugawa, "Preparation of Zirconia-Alumina Fine Powders by Hydrothermal Oxidation of Zr-Al Alloys," Proceedings of the 19th University Conference on Ceramic Science, North Carolina University, Nov. 8–10, 1982.
[6]M. Yoshimura, S. Kikugawa, and S. Sōmiya, "Alumina-Zirconia Fine Powders Prepared by Hydrothermal Oxidation" (in Jap.), *Yogyo-Kyokai-Shi,* **91** [4] 182–88 (1983).
[7]C. T. Lynch; pp 193–216 in Refractory Materials: High Temperature Oxides, Pt. II, Vol. 5. Edited by A. M. Alper. Academic Press, New York, 1970.
[8]M. Yoshimura, S. Kikugawa, and S. Sōmiya, "Preparation of Alpha-Alumina Fine Powders by Hydrothermal Oxidation Method," *J. Jpn. Soc. Powder Powder Metall.,* **30** [5] 207–10 (1983).
[9]H. Toraya, M. Yoshimura, and S. Sōmiya, "Reaction Kinetics in the Hydrothermal Oxidation of Hf," *J. Am. Ceram. Soc.,* **66** [11] 818–22 (1983).
[10]H. Toraya, M. Yoshimura, and S. Sōmiya, "A Computer Program for the Deconvolution of X-ray Diffraction Profiles with the Composite of Pearson Type VII Functions," *J. Appl. Crystallogr.,* **16** [6] 653–57 (1983).
[11]A. J. C. Wilson; pp 37–54 in X-ray Optics, 2d ed. Methuen, London, 1962.
[12]M. Torkar, "Untersuchungen uber Aluminiumhydroxyde und -oxide, 4. Mitt," *Mh. Chem.,* **91**, 658–68 (1960).

Reprinted from Robert F. Davis, Hayne Palmour III and Richard L. Porter (eds), *Emergent Process Methods for High-Technology Ceramics*, Vol. 17 (Plenum Publishing Corporation, 1984) pp. 155–166.

Preparation of Zirconia–Alumina Fine Powders by Hydrothermal Oxidation of Zr–Al Alloys

Shigeyuki Sōmiya, Masahiro Yoshimura and Shinya Kikugawa

Laboratory for Hydrothermal Syntheses
Research Laboratory of Engineering Materials
Tokyo Institute of Technology
4259 Nagatsuta, Midori, Yokohama 227 Japan

ABSTRACT

Fine powders of pure metal oxides can be prepared by hydrothermal oxidation method. This technique can provide the fine powders of homogeneous ZrO_2-Al_2O_3 *mixtures from Zr-Al alloys at temperatures 600–700 °C under 100 MPa pressure for 3 h, according to the reactions,*

$2\ ZrAl_3 + 13\ H_2O = 3\ Al_2O_3 + 2\ ZrO_2 + 13\ H_2$, *and*

$2\ Zr_5Al_3 + 29\ H_2O = 10\ ZrO_2 + 3\ Al_2O_3 + 29\ H_2$.

The average crystallite sizes were 27–35 nm for a-Al_2O_3, *13–24 nm for nonoclinic* ZrO_2 *and 9–12 nm for tetragonal* ZrO_2 *in the products.*

INTRODUCTION

High purity fine powders have become important in modern ceramic science and technology. Some new techniques[1-5] have been developed to prepare high purity ceramics fine powders with submicron to hundreds angstroms size. They include thermal decomposition of metal salts and/or hydroxides derived from inorganic or organometallic compounds, by hydrolysis, precipitation and/or drying, and chemical vapor deposition or physical vapor deposition, etc. Hydrothermal crystallization[6] and spark discharge processes[7] under water have also been used.

Hydrothermal Oxidation Method[8-13]

We have developed a method to prepare fine oxide powders from metal by the reaction with high temperature high pressure water, and called it "hydrothermal oxidation method." Fine grains can be prepared by chemical dividing of metal grains during reactions to form oxide and hydride (in the cases of Hf[11,12] and Zr[8-10]), or hydroxide and oxide (in the case of Al[13]) as follows,

$$Zr + 2\ H_2O = ZrO_2 + 2\ H_2 \tag{1}$$

$$Zr + H_2 = ZrH_2 \tag{2}$$

$$ZrH_2 + 2\ H_2O = ZrO_2 + 3\ H_2 \tag{3}$$

$$Al + 4\ H_2O = Al_2O_3{\cdot}H_2O\ \text{(Boehmite)} + 3\ H_2 \tag{4}$$

$$Al_2O_3{\cdot}H_2O = Al_2O_3 + H_2O \tag{5}$$

This method can provide fine grains of ~24 nm for ZrO_2[9], ~ 33 nm for HfO_2[11] and ~ 110 nm for Al_2O_3,[13] respectively under 100 MPa at temperatures 200 ~ 700°C. The reaction process and grain growth in the hydrothermal oxidation method have been studied in detail in the case of hafnia.[12] It is noteworthy that this method has the advantage to introduce no impurities into the products during reactions because of the reaction between high purity metal and pure water.

Fine-grained Mixed Oxide Powder

Even if fine powders of pure oxide could be prepared, homogeneously mixed (that is, uniformly dispersed) high-purity fine powders are not so easy to prepare. Because impurities might be incorporated into the products during grinding and mixing processes mianly from containers and/or grinding medium if mechanical mixing or blending techniques were used. Some chemical procedures, therefore, such as co-precipitation,[14] co-pyrolysis[15] or co-deposition[16] have been employed to prepare mixed powders from the starting materials of homogeneous mixtures as solutions or gases.

We have tried to prepare homogeneously mixed oxides by the hydrothermal oxidation of alloys. Since an alloy contains different atoms homogeneously mixed in atomic scale, it is expected that the direct oxidation of alloys can produce homogeneously mixed oxides at relatively low temperatures under hydrothermal conditions. Here the hydrothermal oxidation process of Zr-Al alloys is studied to prepare ZrO_2-Al_2O_3 mixed oxides, which are of interest in toughened ceramics.

EXPERIMENTAL PROCEDURE

Preparation of Alloys

Starting materials were Zr powder[a] (99.9% purity involving ~ 2% Hf, < 300 mesh) and Al powder[a] (99.9% purity, 200 ~ 300 mesh). They were weighed in appropriate ratios and mixed thoroughly using a boron carbide mortar and pestle in a glove box under 1 atm nitrogen atmosphere. The mixed metal powder was fused on a cold hearth of water cooled copper in flowing high-purity argon atmosphere by the radiation of a 10 kw Xe arc image furnace.[17] The fused alloy globule about 3 mm in diameter was solidified quickly by removing from the focus of the image furnace.

Hydrothermal Oxidation

The alloy sample was placed in a gold crucible and heated in pressurized (100 ± 5 MPa) water in a Tuttle-Roy type pressure vessel at the rate of 15°C min. After the heat treatment at a certain temperature between 200 and 700°C for 0 ~ 6 h, the pressure vessel was quenched by immersion in cold water. The temperature was measured at the outside wall of the vessel by a Platinel thermocouple[b] (Type K, IEC) calibrated against m.p. of Zn (419.5°C).

Characterization of the Products

All reaction products of the hydrothermal treatments were weighed to check the weight change after drying in air at 120°C for 24 h. Then powder X-ray diffraction[c] was performed for the phase identification and the determination of the average crystallite size (D) of the produced phases. The Scherrer equation,

$$D = 0.9\ \lambda/(\beta \cdot \cos \Theta) \tag{6}$$

was used in this study to calculate the D values based on the diffraction lines, (111) of monoclinic ZrO_2, (111) of tetragonal ZrO_2 and (113) of α-Al_2O_3. The peak width (β) at the half maximum of these peaks were corrected by Warren equation,

$$\beta^2 = \beta^2_{obs} - \beta^2_{inst} \tag{7}$$

using β_{inst} for the (101) line of low quartz as an internal standard.

---

[a] Soekawa Chemicals, Co., Ltd., Tokyo, Japan.
[b] Engelhard Ind. Div. East Newark, NJ, U.S.A.
[c] RU-200 Rigaku Denki Co., Ltd., Tokyo, Japan.

The quantity of each phase in the products was determined by X-ray diffraction using a calibration curve derived from the mixture of reagent grade α-Al_2O_3 and m-ZrO_2 powders.

The products were also examined by scanning electron microscope (SEM[d]) and transmission electron microscope (TEM[e]) with energy dispersive spectroscopy (EDS). The TEM samples were made by the immersion of a collodion film supported on a copper grid in the suspension of sample powder ultrasonically dispersed in water.

RESULTS AND DISCUSSION

Arc image fusion yielded Zr_5Al_3, Zr_2Al_3, $ZrAl_2$ and $ZrAl_3$ phases in the system Zr-Al. The alloy samples of Zr_5Al_3 and $ZrAl_3$ (including trace amounts of free Al) were subjected to hydrothermal oxidation. The weight gain of the fused spherical sample of Zr_5Al_3 was observed above 400°C but it scarcely proceeded at this temperature (Fig. 1). The observed weight gain approached the calculated one, 43.2% according to the reaction,

$$2\ Zr_5Al_3 + 29\ H_2O = 10\ ZrO_2 + 3\ Al_2O_3 + 29\ H_2 \qquad (8)$$

after 6 h at 500°C, 3 h at 600°C or 1 h at 700°C.

Scanning electron microscope revealed the formation of boehmite crystals on the surface of the spherical sample (Fig. 2(b)) after 3 h at 400°C. It caused cracking of the sample. The formation of boehmite and zirconia, and the change of boehmite to corundum at higher temperature brought about the pulverization of the sample into the powder, Fig. 2(c), of the aggregate of the mixed oxide of ZrO_2 and Al_2O_3.

Figure 3 shows that the hydrothermal oxidation of $ZrAl_3$ samples did not proceed below 500°C but became significant at 600°C. The oxidation of samples was almost completed after 1 h at 600°C or 0.5 h at 700°C. The microstructure of the surface and inner part of the samples changed similarly to those for Zr_5Al_3 samples.

A higher temperature was needed for the reaction of the fused alloy samples than the constituent metals, Zr and Al powders, for which a temperature as low as 250°C[10] and 200°C[13], respectively, were sufficient to start the reaction with water under the same conditions. This appeared to be caused by the inactivation of the surface of the samples due to the fusion. When the fused alloy samples were crushed into powder, even though they were such coarse grains as 100

---

[d] JSM T-200, JEOL, Co., Ltd., Tokyo, Japan.
[e] JEM 200CX, JEOL, Co., Ltd., Tokyo, Japan.

μm in diameter, they started to react above 300°C as shown in Fig. 4. This fact suggests that the alloy samples having active surfaces would react quickly with water at relatively low temperatures and yield finer powders.

The reaction products of alloys with water were boehmite, small amounts of K1-Al_2O_3 and monoclinic and tetragonal ZrO_2 below 400°C, and α-Al_2O_3 and monoclinic and tetragonal ZrO_2 above 500°C as seen in Fig. 5. The zirconium hydrides were not formed in the products. The tetragonal ZrO_2 has never been produced in the hydrothermal oxidation of Zr metal alone under the same conditions, so that the formation of t-ZrO_2 is characteristic in the hydrothermal oxidation of Zr-Al alloys. Particularly, the fraction of t-ZrO_2 was higher in early stage of the reaction and at lower temperatures, then it decreased gradually in the prolonged treating time as seen in Fig. 6.

The average crystallite size of each phase increased with the treating time at 600°C and 700°C (Table 1) due to grain growth. The average crystallite size of t-ZrO_2, however, is always smaller than that of m-ZrO_2. This would be one of the reasons why t-ZrO_2 was formed (or stabilized) in the products of the alloy oxidation. In the case of the Zr_5Al_3 samples, the crystallite size of the produced phases was a little larger than those in the case of the $ZrAl_3$ samples, and had similar time dependency.

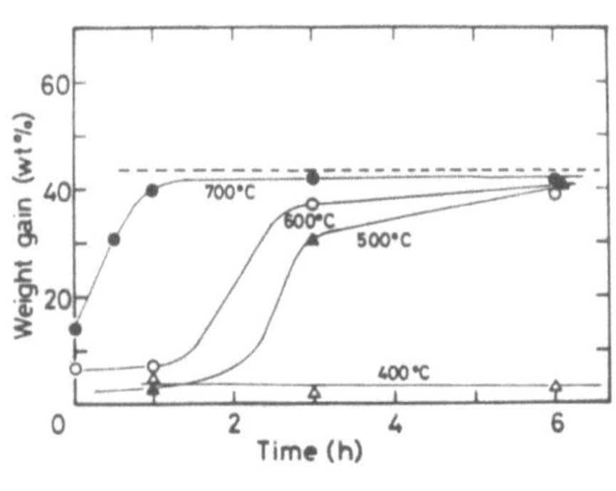

Fig. 1. Weight gain of fused Zr_5Al_3 sample during reactions with 100 MPa water at various temperatures.

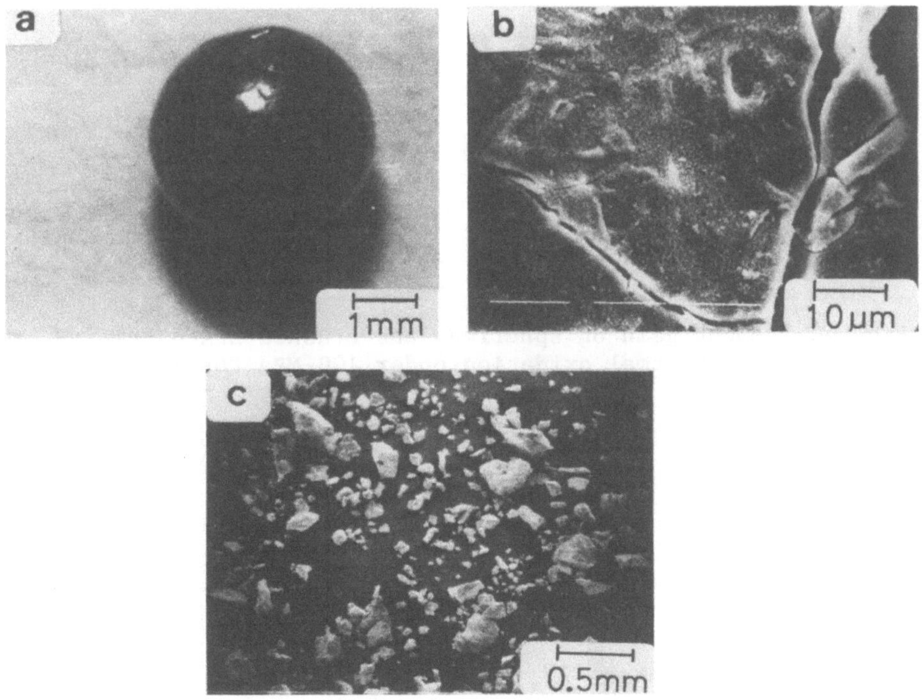

Fig. 2. Scanning electron microphotograph of the Zr_5Al_3 samples under hydrothermal oxidation. (a) Fused spherical alloy sample; (b) surface of partially reacted sample (400°C, 3 h); (c) completely oxidized sample, α-Al_2O_3 + ZrO_2 powder (700°C, 3 h).

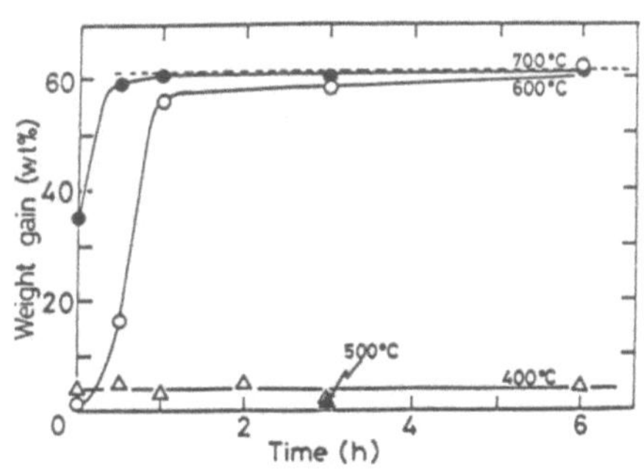

Fig. 3. Weight gain of fused $ZrAl_3$ sample during reactions with 100 MPa water at various temperatures.

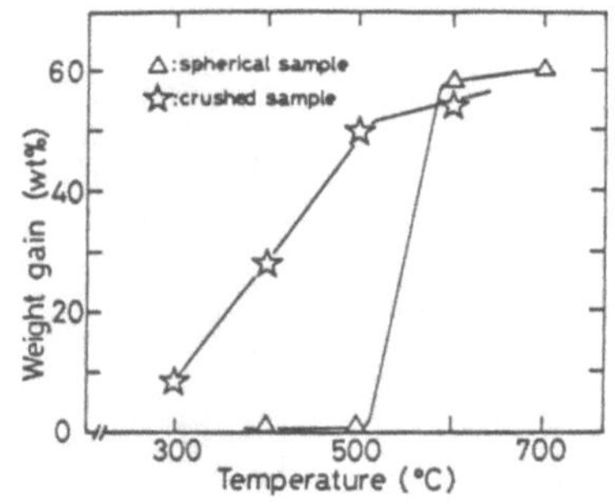

Fig. 4. Weight gain of spherical and crushed $ZrAl_3$ samples by hydrothermal oxidation under 100 MPa for 3 h.

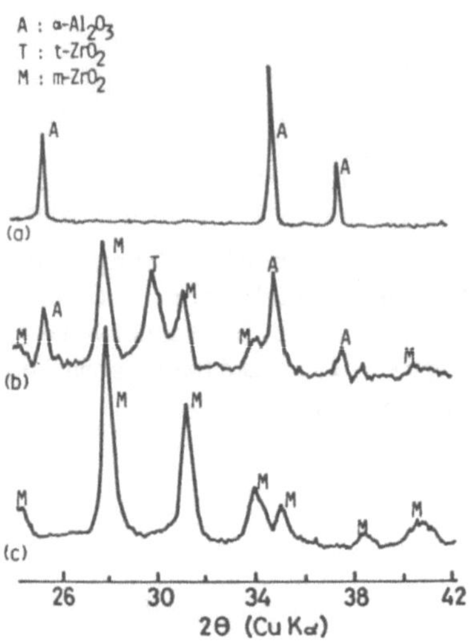

Fig. 5. X-ray powder diffraction patterns of the products prepared at 700°C under 100 MPa for 3 h from: (a) Al and H_2O; (b) $ZrAl_3$ and H_2O; or (c) Zr and H_2O.

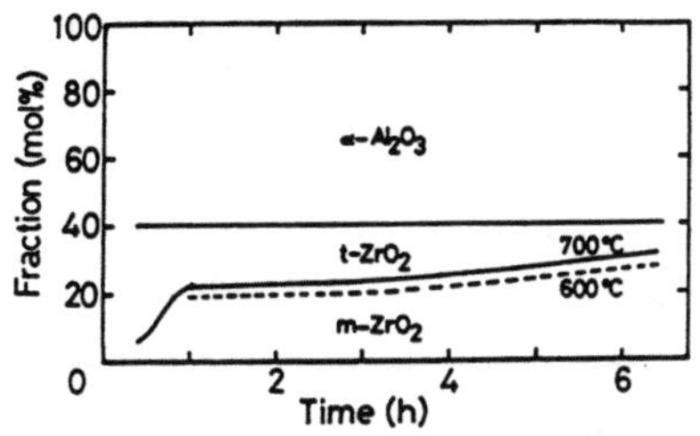

Fig. 6. Change of the contents of ZrO_2 and Al_2O_3 in the products prepared from $ZrAl_3$ and 100 MPa H_2O.

Table 1. The Average Crystallite Size of the Produced Phases with Treating Time of Hydrothermal Oxidation for Fused $ZrAl_3$ Samples.

		Crystallite Size (nm)		
Phase	Temp. (°C)	1 h	3 h	6 h
α-Al_2O_3	600	26.6	27.1	28.3
	700	27.2	30.0	31.2
m-ZrO_2	600	12.6	12.9	15.6
	700	16.9	20.2	21.8
t-ZrO_2	600	9.0	9.0	9.8
	700	11.5	11.7	13.2

The crystallite sizes of ZrO_2 and Al_2O_3 are greatly reduced in the case of the alloy oxidation in comparison with the cases of constituent metal as demonstrated in Table 2. In particular, the crystallite of α-Al_2O_3 is about 5 times less in the alloy oxidation than that in pure Al oxidation. These facts suggest that the growth of the crystallites of ZrO_2 and Al_2O_3 interfered with each other, because Zr and Al atoms mixed in atomic scale in the alloys had to segregate from each other to form ZrO_2 and Al_2O_3 phases separately. Such segregation and crystallite growth processes are difficult at low temperatures, therefore the fine grains of ZrO_2 and Al_2O_3 mix with each other in the products. This is one of the favorable features in the hydrothermal oxidation of alloys to prepare mixed fine powders with homogeneous dispersion.

Table 2. Products and Their Average Crystallite Size of Hydrothermal Oxidation of Metals and Alloys with 100 MPa Water.

Starting Material	Temperature (°C)	Products	Crystallite Size (nm)
Zr	250 - 700	m-ZrO_2	23 - 27
Al	100 - 500	γ-AlOOH	---
	500 - 700	α-Al_2O_3	110 - 200
$ZrAl_3$	600 - 700	m-ZrO_2	13 - 20
		t-ZtO_2	9 - 12
		α-Al_2O_3	27 - 35
Zr_5Al_3	500 - 700	m-ZrO_2	15 - 24
		t-ZtO_2	10 - 12
		α-Al_2O_3	28 - 35

Figure 7 shows the TEM photograph and EDS analysis of the products composed of m-ZrO_2, t-ZrO_2 and α-Al_2O_3. The observed crystallite sizes ranged 9 ~ 35 nm are in good agreement with those calculated by powder X-ray diffraction. The EDS analysis indicated that the area (b) in Fig. 7 contained ZrO_2 and Al_2O_3 particles and that an individual particle was either ZrO_2 or Al_2O_3; for example, the particle (a) in this figure was ZrO_2. These results clearly demonstrate the homogeneous dispersion of the fine particles of ZrO_2 and Al_2O_3 in the products. The EDS analysis showed no detectable Zr contents in Al_2O_3 particles, whereas a small amount of Al (estimated to be ~ 3.5 atom%) was often detected in the ZrO_2 particles as illustrated in Fig. 7(a). Although it is not clear whether this Al_2O_3 is present as a solid solution with ZrO_2 or as a separate phase on the surface and/or in the ZrO_2 particle, this presence of Al_2O_3 would help the formation (or stabilization) of t-ZrO_2 phase. The solid solubility of Al_2O_3 in ZrO_2 phase has not been established yet, but it is suggested[18] at high temperatures around 1900°C. The formation of the t-ZrO_2 phase by the presence of Al_2O_3 was also observed when the mixtures were melted and chilled by a plasma flame.[19] Additional strain (or surface) energy imparted to ZrO_2 grains by Al_2O_3 grains contacting with ZrO_2 seems to be the probable reason for forming (or stabilize) the tetragonal ZrO_2 phase.

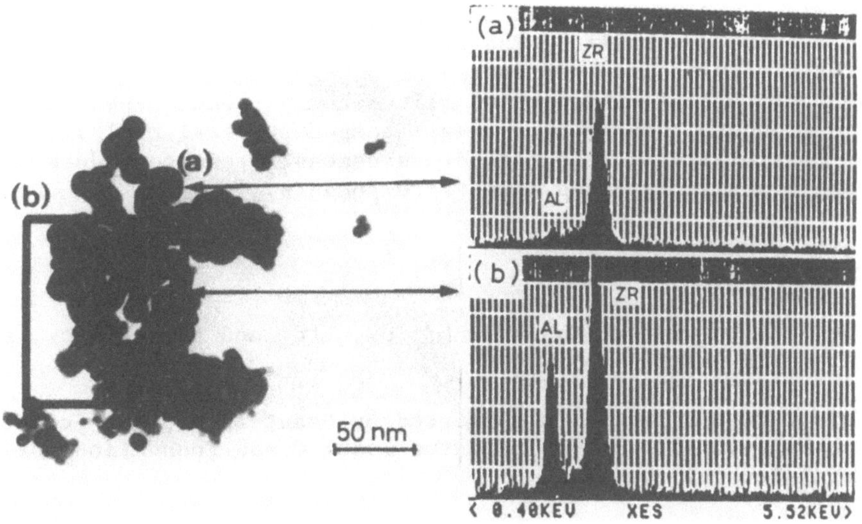

Fig. 7. TEM photograph of the products ($m\text{-}ZrO_2 + t\text{-}ZrO_2 + \alpha\text{-}Al_2O_3$) by hydrothermal oxidation (700°C, 100 MPa, 3 h) of $ZrAl_3$ sample, and EDS analyses of the particle (a), and the area (b).

In addition to the homogeneous mixtures of fine grained ZrO_2 and Al_2O_3 as shown in Fig. 7, larger Al_2O_3 particles have also been formed in the hydrothermal oxidation of Zr-Al alloys. These Al_2O_3 particles have the morphology of hexagonal plate 1 to 5 μm diameter. They appeared to be grown by hydrothermal dissolution/precipitation due to high temperature and high pressure water. The elimination of these larger Al_2O_3 particles is an important study needed in further application of the hydrothermal oxidation method.

CONCLUSION

The hydrothermal oxidation of Zr-Al alloys lead to the following conclusion:

(1) Fused Zr_5Al_3 sample reacted with 100 MPa water above 400°C to form boehmite, then formed $\alpha\text{-}Al_2O_3$, $m\text{-}ZrO_2$ and $t\text{-}ZrO_2$ above 500°C.

(2) Fused $ZrAl_3$ samples reacted similarly but needed above 600°C for the complete oxidation.

(3) The reaction temperature could be reduced below 300°C by crushing the fused alloy samples.

(4) The fine powders consisted of α-Al_2O_3, m-ZrO_2 and t-ZrO_2 having the average crystallite size of 27-35 nm, 13-24 nm and 9-12 nm, respectively, could be prepared at 600-700°C.
(5) The hydrothermal oxidation of alloys could produce the finer oxide powders than that of single component metal could.
(6) The possibility of this method was demonstrated to produce homogeneously dispersed mixed fine powders.

ACKNOWLEDGMENTS

The authors wish to thank JEOL, Co., Ltd. and Hitachi Co., Ltd. for use of TEM and EDS apparatus.

This work was partially supported by Grant-in-Aid for Scientific Research No. 443015, 1980-81, and the Asahi Glass Foundation for Industrial Technology.

REFERENCES

1. D. W. Johnson, Jr., Am. Ceram. Soc. Bull., 60 [2], 221-24, 243 (1981).
2. C. R. Veale, Fine Powders, Preparation, Properties and Uses, Halsted Press Div., Wiley & Sons, NY, 1972.
3. J. L. Pentecost, pp. 1-14 in Ceramic Fabrication Processes, edited by F. F. Y. Wang, Academic Press, NY, 1976.
4. M. Hoch, pp. 13-28 in Science of Ceramics, 9, edited by K. J. DeVries, Nederlandse Keramische Verenging, Noordwijkerhout, The Netherlands, 1977.
5. K. S. Mazdiyasni, Ceram. Int., 8 [2], 42-56 (1982).
6. E. Tani, M. Yoshimura, and S. Somiya, J. Am. Ceram. Soc., 66 [2], in press (1983).
7. T. Shibuya, Yogyo Kyokai Shi, 90 [10], 603-609 (1982).
8. M. Yoshimura and S. Somiya, Am. Ceram. Soc. Bull., 59 [2], 246 (1980).
9. M. Yoshimura and S. Somiya, pp. 453-63 in Science and Technology of Zirconia, Advances in Ceramics, Vol. 3, edited by A. H. Heuer and L. W. Hobbs, The American Ceramics Society, Columbus, 1981.
10. M. Yoshimura and S. Somiya, in Proc. of 5th Int. Round Table Conf. on Sintering, Sept. 7-10, 1981, Portoroz, Yugoslavia, in press.
11. H. Toraya, M. Yoshimura, and S. Somiya, J. Am. Ceram. Cos., 65 [5], C-72 (1982).
12. H. Toraya, M. Yoshimura, and S. Somiya, J. Am. Ceram. Soc., 66 [2], in press (1983).
13. M. Yoshimura, S. Kikugawa, and S. Somiya, submitted to Funtai oyobi Funmatsu Yakin (J. Japan Soc. Powder Metallurgy).

14. P. E. D. Morgan, pp. 67-76 in Processing of Crystalline Ceramics, edited by H. Palmour III, R. F. Davis, and T. M. Hare, Plenum Press, NY, 1978.
15. P. K. Gallagher and F. Schrey, Thermochimica Acta, 1, 465-76 (1970).
16. C. Sheer, S. Korman, D. J. Angier, and R. P. Cahn, pp. 133-52 in Fine Particles, 2nd International Conference, edited by W. E. Kuhn, Electrochemical Society, Princeton, NJ, 1974.
17. M. Yoshimura and S. Somiya, pp. 23-26 in Proc. of the 4th Int. Conf. on Rapidly Quenched Metals, Vol. I, edited by T. Masumoto and K. Suzuki, Japan Institute of Metals, Sendai, 1982.
18. A. M. Alper, pp. 339 in Science of Ceramics, 3, edited by G. H. Steward, Academic Press Inc., London, 1967.
19. A. Krath and H. Meyer, Ber. Deut. Keram. Ges., UU, 61-72 (1965).

Reprinted from *Report of the Research Laboratory of Engineering Materials, Tokyo Institute of Technology*, No. 12 (1987) pp. 59–72.

Formation of TiO_2 and ZnO Powders by Hydrothermal Oxidation of Ti and Zn Metals

By

Masahiro YOSHIMURA*, Hiroyuki OHIRA** and Shigeyuki SŌMIYA*

Abstract

Hydrothermal oxidation of Ti and Zn were studied in a closed system and an open system. By the hydrothermal oxidation of Ti, TiO_2 (rutile) was formed above 450°C (100 MPa, 3 h) with titanium hydride and anatase as intermediate products, which were changed into rutile at higher temperatures. Under these conditions, oxidation completed above 650°C. Also ZnO was formed by the hydrothermal oxidation of Zn above 200°C (100 MPa, 3 h), however, no intermediate products could be detected by XRD. Crystallite sizes of starting Ti and Zn powders were large ($>$ 150 nm) enough, but crystallites of produced rutile and ZnO were 30 – 85 nm and 15 – 125 nm, respectively. Crystal growth of rutile and ZnO was observed in the open system.

1. Introduction

Hydrothermal oxidation method is one of hydrothermal processings (such as hydrothermal systhesis, hydrothermal crystallization and hydrothermal reaction sintering, etc.) which is one of the potential processing in modern and near future ceramic technologies, especially in powder preparation and sintering [1]. As an application of this hydroterminal methods, pure and fine oxide powders can be prepared by the oxidation of metals and alloys with high-temperature and high-pressure water or solutions without contamination compared with other wet chemical methods using metal salts and precipitant(s) in general. Mixed oxide powders dispersed homogeneously can also be prepared if alloys are used as starting materials. By this method, coarse grains of starting materials are divided into finer particles through oxidation, hydrogenation and/or hydration. α–Al_2O_3 [2], monoclinic ZrO_2 [3–5], HfO_2 [6–8] powders, and Al_2O_3–ZrO_2 [9,10] and Al_2O_3–HfO_2 [11] mixed fine powders were prepared.

As for reactions of Ti and Zn with water under hydrothermal condition, Nakayama et al.[12–14] reported on reactions of Ti with high temperature water and steam at 200°–300°C that anatase was formed when much water existed in atmosphere. Moroishi and Shida[15] studied in higher temperature steam (400°–550°C) and reported that only rutile was produced. With high-temperature and high-pressure water, Oota et al.[16] reported rutile formation at 420°–450°C under 20 MPa for 50–528 h. They also reported[17] on oxidation of Zn under hydrothermal condition where it began at

* Professor, Dr. Sci. in Eng.
** Graduate Student, M. Sci. in Eng., Present address; Chichibu Cement Co. Ltd.

350°–400°C and completed at 500°C under 33 MPa in NaF solution (0.7 mol/kg). In the present study, reaction of the hydrothermal oxidation of Ti[21] and Zn[22] were examined as a function of temperature, time and difference in atmosphere.

2. Experimental Procedures

Titanium metal powder (Soekawa Chemicals Co., Ltd., purity 99.9%, < 200 mesh) and zinc metal powder (Koch-Lab., purity 99.999%) were used as starting materials. Specimens for experiments were prepared in (a) a closed system, that is, metal powder and redistilled water were sealed in a gold capsule (I.D. of 2.7 mm and length of 35 mm) in the molar ratio of $H_2O/Ti = 2/1$ and $H_2O/Zn = 1$, or (b) an open system where the powder was set at the bottom of a gold capsule (I.D. of 2.7 or 4.6 mm and length of 20–40 mm) without sealing the top. In the open system, therefore, metals could react with excess water. Then, the capsule was heated in a test-tube-type pressure vessel under the conditions that temperature, pressure and time were 200°–700°C, 10–150 MPa and 0–120 h, respectively.

After the treatment, pressure vessel was quenched in water and solid products in the capsule were examined by X-ray powder diffraction and scanning electron microscopy. Crystallite sizes of the products were determined from the integral breadths of the deconvoluted line profile[7] of (012) reflection of Ti, (220) of $TiH_{1.924}$, (110) of TiO_2 (rutile), (101) of TiO_2 (anatase), (101) or Zn and (102) of ZnO. In this calculation, crystallite was assumed to be spherical and the lattice strain was ignored.

3. Results and Discussion

(A) Hydrothermal Oxidation of Ti

Figures 1 and 2 illustrate variations of products with temperature under 100 MPa for 3 h in the closed and the open systems, respectively. As shown in these figures, Ti changed into titanium hydride (TiH_x, x = 1.924), anatase and rutile above 450°C. Over 600°C, all of Ti and anatase changed into rutile and TiH_x, and rutile was the only products at 700°C. As for the effects of time on these reactions, variations of products at 500°C under 100 MPa in the closed system and the open system are shown in Fig. 3 and 4, respectively. Under this condition, Ti changed into TiH_x, anatase and rutile within

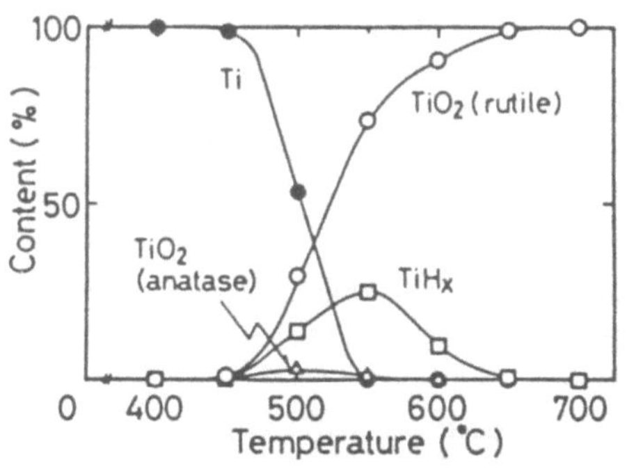

Fig. 1. Variation of amount of products with temperature by the hydrothermal oxidation of Ti in the closed system under 100 MPa for 3 h.

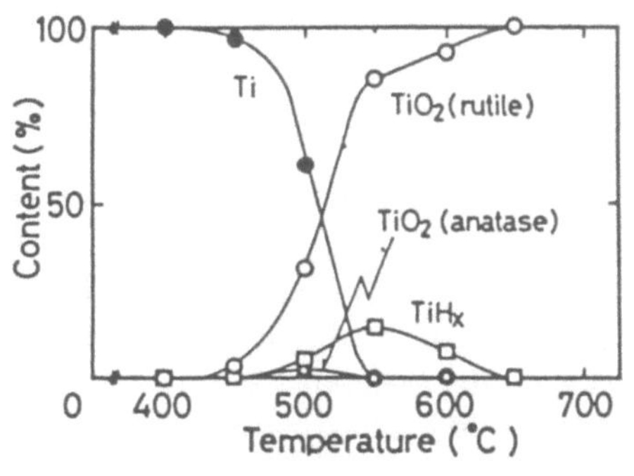

Fig. 2. Variation of amount of products with temperature by the hydrothermal oxidation of Ti in the open system under 100 MPa for 3 h.

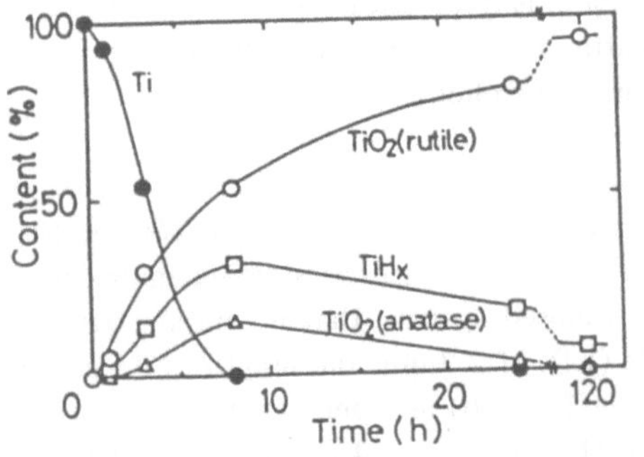

Fig. 3. Variation of amount of products with time by the hydrothermal oxidation of Ti in the closed system at 500°C under 100 MPa.

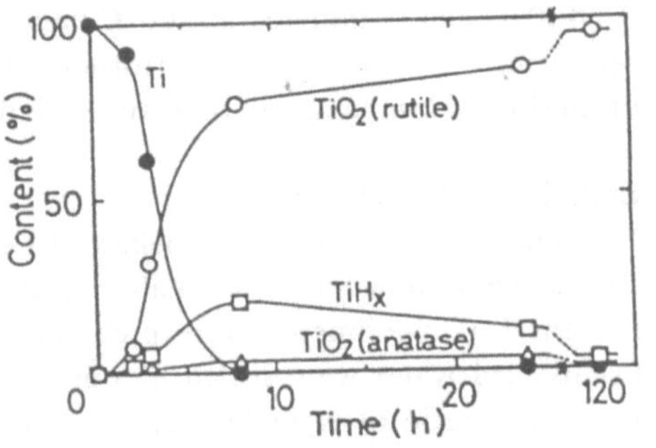

Fig. 4. Variation of amount of products with time by the hydrothermal oxidation of Ti in the open system at 500°C under 100 MPa.

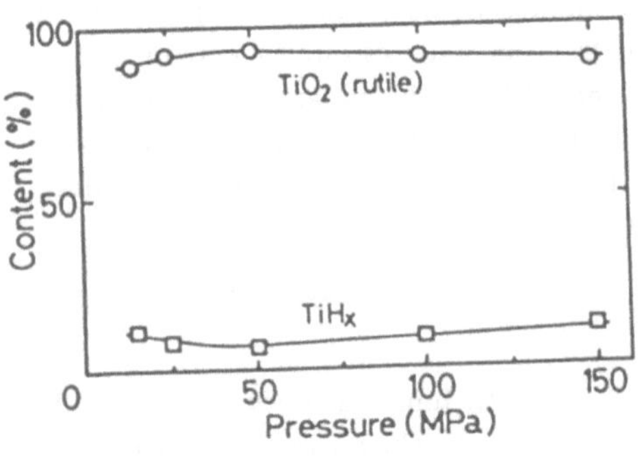

Fig. 5. Variation of amount of products with pressure by the hydrothermal oxidation of Ti in the closed system at 600°C for 3 h.

8 h. After that, reactions from TiH_x and anatase to rutile were continued. Little effects of pressure was observed in both systems (Fig. 5).

From these results, reactions of the hydrothermal oxidation of Ti can be described as following formulas.

$$Ti + 2H_2O \rightarrow TiO_2 + 2H_2 \quad (1)$$

$$Ti + \frac{x}{2} H_2 \rightarrow TiH_x \quad (x = 1.924) \quad (2)$$

$$TiH_x + 2H_2O \rightarrow TiO_2 + (2 + \frac{x}{2}) H_2 \quad (3)$$

At first, surface of Ti is subjected to oxidation (reaction (1)), and then TiH_x is formed by reaction with H_2 produced in the reaction (1). This TiH_x changes into TiO_2 by the reaction with water (reaction (3)).

About the oxidation of Ti under hydrothermal conditions, Oota et al.[16] reported that 8–24% of Ti was oxidized and formed rutile under the conditions of 420°C, 528 h and 450°C, 50 h. Also Moroishi and Shida[15] studied the oxidation of Ti with high temperature steam (450°–550°C) and reported that only rutile was formed on the surface of Ti plate by 24–101 days treatments. These results are in good agreement with those of the present study. On the other hand, Nakayama et al.[12–14] reported that TiO_2 formed by the oxidation with high temperature water (200°C and 300°C) had a double-layered structure which was outer layer of anatase and inner layer of rutile. They further concluded that rutile was formed when little water existed in the atmosphere. Their different results from the present study would be attributed to the differences in the experimental conditions, where their treatments were carried out at much lower temperatures. In addition, they observed the very thin surface of products with the electron diffraction technique, while X-ray powder diffraction technique was applied for the bulk products in the present study.

From Figs. 1–4, it is apparent that there exist some differences between the closed system and the open system. One difference is the amount of TiH_x: in the closed system, TiH_x was formed more than in the open system. This can be attributed to higher hydrogen partial pressure (p_{H_2}) in the closed system than in the open system. In the closed system, hydrogen produced by oxidation of Ti cannot go out from the capsule, while hydrogen leaves from the capsule and p_{H_2} becomes much lower in the open system. Also in reaction (3), high p_{H_2} prevents oxidation of TiH_x, thus as a result, TiH_x was produced in the closed system more than in the open system.

As another difference, it is apparent that more anatase was produced in the closed system than in the open system (Fig. 3 and 4). Oota et al.[16] reported that anatase was formed by the oxidation of TiH_2 under hydrothermal conditions. Therefore anatase could be formed through the oxidation of TiH_x. This corresponds to the results that more anatase was produced in the closed system in which produced more TiH_x. At high temperature, however, formation of anatase was not observed. This is because anatase changed into rutile at about 600°C, at which change of anatase into rutile under 100 MPa took place as reported by Osborn[19].

The comparison in Fig. 3 and 4 indicates the rate of reaction (3) to be relatively lower in the closed system than that in the open system after 8 h. This phenomenon can be explained in two possible reasons. One is decrease of water, that is, water decreased with reaction because limited amount of water was sealed in a closed capsule, so that it causes the rate of reaction decrease. Another is equilibrium state of reaction (3). Since produced hydrogen cannot escape from the capsule in the closed system, p_{H_2} in the capsule increases and reaction (3) may come to equilibrium. To confirm this reason, equilibrium condition of the reaction (3) with hydrogen and water partial pressures (p_{H_2} and p_{H_2O}) was calculated from thermodynamical point of view. For calculation, following formula

$$TiH_2 + 2H_2O \rightarrow TiO_2\ (\text{rutile}) + 3H_2 \tag{4}$$

and total pressure of 100 MPa ($p_{H_2} + p_{H_2O} = 100$ MPa) was applied. Gibbs free energy at T K, ΔG°_T, can be calculated by following equation.

$$\Delta G^\circ_T = (G_{\text{rutile}})^\circ_T + 3(G_{H_2})^\circ_T - (G_{TiH_2})^\circ_T - 2(G_{H_2O})^\circ_T \tag{5}$$

At high pressure, effect of pressure can be calculated as,

$$\Delta G_T = \Delta G^\circ_T + R\,T \ln \frac{(f_{H_2})^3}{(f_{H_2O})^2} \tag{6}$$

where f_{H_2} and f_{H_2O} are fugacities of H_2 and H_2O, respectively. f_{H_2} and f_{H_2O} can be described with fugacity coefficient[20], γ, which is a function of reduced temperature (T_R) and reduced pressure (P_R) as,

$$f_{H_2} = \gamma_{H_2} P_R \ , \quad f_{H_2O} = \gamma_{H_2O} P_R \ .$$

Using these equations, free energy change under total pressure of 100 MPa was calculated and shown in Fig. 6. Also Fig. 7 illustrates the equilibrium condition with the function of total pressure at 800 K. As shown in these figures, only very low p_{H_2O} ($p_{H_2O}/p_{H_2} = 10^{-10}$ at 500°C) is required to complete the oxidation. Therefore, reaction (3) is actually completed without reaching the equilibrium state in the present study, and the reason why the rate of reaction (3) in the closed system decreased can be attributed to the decrease of water.

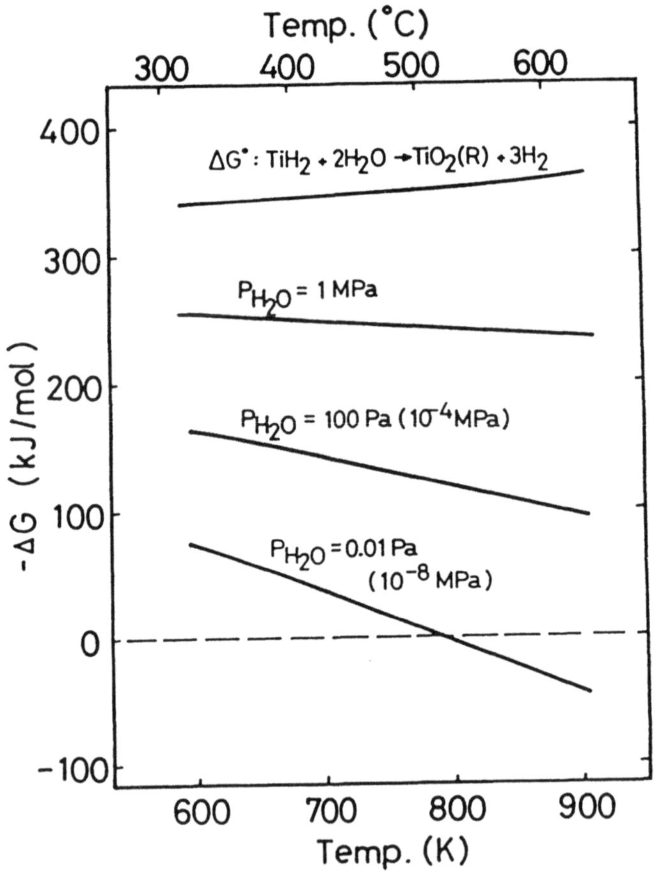

Fig. 6. Free energy change with temperature and water partial pressures for the reaction:
$TiH_2 + 2H_2O \rightarrow TiO_2$ (rutile) $+ 3H_2$
under total pressure of 100 MPa.

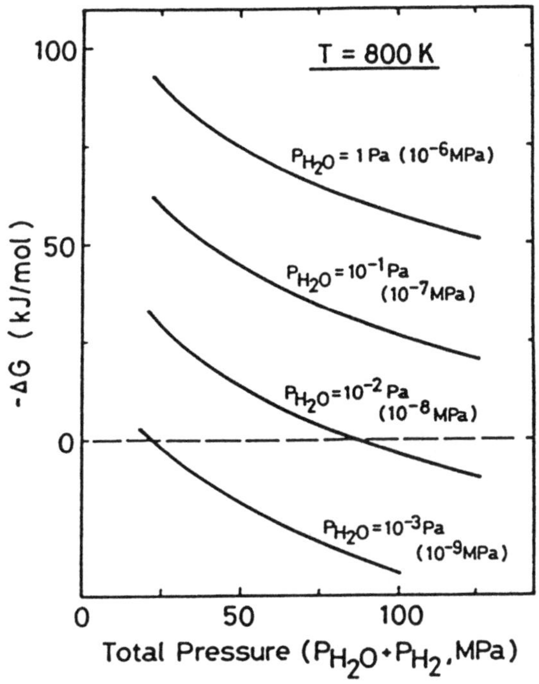

Fig. 7. Free energy change with total pressure and water partial pressures at 800 K for the reaction:
$$TiH_2 + 2H_2O \rightarrow TiO_2 \text{ (rutile)} + 3H_2.$$

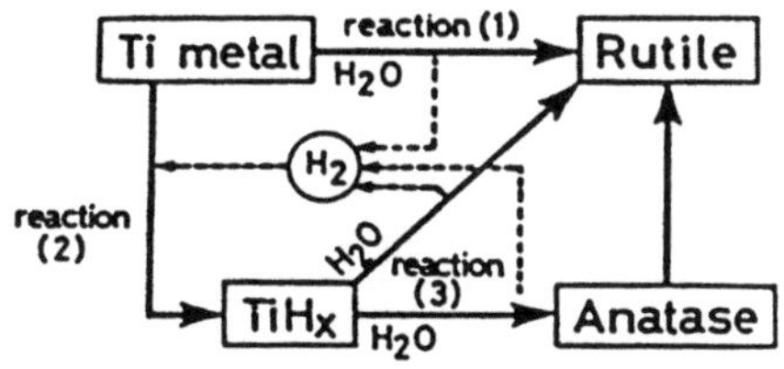

Fig. 8. Schematic diagram of reaction paths of the hydrothermal oxidation of Ti.

From these results, schematic diagram of reaction paths of hydrothermal oxidation of Ti was illustrated in Fig. 8. It can be explained in two paths: one is that rutile is formed directly from Ti (reaction (1)), and another is that it is formed through formation of TiH_x and/or anatase as intermediate products (reaction (2) and (3)).

Figure 9 depicts the surfaces of products by the hydrothermal oxidation of Ti. Starting Ti powder is grain size of $< 100\ \mu m$ and shows crushed-like surface (a). During the oxidation, products are formed on metal surface and grains become roundish ((b) and (c)). At high temperature treatments, surfaces of the products in the open system show the flat plane and/or patterns by crystal growth (d) which was identified as rutile by X-ray powder diffraction, while these rutile of grown crystals and patterns were not observed in the closed system. Because there exists much water enough to act as a solvent

Fig. 9. Scanning electron micrographs of the products by the hydrothermal oxidation of Ti under 100 MPa for 3 h.
(a) starting Ti powder (b) 600°C, closed system
(c) 600°C, closed system (d) 600°C, open system

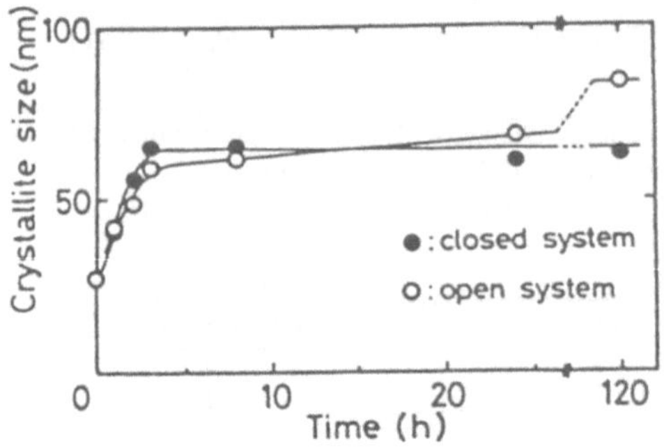

Fig. 10. Variation of crystallite size of rutile with time at 500°C under 100 MPa.

in the open system, these crystals can be grown by a dissolution-precipitation process.

Crystallite size of starting Ti powder determined by X-ray powder diffraction was over 150 nm beyond the limitation by this technique. Crystallite size of TiH_x, however, was less than 20 nm, which shows that Ti was disintegrated during hydrogenation. Crystallite size of rutile was 35–140 nm and it increased with increasing temperature and time. This disintegration process during hydrogenation was also observed in the hydrothermal oxidation of Zr[3–5] and Hf[6–8]. However, crystallite size of rutile was much larger than those of ZrO_2 (24 nm) and HfO_2 (32–34 nm). This difference may be attributed to the difference in solubility in water, that is, rutile crystallite was grown bigger because of its higher solubility in water than ZrO_2 and HfO_2.

Figure 10 illustrates variation of the crystallite size of rutile with time in both systems. Until 3 h, rutile crystallite was grown up to about 65 nm in both systems. In the open system, it kept growing gradually and reached about 85 nm at 120 h, whereas it showed little increase after 3 h in the closed system. Apparently, this result does not agree with the SEM observation, especially in the open system in which rutile crystals were grown and average crystallite size was about 85 nm. That is because not only large crystals on the surface but also crystallites inside of the grains were included for this calculation. Consequently, sharp diffraction by large crystals can be involved in broad peak by a large number of small crystallites, therefore the average crystallite size was much smaller than that expected by the SEM observation.

(B) Hydrothermal Oxidation of Zn

By X-ray diffraction analysis, no intermediate phases could be detected. Therefore reaction of the hydrothermal oxidation of Zn can be described as following formula.

$$Zn + H_2O \rightarrow ZnO + H_2 \tag{7}$$

Figure 11 illustrates proportion of products (Zn and ZnO) with temperature and time in the closed system. At 200°C, Zn was not oxidized very rapidly, so that more than 10% of Zn was remained even after 3 h (100 MPa). Above 300°C, however, oxidation was almost completed within 8 h. Figure 11 clearly shows that rate of reaction (7) decreased when about 70–80% of ZnO was formed. Using the same consideration with the case of Ti, variations of ΔG with temperature and water partial pressure (p_{H_2O}) under total pressure of 100 MPa were calculated as shown in Fig. 12. Figure 13 shows the variation of ΔG with p_{H_2O} and total pressures at 600K. As clearly illustrated in these figures, very low is required to complete the oxidation and, therefore, the decrease of reaction rate shown in Fig. 11 can be attributed to decrease of water. The fact that oxidation was almost completed during heating up to 400°C in the open system also supports this result.

As for oxication of Zn under hydrothermal condition, Oota et al.[17] reported that Zn was oxidized within 5 h at 420°C under about 20 MPa, and that oxidation was completed during heating up to 500°C which took 2 h. The oxidation rate in their results was much lower than that of our results. This may be explained by some differences in

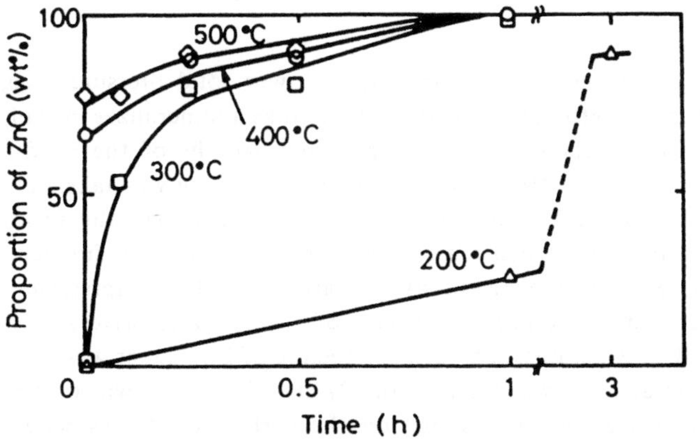

Fig. 11. Variation of proportion of ZnO with temperatures and time by the hydrothermal oxidation of Zn in the closed system under 100 MPa.

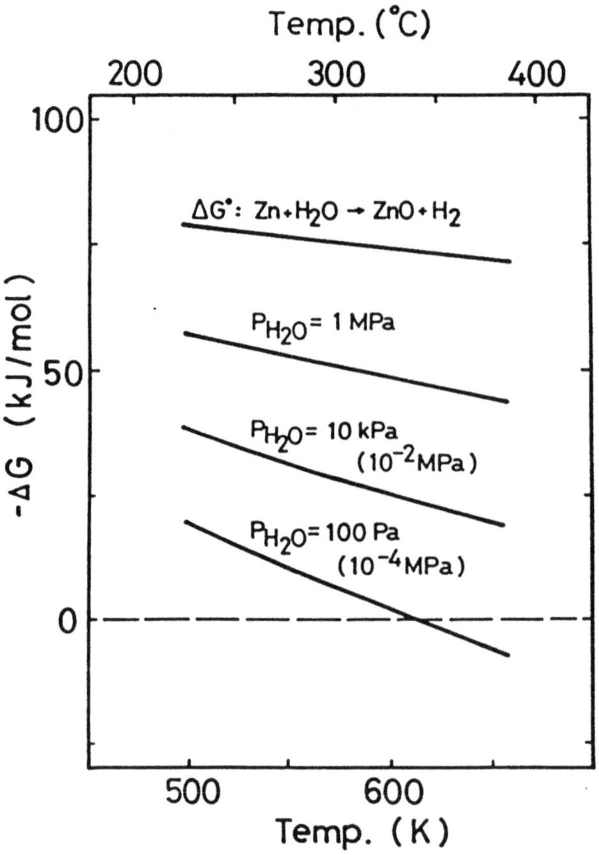

Fig. 12. Free energy change with temperature and water partial pressures under total pressure of 100 MPa for the reaction:

$$Zn + H_2O \rightarrow ZnO + H_2.$$

experimental conditions, for example, their lower total pressure, bigger Zn grains as a starting material and effect of NaF solution used as a mineralizer in their study.

Figure 14 depicts scanning electron micrographs of the products. Grain size of starting Zn powder was 100–200 μm (a), then cracking was taken place on the surface of the product after subjecting to oxidation (b). Although the products were disintegrated into small grains when reaction proceeded rapidly (c), further disintegration was not observed. This disintegration can be attributed to compressive stresses developed in oxide layer during oxidation. Because of these compressive stresses, tensile stresses were produced in Zn grains, and then (1) peeling of ZnO layer was taken place if bonding bettween oxide and Zn was not firm or (2) cracking in Zn was occurred if the bonding was firm and/or surface flaw existed on the surface of Zn, as schematically shown in Fig. 15. Zn surface exposed in these manners could be easily oxidized, therefore, oxidation was proceeded much faster than expected by parabolic law generally expected in the oxidation of metals.

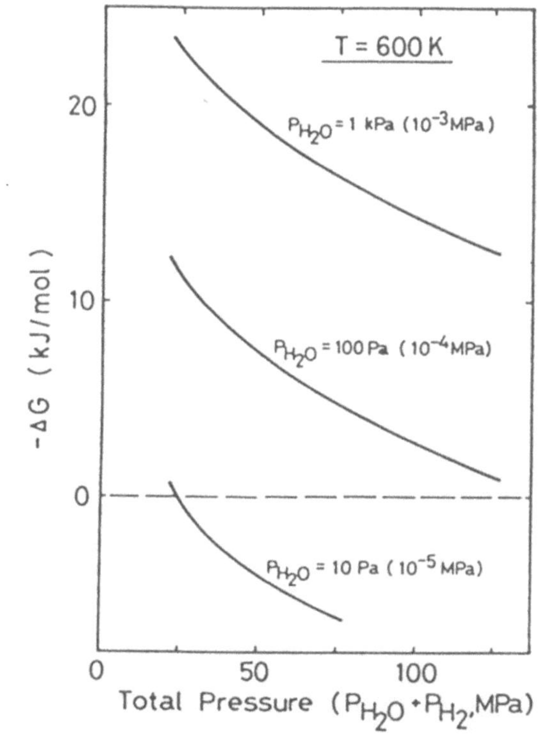

Fig. 13. Free energy change with total pressure and water partial pressures at 600 K for the reaction:

$$Zn + H_2O \rightarrow ZnO + H_2 .$$

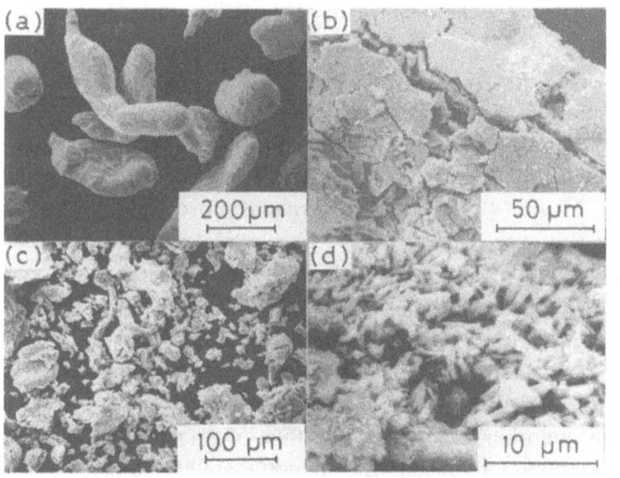

Fig. 14. Scanning electron micrographs of the products by the hydrothermal oxidation of Zn under 100 MPa.
(a) starting Zn powder (b) 200°C, 3h closed system
(c) 400°C, 0h, closed system (d) 400°C, 0h, open system

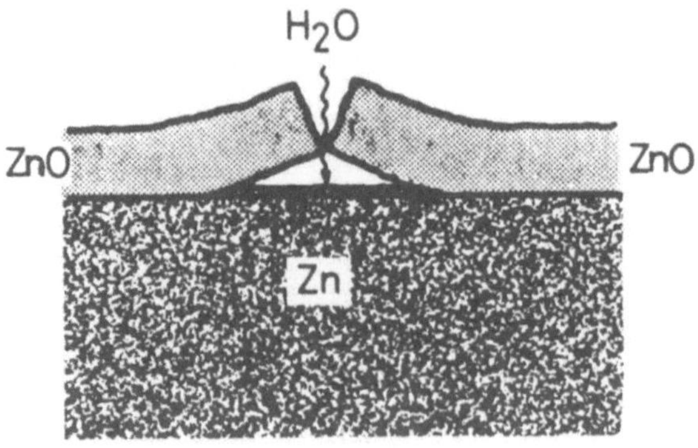

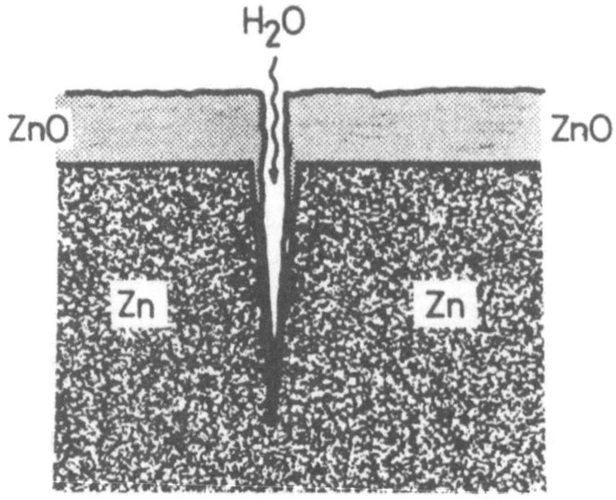

Fig. 15. Schematic illustration of disintegration of ZnO by the hydrothermal oxidation of Zn: (a) Peeling of ZnO (b) Cracking of ZnO.

Variation of crystallite size of ZnO produced in the closed system was illustrated in Fig. 16 as a function of temperature and time. Crystallite size of Zn was beyond the limitation (150 nm). Crystallite size of ZnO produced by 0-hour-treatment, however, was 15–60 nm and it increased with increasing temperature and/or time. Also in the open system, it was 20–125 nm (400°C) and it increased more rapidly than that of the closed system (Fig. 17). In this case, growth of rod-like crystal was also observed (Fig. 14 (d)). Since these crystals could not be observed in the closed system, it could be grown by dissolution-precipitation process similar to the case of rutile crystal growth.

Mechanisms of the hydrothermal oxidation of Ti and Zn were quite different. Crystallite size of Ti decreased with proceeding the oxidation. This fact suggested that oxygen and/or hydrogen diffused into Ti lattice, so that Ti lattice was disintegrated into small crystallites. On the other hand, crystallite size of Zn showed no decrease even at the final stage of the reaction. This means Zn diffused into oxide layer and oxidation took place at ZnO surface.

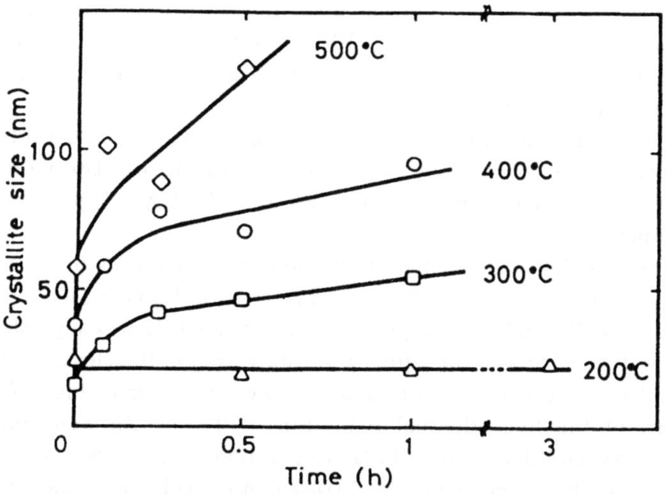

Fig. 16. Variation of crystallite sizes of ZnO with temperatures and time by the hydrothermal oxidation of Zn in the closed system under 100 MPa.

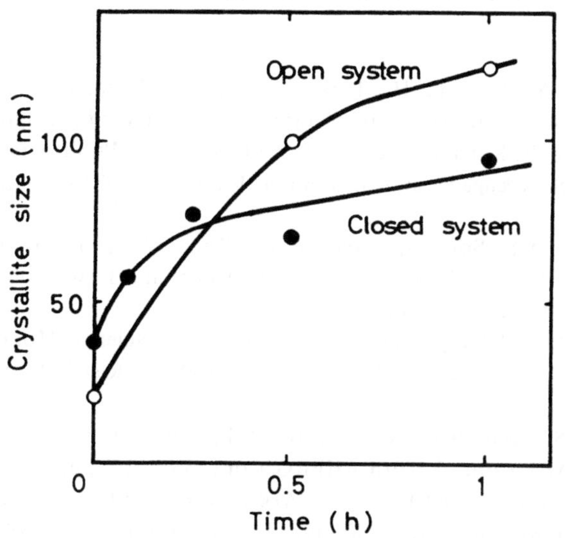

Fig. 17. Comparison of crystallite sizes of ZnO by the hydrothermal oxidation of Zn at 400°C under 100 MPa.

4. Conclusion

Reactions of the hydrothermal oxidation of Ti and Zn were examined in the closed system and the open system. Under 100 MPa for 3 h. Ti was oxidized above 450°C and TiH_x (x = 1.924), anatase and rutile were formed. Then TiH_x and anatase were changed into rutile as these reactions completed at 650°–700°C. During oxidation, large Ti crystallites were disintegrated and produced rutile crystallites of 30–85 nm. Crystallite size of rutile did not increase after 3 h (500°C, 100 MPa) in the closed system, whereas it increased gradually with time. Rutile crystal growth was observed in the open system by dissolution-precipitation process.

ZnO was formed by the hydrothermal oxidation of Zn above 200°C (100 MPa, 3 h) without any intermediate products by X-ray powder diffraction. During oxidation, Zn was disintegrated into much smaller crystallites of ZnO. Crystal growth of ZnO was observed in the open system.

From the thermochemical calculation, equilibrium conditions of the hydrothermal oxidation of Ti and Zn were shown as p_{H_2O}/p_{H_2} of 10^{-10} ($Ti \rightarrow TiO_2$, 500°C) and of 10^{-6} ($Zn \rightarrow ZnO$, 350°C) under total pressure of 100 MPa. It was suggested that these reactions actually does not reach the equilibrium state even in the closed system in which p_{H_2} was high and that the reaction rate became lower at the final stage of the reaction.

Mechanisms of the hydrothermal oxidation of Ti and Zn were quite different. Hydrogen and/or oxygen diffused into Ti lattice, so that Ti lattice was disintegrated into small crystallites. On the other hand, Zn diffused into oxide layer and oxidation took place at ZnO surface. Although mechanisms of disintegration during the hydrothermal oxidation were different in these manners, produced oxide powders had quite small crystallites. Therefore, the hydrothermal oxidation method seems to be effective for preparation of these oxide powders in activated states.

References

1) S. Sōmiya, M. Yoshimura, H. Toraya, E. Tani, M. Suzuki, S. Kikugawa and T. Hattori, "Proc. of Int. Symp. on Ceramic Compornents for Engine," pp. 739–52, Ed. by S. Sōmiya, E. Kanai and K. Ando, KTK Scientific Publishers, Tokyo, 1983.
2) M. Yoshimura, S. Kikugawa and S. Sōmiya, Funtai Oyobi Funmatsu Yakin, **30** [5] 207–10 (1983).
3) M. Yoshimura and S. Sōmiya, "Science and Technology of Zirconia, Advances in Ceramics Vol. 3," pp. 453–63, Ed. by A. H. Heuer and L. W. Hobbs, The American Ceramic Society, Columbus, Ohio, 1983.
4) ibid., "Sintering – Theory and Practice, Materials Science Monograph Vol. 14," pp. 417–22, Ed. by D. Kolar, S. Pejovnic and M. M. Ristič, Elsevier, Amsterdam, 1982.
5) M. Yoshimura, S. Kikugawa and S. Sōmiya, "High Pressure in Research and Industry, Proc. 8th AIRAPT Conf. & 19th EHPRG Conf. Vol. 2," pp. 793–96, Ed. by C-M. Backman, T. Johannison and L. Tegner, Arkitektkopia ISBN, Uppsara, 1982.
6) H. Toraya, M. Yoshimura and S. Sōmiya, J. Amer. Ceram. Soc., **65** [5] C-72 (1982).
7) ibid., **66** [2] 148–50 (1983).
8) ibid., **66** [11] 818–22 (1983).
9) M. Yoshimura, S. Kikugawa and S. Sōmiya, Yogyo-Kyokai-Shi, **91** [4] 182–88 (1983).
10) ibid., "Proc. Conf. Emergent Process Methods for High Technology Ceramics, Materials Science Research Vol. 17," pp. 155–66, Ed. by R.F. Davis, H. Palmour III and R.L. Porter, Plenum Press, New York, 1984.
11) H. Toraya, M. Yoshimura and S. Sōmiya, "Science and Technology of Zirconia II, Advances in Ceramics Vol. 12," pp. 806–15, Ed. by N. Claussen, M. Rühle and

A.H. Heuer, The American Ceramic Society, Columbus, Ohio, 1984.
12) T. Nakayama, H. Yamada, E. Tani and J. Yasui, Nippon Kinzoku Gakkaishi, **27** [7] 318–23 (1963).
13) T. Nakayama and T. Koizumi, ibid., **28** [11] 749–54 (1964).
14) ibid., **31** [4] 380–85 (1967).
15) D. Moroishi and Y. Shida, ibid., **42** [3] 316–23 (1978).
16) T. Oota, I. Yamai and H. Saito, Yogyo-Kyokai-Shi, **87** [7] 375–82 (1979).
17) ibid., **88** [8] 460–68 (1980).
18) H. Toraya, M. Yoshimura and S. Sōmiya, J. Appl. Cryst., **16**, 653–57 (1983).
19) E.F. Osborn, J. Amer. Ceram. Soc., **36** [5] 147–51 (1964).
20) G.M. Barrow, "Physical Chemistry, 4th Ed.," pp. 257, McGraw-Hill, New York, 1979.
21) M. Yoshimura, H. Ohira and S. Sōmiya, Yogyo-Kyokai-shi, **93** [7] 357–63 (1985).
22) ibid., **94** [11], 1186–89 (1986).

Part 3

Hydrothermal Growth

Reprinted from *Collected Abstracts, Fourth International Conference on Crystal Growth*, Tokyo, 29 March 1974, pp. 683–684.

Crystal Growth of Magnetite under Hydrothermal Condition with Hydrogen

S. Hirano* and S. Sōmiya, Res. Lab. Eng. Mat., Tokyo Institute of Technology, Ookayama, Meguro-ku, Tokyo, Japan

The hydrothermal crystal growth of magnetite was investigated under the condition of the coexistence of hydrogen during the growth period without the use of a seed crystal.

The growth experiment was carried out in a hermetically sealed gold capsule containing desired amount of reduced iron and solution which was placed in a cold-seal pressure vessel. The pressure and temperature ranged from 500 to 3000kg/cm^2 and from 200 to 900°C so as to examine the stability of magnetite and the factors found to be important in achieving good crystal growth under the hydrothermal conditions associated with hydrogen. 3m. to 10molality of NaOH, KOH or NH_4Cl solution was used as the solvent in the growth run. The phases and crystals obtained were identified by the X-ray diffraction method with Co Kα radiation.

It was found that magnetite was synthesized even at the temperature as low as 250°C under 1000kg/cm^2 at the controlled atmosphere of the mixture of water and hydrogen. The single crystal of magnetite could be grown spontaneously on the inner wall of the capsule at a nutrient temperature above 400°C and a thermal gradient of approximately 50°C. At the bottom of the capsule, a well sintered magnetite body was formed as a single phase from 350°C to 650°C. The grain size of magnetite was as small as a few microns in the sintered body, depending on the experimental conditions. Wustite crystals were obtained above 650°C. The concentration of the solution remarkably affected the crystallization in the presence of hydrogen. The crystals grown from 3m NaOH or KOH solution did not have any perfect crystal morphology. On the other hand, the crystals grown in the higher concentration of NaOH or KOH such as 5m or 10m showed the good crystallinity and the well developed crystal-faces.

Some dendrite having magnetic property could be grown in NH_4Cl solution. The photograph shows the crystal grown on the inner wall of the gold capsule treated at 550°C for 7 days under 1000kg/cm^2.

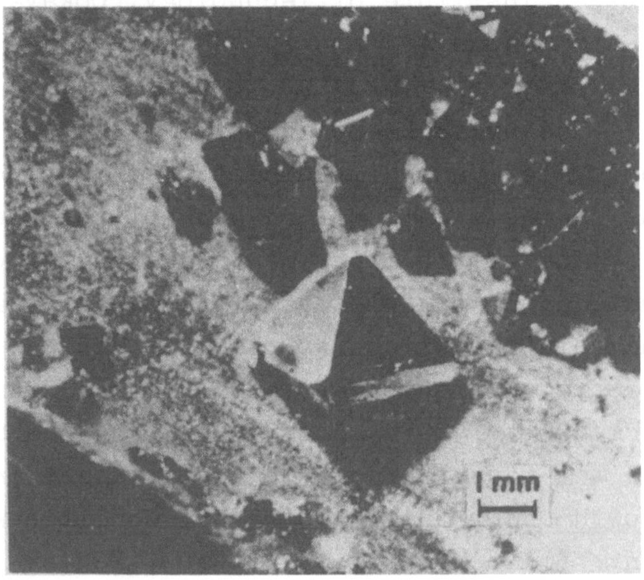

In the present work, the presence of hydrogen found to be essential to make magnetite stable and to grow the single crystal under hydrothermal condition. The hydrothermal method under the coexistence of hydrogen with water revealed the remarkable advantages to grow the oxide crystal which should be stable only in the reduced atmosphere and to obtain the sintered body of a single phase at the lower temperature. This technique is considered to be applied for the synthesis and crystal growth of the materials whose cation valences will be changeable with respect to the condition used.

Reprinted from *Journal of the American Ceramic Society*, **58**(7–8) (1975) 349–350.

Hydrothermal Growth of Brucite Crystals

KAZUO NAKAMURA, SHIN-ICHI HIRANO, and SHIGEYUKI SŌMIYA*

BRUCITE occurs most commonly in nature as an alteration product of periclase; it is also found as a low-temperature hydrothermal vein mineral in serpentinites and in chloritic schists.[1] There have been several studies of the equilibrium between periclase and brucite, e.g. Ref. 2. Previous investigators, however, gave no information on the growth of brucite under hydrothermal conditions. In the present study, single crystals of brucite were grown hydrothermally.

An ordinary test-tube-type hydrothermal vessel was used. The starting periclase powder was prepared by calcining $Mg(NO_3)_2$[†] at 600°C for 24 h. An alkali hydroxide solution[‡] and 20 mg of the periclase powder were inserted into an Au tube 3 or 5 mm in diameter and 35 mm long; the tube was then sealed in a carbon arc. In some cases an Au wire was attached at the top of the capsule as a trap. The solutions used were NaOH and KOH (5 or 15 mol%) and LiOH (3 mol%). The quantity of solution was adjusted so that the pressures inside and outside the capsule could be equalized. The temperatures attained at the top of the capsule were 350°, 450°, 500°, and 550°C; the temperature at the bottom was 25°C higher than that at the top in every case. Pressures of 500, 1000, and 2000 kg/cm² were used. After a growth run of 7 or 10 days, the vessel was quenched into cold water. The capsule removed from the vessel was cut open, and the crystals were washed with redistilled water and observed with an optical microscope.[§] For observations in reflected light, the surface of the crystal was coated with Au in vacuum.

Received February 27, 1975; revised copy received May 28, 1975.
The writers are with the Tokyo Institute of Technology, Tokyo, Japan.
*Member, the American Ceramic Society.
†Prepared by dissolving Mg metal >99.95% pure, Furukawa Magnesium Co., Ltd., Tokyo, Japan) in HNO_3.
‡Prepared from reagent-grade hydroxide (Wako Pure Chemical Industry, Ltd., Osaka, Japan) and water redistilled in a quartz apparatus.
§Vanox, Olympus Optical Co., Ltd., Tokyo, Japan.

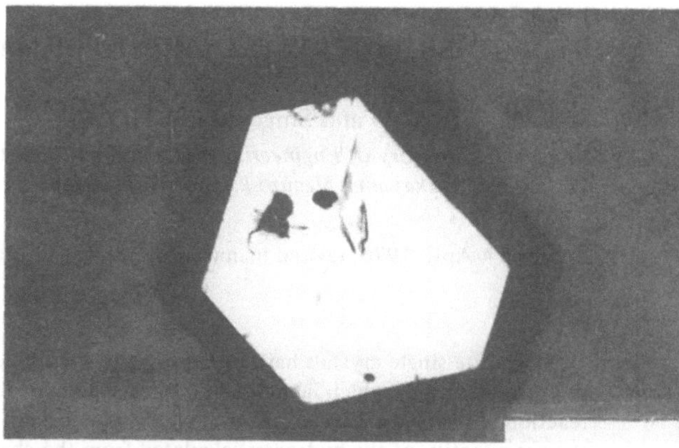

Fig. 1. Photomicrograph of brucite single crystal grown in 5 mol% KOH at 450°C and 500 kg/cm² for 7 days. Crystal is ≈0.04 mm thick. Bar=0.1 mm.

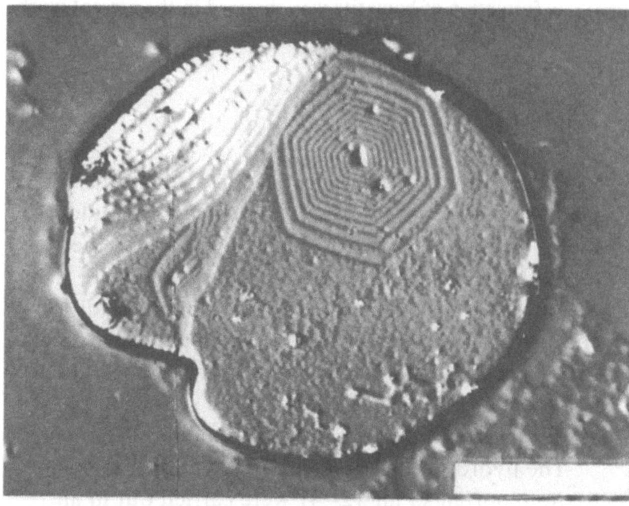

Fig. 2. Photomicrograph of hexagonal spiral on (0001) plane of brucite grown in 3 mol% LiOH at 450°C ar d 1000 kg/cm² for 10 days. Bar=0.05 mm.

The crystals trapped at the upper Au wire in the experiment in which the wire was used were well-developed transparent plates with the (0001) plane. The crystals grown in NaOH or KOH solution were thick hexagonal plates with straight edges (Fig. 1). On the other hand, the crystals grown in LiOH solution were thin with rounded edges. The growth rate normal to the *c* axis ($R_{\perp c\ \mathrm{axis}}$) was greater than that parallel to the *c* axis ($R_{\parallel c\ \mathrm{axis}}$) in all the solutions, but the ratio of these growth rates differed, depending on the solution, i.e.

$$\left(\frac{R_{\perp c\ \mathrm{axis}}}{R_{\parallel c\ \mathrm{axis}}}\right)_{\mathrm{LiOH}} > \left(\frac{R_{\perp c\ \mathrm{axis}}}{R_{\parallel c\ \mathrm{axis}}}\right)_{\mathrm{NaOH\ or\ KOH}}$$

Therefore, the crystals grown in NaOH or KOH solution represent the crystal habit, whereas those grown in LiOH solution do not.

The surfaces, which were observed primarily by Nomarsky-type differential interferometry, exhibited various types of spiral patterns. Figure 2 shows a hexagonal spiral originating from a single dislocation on the (0001) plane of brucite grown in LiOH solution. In addition to this type of spiral, a single circular spiral, closed loops from a pair of dislocations with opposite sign, and many spiral patterns originating from a group of dislocations were observed. Thus, it appears that brucite grows under hydrothermal conditions by a spiral growth mechanism.

[1] W. A. Deer, R. A. Howie, and J. Zussman, Rock Forming Minerals, Vol. 5. Longmans, Green and Co., Ltd., London, 1962; pp. 89–92.
[2] (*a*) D. M. Roy and Rustum Roy, "A Redetermination of Equilibria in the System $MgO-H_2O$ and Comments on Earlier Works," *Amer. J. Sci.*, **255**[10] 573–82 (1957).
(*b*) J. W. Meyer and J. C. Yang, "Some Observations in the System $MgO-H_2O$," *ibid.*, **260** [11] 707–17 (1962).

Reprinted from *Journal of Crystal Growth*, **35** (1976) 273–278.

HYDROTHERMAL CRYSTAL GROWTH OF MAGNETITE IN THE PRESENCE OF HYDROGEN

Shin-ichi HIRANO and Shigeyuki SŌMIYA

Research Laboratory of Engineering Materials and Department of Materials Science and Engineering, Tokyo Institute of Technology, Ookayama, Meguro-ku, Tokyo 152, Japan

Received 4 April 1976; revised manuscript received 11 May 1976

Magnetite single crystals have been grown by spontaneous nucleation on the inner wall of a gold capsule under hydrothermal conditions which included hydrogen. The hydrogen for the control of the reduced atmosphere was supplied by the reaction of charged metallic iron with water. The leakage of the hydrogen through a gold capsule was found to be 2% of the net quantity of the hydrogen calculated from the charged iron at 550°C for 21 days under 1000 kg/cm^2. The leakage was increased due to the grain growth of the wall of the gold capsule with the increases of the growth period and the concentration of solution used. The {110} faces of magnetite were found to predominate over the normal {111} faces as the concentration of the hydrogen was increased. Optimum conditions to grow bulky single crystals of magnetite were as follows: 10m NaOH as a solution; crystallization temperature 500°C; percent fill of free volume, about 55%; ΔT, 50°C. No formation of hematite was observed in the present work. Dendritic magnetite crystals were grown in 5m NH_4Cl solution at 500°C under 1000 kg/cm^2.

1. Introduction

It has been of great importance to establish a simple method to control the redox environment under the hydrothermal condition for the growth of the oxide with the desired chemical composition where the cation valence is changeable by the growth condition employed. There have, however, been published few papers on the hydrothermal crystal growth of oxides under such conditions except for the studies of phase equilibria from the view point of geology [1].

The hydrothermal crystal growth studies of magnetite reported so far [2–4] were carried out in an unlined or silver-lined autoclave which is permeable to hydrogen, so that mixtures of magnetite and hematite have been obtained due to the failure of maintaining the oxygen partial pressure within the magnetite stability field during the growth run.

In the present work, the hydrothermal crystal growth of stoichiometric magnetite was investigated as an example to grow oxides of lower cation valences under the controlled condition with coexistence of hydrogen formed by the reaction of iron and supercritical water without the use of a seed crystal. The leakage of hydrogen through a gold capsule was checked under various temperatures and concentrations of solution. The effect of the coexisting hydrogen gas on the morphology of the grown magnetite crystal was also studied to deduce the morphological aspect of natural magnetite crystals.

2. Experimental

The starting iron (analytical grade, purity >99%, Merck, Germany) consisted of reduced iron particles of spherical shape with an average size of about 7 μm. Analysis of the iron powder reported by the supplier was as follows: soluble part in water <0.01%, insoluble part in sulfuric acid <0.1%, S < 0.05%, N < 0.01%, Cu < 0.002%, Zn < 0.005%, Mn < 0.01% and As < 0.0002%. Water was prepared by re-distillation with a quartz distillator of water which had been obtained through an ion exchanger from distilled water. Sodium hydroxide, potassium hydroxide and ammonium chloride were extra reagent grade.

The following reaction was employed to form magnetitite and also to control the oxygen partial pressure:

$$3Fe + 4H_2O \rightarrow Fe_3O_4 + 4H_2 \, .$$

Weighed amounts of iron powder, pure water and solution were sealed hermetically without any seed crystal into a gold capsule, 5.0 mm in outer diameter, 0.15 mm thick and 50 mm in length. The typical amounts of these substances for an experiment at the crystallization zone of 500°C under 1.000 kg/cm^2 were as follows: 111.7 mg iron, 48.0 mg pure water and 330 μl 10m NaOH solution; 223.4 mg iron, 96.0 mg pure water and 330 μl 10m NaOH solution; and 335.1 mg iron, 144.0 mg pure water and 330 μl 10m NaOH solution. The amounts of both iron and pure water were changed to keep the quantity of solution constant when the other level of the concentration of hydrogen was required. The sealed capsule was placed in a test-tube type pressure vessel of Stellite 25, 6.5 mm in inner dimaeter, and subjected to the hydrothermal reaction and growth under the desired pressure and temperature in a furnace with temperature gradient. The growth experiments were thus carried out at temperatures from 250°C to 700°C under pressures up to 2000 kg/cm^2. As a solvent, sodium hydroxide or potassium hydroxide of the concentrations up to 20 molality or ammonium chloride solution up to 5 molality was employed. The temperature gradient adopted was from 3 to 15°C/cm. Pressure was applied first to a desired value for the growth run through a capsule with water and the temperature was then raised at a heating rate of about 10°C/min under constant pressure. The pressure was measured with a calibrated Heise gauge and the temperature was measured with a potentiometer P-31 (Yokogawa Elec. Works, Ltd. Tokyo) connected to three platinel thermocouples attached on the outside wall of the pressure vessel at three positions corresponding to top, middle and dissolution zone of the capsule. The thermocouples were calibrated against the melting points of gold (1,064.4°C) and zinc (419.5°C). After a growth period, the vessel was quenched into cold water. Then the amount of hydrogen which remained in the capsule was measured by weighing the capsule dried beforehand with acetone before and after puncturing with a thin needle and compared with that calculated from the amount of the starting iron. The crystals formed in the capsule were observed with an optical microscope (Vanox, Olympus Optical Co. Ltd., Tokyo, Japan) and a scanning electron microscope (HSM-2, Hitachi Ltd., Tokyo, Japan) and also identified with a X-ray diffractometer (JDX-7F, JEOL Co., Ltd., Tokyo, Japan) using Fe-filtered CoKα or CuKα radiation through a graphite monochromator. Some of the crystals thus grown were pulverized into powder in order to measure the precise unit cell parameter. The calculated lattice parameters from higher angle peaks above 60° in θ corrected with an internal standard of hydrothermally grown calcium fluoride beforehand defined with 99.999% silicon, were plotted against $\cos^2\theta$ and were extrapolated to $\cos^2\theta = 0$.

3. Results and discussion

The reaction of the starting iron with water to form magnetite was found to occur even at a temperature as low as 250°C and complete within 1 h after reaching the temperature above 400°C under 1000 kg/cm^2. The leakage of hydrogen through a gold capsule was checked in order to confirm whether the reaction had completed and the growth run could be performed in the reduction condition required. Fig. 1 shows the percent leakage of hydrogen during a growth run, where the ratio of the amount of hydrogen loss to that theoretically calculated from the quantity of the starting iron powder charged was plotted against the growth period under various conditions of temper-

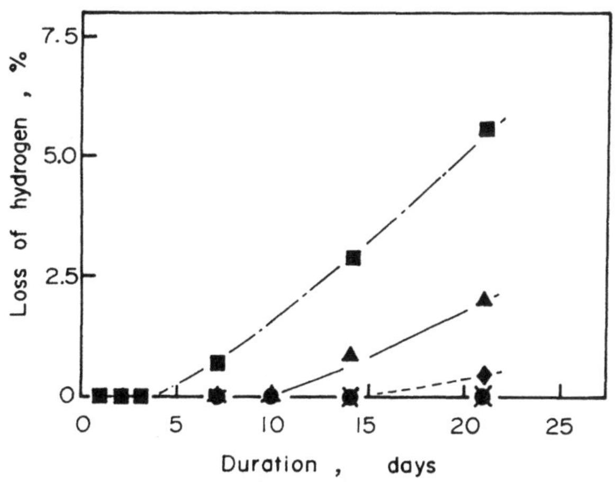

Fig. 1. Leakage of hydrogen (in percent) as a function of time, theoretical amount of hydrogen formed of 8 m-mole in 330 μl of NaOH solution under 1000 kg/cm^2: (■) 600°C, 10m NaOH soln; (▲) 550°C, 10m NaOH soln; (●) 500°C, 10m NaOH soln; (×) 400°C, 10m NaOH soln: (♦) 600°C, 2.5m NaOH soln.

ature and concentration of NaOH solution. In the experiment for the measurement of the leakage, iron powder of 335.1 mg, 144.0 mg of pure water and 330 μl of NaOH solution were charged in each capsule, therefore 16.0 mg of hydrogen should be formed. The results revealed that the higher the growth temperature and the concentration of the solution, the greater the leakage of hydrogen through the capsule wall was observed to take place above 550°C under 1000 kg/cm^2, while no leakage of hydrogen was detected in the case of pure water as solution. With 10m NaOH solution, the leakage of hydrogen amounted to about 2% of the total quantity of hydrogen formed and 6% at 550°C and at 600°C, respectively, for 21 days under 1000 kg/cm^2. Such a leakage of hydrogen was considered to be responsible for the diffusion through the wall of the gold capsule due to its grain growth which was observed microscopically. No measurable amount of leakage was detected within 10 days at 550°C in 10m NaOH solution. Thus most of the growth runs were carried out for the growth period up to 10 days in this work.

Magnetite crystals could be synthesized even at a temperature as low as 250°C under 1000 kg/cm^2 in NaOH, KOH and NH_4Cl solution. Single crystals of magnetite visible even to the naked eye were grown spontaneously on the inner wall of the capsule at the crystallization zone above 350°C and the temperature difference of 50°C. No formation of hematite could be detected in all of the present experiments, even if the growth run was carried out at 600°C in 10m NaOH solution for 21 days under 1000 kg/cm^2. It turns out that the conditions employed in the present study using a gold capsule were well within the magnetite stable field reported by Eugster [5]. The concentration of the solution influenced remarkably the crystal growth of magnetite. The size of the crystals grown in the solution of 2.5m NaOH or KOH only reached about 1.5 mm at the crystallization zone of 550°C with about 50°C of temperature difference from the dissolution zone under 1000 kg/cm^2 for days. On the other hand, crystals grown in the solution of higher concentration such as 10m NaOH showed good crystallinity and were about 3 mm with well-developed crystal faces under the same conditions. In order to investigate the growth feature in a capsule, the scanning electron microscopic observations were made on cut-opened capsule surfaces. Fig. 2 shows the scanning electron micrographs of the crystals grown on the inside wall at different position, of the gold capsule in 10m NaOH solution with 5.3 m-mole of hydrogen for 1 day under 1000 kg/cm^2 at 550°C of the dissolution zone with about 12.5°C/cm of the temperature gradient. The features as shown in fig. 2 were normally recognized regardless of the concentration of NaOH or KOH solution. Such a sintering phenomenon of magnetite at the bottom of the capsule as shown in fig. 2a led to the development of a new method to sinter the oxide with the lower cation valence and the desired chemical composition at lower temperature as low as 550°C under hydrothermal conditions [6]. All of the crystals in the zone b (ΔT: about 10°C) showed a characteristic of "the mosaic" as illustrated in fig. 2b, which with the longer growth period tended to develop the faces of the low indices and continuously grew to the crystal shown in fig. 2c. The crystals in the zone c, as shown in fig. 2c, had well-developed flat {111} faces with stepped side faces. The zone d (ΔT: about 50°C) was the most suitable position for efficient nucleation and growth of bulky crystals as shown in fig. 2d.

The optimum conditions to grow bulky crystals with well-defined crystal faces were as follows: Crystallization zone 500°C; dissolution zone 550°C; pressure 1000 kg/cm^2; filling degree of free volume, about 55%; and solvent 10m NaOH.

Fig. 3 shows a representative magnetite crystal grown under such conditions for 7 days. All of the crystals thus grown were stuck so strongly on the inside wall of the gold capsule that they hardly could be removed. The phenomenon was considered to reflect that the nucleation might have taken place on a preferred position of the gold capsule surface. Figs. 4a and 4b show the surface figures on the (111) face of the crystal shown in fig. 3, taken by a differential interference method and a reflection microscopic method with the tilted illumination, respectively. The surface structure showed a roughness on the order of 1 μm and then no evidence of spiral growth.

Hydrogen was found to play a very important role in the morphology of the grown crystal as well as in the control of the atmosphere under hydrothermal conditions. Figs. 5a, 5b and 5c are the representative photographs of the magnetite crystals grown with the various amounts of hydrogen such as 2.6 m-mole, 5.3 m-mole and 8.0 m-mole, respectively, with 330 μl

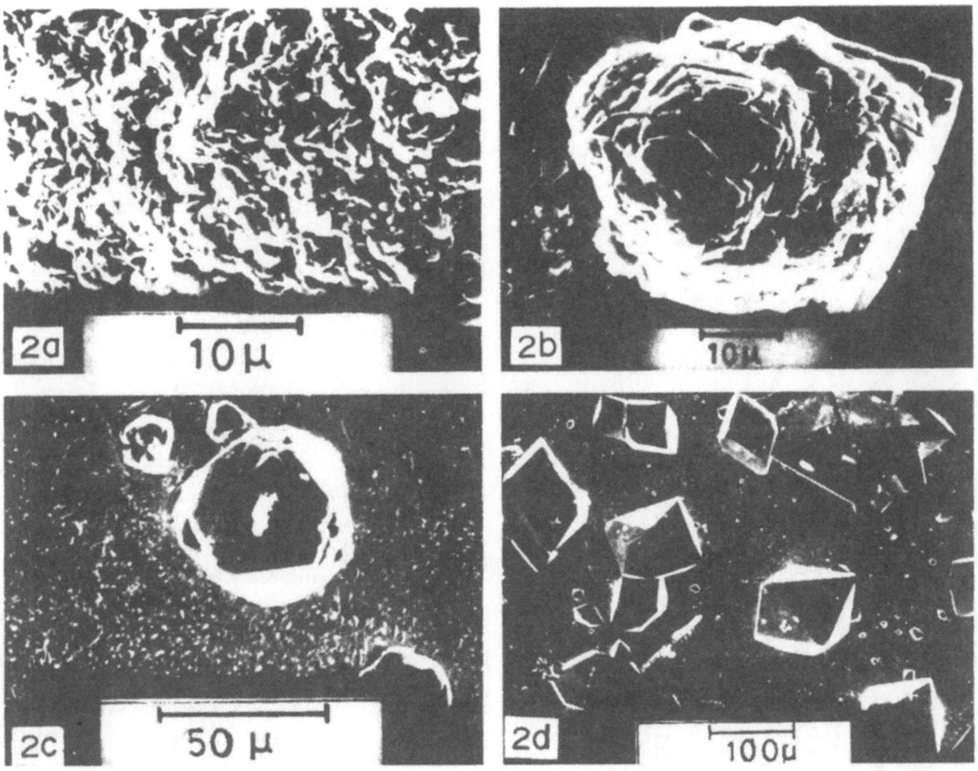

Fig. 2. Scanning electron micrographs of magnetite crystals on inner wall of cut-opened capsule at different position (at temperature as below); 10m NaOH, 1 day, 1000 kg/cm^2: (a) 550°C bottom of capsule); (b) 540°C; (c) 525°C; (d) 500°C (top of capsule).

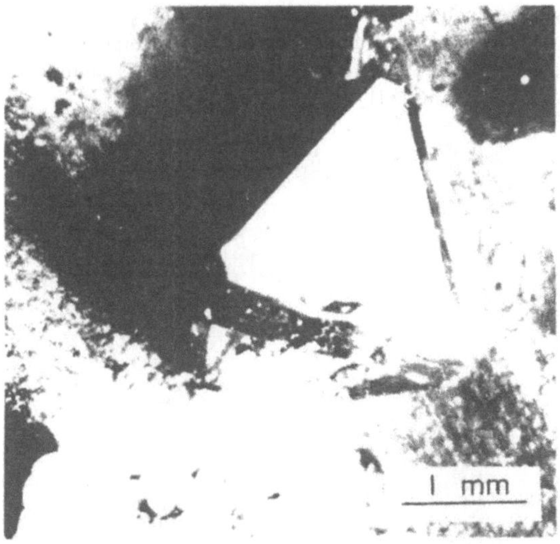

Fig. 3. Magnetite crystal grown at crystallization zone temperature of 500°C, ΔT = 50°C, under 1000 kg/cm^2, in 10m NaOH for 7 days.

of 10m NaOH solution at a crystallization temperature of 500°C under 1000 kg/cm^2. The higher the concentration of the coexisting hydrogen in the gold capsule, the more remarkable the development of the {110} faces becomes, although the exact oxygen partial pressure could not be calculated from the data of the amount of hydrogen formed in 10m NaOH solution because of the lack of information on the behaviour of hydrogen in NaOH solution. Finally, the crystals grown in the presence of 8.0 m-mole hydrogen in 330 μl of 10m NaOH solution tended to be bounded only by {110} faces.

The lattice parameters of the magnetite crystals thus grown were 8.3962 ± 0.0003 Å, irrespective of the concentration of coexisting hydrogen and solution, which corresponds to the value of the stoichiometric composition of magnetite reported by Basta [7]. Therefore, the stoichiometry of magnetite was considered not to have any direct connection with the

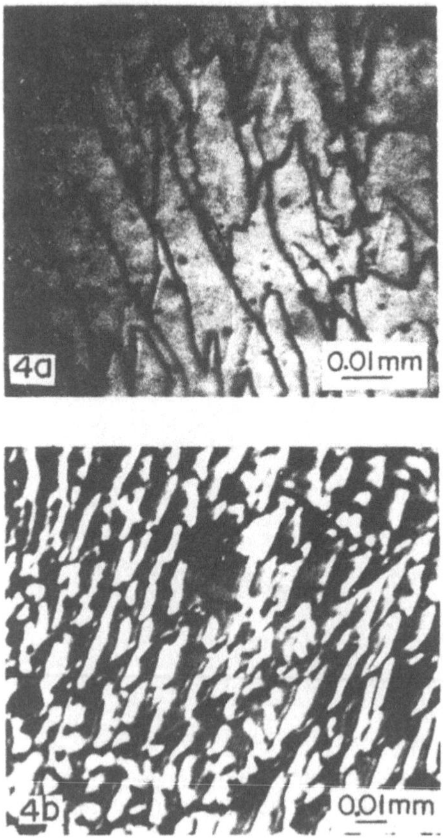

Fig. 4. (111) surface figures: (a) differential interference micrograph; (b) reflection micrographs taken with tilted illumination.

change in morphology. Such an effect of hydrogen concentration under hydrothermal conditions has also been observed in the hydrothermal growth experiments of fayalite in the authors' laboratory [8]. Another series of experiments would be needed to find out a clue to the exact reason for the morphological effect.

It was found that the morphology of magnetite crystals could be changed with changes in the kind and the concentration of solution as well as the amount of hydrogen. Small magnetite single crystals of typical octahedral habit with {111} faces were also grown under the conditions of coexistence with 5.3 m-mole of hydrogen in 1m NH_4Cl solution of 330 μl with the crystallization zone at 500°C and ΔT = 50°C. On the other hand, dendritic crystals of magnetite growing in the ⟨100⟩ direction as shown in fig. 6 were

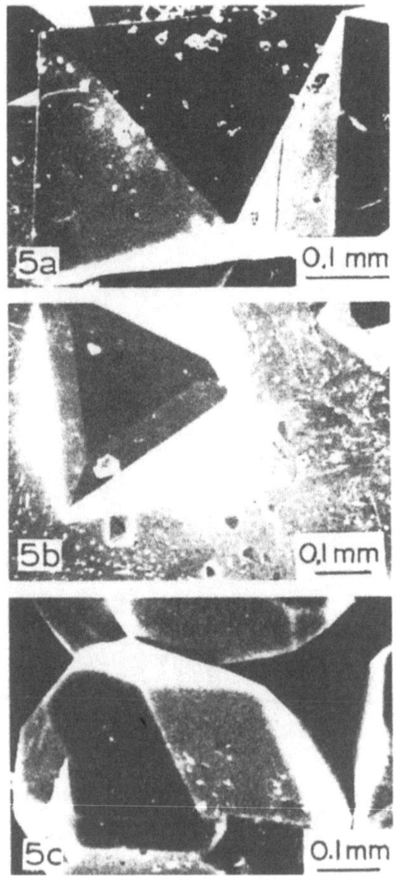

Fig. 5. Photomicrographs of magnetite crystals grown at various hydrogen concentrations in 300 μl of 10m NaOH solution under 1000 kg/cm^2 for 3 days, crystallization temperature 500°C, ΔT = 50°C (a) in 2.6 m-mole H_2, (b) in 5.3 m-mole H_2, and (c) 8.0 m-mole H_2.

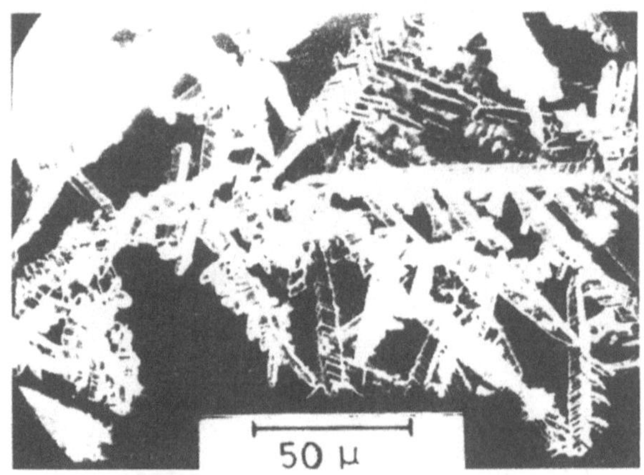

Fig. 6. Dendritic magnetite crystals grown in 5 m-NH_4Cl solution in 5.3 m-mole H_2, crystallization temperature 500°C, ΔT = 50°C under 1000 kg/cm^2 for 3 days.

observed by maintaining the other conditions constant except for raising the concentration of NH_4Cl solution to 5m.

As for the material of a container in which the hydrothermal growth experiments were carried out, glassy carbon products were found to be the most suitable material at temperatures up to 600°C for a longer growth run, which could stand against the attack by the concentrated solutions such as a 10m NaOH [9].

4. Conclusions

The hydrothermal crystal growth method in the presence of hydrogen was found to be capable of the growth of stoichiometric magnetite single crystals without any formation of hematite and also to be applicable to the single crystal growth of the oxide of lower cation valence. The gold capsule material maintained the oxygen partial pressure well within the magnetite stable field even for a long period of 21 days at 600°C in 10m NaOH solution. The leakage of hydrogen through the gold capsule depended remarkably on temperature, duration and concentration of solution such that the leakage of hydrogen was raised to the detectable amount at temperatures above 550°C in 10m NaOH, that is about 6% of hydrogen at 600°C for 21 days in 10m NaOH solution under 1000 kg/cm^2 with a gold capsule, 5.0 mm in outer diameter, 0.15 mm thick and 50 mm in length, presumably due to the increased grain growth of the gold capsule wall, while no leakage was detected in pure water under the same conditions. The coexisting hydrogen during a growth run played an important role in the morphology of the grown magnetite as well as in the control of the reduction condition. The more remarkable was the development of {110} faces which were observed, the higher was the hydrogen concentration in the growth atmosphere. Dendritic crystals of magnetite growing in ⟨100⟩ directions were formed in 5m NH_4Cl solution at a crystallization zone temperature of 500°C and $\Delta T = 50°C$ under 1000 kg/cm^2.

Acknowledgments

The authors are grateful to Dr. R.A. Laudise (Bell Labs, NJ, USA), Professor P. Hartman (State University of Leiden, The Netherlands), and I. Sunagawa (Tohoku University, Sendai, Japan) for their encouragements and valuable comments for this work.

References

[1] H.P. Eugster, J. Chem. Phys. 26 (1957) 1760.
[2] N. Yu. Ikornikova, Dokl. Akad. Nauk SSSR 130 (1960) 610.
[3] J. Koenig, reported by R.A. Laudise and J.W. Nielsen, in: Solid State Physics, Vol. 12, Eds. F. Seitz and D. Turnbull (Academic Press, New York, 1961) p. 149.
[4] E.D. Kolb, A.J. Caporaso and R.A. Laudise, J. Crystal Growth 19 (1973) 242.
[5] H.P. Eugster, reported by H.P. Eugster, in: Research in Geochemistry, Vol. 1, Ed. P.H. Abelson (Wiley, New York, 1959) p. 397.
[6] S. Hirano and S. Somiya, J. Am. Ceram. Soc., to be published.
[7] E.Z. Basta, Mineral. Mag. 31 (1957) 431.
[8] S. Hirano, Y. Iwai and S. Somiya, in preparation.
[9] S. Hirano and S. Somiya, J. Ceram. Soc. Japan, to be published.

Reprinted from *Materials Research Bulletin*, **11** (1976) 1023–1030.

Crystal Growth of Sodium Iron Titanium Bronze Compounds under Hydrothermal Conditions

Shin-ichi Hirano, M.G.M.U. Ismail* and Shigeyuki Sōmiya
Research Laboratory of Engineering Materials and Department of Materials Science and Technology, Tokyo Institute of Technology
Ookayama, Meguro-ku, Tokyo 152, Japan

(Received June 1, 1976; Communicated by R. C. DeVries)

ABSTRACT

Single crystals of four types of oxides consisting of sodium iron and titanium were grown hydrothermally under reduced conditions starting with a mixture of TiO_2 (rutile) and metallic Fe powder in the ratio 5:1 (weight basis) at 650°C and 1000kg/cm^2 in 10m NaOH solution for different durations. Formation of single crystals having an orthorhombic cell was observed after one week. In longer runs under the same conditions the initial phase undergoes transformation to form crystals having hexagonal and orthorhombic cells and finally to monoclinic freaudenbergite ($Na_2O\cdot Fe_2O_3\cdot 7TiO_2$). From Mössbauer and EPMA data, the common chemical formula for these new phases can be written as $(Na_xFe_y^{2+})\cdot[Ti_pFe_q^{2+}Fe_r^{3+}O_{2-\delta}]_n$.

Introduction

In the course of investigation on the stability of ilmenite in NaOH solution under hydrothermal conditions a new type of nonstoichiometric oxide consisting of sodium, iron and titanium was found. In order to synthesize this nonstoichiometric compound under hydrothermal conditions, experiments were carried out starting with a mixture of TiO_2 (rutile) and Fe using NaOH solution. Formation of three new phases were observed, and the investigations suggested that these nonstoichiometric compounds were to be sodium iron titanium bronzes. The existence of alkali titanium bronze has been demonstrated by Anderson and Wadsley(1). From the structural analysis of $Na_xTi_4O_8$, an alkali metal titanium dioxide bronze, they showed that the presence of Na as an impurity or guest atom may be necessary to form the host lattice. Mumme and Reid(2) also reported the existence of $Na_xFe_xTi_{2-x}O_4$ which shows only the presence of Fe^{3+} ion, and the structure of this compound was identical to

*Present address; Ceylon Institute of Scientific and Industrial Research, P.O. Box 787, Colombo 7, Sri Lanka.

$NaFeTiO_4$. In the present study, the nonstoichiometric alkali iron titanates synthesized under hydrothermal conditions, showed the presence of both Fe^{2+} and Fe^{3+} ions in different coordinations. This paper presents the crystal growth of these compounds under hydrothermal conditions and their crystal data.

Experimental

Preparation of starting material

The conversion of TiO_2(anatase) to TiO_2(rutile) was carried out at 650°C and 1000kg/cm^2 for one week duration, under hydrothermal conditions. The presence of only rutile was confirmed by taking the X-ray pattern of the converted samples. The TiO_2(rutile) and powdered metallic Fe were weighed out in proper ratio and mixed together. In the case of metallic Ti and Fe powders the proper ratio to be mixed was determined by considering the reaction of Ti with water to form TiO_2 under hydrothermal conditions. The mixture was transferred to a gold capsule of 2.7mm in inner diameter, 0.15mm wall thickness and 35mm in length, containing a known amount of NaOH solution. The capsule was sealed and tested for the leakage and weighed. The capsules were then loaded into a Roy and Tuttle(3) test tube type hydrothermal apparatus and the pressure was brought up to the desired value. Then the temperature was increased gradually by taking care to prevent overshooting of pressure, up to the operating temperature. A platinel thermocouple calibrated at the melting point of gold (1064.4°C) was attached to the outside of the pressure vessel for the measurement of temperature. The temperature and pressure were controlled within ±3°C and ±20kg/cm^2 of the desired value. At the completion of experiment the vessel was quenched into cold water. The capsule was weighed for observation of loss in weight. The final products were examined by the following procedures.

Microscopic examination

An optical microscope of "Vanox" (Olympus optical Co. Ltd., Japan) was used in this investigation. The crystals grown were washed with ethanol and observed under the optical microscope for growth habits, patterns, twins, etc.

X-ray data

The phase identification and the powder patterns of the final products were carried out using a JDX-7F (JEOL Co., Japan) X-ray diffractometer with a graphite monochromator, employing Cu Kα radiation. Single crystal oscillation and Weissenberg photographs were taken for all four phases employing Ni filtered Cu K_α radiation. The structure, space group and the unit cell constants were determined from Weissenberg photographs.

Mössbauer spectra

The Mössbauer spectra of the phase III was recorded using ^{57}Co in a Cu source and a combination of a constant acceleration electromagnetic drive with a 400 channel pulse-height analyser operated in time mode. The estimation of Fe^{2+}/Fe^{3+} was done by computer simulation of the spectra. The coordination of Fe^{2+} and Fe^{3+} ions to the ligands was determined by calculation of isomer shift and quadrupole splittings of the doublets. The isomer shift is expressed relative to stainless steel.

Chemical analysis

The single crystals of the new phases were selected for oscillation photographs along different axes and used for EPMA analysis. The beam used was about 1μ in diameter and by computer simulation the Fe^{2+} to Fe^{3+} ratio was counted. The atomic ratios of Na^+, Ti^{4+}, Fe^{2+} and Fe^{3+} were determined by fixing the oxygen as 4.

Results and Discussion

Crystal growth of phase I

Starting with a mixture of TiO_2 (rutile) and metallic Fe powder in the ratio 5:1 (weight basis), runs were carried out at temperatures between 500°C and 800°C under 1000kg/cm^2 for durations of one week in 10m NaOH solution. Formations of different type of phases were found at temperatures above 675°C. Black single crystals with a metallic lustre were formed in the low temperature region of the capsule at 650°C. There were crystals of prismatic, hexagonal prismatic and needle like habits also. The growth of these crystals was considered to take place by the process of layer by layer mechanism, and <001> direction was observed as the fastest growing direction. This phase will be referring as phase I in the following text of this paper. Figure 1 shows a photomicrograph of such a crystal.

Crystal growth of phase II

In longer runs under same conditions, formation of crystals possessing an entirely different habit was observed. When the duration was 8 days, crystals of rhombohedral habit were grown in the low temperature region of the capsule. The oscillation and Weissenberg studies on the crystals selected randomly, suggested that there are two types of crystals having hexagonal and orthorhombic unit cells. Figure 2 shows the habit of the crystal having a hexagonal unit cell. This phase will be referred to as phase II.

Crystal growth of phase III

After 10 days under the same temperature and pressure as in the case of phase I, formation of single crystals of phase III compound as single phase with orthorhombic unit cell was observed. Figure 3 shows the photomicrograph of the phase III single crystal. The fastest growing direction was <001>. The growth of these crystals was found to have occurred by the mechanism of layer by layer. Twinning of crystals were also observed along <101>. This phase will be referred to as phase III.

Crystal growth of freaudenbergite

Under the same temperature and pressure in 10m NaOH solution, formation of crystals with a monoclinic cell was observed, when the duration was 21 days. This compound was identified as freaudenbergite (Figure 4).

Use of metallic Fe and Ti as the starting materials

Metallic Fe and Ti powders were mixed in the ratio 1:4 (weight basis) and the runs were carried out at different temperatures to study the possibilities of growing the same type of crystals as those grown when TiO_2 and Fe were used as the starting materials. At 700°C and 1000kg/cm^2, single crystals of phase I were also grown in 10m NaOH solution in 5 days. Single crystals of phase II were grown with single crystals of phase III in 6 days under the same experimental conditions. Further increase of duration of treatment to 8 days served to grow only the phase III single crystals. The single crystals of phase I, II and III grown in 10m NaOH solution starting with the metallic Ti and Fe powders showed the same crystal habits as when TiO_2 and Fe were the starting materials. Runs carried out to investigate the possibilities of growing single crystals of phase III, by starting with the phase I compound and 1m NaOH solution at 650°C and 1000kg/cm^2, gave the single crystals of the phase III in one week of duration. The crystals grown showed the same type of crystal habit as in the former experiments. Single crystals of the phase III also were grown at 700°C and 1000kg/cm^2 in 5m NaOH solution for one week of duration, by starting with TiO_2 and Fe in the ratio 5:1 (weight basis).

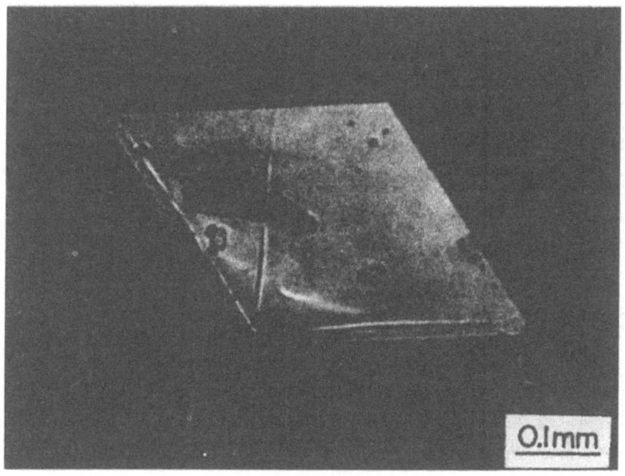

FIG. 1
Single crystal of phase I

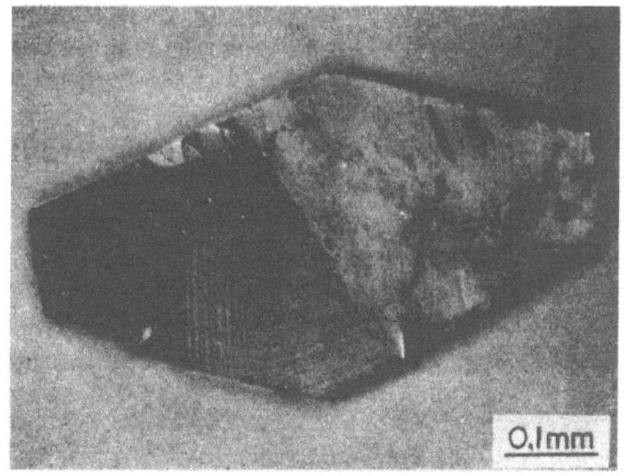

FIG. 2
Phase II single crystal

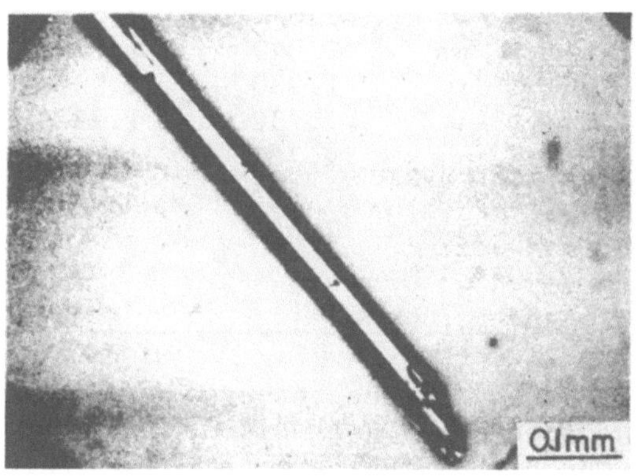

FIG. 3
Phase III single crystal

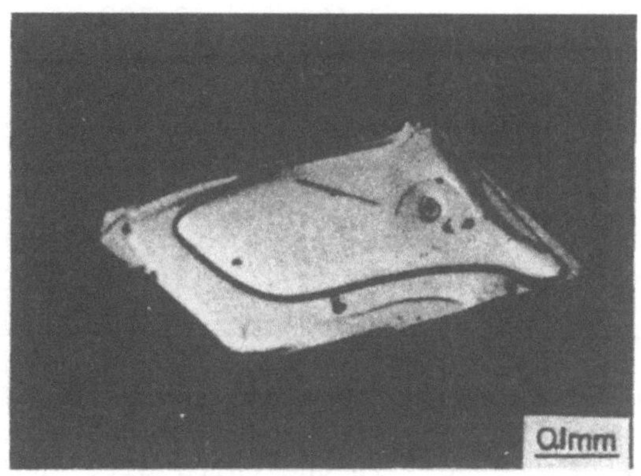

FIG. 4
Freaudenbergite single crystal

X-ray studies

The oscillation and Weissenberg photographs of single crystals of the phases I, II and III were taken along different axes. The cell constants and the possible space groups were determined from the Weissenberg photographs. The crystal data are given in Table 1. The X-ray powder diffraction data of phases I and III are given in Table 2.

Mössbauer study

The Mössbauer spectra of the phase III is shown in Figure 5. The pattern consisted of three kinds of doublets, one of which corresponded to Fe^{3+}, and the other two doublets were due to the two kinds of sites of Fe^{2+}. Considering these results, the Fe^{2+} and Fe^{3+} ions were assigned to three different sites as given by the formula $(Na_xFe_y^{2+})[(Ti_pFe_q^{2+}Fe_r^{3+})O_{2-\delta}]_n$. The computed isomer shift and quadrupole splitting of the three sites on the spectra are given in Table 3. The isomer shift of Fe^{3+} is diagnostic of low spin Fe^{3+}.

TABLE 1

Crystal data of new phases

Phase	Structure	Space group	a Å	b Å	c Å
I	Orthorhombic	Pbnm	9.13	10.82	2.93
II	Hexagonal	P6/mmm	5.093	5.093	14.099
III	Orthorhombic	Pcaa	8.90	14.21	5.02

TABLE 2

Diffraction data of the phases I and III.

Phase I				Phase III			
h k ℓ	I	d_{obs}(Å)	d_{calc}(Å)	h k ℓ	I	d_{obs}(Å)	d_{calc}(Å)
0 2 0	w	5.411	5.411	1 2 1	s	3.726	3.722
1 2 0	vs	4.656	4.649	2 2 1	s	3.014	3.017
2 0 0	vs	4.559	4.566	0 1 2	vw	2.470	2.469
2 2 0	w	3.491	3.491	2 5 0	vw	2.395	2.396
1 0 1	w	2.787	2.789	1 2 2	vw	2.287	2.286
1 1 1	vs	2.699	2.700	3 4 1	vs	2.072	2.071
3 2 0	vs	2.647	2.647	0 4 2	ms	2.049	2.049
1 2 1	s	2.478	2.479	2 6 1	w	1.930	1.930
3 3 0	vw	2.324	2.325	2 5 2	w	1.732	1.732
3 0 1	vw	2.111	2.111	0 6 2	vw	1.722	1.722
4 2 0	w	2.103	2.102	4 3 2	vw	1.575	1.576
2 4 1	w	1.823	1.823	1 3 3	vw	1.551	1.552
2 6 0	w	1.678	1.677				
5 0 1	vw	1.549	1.550				

This indicates the covalent character of Fe^{3+} ion. Hence the assignment of Fe^{3+} ion for Ti sites agrees with the results obtained. The isomer shift value of the Fe^{2+} in the third site shows the high spin character. But the isomer shift value obtained for Fe^{2+} in the second site has not been clearly understood. This value is an intermediate value between high and low spin Fe^{2+} and may be an indication of the partial delocalization of electrons By setting the Fe^{2+} to Fe^{3+} ratio at 1:1 and by assignment of 73% of the total Fe^{2+} in the second site (Ti site) and 27% in the third site (Na site), the fittings to the observed spectra were carried out by the computer simulation. The computed R value for the fittings was 3%, and this shows the assignments of Fe^{2+} and Fe^{3+} to different sites were in good agreement with the suggested formula. The Fe^{2+} to Fe^{3+} ratio was also determined by EPMA method and it was 1:1. Hence the value determined for Fe^{2+} to Fe^{3+} ratio by Mössbauer spectra was in good agreement with the EPMA data. Taking Mössbauer and EPMA data into consideration, the possible chemical formula for phase III was written as,

$$(Na_{0.023}Fe^{2+}_{0.17})[(Ti_{1.21}Fe^{2+}_{0.45}Fe^{3+}_{0.62})O_{4-\delta}].$$

Chemical analysis and the possible chemical formula

The chemical analysis of phase I, II and III were carried out using EPMA method. Single crystals of the three phases were selected by checking with oscillation photographs. The Fe^{2+} to Fe^{3+} ratio was deduced by the computer simulation and the appropriate Ti and Na contents were determined by fixing

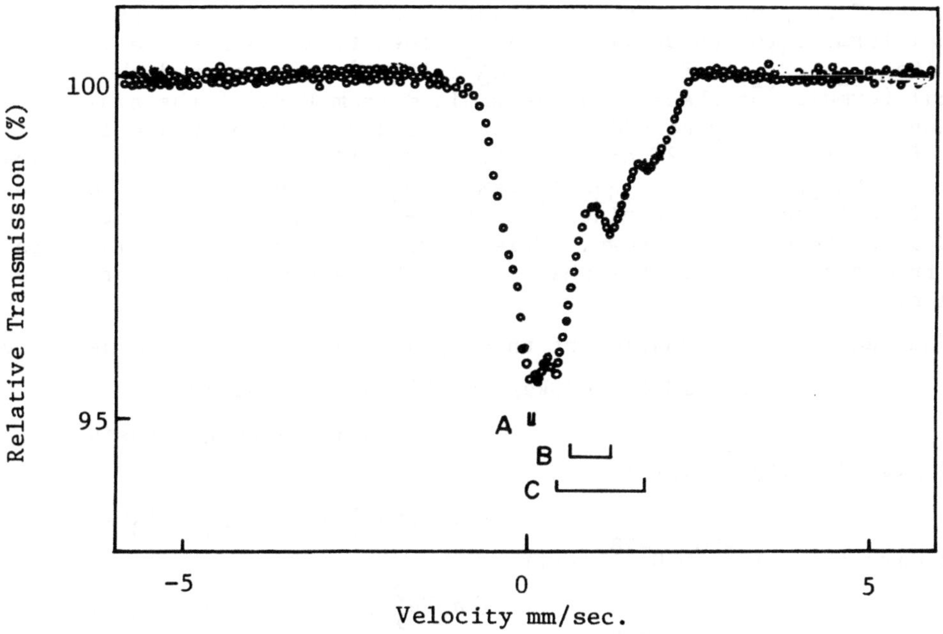

FIG. 5

Mössbauer spectra of Phase III
at room temperature (295°K)
A: Fe^{3+}(Ti site), B: Fe^{2+}(Ti site)
C: Fe^{2+}(Na site)

TABLE 3

Mössbauer data for phase III

	Isomer shift (mm/sec)	Quadrupole splitting (mm/sec)
Fe^{3+}(Ti site)	0.01692	0.1692
Fe^{2+}_1(Ti site)	0.86293	0.6768
Fe^{2+}_2(Na site)	1.08852	1.41

the oxygen content at 4. It was observed that phase I has a higher concentration of Na than phase II or III. The single crystals of phase II examined showed a variation of Na content from edge to center within one crystal. The fact may suggest that the phase II might be metastable. The samples which were treated for longer than 7 days at 650°C and 1000kg/cm^2 showed a decrease in weight of samples after treatment. This could have occurred due to the grain growth of the capsule material in long runs in concentrated NaOH solution at 650°C and 1000kg/cm^2, permitting hydrogen to diffuse out through the wall of the capsule. The leakage of hydrogen from the capsule increases the oxygen partial pressure. The rough estimation of the oxygen partial pressure for formations of the four phases were made by referring to the data given by Eugster(3). The Mössbauer study of the phase III showed the presence of Fe^{2+} and Fe^{3+} in the ratio 1:1 and this was in good agreement with the EPMA data. The phase III was formed at higher oxygen partial pressure than phase I or II. Furthermore on longer duration of treatment, the formation of freaudenbergite ($Na_2O \cdot Fe_2O_3 \cdot 7TiO_2$) was observed. This indicates the decrease of Fe^{2+} content with increase of duration of treatment. Taking these facts into consideration, it can be said that the ratio of the content of Fe^{2+} to that of Fe^{3+} in phase I or phase II should be higher than in phase III. The chemical formula for phase III was determined from Mössbauer and EPMA studies to be $(Na_{0.023}Fe^{2+}_{0.17})[(Ti_{1.21}Fe^{2+}_{0.45}Fe^{3+}_{0.62})O_{4-\delta}]$ as mentioned above. The chemical formula for phase I was determined from EPMA of the single crystal of phase I and also by considering the Mössbauer data of phase III. The ratio of 3:1 for Fe^{2+}/Fe^{3+} obtained from EPMA fits well with the following formula: $(Na_{0.6}Fe^{2+}_{0.1})[(Ti_{1.55}Fe^{2+}_{0.3}Fe^{3+}_{0.13})O_{4-\delta}]$. The chemical formula for phase II could not be determined due to the variation of Na content from 0.6 to 0.023 in the crystals grown. This shows that the phase II is a transient phase from the phase I to the phase III or has a broad region of solid solutions.

The sequence of formation of these phases are represented as follows,

at 650°C, 1000kg/cm^2, 10m NaOH solution

	TiO_2 + Fe (5:1) weight basis
7 days ($Po_2 \lesssim 10^{-20}$ atm.)	↓
	$(Na_{0.6}Fe^{2+}_{0.1})[(Ti_{1.55}Fe^{2+}_{0.3}Fe^{3+}_{0.13})O_{4-\delta}]$
8 days (Po_2; $10^{-20} \sim 10^{-18}$ atm.)	↓
	$(Na_xFe^{2+}_y)[(Ti_rFe^{2+}_sFe^{3+}_t)O_{4-\delta}]$
10 days (Po_2; $10^{-18} \sim 10^{-14}$ atm.)	
	$(Na_{0.023}Fe^{2+}_{0.17}[(Ti_{1.21}Fe^{2+}_{0.45}Fe^{3+}_{0.62})4-\delta]$
21 days ($Po_2 \gtrsim 10^{-10}$ atm.)	↓
	$Na_2O \cdot Fe_2O_3$ $7TiO_2$

Conclusion

The single crystals of phases I, II and III were black with a metallic lustre. The decrease of the total amount of ions in Na site with the order of formation is considered to cause the transformation of phase I through phase III. The phase I single crystals showed entirely different types of habits compared with phase II or III. The Mössbauer spectra of phase III showed the presence of two Fe^{2+} sites and one Fe^{3+} site. The chemical formulae for the phases were deduced by considering the Mössbauer and EPMA data, but further Mössbauer studies at low temperatures are necessary for final confirmation of the Fe^{2+} and Fe^{3+} sites. The X-ray structural analysis of the phase I, II and III are in progress. Taking the results obtained for phases I, II and III into consideration, it can be concluded that these compounds belongs to the "oxide bronzes" category. The electrical conductivity and magnetic susceptibility measurements are also to be performed for the clarification of the properties, because the compounds seem to have characteristics associated with the coexistence of ions and electrons.

Acknowledgements

The authors wish to express their gratitude to Associate Professor F. Marumo and Dr. K. Tanaka of the same institute for the advises given in the determination of the structure of the new phases, Dr. T. Okada of the Institute of Physical and Chemical Research, Saitama, Japan, for taking the Mössbauer spectra, and to Dr. K. Okuda of Geological Survey of Japan for providing the EPMA equipment.

M.G.M.U.I. is indebted to the Ministry of Education of Japan scholarship programme for providing this opportunity to study in Japan.

References

1. S. Anderson and A. D. Wadsley, Acta Cryst. 15, 201 (1962).
2. W. G. Mumme and A. F. Reid, Acta Cryst. B 24, 625 (1968).
3. H. P. Eugster, J. Chem. Phys. 26, 1760 (1957).

Reprinted from *American Ceramic Society Bulletin*, **56**(5) (1977) 513–514.

Hydrothermal Growth of Y_2O_3-Stabilized Cubic ZrO_2 Crystals

KAZUO NAKAMURA, SHIN-ICHI HIRANO and SHIGEYUKI SŌMIYA*

Research Lab of Engineering Materials,
Tokyo Institute of Technology, Tokyo, Japan

Since cubic ZrO_2 stabilized with CaO, Y_2O_3, or MgO has a high liquid formation temperature (>2600°C) and high oxygen-ion conductivity at high temperatures, it is useful for refractories, oxygen partial pressure sensors and electrodes.

The synthesis of monoclinic ZrO_2 single crystals by the flux method[1] and the hydrothermal method[2] has been widely investigated. There are, however, few reports concerning the growth of stabilized cubic ZrO_2 single crystals. Reed[3] and Goldsmith et al.[4] synthesized CaO-stabilized ZrO_2 crystals using a modified Verneuil-type apparatus, but these crystals appeared to contain many cracks resulting from the large thermal stresses.

The present workers have reported a single-phase region of cubic ZrO_2 solid solution in the system ZrO_2-Y_2O_3 under hydrothermal conditions at 1000 kg/cm².[5] In the present work, the possibility of hydrothermal growth of Y_2O_3-stabilized ZrO_2 single crystals was confirmed.

Zirconia was prepared by calcining Hf-free $ZrCl_2$. Zirconia powder containing 20 mol% Y_2O_3 was formed into thin tablets and sintered at 1400°C for 24 h in air, crushed coarsely and sealed in an Au or Pt capsule 3 mm in diameter and 12 mm long having 1 or 2 holes 1 mm in diameter. This capsule was placed in the dissolution zone of an Au or Pt capsule 5 mm in diameter and 50 mm long and then immersed in 15 mol% NaOH solution. In some cases, Au or Pt wire was attached to the top of the capsule to control nucleation of the ZrO_2 crystal. The capsule was treated in an ordinary test-tube-type hydrothermal vessel for 1 week. The temperature of the growth zone at the top of the capsule was 650°, 700°, or 750°C; the temperature at the bottom was 50°C higher than that at the top in every case. The pressure for the growth run was 1000 or 1500 kg/cm². Only stabilized cubic ZrO_2 solid solution was stable under all the conditions used. Largest crystals obtained were 0.5 mm.

A crystal (Fig. 1) grown on the Au wire by heterogeneous nucleation with the pressure vessel vertical had a subhedral form. Homogeneous nucleation and growth also could occur in the hydrothermal solution, but the crystals thus grown did not have euhedral forms. Redissolution of crystals which fell to the higher temperature zone might be responsible for this behavior. Thus, the vessel was placed in a horizontal position to prevent crystals grown in the lower temperature zone of the solution from falling into the higher temperature zone. In this case, octahedral crystals bounded by {111} faces were obtained (Fig. 2).

In addition to {111} faces, small {100} faces were also observed in some cases. When the growth temperature was 650°C, the type of morphology shown in Fig. 3 was observed. This crystal is essentially composed of {100} cubic faces with a few small blocks bounded by {110} faces like a roof; it seems to be formed of small, thin plates. This type of crystal may change in morphology from cubic to octahedral during growth.

When a pressure of 1000 kg/cm² was used, needlelike crystals

*Member, the American Ceramic Society.

Received September 10, 1976.

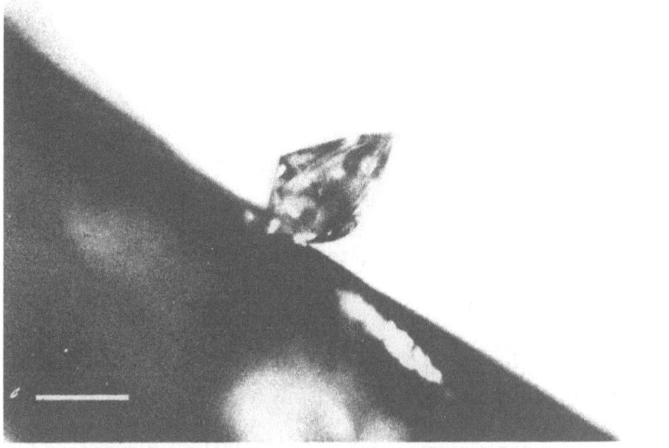

Fig. 1. Photomicrograph of stabilized cubic ZrO_2 grown on Au wire at 750°C and 1500 kg/cm² (bar = 100 μm).

Fig. 2. SEM of stabilized cubic ZrO_2 with octahedral form grown at 650°C and 1500 kg/cm² (bar = 100 μm).

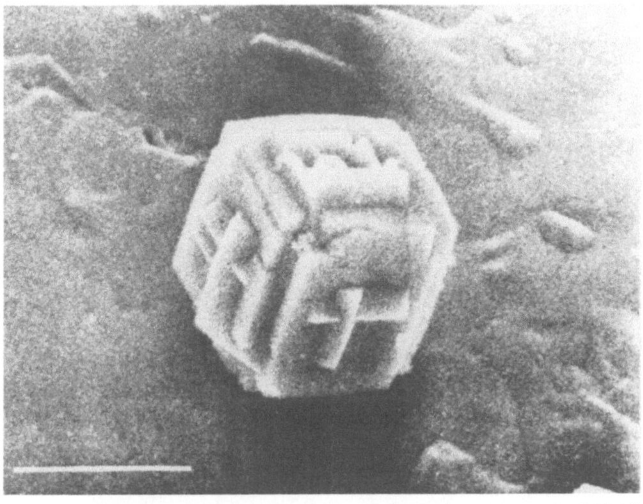

Fig. 3. SEM of stabilized cubic ZrO_2 with cubic form grown at 650°C and 1500 kg/cm² (bar = 10 μm).

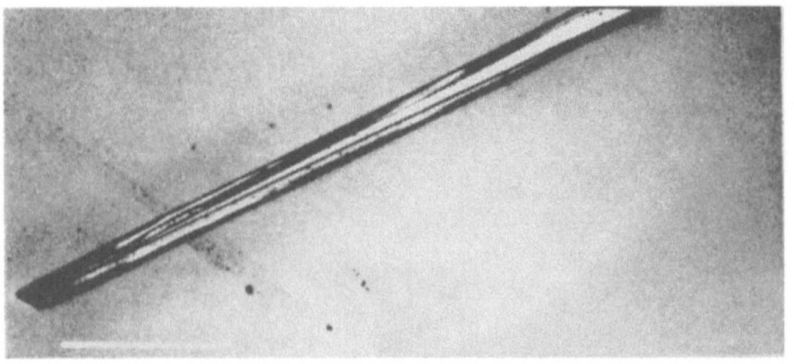

Fig. 4. SEM of needlelike stabilized cubic ZrO_2 grown at 700°C and 1000 kg/cm² (bar = 1 mm).

(Fig. 4) were often obtained along with bulky octahedral crystals. The growth direction was [1 10], and this type of crystal is normally found in bundles with thinner crystals.

From these experiments, a pressure of 1500 kg/cm² was found to be more suitable than 1000 kg/cm², because the solution is denser at 1500 kg/cm² than at 1000 kg/cm². It is concluded that Y_2O_3-stabilized cubic ZrO_2 crystals can be grown hydrothermally in NaOH solution and the nutrient transported from the hot zone to the cold zone in this solution, in contrast to the growth of pure ZrO_2 in HF or NH_4F solution.

References

[1] (*a*) V. A. Kuznetzov, "Crystallization of Titanium, Zirconium and Hafnium Oxides and Some Titanate and Zirconate Compounds under Hydrothermal Conditions," *J. Cryst. Growth,* **3, 4,** 405–10 (1968).

(*b*) Y. Fujiki, T. Mitsuhashi and Y. Suzuki, "Hydrothermal Growth and Etching of ZrO_2 Crystals," *J. Am. Ceram. Soc.,* **55** [4] 223–24 (1972).

[2] (*a*) A. B. Chase and J. A. Osmer, "Growth of Single Crystals of ZrO_2 and HfO_2 from PbF_2," *Am. Mineral.,* **51** [11,12] 1808–11 (1966).

(*b*) Y. Fujiki, "Growth of ZrO_2 Single Crystals Using V_2O_5-NaF Flux," *Yogyo Kyokai Shi,* **82** [7] 402–03 (1974).

[3] T. B. Reed, "Growth of Refractory Crystals Using the Induction Plasma Torch," *J. Appl. Phys.,* **32** [12] 2534–35 (1961).

[4] M. Kestigian, G. J. Goldsmith and M. Hopkins, "High Intensity Carbon-Arc-Image Furnace and Its Application to Single Crystal Growth of Refractory Oxides," *J. Electrochem. Soc.,* **111** [2] 260–62 (1964).

[5] K. Nakamura, S. Hirano and S. Somiya, "The System ZrO_2-Y_2O_3 under Hydrothermal Conditions at 1000 kg/cm²," *Yogyo Kyokai Shi,* **83** [12] 570–74 (1975). □

Reprinted from *Research Reports of Nagaoka Technical College*, **14**(4) (1978) 89–96.

Hydrothermal Growth of Fayalite Crystals****

Shigeyuki SŌMIYA*, Shin-ichi HIRANO**

and

Yutaka IWAI***

(Received October 19, 1978)

ABSTRACT

Fayalite single crystals without any detectable contamination of the ferric ion could be hydrothermally grown under the controlled reducing conditions.

The adjustment method of the reduced atmosphere established in the present work was found to be applicable for keeping the hydrothermal condition within the stability field of the ferrous ion. The morphology of the grown fayalite crystal was affected by the grown temperature, the concentration of NaOH solution and the oxygen fugacity. The lower the growth temperature, the more remarkable become the developments of {021} and {110} faces and the diminishment of the {100} and {111} faces. It was also found the same effects of the concentration of NaOH and oxygen fugacity.

The results obtained in this experiment will be useful for the growth of larger and highly qualified fayalite crystals with a large-sized pressure vessel.

INTRODUCTION

Fayalite, Fe_2SiO_4, is the iron-rich end member in the olivine series of silicates, of which most part of upper mantle of the earth consists. Thus one of the most important minerals for geophysical research. Silicates of ferrous oxides have been the subject of numerous investigation of geologists and chemists.

To grow crystals containing transition elements such as Fe_2SiO_4 has been difficult. So, it is also very important to develop the technique which should be established for the application of synthesizing lower oxide materials.

N.L. Bowen and J.F. Schairer[(1)] synthesized fayalite on FeO–SiO_2 system, however, large single crystals were not obtained. T.J. Shankland[(2)] grew large single crystals of colourless forsterite (Mg_2SiO_4) crystals by flame fusion method. He[(3)] also extended his research also for growing olivine crystals doped with more than about 0.3m/o iron and obtained only poor result due to the partial alteration of Fe^{2+} to Fe^{3+}. By the Czochralski

* Laboratory for Hydrothermal Synthesis, Research Laboratory of Engeneering Materials, Tokyo Institute of Technology, Meguro-ku, Tokyo 152.

** Synthetic Crystal Research Laboratory, Faculty of Engineering, Nagoya University, Nagoya.

*** Department of Industrial Chemistry, Nagaoka Technical College, Nagaoka-shi, Niigata 940.

**** This work was done at the Laboratory for Hydrothermal Synthesis. Tokyo Institute of Technology.

method, H. Takei[4] grew large single crystals of forstelite and also intended to grow olivine crystals with ferrous ion under lower oxygen partial pressures. He reported that the results of growing crystals with a tan colour was responsible for the inclusion of the considerable amount of ferric ion and precipitates of γ-iron (due to the fairue of the homogenization of the reduced atomosphere). He[4,5] reported on the difficulties of growing olivine crystals by Czochralski technique:

i) Strain-free single crystals were difficult to be pulled, since the temperature gradient above the melt was very steep owing to the high melting point of these compounds.

ii) Durable materials of the crusible were hardly available, being the melts chemically active.

iii) It was difficult to keep iron in ferrous state without partial alteration of Fe^{2+} to Fe^{3+} owing to the instability of ferrous state under the experimental conditions. Especially, fayalite compound, Fe_2SiO_4, melts congluently under the reducing condition desired as shown in the equilibrium diagram of FeO–Fe_2O_3–SiO_2 descrived by E.F. Osborn and A. Muan.[6]

Then, none of the attempts to grow ferric ion-free single crystals of olivine by melt growth gave favorable results. Large single crystals of fayalite obtained by H. Takei[5] using travelling solution zone melting (TSZM) method was tan coloured suggesting a contamination of ferric ion.

The authers have already reported that the transparent fayalite crystals free from ferric ion could be successfully grown hydrothermally under the reducing conditions[7] using the same method for the hydrothermal grouth of magnetite.[8] This method led to the development of the way to grow fayalite crystals of high quality sufficiently large for the physical measurments.

In this paper, the effects of growth temperature, kind of solution and the oxygen fugacity during growth on varriance of crystal morphology and colour have been described.

EXPERIMENTAL

A gold capsule (outer diamater of 5 mm, inner diamater of 4.6mm and 50mm long) containing starting materials and minerlizer solution was placed in a test-tube type cold seal pressure vessel and subjected to the specified temperatures and pressures. The pressure was measured Heise bourdon tube pressure gauge calibrated by the dead-weight method and the temperature was measured at the outside wall of the vessel using a caliblated platinel thermocouple. The accuracies of the measurments of pressure and temperature were within $\pm 20 kg/cm^2$ and $\pm 2°C$, respectively.

Metallic iron powder (analytical grade, purity>99%, Merck, Germany) and metallic silicon powder (purity>99.999%, Shin-etsu Chem. Co., Japan) or amorphous silica (purity>99%, Kanto Chem. Co., Japan) were usued as starting materials. To control the conditions to be reduced these metallic powders were reacted with the super critical water to form the oxides and hydrogen. The reactions were as follows.

$$2\,Fe + SiO_2 + 2\,H_2O \longrightarrow Fe_2SiO_4 + 2\,H_2 \qquad (1)$$

$$2\,Fe + Si + 4\,H_2O \longrightarrow Fe_2SiO_4 + 4\,H_2 \qquad (2)$$

The stoichiometric composition of the starting materials, pure water necessary to form form fayalite and the solution were sealed harmetically without any seed crystal in the gold capsule. The amount of the solution was adjusted so that the pressures in the inside and outside of the capsule were balanced.

The hydrothermal conditions employed in the present work were as follows: the growth temperature, up to 800°C; the temperature difference between the growth zone and the dissolution zone, 25°C or 50°C; the pressure, 1,500kg/cm²; mineralizer solution, sodium hydroxide of the concentrations up to 10 molality; sodium chloride, up to 5.0 molality or certain mixtures of sodiun hydroxide and sodium chloride solutions.

After a desired growth period, the vessel was quenched into cold water. The crystals thus grown in the capsule were observed with an optical microscope (Vanox, Olympus Optical Co. Ltd., Tokyo), and also identified with X-ray diffractomater (JDF-7F, JEOL Co. Ltd., Tokyo).

RESULTS AND DISCUSSIONS

Fayalite single crystals could be grown under hydrothermal conditions at any temperatures from 400° to 800°C. The single crystals of about 3 mm long grew in the gold capsule under the conditions of 650°C and 1,500 kg/cm² for 3 days of growth period with NaCl solutions, NaOH solutions or mixture of them.

The typical growth appearnces in the capsule are shown in Fig 1. Crystals nucleated spontaneously on the inner wall of the gold capsule at the growth zone and then grew in a group of crystals as illustrated in Fig. 1—a. Figs. 1—b and 1—c represent ones separated from the group of crystals grown under the proper conditions at 400°C and 1,500 kg/cm² in 1.0m NaOH solution for 7 days, and at 450°C and 1,500kg/cm² in 2.5m NaOH solution for 7 days, respectively.

The crystal morphology was found to be influenced by the growth temperature,

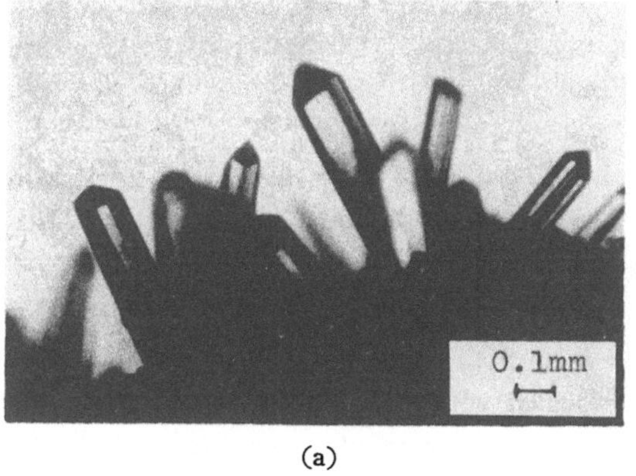

(a)

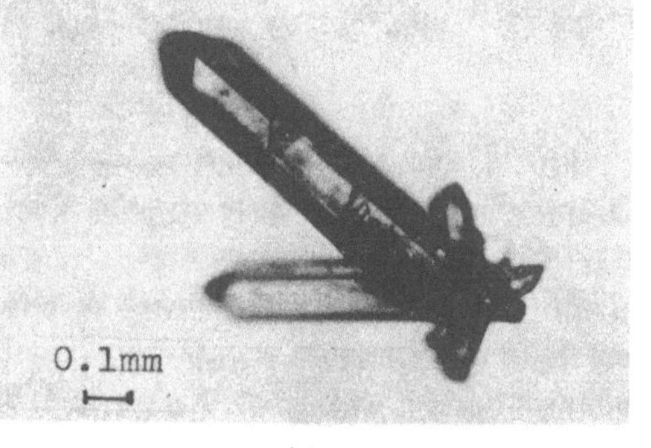

(b)

(c)

Fig. 1. Single crystals of fayalite grown hydrothermally. (a) and (b) At 400°C and 1,500kg/cm² in 1.0m NaOH soln. (c) At 450°C and 1,500kg/cm² in 2.5m NaOH soln.

the kind of solution and the oxygen fugacity. The morphology change was described by the comparison of the development of crystal face. The Miller indices were determined applying X-ray Lawe method. Schematic figures of fayalite crystals were illustrated in Fig. 2.

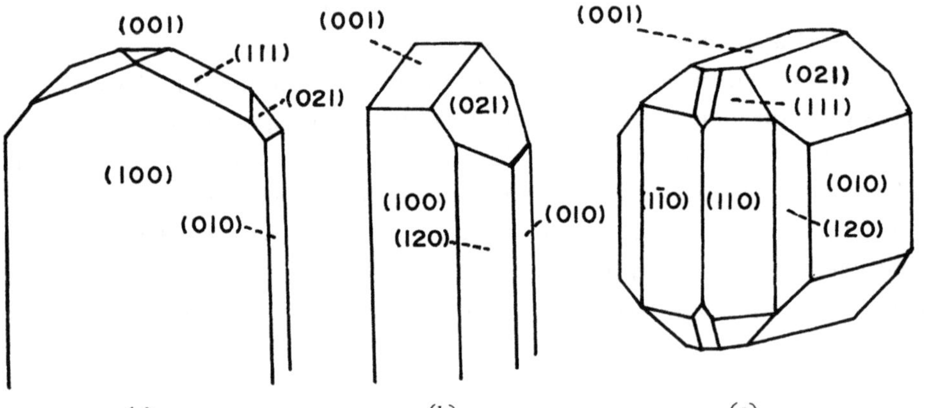

(a) (b) (c)

Fig. 2. Schimatic figures of fayalite single crystals.· (a) Platy crystal. (b) Needle like crystal. (c) Bulky crystal.

Fig. 3. shows the change in morphology with the growth temperature. At temperatures below 600°C, most crystals had the platy shape bounded mainly by {110}, {111} and {010} as shown in Figs. 3—a and 3—b. As the growth temperature was raised, {100} and {111} faces tended to become minor, whereas the development of {021} and {110} faces become remarkable (Figs. 3—c and 3—d). The same results were obtained in 1.0 and 2.5m NaOH solutions, respectively. In NaOH solutions of the concentration above 2.5m, serpentine or mica group of crystals were formed at the temperatures below 500°C by the substitution of Fe^{2+} and Mg^{2+} for Fe^{3+} and Al^{3+}, respectively. Under the hydrothermal conditions of very low oxygen fugacity, serpentine group of crystals were found to be more stable ferrous ion in Mg^{2+}.

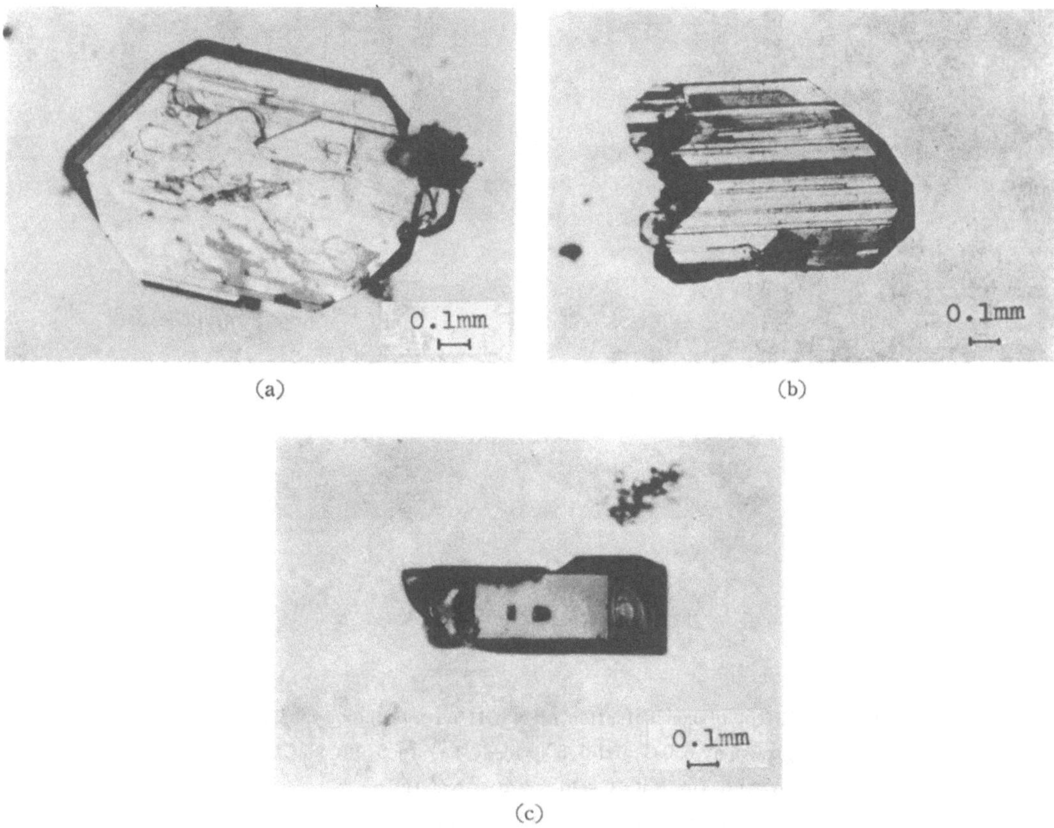

Fig. 3. Effect of temperature on morphology of fayalite crystals at 1,500kg/cm² in 5.0m NaCl soln. (a) At 600°C. (b) At 700°C. (c) At 800°C.

The effects of the concentrstion of NaOH solution on the morphology and colour of fayalite crystal grown were also found to be very remarkable as shown in Fig. 4. The addition of NaOH solution gave rise to the development of the {021} and {110} faces and brought the morphological importance of the {001} and {111} faces comparable. Then the shape of fayalite crystal was changed from platy to bulky or needle-like. It was also found that the colour of crystals grown varied with the kind of solution used. Light yellow single crystals of fayalite were obtained in case of growing in 2.5 and 5.0m NaCl solutions. In the mixture solutions containing NaOH, however, single crystals mainly produced with brown or tan colour. It was probably due to the partial alteration of Fe^{2+} to Fe^{3+}.

The effects of the oxygen fugacity on the morphology of grown fayalite crystals were examined with varying the amount of hydrogen formed in the capsule and keeping

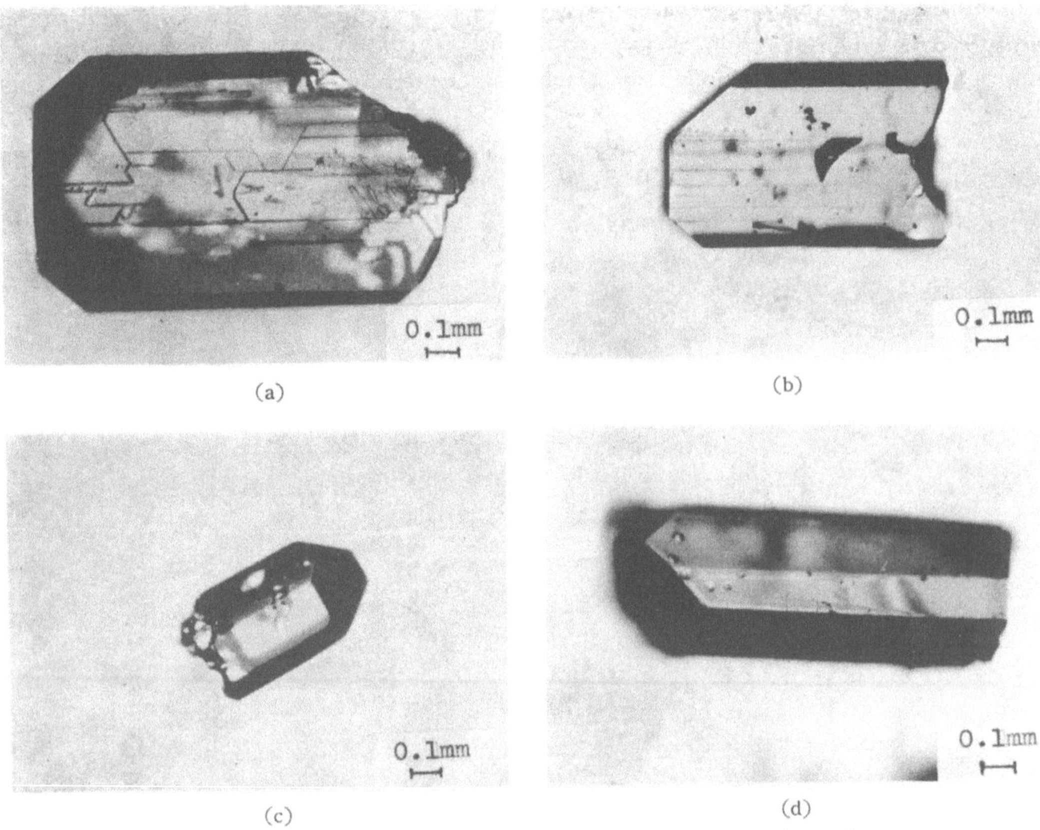

Fig. 4. Effect of concentration of NaOH on morphology of fayalite grown at 650°C and 1,500kg/cm². (a) In 5.0m NaCl soln.
(b) +0.1m NaOH soln (5 m/o NaOH).
(c) +0.5m NaOH soln (10m/o NaOH).
(d) +1.0m NaOH soln (20m/o NaOH).

the other variables constant. Since the acurate value of the oxygen fugacity in the hydrothermal solution has not been measured owing to the lack of the basic data, the estimated amount of hydrogen formed by the reaction of the metallic starting substances with supercritical water was used instead of oxygen fugacity of each sample. Fig. 5. shows the effects of oxygen fugacity (the amounts of hydrogen present in the growth atomosphere) on the morphology of fayalite crystal grown at 650°C and 1,500 kg/cm² for 3 days. (Hydrogen partial pressures in the capsules were estimated to be 180 atm in weakly reducing condition and 360 atm in highly reducing condition at 650°C. But, in this experiment, the pressure of hydrogen in the capsule was supposed to include large deviation owing to the leakage of hydrogen out of thecapsule during the growth period.) In the lower oxygen fugacities colourless platy fayalite crystals grew with the shape bounded mainly by {100}. With increasing the oxygen fugacity, {021} and {110} faces developed to be comparable to {100} and {111} faces, and the crystals tended to have the bulky or needle–like shape. The colour of crystals was also altered by change of the growth condition. The platy crystals grown under the low oxygen fugacity showed a colour varying colonrless or lightgreen to yellow. And deep brown crystals in the form of bulk were yielded in the high oxygen fugacity. These results were considered to be owing to the partial alteration of Fe^{2+} to Fe^{3+} in the high oxygen fugacity.

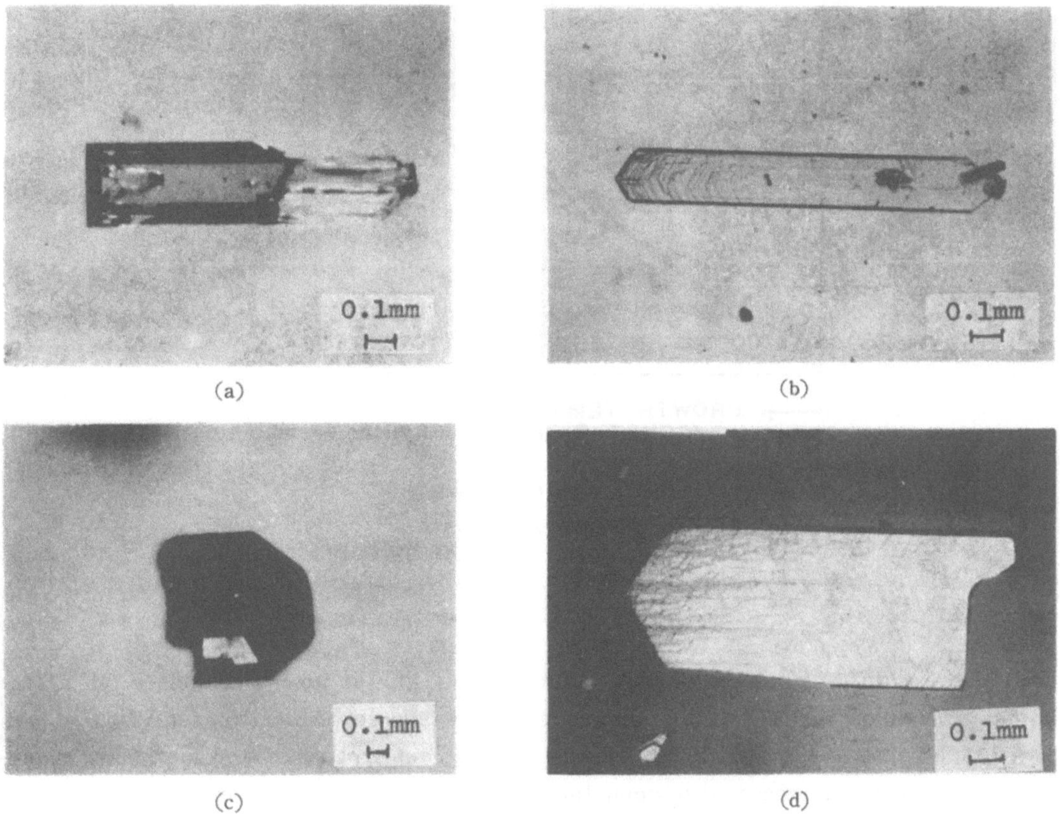

Fig. 5. Effect of oxygen fugacity on morphology of fayalite crystals. grown at 650°C and 1,500kg/cm². (a) In 2.5m NaOH under weakly reducing conditin. (b) Under highly reducing condition. (c) In 2.5m NaCl soln. under weakly reducing condition. (d) Under highly reducing condition.

Fig. 6. shows the effects of the growth temperature, the concentration of NaOH solution and the oxygen fugacity on the morphology of the grown fayalite crystals as a schematic diagram on the morphological importance of the developed crystal face. Most remarkable phenomena were found to be the intensive developments of {021} and {110} faces, and the reverse diminishment of {111} and {100} faces with increasing the growth temperature, the concentration of NaOH solution and the oxygen fugacity increased. These results can be applicable for the determination of the growth conditions according to the demand of the certain shape of crystal.

The attack of the gold capsule wall by concentrated hydrothermal solution was

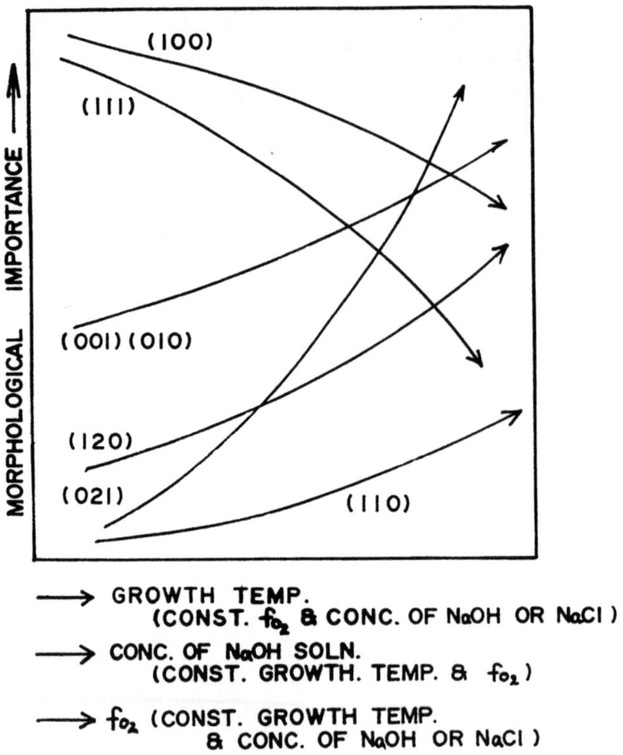

Fig. 6. Schimatic diagram on the effects of tempreture, concentration of NaOH, and oxygen fugacity.

found to cause the grain growth and the corrosion at the grain boundary of gold, leading to the partial leakage of hydrogen gas especially at higher temperatures. Thus it was understood that the influence of the growth temperature on the morphology was exaggarated by the increase in oxygen fugacity due to the partial leakage of hydrogen.

SUMMARY

The highly qualified fayalite single crystals without any detectable amount of ferric ion could be grown by the hydrothermal method under the reduced condition established in the present study. The size of grown crystals was small for the physical measurments. The method developed in this experiment, however, would be useful for the growth of large fayalite crystals by using a large sized pressure vessel.

REFERANCES

(1) N. L. Bowen and J. F. Schairer : *Am. J. Sci.*, 5 th Ser., 24, 177 (1932)
(2) T. J. Shankland and K. Hemmenway : *Am. Min.*, 48, 200 (1963)
(3) T. J. Shankland and K. Hemmenway : *Am. Ceram. Soc. Bull.*, 46 1160 (1967)
(4) H. Takei : *Hōseki Gakkai Shi*, 1 (2), 7 (1974) Japanese.
(5) H. Takei : *Nihon Kessyō Seichō Gakkai Shi*. 4 (3), 29 (1977), Japanese.
(6) E. F. Osborn and A. Muan : *'Phase Diagram of Oxide Systems' Am. Ceram. Soc*,. (1960)
(7) S. Hirano, Y. Iwai, S. Sōmiya and S. Saitō : *Proc. 6th Int'l Conf. on High Pressure*. (1977)
(8) S. Hirano and S. Sōmiya : *J. Crys. Grow.*, 35, (3), 273 (1976)

Reprinted from *High-Pressure Science and Technology*, Vol. 1, Sixth AIRAPT Conference (Plenum Press, 1979) pp. 970–976.

Hydrothermal Growth of Fayalite Crystals

S. Hirano, Y. Iwai, S. Sōmiya, and S. Saito

Tokyo Institute of Technology
Tokyo, Japan

INTRODUCTION

Fayalite, Fe_2SiO_4, is the iron-rich end member of the olivine series, which constitutes most of the upper mantle of the earth. It is thus a most important mineral with respect to geophysical science.

Silicates of ferrous oxides have been the subject of numerous investigations of importance to geologists. Bowen and Schairer [1] synthesized fayalite in their study of the $FeO-SiO_2$ system. Flaschen and Osborn [2] completed a hydrothermal study on the system $FeO-SiO_2-H_2O$ at low oxygen partial pressures and also synthesized fayalite. However, no high-perfection single crystals of fayalite free of ferric iron have been grown, because of the difficulty of controlling the valence state of iron. Shankland [3] grew large single crystals of colorless forsterite, Mg_2SiO_4, by the flame fusion method. He then extended his research to growing olivine crystals doped with more than about 0.3 mol % iron, but obtained only partly satisfactory results due to the partial alteration of Fe^{2+} to Fe^{3+}. By the Czochralski method, Takei [4] grew large single crystals of forsterite and also intended to grow olivine crystals with ferrous iron under low oxygen partial pressure. He reported the growth of crystals with a tan color due to the inclusion of considerable amounts of ferric ion and precipitates of γ-iron due to the failure to obtain a homogeneous reduced atmosphere.

In the present study, transparent fayalite crystals free of ferric ion were successfully grown hydrothermally under the reduced conditions by adjusting the iron valance, using the same method used in the hydrothermal growth of magnetite [5]. This led to the development of a method for growing high quality fayalite crystals of sufficient size for physical measurements.

EXPERIMENTAL

A gold capsule, 5 mm in outer diameter, 4.6 mm in inner diameter and 50 mm in length, containing starting materials and mineralizer solution, was set in a test-tube type cold seal pressure vessel and subjected to the desired temperature and pressure. The pressure was measured with a Heise bourdon tube pressure gauge calibrated by the dead-weight method, and the temperature was measured at the outside wall of the vessel using a platinel thermocouple calibrated against the melting point of gold. The accuracies of pressure and temperature were within $\pm$ 20 kg/cm^2 (1.8 x 10^6 Pa) and $\pm$ 2°C.

Metallic iron powder (analytical grade, purity > 99%, Merck, Germany) and silicon powder (purity > 99.999%, Shinetsu Chem. Co., Japan) or amorphous silica (purity > 99%, Kanto Chemical Co., Japan) were used as the starting materials. These metallic powders react with supercritical water to form oxides and hydrogen sufficient to produce a reducing atmosphere. That is, the following reaction was employed to form fayalite and also control the oxygen fugacity,

$$2Fe + SiO_2 + 2H_2O \rightarrow Fe_2SiO_4 + 2H_2 \qquad (1)$$

$$2Fe + Si + 4H_2O \rightarrow Fe_2SiO_4 + 4H_2 \qquad (2)$$

Stoichiometric amounts of the starting materials and pure water necessary to form fayalite and solution were sealed hermetically without any seed crystal into the gold capsule. The amount of solution was adjusted so that the pressures inside and outside the capsule would be balanced.

The hydrothermal conditions employed in the present work were as follows; the growth temperature, up to 800°C (1073 K); the temperature difference between the growth zone and the dissolution zone, 25 or 50°C; the pressure 1500 kg/cm^2 (1.47 x 10^8 Pa); mineralizer solution was sodium hydroxide with concentrations up to 10 molar, sodium chloride up to 5.0 molar or certain mixtures of sodium hydroxide and sodium chloride solutions.

After a desired growth period, the vessel was quenched in cold water. The crystals grown in the capsule were observed with an optical microscope and identified by x-ray diffraction. The oxidation states of iron in the grown crystal were measured by the Mössbauer spectroscopic method with Co^{57} in Pd and sodium nitroprusside, $Na_2Fe_2(CN)_5NO \cdot 2H_2O$ as a reference.

RESULTS AND DISCUSSION

Fayalite single crystals could be grown under hydrothermal conditions at temperatures from 400 to 800°C. Single crystals, about 3 mm in length, were grown in the gold capsule at 650°C and 1500 kg/cm^2 for 3 days in NaCl solution, NaOH solutions, or mixture of these solutions.

The crystal morphology was found to be influenced phenomenologically by the growth temperature, the type of solution, and the oxygen fugacity. Figure 1 shows the change in morphology with the growth temperature. At temperatures below 600°C, most crystals were plate-shaped bounded mainly by {100}, {111}, and {010} as shown in Figs. 1a and 1b. As the growth temperature was raised, the {100} and {111} faces tended to become smaller, whereas the development of {021} and {110} faces was remarkable (see Figs. 1c and 1d). Similar results were also obtained in 1.0 or 2.5 molar NaOH solution. In a NaOH solution with a concentration above 2.5 molar, serpentine or a member of the mica group was formed at temperatures below 500°C by the substitutions of Fe^{2+} and Fe^{3+} for Mg^{2+} and Al^{3+}, respectively. Under the hydrothermal condition of especially low oxygen fugacity (0.005 mole hydrogen in 0.35 ml of 2.5 molar NaCl or 2.5 molar NaOH solution), serpentine crystals were found to be more stable at lower temperatures by the substitution of stable ferrous ion for Mg^{2+}.

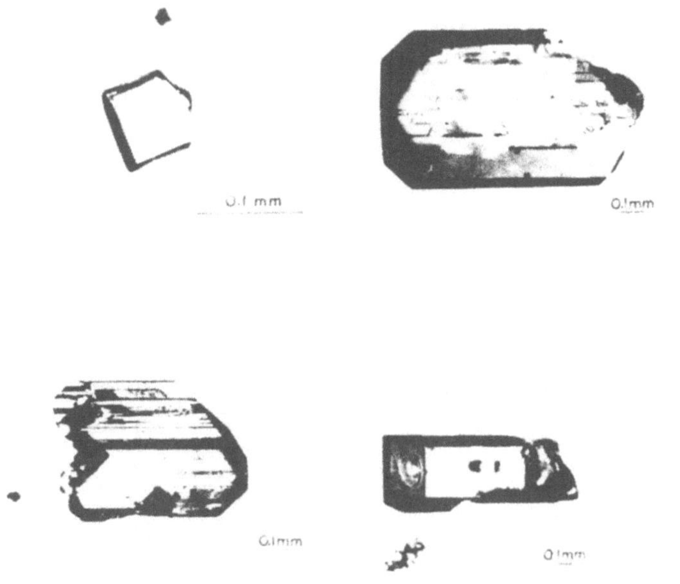

Fig. 1. Changes in the morphology of fayalite crystals with growth temperature (in 5 molar NaCl solution at 1500 kg/cm^2 for 3 days). Legend: (a) at 500°C; (b) at 650°C; (c) at 700°C; and (d) at 800°C.

The effect of the concentration of NaOH solution on the morphology of fayalite was also found to be remarkable, as shown in Fig. 2. The addition of NaOH to the NaCl solution led to the development of the {021} and {110} faces and to also increase the morphological importance of the {001} and {111} faces. Thus, the addition of NaOH changes the morphology of fayalite from platelets to needles.

The effect of the oxygen fugacity on the morphology of fayalite was examined by keeping the other variables constant except for the amount of hydrogen formed in the capsule. Since the actual value of the oxygen fugacity in hydrothermal solution is not available due

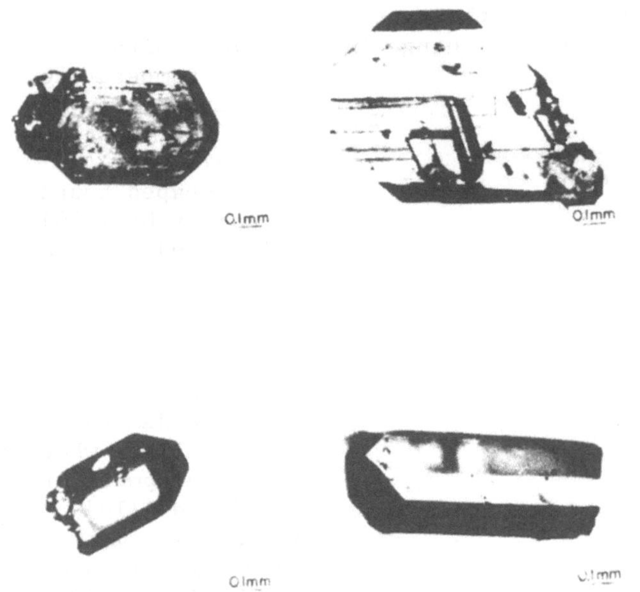

Fig. 2. Changes in the morphology of fayalite crystals prepared at 650°C and 1500 kg/cm^2 for 3 days. Legend: (a) in 5 molar NaCl solution; (2) in mixed (4.9 molar NaCl/0.1 molar NaOH) solution; (c) in mixed (4.5 molar NaCl/0.5 molar NaOH) solution; and (d) in mixed (4.0 molar NaCl/1.0 molar NaOH) solution.

to the lack of basic data, the amount of hydrogen formed by the reaction of the metallic starting substances with supercritical water was changed for comparison with the reducing environment in the present work. Figure 3 shows the effect of the oxygen fugacity (the amount of hydrogen present in the growth atmosphere) on the morphology of fayalite crystals grown at 650°C and 1500 kg/cm^2 for 3 days. At the lower oxygen fugacity, colorless plate-shaped fayalite crystals grew with the shape bounded mainly by {100}. With an increase in the oxygen fugacity, {021} and {110} faces developed to the extent comparable to {100} and {111} faces, and the crystals tended to have a bulky or needle shape.

In Fig. 4, the effects of the growth temperature, the concentration of NaOH solution, and the oxygen fugacity on the morphologies of the grown fayalite crystals are schematically summarized. Most notable are the development of {021} and {110} faces and the decrease in importance of the {111} and {100} faces as the magnitude of the growth temperature, the concentration of NaOH solution and the oxygen fugacity increased. The results may be applied to determine the growth conditions of a crystal.

The attack on the wall of the gold capsule by the concentrated hydrothermal solution was found to cause grain growth of the gold, which leads to partial leakage of hydrogen gas especially at higher temperatures. Thus, it should be understood that the phenomenon of the influence of the growth temperature on the morphology was exaggerated by the increase in the oxygen fugacity due to the partial leakage of hydrogen.

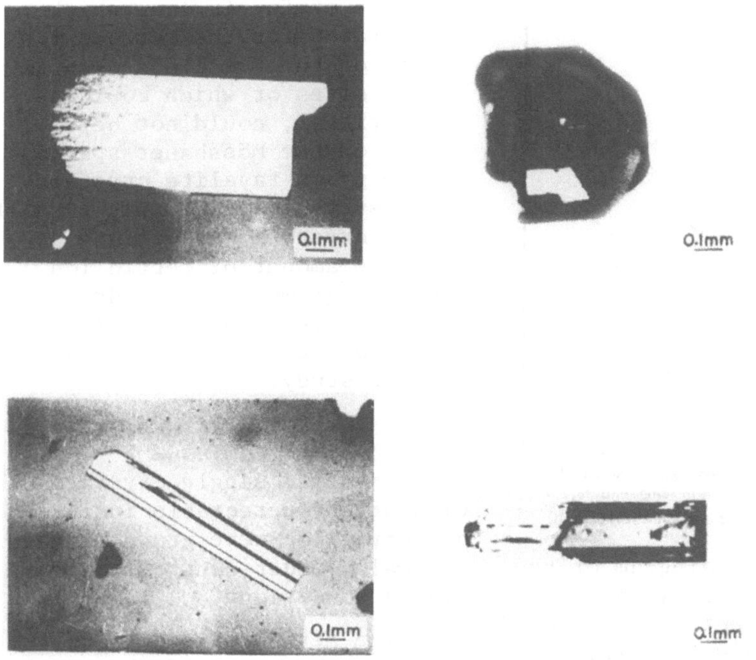

Fig. 3. Changes in the morphology of fayalite crystals prepared at 650°C and 1500 kg/cm^2 for 3 days with an amount of hydrogen in growth capsule of (a) 0.0025 mole H_2 in 0.35 ml of 2.5 m NaCl solution; (b) 0.00125 mole H_2 in 0.35 ml of 2.5 m NaCl solution; (c) 0.00125 mole H_2 in 0.35 ml of 2.5 m NaOH solution; and (d) 0.00125 mole H_2 in 0.35 ml of 2.5 m NaOH solution.

The typical growth appearances in the capsule are shown in Fig. 5. Crystals nucleated spontaneously on the inner wall of the gold capsule at the growth zone and then grew in a group of crystals as illustrated in Fig. 5a. Figures 5b and 5c represent individual crystals separated from the group of crystals grown at 650°C and 1500 kg/cm^2 in 5 molar NaCl solution for 3 days and at 450°C and 1500 kg/cm^2 in 2.5 molar NaOH solution for 3 days, respectively. The size of the crystals grown in the present work is hardly sufficient for physical measurements under very high pressure, but the basic data obtained in this study will be useful for growing larger fayalite crystals in a larger pressure vessel of glassy carbon now under construction.

Figure 6 shows a typical surface microtopograph on a (100) face of a fayalite crystal. This kind of pattern was characteristic of the (100) face of the fayalite crystals grown at a relatively low oxygen fugacity and a lower concentration of NaOH solution.

The Mössbauer spectroscopic method was applied to the fayalite crystals in order to observe the oxidation state of the iron. Figure 7 shows the spectrum taken at 25°C on a fayalite crystal grown at 650°C under 1500 kg/cm^2 in 2.5 molar NaCl solution for 3 days from iron and silicon powders as the starting materials. The

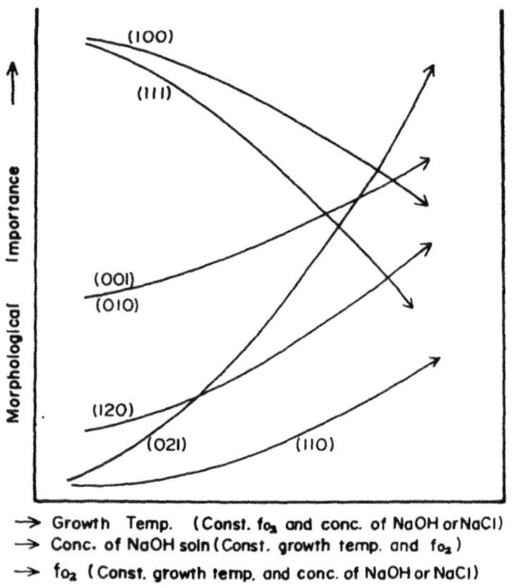

Fig. 4. Schematic diagram of the morphological changes with growth temperature, concentration of NaOH solution and oxygen fugacity.

presence of ferric ion, which substitutes for the ferrous ion in one of the two kinds of octahedral sites or which could be interstitial, could not be detected on the Mössbauer spectra of the grown fayalite crystals. The results indicate that fayalite single crystals without any detectable amount of ferric ion could be grown by the hydrothermal method under the reduced conditions established in the present study.

SUMMARY

Fayalite single crystals without any detectable inclusion of ferric ion were grown hydrothermally under controlled, reducing conditions.

The method of adjusting the reducing character of the atmos-

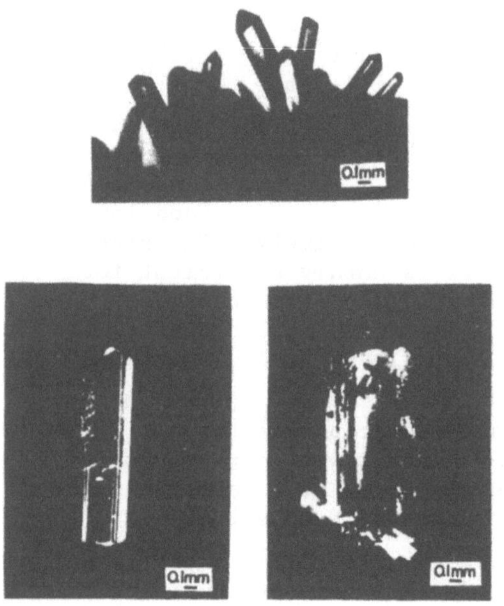

Fig. 5. Representative synthetic fayalite crystals. (a) Growth appearance in capsule (at 400°C under 1500 kg/cm ^{2} in 1 molar NaOH solution for 3 days); (b) Crystal grown at 650°C under 1500 kg/cm ^{2} in 5 molar NaCl solution for 3 days; and (c) Crystal grown at 450°C under 1500 kg/cm^2 in 2.5 molar NaOH solution for 3 days.

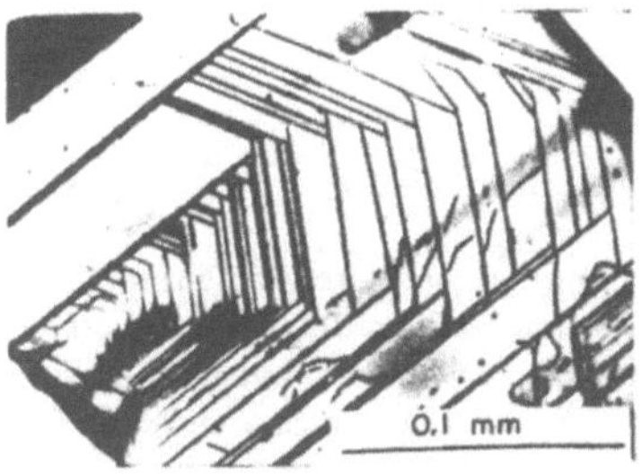

Fig. 6. Typical surface microtopograph on {100} face of fayalite crystal grown at lower oxygen fugacity.

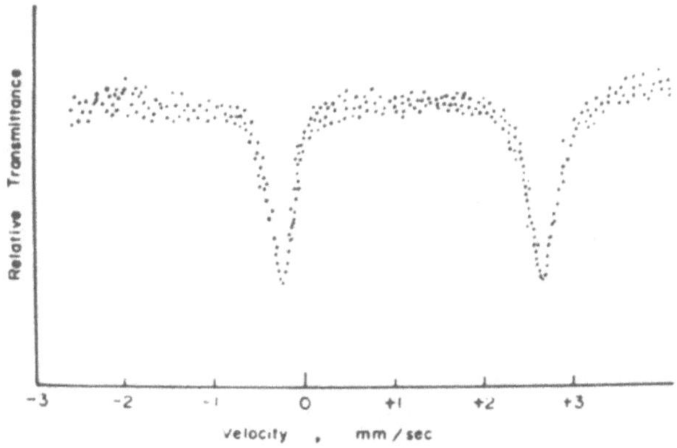

Fig. 7. Mössbauer spectrum of fayalite crystal grown at 650°C under 1500 kg/cm^2 for 3 days in 2.5 molar NaCl solution.

phere established in the present work, was found to be successful for keeping the hydrothermal growth conditions within the stability field of the divalent iron. The morphology of the fayalite crystals was markedly affected by the growth temperature, the concentration of NaOH solution, and the oxygen fugacity. The lower the growth temperature, the concentration of NaOH solution, and the oxygen fugacity, the more significant the development of {021} and {110} faces and, conversely, the decreasing importance of the {100} and {111} faces. Mössbauer spectra indicated that no detectable amount of ferric ion was included in the grown fayalite crystals. The data obtained in the present study should be useful with respect to future experiments on the growth of large fayalite crystals in a larger pressure vessel constructed of glassy carbon.

ACKNOWLEDGMENT

The authors are grateful to T. Fujii of the Fundamental Research Laboratory of Nippon Steel Company, Japan, for his kind support and use of the Mössbauer spectrometer.

REFERENCES

1. N. L. Bowen and J. F. Schairer, Am. J. Sci. 24, 176 (1932).
2. S. S. Flaschen and E. F. Osborn, Econ. Geol. 52, 923 (1957).
3. T. J. Shankland and K. N. Hemmenway, Am. Min. 48, 208 (1963).
   T. J. Shankland, Am. Ceram. Soc. Bull. 46, 1160 (1967).
4. F. Takei, J. Gemmol. Soc. Japan 1, 59 (1974).
5. S. Hirano and S. Somiya, J. Crystal Growth 35, 273 (1976).

Reprinted from *Journal of Crystal Growth*, **52** (1981) 837–842.

HYDROTHERMAL GROWTH OF QUARTZ CRYSTALS IN NaCl SOLUTION

Masahiro HOSAKA and Sadao TAKI

Institute of Inorganic Synthesis, Faculty of Engineering, Yamanashi University, Miyamae-7, Kofu, Japan

It has been shown that quartz crystals can be grown in NaCl solution by the hydrothermal temperature gradient method, and it was possible to produce large crystals after a few months. The effect of the various experimental conditions on the growth rates was investigated and observations of morphology and infrared absorption spectra were made. Two of the outstanding differences between quartz crystals grown in NaCl solution and those grown in alkaline solutions were that contrary to NaOH in NaCl striations were observed on the prism face and that the relative growth rates for the R and r faces were $R > r$.

1. Introduction

On the basis of analytical data for fluid inclusions in natural quartz, it is evident that quartz-bearing solutions in nature contain Na^+, Ca^{2+}, K^+, Cl^-, HCO_3^-, SO_4^{2-}, ions, and many others [1–4]. Among these, Na^+, K^+ and Cl^- are the major components. On the whole, Na^+ ions are more prevalent than K^+ ions. For example, Sushcevskaya [4] obtained the following results concerning the fluid inclusions in quartz from Myao–Chan tin ore deposits: Na^+ 30.50 g/l; Ca^{2+} 3.74 g/l; K^+ 11.43 g/l; Li^+ 0.04 g/l; Cl^- 61.93 g/l; HCO_3^- 3.66 g/l and F^- 2.56 g/l. Roedder [5] showed that the salinity corresponded to NaCl solution equivalent concentration ranging from zero to about 25 wt% in the case of the fluid inclusions in quartz from non-metalliferous occurrences, deposits of the Mississippi Valley type, and other ore deposits. Lisitsin [1] and Takenouchi [6] reported that the salinity of the ore-forming solution for quartz deposits occasionally ranged from 30 to 40 wt%. In general, the salinity ranges from practically zero to 40 wt% as NaCl.

In contrast, synthetic quartz usually has been grown in alkaline solutions such as NaOH or Na_2CO_3. Compared with natural quartz, the synthetic material displays differences in morphology and other features. Specifically, no straitions can be observed on the prism faces of synthetic quartz as is seen on the same faces of the natural crystals. The differences observed are probably caused by the difference of the growth solution mentioned above. In order to verify this hypothesis, NaCl solutions were used as a solvent in our experiments. Although this solvent has been used in the growth of α-quartz using fused quartz as a source under isothermal conditions [7], no study has ever been reported where NaCl solution was employed in the synthesis of quartz crystals by a temperature gradient method. In our experiments with the temperature gradient method, we have been able to show that quartz crystals can be successfully grown in NaCl solutions, and have been able to produce large crystals after a few months. The effect of NaCl solution on growth rate, and the observations of morphology and infrared absorption spectra were also studied and are described below.

2. Experimental method

Using a temperature gradient method, the hydrothermal growth of quartz crystals was carried out in platinum capsules (10 mm ID × 210 mm length and 22 mm ID × 290 mm length) placed in autoclaves of different sizes, made of stellite No. 25, with a cold-cone seat closure or flat plate closure. The seeds used were X-cut($11\bar{2}0$), R-cut($10\bar{1}1$), r-cut($01\bar{1}1$), m-cut($10\bar{1}0$), AT-cut plates, and Y-bar. Each seed was etched in HF solution for a few minutes to remove surface damage caused by cutting and polishing. The seeds were suspended in the upper region, the growth region, and natural quartz as a source material was placed in the bottom of the capsule, the dissolving region. The concentrations of NaCl employed were from 2 to 30 wt%. The platinum capsule was filled to the fraction of its free volume with NaCl solution, and the space between the capsule and the autoclave was also filled to the same fraction. The growth temperatures varied from 350 to 500°C. The temperature gradient between the growth and the dissolving regions was held at either 25 or 50°C. The baffle was 30% open in all experiments. Temperatures were measured with chromel–alumel thermocouples attached to the skin of the autoclave. The growth period lasted from a few days to 2–3 months.

The surface topography of quartz crystals was observed by ordinary optical, differential interference contrast and scanning electron microscopy, after coating samples with a thin film of gold.

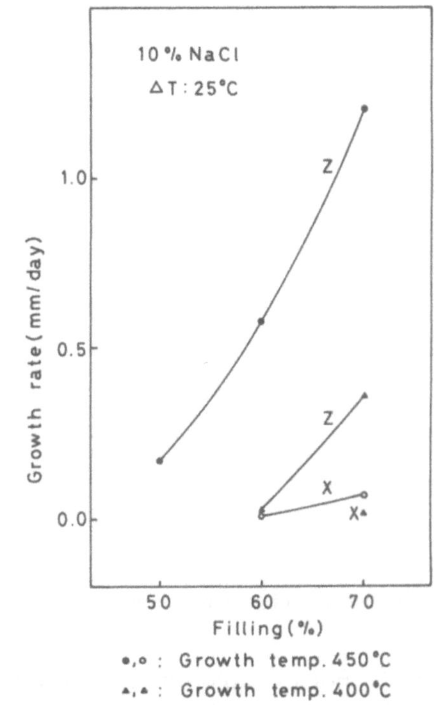

Fig. 1. Dependence of growth rates on degree of fill.

3. Results and discussion

3.1. Growth rates

The dependence of the growth rates on the degree of fill in the range from 50 to 70% is shown in fig. 1. The growth rates decreased with a decrease in the percent fill, and the seeds suspended in the growth region were dissolved when fills below approximately 50% were used. Hence, 70% fill, which gave the fastest growth rates in the range studied, was used in all the quartz preparation experiments made in this research.

The effect of NaCl solution concentration on growth rates is shown in fig. 2. Growth rates for both *Z* and *X* faces increased with the concentration of NaCl up to 10 wt%, and decreased for concentrations above 10 wt%.

The dependence of the growth rates for the *Z*, *X*, ***R***, and *r* faces on temperature is shown in figs. 3 and 4. Growth rates showed a conspicuous increase with an increase in growth temperatures above 400°C. The fastest-growing face under all conditions was the *Z* plane. The growth rate for the *m* face was too slow to show any quantitative dependence on growth temperature. Growth rates for all the faces studied were quite small

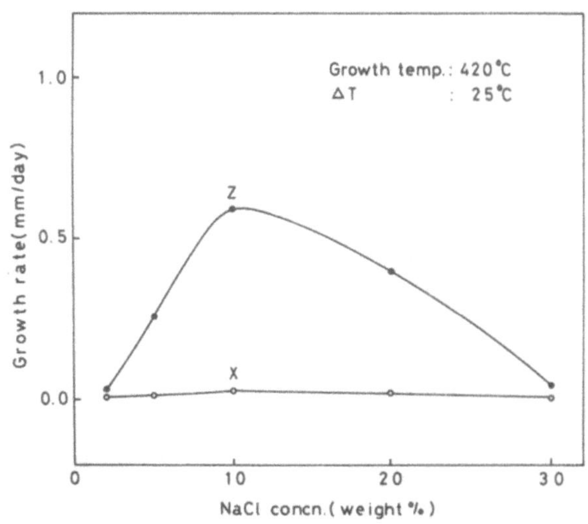

Fig. 2. Effect of concentration of NaCl solution on growth rates.

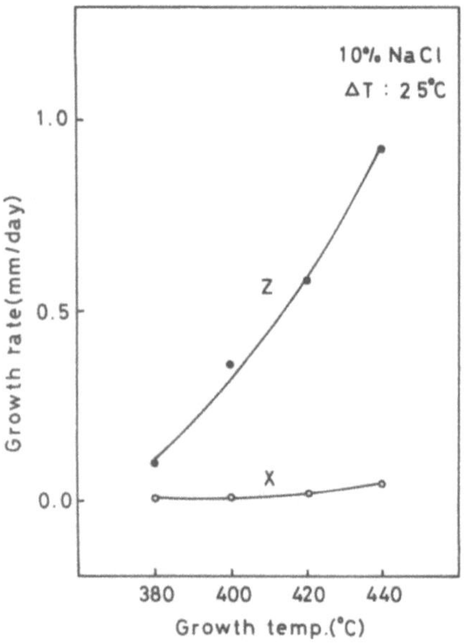

Fig. 3. Dependence of growth rates on temperatures.

below 380°C. For example, the growth rates for the Z and X faces under the following conditions: growth zone 350°C, dissolving zone 375°C, and concentration of NaCl 10 wt%, were 0.15 and 0.01 mm/day, respectively.

The energies of activation for each face of the Z, X, $\boldsymbol{R}$, and $\boldsymbol{r}$ were 21,000, 31,000, 26,000, and 38,000 cal/mole when the temperature gradient was 25°C. The energies of activation were nearly identical to those obtained in the case of alkaline solutions [8, 9].

The growth rates for the $\boldsymbol{R}$ and $\boldsymbol{r}$ faces were of the order of 10^{-2} mm/day as shown in fig. 4. They are thus from one tenth to one hundredth of the growth rates of quartz crystals grown in alkaline solutions, when the same autoclave and furnace were used. For example, in 1N NaOH solution, the growth rates for the $\boldsymbol{R}$ and $\boldsymbol{r}$ faces were 0.58 and 1.34 mm/day, respectively, with the growth region at 400°C and the dissolving region at 425°C. The relative growth rates for the $\boldsymbol{R}$ and $\boldsymbol{r}$ faces grown in NaCl solution were: $\boldsymbol{R} > \boldsymbol{r}$, which were contrary to our experimental results for synthetic growth in alkaline solutions which showed $\boldsymbol{r} > \boldsymbol{R}$. The difference in the growth rates between the $\boldsymbol{R}$ and $\boldsymbol{r}$ faces was reduced somewhat at a temperature gradient of 50°C (fig. 4b) in comparison with a temperature gradient 25°C (fig. 4a).

The growth rates for the X face were also slow. For example, the growth rate for the total of both sides of an X-cut seed was 0.05 mm/day at a growth temperature 440°C in fig. 3. The

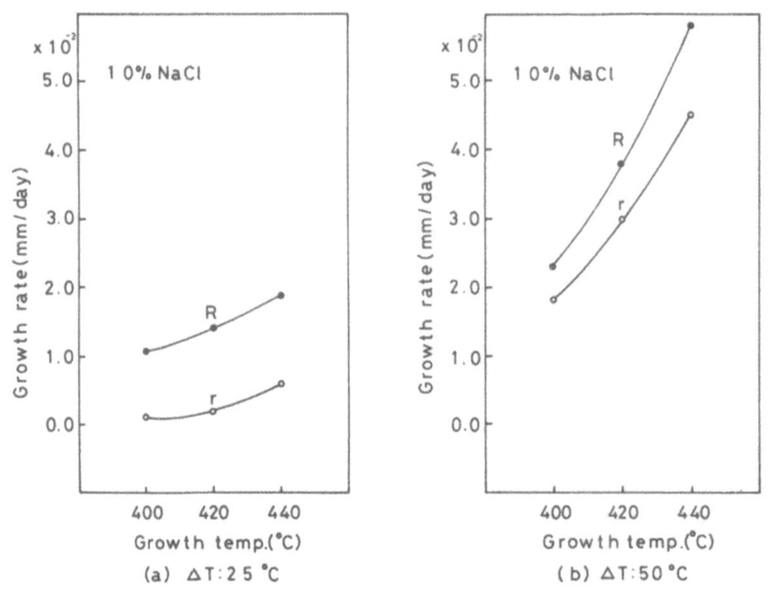

Fig. 4. Dependence of growth rates on temperatures.

ratios of the growth rates for the $-X$ face to the growth rate for the $+X$ face, i.e. V_{-x}/V_{+x}, were 0.3–0.5. This ratio is fairly close to those for quartz crystals grown in alkaline solutions.

3.2. Characterization of grown crystals

Quartz crystals grown on Y-bar seeds had flat shapes well-developed in the direction of the Z axis, because the growth rate for the Z face was much faster than that for the X face as shown in fig. 2.

Fig. 5 shows some transparent crystals, consisting only of the *R*, *r*, and *m* faces, grown on X-cut seed plates. Fig. 6 similarly shows a transparent crystal grown on an AT-cut seed plate. Such flawless quartz crystals were always obtained with X-cut, R-cut, r-cut, and AT-cut seeds. Crevice flawing, in which the cross-section normal to the Z axis was triangular in shape, was frequently observed on the basal faces which exhibited fairly rough growth. This "crevice flawing" is probably caused by constitutional supersaturation [10] due to the large concentration gradient, caused by the relatively high growth rate, which occurs at the crystal–liquid interface.

Fig. 7 shows the surface topography of prism faces of quartz crystal grown on a Z-bar seed ($0.9 \times 1.65 \times 15$ mm along X, Y, and Z axes). Striations clearly run parallel to the intersection of the prism faces with the rhombohedron, i.e., perpendicular to the Z axis. The striations are similar to those seen in natural quartz, and the crystals are in marked contrast to those grown in alkaline solutions, where no striations occur at all, and where rectangular or hexagonal growth pyramids are usually observed on the surface [11, 12]. Fig. 8 shows photographs of the rhombohedrons taken from the Z axis direction of the quartz crystal of fig. 7.

As is generally observed, in alkaline solution (fig. 8a), as the growth proceeds, the *r* faces become smaller because of the faster-growing rates for the *r* faces in comparison with the growth rates for the *R* faces. The *r* faces finally disappear and only the three *R* faces remain. In

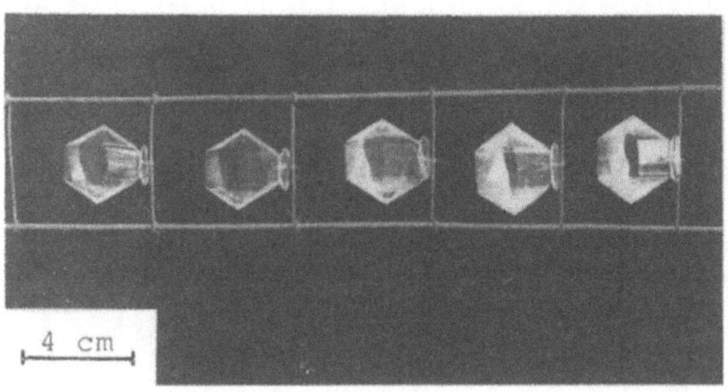

Fig. 5. Transparent crystals grown on X-cut seeds.

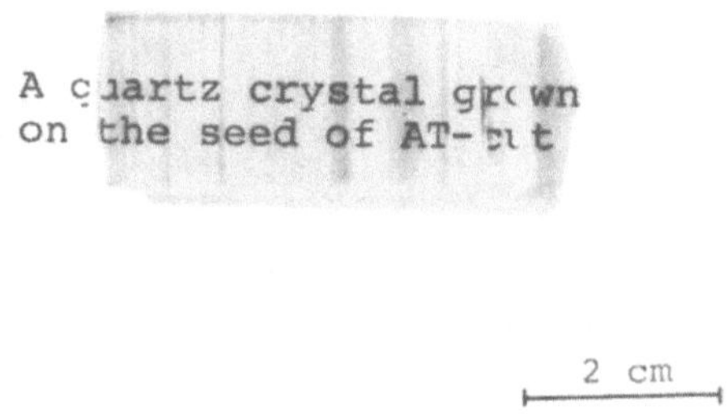

Fig. 6. Transparent crystal grown on AT-cut seed.

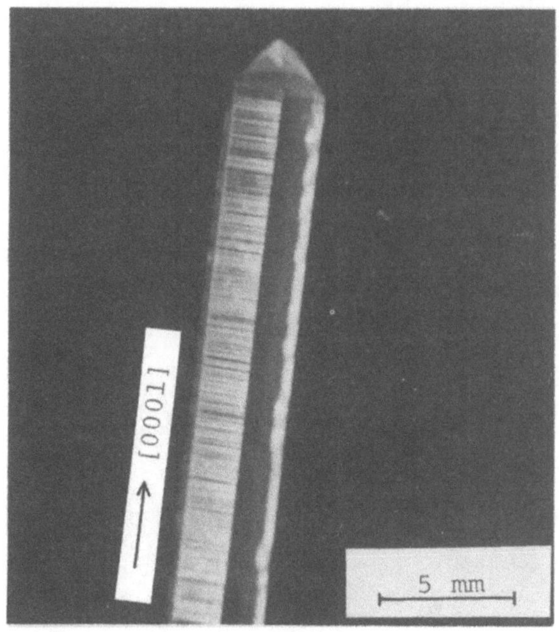

Fig. 7. Surface topography of prism faces grown on a Z-bar seed.

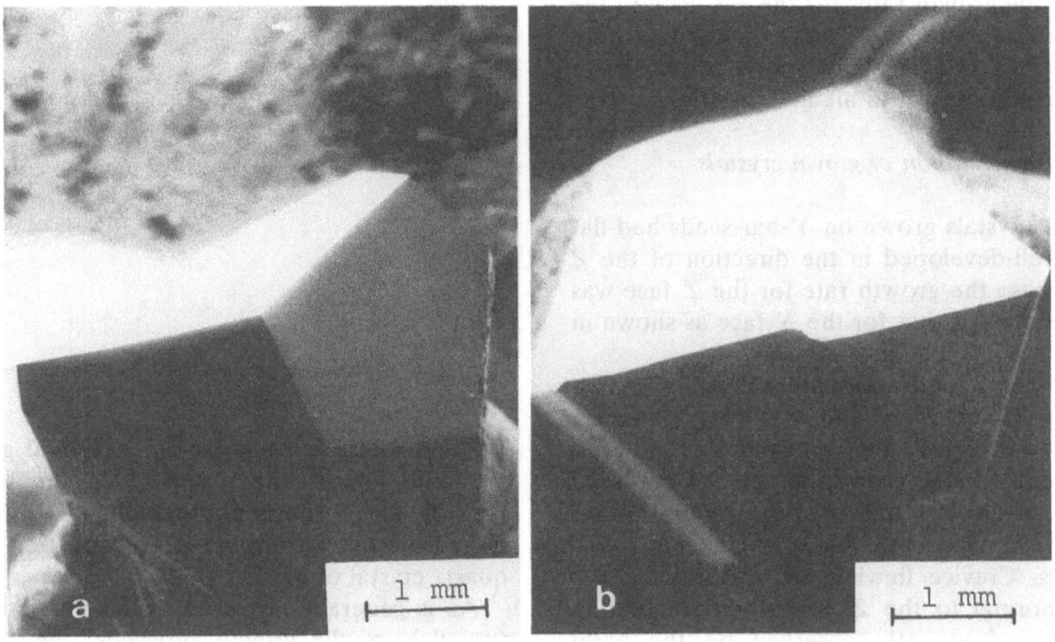

Fig. 8. Photographs of rhombohedrons taken from the Z axis direction of quartz crystals grown in: (a) NaOH solution; (b) NaCl solution.

NaCl solution (fig. 8b), however, the R and r faces, i.e., the six rhombohedrons, were observed even after a growth period of approximately one month since the difference in growth rates for the R and r faces in NaCl solution was small. Fig. 9 (NaCl growth) shows the growth surface of the rhombohedron where wedge-shaped patterns are arranged regularly together with growth hillocks. The height of the respective wedge-shaped patterns was about 1000–6000 Å. These unique growth patterns associated with NaCl growth were frequently observed with growth rates above 0.5×10^{-2} mm/day, but with growth rates of approximately 0.1×10^{-2} mm/day, only growth hillocks with flat regions between steps were observed (fig. 10).

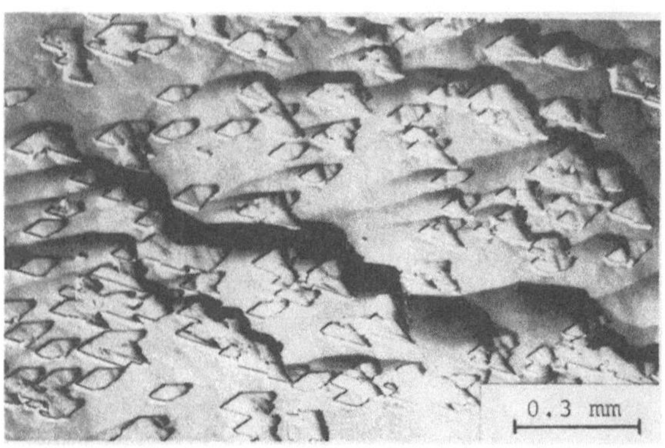

Fig. 9. Growth surface of the rhombohedron with wedge-shaped patterns.

Fig. 10. Growth surface of the rhombohedron with growth hillocks.

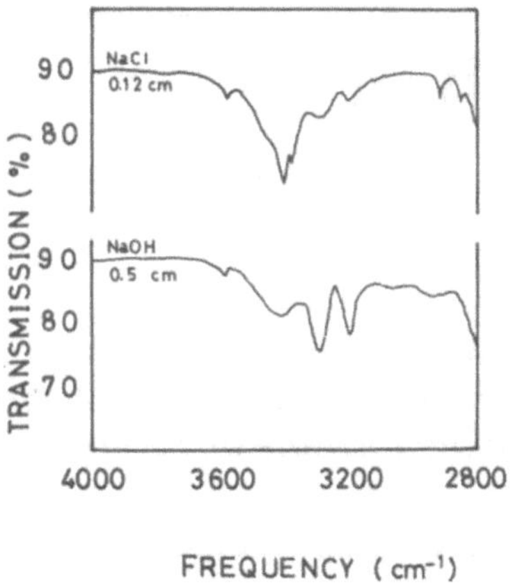

Fig. 11. 2800–4000 cm^{-1} absorption spectra of quartz crystals grown in NaCl and in NaOH solutions.

3.3. *Infrared absorption spectra*

Fig. 11 shows the 2800–4000 cm^{-1} absorption spectra of quartz crystals grown in NaCl and in NaOH solution measured at room temperature using JASCO DS-701G Infrared Spectrophotometer. Infrared absorption spectra are seen at 3200, 3300, 3400, and 3585 cm^{-1} in the case of NaOH growth in agreement with past studies [13, 14]. However, in the case of NaCl growth, the absorption peaks were observed at 3205, 3305, 3385, 3405, and 3585 cm^{-1}. The absorption at 3405 cm^{-1} in NaCl grown synthetic quartz was very sharp, compared with that from NaOH solution in which the absorption at 3400 cm^{-1} is very broad. The shape of the absorption spectra in the 3405 cm^{-1} region bore some resemblance to that of one kind of natural quartz; this was discovered through our experiments with natural quartz.

4. Conclusions

Using the temperature gradient method, flawless transparent quartz crystals were hydrothermally grown in NaCl solutions, except when Z-cut seed plates were used, in that case, flawed growth resulted. NaCl solution produced differences in growth rates, morphology and infrared absorption spectra, compared with alkaline solutions. Important observations differing from $(OH)^-$ grown quartz include: (1) Striations were always observed on the prism face, and they corresponded to that of natural quartz. (2) The relative growth rates for the R and r faces were: $R > r$, (3) IR absorption spectra showed differences from those observed for $(OH)^-$ grown quartz in the 3200–3585 cm^{-1} region with certain features similar to natural quartz. Thus, this work lends evidence to the belief that natural quartz is frequently formed in NaCl solutions.

References

[1] A.E. Lisitsin and S.V. Malinko, Geokhimia 9 (1961) 789.
[2] R.O. Rye and J. Haffty, Econ. Geol. 64 (1969) 629.
[3] T. Miyazawa and M. Enjoji, J. Japan Geothermal Energy Assoc. 34 (1972) 60.
[4] T.M. Sushchevskaya and G.F. Ivanova, Geokhimia 9 (1967) 1099.
[5] E. Roedder, Econ. Geol. 58 No. 2 (1963) 167.
[6] S. Takenouchi, Mining Geol. 12 (1962) 294.
[7] J.F. Corwin and A.C. Swinnerton, J. Am. Chem. Soc. 73 (1951) 3598.
[8] R.A. Laudise, J. Am. Chem. Soc. 81 (1959) 562.
[9] F. Muto and S. Taki, Rept. Fac. Eng. Yamanashi Univ. 13 (1963) 152.
[10] R.A. Laudise, in: Crystal growth, Ed. P. Hartman (North-Holland, Amsterdam, 1973) p. 162.
[11] S. Taki, Kagaku 30 (1960) 206.
[12] M.S. Joshi and A.S. Vagh, J. Appl. Phys. 37 (1966) 315.
[13] A. Kats, Philips Res. Rept. 17 (1962) 133.
[14] A. Kats, Philips Res. Rept. 17 (1962) 201.

Reprinted from *Journal of Materials Science*, **16** (1981) 813–816.

Hydrothermal crystal growth of perovskite-type fluorides

SHIGEYUKI SŌMIYA, SHIN-ICHI HIRANO*, MASAHIRO YOSHIMURA, KAZUMICHI YANAGISAWA
Laboratory for Hydrothermal Syntheses, Research Laboratory of Engineering Materials and Department of Material Science and Engineering, Tokyo Institute of Technology, Nagatsuta, Midori, Yokohama 227, Japan

Single crystals of perovskite-type fluorides were grown from potassium fluoride and di-valent metal chloride solutions by a hydrothermal method under a temperature gradient, at maximum temperature 600° C and pressure 98 MPa. Single-crystal cubes of $KMnF_3$, $KFeF_3$, $KCoF_3$ and $KZnF_3$, of cube edge sizes ranging from 0.2 to 2.0 mm, were grown. Crystals of $KNiF_3$ grew in rectangular prisms and those of $KCdF_3$ in anhedral form. Single crystals of $KCuF_3$ were grown in pure water from co-precipitated $KCuF_3$ powder. $KMgF_3$ crystals were not grown by this method.

1. Introduction

Perovskite-type fluorides have very interesting magnetic and optical properties [1–8], and several methods have been employed to grow single crystals suitable for measurements of their properties [1–15].

Leckebusch [9] grew small crystals of $KMnF_3$ by a gel method. Hirakawa *et al.* [1] grew $KCuF_3$ crystals from aqueous HF solution by evaporating the solvent. However, crystals grown from aqueous solution are seldom stoichiometric [2].

Garrard *et al.* [10] grew crystals of $KCoF_3$ and $KFeF_3$ by a flux method. Nitrogen gas prevented oxidation of the fluorides. The starting mixture was heated to 500° C with a small amount of NH_4HF_2 for purification and dehydration.

Both Bridgeman–Stockbargar and Czochralski methods have been used to grow crystals of perovskite-type fluorides [4–8, 13, 14]. The starting materials had to be purified and dehydrated before use and both were carried out in an atmosphere of nitrogen or inert gas.

The hydrothermal method [15] has been developed to grow single crystals of $KCoF_3$ and $KFeF_3$, and is reported here for the growth of other perovskite-type fluorides.

2. Experimental procedure

For the production of perovskite-type fluorides, the following reaction was used:

$$MCl_2 + 3KF \xrightarrow{H_2O} KMF_3 + 2KCl, \qquad (1)$$

where M is di-valent metal ion. KMF_3 produced by this reaction is somewhat soluble in the resulting KCl solution under hydrothermal conditions and may be transported by a temperature gradient to crystallize in the low temperature region.

The starting materials were KF, $MnCl_2 \cdot 4H_2O$, $FeCl_2 \cdot nH_2O$, $CoCl_2$ (anhydrous), $NiCl_2 \cdot 6H_2O$, $CuCl_2 \cdot 2H_2O$, $CdCl_2 \cdot 2.5H_2O$ (all of reagent grade from Wako Pure Chemical Industries Co., Ltd, Tokyo, Japan) $ZnCl_2$ (anhydrous) and $MgCl_2 \cdot 6H_2O$ (of reagent grade from Koso Chemical Co., Ltd, Tokyo, Japan). Water was twice distilled.

Aqueous solutions of potassium fluoride of three concentrations were used: concentrated (35 g/65 g H_2O, 9.27 m), medium (40 g/100 g H_2O, 3.07 m) and dilute (15 g/85 g H_2O, 3.07 m). Potassium fluoride solution was injected into a gold capsule of outside diameter 3.0 mm, of thickness 0.2 mm and of length 50 mm. The amount of solution was calculated to balance the pressure outside and inside the capsule at the growth tem-

*Present address: Synthetic Crystal Research Laboratory, Faculty of Engineering, Nagoya University, 1 Furo-cho, Chikusa, Nagoya 464, Japan.

TABLE I The best experimental conditions for crystal growth of perovskite-type fluorides at a pressure of 98 MPa

KMF_3	Solution (m)	Growth temperature (° C)	Temperature difference (° C)	Growth period (days)	Results
$KMnF_3$	Dil*	500	20	4	Partly aggregated. Pink cubes with smooth surfaces (see Fig. 1). Largest is of edge size 1 mm.
$KFeF_3$	Dil	500	20	7	Pale yellow cubic crystals of $KFeF_3$ of edge size up to 2 mm.
$KCoF_3$	Dil	500	20	7	Partly aggregated. Cubic or rectangular prism form.
$KNiF_3$	Dil	600	20	7	Aggregates of rectangular prism crystals. Largest is of length 0.5 mm (see Fig. 2).
$KCuF_3$	Med*	500	20	2	A few plate-like crystals of $KCuF_3$. Largest is 0.5 mm × 0.5 mm × 0.1 mm. Also produced are crystals of CuO and Au.
$KZnF_3$	Dil	500	20	4	A few cubic $KZnF_3$ crystals with rough surfaces. Largest is of edge size 1 mm (see Fig. 3).
$KCdF_3$	Dil	500	20	4	A transparent $KCdF_3$ crystal without any particular faces.
$KMgF_3$	Med	500	50	4	Suspended. Very small crystals of sizes from 0.02 to 0.05 mm (see Fig. 5).

*Dil: 3.07 m KF solution; Med: 6.89 m KF solution.

perature. The equivalent amounts of chlorides according to Equation 1 were added to the solution. For $KFeF_3$, the calculated amount of solid KF was added to $FeCl_2$ solutions, which were prepared from a saturated solution at 10° C. The capsule was welded shut with an electric arc.

The sealed capsule containing the reactants was put into a Stellite 25 test-tube-type pressure vessel. These were heated in a temperature gradient furnace. Temperatures were measured on the wall of the pressure vessel by two Platinel (Engelhard) thermocouples at two positions, corresponding to the top and the bottom of the capsule. The temperature at the top, that is at the growth temperature zone, was adjusted to 400, 500 or 600° C. The temperature difference between the growth temperature and the dissolution temperature was either 20 or 50° C higher. The pressure was adjusted to 98 MPa (1000 kg cm^{-2}). The duration of the crystal growth experiments was 2, 4 or 7 days.

At the conclusion of a run, the pressure vessel was quenched in cold water. The capsule was cut apart with a knife, the contents were washed in water, and identified from X-ray powder diffraction study patterns and from microscopic observations.

3. Results and Discussion

3.1. Crystal growth

The composition corresponding to Equation 1 was used in the growth experiments. The conditions are summarized in Table I.

Crystals of $KMnF_3$, $KFeF_3$, $KCoF_3$, $KNiF_3$ and $KZnF_3$ were aggregated when the temperature difference between the growth temperature and the dissolution temperature was 50° C and the concentration of the KF solution was 9.27 or 6.87 m. The morphology was improved by reducing the temperature difference to 20° C and reducing the concentration of KF solution to 3.07 m (see Figs 1 and 2). These results indicate that the number of crystals grown decreases in the dilute solution with a small temperature gradient. Increase of the growth temperature to 600° C was effective to grow single crystals, especially for $KNiF_3$ (see Fig. 3).

Growth experiments on $KCuF_3$ gave only a small yield of $KCuF_3$ crystals and crystals of cupric oxide were also formed in the low tempera-

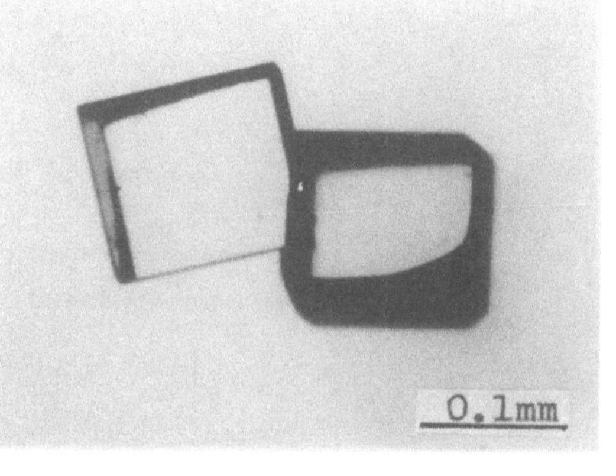

Figure 1 $KMnF_3$ single crystals.

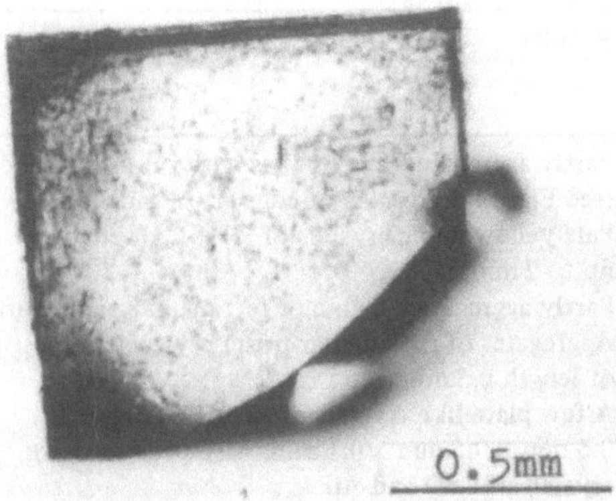

Figure 2 $KZnF_3$ single crystal.

Figure 4 $KCuF_3$ single crystal.

ture region. The gold capsule was corroded at the dissolution zone and gold crystals grew at the top of the capsule. A few larger single crystals of $KCuF_3$, however, were grown from co-precipitated $KCuF_3$ using pure water instead of KCl solutions as a solvent. Fig. 4 shows the product.

Anhedral crystals of $KCdF_3$ were formed under any of the conditions tried in this work. $KMgF_3$ crystals were suspended in the solution but were not grown by transportation to the low temperature region (see Fig. 5).

3.2. Effect of the starting composition on phases synthesized by the hydrothermal reaction

The starting compositions studied were, corresponding to (a) Equation 1, $KF/MCl_2 = 3$; (b) KF excess, $KF/MCl_2 = 6$; (c) MCl_2 excess, $KF/MCl_2 = 1.5$. The experimental conditions and the phases produced in these experiments are summarized in Table II.

Some by-products were occasionally produced in these reactions. The amounts of oxides produced by the hydrothermal reaction were in the following order: manganese oxides (none) $< \gamma$-Fe_2O_3 (maghemite) $\simeq CoO < NiO \ll CuO$ (in accordance with atomic numbers). This suggests that the stability of the fluorides KMF_3 (where M = Mn, Fe, Co, Ni and Cu) under the hydrothermal condition tends to decrease with decreasing numbers on unpaired d-electrons.

With addition of excess KF, crystals of $KMnF_3$, $KFeF_3$, $KCoF_3$ and $KZnF_3$ were well-shaped in cubic or rectangular prism form of edge sizes up to 2 mm. An excess of di-valent metal chlorides seemed to be less effective in promoting crystal growth.

4. Conclusion

Crystals of KMF_3 (where M = Mn, Fe, Co, Ni, Cu, Zn and Cd) were grown by the hydrothermal method under pressure of 98 MPa at 500° C, with a temperature gradient of 20° C, using 3.07 m KF solution. Single crystals of $KCuF_3$ were grown in

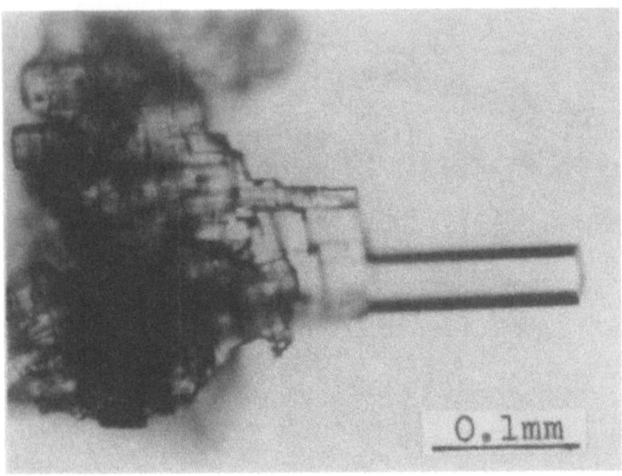

Figure 3 $KNiF_3$ crystals.

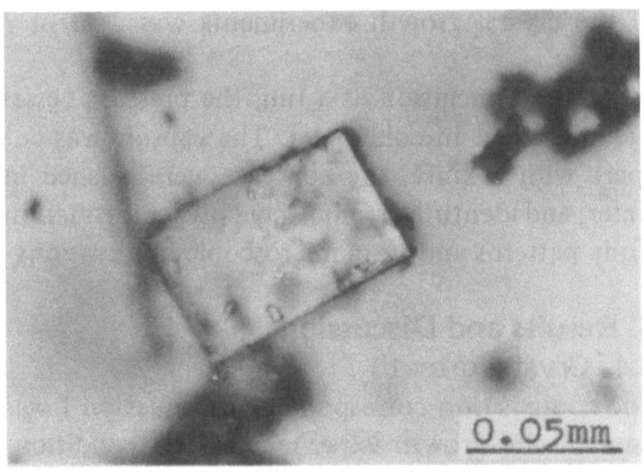

Figure 5 $KMgF_3$ crystals.

TABLE II Solid phases produced by the hydrothermal reaction with different starting compositions at a pressure of 98 MPa and 500° C in 7 days with a temperature difference of 20° C, using 6.89 M KF solution

KMF_3	MCl_2 excess $KF/MCl_2 = 1.5$	Equivalent (Equation 1) $KF/MCl_2 = 3$	KF excess $KF/MCl_2 = 6$
Mn	$KMnF_3 + MnF_2$	$KMnF_3$	$KMnF_3$
Fe	$KFeF_3 + \gamma\text{-}Fe_2O_3$*	$KFeF_3 + \gamma\text{-}Fe_2O_3$*	$KFeF_3 + \gamma\text{-}Fe_2O_3$*
Co	$KCoF_3 + U_1$	$KCoF_3 + CoO$*	$KCoF_3 + CoO$*
Ni	U_2†	$KNiF_3 + NiO$*	$KNiF_3$† + NiO*
Cu	$KCuF_3$† + CuO†	$KCuF_3 + CuO$	CuO
Zn	$KZnF_3$	$KZnF_3$	$KZnF_3$
Cd	$KCdF_3$ + CdOHF	$KCdF_3$ + CdOHF	$KCdF_3$
Mg	MgF_2†	$KMgF_3$† + MgF_2†	$KMgF_3$†

*Precipitated.
†Suspended in the solution.
All others are transported to the top of the capsule. U_1 and U_2 are unidentified phases.

pure water using co-precipitated $KCuF_3$ powder. The infra-red spectra of $KCoF_3$ showed no evidence of the presence of hydroxyl ions [15]. Thus, the uptake of hydroxyl ions by crystals of perovskite-type fluorides appears to be negligible. The amount of oxide by-products increased with the decrease of unpaired 3d-electrons from Mn to Cu. Addition of excess KF promoted growth of crystals of $KMnF_3$, $KFeF_3$, $KCoF_3$ and $KZnF_3$ of cube edge size up to 2 mm. The increase of the growth temperature was also effective in growing crystals of $KNiF_3$. These results show the possibility of growing larger and higher quality single crystals of perovskite-type fluorides by a hydrothermal technique using KF solution at high temperatures above 600° C in a larger container, such as an autoclave.

Acknowledgement

The authors thank Dr F. P. Glasser for his critical reading of the manuscript and his helpful advice.

References

1. K. HIRAKAWA, K. HIRAKAWA and T. HASHIMOTO, *J. Phys. Soc. Japan* **15** (1960) 2063.
2. K. KNOX, *Acta Cryst.* **14** (1961) 583.
3. Y. SUEMUNE and H. IKAWA, *J. Phys. Soc. Japan* **19** (1964) 1686.
4. W. W. HOLLOWAY and M. KESTIGIAN, *J. Chem. Phys.* **45** (1966) 639.
5. K. GESI, J. D. AXE and G. SHIRANE, *Phys. Rev.* **B5** (1972) 1033.
6. E. S. SVENSSON, W. J. L. BUYERS, T. M. HOLDEN, R. A. COWLEY and R. W. H. STEVENSON, *Canadian J. Phys.* **47** (1969) 1983.
7. R. H. ROSE, J. E. RHOADS and L. E. HALLIBURTON, *Phys. Rev.* **B14** (1976) 3583.
8. H. TAKEUCHI, K. HORAI and M. ARAKAWA, *J. Phys. Soc. Japan* **46** (1979) 18.
9. R. LECKEBUSCH, *J. Crystal Growth* **23** (1974) 74.
10. B. J. GARRARD, B. M. WANKLYN and S. H. SMITH, *ibid.* **22** (1974) 169.
11. B. M. WANKLYN, *ibid.* **5** (1969) 279.
12. M. SAFA, B. K. TANNER, B. J. GARRARD and B. M. WANKLYN, *ibid.* **39** (1977) 243.
13. R. LECKEBUSCH, *Fortshr. Miner.* **55** (1977) 9.
14. D. SKRZYPEK, P. JAKUBOWSKI, A. RATUSZNA and A. CHEZKOWSKI, *J. Crystal Growth* **48** (1980) 475.
15. S. HIRANO, K. YANAGISAWA and S. SŌMIYA, *ibid.*, to be published.

Received 15 May and accepted 2 September 1980.

Reprinted from *Journal of Crystal Growth*, **53** (1981) 542–546.

HYDROTHERMAL GROWTH OF QUARTZ CRYSTALS IN KCl SOLUTION

Masahiro HOSAKA and Sadao TAKI

Institute of Inorganic Synthesis, Faculty of Engineering, Yamanashi University, Miyamae-7, Kofu, Japan

Received 31 October; manuscript received in final form 5 January 1981

Quartz crystals were hydrothermally grown in KCl solutions in the range 350–500°C, and it was possible to produce large crystals after a few months. The best conditions to obtain sufficient growth rates were found to be: growth temperatures above 400°C and 10 wt% KCl solution. Striations were observed on the prism face and the relative growth rates were $R > r$. Furthermore, the solubility of quartz was measured in 2, 10, and 30 wt% KCl solutions in the range 400–440°C. It was found that the solubility obeyed the Van 't Hoff equation with a heat of solution of 7,600, 6,900, and 8,000 cal/mole, respectively, in 2, 10, and 30 wt% KCl solutions.

1. Introduction

In a previous report [1], it was elucidated that the liquid constituent of fluid inclusions in natural quartz contained mainly Na^+, K^+, and Cl^- ions, and the total salinity; i.e. NaCl equivalent concentration, ranged from practically zero to over 40 wt%. It is evident that Na^+ ions are more prevalent than K^+ ions, because the atomic ratio of the K^+ ions to the Na^+ ions varies from approximately 0.02 to 0.5 [2,3] which is very wide range.

From the point of view of growing quartz crystals under conditions close to the growth environment of natural quartz, hydrothermal growth of quartz crystals in NaCl solutions has already been reported [1], and some differences were observed in comparison with those grown in alkaline solutions. In the present experiment, quartz crystals were grown in KCl solutions, and again we were able to produce large crystals in the same manner as those grown in NaCl solutions. The effect of KCl solution on growth rates and the observations of morphology and infrared absorption spectra, and the effect on features of crystals were examined and are described below.

2. Experimental method

The hydrothermal growth of quartz crystals was carried out in platinum capsules (10 mm ID × 210 mm length and 22 mm ID × 190 mm length) placed in autoclaves of different sizes, made of stellite No. 25, with a cold-cone seat closure or flat plate closure. The seeds used were *X*-cut ($11\bar{2}0$), *m*-cut ($10\bar{1}0$), *R*-cut ($10\bar{1}1$), *r*-cut ($01\bar{1}1$), and *Y*-bar. The concentrations of KCl employed were from 2 to 30 wt%. The growth temperatures varied from 350–500°C. The degree of fill was 70% in all experiments. Other experimental conditions were the same as those for NaCl growth.

The measurement of the solubility of quartz crystals in KCl solutions was carried out in an autoclave of small size (V = 23.8 ml). A quartz crystal of *Y*-bar (2 × 2.3 × 150 mm along *X*, *Y*, and *Z* axes) was put in the autoclave and it was filled to 70% with KCl solution of definite concentration. The autoclave was finally closed with a cap, and it was set in a furnace. After the temperature of the autoclave reached a state of equilibrium at a desired temperature, it was kept for 2 days at a temperature gradient of 25°C and for an additional day under isothermal condition. After 3 days, the autoclave was quenched in water, and the weight loss of the *Y*-bar was measured. The solubility was measured in the respective concentrations of 2, 10, and 30 wt% KCl solutions in the range 400–440°C.

3. Experimental results and discussion

3.1. Results of solubility measurement

Fig. 1 shows the solubility of quartz crystals obtained in KCl solutions with the solubility data obtained in NaOH or Na_2CO_3 solutions [4], becoming fundamental data for the growth of quartz crystals. The solubility showed the largest value when the concentration of KCl solution was 10 wt%. The respective solubility for 2, 10, and 30 wt% increased almost proportionally with the increase of temperatures. Each solubility was below about one tenth of those obtained in alkaline solutions. Fig. 2 shows the relation between the reciprocal of absolute temperatures and the logarithm of solubility. The heat of solution of quartz in 2, 10, and 30 wt% KCl solutions was found to be 7600, 6900 and 8000 cal/mole, respectively. They approached 9500 cal/mole obtained in 0.5N NaCl at 70% fill in the range from 330 to 390°C [4]. But they were very large compared with 900 cal/mole (0.5N NaOH, 70% fill) and 5200 cal/mole (5% Na_2CO_3, 70% fill), which were calculated by Taki [4] from the data for NaOH and Na_2CO_3 plotted in fig. 1.

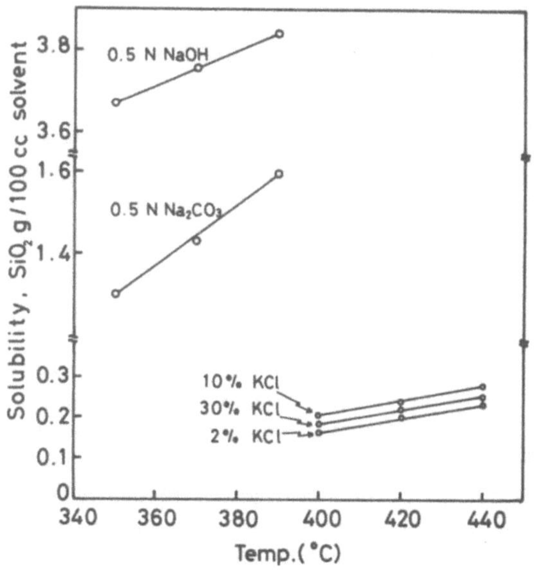

Fig. 1. Dependence of solubility of quartz on temperatures at 70% fill.

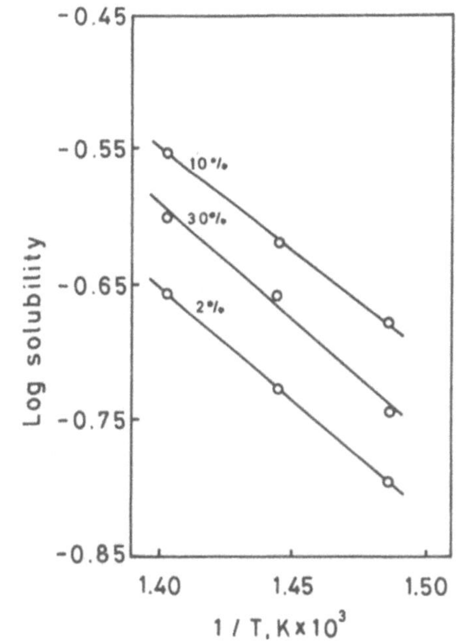

Fig. 2. Relation between the reciprocal of absolute temperatures and the logarithm of solubility.

3.2. Growth rates

Fig. 3 shows the dependence of the growth rates for *Z* and *X* faces grown on *Y*-bar seeds on the concentrations of KCl solution. The growth rates for both *Z* and *X* faces indicated the maximum value in

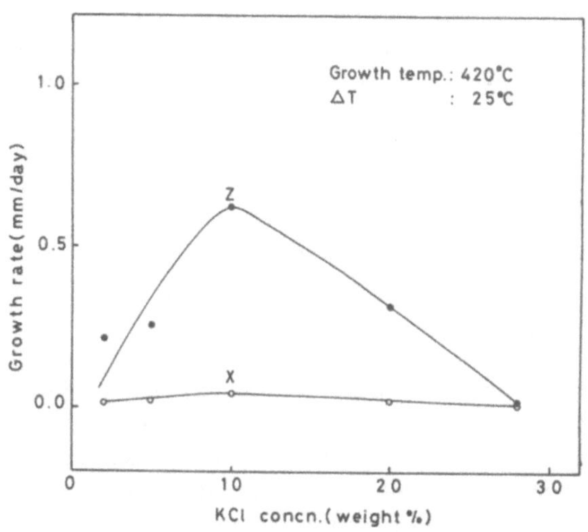

Fig. 3. Effect of concentration of KCl solution on growth rates.

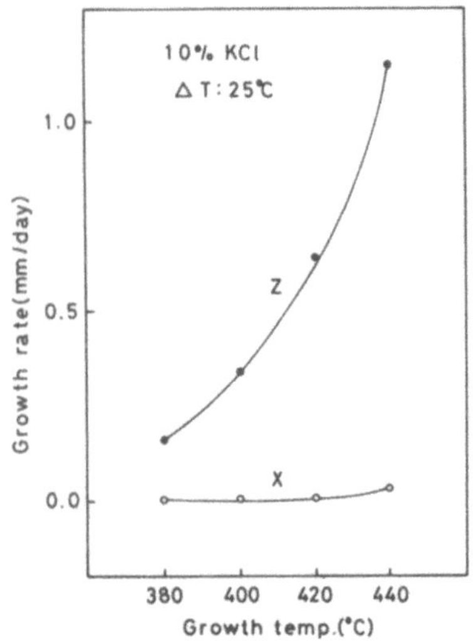

Fig. 4. Dependence of growth rates on temperatures.

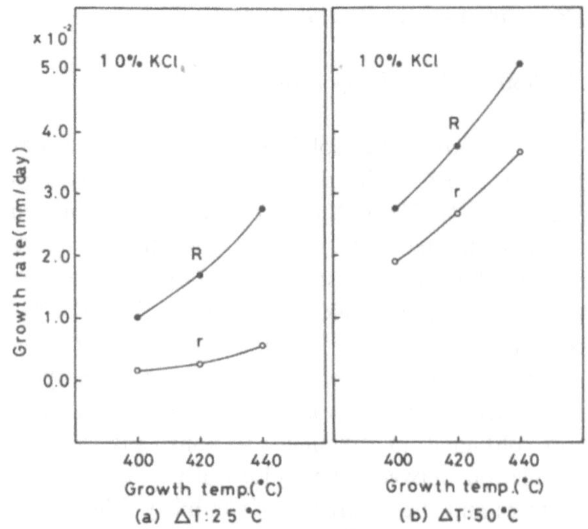

Fig. 5. Dependence of growth rates on temperatures.

the vicinity of 10 wt% KCl solution, showing the same tendency of curve as obtained in the growth of quartz crystals in NaCl solutions. It is thought that there is correlation between the growth rates and the dissolution quantity of the nutrient quartz, since the dissolution quantity of quartz crystals was the largest for 10 wt% KCl solution in the range of the studied concentrations, as shown in fig. 1. Fluid inclusions in natural quartz rarely contain 40–50 wt% NaCl equivalent concentration. Accordingly, when quartz crystals were grown in such high concentrations of KCl solution, it was found that the growth of quartz crystals was barely able to be noticed after a long growth period because of the very slow growth rates.

Fig. 4 shows the dependence of the growth rates for the *Z* and *X* faces on growth temperatures. The growth rates for the *Z* faces increased remarkably with a rise in growth temperatures, but those for the *X* faces were much smaller. A striking difference was that the ratio of the growth rates for the *Z* face to the growth rates for the *X* face; i.e. V_z/V_x, was more than about 8, and was extremely large compared with those of quartz crystals grown in alkaline solutions. Fig. 5 shows the dependence of the growth rates for the *R* and *r* faces. The growth rates were as slow as those of quartz crystals grown in NaCl solutions and they were from one tenth to one hundredth of the growth rates observed in the case of alkaline solutions. It is noteworthy that the growth rate for the *R* face is larger than that for *r* face.

The energies of activation for each face of the *Z*, *X*, *R*, and *r* faces were 22,000, 44,000, 24,000, and 31,000 cal/mole when the temperature gradient was 25°C. These energies were nearly identical to those obtained in NaCl solution [1] and to those obtained in NaOH and Na_2CO_3 solutions [5,6].

4. Characterization of grown quartz crystals

Quartz crystals grown on *Y*-bar seeds were obtained as flat shapes well-developed in the *Z* direction because of much faster growth rates for the *Z* faces in comparison with those for the *X* faces. Six rhombohedral faces were observed in the observation from the *Z* direction of quartz crystals grown on *Z*-bar seeds, as estimated in fig. 5b, when the temperature gradient was 50°C. On the contrary, in the case of growth in alkaline solutions, only the three rhombohedral faces remain as the growth proceeds, because the difference between the growth rates for the *R* faces and the growth rates for the *r* faces generally is large. Crevice flawing was frequently observed in the basal face, and fluid inclusions were sometimes formed in the *Z* region. Fig. 6 shows some crevice flawing which emerged on the *Z* surface. It was triangular and different from the flat-shaped crevice

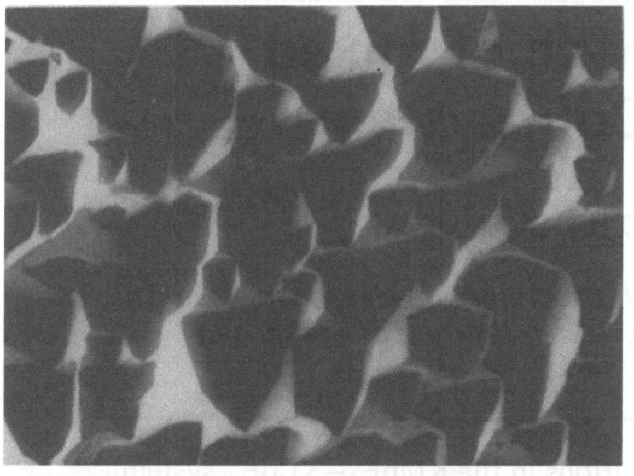

Fig. 6. SEM micrograph of crevice flawing on Z surface. Magn. ×105.

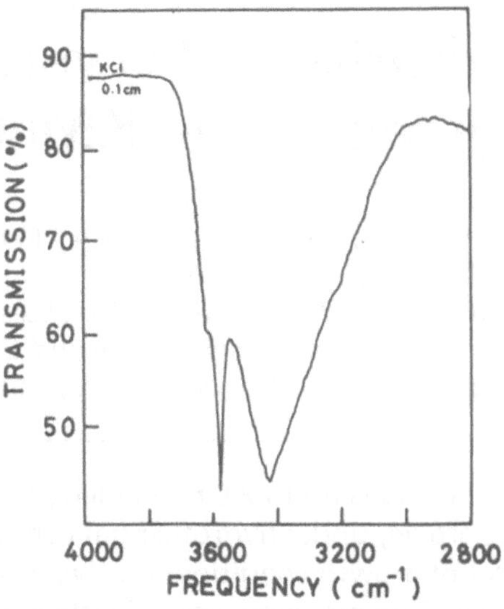

Fig. 7. 2800–4000 cm^{-1} absorption spectra of quartz crystal grown in KCl solution.

flawing found between cobbles on the Z surface of quartz crystals grown in alkaline solutions. Such flawing was frequently seen from the beginning stage of growth when temperatures were above 400°C. But when the growth temperatures were below 400°C and the growth rates were very slow, flawless growth was observed in the range of 1–3 mm from the boundary between the seed and and the growth region. This flawless region remained for a growth period of 5–7 days. The region of the R, r, and X direction was transparent with the naked eye. Striations were observed on the prism faces, and they clearly run parallel to the intersection of the prism faces with the rhombohedral faces. $+X$ surface was very smooth and lustrous, but $-X$ surface was lusterless and somewhat rough. These characteristics were contrary to those of $+X$ and $-X$ surfaces of quartz crystals grown in alkaline solutions. $+X$ and $-X$ regions were colored by Co-γ radiation in the case of quartz crystals grown in alkaline solutions, whereas quartz crystals grown in KCl solutions were hardly colored even by the irradiation of 5.4×10^6 Roentgen. It is thought that either the concentrations of aluminium are lower than those incorporated in quartz crystals grown in alkaline solutions of aluminium incorporates intersticially in the crystals [7,8].

Fig. 7 shows the infrared absorption spectra measured at room temperature by a JASCO DS-701G Spectrophotometer. The absorption spectra due to OH vibration were observed at 3430, 3585, and 3610 cm^{-1}, and the respective absorptions had very large extinction coefficients. The absorptions due to OH vibration were more than those of quartz crystals grown in NaCl, NaOH, and Na_2CO_3 solutions, and the OH background absorption in the 3430 cm^{-1} region was very great.

5. Conclusions

Quartz crystals were hydrothermally grown in KCl solution, as well as in NaCl solution. From the experimental results, it was apparent that flawless crystals were obtained except for Z direction, and growth temperatures needed to be raised above approximately 400°C to obtain sufficient growth rates to compared with usual alkaline growth, and growth rates showed the largest value at 10 wt% KCl solution. Some differences observed, as compared with growth in alkaline solutions, are as follows: (1) The growth rate for the R face was larger than that for r face. (2) Striations were observed on the prism faces of quartz crystals grown in KCl solution. Particularly, the striations were similar to those observed on the same faces of quartz crystals grown in NaCl solution and to those found in natural quartz.

References

[1] M. Hosaka and S. Taki, J. Crystal Growth 51 (1981) 640.
[2] M. Enjoji, Sci. Rept. Tokyo Educ. Univ. Sec. C11 No. 106 (1972) 79.
[3] T.M. Sushchevskaya and G.F. Ivanova, Geokhimia 9 (1967) 1099.
[4] S. Taki and K. Yamada, Rept. Fac. Educ. Yamanashi Univ. 12 (1961) 146.
[5] R.A. Laudise, J. Am. Chem. Soc. 81 (1959) 562.
[6] F. Muto and S. Taki, Rept. Fac. Eng. Yamanashi Univ. 13 (1963) 152.
[7] A.J. Cohen, J. Phys. Chem. Solids 13 (1960) 321.
[8] G. Lehmann and H.U. Bambauer, Angew. Chem. Intern Ed. 12 No. 4 (1973) 283.

Reprinted from *Journal of Materials Science*, **17** (1982) 177–182.

Hydrothermal crystal growth of mixed crystals in the $KMnF_3$–$KCoF_3$ system

KAZUMICHI YANAGISAWA, MASAHIRO YOSHIMURA, SHIGEYUKI SŌMIYA
Laboratory for Hydrothermal Syntheses, Research Laboratory of Engineering Materials, Department of Materials Science and Engineering, Tokyo Institute of Technology, Nagatsuta, Midori, Yokohama 227, Japan

Single crystals up to 2.0 mm in length of solid solutions in the $KMnF_3$–$KCoF_3$ system were grown, under hydrothermal conditions from 400 to 600° C under 100 MPa. The effect of growth conditions on shape and surface structure of the single crystals was studied. The measurements of lattice parameter and magnetic susceptibility demonstrated that the composition of the crystals was the same as that of the starting mixture of $KMnF_3$ and $KCoF_3$ powders. The distribution of Mn^{2+} and Co^{2+} ions in a single crystal of the solid solution was investigated and it appeared to be homogeneous.

1. Introduction

Mixed crystals in the $KMnF_3$–$KCoF_3$ system are of interest because of their structural and magnetic properties [1–6]. Single crystals have been grown by several methods [1–7]: Svensson *et al.* [2] grew single crystals by the Stockbarger method and Skrzypek *et al.* [7] used the Bridgman method. In these methods, anhydrous pure fluorides were used as starting materials and these had to be melted down at temperatures above 1000° C. In order to prevent oxidation of fluorides, these methods require an argon or nitrogen atmosphere. The crystals had different compositions, far from those of the starting materials, because the solid crystallized along the solidus line, not the liquidus line.

The flux method [3, 6] has also been used to grow these single crystals but little information has been given. This method might have problems similar to the Stockbarger and Bridgman methods.

The hydrothermal method has been used to grow single crystals of the perovskite-type fluorides without particular difficulties [8, 9]. The uptake of OH^- ion in the crystal seems to be negligible because the crystal of $KCoF_3$ produced by this method showed no OH absorption in the infra-red spectrum [9]. This method has been developed to grow crystals of mixed perovskite-type fluorides in the present work.

2. Experimental procedure

Chemicals used were guaranteed reagent-grade KCl, KF, $MnCl_2 \cdot 4H_2O$ (Wako Pure Chemical Ind. Ltd, Tokyo, Japan) and $CoCl_2 \cdot 6H_2O$ (Kanto Chemical Co. Inc., Tokyo, Japan). $KMnF_3$ and $KCoF_3$ powders were separately synthesized by co-precipitation from KF and divalent metal chloride solutions. The precipitates were filtered, washed with water and dried at 110° C in air for 24 h. They were confirmed to be pure $KMnF_3$ and $KCoF_3$, respectively, by X-ray powder diffraction. Thermogravimetric analysis showed that the moisture content in the powders was within 0.1%. Powders were mixed by a wet blending method with ethanol to get the desired compositions. The mixed powder was moulded and isostatically pressed at 200 MPa into rods 1.5 mm in diameter.

The solvents used were aqueous solutions of KCl with 1.0, 2.0 and 4.0 molality and of KF with 2.57 molality. The calculated amount of solution, by which the inside pressure could be balanced with the outside pressure on the capsule at the growth temperature, was injected into a gold capsule, 3.0 mm in outer diameter, 0.15 mm in thickness and 50 mm in length. A few rods of the mixed powder, about 70 mg, were put into the solution. The capsule was closed by welding with an electric arc.

The sealed capsule was put into a test-tube-type

pressure vessel of Stellite 25 and heated at 25° C min^{-1} in an electric furnace. Temperatures were measured on the wall of the pressure vessel by two platinel thermocouples calibrated against the melting point of zinc (419.5° C), at two positions corresponding to the top and the bottom of the capsule. The temperature at the top, that is the growth temperature, T_g, was adjusted to 400, 500 and 600° C by an electric controller. The temperature at the bottom, that is the dissolution temperature, T_d, was adjusted to be 20 or 50° C higher than at the top by using the temperature gradient in the furnace. The pressure was measured by a calibrated Heise pressure gauge and adjusted to 100 MPa. The duration of crystal growth experiments was 4 or 7 days.

At the completion of the experiments, the vessel was quenched in cold water. The capsule was cut open to reveal the contents, which were washed with water. The shape and surface structure of the crystals were observed with an optical microscope (VANOX, Olympus Optic Ind. Ltd, Tokyo, Japan) and a scanning electron microscope (SEM) (JSM-T200, JEOL Ltd, Tokyo, Japan), and the crystals were identified by X-ray powder diffraction (RU-200, Rigaku Denki Co. Ltd, Tokyo, Japan). Lattice parameters were calculated from high-angle diffraction lines with a slow scan rate (0.25° min^{-1}), using Si (99.999% pure) as an internal standard. The compositional change of the solid solution was checked by the measurement of the lattice parameter. This method allows changes within 1 wt% to be detected, which is more precise than electron probe micro-analysis (EPMA). The magnetic susceptibility was measured by a Faraday-type magneto-balance (MB-1A, Shimadzu Seisakusho Ltd, Kyoto, Japan). The distribution of metal ions in the single crystals was measured by EPMA (EMX 11 A, Shimadzu Seisakusho Ltd, Kyoto, Japan).

3. Results and discussion

3.1. Hydrothermal reaction in the $KMnF_3$–$KCoF_3$ system

The rate of the hydrothermal reaction in the $KMnF_3$–$KCoF_3$ system was studied by quenching samples of 50 mol% $KMnF_3$ after heating in 2 M-KCl solution at 500° C under 100 MPa for 0, 1, 3, 6, 12 or 24 h.

The sample quenched as soon as the temperature reached 500° C showed a diffuse diffraction peak (Fig. 1b) between those of $KMnF_3$ and

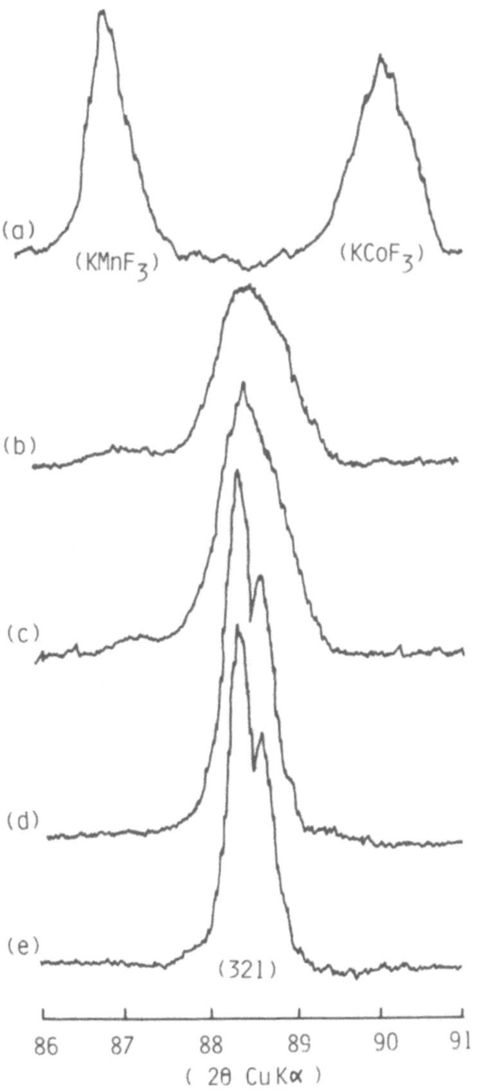

Figure 1 Peak profile of (3 2 1) diffraction line of $KMn_{0.5}Co_{0.5}F_3$ synthesized at 500° C in 2 M KCl solution under 100 MPa. (a) Starting material, (b) after 0 h, (c) after 1 h, (d) after 12 h, (e) crystals grown for 7 days.

$KCoF_3$. It showed that $KMnF_3$ and $KCoF_3$ powders had already reacted and produced solid solutions between them during heating up to 500° C. As the duration of heating at 500° C was increased, the peak of the solid solution became gradually sharper. The solid solutions were considered to have homogeneous compositions after 12 h (Fig. 1d). No oxides were detected by X-ray powder diffraction. It is probable that the compositions of the solid solutions produced by this hydrothermal method are essentially equal to the starting compositions, because the solubilities of

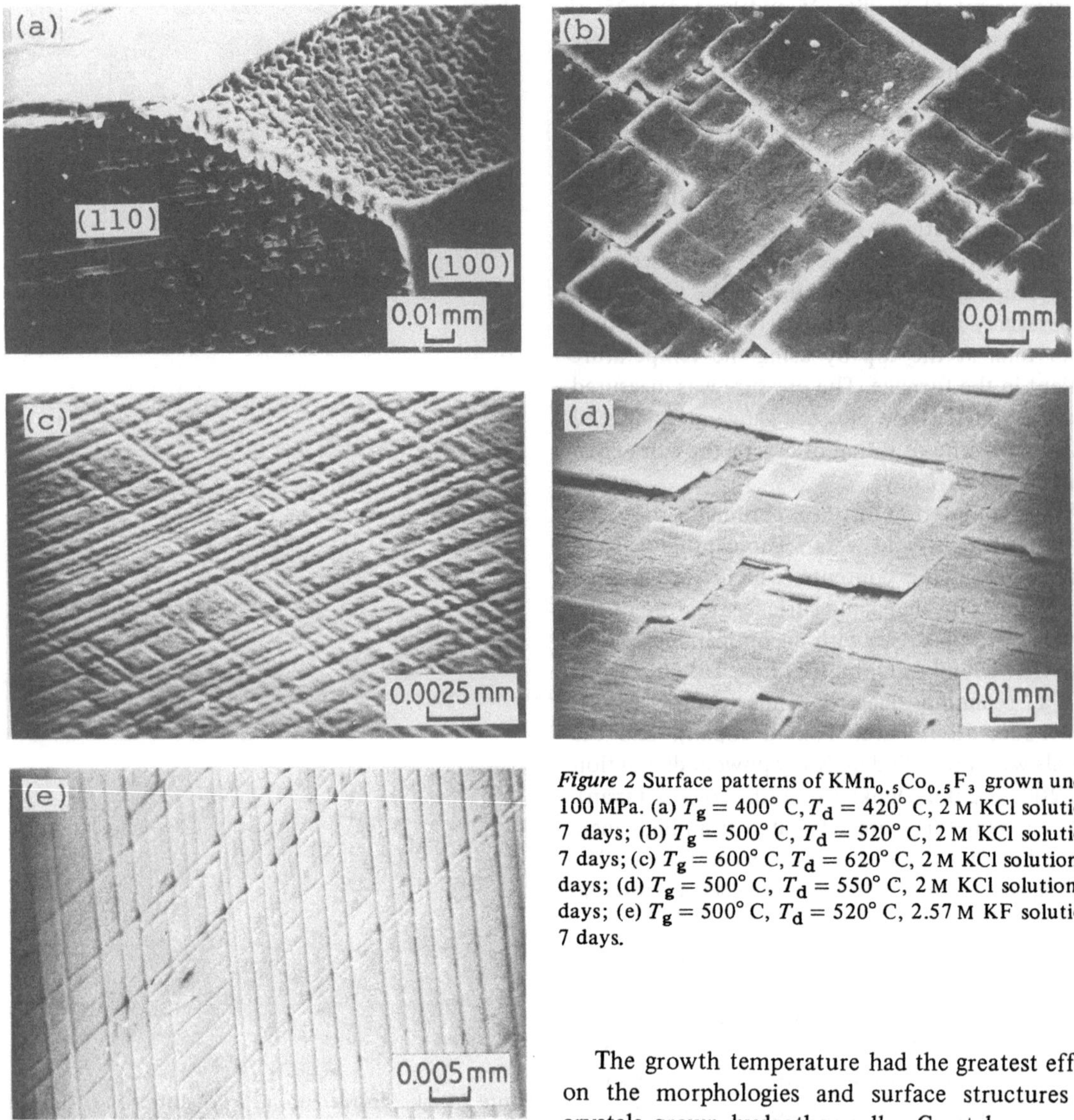

Figure 2 Surface patterns of $KMn_{0.5}Co_{0.5}F_3$ grown under 100 MPa. (a) $T_g = 400°$ C, $T_d = 420°$ C, 2 M KCl solution, 7 days; (b) $T_g = 500°$ C, $T_d = 520°$ C, 2 M KCl solution, 7 days; (c) $T_g = 600°$ C, $T_d = 620°$ C, 2 M KCl solution, 7 days; (d) $T_g = 500°$ C, $T_d = 550°$ C, 2 M KCl solution, 4 days; (e) $T_g = 500°$ C, $T_d = 520°$ C, 2.57 M KF solution, 7 days.

$KMnF_3$ and $KCoF_3$ are small in comparison with the amount of undissolved solid under hydrothermal conditions.

3.2. Crystal growth

Mixed crystals of $K(Mn, Co)F_3$ were grown at the top of the capsule under hydrothermal conditions between 400 and 600° C under 100 MPa. Mixed crystals of $KMn_{0.5}Co_{0.5}F_3$ (Fig. 1e) grown for 7 days shows a diffraction peak at the same position as that of the solid solution (Fig. 1d) produced by the hydrothermal reaction under the same conditions. This result revealed that the mixed crystals had the same compositions as the starting compositions.

The growth temperature had the greatest effect on the morphologies and surface structures of crystals grown hydrothermally. Crystals grown at 400° C had (100) and (110) surfaces (Fig. 2a). Small and rough (110) surfaces were observed on the edges of cubic or rectangular prism crystals. Their growth rate is probably faster than (100) surfaces. Crystals grown above 500° C were surrounded only (100) surfaces with chequered patterns (Fig. 2b, c). Growth patterns and difference of height on surfaces of crystals grown at 600° C (Fig. 2c) were much smaller than at 500° C (Fig. 2b).

The increase of the temperature gradient caused rough surfaces (Fig. 2d). Increasing the concentration of KCl solution used also had similar effects. When KF solution was used as a solvent, the crystals had flatter surfaces with linear patterns (Fig. 2e).

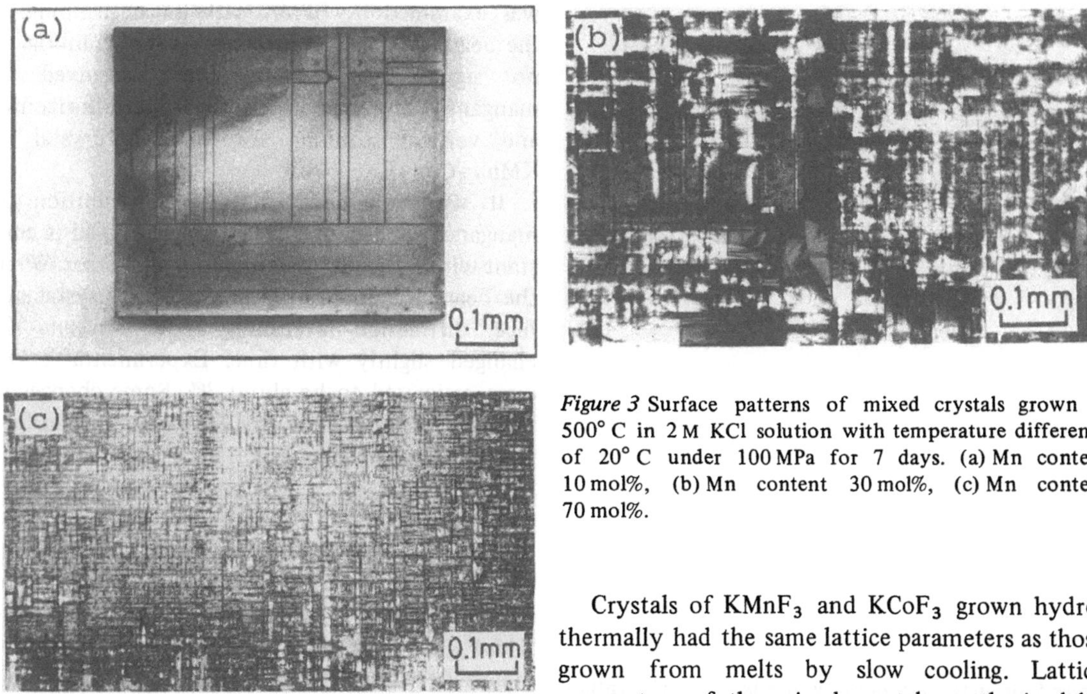

Figure 3 Surface patterns of mixed crystals grown at 500° C in 2 M KCl solution with temperature difference of 20° C under 100 MPa for 7 days. (a) Mn content 10 mol%, (b) Mn content 30 mol%, (c) Mn content 70 mol%.

Crystals with smooth surfaces were grown at the high growth temperature with a small gradient in diluted KCl solution or in KF solution. The surface structures were considered to be affected by the crystal growth rate, which depended on the supersaturation at the surface and on the deposition rate. The increased growth temperature gave an increased deposition rate. The decrease of the gradient resulted in decrease of the supersaturation. The solubility of K(Mn, Co)F_3 seemed to be lower in the diluted KCl solution or in KF solution. The low solubility might result in smaller supersaturation. It was concluded that the smooth crystal surfaces were achieved when the deposition rate was larger than the feed rate by the supersaturation.

Growth patterns on the crystal surfaces changed with starting compositions. Crystals with 10 mol% $KMnF_3$ had a few linear lines on the surface (Fig. 3a). As the concentration of manganese increased, the growth patterns became gradually smaller (Fig. 3b, c).

3.3. Characterization of crystals

3.3.1. *Lattice parameter*

Lattice parameters of hydrothermally grown crystals are shown against composition in Fig. 4. The lattice parameter–composition curve deviates slightly upward from Vegard's law.

Crystals of $KMnF_3$ and $KCoF_3$ grown hydrothermally had the same lattice parameters as those grown from melts by slow cooling. Lattice parameters of the mixed crystals synthesized by cooling melts quickly are shown with crosses in Fig. 4. They too showed the slight deviation from Vegard's law.

3.3.2. *Magnetic susceptibility*

Magnetic susceptibilities of mixed crystals synthesized under hydrothermal condition are shown in Fig. 5. The temperature of the maximum value of the susceptibility changes from 80 to 125 K with the decrease of manganese contents. The values of susceptibility measured in this work were identical with those reported previously [6, 10]. This means that the hydrothermal method as well as the flux

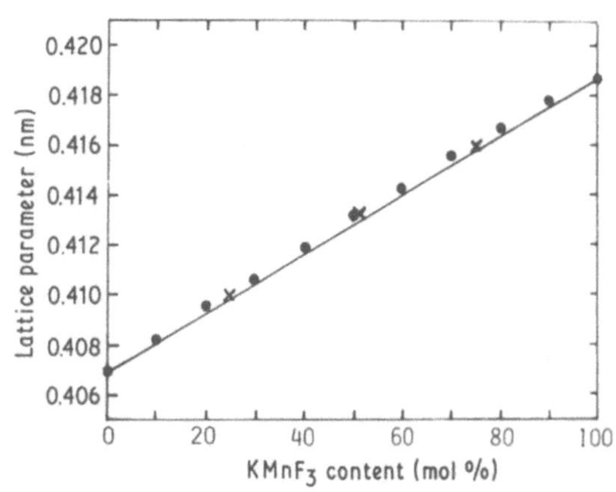

Figure 4 Lattice parameters of mixed crystal grown at 500° C in 2 M KCl solution under 100 MPa for 7 days.

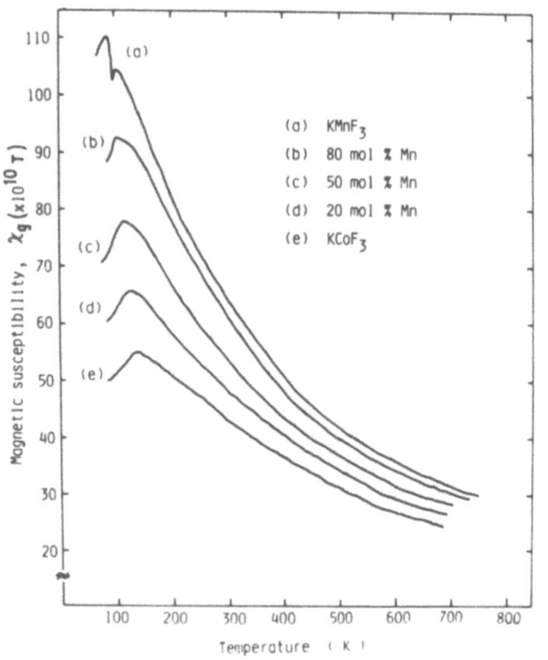

Figure 5 Magnetic susceptibility of mixed crystals synthesized at 500° C in 2 M KCl solution under 100 MPa for 1 day.

method [6] can be employed to synthesize mixed crystals in the $KMnF_3$–$KCoF_3$ system.

3.3.3. *Distribution of metal ions*

The distribution of metal ions in the single crystal was examined by EPMA with linear scanning of the beam on polished surfaces. As shown in Fig. 6, no significant variations were observed in manganese or cobalt intensities in both horizontal and vertical scanning for a single crystal of $KMn_{0.5}Co_{0.5}F_3$.

It was concluded that the distribution of manganese and cobalt in the single crystal is constant within the limits of experimental error. When the beam was fixed on a point of the crystal surface, diffraction intensities of K, Mn and Co changed slightly with time. Experimental errors were estimated to be about 3%. Some changes in intensity resulted from small cracks on the surface made during polishing.

4. Conclusions

The hydrothermal method has been applied to the growth of single crystals of solid solutions in the $KMnF_3$–$KCoF_3$ system. The co-precipitated $KMnF_3$ and $KCoF_3$ powders, used as starting materials, reacted rapidly under hydrothermal condition at 500° C and produced solid solutions. Crystals grown hydrothermally had the same compositions as the starting mixtures. The formation of continuous solid solutions at 500° C was revealed by the measurement of lattice parameters and magnetic susceptibilities. The results of EPMA showed homogeneous distribution of metal ions in the single crystals.

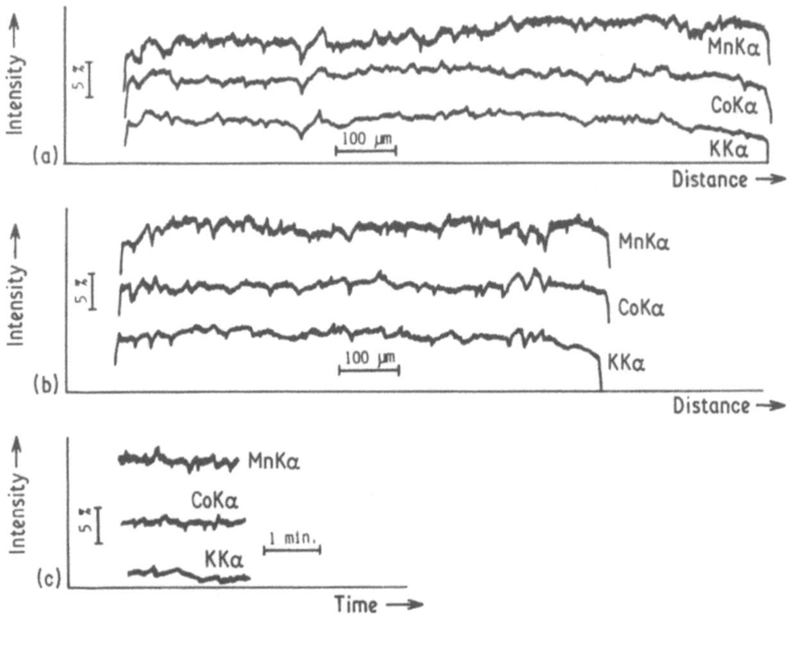

Figure 6 EPMA result for a single crystal of $KMn_{0.5}Co_{0.5}F_3$ grown at 500° C in 2 M KCl solution under 100 MPa for 7 days. (a) Horizontal scan, (b) vertical scan, (c) steady beam.

References

1. W. W. HOLLOWAY and M. KESTIGIAN, *J. Appl. Phys.* **38** (1967) 1480.
2. E. C. SVENSSON, W. J. L. BUYERS, T. M. HOLDEN, R. A. COWLEY and R. W. H. STEVENSON, *Can. J. Phys.* **47** (1969) 1983.
3. M. A. VINNIK and L. N. SELEZNEVA, *Sov. Phys. Cryst.* **14** (1970) 14.
4. A. CHEŁKOWSKI, P. JAKUBOWSKI, D. KRASKA, A. RATUSZNA and W. ZAPART, *Acta Phys. Pol.* **A47** (1975) 347.
5. A. RATUSZNA, A. PIETRASZKO, A. CHEŁKOWSKI and K. ŁUKASZEWCZ, *Phys. Stat. Sol.* **A54** (1979) 739.
6. T. HASHIMOTO, *J. Phys. Soc. Japan* **18** (1963) 1140.
7. D. SKRZYPEK, P. JAKUBOWSKI, A. RATUSZNA and A. CHEŁKOWSKI, *J. Cryst. Growth* **48** (1980) 475.
8. S. SŌMIYA, S. HIRANO, M. YOSHIMURA and K. YANAGISAWA, *J. Mat. Sci.* **16** (1981) 813.
9. S. HIRANO, K. YANAGISAWA and S. SŌMIYA, *J. Cryst. Growth* to be published.
10. K. HIRAKAWA, K. HIRAKAWA and T. HASHIMOTO, *J. Phys. Soc. Japan* **11** (1960) 2063.

Received 19 March
and accepted 18 June 1981

Reprinted from *Journal of Materials Science Letters*, **1** (1982) 461–462.

Crystallization and crystal growth of CeO_2 under hydrothermal conditions

E. TANI, M. YOSHIMURA, S. SŌMIYA
Research Laboratory of Engineering Materials, and Department of Materials Science and Engineering, Tokyo Institute of Technology, Nagatsuta, Yokohama, 227 Japan

CeO_2 crystals have previously been grown by the flux method [1–4] due to their high melting point and high vapour pressure [5] at high temperatures. The flux method requires high temperatures and generally brings about greater contamination of the flux than does the hydrothermal method. This study deals with the effect of several mineralizers on the crystallization and the crystal growth of CeO_2 under hydrothermal conditions.

The starting material was hydrous ceria precipitate from $Ce(NO_3)_3 \cdot 6H_2O$ (99.9%, Nissan Kigenso Co., Ltd, Osaka, Japan) solution with 3 N NH_4OH*. The precipitate was filtered and dispersed in distilled water, then filtered again for washing. After repeating this washing procedure four times, it was dried at 120° C for 4 days. The water content of the hydrous ceria could be presented by the formula of $H_2O/CeO_2 \doteqdot 1$ (molar ratio) by TG–DTA (CN8076El, Rigaku Denki Co., Ltd, Tokyo, Japan).

Distilled water or a solution of KF* (8 wt %), LiCl† (15 wt %), Libr‡ (10 wt %), K_2CO_3‡ (10 wt %), $LiNO_3$* (10 wt %), Li_2SO_4* (10 wt %) and NaOH* (30 wt %) was used as a mineralizer (all reagent grade). These mineralizers were selected on the basis of our previous results [6].

A Pt or Au tube (2.7 mm i.d., 0.15 mm thick and 32 mm long) filled with the hydrous ceria and a mineralizer solution was treated under 100 MPa at 500 to 700° C for either 1 or 3 days using a test-tube type pressure vessel of Stellite 25 (6.5 mm i.d.). After this hydrothermal treatment, the vessel was quenched in cold water. The products were washed in distilled water and examined using an X-ray powder diffractometer (RU-200, Rigaku Denki Co. Ltd, Tokyo, Japan), and a scanning electron microscope (JSM-01BM, JEOL Co. Ltd, Tokyo, Japan). The crystallite sizes of the phases were calculated from the Scherrer formula, using the (1 1 1) diffraction peak of CeO_2.

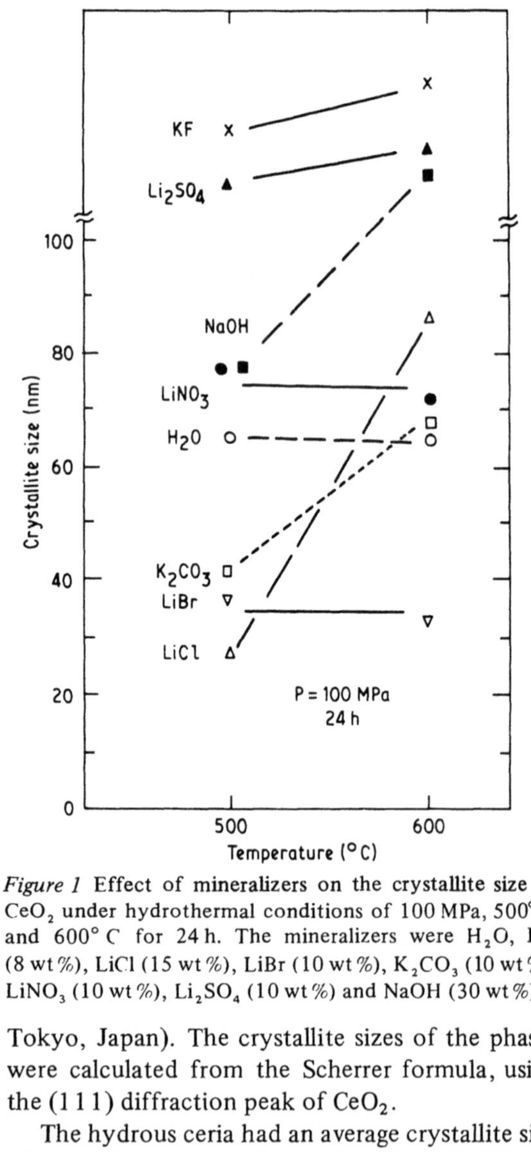

Figure 1 Effect of mineralizers on the crystallite size of CeO_2 under hydrothermal conditions of 100 MPa, 500° C and 600° C for 24 h. The mineralizers were H_2O, KF (8 wt %), LiCl (15 wt %), LiBr (10 wt %), K_2CO_3 (10 wt %), $LiNO_3$ (10 wt %), Li_2SO_4 (10 wt %) and NaOH (30 wt %).

The hydrous ceria had an average crystallite size of about 9 nm. Fig. 1 shows the effect of different

*Wako Pure Chemical Industry Co. Ltd, Tokyo, Japan.
†Kanto Chemical Co. Inc., Tokyo, Japan.
‡Koso Chemical Co. Ltd, Tokyo, Japan.

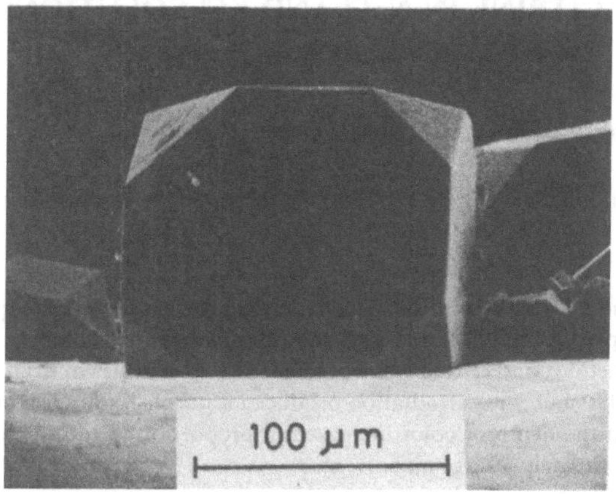

Figure 2 Scanning electron micrograph of CeO_2 crystals grown from hydrous ceria under 100 MPa at 700° C for 3 days with 8 wt % KF solution.

mineralizers on the crystallite size of CeO_2 treated hydrothermally from hydrous ceria under 100 MPa at 500 and 600° C for 24 h. KF and Li_2SO_4 solutions considerably enhanced the growth of CeO_2 crystallites up to above 100 nm at 500 and 600° C.

This result revealed the KF solution to be optimum for the growth of CeO_2 crystals. Therefore, the crystal growth of CeO_2 was carried out with the 8 wt % KF solution.

Fig. 2 shows CeO_2 crystals grown under 100 MPa at 700° C ($\Delta T = 50°$ C; the temperature difference between the lower part of the Pt or Au tube and the upper part) for 3 days. The crystals were transported to the top of the Pt tube and grew up to 100 μm in size and showed a cubic shape with apparent (1 0 0) and (1 1 1) faces. The EDS analysis showed only Ce in the crystals, but the mass spectrometric analysis showed the presence of K in the order of 100 p.p.m. and F in the order of 10 p.p.m. in the products.

Acknowledgements

This research was supported by Grant-in-Aid for Scientific Research No. 443015 (1979–1980) and the Iwatani Naoji Foundation, 1980. The authors thank JEOL Co. Ltd. For providing the opportunity to use the EDS.

References

1. C. B. FINCH and G. WAYNE CLARK, *J. Appl. Phys.* 37 (1966) 3910.
2. W. H. GRODKIEWICZ and D. J. NITTI, *J. Amer. Ceram. Soc.* 49 (1966) 576.
3. R. C. LINARES, *J. Phys. Chem. Solids* 28 (1967) 1285.
4. B. M. WANKLYN, *J. Cryst. Growth* 5 (1969) 219.
5. A. ROUANET, *C.R. Acad. Sci., Paris* 266C (1968) 908.
6. E. TANI, M. YOSHIMURA and S. SŌMIYA, *Yogyo Kyokai Shi* 90 (1982) 195.

Received 17 November 1981 and accepted 27 May 1982

Reprinted from *Journal of Crystal Growth*, **64** (1983) 572–576.

HYDROTHERMAL GROWTH OF AMETHYST AND CITRINE IN NaCl AND KCl SOLUTIONS

Masahiro HOSAKA * and Sadao TAKI

Institute of Inorganic Synthesis, Faculty of Engineering, Yamanashi University, Miyamae-7, Kofu, Japan

Received 28 July 1983; manuscript received in final form 16 September 1983

Natural quartz grows from solutions containing mainly Na^+, K^+ and Cl^- ions. In this experiment, the growth of synthetic quartz under the same growth conditions as those of natural quartz was attempted using NaCl and KCl solutions as a growth solution. As a result, colored quartz equivalent in quality and color tone to natural amethyst and citrine was obtained. The quartz grown from *r*-cut and *R*-cut plate seeds with iron added to the growth solution was colorless but γ-ray irradiation produced a purplish tint. The quartz grown from *X*-cut plate and *Y*-bar seeds had a yellowish tint. Although the depth of color increased in proportion to the amount of iron, it reached close to saturation with an iron concentration of approximately 90 μg/ml solvent.

1. Introduction

Amethyst, smoky quartz, citrine, and rose quartz represent colored quartz existing in nature. Natural quartz grows in a solution containing mainly the Na^+, K^+ and Cl^- ions, and colored quartz is a product in which impurities contained in a solution were incorporated during the process of crystal growth. The coloration mechanism of colored quartz has been fairly well explicated in only relatively recent years. It is known that citrine occurs in the presence of substitutional Fe^{3+} [1], yellow quartz in the presence of interstitial Fe^{3+} [2], amethyst in the presence of interstitial and substitutional Fe^{3+} [3,4], and smoky quartz in the presence of substitutional Al^{3+} [5]. Amethyst and smoky quartz [6,7] are decolorized by heating and colored by irradiation. For artificial synthesis of colored quartz, an alkaline solution has been mainly used as a growth solution. Amethyst [8], citrine and green quartz [9] were synthesized using K_2CO_3 solution and KOH solution with added Fe. Blue quartz was obtained in NaOH solution by doping with Co^{2+} [10]. Amethyst was also obtained in NH_4F solution by doping Fe [11]. However, during the growth process using an alkaline solution, the color tone undergoes unstable changes depending on the growth conditions such as impurities used.

In this experiment, synthesis of amethyst and citrine was attempted with growth conditions as close to the growth of natural quartz as possible using NaCl and KCl solutions with iron as an impurity. Synthetic amethyst and citrine of the color tone almost equal to natural colored quartz were obtained.

2. Experimental method

The hydrothermal growth of colored quartz crystals was carried out using a platinum capsule (10 mm ID × 210 mm length) placed in an autoclave. *X*-cut $(11\bar{2}0)$, *R*-cut $(10\bar{1}1)$ and *r*-cut $(01\bar{1}1)$ plate and *Y*-bar seeds were used. Synthetic quartz crystals were used as a nutrient. Fe or α-Fe_2O_3 powder was used as an additive in concentrations of 1–1000 μg/ml solvent. As a growth solution, 10 wt% NaCl solution and 10 wt% KCl solution were used at a fill rate of 70%. The growth period was 5–20 days. The growth temperatures varied from 400 to 450°C. The temperature gradient was held at either 25 or 50°C. The crystals grown were

* Present address: Institute of Gemology and Jewelry Arts of Yamanashi Prefecture, Tokojicho 1955-1, Kofu, Japan.

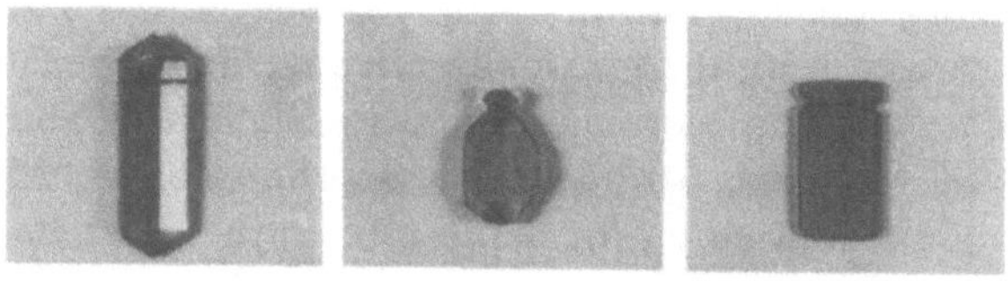

Fig. 1. Amethyst and citrine grown in NaCl solution with iron added. Left: citrine grown on *X*-cut seed. Middle: citrine grown on *Y*-bar seed. Right: amethyst grown on *r*-cut seed.

colored by irradiation with γ-rays at a dose of 2×10^6–15×10^6 roentgen using Co^{60}, and evaluated macroscopically as well as spectrophotometrically using a Shimazu spectrophotometer (Model MPS-50L).

3. Results and discussion

Under these experimental conditions, transparent high-quality crystals were obtained from the NaCl and KCl solutions containing Fe and α-Fe_2O_3. Fig. 1 shows amethyst and citrine grown in NaCl solution. The growth rates of the *X*-face, *R*-face and *r*-face were approximately 0.1, 0.01 and 0.01 mm/day, respectively. The growth characteristics such as the growth pattern obtained in this experiment were the same as previously reported [12–14]. When iron was added, the crystals grown from the *R*-cut and *r*-cut plate seeds were colorless, and the crystals grown from the *X*-cut plate and *Y*-bar seeds were yellow.

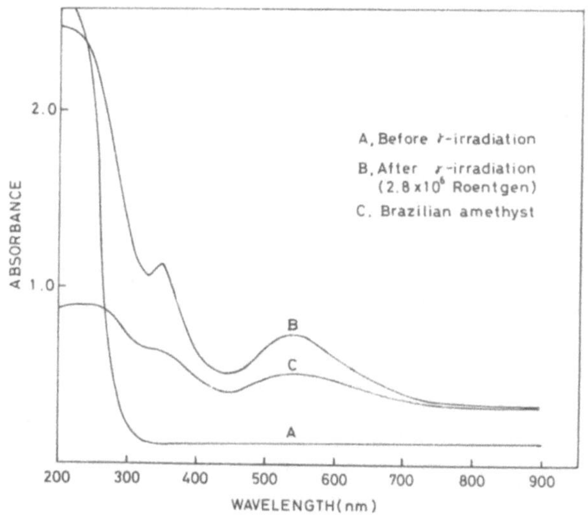

Fig. 2. Absorption spectra for synthetic and Brazilian amethyst samples.

3.1. Amethyst

The crystals grown from the *R*-cut and *r*-cut plate seeds were colorless and transparent, despite an iron content when grown. However, γ-ray irradiation changed them into amethyst of a beautiful purplish tint. Fig. 2 shows the absorption characteristics of these crystals before and after γ-irradiation in comparison with natural amethyst. Although not very distinct in the figure, the specimen before γ-ray irradiation shows a large absorp-

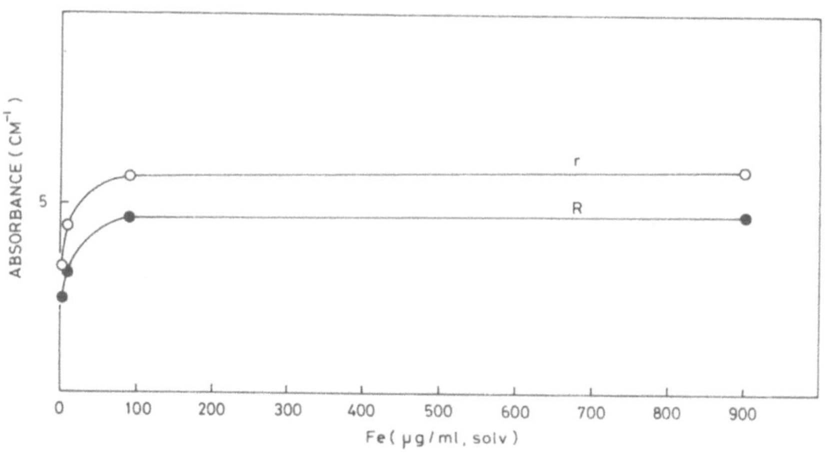

Fig. 3. Relation between the concentration of added Fe and absorbance at 545 nm.

tion at 225 nm and a broad absorbance zone at 270 nm. The absorbance at 225 nm is due to the presence of Fe^{3+} [4] and this absorbance was found to grow larger with an increase of the iron concentration in the solution. As seen in the figure, by irradiation with γ-rays new absorption appeared at 350 and 545 nm. Conversely, there appeared a tendency of slight diminution of absorption in the vicinity of 220 nm. The purplish tint of amethyst comes from the amethyst center at 545 nm. There was no difference at all in the absorption characteristics between synthetic amethyst and natural amethyst as shown in fig. 2. The depth of the purplish tint depended on the amount of iron mixed in the growth solution and the dose of γ-ray irradiation. Fig. 3 shows the relation between the amount of Fe and the absorbance at 545 nm.

Even with a very small amount (0.9 μg/ml solvent; added amount: 0.01 mg) of iron, a distinct purplish tint was obtained and the purplish color deepened with an increase in iron. However, coloring reached saturation at an iron content of above 90 μg/ml solvent. In γ-ray irradiation, 2×10^6 roentgen were the maximum effective dose. When a comparison was made between the NaCl and KCl solutions, neither of the solutions produced any effect at all on the depth of coloring. Furthermore, when a comparison was made between Fe and α-Fe_2O_3, Fe produced a better effect on the depth of coloring.

Although the *Z* growth region of *Y*-bar synthetic quartz was colored with a slightly yellowish tint, it changed to a purplish tint by irradiation with γ-rays. Yellowish coloration prior to γ-ray irradiation suggests the presence of a high proportion of substitutional Fe^{3+}. However, the change into purple by irradiation indicates the presence of interstitial Fe^{3+}. It seems, therefore, that this ion was incorporated into the tunnel extending in a *Z*-axial direction which was formed by spiral linkage of SiO_4 tetrahedrons.

Amethyst is decolorized by heat treatment [15]. Fig. 4 shows the changes in the absorbance characteristics by heat treatment of amethyst grown from the *r*-cut plate seed with an iron concentra-

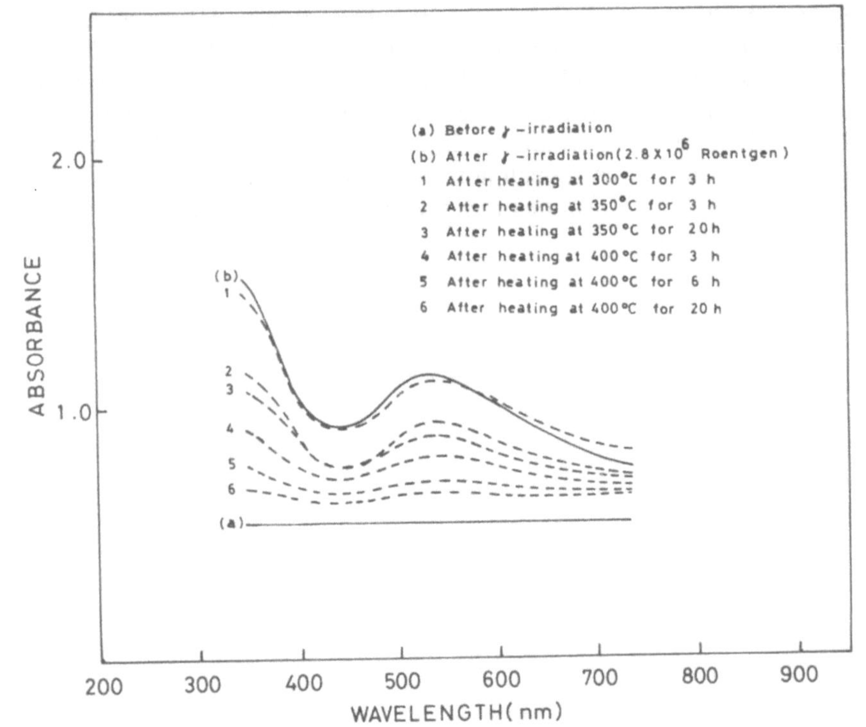

Fig. 4. Absorption spectra for synthetic amethyst before and after heating.

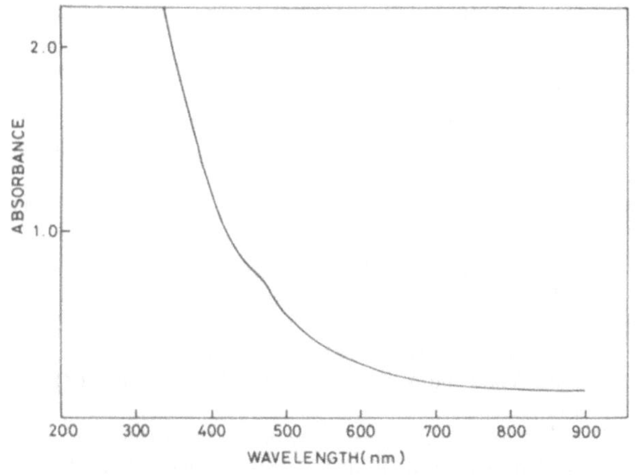

Fig. 5. Absorption spectrum for synthetic citrine.

tion of 92.6 μg/ml solvent. Decolorization was slow at 300°C. But the decolorization rate increased at 400°C and complete decolorization was obtained after approximately 30 h of heating. The decolorized appearance of synthetic amethyst was exactly the same as that of natural amethyst. Reirradiation with γ-rays at a dose of 2.8×10^6 roentgen of decolorized amethyst restored completely the original color and absorbance.

3.2. Citrine

The absorption spectrum of synthesized citrine is shown in fig. 5. The absorbance at the near-ultraviolet zone by Fe^{3+} was considerably large. A small absorption was also observed at 470 nm. Fig. 6 shows the relation between the amount of Fe added and the absorbance at 470 nm. Even in the presence of a very small amount (0.9 μg/ml solvent) of iron, the crystals showed distinctly a yellowish tint. The tint, however, increased with an increase of the iron content in the solution. However, at an iron concentration above 92.6 μg/ml solvent, the increase in absorbance considerably slackened. At an iron concentration of 926.7 μg/ml solvent, the color was turned into dark yellow. The dark yellow color may be attributable to the presence of interstitial Fe^{3+}, besides substitutional Fe^{3+}, contained in the high iron concentration. Addition of α-Fe_2O_3 in the growth solution produced similar effects to Fe but coloring was slightly less deep. At the same iron concentrations, the NaCl solution produced crystals of somewhat deeper color than the KCl solution.

Irradiation with γ-rays of synthesized citrine had no effect on the color. A high dose of 15×10^6 roentgen produced only a slight brownish tone. Heat treatment also did not cause any change in color and the absorption characteristics underwent no change even by heating at 500°C.

4. Conclusion

In order to obtain colored crystals under growth conditions as close to the growth of natural quartz

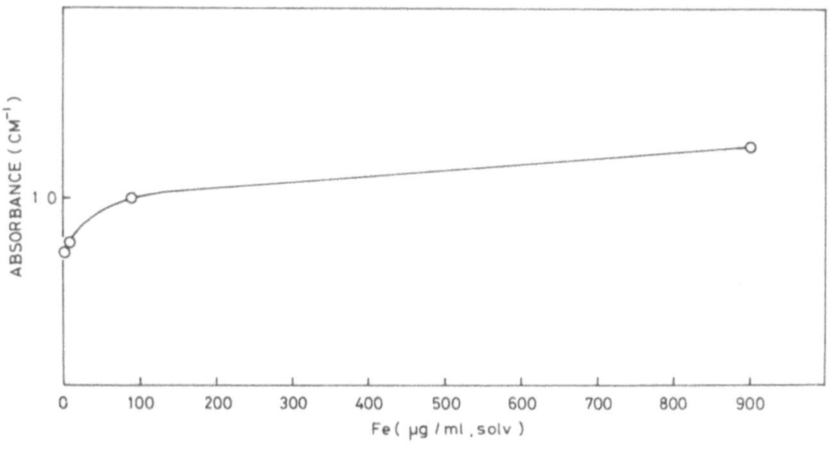

Fig. 6. Relation between the concentration of added Fe and absorbance at 470 nm.

as possible, the hydrothermal growth of quartz was undertaken using 10 wt% NaCl and 10 wt% KCl solutions with iron as an impurity. Citrine was synthesized from *X*-cut plate and *Y*-bar seeds. Amethyst was obtained by irradiation with γ-rays of the transparent crystals grown from *R*-cut and *r*-cut plate seeds. The color tone was as excellent as high-quality natural quartz. The color depth increased with an increase of iron concentration in the growth solution, but at a concentration above 100 μg/ml solvent, the color did not increase its depth.

References

[1] T.I. Barry and W.J. Moore, Science 144 (1964) 289.

[2] G. Lehmann and W.J. Moore, J. Chem. Phys. 44 (1966) 1741.

[3] G. Lehmann and H.U. Bambauer, Angew. Chem. Intern. Ed. Engl. 12 (1973) 283.

[4] A.J. Cohen, Am. Mineralogist 59 (1974) 719.

[5] C.S. Brown and L.A. Thomas, J. Phys. Chem. Solids 13 (1960) 337.

[6] K. Nassau and B.E. Prescott, Phys. Status Solidi (a) 29 (1975) 659.

[7] K. Nassau and B.E. Prescott, Mineral. Mag. 41 (1977) 301.

[8] L.I. Tsynober and L.G. Chentsova, Kristallografiya 4 (1959) 633.

[9] A.A. Ballman, Am. Mineralogist 46 (1961) 439.

[10] D.L. Wood and A.A. Ballman, Am. Mineralogist 51 (1966) 216.

[11] V.S. Balitsky, J. Crystal Growth 41 (1977) 100.

[12] M. Hosaka and S. Taki, J. Crystal Growth 52 (1981) 837.

[13] M. Hosaka and S. Taki, J. Crystal Growth 53 (1981) 542.

[14] M. Hosaka and S. Taki, J. Crystal Growth 55 (1981) 363.

[15] G. Lehman and W.J. Moore, Science 152 (1966) 1061.

Reprinted from *Journal of Materials Science Letters*, **3** (1984) 1097–1098.

Hydrothermal growth of La^{3+}-doped SrF_2 single crystal

M. YOSHIMURA, K. J. KIM, S. SŌMIYA
Research Laboratory of Engineering Materials, and Department of Materials Science and Engineering, Tokyo Institute of Technology, 4259, Nagatsuta, Midori, Yokohama, 227 Japan

Both pure and doped (with rare earth or other inorganic ions) alkali earth fluoride single crystals have widely been grown by the Bridgmann–Stockbarger [1, 2], Czochralski [3, 4], and melting [5] methods. These methods require an HF atmosphere and high temperature due to their high melting points. These requirements have often brought about the contamination from containers and/or the atmosphere. We reported possible growth of some fluoride single crystals under hydrothermal conditions at low temperatures such as 400 to 700° C [6, 7]. In the present work, we have studied the possibility of hydrothermal growth of La^{3+}-doped SrF_2 single crystals.

The suitabilities of several solutions were studied as a mineralizer for hydrothermal crystal growth. In the equilibrium studies of the system SrF_2–LaF_3 at 500 to 750° C under hydrothermal conditions using two types of starting samples, where one sample was the mixed sample of 99.9% SrF_2 (Tri Chemical Lab. Inc., Sagamihara, Japan), and 99.9% LaF_3 (Shin-Etsu Chemical Co. Ltd., Tokyo, Japan), and the other was a sample from the mixed samples fired at 1300° C for 4 h in argon, a good mineralizer led to the same products from both samples. The 10 wt % solutions of LiCl*, LiBr†, $LiClO_4$‡, NaCl§, or Na_2SO_4† (all reagent grade) produced the identical SrF_2 solid solution with the same lattice parameter [8].

The sealed capsule of platinum (2.7 mm inner diameter, 0.5 mm thick, and 32 mm long) filled with the mixed sample (10 or 20 wt % LaF_3) and a mineralizer solution, was heated in a Roy and Tuttle-type [9] (stellite 25; 6.5 mm inner diameter) hydrothermal apparatus, and quenched into water. The hydrothermal runs were carried out at 500 to 700° C with a temperature gradient ($\Delta T = 5$, 10 or 15° C between the top and the bottom of the capsule) for 1 to 5 days under 100 MPa.

The crystals grown by transportation at the top of the capsule were examined by an X-ray diffractometer (XRD, RU-200, Rigaku Denki Co. Ltd., Tokyo, Japan), a scanning electron microscope (SEM, HITACH H-700H, HITACHI Co. Ltd., Hitachi, Japan., or SED-880, Daini Seikosha Co. Ltd., Tokyo, Japan) analysis. The EDS could not detect Li^+ or other contamination in the products using an LiCl solution. It also indicated the lowering of the ratio of La/Sr in crystals ($Sr_{1+x}La_xF_{2+x}$; $x = 0.05$ to 0.1) from the starting powder ($x =$ 0.07 to 0.14).

Fig. 1 shows the effect of some mineralizers on the hydrothermal growth of $Sr_{1+x}La_xF_{2+x}$ crystals at 600° C ($\Delta T = 10$° C) for 2 days under 100 MPa. In the case of 10 wt % LiCl and LiBr solutions,

Figure 1 SrF_2 (10 wt % LaF_3) crystals grown in various 10 wt % solutions at 600° C ($\Delta T = 10$° C) for 2 days under 100 MPa.

*Kanto Chemical Co. Inc., Tokyo, Japan.
†Koso Chemical Co. Ltd., Tokyo, Japan.
‡ Nakarai Chemical Co. Ltd., Kyoto, Japan.
§Wako Pure Chemical Indus. Ltd., Osaka, Japan.

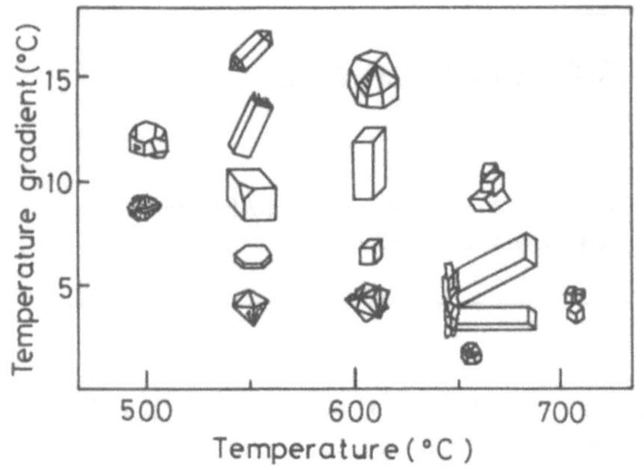

Figure 2 Crystal morphology and sizes of $Sr_{1+x}La_xF_{2+x}$ in 10 wt % LiCl solution at various temperatures for 2 days under 100 MPa.

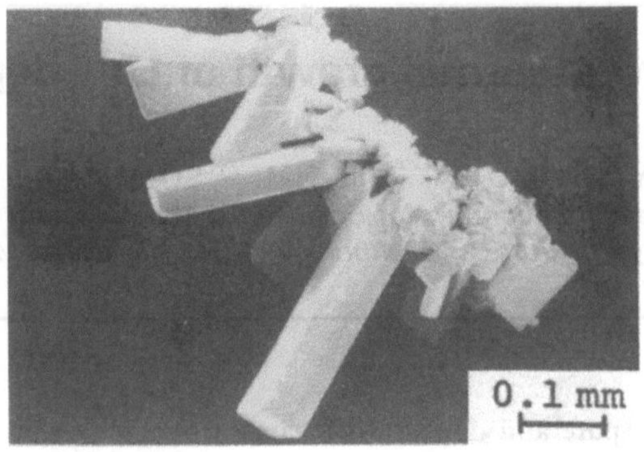

Figure 3 SrF_2 (20 wt % LaF_3) crystals grown in 10 wt % LiCl solution at 650° C ($\Delta T = 5°$ C) for 2 days under 100 MPa.

$Sr_{1+x}La_xF_{2+x}$ was crystallized into penetration twinning crystals about 50 μm in size. In the case of 10 wt % NaCl or Na_2SO_4 solutions, only small octahedral crystals were obtained of about 20 μm size. This result revealed the LiCl or LiBr solution to be preferable for the growth of $Sr_{1+x}La_xF_{2+x}$ crystals. Therefore, the hydrothermal growth of $Sr_{1+x}La_xF_{2+x}$ crystals was carried out with the 10 wt % LiCl solution. Fig. 2 shows the size and morphology of $Sr_{1+x}La_xF_{2+x}$ crystals as functions of temperature and temperature gradient in the hydrothermal conditions. The formation region of crystals having a characterized morphology and size, from small octahedra, (hexagon), cubic with twinning, cubic and columnar, to small aggregated cubic, seemed to array in a band from low temperature with a small temperature gradient to high temperature with a large temperature gradient. Optimal conditions to grow larger crystals appeared to be in the regions around 600° C, $\Delta T = 10°$ C and 650° C, $\Delta T \doteq 5°$ C. The largest crystals grown by spontaneous nucleation in the present study were about 0.2 mm in length as shown in Fig. 3. They have a columnar morphology with (1 0 0) faces. A larger crystal will be grown by using a seed and a larger container. Such a study is now in progress.

References

1. D. C. STOCKBARGER, *J. Opt. Soc. Amer.* **39** (1949) 731.
2. A. ROOS, *Mater. Res. Bull.* **18** (1983) 405.
3. K. NASSAU, *J. Appl. Phys.* **32** (1961) 1820.
4. H. GUGGENHEIM, *ibid.* **34** (1963) 2482.
5. D. A. JONES and W. A. SHAND, *J. Cryst. Growth.* **2** (1968) 316.
6. S. SŌMIYA, S. HIRANO, M. YOSHIMURA and K. YANAGISAWA, *J. Mater. Sci.* **16** (1981) 813.
7. K. YANAGISAWA, M. YOSHIMURA and S. SŌMIYA, *ibid.* **17** (1982) 177.
8. M. YOSHIMURA, K. J. KIM and S. SŌMIYA, *Yogyo-Kyokai-Shi (J.Ceram. Soc. Jpn.)* submitted.
9. R. ROY and O. F. TUTTLE, "Physics and Chemistry of the Earth 1" (Pergamon Press, London, New York, 1956) p. 138.

Received 16 April
and accepted 31 May 1984

Reprinted from *Journal of Crystal Growth*, **71** (1985) 277–279.

LETTER TO THE EDITORS

CRYSTAL GROWTH OF YTTRIA STABILIZED ZIRCONIA (YSZ) UNDER HYDROTHERMAL CONDITIONS

Masahiro YOSHIMURA, Takehiro HIUGA and Shigeyuki SŌMIYA

Research Laboratory of Engineering Materials and Department of Materials Science, Tokyo Institute of Technology, 4259 Nagatsuta, Midori, Yokohama 227, Japan

Received 15 May 1984; manuscript received in final form 27 October 1984

Yttria stabilized zirconia (YSZ) crystals were grown on seed crystals under hydrothermal conditions. Using 30 wt% solution of NaOH, one of the best mineralizers, an epitaxial deposition of YSZ 10 μm thick was attained at 700°C with $\Delta T = 10$°C under 140 MPa for 24 h.

Yttria stabilized zirconia (YSZ) crystals have generally been synthesized by skull melting methods [1]. Recently, the synthesis of YSZ crystals by using flux methods [2], floating zone methods [3] and hydrothermal methods [4,5] has been reported. Often, the crystallinity and quality of products grown by hydrothermal methods are better than the quality of materials grown at higher temperatures. In the crystal growth of YSZ, however, the hydrothermal method until now has only produced spontaneous nucleation. We report here the hydrothermal crystal growth of YSZ on a seed crystal.

The seed crystal was a commercial YSZ single crystal synthesized by a skull melting method containing 14 mol% Y_2O_3. It was polished and annealed at 1400°C for 4 h in air. As a feed for the crystal growth, ZrO_2–Y_2O_3 mixed powder containing 14 mol% Y_2O_3 was prepared by 6N NH_4OH co-precipitation of a $ZrOCl_2 + Y(NO_3)_3$ solution. The precipitate was filtered and washed with distilled water and then dried at 125°C for 24 h. Solutions of NaOH (15, 30 wt%), KOH (30 wt%), K_2CO_3 (30 wt%), and NaF (30 wt%) were used as mineralizers. According to our previous investigation [6], YSZ crystals do not destabilize and/or react with these basic solutions. YSZ dissolves and recrystallizes quite clearly from these mineralizers although a small amount of monoclinic ZrO_2 was also observed to precipitate in some cases.

A Pt or Au capsule (2.7 mm ID and 30 mm long) was filled with a mineralizer solution, feed powder was added and then a seed crystal was suspended at the top of the capsule with a Pt wire. After sealing the capsule by welding with an electric arc, the capsule was treated hydrothermally in a test-tube type pressure vessel for 24 h. The temperature of the growth zone at the top of capsule was in the range between 600 and 700°C. The temperature of the dissolution zone at the bottom was 10°C higher ($\Delta T = 10$°C) than that of the top in all cases. YSZ has a positive temperature coefficient for dissolution in all mineralizers used. The pressure for the growth run was between 100 and 140 MPa. After the hydrothermal treatment, the pressure vessel was quenched in cold water. The seed crystal was washed in distilled water and the growth on the seed was analyzed by energy dispersive spectroscopy (SED-880, Daini Seikosha Co. Ltd., Tokyo, Japan). The growth was characterized as YSZ because of the detection of both Zr and Y in the precipitate. The amount of growth of YSZ was determined by a measurement of the weight increase of the seed crystal.

The precipitation of YSZ crystal was observed in all of the mineralizer solutions under the hydrothermal conditions of 600°C, $\Delta T = 10$°C, 100 MPa, 24 h. In the case of the NaOH solution,

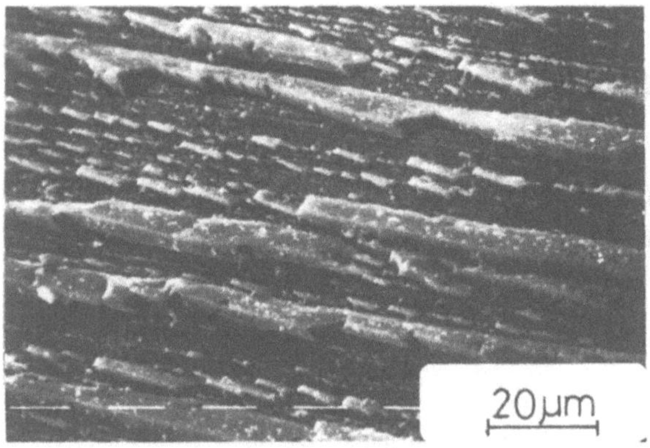

Fig. 1. SEM of YSZ crystals on seed grown hydrothermally at 600°C ($\Delta T = 10$°C) under 100 MPa for 24 h with 15 wt% NaOH solution.

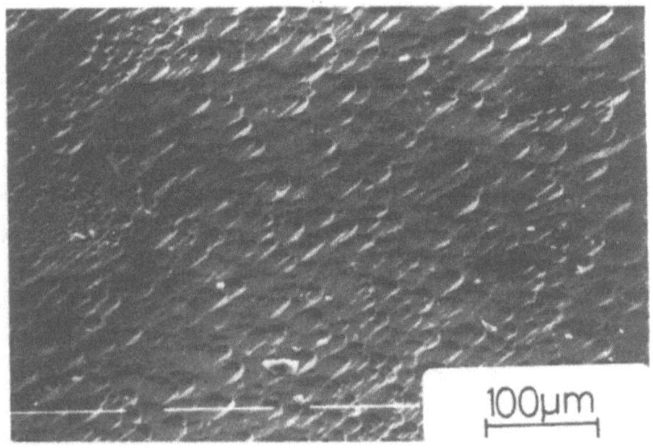

Fig. 3. SEM of YSZ crystal on seed grown hydrothermally at 700°C ($\Delta T = 10$°C) under 140 MPa for 24 h with 30 wt% NaOH solution.

more growth was observed than with the other solutions. The precipitated YSZ crystal was shaped like a terrace along the polishing flaws of the surface of the seed crystal and covered it incompletely. Fig. 1 shows a scanning electron micrograph of YSZ crystals precipitated using a 15 wt% NaOH solution as a mineralizer. The conditions were 600°C $\Delta T = 10$°C, 100 MPa, 24 h.

When the co-precipitated powder containing 14 mol% Y_2O_3 was used as a feed, a slight amount of precipitation of (monoclinic) m-ZrO_2 was also observed in each of the mineralizer solutions. Apparently, the solubility of the Y_2O_3 is larger than that of the ZrO_2 component in hydrothermal solutions. To alleviate this problem, a small amount of

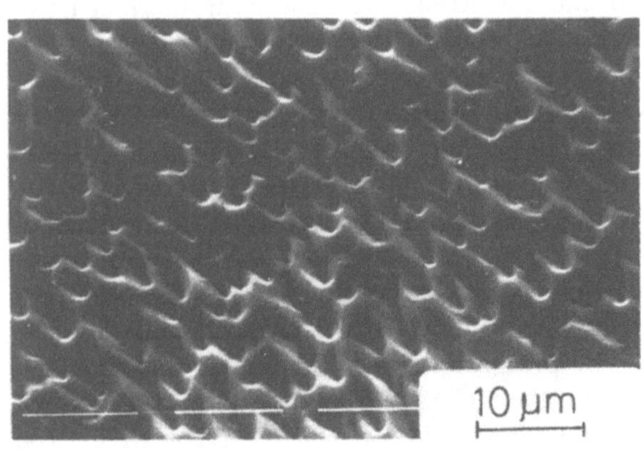

Fig. 2. SEM of YSZ crystal on seed grown hydrothermally at 700°C ($\Delta T = 10$°C) under 140 MPa for 24 h with 15 wt% NaOH solution.

Y_2O_3 powder was added to the 14 mol% Y_2O_3 co-precipitates powder so that the total content of Y_2O_3 in the mixed powder was 18 mol%. In these case, no precipitation of m-ZrO_2 was observed. The crystal growth of YSZ was carried out with the mixed powder containing 18 mol% Y_2O_3. Fig. 2 shows YSZ crystals on the seed grown at 700°C ($\Delta T = 10$°C) under 140 MPa for 24 h with 15 wt% NaOH solution. Under these conditions, the YSZ crystal uniformly covered the entire surface of the seed crystal. We attribute this improvement to the suppression of m-ZrO_2 and to the higher growth temperature.

A back-reflection Laue pattern confirmed that the material grown at 700°C was single crystal YSZ.

Increasing the concentration of NaOH from15 wt% to 30 wt% yielded a smoother surface on the growth YSZ crystals as shown in fig. 3. This also brought about increase YSZ deposition on the seed, about 10 μm thickness in this case for a period of 1 day. Although these growth rates are not rapid enough to contemplate bulk growth they have encouraged us to begin a program of epitaxial growth. These results constitute the fruit of successful seeded growth of YSZ crystals under hydrothermal conditions.

The author would like to thank Professor T. Mori for the work on Laue reflection and the Iwatani Naoji Foundation for financial support.

References

[1] V.I. Aleksandrov, V.V. Osiko, A.M. Prokhorov and V.M. Tatarintsev, in: Current Topics in Material Science, Vol. 1, Ed. E. Kaldis (North-Holland, Amsterdam, 1978) pp. 421–480.

[2] M. Yokoyama, T. Ota and I. Yamai, The Synthesis of Stabilized Zirconia Single Crystal by a Flux Method, in: Proc. 28th Symp. on Artificial Minerals, Kyoto, Japan, 1983, B-3, p. 57.

[3] T. Yamakawa, O. Sakurai, K. Uematsu, N. Mizutani and M. Kato, Growth of Yttria Stabilized Zirconia Crystals by Floating-Zone Method Using an Arc-Imaging Furnace, in: Proc. Ann. Meetings Ceram. Soc. Japan, Kyoto, Japan, 1982, B-5, p. 18.

[4] K. Nakamura, S. Hirano and S. Sōmiya, Am. Ceram. Soc. Bull. 56 (1977) 513.

[5] H. Nishizawa, N. Yamasaki, K. Matsuoka and H. Mitsushio, J. Am. Ceram. Soc. 65 (1982) 343.

[6] M. Yoshimura, T. Hiuga and S. Sōmiya, Hydrothermal Corrosion of Yttria Stabilized Zirconia (YSZ) crystals, in: Proc. Meeting of Zirconia Ceramics '83, T.I.T., Yokohama, Japan, December 2, 1983, Zirconia Ceramics 2 (in Japanese), Ed. S. Sōmiya and M. Yoshimura (Uchida Rokakuho, Tokyo, 1984) pp. 137–147.

Reprinted from *Chemistry Letters* (1985) pp. 1613–1616.

Hydrothermal Crystal Growth of Calcite in NaCl and KCl Solutions

Shin-ichi HIRANO and Kō-ichi KIKUTA

Department of Applied Chemistry, Faculty of Engineering, Nagoya University, Furocho, Chikusa-ku, Nagoya 464

Hydrothermal growth of calcite single crystal was studied by using NaCl and KCl solutions which have been considered to be naturally existing as the saline mineralizer. Both of NaCl and KCl solutions were found to be the effective mineralizer for the growth of calcite crystal of optical grade. Optically clear single crystals of calcite could be grown from each chloride solution.

Calcite ($CaCO_3$) has been used as an optical material in microscopes and has been recently watched with keen interest for the application to laser devices. The high quality single crystal of calcite for optical use, however, has been parched in nature. So, it becomes urgent to grow the optical grade of calcite single crystals.

Many artificial methods were attempted to grow the single crystal of calcite.[1,2] The decomposition of calcite at high temperature makes it difficult to grow the high quality crystals. The hydrothermal growth method is one of the attractive techniques to suppress the dissociation of CO_2 and to lower the strain induced by the thermal stress if the inclusion of the solution is avoided.

In the system of hydrothermal synthesis, the rule to select the solvent has not been defined. In the first place, it is necessary to examine whether single crystals can be grown without any formation of second phase or compound. Some solutions containing Na^+ or K^+ were employed as the solvent to elucidate the effect of the mineralizer. All runs were carried out in a sealed gold capsule (5 mm O.D, 70 mm in length) set in a cone-in-cone type of the hydrothermal reaction vessel.

The crystallized product at 500 °C under 100 MPa with 3 mol kg^{-1} Na_2CO_3 solution was not calcite but unknown compound. Its XRD datum does not agree with those of pirssonite and gaylussite, which were confirmed by Balascio et al.[3]

The second item to be considered is the morphology and perfection of the crystals grown under the hydrothermal condition. In contrast to Na_2CO_3 solution, the experimental run in NaOH or KOH hydrothermal solution did not give any other phase except calcite. The grown crystals, however, showed dendritic form (Fig. 1), which were not acceptable for the optical application.

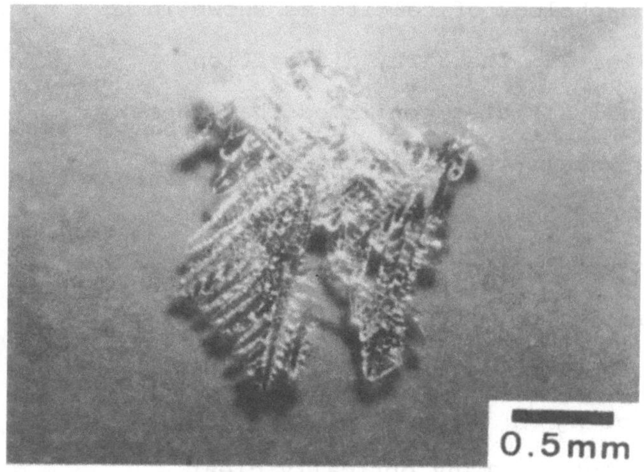

Fig. 1. Photomicrograph of calcite crystal grown from 2 mol kg^{-1} NaOH sol.

As stated above, it turns out from these data that the attention has to be paid for anion species as well as the cation species. The kinds of ions in solvent are the strong influencing factors for the stable phase formation, growth rate and the morphology of the grown crystal.

NaCl and KCl aqueous solutions have been well known as the main component of the naturally existing mineralizer. The NaCl solution must play an important role for the growth of natural calcite single crystals as well known natural saline mineralizer.[4]

The solubility dependence of $CaCO_3$ on the concentration of some hydrothermal chloride solutions were reported by N.Yu. Ikornikova.[5] He reported that the solubility in NaCl solution is much smaller than those in other chloride solutions such as $CaCl_2$ and LiCl.

Ikornikova tried to grow calcite crystals with seed, but did not report the experimental details and results.[6] In the present work, we focused on the elucidation of the NaCl solution as the mineralizer for the growth of calcite single crystal. KCl solution was also employed as the hydrothermal solution for the growth of calcite crystals and evaluation of the calcite stability in the solution.

A preliminary growth run was begun with saturated NaCl solution (6.1 mol

kg^{-1}) at 470 °C (nutrient temperature) under 75 MPa for 7 days.

The single crystals of the size 1 mm could be grown spontaneously on the wall of the gold capsule. The grown single crystals were bounded by typical {10$\bar{1}$1} rhombohedral faces, which is the characteristic morphology of calcite as shown in Fig. 2. The X-ray diffraction analysis also confirmed that these crystals were only calcite. In order to grow more perfect single crystal of calcite without any inclusion, the concentration of NaCl solution had to be lowered to 3 mol kg^{-1}. With 3 mol kg^{-1} NaCl solution, many optically clear calcite single crystal grew at the inner top of the gold capsule at 450 °C under 50 MPa for 7 days. (Fig. 3).

The growth experiment with saturated KCl solution (4.5 mol kg^{-1}) at 470 °C and 75 MPa also could give polyhedral single crystals as shown in Fig. 4. The difference of growth rate on crystallographic plane gave some suggestions when a seed crystal is cut to orient to suitable facet. {10$\bar{1}$1} cleavage crystal is easy to prepare but growth rate on the plane was obviously slow.

In this work, it turned out

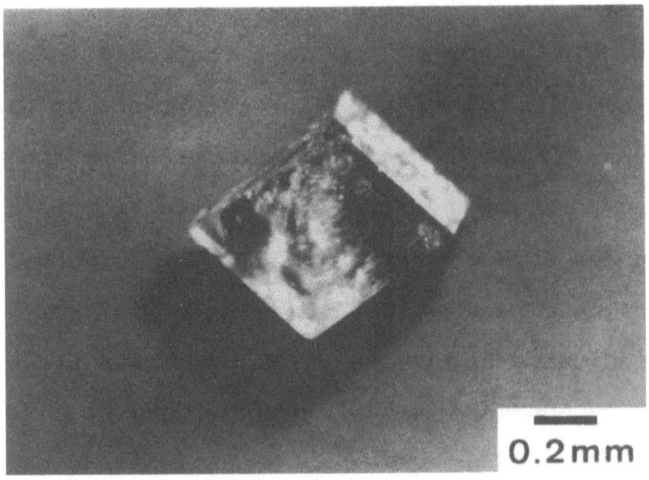

Fig. 2. Photomicrograph of calcite crystal grown from sat. NaCl sol.

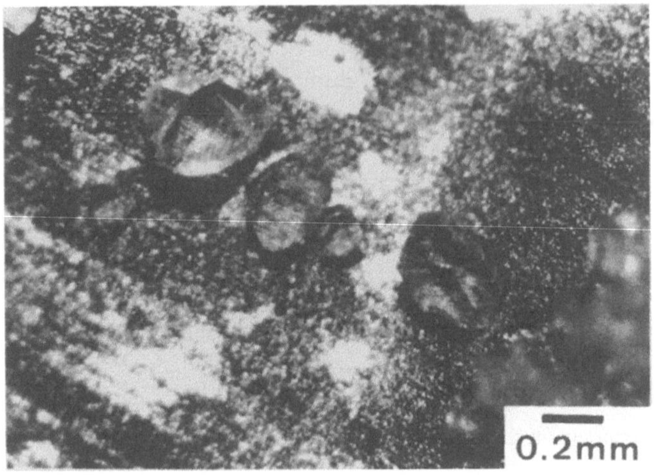

Fig. 3. Photomicrograph of calcite crystal grown from 3 mol kg^{-1} NaCl sol.

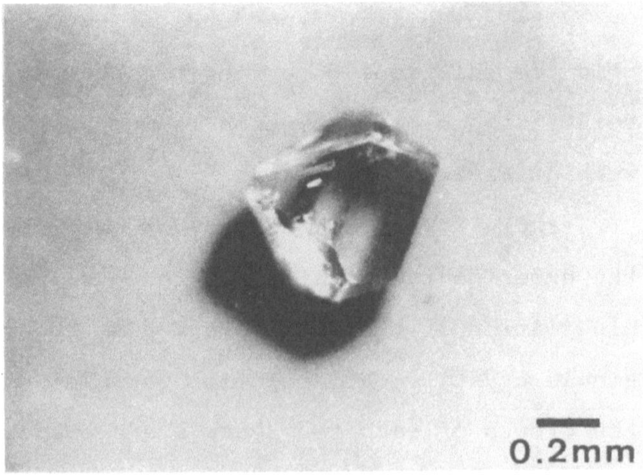

Fig. 4. Photomicrograph of calcite crystal grown from sat. KCl sol.

both NaCl and KCl solutions were employable for hydrothermal synthesis of calcite single crystals. The fact that calcite could be grown in these chloride solutions may support the occurrence of natural single crystal such as Iceland spar.[4)] It is important to determine the solubility of calcite as a function of temperature in order to investigate the behavior of chemical species exsisting under hydrothermal condition and to grow large single crystals. The solubility of calcite was measured in the 3 mol kg^{-1} NaCl solution under 100 MPa. The result indicates an obvious increase in its solubility above 400 °C, and the amount of calcite dissolved under these conditions rises up to 2 g/l at 500 °C, The value of solubility at 500 °C is enough to grow calcite single crystal at the significant growth rate in the industrial scale. On the basis of Arrhenius relationship, we calcutated ΔH to be about 46 kJ/mol. Because KCl solution was found to be one of the useful hydrothermal solvents, further detailed research is also in progress to determine the solubility at various temperature.

References

1) J.F. Balascio and W.B. White, J. Crystal Growth, 23, 101 (1974).
2) C. Belin, J.J. Brissot, and R.E. Jesse, J. Crystal Growth, 13/14, 597 (1972).
3) J.F. Balascio and W.B. White, Mater. Res. Bull., 7, 1461 (1972).
4) L.G. Berry, B. Mason, "Mineralogy," W.H. Freeman and Company, San Francisco (1959), p. 406.
5) N.Yu. Ikornikova, Soviet Phys. Cryst., 5, 726 (1961).
6) N.Yu. Ikornikova, Growth of Crystals, 3, 297 (1962).

(Received July 22, 1985)

Reprinted from *Journal of Crystal Growth*, **78** (1986) 408–412.

HYDROTHERMAL CRYSTAL GROWTH OF $BaPb_{1-x}Bi_xO_3$ ($0 \leqslant x \leqslant 0.30$)

Shin-ichi HIRANO and Seiji TAKAHASHI

Department of Applied Chemistry, Faculty of Engineering, Nagoya University, Furocho, Chikusa-ku, Nagoya 464, Japan

Received 14 February 1986; manuscript received in final form 15 June 1986

$BaPb_{1-x}Bi_xO_3$ ($0 \leqslant x \leqslant 0.30$) single crystals with a very sharp transition to the superconducting state were grown hydrothermally at temperatures below 450 °C. The optimum conditions to grow bulk single crystals of $BaPb_{1-x}Bi_xO_3$ in KCl solution were as follows: 4.5m KCl solution as a solvent; temperatures at the dissolution zone in a gold capsule, 400–450 °C; temperature difference ΔT, 20–30 °C; pressure, 1000 kg/cm^2. The composition of the grown crystals deviated slightly on the lean side of the Bi content from that of the original nutrient oxide particles. It was found that exact control of the composition of the grown crystals could be achieved by using calcined $BaPb_{1-x}Bi_xO_3$ with an adjusted composition x of perovskite structure starting material. The superconducting transition temperature, T_c, changed with the composition x. Single crystals with $x = 0.24$ showed a T_c of 11.7 K, and a transition temperature width of 1.8 K, which is the narrowest value reported so far.

1. Introduction

Crystalline $BaPb_{1-x}Bi_xO_3$ has been found by Sleight et al. [1] to show the superconducting character in the narrow composition region of $0.05 \leqslant x \leqslant 0.30$. The superconducting transition temperature, T_c, is about 13 K when $x = 0.25$. This is the highest T_c for a superconductor not involving a transition element. Thanh et al. [2] reported the electrical resistivity, Hall coefficient and upper critical magnetic field for compositions in this system using hot-press-sintered specimens. Then resistivity measurements indicated semi-metallic conductivity in the composition range $x < 0.35$, and semiconductivity when $x > 0.35$. The carrier density, which was calculated from the Hall coefficient, varied with composition. The carrier concentration was closely related to the change in T_c with x. The value of the carrier density reported in ref. [2] was exceptionally low for superconductors with a high T_c. The Fermi energy E_F and the density of states $N(E_F)$ were estimated by Tani et al. [3], using Thanh's results for the carrier density. Their results also revealed that $N(E_F)$ varied with x and they reported an absolute value of $N(E_F)$ of about 10^{21} states/eV · cm^3 which is one to two orders of magnitude smaller than that of high T_c superconductors. These data suggest that the superconductivity of this system may be attributable to plasmon and exciton effects, and not explained by the BCS theory. Results were obtained using polycrystalline specimens. Khan et al. [4] pointed out the problem that the transition from the normal to the superconducting state could not be fully completed on polycrystalline specimens with $T_c =$ 12.2–6 K. Uncertainties in T_c due to polycrystalline ceramics were also reported by Ito et al. [5] on the basis of the fact that the specific heat anomaly at T_c could not be observed on such samples. These problems have been understood to be caused by the inhomogeneity of polycrystalline specimens or grain boundary effects.

Demands for the growth of homogeneous and well characterized single crystal of $BaPb_{1-x}Bi_xO_3$ have increased because of the need to characterize the intrinsic superconducting features and to use the crystal as a sensitive detector in the infrared region. The low carrier densities characteristics of this oxide superconductor make single crystals applicable for optical devices, especially in the infrared region. $BaPb_{1-x}Bi_xO_3$ single crystals have been grown by the flux method above 900°C [6–10]. The superconducting transition of crystals grown by the flux method, however, was unfortunately broad and the transition temperature width, ΔT_c, was as wide as about 5 K. $BaPb_{1-x}Bi_xO_3$ crystals might be likely to undergo a structural transition and exhibit oxygen difficiency at temperatures above 500°C. The growth of single crystals probably would be required to be carried out below 500°C.

In this paper, the hydrothermal growth of $BaPb_{1-x}Bi_xO_3$ single crystal was performed at temperatures below 450°C to prepare superconducting oxide.

2. Experimental

The control of the composition of grown single crystals was found to be rather difficult when starting with the required mixture of $BaPbO_3$ and $BaBiO_3$ as the nutrient. Crystalline $BaPb_{1-x}Bi_xO_3$ powders with an adjusted x value had to be prepared as the nutrient beforehand for hydrothermal growth as mentioned below. Stoichiometric amounts of $BaCO_3$, PbO_2 and Bi_2O_3 powders (reagent grade) were weighed so as to give the desired composition, mixed throughly, and then calcined for 24 h at 850°C in an oxygen atomosphere. This calcination was repeated twice to complete the synthesis of polycrystalline starting materials. The prepared powders were identified by X-ray diffractometry (XRD). Atomic absorption spectroscopy (AAS) was employed for the determination of the exact composition of $BaPb_{1-x}Bi_xO_3$ starting materials and grown crystals.

Weighed amounts of the starting materials as the nutrient and KCl solution as the solvent were sealed hermatically into a gold capsule, 5.0 mm in outer diameter, 0.15 mm thick and 66 mm in length. The sealed capsule was placed in a test tube-type pressure vessel and was subjected to hydrothermal treatment with a temperature gradient. The hydrothermal growth experiments were carried out at temperatures of 350–450°C at 1000 kg/cm^2 for 7–14 days without any seed crystal. A KCl solution of 4.5 molarity was employed as one of the solvents on the basis of the results of preliminary experiments. A temperature gradient of 3–5°C/cm was used. The temperature in the vessel was measured with a sheathed CA thermocouple, which was calibrated with a standard thermocouple, inserted in the pressure vessel. After a hydrothermal growth run, the pressure vessel was quenched in cold water.

Crystals grown in capsules were observed under an optical microscope and identified with XRD. The chemical composition was measured by atomic absorption spectroscopy on a solution of crystals dissolved in hydrochloric and nitric acids. The superconducting transition was measured using the Meissner effect measuring the transition fraction from the normal to the Meissner state with SQUID equipment.

3. Results and discussion

3.1. Growth of $BaPb_{1-x}Bi_xO_3$ single crystals ($0 \leqslant x \leqslant 0.30$)

$BaPb_{1-x}Bi_xO_3$ single crystals could be successfully grown by the hydrothermal method at temperatures of 400–450°C. The crystals were mainly grown on the inner wall of the upper part (growth zone) of gold capsules. The conditions of the growth of $BaPb_{1-x}Bi_xO_3$ crystal are summarized in table 1. A growth temperature of 350°C was too low to dissolve and crystallize the nutrient in KCl solution. It was found that the optimum condition to grow large single crystal changed with the composition of the oxide system. In the case of growth using the nutrient with $x = 0$, a few large and euhedral crystals were spontaneously grown at 400°C. On the other hand, at 450°C, many small crystals with a rough surface were grown. It turns out that the solubility of $BaPbO_3$ in KCl hydrothermal solution is strongly dependent on the temperature between 400 and 450°C. The optimum temperature to grow a large and euhedral crystal of $BaPbO_3$ with a rather smooth surface was 400°C in 4.5 molarity KCl solution at 1000 kg/cm^2. With increase in x of $BaPb_{1-x}Bi_xO_3$, a higher growth temperature was required to grow large and euhedral single crystals. Growth temperatures above 400°C were needed to grow single crystals with $x = 0.30$. These phenomena indicate that the solubility of $BaPb_{1-x}Bi_xO_3$ in 4.5m KCl

Table 1
Results of hydrothermal experiments (in 4.5m KCl solution at 1000 kg/cm^3 for 7–14 days)

Temperature (°C)	Composition x of nutrient					
	0	0.10	0.20	0.25	0.30	0.35
350	a	a	a	a	a	a
400	b	b	b	b	b	c
450	d	d	d	d	d	b

(a) No crystal in growth zone of gold capsule.
(b) A few large crystals in growth zone of gold capsule.
(c) A few small crystals in growth zone of gold capsule.
(d) Many small crystals in growth zone of gold capsule.

solution under hydrothermal conditions changed with the composition of this system. Measurement of solubility is in progress.

An important observation also was the shift of the composition of the grown crystals from that of the starting nutrient oxide. This relation is shown in fig. 1. The composition of the grown crystal was shifted to the Bi lean side from the composition of the original nutrient. Fortunately the linear relationship between the compositions of the nutrient and the grown crystal, which was very reproducible, makes it easy to adjust the exact composition of the grown single crystal. Therefore, the composition of the grown crystal can be controlled by selecting beforehand the proper composition of the nutrient oxide prepared by ordinary ceramics processing with an increase of the composition to Bi rich side by a factor of 1.2.

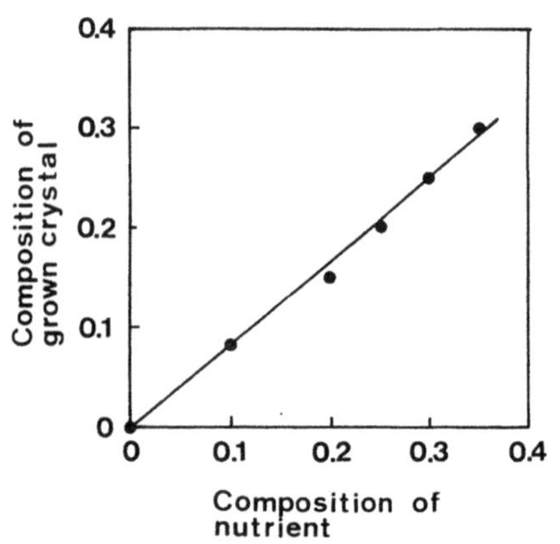

Fig. 1. Relation between compositions of nutrient and grown crystal.

Microphotographs of the grown crystals are shown in fig. 2. The maximum size of the grown crystal was about $3 \times 3 \times 1$ mm^3 by the spontaneous nucleation process with a gold capsule of 4.7 mm in inner diameter. The color of the crystal with $x = 0$ is black and turns into bronze as the composition x increases to 0.30.

3.2. Properties of grown single crystal

The superconducting property of the grown single crystal with $x = 0.24$ was measured using the Meissner effect. Fig. 3 shows the temperature–susceptibility curve under 1.6 G. The crystal began to transform abruptly to the Meissner state at 11.7 K and reached 100% of the superconducting state at 9.9 K. The result shows that the superconducting temperature T_c of the grown crystal with $x = 0.24$ was 11.7 K and the transition width ΔT_c was 1.8 K. This transition width was the narrowest and the transition to the superconducting state was the sharpest among the single crystals grown so far.

The composition of the grown single crystal with $x = 0.24$ was carefully analyzed relative to stoichiometry by the atomic absorption spectroscopic method. Oxygen defficiency of the crystal was undetectable for $Ba_{1.00}Pb_{0.76}Bi_{0.24}O_{3.0}$. Inhomogeneity of the atomic distribution could not be observed with electron microprobe analysis. Therefore, it was considered that the single crystals grown hydrothermally were very homogeneous. Thus hydrothermal method seems to be the most promising method for the growth of homogeneous $BaPb_{1-x}Bi_xO_3$ single crystals.

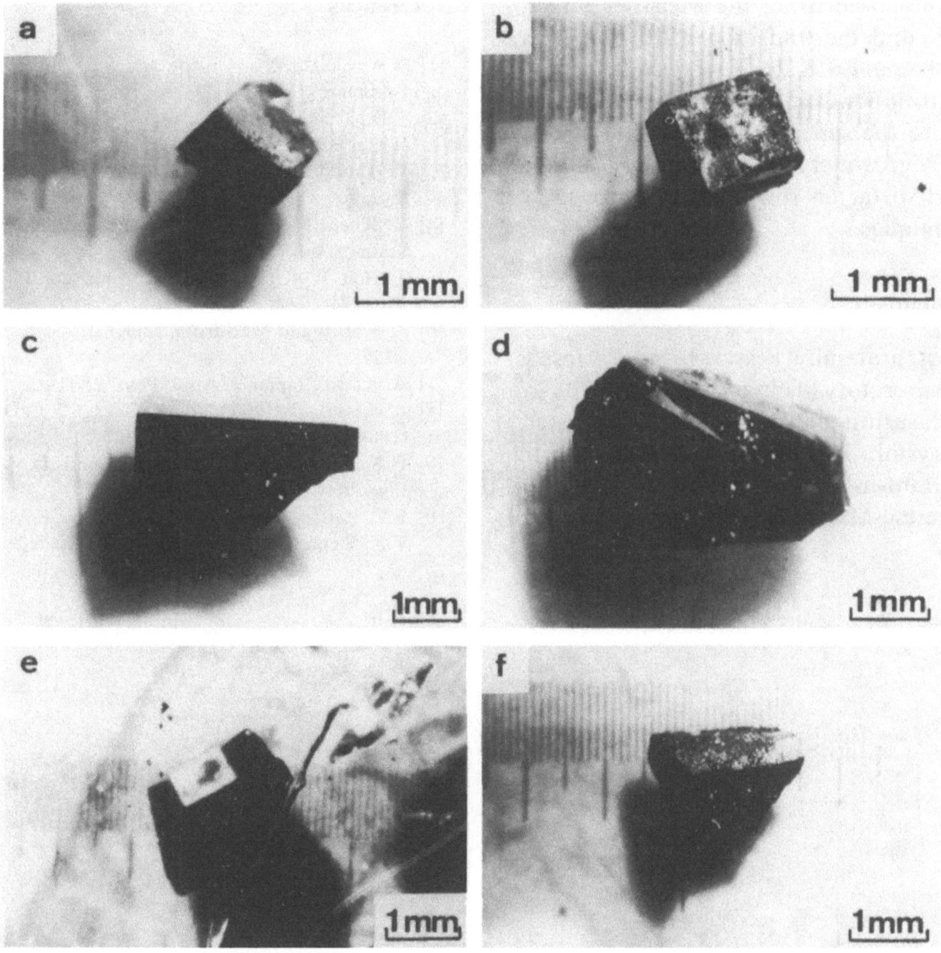

Fig. 2. Microphotographs of grown crystals with: (a) $x = 0$; (b) $x = 0.08$; (c) $x = 0.15$, (d) $x = 0.20$; (e) $x = 0.24$; (f) $x = 0.30$.

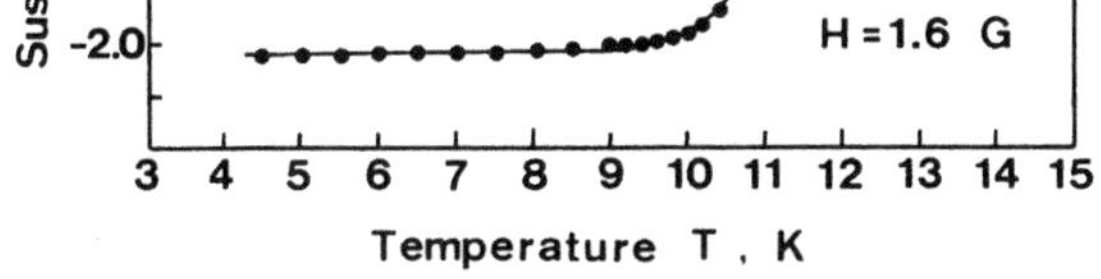

Fig. 3. Temperature–susceptibility curve of the grown crystal with $x = 0.24$.

4. Conclusions

$BaPb_{1-x}Bi_xO_3$ $(0 \leqslant x \leqslant 0.30)$ single crystals could be successfully grown by the hydrothermal method in 4.5m KCl solution at temperatures of 400–450 °C, and 1000 kg/cm^2 in 7–14 days.

The composition of the grown crystal was poorer in Bi than the nutrient by a factor of 0.8. Therefore, the composition of the grown crystals can be controlled readily by selecting the appropriate composition of starting nutrient prepared by ceramics processing of $BaPb_{1-x}Bi_xO_3$ of perovskite structure.

The superconducting transition temperature T_c of the single crystals grown by this hydrothermal method was measured using the Meissner effect. T_c was 11.7 K and the transition width ΔT_c was very narrow, being 1.8 K for a crystal with $x =$ 0.24. The transition width was the narrowest and the transition to the superconducting state of this hydrothermally grown crystal was sharpest among those reported so far on single crystals grown by the other techniques.

Acknowledgements

The authors gratefully acknowledge Professor S. Tanaka's Laboratory (Tokyo Uinversity, Tokyo, Japan) for measuring the superconducting properties of the crystal. This work was supported in part by a Grant-in-Aid for Scientific Research from the Japanese Ministry of Education, Science, and Culture.

References

[1] A.W. Sleight, J.L. Gillson and P.E. Bierstedt, Solid State Commun. 17 (1975) 27.
[2] T.D. Thanh, A. Koma and S. Tanaka, Appl. Phys. 22 (1980) 205.
[3] T. Tani, T. Itoh and S. Tanaka, J. Phys. Soc. Japan 49 Suppl. A, (1980) 309.
[4] Y. Khan, K. Nahm, M. Rosenberg and H. Willer, Phys. Status Solidi (a) 39 (1977) 79.
[5] T. Itoh, K. Kitazawa and S. Tanaka, J. Phys. Soc. Japan 53 (1984) 2668.
[6] A. Katsui and M. Suzuki, Japan. J. Appl. Phys. 21 (1982) L157.
[7] A. Katsui, Japan. J. Appl. Phys. 21 (1982) L553.
[8] A. Katsui, Y. Hidaka and H. Takagi, J. Crystal Growth 66 (1984) 228.
[9] G.K. Wertheim, J.P. Remeika and D.N.E. Buchanan, Phys. Rev. B26 (1982) 2120.
[10] S.V. Zaitsev-Sotov, A.V. Kuznetsov, E.A. Protasov and V.N. Stepankin, Soviet Phys.-Solid State 26 (1984) 1928.

Reprinted from *Journal of Crystal Growth*, **79** (1986) 223–226.

SOLUBILITY AND HYDROTHERMAL GROWTH OF CALCITE SINGLE CRYSTAL IN NITRATE SOLUTIONS

Shin-ichi HIRANO and Kō-ichi KIKUTA

Department of Applied Chemistry, Faculty of Engineering, Nagoya University, Furo-cho, Chikusa-ku, Nagoya 464, Japan

The hydrothermal synthesis of calcite single crystals was studied using strong electrolyte solutions mineralizers. Alkali nitrate solutions ($NaNO_3$, KNO_3) were found to be effective solvents. The solubility of calcite in 3m (molarity) $NaNO_3$ solutions was measured by the weight loss method and was around 2.2 g/l at 450 °C and 100 MPa, which was twice as high as that in 3m NaCl solution. The (0001) plane tended to develop in $NaNO_3$ hydrothermal solutions at higher temperatures, above 500 °C, while rhombohedral $(10\bar{1}1)$ planes developed predominately on crystals grown below 500 °C.

1. Introduction

Calcite single crystals, known as Iceland spar, have been used as an optical material for a long time. Recently, the demand for optically clear calcite single crystals has been increasing, with the development of the laser devices such as an optical isolator, since natural resources have been depleted.

Many attempts have been made to grow calcite single crystals [1–3]. The quality of the grown crystals, however, was not acceptable for optical applications owing to the large number of dislocations and impurities introduced from the flux.

The hydrothermal method generally has the advantage of suppressing the dissociation of CO_2 and yielding high quality crystals with less thermal strain, in a homogeneous ambient, at relatively low temperatures. It is well known that Iceland spar grew in hydrothermal solutions in nature [4].

Ikornikova reported the hydrothermal synthesis of calcite single crystal in NaCl and other chloride solutions [5,6]. Growth experiments with alkali carbonate solutions have been carried out by Balascio and White [7] and Kinloch et al. [8]. The authors reported that the euhedral clear calcite single crystals could be grown in NaCl and KCl hydrothermal solutions [9].

In the present work, nitrate solutions were found to be effective solvents for calcite growth. In order to understand and improve the hydrothermal method for the growth of calcite, the solubility and the phase stability of calcite were studied in hydrothermal nitrate solution, compared with NaCl solution.

2. Experimental

Both reagent grade calcite powder (Wako Pure Chemical Industries, Ltd.) and highly pure natural single crystals were used as starting materials. In this work, an hydrothermal pressure vessel of the cold seal type ($\frac{1}{4}$inch ID) was employed, with a gold capsule as the specimen container.

The temperature was controlled by a calibrated chromel-alumel thermocouple set in the pressure vessel. The pressure was measured by a calibrated Heise gauge. The solubility of calcite was determined in the gold capsule (5.0 mm OD, 4.7 mm ID, 50 mm in length) with cleaved Iceland spar crystals about 3 mm in size. Crystals and 3m $NaNO_3$ solvent were sealed in the capsule by welding and then placed in the pressure vessel. Solubility runs were maintained, under each condition, for 3 days, which had been confirmed previously to be long enough bring the system to equilibrium. Following a run the pressure vessel was quenched with cold water. A stainless steel filler rod was also placed in the space above the capsule so as to reduce convection and achieve thermal uniformity of the capsule.

The phase of the crystals, after a run, was identified by X-ray diffraction technique and the morphology of the grown crystals was observed by optical microscopy. The grown crystals were also characterized by infra-red spectroscopy.

3. Results and discussion

3.1. Selection of hydrothermal solvent and stability of calcite

The selection and comparison of the hydrothermal solvents were made on the basis of the stability and morphology of the calcite crystals. Experiments were carried out at constant conditions of 500°C and 100 MPa for 3 days with reagent grade powder of calcite. The results are shown in table 1. X-ray diffraction analysis of the crystals grown in these solutions showed that calcite did not change to any other phase or compound except when the solution used was 3m Na_2CO_3. In this case, the calcite changed to an unknown compound which was neither pirssonite nor gaylussite, as reported by Balascio and White [7].

Dendritic crystals were grown in NaOH and KOH solutions. It was very difficult to grow the bulk crystals bounded by flat surfaces in these alkali hydroxide solutions. The results shown in table 1 indicate that the anion group in the salt solution is a more dominant factor than is the type of cation in the solvent used for the growth of calcite single crystals.

Table 1
Results of hydrothermal treatment of calcite in several solutions [a)]

Solution	Crystal form	XRD
3m NaCl	Polyhedral	Calcite
2m KCl	Polyhedral	Calcite
3m $NaNO_3$	Polyhedral	Calcite
2m KNO_3	Polyhedral	calcite
2m NaOH	Dendritic	Calcite
2m KOH	Dendritic	Calcite
3m Na_2CO_3	–	Unknown compound

[a)] 500°C, 100 MPa, 3 days.

It has been confirmed that both alkali chloride and alkali nitrate solutions were useful for growing calcite crystals as the single phase. Nitrate solutions showed a higher solubility for calcite and were therefore selected as being superior to chloride solutions.

3.2. Solubility of calcite in 3m $NaNO_3$ solutions

The solubility of calcite in 3m $NaNO_3$ solution was measured as a function of temperature under 100 MPa by the weight loss method (fig. 1). The solubility of calcite in 3m NaCl is also given in this figure for comparison. The solubility in 3m $NaNO_3$ increased abruptly with increasing temperature and it amounted to 2.2 g/l at 450°C. This solubility is acceptable for the growth of calcite single crystal by the hydrothermal technique. Since the value of the solubility in 3m $NaNO_3$ is about twice that for 3m NaCl, $NaNO_3$ solutions were judged to be suitable for the growth at lower temperatures.

Fig. 2 shows the plot of log S (solubility) versus $1/T$. The data show that the Van 't Hoff law is not obeyed in this system. This fact suggests that

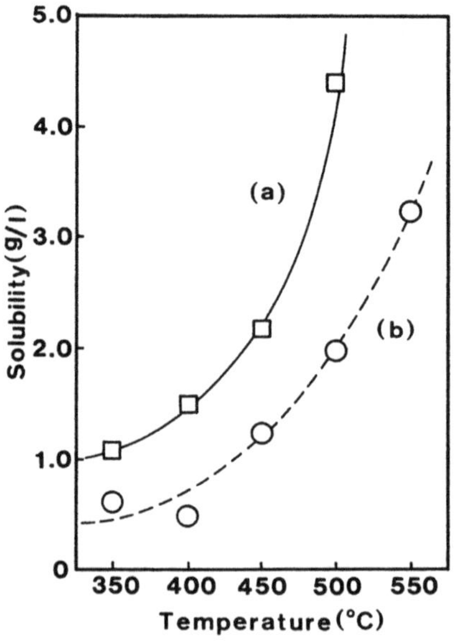

Fig. 1. Solubility of calcite in 3m $NaNO_3$ solution as a function of temperature: (a) in 3m $NaNO_3$ solution (□); (b) in 3m NaCl solution (○).

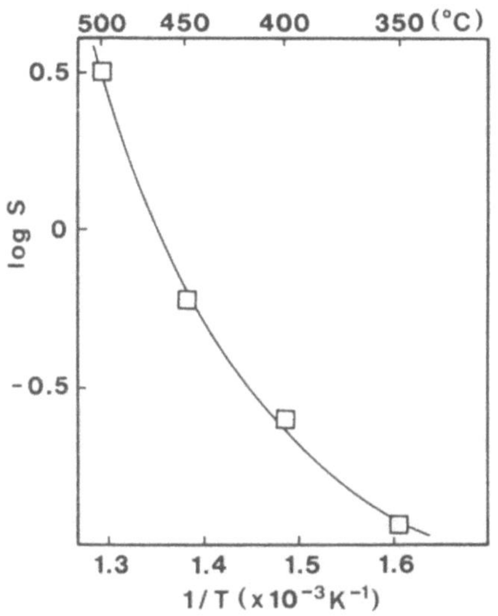

Fig. 2. Log S (solubility) of $CaCO_3$ as a function of $1/T$ in 3m $NaNO_3$.

the state of the dissolved species may change with increasing temperature.

3.3. Morphology and IR spectra of grown calcite single crystals

Some growth experiments were carried out with highly pure Iceland spar as the starting material. Single crystals to 0.6 mm in size could be prepared spontaneously on the inner wall of the capsule at 400°C and 100 MPa for 10 days (fig. 3). The crystals were mainly bounded by $\{10\bar{1}1\}$ faces. The development of smooth $\{10\bar{1}1\}$ faces lead to a lowering of the growth rate. For example, the average growth rate on the $(10\bar{1}1)$ seed crystal was about 0.006 mm/day at 400°C. To grow a large single crystal, it is necessary to employ a seed crystal with the correct orientation, such as {0001} and $\{11\bar{2}0\}$ faces. The crystal growth experiments above 500°C revealed that the crystals were grown with well developed {0001} faces (fig. 4). This tendency in the morphological changes of the grown crystal agree with those in naturally occurring Iceland spar. Sunagawa showed that the plate-like {0001} calcite crystal might occur at elevated temperatures in the first stage of the precipitation of Iceland spar [10]. It has been

Fig. 3. Photomicrograph of a calcite crystal grown in 3m $NaNO_3$ solution at 400°C and 100 MPa for 10 days.

recognized generally that the variation of the morphology is closely related to the degree of supersaturation, impurities and the dissolved species, etc. Further detailed studies are now in progress to elucidate these problems.

Infra-red spectra of the grown single calcite crystals showed that they were identical to highly perfect natural calcite crystals with a stretching vibration around 3500 cm^{-1}, due to trace OH^- (fig. 5). The up-take of OH^- in these hydrothermally grown crystals is not considered to be serious, compared with high quality natural calcite single crystals which are useful for optical devices.

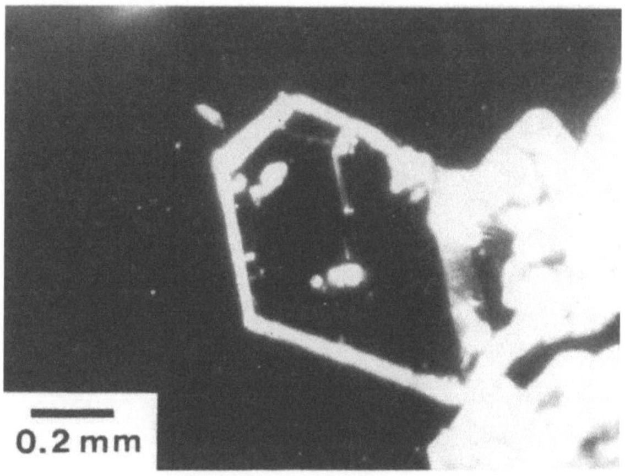

Fig. 4. Photomicrograph of calcite crystals grown in 3m $NaNO_3$ solution at 550°C and 100 MPa for 5 days.

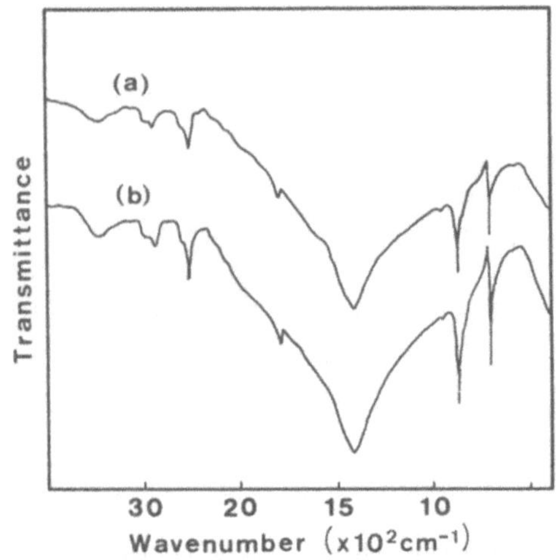

Fig. 5. IR spectra of calcite single crystals: (a) hydrothermally grown crystal in $NaNO_3$ solution; (b) natural Iceland spar crystal.

4. Conclusion

Both $NaNO_3$ and KNO_3 solutions were found to be acceptable as hydrothermal solvents for growing polyhedral calcite single crystals, without formation of second phase inclusions under rather easier conditions than for NaCl solution.

Plate-like single crystals, bounded by {0001} faces, could be prepared at elevated temperatures above 500°C, which indicates good agreement with the mineralogical appearance of natural Iceland spar.

The characteristic IR spectra of the grown single crystals of calcite were identical to those for high quality natural Iceland spar, thus showing a comparable trace uptake of OH^- in the grown crystals.

References

[1] P.M. Gruzensky, in: Crystal Growth, Ed. H.S. Peiser (Pergamon, Oxford, 1967) p. 365.

[2] J.F. Balascio and W.B. White, J. Crystal Growth 23 (1974) 101.

[3] C. Belin, J. Crystal Growth 34 (1976) 341.

[4] L.G. Berry and B. Mason, Mineralogy (Freeman, San Francisco, 1959).

[5] N.Yu. Ikornikova, Soviet Phys.-Cryst. 5 (1961) 726.

[6] N.Yu. Ikornikova, Growth of Crystals 3 (1962) 297.

[7] J.F. Balascio and W.B. White, Mater. Res. Bull. 7 (1972) 1461.

[8] D.R. Kinloch, D.F. Belt and R.C. Puttbach, J. Crystal Growth 24/25 (1974) 610.

[9] S. Hirano and K. Kikuta, Chem. Letters (1985) 1613.

[10] I. Sunagawa, N. Morimoto and A. Miyashiro, Kobutsugaku (Iwanami-shoten, Tokyo, 1975).

Part 4

Crystallization

Reprinted from *Journal of the American Ceramic Society*, **64**(12) (1981) C-181.

Hydrothermal Preparation of Ultrafine Monoclinic ZrO_2 Powder

EIJI TANI, MASAHIRO YOSHIMURA,* AND SHIGEYUKI SŌMIYA*

Research Laboratory of Engineering Materials, and the Department of Materials Science and Engineering, Tokyo Institute of Technology, Nagatsuta, Yokohama, 227 Japan

Ultrafine powder of single-phase monoclinic ZrO_2 was prepared by hydrothermal treatments of amorphous hydrated zirconia with 8 wt% KF solution under 100 MPa at 200° to 500°C for 24 h. The process yielded well-crystallized particles 16 nm and 22 nm in size at 200° and 500°C, respectively.

ULTRAFINE ZrO_2 powder with a particle size of <30 nm has been intensively sought as a raw material for modern zirconia ceramics. Such powders have been prepared by thermal decomposition of amorphous hydrated zirconia[1-6] and zirconium salts,[2,5,7] or by vapor reaction.[8] In these methods, however, tetragonal ZrO_2 was produced in addition to monoclinic ZrO_2 at low temperatures. The stabilization of the tetragonal ZrO_2 has been considered to result from particle size,[2,8] impurities,[1,6,7] topotactic crystallization,[3,5] and/or strain,[6] however is not completely understood.

Mitsuhashi *et al.*[6] treated amorphous hydrated zirconia with distilled water at 98 MPa from 215° to 245°C for 10 min to 22 h and obtained ≈80% tetragonal ZrO_2. The powders obtained by the hydrothermal treatments have been strain-free, uniform single-domain particles with an average crystallite size of 10 to 30 nm.

The authors have studied the hydrothermal treatment of amorphous hydrated zirconia to prepare uniform ultrafine ZrO_2 powder. Pure ultrafine monoclinic ZrO_2 powder has been prepared by this method using a KF solution as a mineralizer.

The starting material was amorphous hydrated zirconia precipitated from Hf-free $ZrCl_4$* solution with NH_4OH, then washed with distilled water and dried at 120°C for 48 h. A Pt or Au tube (3 mm ID, 0.15 mm thick, and 32 mm long) filled with the starting material and 8 wt% KF solution was treated under 100 MPa at 200° to 600°C for 24 h using a test tube type pressure vessel of Stellite 25, (6.5 mm ID). After these hydrothermal treatments, the vessel was quenched in cold water. The products were washed in water and examined with an X-ray powder diffractometer.† The crystallite sizes were calculated from the Scherrer formula using the (111) diffraction peak of monoclinic ZrO_2 in the products; $D=0.9\ \lambda(\beta\ \cos\ \theta)^{-1}$, where D is the crystallite size, λ the wavelength of X-ray ($CuK\alpha$), β the corrected half-width using low quartz as a standard, and θ is the diffraction peak angle.

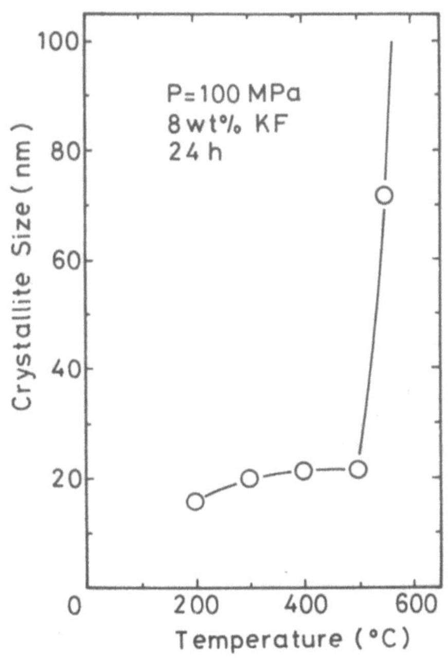

Fig. 1. Variation with temperature of particle size of monoclinic ZrO_2 produced by hydrothermal treatment under 100 MPa for 24 h using KF solution.

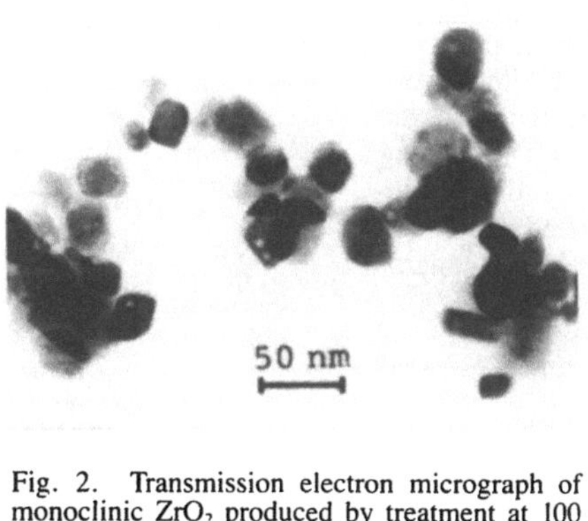

Fig. 2. Transmission electron micrograph of monoclinic ZrO_2 produced by treatment at 100 MPa and 500°C for 24 h using KF solution.

The product was confirmed to be monoclinic ZrO_2 phase by X-ray powder diffraction. Potassium and other cationic impurities were not detected in the products by energy dispersive spectrometry. Figure 1 shows that the crystallite size of monoclinic ZrO_2 produced was ≈16 nm at 200°C, and it increased gradually up to ≈22 nm at 500°C. Abrupt increase in the crystallite size has been observed above this temperature, i.e. 70 nm at 550°C and >100 nm at 600°C. This phenomenon corresponds to the fact that crystal growth of ZrO_2 in KF solution appears to be significant above 600° to 650°C as reported by Kuznetzov.[9]

Figure 2 shows the transmission electron micrograph‡ of monoclinic ZrO_2 formed by the hydrothermal treatment under 100 MPa at 500°C for 24 h. Electron diffraction indicated each particle to be crystalline. The figure shows particles to be uniform in shape and in size (≈20 nm) which agrees with the crystallite size obtained by X-ray diffraction.

Use of the KF solution as a mineralizer yields only monoclinic ZrO_2 by hydrothermal treatments. Use of H_2O or LiCl solutions, however, produced tetragonal ZrO_2 as well as monoclinic ZrO_2 under the same conditions. An investigation on the stabilization of tetragonal ZrO_2 under hydrothermal conditions is in progress.

CONTRIBUTING EDITOR—P. K. GALLAGHER

Received June 24, 1981; revised copy received August 6, 1981.

Supported by Grant-in-Aid for Scientific Research, No. 443015, and by Iwatani Naoji Foundation.

*Member, the American Ceramic Society.

*Hf-free $ZrCl_4$, (Sn–680 ppm, Fe–620 ppm, Hf<70 ppm, Al<30 ppm), Nippon Mining Co., Ltd., Tokyo, Japan.

†RU-200, Rigaku Denki Co., Ltd., Tokyo, Japan.

Acknowledgment

The authors thank JEOL Co., Ltd. for providing the opportunity to use the transmission electron microscope.

References

[1] R. Cyprès, R. Wollast, and J. Raucq, "Contribution on the Polymorphic Conversion of Pure Zirconia," *Ber. Deut. Keram. Ges.*, **40** [9] 527–32 (1963).

[2] R. C. Garvie, "Occurrence of Metastable Tetragonal Zirconia as a Crystallite Size Effect," *J. Phys. Chem.*, **69** [4] 1238–43 (1965).

[3] J. Livage, K. Doi, and C. Mazières, "Nature and Thermal Evolution of Amorphous Hydrated Zirconium Oxide," *J. Am. Ceram. Soc.*, **51** [6] 349–53 (1968).

[4] G. Katz, "X-ray Diffraction Powder Pattern of Metastable Cubic ZrO_2," *ibid.*, **54** [10] 531 (1971).

[5] Y. Murase, E. Kato, and H. Matsumoto, "Crystallization and Phase Transformation of Low Temperature Form Zirconia Prepared from Different Starting Materials," *Nippon Kagaku Kaishi*, **1972**, No. 12, 2329–36.

[6] T. Mitsuhashi, M. Ichihara, and U. Tatsuke, "Characterization and Stabilization of Metastable Tetragonal ZrO_2," *J. Am. Ceram. Soc.*, **57** [2] 97–101 (1974).

[7] E. D. Whitney, "Kinetics and Mechanism of the Transition of Metastable Tetragonal to Monoclinic Zirconia," *Trans. Faraday Soc.*, **61** [9] 1991–2000 (1965).

[8] Y. Suyama, T. Mizobe, and A. Kato, "ZrO_2 Powders Produced by Vapor Phase Reaction," *Ceramurgia Int.*, **3** [4] 141–46 (1977).

[9] V. A. Kuznetzov, "Crystallization of Titanium, Zirconium and Hafnium Oxides and Some Titanate and Zirconate Compounds Under Hydrothermal Conditions," *J. Cryst. Growth*, **3,4**, 405–10 (1968). □

‡JEM-100CX, JEOL Co., Ltd., Tokyo, Japan.

Reprinted from *Journal of the American Ceramic Society*, **65**(7) (1982) 343–346.

Crystallization and Transformation of Zirconia Under Hydrothermal Conditions

H. NISHIZAWA, N. YAMASAKI, AND K. MATSUOKA

Department of Chemistry, Faculty of Science

H. MITSUSHIO

Department of Geology, Faculty of Science, Kochi University, Kochi, Japan 780

During constant-rate heating to 350°C in concentrated NaOH solutions, cubic ZrO_2 crystallized at ≈120°C from hydrated amorphous ZrO_2; these fine cubic ZrO_2 particles abruptly changed into needlelike monoclinic ZrO_2 single crystals at 300°C. Crystallization and phase transformation were studied by XRD, TEM, and EPMA. Cubic ZrO_2 appears to crystallize via collapse of the $ZrO_2{\cdot}nH_2O$ structure and subsequent slight rearrangement of the lattice. The abrupt formation of monoclinic ZrO_2 was considered to result when the very fine cubic ZrO_2 particles coagulated in a highly oriented fashion.

I. Introduction

THE kinetics and mechanism of solid-state crystallization and transformation in ZrO_2 at low temperatures have been well studied.[1–7] Less attention has been paid to the relation between the polymorphs under aqueous hydrothermal conditions. Cypres *et al.*[1] proposed that OH^- or other anion impurities stabilize the tetragonal form at room temperature, but Clearfield[8] obtained monoclinic ZrO_2 by refluxing hydrated amorphous ZrO_2 in distilled water. Mitsuhashi *et al.*[9] found that tetragonal ZrO_2, plus a small amount of the monoclinic form, crystallized abruptly at 235°±10°C in distilled water under 100 MPa. The single-domain tetragonal particles obtained were strain-free and transformed easily to the monoclinic phase. These investigators concluded that one single-domain tetragonal crystal transforms martensitically to one single-domain monoclinic crystal in fine ZrO_2 particles.

Recently, the present workers obtained isothermal transformation curves for ZrO_2 at 250° to 400°C under hydrothermal conditions. The experimental results indicate that this transformation occurs not through dissolution and precipitation but via solid-state lattice deformation. In the present work, the effects of mineralizing agents, e.g. NaOH, on crystallization and transformation were investigated. Constant-rate heating and rapid quenching were used to observe abrupt crystallization and transformation.

II. Experimental Procedure

Amorphous hydrated ZrO_2 precipitated from $ZrOCL_2{\cdot}8H_2O$* solution with NH_3* was washed repeatedly with distilled water until the wash water did not become turbid when 1*M* $AgNO_3$ was added, then dried with an ir heater. The pulverized sample was inserted into a microautoclave† with 5 cm^3 of aqueous NaOH‡ and reacted at temperatures up to 350°C at the autogeneous pressure§ of the autoclave. The heating rate was controlled at 6°C/min from 50° to 350°C. After the desired temperature was reached, the autoclave was quenched rapidly to room temperature by immersing it in ice water. The products were washed several times with distilled water, oven-dried at 60°C, and identified by X-ray diffractometry¶ using 99.99% Si powder as an internal standard.

The monoclinic, tetragonal, and cubic forms were detected. The X-ray diffraction (XRD) patterns showed broad diffuse lines, indicating a very fine crystallite size. Despite the broadened lines, a very complete pattern was observed. The lines from $2\Theta=72°$ to 76° were used to distinguish between the tetragonal and cubic forms (Fig. 1). The effective crystallite size of the zirconia was calculated from Scherrer's equation with Warren's correction, using the half-widths of the cubic (tetragonal) (111) and monoclinic $(11\bar{1})$ reflections.[10] This method reportedly gave a precision of ±10% at 10 to 100 nm for a related series of samples, with an absolute accuracy of ±25%.[4]

The hydrothermally prepared samples generally consist of particles with a single-domain structure.[9] The samples were ultrasonically dispersed in acetone for a few seconds and poured into distilled water. A small drop was allowed to dry on a collodion film on a Cu-mesh screen for TEM examination.‖ Some collodion films were supported by evaporated carbon. The domain size of the ZrO_2 particles, as estimated by electron microscopy, was perfectly consistent with the calculated effective crystallite size. The percent of cubic (tetragonal) phase in the sample was calculated from the relative intensities of the two monoclinic peaks ($11\bar{1}$ and 111) and the cubic (tetragonal) peak (111)[11]; a similar approach was used by Mitsuhashi *et al.*[9] for tetragonal ZrO_2 powder.

‖JEM-100U, Japan Electron Optics Co., Ltd., Tokyo, Japan.

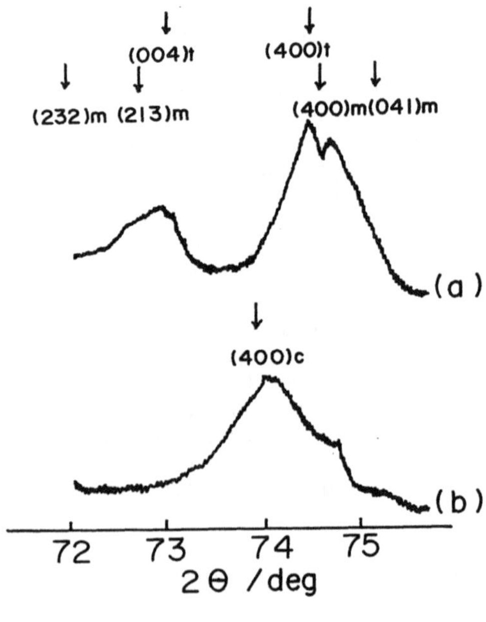

Fig. 1. X-ray diffraction patterns of samples hydrothermally treated (*a*) in pure water to 350°C (60% tetragonal and 40% monoclinic) and (*b*) in 1 molal NaOH to 210°C (100% cubic). Diffraction peaks marked by arrows are indexed on the basis of $a=0.5148$ nm, $b=0.5203$ nm, $c=0.5315$ nm, and $\beta=99°23'$ for the monoclinic form, $a=0.509$ nm, $c=0.518$ nm for the tetragonal form, and $a=0.5119$ nm for the cubic form.

Received June 15, 1981, revised copy received January 25, 1982.

*Guaranteed reagents with metallic impurities of 0.004 and 0.0003%, respectively; Wako Pure Chemical Industry, Ltd., Osaka, Japan.

†Lined with Hastelloy-C, Cabot Corp., Kokomo, IN; capacity 8 cm^3.

‡Prepared from reagent-grade sodium hydroxide (Wako Pure Chemical Industry, Ltd.) and redistilled water as concentrated (32 g/200 g H_2O, 4.0 molal) and medium (8 g/200 g H_2O, 1.0 molal) solutions.

§6 and 16 MPa at 250° and 350°C, respectively.

¶Ni-filtered Cu radiation.

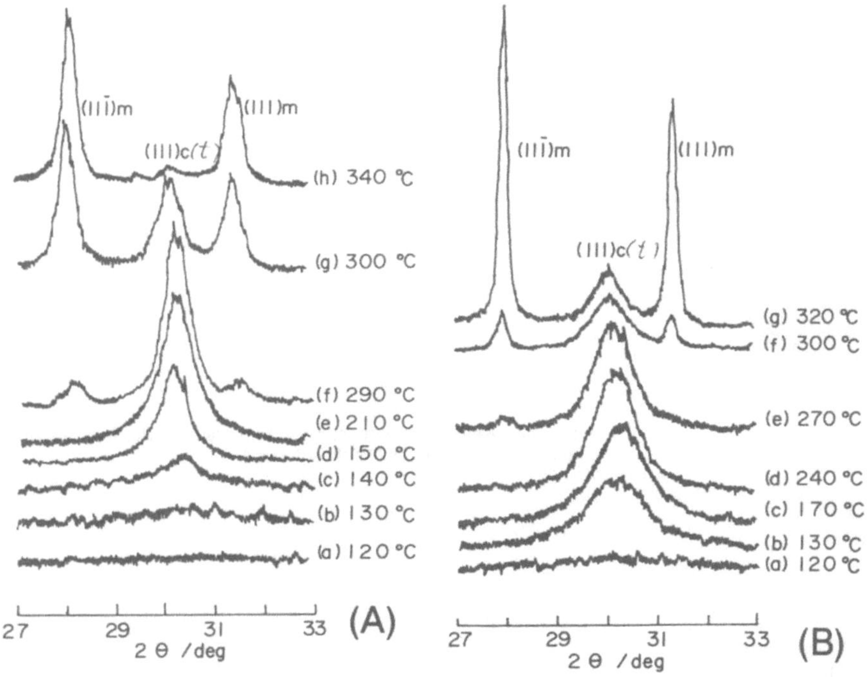

Fig. 2. X-ray diffraction patterns of samples hydrothermally treated in (*A*) 1 molal NaOH solution and (*B*) 4 molal NaOH solution up to temperatures indicated.

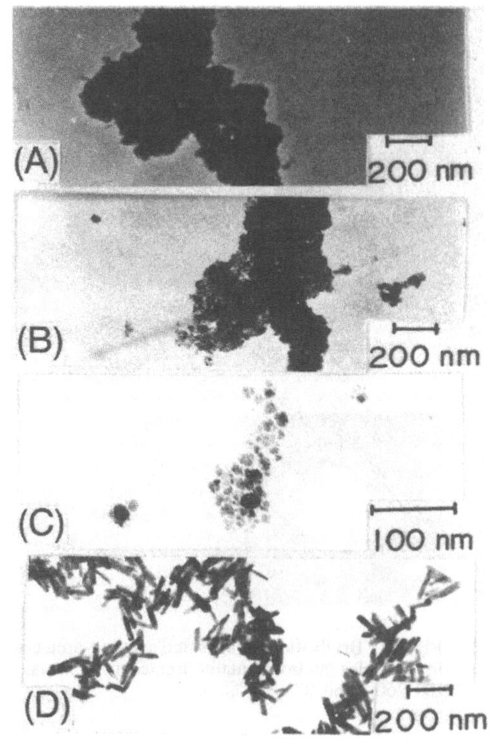

Fig. 4. Bright-field images of samples hydrothermally treated in 1 molal NaOH up to (*A*) 120°, (*B*) 130°, (*C*) 240°, and (*D*) 350°C.

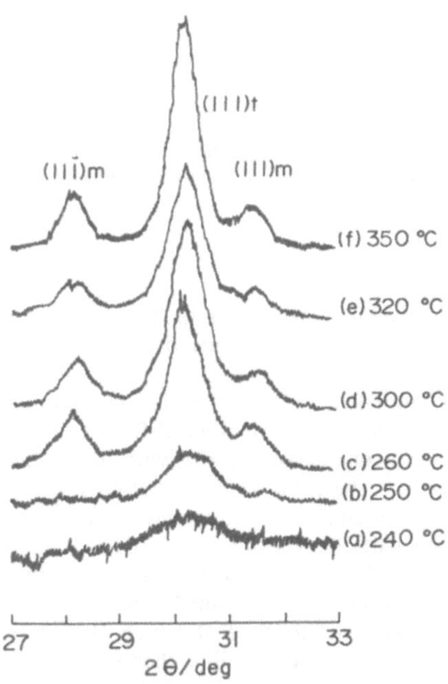

Fig. 3. X-ray diffraction patterns of samples hydrothermally treated in pure water up to temperatures indicated.

The Na^+ content of the powder was determined semiquantitatively by EPMA**; an albite single crystal†† was used as the standard.

III. Results

(1) *X-Ray Diffraction Patterns*

X-ray diffraction patterns for samples hydrothermally treated in 1 and 4 molal NaOH up to 120° to 340°C are shown in Fig. 2. The first phase crystallized, beginning at 130°C, was identified as cubic ZrO_2, with lattice constant 0.5119 nm. Monoclinic ZrO_2 first appeared at ≈290°C, where the amount of cubic ZrO_2 began to decrease.

Several investigators[12–14] have reported the existence of metastable cubic and tetragonal ZrO_2 at temperatures lower than those at which the stable modifications exist. Katz,[14] in studying the dehydration of $Zr(OH)_4$, observed that the metastable cubic phase (lattice constant 0.509 nm) appeared at 285°C and transformed to the metastable tetragonal form at 300°C; the stable monoclinic form began to appear together with the metastable tetragonal form at 320°C and only the monoclinic form was detected at >600°C.

In the present case, crystallization was observed at temperatures (130° and 140°C) much lower than those at which ZrO_2 crystallizes in air. The effective crystallite size of the cubic and monoclinic ZrO_2 obtained is essentially unchanged during crystallization and transformation. As the concentration of the NaOH solution is increased, the crystallite size of the cubic ZrO_2 (which appears first) decreases and that of the monoclinic ZrO_2 obtained after transformation increases. Single-phase cubic or monoclinic ZrO_2 can be obtained in the range 130° to 350°C. Complete crystallization occurs in pure water at 260°C, producing a phase ratio tetragonal: monoclinic of 60:40, which is practically unchanged up to 350°C (Fig. 3). A single-phase product could not be obtained in 0.7 molal NaOH in the range 260° to 400°C.

(2) *Electron Microscopy*

Electron microscopy shows that samples hydrothermally treated to 130°C in 1 molal NaOH contain aggregates of fine ZrO_2 particles, whereas those treated at lower temperatures appear amorphous and nearly identical to the starting gel (Fig. 4(*A*) and (*B*)). Crystallization appears to proceed via collapse of the amorphous gel, resulting in grains with a mosaic structure. The cubic ZrO_2 particles are essentially unchanged in size until the monoclinic phase appears (Fig. 4(*C*)). Figure 4(*D*) shows single-crystal monoclinic ZrO_2 obtained after complete transformation. These needle-like crystals have a uniform crystallite size (150×30 nm).

Figure 5 shows bright-field images and selected-area electron diffraction patterns of powders obtained during the cubic-to-monoclinic transformation in 4 molal NaOH; schematic electron

**JXA-5A, Japan Electron Optics Co., Ltd..
††Containing 11.49 wt% Na as Na_2O.

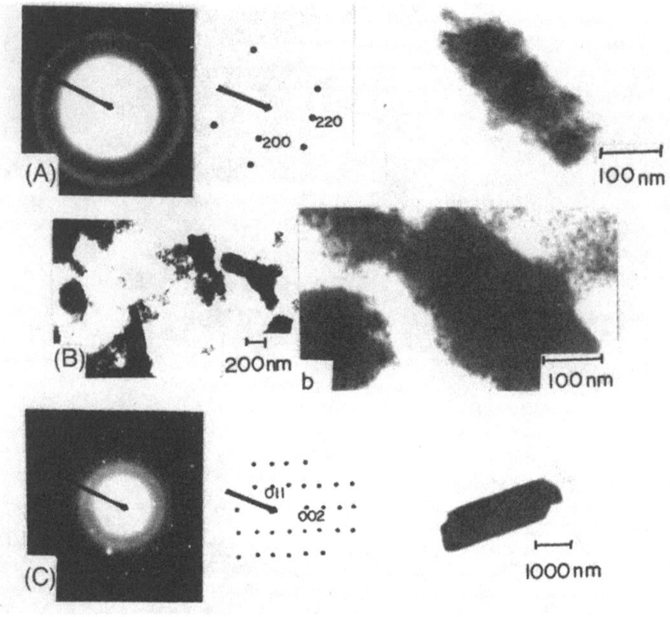

Fig. 5. Bright-field images and selected-area electron diffraction patterns for samples hydrothermally treated in 4 molal NaOH up to (*A*) 150°, (*B*) 260°, and (*C*) 350°C.

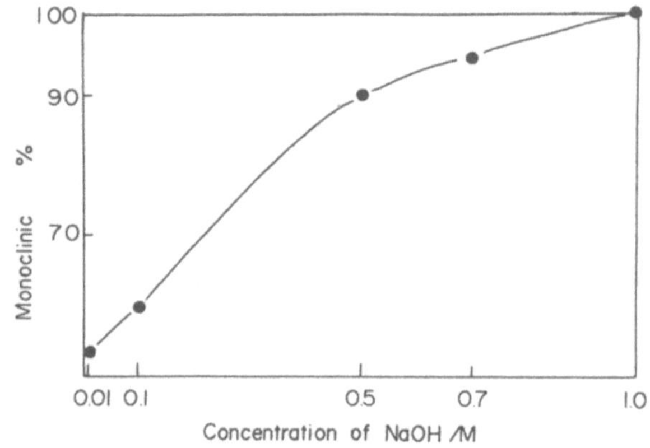

Fig. 6. Amount of monoclinic ZrO_2 obtained after hydrothermal treatment to 350°C in NaOH solutions.

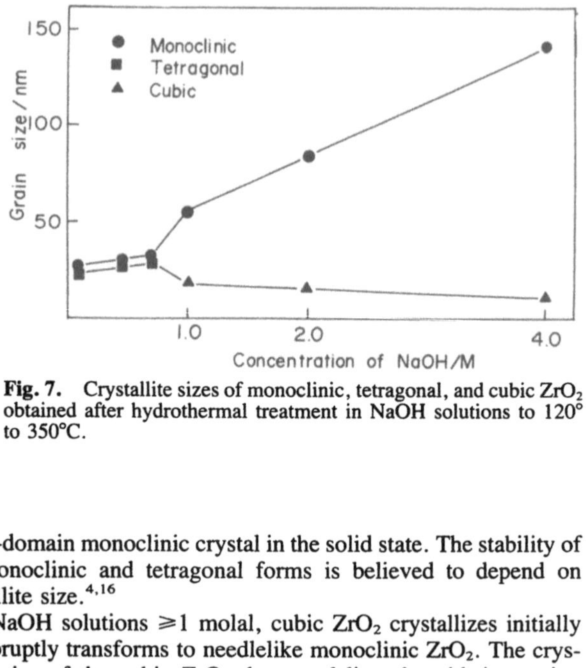

Fig. 7. Crystallite sizes of monoclinic, tetragonal, and cubic ZrO_2 obtained after hydrothermal treatment in NaOH solutions to 120° to 350°C.

diffraction patterns were drawn without the halo apparent in (*A*) and (*C*), which may have resulted from the carbon film. Some spots were indexed on the basis of $a = 0.5119$ nm for cubic ZrO_2 and $a = 0.5148$, $b = 0.5203$, $c = 0.53156$ nm, $\beta = 99°23'$ for monoclinic ZrO_2. Figure 5(*A*) represents a sample for which the X-ray powder diffraction pattern revealed only cubic ZrO_2 (crystallite size 10 nm); the electron diffraction pattern contains the spots of single-crystal cubic ZrO_2, suggesting that many small cubic ZrO_2 particles are arranged in a highly oriented manner. Figure 5(*B*) shows a mixture of cubic and monoclinic ZrO_2. The monoclinic single crystals are surrounded by very fine cubic particles. Cubic ZrO_2 transforms to the monoclinic form simultaneously with monoclinic crystal growth. Figure 5(*C*) represents a needlelike single crystal of monoclinic ZrO_2; the growth direction is [001].

IV. Discussion

Amorphous ZrO_2 transforms into the crystalline form at 130° to 140°C under hydrothermal treatment. The gross morphology of the amorphous reactant is preserved in the cubic ZrO_2 product. EPMA indicated that Na^+ is introduced into the amorphous ZrO_2 during crystallization. As the concentration of the NaOH solution is increased, the crystallite size of the cubic ZrO_2 decreases. These facts suggest that Na^+ (and OH^-) adsorbed on the amorphous reactant rupture the Zr–O–Zr bridges, thereby collapsing the structure and leading to dehydration. Single-crystal cubic ZrO_2 seems to be formed by slight structural rearrangement of the reactant lattice. Whitney[15] observed such behavior during the dehydration of hydrous zirconia in air. The absence of the monoclinic phase during the first stages of crystallization may be due to the close relation between the developing cubic ZrO_2 and the amorphous reactant or to the small crystallite size (11 to 18 nm).[4,9]

In NaOH solutions ≤ 0.7 molal, tetragonal ZrO_2 appeared together with monoclinic ZrO_2 in the initial stages of crystallization, and both forms remained up to 350°C, as in pure water. The amount of residual tetragonal ZrO_2 decreases with increasing NaOH concentration, becoming zero in 1 molal solution (Fig. 6). The tetragonal crystallite size seems to increase slightly with increasing NaOH concentration up to 25 nm (Fig. 7). Monoclinic ZrO_2 crystallites are very similar in size to the coexisting tetragonal ZrO_2 crystallites (Fig. 7) and never grow to needlelike single crystals like those shown in Fig. 5(*C*). In this case, one single-domain tetragonal crystal must transform instantaneously to one single-domain monoclinic crystal in the solid state. The stability of the monoclinic and tetragonal forms is believed to depend on crystallite size.[4,16]

In NaOH solutions ≥ 1 molal, cubic ZrO_2 crystallizes initially and abruptly transforms to needlelike monoclinic ZrO_2. The crystallite size of the cubic ZrO_2 decreased linearly with increasing NaOH concentration (Fig. 7). The cubic ZrO_2 crystal grows very little prior to transformation. Such a very fine crystal is considered to behave like a colloidal particle. Hauser and LeBeau,[17] who investigated the properties of films formed by gradually extracting water from bentonite sols of very small and uniform particle size, observed that a concentration was reached at which Brownian motion ceased and the particles tend to align themselves. On further removal of water the particles appeared to snap into position and form "crystallites" of highly anisometric shape. The films that formed were chemically similar to mica and possessed some of its properties, including its structural characteristics as revealed by XRD. In the present case, the possibility of monoclinic ZrO_2 crystal growth from colloidal cubic ZrO_2 dispersions (not from true solution) is possible. It is plausible that cubic ZrO_2 particles with small crystallite sizes form the larger monoclinic ZrO_2 single crystals (Fig. 7) because it is easier for smaller particles to migrate and align themselves. Sodium ions incorporated into cubic ZrO_2 during crystallization are excluded during the transformation to the monoclinic form. The role of Na^+ in the abrupt formation of the needlelike single crystals is still under study.

Acknowledgments: The writers thank H. Tebika, S. Yoshikura, S. Mizuta, and T. Kariya for help in TEM and EPMA.

References

[1]R. Cypres, R. Wollast, and J. Raucq, "Polymorphic Conversion of Pure Zirconia," *Ber. Deut. Keram. Ges.*, **40** [9] 527–32 (1963).

[2]G. L. Clark and D. H. Reynolds, "Chemistry of Zirconium Dioxide. X-ray Studies," *Ind. Eng. Chem.*, **29** [6] 711–15 (1937).

[3]E. D. Whitney, "Kinetics and Mechanism of Transition of Metastable Tetragonal to Monoclinic Zirconia," *Trans. Faraday Soc.*, **61** [9] 1991–2000 (1965).

[4]R. C. Garvie, "Occurrence of Metastable Tetragonal Zirconia as a Crystallite Size Effect," *J. Phys. Chem.*, **69** [4] 1238–43 (1965).

[5]Yu. M. Polezhaev, "Low-Temperature Cubic and Tetragonal Form of Zirconium Dioxide," *Zh. Fiz. Khim.*, **41** [11] 2958–59 (1967).

[6]K. S. Mazdiyasni, C. T. Lynch, and J. S. Smith, "Preparation of Ultra-High-Purity Submicron Refractory Oxides," *J. Am. Ceram. Soc.*, **48** [7] 372–75 (1965).

[7]J. E. Bailey, D. Lewis, Z. M. Librant, and L. J. Porter, "Phase Transformations in Milled Zirconia," *Trans. J. Brit. Ceram. Soc.*, **71** [1] 25–30 (1972).

[8]A. Clearfield, "Crystalline Hydrous Zirconia," *Inorg. Chem.*, **3** [1] 146–48 (1964).

[9]T. Mitsuhashi, M. Ichihara, and U. Tatsuke, "Characterization and Stabilization of Metastable Tetragonal ZrO_2," *J. Am. Ceram. Soc.*, **57** [2] 97–101 (1974).

[10]H. P. Klug and L. E. Alexander, X-Ray Diffraction Procedures, Ch. 9. Wiley & Sons, New York, 1954.

[11]R. C. Garvie and P. S. Nicholson, "Phase Analysis in Zirconia Systems," *J. Am. Ceram. Soc.*, **55** [6] 303–305 (1972).

[12]K. S. Mazdiyasni, C. T. Lynch, and J. S. Smith, "Metastable Transition of Zirconium Oxide Obtained from Decomposition of Alkoxides," *J. Am. Ceram. Soc.*, **49** [5] 286–87 (1966).

[13]R. Ruh and T. J. Rockett, "Proposed Phase Diagram for the System ZrO_2," *J. Am. Ceram. Soc.*, **53** [6] 360 (1970).

[14]G. Katz, "X-Ray Diffraction Powder Pattern of Metastable Cubic ZrO_2," *J. Am. Ceram. Soc.*, **54** [10] 531 (1971).

[15]E. D. Whitney, "Observation on the Nature of Hydrous Zirconia," *J. Am. Ceram. Soc.*, **53** [12] 697–98 (1970).

[16]R. C. Garvie, "Stabilization of the Tetragonal Structure in Zirconia Microcrystals," *J. Phys. Chem.*, **82** [2] 218–24 (1978).

[17]E. A. Hauser and D. S. LeBeau, "Gelation and Film Formation of Colloidal Clays," *J. Phys. Chem.*, **42**, 961–69 (1938).

Reprinted from *Journal of Materials Science Letters*, **1** (1982) 461–462.

Crystallization and crystal growth of CeO_2 under hydrothermal conditions

E. TANI, M. YOSHIMURA, S. SŌMIYA
Research Laboratory of Engineering Materials, and Department of Materials Science and Engineering, Tokyo Institute of Technology, Nagatsuta, Yokohama, 227 Japan

CeO_2 crystals have previously been grown by the flux method [1–4] due to their high melting point and high vapour pressure [5] at high temperatures. The flux method requires high temperatures and generally brings about greater contamination of the flux than does the hydrothermal method. This study deals with the effect of several mineralizers on the crystallization and the crystal growth of CeO_2 under hydrothermal conditions.

The starting material was hydrous ceria precipitate from $Ce(NO_3)_3 \cdot 6H_2O$ (99.9%, Nissan Kigenso Co., Ltd, Osaka, Japan) solution with 3 N NH_4OH*. The precipitate was filtered and dispersed in distilled water, then filtered again for washing. After repeating this washing procedure four times, it was dried at 120° C for 4 days. The water content of the hydrous ceria could be presented by the formula of $H_2O/CeO_2 \doteqdot 1$ (molar ratio) by TG–DTA (CN8076El, Rigaku Denki Co., Ltd, Tokyo, Japan).

Distilled water or a solution of KF* (8 wt %), LiCl† (15 wt %), Libr‡ (10 wt %), K_2CO_3‡ (10 wt %), $LiNO_3$* (10 wt %), Li_2SO_4* (10 wt %) and NaOH* (30 wt %) was used as a mineralizer (all reagent grade). These mineralizers were selected on the basis of our previous results [6].

A Pt or Au tube (2.7 mm i.d., 0.15 mm thick and 32 mm long) filled with the hydrous ceria and a mineralizer solution was treated under 100 MPa at 500 to 700° C for either 1 or 3 days using a test-tube type pressure vessel of Stellite 25 (6.5 mm i.d.). After this hydrothermal treatment, the vessel was quenched in cold water. The products were washed in distilled water and examined using an X-ray powder diffractometer (RU-200, Rigaku Denki Co. Ltd, Tokyo, Japan), and a scanning electron microscope (JSM-01BM, JEOL Co. Ltd, Tokyo, Japan). The crystallite sizes of the phases were calculated from the Scherrer formula, using the (1 1 1) diffraction peak of CeO_2.

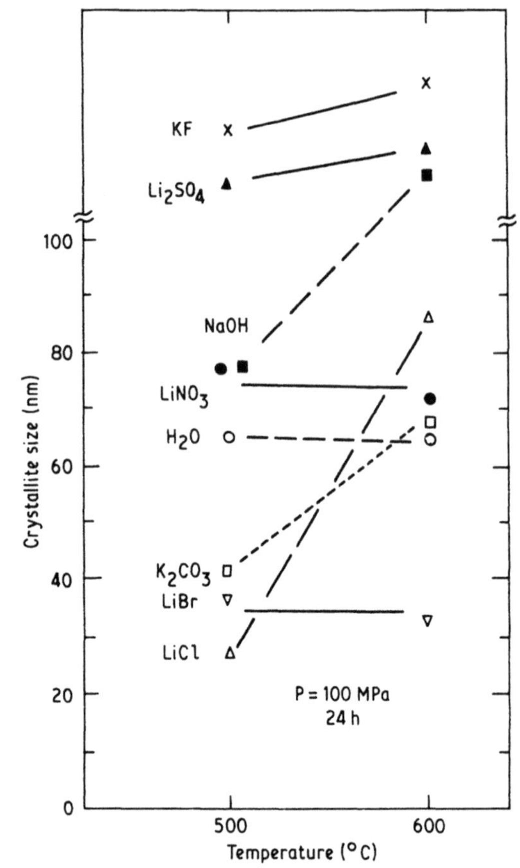

Figure 1 Effect of mineralizers on the crystallite size of CeO_2 under hydrothermal conditions of 100 MPa, 500° C and 600° C for 24 h. The mineralizers were H_2O, KF (8 wt %), LiCl (15 wt %), LiBr (10 wt %), K_2CO_3 (10 wt %), $LiNO_3$ (10 wt %), Li_2SO_4 (10 wt %) and NaOH (30 wt %).

The hydrous ceria had an average crystallite size of about 9 nm. Fig. 1 shows the effect of different

* Wako Pure Chemical Industry Co. Ltd, Tokyo, Japan.
† Kanto Chemical Co. Inc., Tokyo, Japan..
‡ Koso Chemical Co. Ltd, Tokyo, Japan.

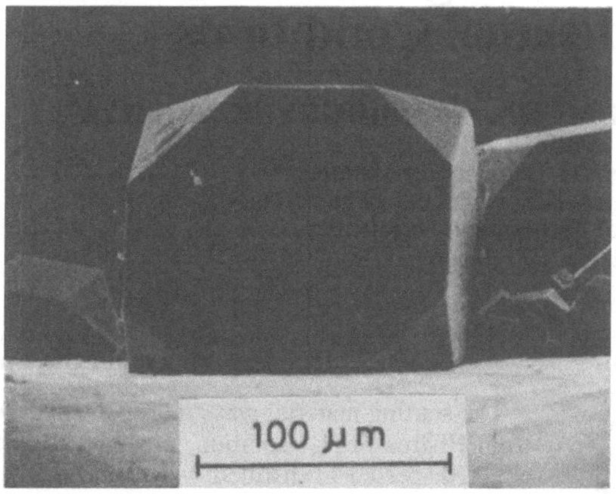

Figure 2 Scanning electron micrograph of CeO_2 crystals grown from hydrous ceria under 100 MPa at 700° C for 3 days with 8 wt % KF solution.

mineralizers on the crystallite size of CeO_2 treated hydrothermally from hydrous ceria under 100 MPa at 500 and 600° C for 24 h. KF and Li_2SO_4 solutions considerably enhanced the growth of CeO_2 crystallites up to above 100 nm at 500 and 600° C.

This result revealed the KF solution to be optimum for the growth of CeO_2 crystals. Therefore, the crystal growth of CeO_2 was carried out with the 8 wt % KF solution.

Fig. 2 shows CeO_2 crystals grown under 100 MPa at 700° C ($\Delta T = 50°$ C; the temperature difference between the lower part of the Pt or Au tube and the upper part) for 3 days. The crystals were transported to the top of the Pt tube and grew up to 100 μm in size and showed a cubic shape with apparent (100) and (111) faces. The EDS analysis showed only Ce in the crystals, but the mass spectrometric analysis showed the presence of K in the order of 100 p.p.m. and F in the order of 10 p.p.m. in the products.

Acknowledgements

This research was supported by Grant-in-Aid for Scientific Research No. 443015 (1979–1980) and the Iwatani Naoji Foundation, 1980. The authors thank JEOL Co. Ltd. For providing the opportunity to use the EDS.

References

1. C. B. FINCH and G. WAYNE CLARK, *J. Appl. Phys.* 37 (1966) 3910.
2. W. H. GRODKIEWICZ and D. J. NITTI, *J. Amer. Ceram. Soc.* 49 (1966) 576.
3. R. C. LINARES, *J. Phys. Chem. Solids* 28 (1967) 1285.
4. B. M. WANKLYN, *J. Cryst. Growth* 5 (1969) 219.
5. A. ROUANET, *C.R. Acad. Sci., Paris* 266C (1968) 908.
6. E. TANI, M. YOSHIMURA and S. SŌMIYA, *Yogyo Kyokai Shi* 90 (1982) 195.

Received 17 November 1981
and accepted 27 May 1982

Reprinted from *Journal of the American Ceramic Society*, **66**(1) (1983) 11–14.

Formation of Ultrafine Tetragonal ZrO_2 Powder Under Hydrothermal Conditions

EIJI TANI, MASAHIRO YOSHIMURA,* and SHIGEYUKI SŌMIYA*

Research Laboratory of Engineering Materials
and Department of Materials Science and Engineering,
Tokyo Institute of Technology, Nagatsuta,
Yokohama, 227 Japan

Ultrafine tetragonal ZrO_2 powder was prepared by hydrothermal treatment at 100 MPa of amorphous hydrous zirconia with distilled water and LiCl and KBr solutions. The resulting powder consisted of well-crystallized particles; at 200°C, the particle size was 16 nm and at 500°C, 30 nm. Under hydrothermal conditions tetragonal ZrO_2 appears to crystallize topotactically on nuclei in the amorphous hydrous zirconia.

I. Introduction

THERMAL decomposition of amorphous hydrous zirconia,[1–8] zirconium alkoxides,[9,10] and zirconium salts,[1,4,7,11] ball-milling of monoclinic ZrO_2,[12] and vapor-phase reactions[13] are known to produce tetragonal ZrO_2, in addition to monoclinic ZrO_2, at low temperatures. However, the stabilization of tetragonal ZrO_2 at low temperatures is as yet not completely understood. Cyprès *et al.*[2] and others[3,11] have attributed the stabilization of tetragonal ZrO_2 to OH^-, SO_4^{2-}, and other anionic impurities, whereas Garvie[4,14] and others[9,12,13] have explained it in terms of a particle size effect, on the basis that the surface energy of tetragonal ZrO_2 is lower than that of the monoclinic form. Livage *et al.*[5] and Murase *et al.*[7] considered the stabilization to be related to the structural similarity between amorphous and tetragonal ZrO_2. Garvie[14] and Mitsuhashi *et al.*[8] pointed out the effect of strain in the stabilization of tetragonal ZrO_2. In these studies, however, several effects appeared to occur simultaneously. Therefore, it is difficult to determine which mechanisms are actually operative in the formation of tetragonal ZrO_2.

Previous studies[8,15] have demonstrated that the effects of strain and of inhomogeneous particle size in products can be eliminated under hydrothermal conditions. Mitsuhashi *et al.*,[8] who treated amorphous hydrous zirconia with distilled water at 98 MPa at 215° to 245°C for 10 min to 22 h, obtained ≈80% tetragonal ZrO_2. Tani *et al.*[15] obtained ultrafine single-phase monoclinic ZrO_2 powder by hydrothermal treatment of amorphous hydrous zirconia with a KF solution at 100 MPa from 200° to 500°C for 24 h. Under hydrothermal conditions, crystallization processes and products can be changed radically by mineralizers.

The effect of mineralizers on crystallization has been studied[16] in detail in the system ZrO_2-CeO_2 under hydrothermal conditions. Crystallization of solid solutions in this system proceeded by dissolution/precipitation and by structural rearrangement. KF and NaOH solutions and LiCl and KBr solutions typically enhanced the dissolution/precipitation and structural rearrangement mechanisms, respectively.

In this report, formation of tetragonal and monoclinic ZrO_2 by hydrothermal treatments of amorphous hydrous zirconia is described; the mechanism for the occurrence of tetragonal ZrO_2 is discussed.

Received October 30, 1981; revised copy received April 14, 1982; approved August 8, 1982.
Supported by Grant-in-Aid for Scientific Research No. 443015 (1979–80) and the Iwatani Naoji Foundation, 1980.
*Member, the American Ceramic Society.

II. Experimental Procedure

The starting material was amorphous hydrous zirconia prepared from Hf-free $ZrCl_4$* solution, precipitated with 3*N* NH_4OH,† filtered, dispersed in distilled water, and refiltered. After four dispersion and filtration processes, the precipitate was dried at 120°C for 48 h. The water content of this material was determined by TG-DTA.‡ Precipitates calcined at 240°, 420°, 620°, or 820°C in air for 60 min at a heating rate of 10°C/min were also used as starting materials. Distilled water and solutions of 8 wt% KF†; 30 NaOH†; 7, 15, and 30 LiCl§; and 10 KBr† were used as mineralizers. A Pt or Au tube 2.7-mm ID, 0.15 mm thick, and 32 mm long was filled with the starting material and solution and treated under 100 MPa at 200° to 600°C for 24 h using a test-tube-type pressure vessel of Co-Cr-W alloy¶ (6.5-mm ID); the vessel was then quenched into cold water. The products were washed in distilled water, dried, and examined by X-ray powder diffractometry,** transmission electron microscopy,†† and energy dispersion spectrometry.††

Crystallite sizes were calculated using the (111) diffraction peak of monoclinic and tetragonal ZrO_2 from the Scherrer formula,[17] $D=0.9\lambda(\beta\cos\theta)^{-1}$, where D is the crystallite size, λ the wavelength ($CuK\alpha$), β the corrected half-width obtained using the (101) peak of low quartz as the standard and the Warren formula,[17] and θ the diffraction peak angle.

The tetragonal ZrO_2 content was estimated from the relative intensities of the (111) peaks of tetragonal and monoclinic ZrO_2.

III. Results and Discussion

The starting material dried at 120°C was amorphous by X-ray diffraction; TG-DTA gave an H_2O/ZrO_2 ratio of ≈2 for this material.

(1) Formation of Ultrafine Tetragonal ZrO_2 Powders

The phases and crystallite sizes produced by hydrothermal treatment at 300°C for 24 h with several mineralizers are shown in Table I. Both tetragonal and monoclinic ZrO_2 formed when H_2O and LiCl and KBr solutions were used, whereas only the monoclinic phase was produced with KF and NaOH solutions. The same results were previously obtained with KF solution at treatment

*Sn 680 ppm, Fe 620, Hf<70, Al<30; Nippon Mining Co., Ltd., Tokyo, Japan.
†Guaranteed reagent, Wako Pure Chemical Industry Co., Ltd., Tokyo, Japan.
‡Model CN8076E1, Rigaku Denki Co., Ltd., Tokyo, Japan.
§Guaranteed reagent, Kanto Chemical Industry Co., Inc., Tokyo, Japan.
¶Stellite 25, Union Carbide Corp., New York, NY.
**RU-200, Rigaku Denki Co., Ltd.
††JEM-100CX, JEOL Co., Ltd., Tokyo, Japan.

Table I. Phases Present and Crystallite Size of Hydrothermally Treated* Products

Mineralizer	Avg crystallite size of products (nm)	
	Tetragonal ZrO_2	Monoclinic ZrO_2
KF (8 wt%)	ND	16 nm
NaOH (30 wt%)	ND	40 nm
H_2O	15 nm	17 nm
LiCl (15 wt%)	15 nm	19 nm
KBr (10 wt%	13 nm	15 nm

*100 MPa, 300°C, 24 h.

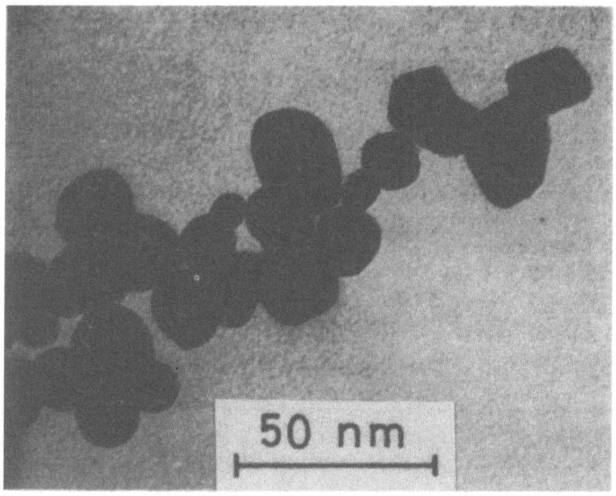

Fig. 1. Transmission electron micrograph of tetragonal and monoclinic ZrO_2 powders produced by hydrothermal treatment at 100 MPa at 300°C for 24 h using 15 wt% LiCl solution.

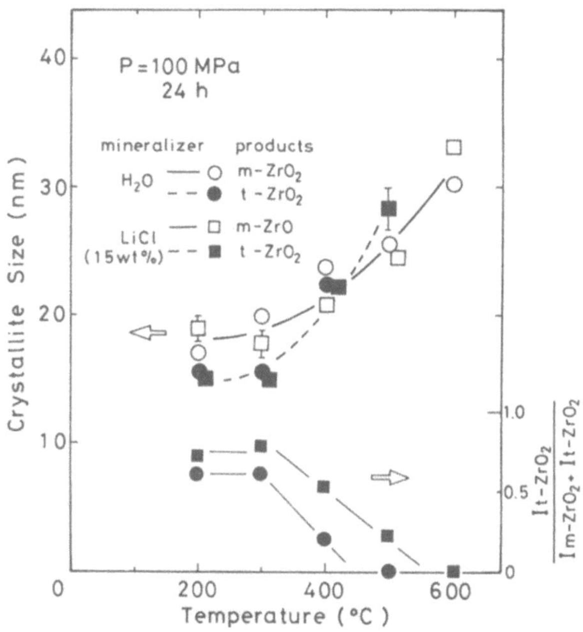

Fig. 2. Variation with temperature of crystallite size and tetragonal ZrO_2 contents in products of hydrothermal treatment at 100 MPa for 24 h.

temperatures up to 600°C.[15] The crystallization process clearly depends on the mineralizer solution. The crystallite size, however, was almost the same (≈17 nm) for the monoclinic and tetragonal phases in all products except those obtained with NaOH solution.

Figure 1 shows fine monoclinic and tetragonal ZrO_2 particles formed by hydrothermal treatment at 100 MPa and 300°C for 24 h with 15 wt% LiCl solution. The particles are well shaped and isolated from each other. The pure monoclinic ZrO_2 particles produced by hydrothermal treatment of the same starting materials with 8 wt% KF solution at 200° to 500°C were similarly shaped and isolated.[15] Thus, hydrothermal treatment yields well-shaped, isolated fine particles, in contrast to the aggregated particles resulting from calcination of amorphous hydrous zirconia in air.

(2) Effect of Particle Size on the Formation of Tetragonal ZrO_2

Figure 2 shows the variation with temperature of crystallite size and tetragonal ZrO_2 content for the products of hydrothermal treatment for 24 h with H_2O and 15 wt% LiCl solution. The crystallite sizes of monoclinic and tetragonal ZrO_2 were ≈18 and ≈15 nm, respectively, below 300°C and increased with treatment temperature above 300°C. Monoclinic ZrO_2 crystallites were ≈25 nm in diameter at 500°C; tetragonal ZrO_2 was not produced when H_2O was used, but 30-nm tetragonal crystallites resulted with LiCl solution at 500°C. At 600°C, monoclinic ZrO_2 crystallites ≈32 nm in size were found; no tetragonal ZrO_2 was produced with either solution. Therefore, the maximum crystallite size of tetragonal ZrO_2 produced by hydrothermal treatment was ≈30 nm.

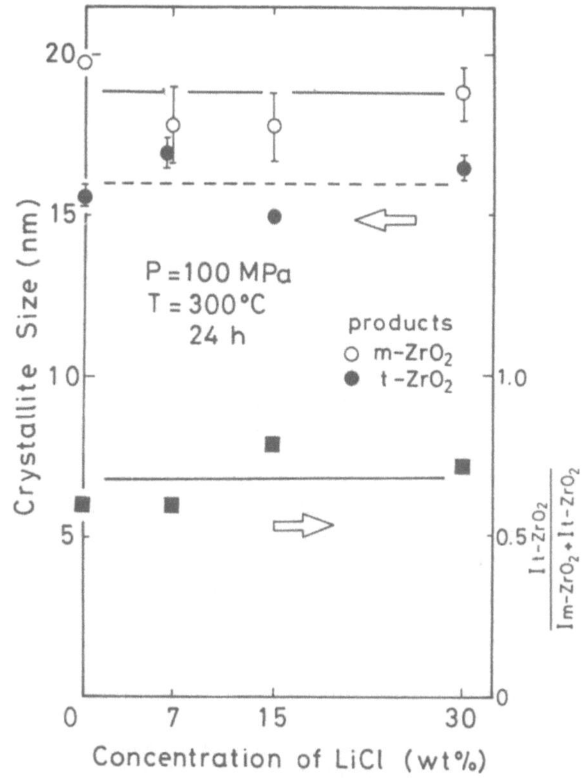

Fig. 3. Variation with LiCl concentration of crystallite size and tetragonal ZrO_2 content in products of hydrothermal treatment under 100 MPa at 300°C for 24 h.

Garvie[4,14] and others[9,12,13] attributed the stabilization of tetragonal ZrO_2 to a particle size effect, i.e. smaller grains can be stabilized in the tetragonal form because their surface energy is much less than that of the monoclinic form. If this were true, the critical particle size would be ≈30 nm (Fig. 2). However, single-phase monoclinic ZrO_2 with a crystallite size of ≈16 nm was produced with KF solution (Table I); this is obviously smaller than the proposed critical particle size. In addition, crystallite size was almost the same for tetragonal and monoclinic ZrO_2 formed with H_2O, LiCl, and KBr solutions (Table I). This evidence clearly indicates that particle size is not a primary factor in the stabilization of tetragonal ZrO_2.

(3) Effect of Impurities on the Formation of Tetragonal ZrO_2

Cyprès *et al.*[2] and others[3,11] attributed the stabilization of tetragonal ZrO_2 to OH^- or other anionic impurities. The contribution of the OH^- ion might be enhanced under hydrothermal conditions. However, only monoclinic ZrO_2 was produced by the hydrothermal treatment of amorphous hydrous zirconia with KF and NaOH solutions, whereas both tetragonal and monoclinic ZrO_2 were produced under the same hydrothermal conditions with H_2O, LiCl, and KBr solutions. This result cannot be explained in terms of a contribution of the OH^- ion to stabilization of tetragonal ZrO_2.

To check the stabilizing effect of LiCl, the starting materials were treated with LiCl solutions of varied concentration at 300°C for 24 h. Figure 3 shows that the crystallite size and tetragonal ZrO_2 content were almost independent of LiCl concentration. Chlorine and other impurities were not detected by energy dispersion spectrometry in the products of hydrothermal treatment with 15 wt% LiCl solution.

Therefore, anionic impurities cannot be considered an important factor in the formation of tetragonal ZrO_2. Furthermore, on the basis of results obtained by vapor-phase reaction of $ZrCl_4$ and O_2, Suyama *et al.*[13] concluded that the stability of the tetragonal phase cannot be explained by impurity effects.

(4) Effect of Strain on the Formation of Tetragonal ZrO_2

Strain-free particles of both monoclinic ZrO_2 (produced in a KF solution[15]) and tetragonal ZrO_2 (produced hydrothermally in H_2O[8] and LiCl solutions) have been isolated. This result indicates that strain is not a main reason for the formation of tetragonal ZrO_2 under hydrothermal conditions.

(5) Topotactic Crystallization

The phases and crystallite sizes of hydrothermal products depend not only on mineralizers but also on the temperature of calcination of the starting hydrous zirconia. The ZrO_2 phases present after calcination of amorphous hydrous zirconia for 60 min in air are shown in Fig. 4*(A)*. X-ray diffraction showed that samples calcined below 240°C contained only the amorphous phase; above 420°C, the tetragonal and monoclinic phases appeared, whereas at 820°C only the monoclinic phase was present. These results agree well with previous studies.[5,6] The phases produced by hydrothermal treatment (100 MPa at 300°C for 24 h with 15 wt% LiCl solution) of these calcined samples clearly differed from the calcined phases (Fig. 4*(B)*). The samples calcined below 240°C, which were amorphous, produced both tetragonal and monoclinic ZrO_2. However, the samples calcined above 420°C, which contained no amorphous phase, yielded only monoclinic ZrO_2; tetragonal ZrO_2 appears to be produced by hydrothermal treatment only when amorphous zirconia is present.

Distribution functions obtained by X-ray and neutron diffraction have shown[5] that the atoms in amorphous zirconia are not distributed at random; certain preferred interatomic spacings exist. These interatomic spacings, characteristic of short-range order, correspond to distances observed in tetragonal ZrO_2. Thus, amorphous zirconia contains, in short-range-ordered regions, tetragonal ZrO_2 nuclei which can grow during crystallization under hydrothermal conditions.

It has been proposed that crystallization under hydrothermal conditions proceeds by dissolution/precipitation and structural rearrangement.[15] Crystallite growth by dissolution/precipitation would result in increased crystallite size after prolonged hydrothermal treatment. With NaOH solution, larger monoclinic ZrO_2 crystallites resulted after hydrothermal treatment at 300°C for 24 h, but with 15 wt% LiCl solution, the crystallite size increased only slightly compared to that of samples calcined above 420°C in air (Fig. 5). The amorphous phase produced by calcination below 240°C produced crystallites ≈17 nm in size (smaller than the critical size to form tetragonal ZrO_2 on hydrothermal treatment). Furthermore, the crystallite sizes of the monoclinic and tetragonal phases produced hydrothermally for 168 h were almost the same as those produced in 24 h at 500°C with 15 wt% LiCl solution. These results suggest that the dissolution/precipitation mechanism is not effective in the crystallization of zirconia in the presence of LiCl solutions.

Therefore, when LiCl solutions are used, tetragonal ZrO_2 appears to form by structural rearrangement rather than the dissolution/precipitation mechanism observed in the case of solid solutions in the system ZrO_2-CeO_2.[15] However, in that system, only the monoclinic phase was produced (in addition to CeO_2) by the dissolution/precipitation mechanism; the tetragonal phase resulted from structural rearrangement.

Consequently, the mechanism of tetragonal ZrO_2 formation is thought to be topotactic crystallization on the nuclei in the amorphous zirconia. This phenomenon seems to follow Ostwald's step rule,[18] which states that when a substance can exist in more than one modification, the least stable is formed first and changes ultimately into the more stable. Amorphous zirconia might be considered to be the higher modification; therefore, the tetragonal phase can be formed as a more stable one. The fact that the calcined samples which had been crystallized produced no tetragonal ZrO_2 after hydrothermal treatment supports this hypothesis.

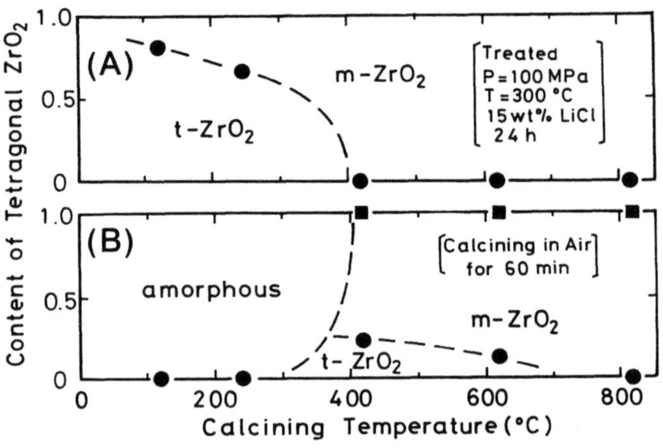

Fig. 4. Tetragonal ZrO_2 content after *(A)* hydrothermal treatment (100 MPa, 300°C for 24 h in 15 wt% LiCl solution) of amorphous hydrous zirconia which was first calcined at temperatures indicated and *(B)* calcination of amorphous hydrous zirconia for 60 min at varied temperatures in air.

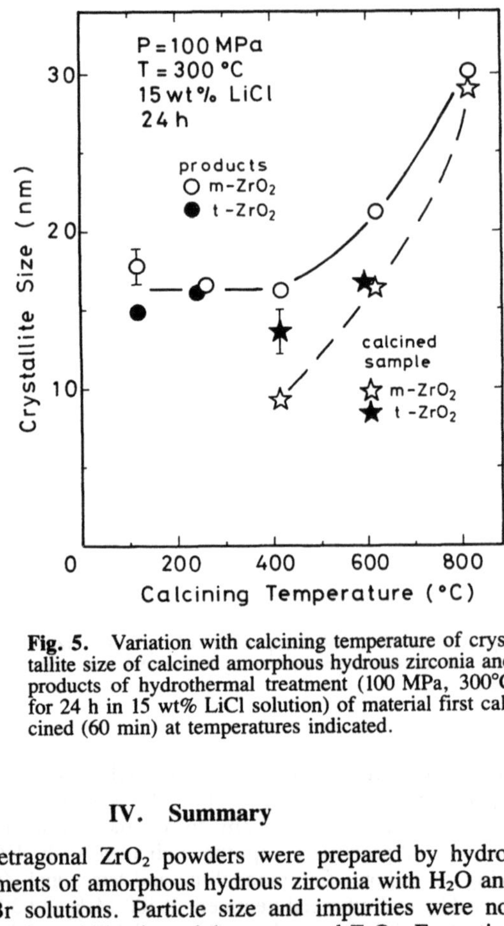

Fig. 5. Variation with calcining temperature of crystallite size of calcined amorphous hydrous zirconia and products of hydrothermal treatment (100 MPa, 300°C for 24 h in 15 wt% LiCl solution) of material first calcined (60 min) at temperatures indicated.

IV. Summary

Ultrafine tetragonal ZrO_2 powders were prepared by hydrothermal treatments of amorphous hydrous zirconia with H_2O and LiCl and KBr solutions. Particle size and impurities were not primary factors in stabilization of the tetragonal ZrO_2. Formation of tetragonal ZrO_2 is closely related to the existence of amorphous zirconia. Topotactic crystallization on nuclei in the amorphous zirconia is proposed as the mechanism of tetragonal ZrO_2 formation under hydrothermal conditions.

Acknowledgments: The authors thank JEOL Co., Ltd. for providing the opportunity to use the transmission electron microscope and Nippon Mining Co., Ltd. for providing the $ZrCl_4$.

References

[1]G. L. Clark and D. H. Reynolds, "Chemistry of Zirconium Dioxide. X-Ray Diffraction Studies," *Ind. Eng. Chem.*, **29** [6] 711–15 (1937).

[2]R. Cyprès, R. Wollast, and J. Raucq, "Contribution on the Polymorphic Conversion of Pure Zirconia," *Ber. Dtsch. Keram. Ges.*, **40** [9] 527–32 (1963).

[3]A. Clearfield, "Crystalline Hydrous Zirconia," *Inorg. Chem.*, **3** [1] 146–48 (1964).

[4]R. C. Garvie, "Occurrence of Metastable Tetragonal Zirconia as a Crystallite Size Effect," *J. Phys. Chem.*, **69** [4] 1238–43 (1965).

[5]J. Livage, K. Doi, and C. Mazières, "Nature and Thermal Evolution of Amorphous Hydrated Zirconium Oxide," *J. Am. Ceram. Soc.*, **51** [6] 349–53 (1968).

[6]G. Katz, "X-Ray Diffraction Powder Pattern of Metastable Cubic ZrO_2," *J. Am. Ceram. Soc.*, **54** [10] 531 (1971).

[7]Y. Murase, E. Kato, and H. Matsumoto, "Crystallization and Phase Transformation of Low Temperature from Zirconia Prepared from Different Starting Materials," *Nippon Kagaku Kaishi*, **1972**, No. 12, 2329–36.

[8]T. Mitsuhashi, M. Ichihara, and U. Tatsuke, "Characterization and Stabilization of Metastable Tetragonal ZrO_2," *J. Am. Ceram. Soc.*, **57** [2] 97–101 (1974).

[9]K. S. Mazdiyasni, C. T. Lynch, and J. S. Smith, "Preparation of Ultra-High-Purity Submicron Refractory Oxides," *J. Am. Ceram. Soc.*, **48** [7] 372–75 (1965).

[10]K. S. Mazdiyasni, C. T. Lynch, and J. S. Smith, "Metastable Transitions of Zirconium Oxide Obtained from Decomposition of Alkoxides," *J. Am. Ceram. Soc.*, **49** [5] 286–87 (1966).

[11]E. D. Whitney, "Kinetics and Mechanism of the Transition of Metastable Tetragonal to Monoclinic Zirconia," *Trans. Faraday Soc.*, **61** [9] 1991–2000 (1965).

[12]J. E. Bailey, D. Lewis, Z. M. Librant, and L. J. Porter, "Phase Transformations in Milled Zirconia," *Trans. J. Br. Ceram. Soc.*, **71**, 25–30 (1972).

[13]Y. Suyama, T. Mizobe, and A. Kato, "ZrO_2 Powders Produced by Vapor Phase Reaction," *Ceramurgia Int.*, **3** [4] 141–46 (1977).

[14]R. C. Garvie, "Stabilization of the Tetragonal Structure in Zirconia Microcrystals," *J. Phys. Chem.*, **82** [2] 218–24 (1978).

[15]E. Tani, M. Yoshimura, and S. Sōmiya, "Hydrothermal Preparation of Ultrafine Monoclinic ZrO_2 Powder," *J. Am. Ceram. Soc.*, **64** [12] C-181 (1981).

[16]E. Tani, M. Yoshimura, and S. Sōmiya, "Effect of Mineralizers on the Crystallization of the Solid Solutions in the System ZrO_2-CeO_2 under Hydrothermal Conditions," *Yogyo Kyokai Shi*, **90** [4] 195–201 (1982).

[17]B. D. Cullity; pp. 261–63 in Elements of X-Ray Diffraction, Addison-Wesley, Reading, MA, 1967.

[18]J. Grant; p. 479 in Hackh's Chemical Dictionary, McGraw-Hill, New York, 1972.

Reprinted from *Report of the Research Laboratory of Engineering Materials, Tokyo Institute of Technology*, No. 8 (1983) 47–53.

Hydrothermal Crystallization and Crystal Growth of Un-doped and ZrO_2-doped CeO_2

By
Eiji TANI*, Masahiro YOSHIMURA**
and Shigeyuki SŌMIYA***

Abstract

The crystallization of un-doped and ZrO_2-doped CeO_2 was studied from the precipitated hydrous ceria with KF, LiCl, LiBr, K_2CO_3, $LiNO_3$, Li_2SO_4 and NaOH solutions under 100 MPa at 500 to 600°C for 24 h. The KF and Li_2SO_4 solutions enhanced effectively the growth of the crystallite of un-doped CeO_2. The LiCl, LiBr and Li_2SO_4 solutions, whereas, enhanced the growth of the crystallite of ZrO_2-doped CeO_2. The calcination of the precipitated hydrous ceria contributed to the hydrothermal growth of the crystallite of ZrO_2-doped CeO_2 with LiCl solution. The single crystals of un-doped CeO_2 (120 μm) and ZrO_2-doped CeO_2 (200 μm) were obtained by the hydrothermal treatments with KF and LiCl solutions, respectively.

1. Introduction

Ceria (CeO_2) has a cubic fluorite structure as well as the highest-temperature form of ZrO_2, so that it makes wide solid solutions with ZrO_2 at high temperatures. The solid solutions are of interest for high electronic conductors[1] or stabilizers of toughening ceramics.[2] We have studied the crystallization[3] and the phase relations[4,5] of these solid solutions under hydrothermal conditions.

CeO_2 crystals were previously grown only by the flux method due to its high melting point and high vapor pressure[6] at high temperatures. The flux used was $Li_2O \cdot 2WO_3$,[7] PbF_2 or BiF_3 with small amount of B_2O_3,[8] PbF_2-B_2O_3,[9] or PbF_2.[10] The flux method needs relatively high temperatures and generally brings about the contanination from the flux as compared to hydrothermal methods. Therefore, we have tried to grow CeO_2 and ZrO_2-doped CeO_2 crystals using hydrothermal methods. It has been found that the crystallization process of the ZrO_2-CeO_2 solid solutions depends mainly on the anions of solute in the solution used as a mineralizer under hydrothermal conditions.[3]

This paper deals with the effect of several mineralizers on the crystallization and the crystal growth of un-doped and ZrO_2-doped CeO_2 under hydrothermal conditions. The calcination effect of the precipitated hydrous ceria was also studied in the growth of ZrO_2-doped CeO_2 under hydrothermal conditions.

---

* Graduate Student: Present Address, Government Industrial Research Institute of Kyūshū, Tosu, Saga 841
** Associate Professor, Dr. Sci. in Eng.
*** Professor, Dr. Sci. in Eng.

2. Experimental

The starting material for un-doped CeO_2 was hydrous ceria precipitated from $Ce(NO_3)_3 \cdot 6H_2O$ (99.9%, Nissan Kigenso Co., Ltd., Osaka, Japan) solution with 3N-NH_4OH. The hydrous ceria was filtered, then dispersed again in distilled water for washing, and then filtered again. After four times of washing and filtration treatments, they were dried in air at 120°C for 4 days.

The starting material for ZrO_2-doped CeO_2 was hydrous zirconia-ceria solid solutions (containing 40 mol% CeO_2) coprecipitated from the mixture of $Ce(NO_3)_3 \cdot 6H_2O$ and Hf-free $ZrCl_4$ (Sn; 680 ppm, Fe; 620 ppm, Hf; < 70 ppm, Al; < 30 ppm, Nippon Mining Co., Ltd., Tokyo, Japan) solutions with 3N-NH_4OH. The hydrous solid solutions were filtered, then dispersed again in a 0.25N-NH_4OH solution for washing, and then filtered again. After four times of washing and filtration treatments, they were dried in air at 120°C for 7 days. Some samples were heated at a heating rate of 10°C/min and calcined at certain temperatures up to 820°C for 60 min.

The water content of the staring materials were determined by TG-DTA (Thermoflex, Rigaku Denki Co., Ltd., Tokyo, Japan). Distilled water or the solutions of KF* (8 wt%), LiCl** (15 wt%), LiBr*** (10 wt%), K_2CO_3*** (10 wt%), $LiNO_3$* (10 wt%), Li_2SO_4* (10 wt%) and NaOH* (30 wt%) were used as mineralizer solutions (all of reagent grade; *Wako Pure Chemical Industry Co., Ltd., Tokyo, Japan; **Kanto Chemical Co., Inc., Tokyo, Japan; ***Koso Chemical Co., Ltd., Tokyo, Japan). They were selected as typical mineralizers containing different anions effective on the crystallization[3)] of the ZrO_2-CeO_2 solid solutions.

A Pt or Au tube (2.7 mm ID, 0.15 mm thick and 32 mm long) filled with the starting material and a mineralizer solution was treated under 100 MPa at 500 to 700°C for 1 or 3 days, using a test-tube-type pressure vessel of Stellite 25 (6.5 mm ID). After these hydrothermal treatments, the vessel was quenched in cold water. The products were washed in distilled water and examined by X-ray powder diffractometer (RU-200, Rigaku Denki Co., Ltd., Tokyo, Japan), scanning electron microscope (JSM-T200, JEOL Co., Ltd., Tokyo, Japan) and energy dispersive spectroscopy (JEM-100CX, JEOL Co., Ltd., Tokyo, Japan).

The crystallite sizes of the produced phases were calculated from the Scherrer formula, using (111) diffraction peak of CeO_2; $D = 0.9\,\lambda\,(\beta \cdot \cos\theta)^{-1}$, where D is the crystallite size; λ the wavelength of X-rays (CuKα); β the corrected half-width using (101) peak of low quartz as a standard and the Warren formula; and θ the diffraction peak angle.

3. Results and Discussion

The starting materials for un-doped and ZrO_2-doped CeO_2 showed the X-ray patterns of a fluorite-type structure. Their average crystallite size were about 9 and 5 nm for un-doped and ZrO_2-doped CeO_2, respectively. The water contents of these materials could be presented by the formula of $CeO_2 \cdot H_2O$ and $3ZrO_2 \cdot 2CeO_2 \cdot 2H_2O$ by TG-DTA.

3.1 Crystallization

A) Un-doped CeO_2

Figure 1 (a) shows the variation with mineralizers of the crystallite size of un-doped CeO_2 treated hydrothermally at 500 and 600°C for 24 h. The KF and Li_2SO_4 solutions considerably enhanced the growth of CeO_2 crystallite more than 100 nm at 500 and

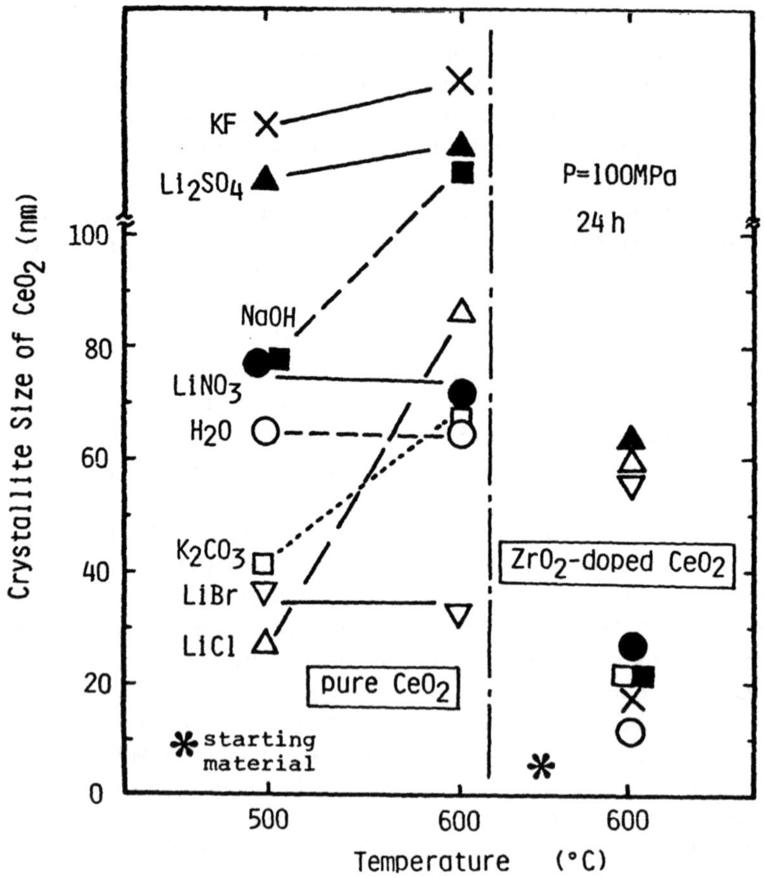

Fig. 1 Effect of mineralizers on the growth of CeO_2 crystallite under hydrothermal conditions of 100 MPa at 500° or 600°C for 24 h.

and 600°C. The NaOH solution grew the crystallite of CeO_2 to about 80 nm at 500°C and more than 100 nm at 600°C. These solutions seemed to dissolve CeO_2 considerablly under hydrothermal conditions studied. The other solutions did not grow the crystallite of CeO_2 above 100 nm at 500 and 600°C probably due to less solubilities of CeO_2.

This result revealed the KF and Li_2SO_4 solutions to be suitable for the growth of un-doped CeO_2 crystals.

B) ZrO_2-doped CeO_2

The products after hydrothermal treatments of the ZrO_2-doped CeO_2 samples at 600°C for 24 h were identified to be CeO_2 s.s. (~98 wt% CeO_2), monoclinic ZrO_2 s.s. (~97 wt% ZrO_2) and/or ZrO_2-rich s.s. phases by X-ray diffraction.[3] The crystallite of CeO_2 s.s. was grown to about 60 nm with Li_2SO_4, LiBr and LiCl solutions at 600°C (Fig. 1 (b)). Whereas, the other solutions did not contribute greatly to the growth of the crystallite of CeO_2 s.s. The KF solution which was effective for the growth of un-doped CeO_2 did not contribute to the growth of ZrO_2-doped CeO_2.

The effect of mineralizers on the crystallization of un-doped and ZrO_2-doped

CeO_2 was not identical. This different effect of mineralizers to the growth of ceria crystallites was considered to be caused by the fact that the dissolution rate of CeO_2 phase was influenced by that of ZrO_2 phase. For example, CeO_2 component would be less dissolved when a mineralizer solution dissolved ZrO_2 component selectively. The crystallization of ZrO_2-CeO_2 solid solutions under hydrothermal conditions consists of the dissolution/precipitation and/or the structural rearrangement processes.[3] LiCl and LiBr solutions, which seemed to have less solubilities of CeO_2 and ZrO_2 components, contributed the structural rearrangement process in the crystallization of ZrO_2-CeO_2 solid solutions.[3]

The calcined coprecipitated samples were treated hydrothermally with LiCl solution. The coprecipitated samples for ZrO_2-doped CeO_2 were calcined up to 820°C for 60 min in air, then the crystallite grew up only from 5 nm to 9 nm (Fig. 2). However, the crystallite of CeO_2 grew up more than 100 nm after hydrothermal treatments at 500°C for 24 h when they had been calcined above 620°C in air. The calcination effect was significant for the crystallite growth of CeO_2 phase.

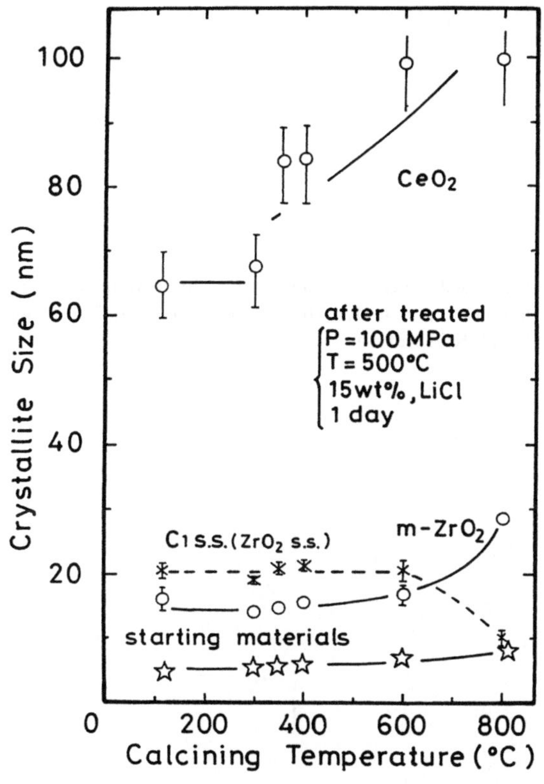

Fig. 2 Effect of calcination of starting CeO_2-ZrO_2 solid solutions on the growth of CeO_2, m-ZrO_2 and ZrO_2 s.s. crystallites after hydrothermal treatments at 500°C under 100 MPa for 1 day with 15 wt% LiCl solution.

3.2 Crystal Growth

The crystal growth of un-doped and ZrO_2-doped CeO_2 was carried out with KF and LiCl solutions, respectively.

Figure 3 shows un-doped CeO_2 crystals grown from hydrous ceria under 100 MPa at 700°C ($\Delta T = 50°C$; the difference of temperature from the lower part of Pt or Au capsule to the upper one) for 3 days with KF solution. The crystals were transported to the top of the capsule and grew up to about 120 μm in size and showed cubic shape with apparent (100) and (111) faces.

Figure 4 shows un-doped CeO_2 crystals grown from the hydrous ceria (calcined at 820°C for 60 min in air) treated under 100 MPa at 600°C ($\Delta T = 20°C$) for 3 days with

Fig. 3 Un-doped CeO_2 crystals grown hydrothermally from hydrous ceria under 100 MPa at 700°C for 3 days with 8 wt% KF solution.

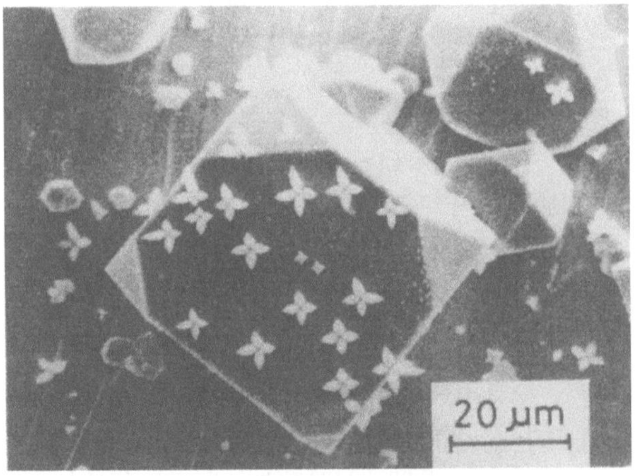

Fig. 4 Un-doped CeO_2 crystals with secondary crystals from calcined hydrous ceria under 100 MPa at 600°C for 3 days with 8 wt% KF solution.

KF solutions. These crystals grew up to 50μm in size with the (100) and (111) faces. On the (100) surfaces of the crystals many flower-like secondary crystals which have four petals were observed. Since the petals seemed to grow along the [100] direction of the crystals, they might be due to epitaxial crystallization.

The flower-like secondary crystals with three petals were also observed on the (111) face of the un-doped CeO_2 crystals grown from hydrous ceria under 100 MPa at 700°C (ΔT = 50°C) for 3 days with KF solution.

Figure 5 shows ZrO_2-doped CeO_2 crystals grown from the hydrous zirconia-ceria s.s. (calcined at 820°C for 60 min) treated under 100 MPa at 600°C (ΔT = 15°C) for 3 days with LiCl solution. The crystals grew up to 200 μm in size and had apparent (100) and (111) faces.

The ZrO_2 content of these crystals could not be determined by energy dispersive spectroscopic analysis, but they should contain 1.5~2 mol% ZrO_2, according to the phase diagram[4),5)] established under hydrothermal conditions.

Fig. 5 ZrO_2-doped CeO_2 crystal grown hydrothermally under 100 MPa at 600°C (ΔT = 15°C) for 3 days with 15 wt% LiCl solution.

4. Summary

The effect of mineralizers on the crystallization of un-doped and ZrO_2-doped CeO_2 was not identical. KF and Li_2SO_4 solutions enhanced the growth of un-doped CeO_2 crystallite. Whereas, Li_2SO_4, LiBr and LiCl solutions contributed to the growth of ZrO_2-doped CeO_2 crystallite. The calcination of the coprecipitated zirconia-ceria solid solutions resulted in the growth of ZrO_2-doped CeO_2 crystallite after hydrothermal treatments. Un-doped CeO_2 crystals (120μm) were obtained from hydrous ceria under 100 MPa at 700°C for 3 days with 8 wt% KF solution. On the other hand, ZrO_2-doped CeO_2 crystals (200μm) were grown from the calcined hydrous zirconia-ceria solutions with 15 wt% LiCl solution under 100 MPa at 600°C for 3 days.

References

1) S. F. Pal'guev and Z. S. Volchenkova: Russ. J. Phys. Checm., **34** (1960) 211.
2) P. F. Becher: J. Am. Ceram. Soc., **64** (1981) 37.
3) E. Tani, M. Yoshimura and S. Sōmiya: Yogyo Kyokai Shi, **90** (1982) 195.
4) M. Yoshimura, E. Tani and S. Sōmiya: Solid State Ionics, **3/4** (1981) 477.
5) E. Tani, M. Yoshimura and S. Sōmiya: J. Am. Ceram. Soc., **66** [7], in press (1983).
6) A. Rouanet: C. R. Acad. Sci., Paris **266C** (1968) 908.
7) C. B. Finch and G. Wayne Clark: J. Appl. Phys., **37** (1966) 3910.
8) W. H. Grodkiewicz and D. J. Nitti: J. Am. Ceram. Soc., **49** (1966) 576.
9) R. C. Linares: J. Phys. Chem. Solids, **28** (1967) 1285.
10) B. M. Wanklyn: J. Crystal Growth, **5** (1969) 219.

Reprinted from *Report of the Research Laboratory of Engineering Materials, Tokyo Institute of Technology*, No. 9 (1984) 53–64.

Fine Zirconia Powders by Hydrothermal Processing

By
Masahiro YOSHIMURA* and Shigeyuki SŌMIYA**

Abstract

Fine zirconia powders could be prepared by two hydrothermal techniques, (a) Hydrothermal crystallization of amorphous zirconia, and (b) Hydrothermal oxidation of Zr metal, both recently developed in our laboratory.

In the hydrothermal crystallization, ultrafine monoclinic ZrO_2 crystals with 16 nm of particle size were prepared using KF solutions at 200°C under 100 MPa for 24 h, whereas ultrafine tetragonal ZrO_2 (15 nm) and monoclinic ZrO_2 (19 nm) crystals were obtained with LiCl solutions at 300°C under 100 MPa for 24 h. The ZrO_2 particles were well shaped and isolated from each other. Metastable tetragonal ZrO_2 was formed only from amorphous phase.

Hydrothermal oxidation of Zr metal by pure water yielded only monoclinic ZrO_2 with 24 nm of average crystallite size above 400°C under 100 MPa for 3 h. When Ca solutions were used instead of H_2O, fine cubic zirconia grains were partially formed, in addition to monoclinic zirconia particularly in the case of $Ca(NO_3)_2$ solutions with excess an amount of Zr metal at high temperatures, 700°C. This cubic zirconia seemed to be stabilized both by calcium and by nitrogen. $Y(NO_3)_3$ or NH_4NO_3 solution also yielded a small amount of cubic zirconia phase in the products.

1. Introduction

Fine powders have become important in recent sciences and technologies of ceramics, chemistry, physics and biology. The preparation methods of fine powders can be divided into two categories; (A) Breakdown (size reduction) and (B) Build-up (size increase) methods.[1] The breakdown processes where large grains are broken physically (mechanically) into small grains by crushing, grinding or milling are inevitable the incorporation of inhomogeneous strain, size distribution and/or impurities from containers or grinding media. New techniques[2,3] recently developed to prepare fine powders are, therefore, mostly build-up methods where fine particles are made from much smaller particles such as clusters, molecules, ions or atoms. The dispersed particles in medium as gas, vapor, solvent and/or melt are condensed physically by deposition or solidification, or by chemical reactions; burning, decomposition, precipitation, hydrolysis etc. or by solvent removal by drying or extraction etc. The shape, size, crystalline system

* Associate Professor, Dr. Sci. in Eng.
** Professor, Dr. Sci. in Eng.

and/or crystallinity of the particles can be controlled by the controle of the condensation rate and growth rate. The hydrothermal method described here, has similar features to these liquid and gas methods because of the use of high temperature-high pressure solutions and/or vapors as a dispersed medium.

In the present report, we describe a build-up method:

(1) Hydrothermal crystallization[4,5] and a breakdown method:

(2) Hydrothermal oxidation.[6–14] In the latter, chemical breakdown pluverizes the grains to fine particles without the incorporation of impurities and agglomeration during reaction. Particularly, this method using the reaction between high-pure metal and pure water has the merit of no contamination from solvent or precipitant needed in other solution (wet) methods.

2. Hydrothermal Crystallization Method[4,5]

2.1 Experimental Procedure

The starting material was amorphous hydrous zirconia prepared from Hf-free $ZrCl_4$ (Sn: 680, Fe: 620, Hf: < 70, Al: < 30, ppm, Nippon Mining Co., Ltd., Tokyo) solution with 3N NH_4OH, filtered, dispersed in distilled water and refiltered. After four dispersion and filtration processes, the precipitate was dried at 120°C for 48 h. The water content of this material was determined by TG-DTA (Thermoflex, Rigaku Denki Co., Ltd., Tokyo). The precipitates calcined at 240°, 420°, 620° and 820°C in air for 60 min at a heating rate of 10°C/min were also used as starting materials. Distilled water and solutions of 8 wt% KF, 30 NaOH, 7, 15 and 30 LiCl, and 10 KBr were used as mineralizers. A Pt or Au tube 2.7 mm ID, 0.15 mm thick and 32 mm long was filled with the starting material and solution, and treated under 100 MPa at 200° to 600°C for 24 h using a test-tube-type pressure vessel. The temperature was measured by two Platinel (Au-Pd, Type K) thermocouples calibrated against the melting point of Zn (419.5°C) and controlled within ±3°C during a run. The pressure was measured by a calibrated Heise bourdon gauge and controlled within ±0.2 MPa. The vessel was then quenched into cold water. The products were washed in distilled water, dried and examined by X-ray powder diffractometry (RU-200, Rigaku Denki Co., Ltd., Tokyo), transmission electron microscopy and energy dispersive spectroscopy (JEM-100CX, JEOL Co., Ltd., Tokyo).

Crystallite sizes were calculated using the (111) diffraction peak of monoclinic and tetragonal ZrO_2 from the Scherrer formula, $D = 0.9\lambda/(\beta \cos\theta)$, where D is the crystallite size, λ the wave length (CuKα), β the corrected half width obtained using the (101) peak of low quartz as the standard and the Warren formula, and θ the diffraction peak angle. The tetragonal ZrO_2 content was estimated from the relative intensities of the (111) peaks of tetragonal and monoclinic ZrO_2.

2.2 Results and Discussion

The starting material dried at 120°C was amorphous (Fig. 1) by X-ray diffraction, and TG-DTA gave H_2O/ZrO_2 ratio of ≈2 for this material. This amorphous material was crystallized in monoclinic and/or tetragonal ZrO_2 by hydrothermal treatments. Under the hydrothermal conditions of 100 MPa, 300°C for 24 h, monoclinic ZrO_2 only formed in KF or NaOH solutions as shown in Fig. 2. In the case of KF solutions, only monoclinic ZrO_2 was obtained at temperatures of 200°~600°C. The formed monoclinic ZrO_2 particles are homogeneous, non-agglomerated microcrystals with the diameter of ~20 nm (Fig. 3) by TEM observations.

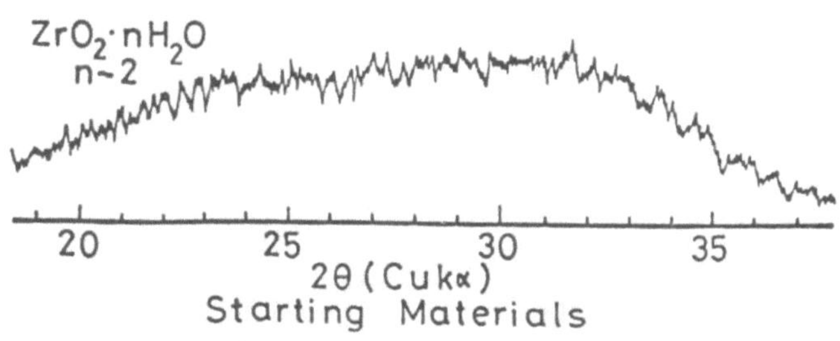

Fig. 1. X-ray diffraction pattern of amorphous hydrous zirconia used as a starting material of hydrothermal crystallization, $ZrO_2 \cdot nH_2O$ ($n \approx 2$).

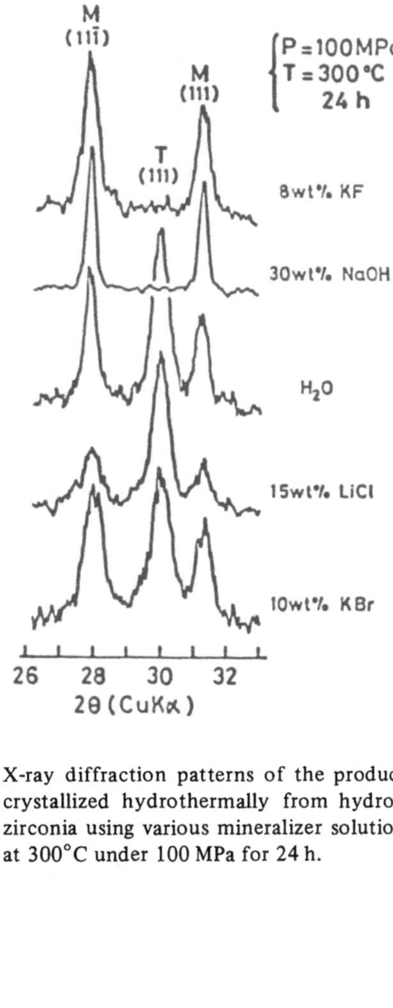

Fig. 2. X-ray diffraction patterns of the products crystallized hydrothermally from hydrous zirconia using various mineralizer solutions at 300°C under 100 MPa for 24 h.

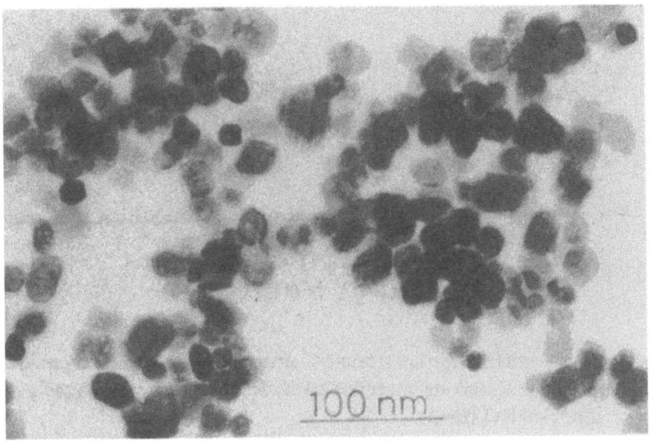

Fig. 3. Transmission electron micrograph of monoclinic ZrO_2 produced by hydrothermal crystallization at 400°C under 100 MPa for 24 h using 8 wt% KF solution.

The observed particle size is in good agreement with the crystallite size calculated from the half width of the X-ray diffraction peak; 20 nm at 300°C, 21 nm at 400°C as indicated in Fig. 4. This shows that the hydrothermally crystallized particles are single-grain microcrystals isolated each other, in contrast to agglomerated particles resulting from calcination of amorphous hydrous zirconia in air. Figure 4 shows that the crystallite size of monoclinic ZrO_2 increased slightly with the increase of treating temperature from 200° to 500°C, but increased rapidly above 500°C; i.e. 70 nm at 550°C, $\gg$ 100 nm at 600°C. This phenomenon appears to be caused by the fact that the solubility of zirconia increased rapidly in KF solution and consisting with the fact that the growth rate of monoclinic ZrO_2 crystal became significant above 600°C.[15]

In the NaOH solution, the crystallite size of monoclinic ZrO_2 was grown to 40 nm at even 300°C because of larger solubility than in KF solutions as indicated in Table 1.

On the other hand, tetragonal ZrO_2 was crystallized in addition to small amount of monoclinic ZrO_2 in the mineralizer solutions of H_2O, LiCl and KBr as indicated in Table 1 and Fig. 2. The ratio of tetragonal ZrO_2/monoclinic ZrO_2 varied with the species and concentrations of the solutions and with treating temperatures, but single-phase tetragonal ZrO_2 could not be obtained under the hydrothermal conditions studied. Figure 5 shows the tetragonal (70%) and monoclinic (30%) ZrO_2 particles formed at 300°C in 15 wt% LiCl solution under 100 MPa for 24 h. The average crystallite size was calculated to be 15 nm for tetragonal ZrO_2 and 19 nm for monoclinic ZrO_2 in this product as seen in Table 1. The particles in Fig. 5 seems to be single crystalline particles of tetragonal or monoclinic ZrO_2 because of the agreement between the particle size and the crystallite size.

The formation mechanism of tetragonal ZrO_2 particle has been studied in detail[5,16] and concluded that particle size and impurities are not primary factors in the stabilization of tetragonal ZrO_2. The formation of tetragonal ZrO_2 is closely related to the existence of amorphous ZrO_2.[5] A kinetic formation mechanism seems to be responsible to form tetragonal ZrO_2 from amorphous zirconia.

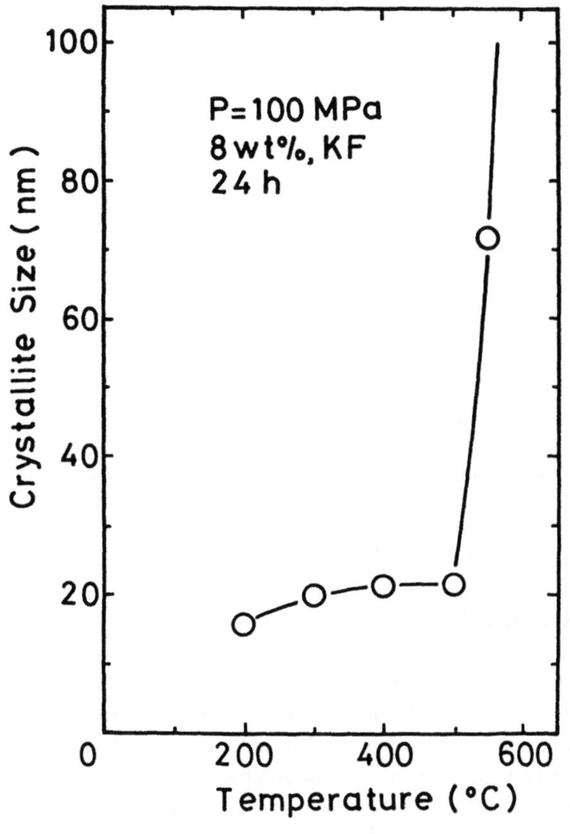

Fig. 4. Variation with temperature of crystallite size of monoclinic ZrO_2 produced by hydrothermal crystallization under 100 MPa for 24 h using 8 wt% KF solution.

Table 1. Phases Present and Crystallite Size of Hydrothermally Crystallized Products at 100 MPa for 24 h.

Mineralizer	Temp (°C)	Average crystallite size (nm)	
		Tetragonal ZrO_2	Monoclinic ZrO_2
KF (8 wt%)	200	Not detected	16 nm
KF (8 wt%)	300	Not detected	20 nm
NaOH (30 wt%)	300	Not detected	40 nm
H_2O	300	15 nm	17 nm
LiCl (15 wt%)	300	15 nm	19 nm
KBr (10 wt%)	300	13 nm	15 nm

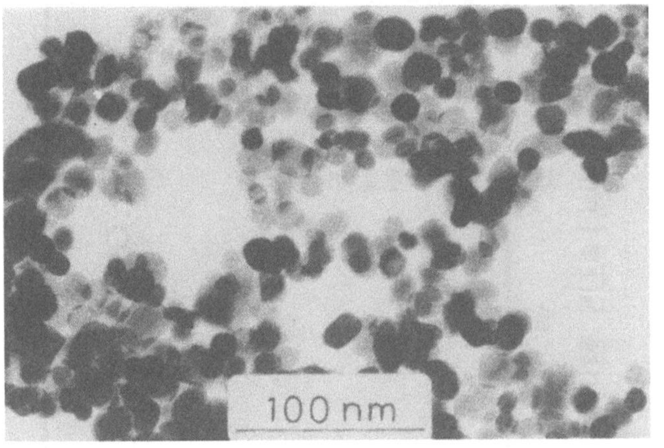

Fig. 5. Transmission electron micrograph of tetragonal (70%) and monoclinic (30%) ZrO_2 powders produced by hydrothermal crystallization at 300°C under 100 MPa for 24 h using 15 wt% LiCl solution.

3. Hydrothermal Oxidation Method

3.1 Experimental

The starting material was zirconium metal powder, 99.9% including ≈2% Hf, consisting of particles of 1 to 10μm in diameter. This zirconium and redistilled water after deionization were hermetically sealed in a Pt capsule 2.7 mm ID and 35 mm long. The sealed capsule was heated in a test-tube-type vessel at temperatures up to 700°C under the pressure of 100 MPa. After the heat treatment for a desired period, the vessel was quenched into cold water, then the capsule was weighed to check leakages. The solid products were examined by X-ray powder diffractometry and electron microscopy.

To study the direct formation of stabilized zirconia fine powders, Ca- or Y-containing solutions were used as an oxidation agent instead of pure water. They were 59 wt% $Ca(NO_3)_2{\cdot}4H_2O$, 41 wt% $Y(NO_3)_3{\cdot}6H_2O$ solutions. $Ca(OH)_2$ or NH_4NO_3 solution was also used as a reaction agent.

3.2 Results and Discussion

3.2.1 Reaction between zirconium and pure water[6–9]

Under the pressure of 100 MPa, the reaction between Zr and H_2O has been observed above 250°C. (eq. (1))

$$Zr + 2\,H_2O \rightarrow ZrO_2 + 2\,H_2 \qquad (1)$$

The solid products in the capsule at 300°C consist of monoclinic zirconia (m-ZrO_2), zirconium hydride (ZrH_2) and residual zirconium metal (Zr), which make a popcorn-like grains as seen in Fig. 6. Zirconium hydride is formed by the reaction between Zr and produced H_2 as eq. (2),

$$Zr + H_2 \rightarrow ZrH_2. \qquad (2)$$

As the hydride reacts with water as eq. (3)

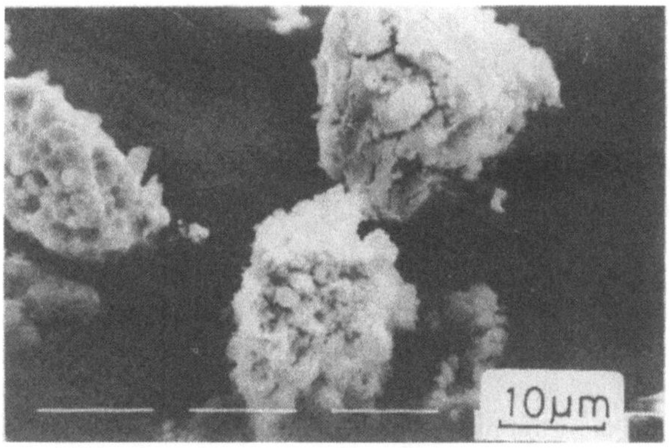

Fig. 6. Scanning electron micrograph of popcorn-like grains produced by hydrothermal oxidation of Zr with H_2O at 300°C under 100 MPa for 3 h.

$$ZrH_2 + 2\,H_2O \rightarrow ZrO_2 + 3\,H_2, \tag{3}$$

sole monoclinic zirconia is the final product above 400°C. During the oxidation and hydridation, zirconium grains are divided into smaller particles of 20~30 nm, Fig. 7, which are in good agreement with the average crystallite size (24 nm) calculated from the X-ray diffraction peak profile. No tetragonal zirconia was formed by this hydrothermal oxidation process, even though the crystallite size (≈24 nm) is smaller than so-called critical crystallite size (≈30 nm)[17] for metastable tetragonal zirconia. This result is consistent with the results of the hydrothermal crystallization previously mentioned, where metastable tetragonal zirconia might be formed from amorphous zirconia by a kinetic formation mechanism.

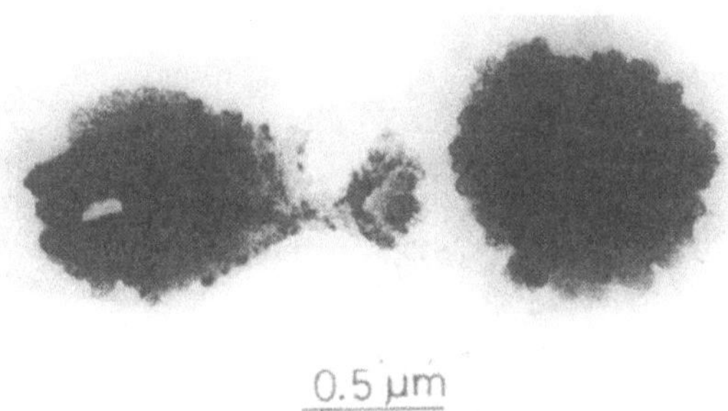

Fig. 7. Transmission electron micrograph of m-ZrO_2 grains produced by hydrothermal oxidation of Zr with H_2O at 400°C under 100 MPa for 3 h.

Hydrothermal oxidation of hafnium (Hf) has also been studied[10,11,14] and revealed that the hydridation process was mainly responsible to pluverize the grains to fine particles during reactions.

3.2.2 Reaction between zirconium and solutions containing stabilizer[8,9]

In the hydrothermal oxidation of zirconium metal with stabilizer agents, it was expected that all water in the capsule would react with zirconium to yield oxides as follows:

$$0.15\ Ca(OH)_2 + 0.85\ Zr + 1.55\ H_2O \rightarrow Zr_{0.85}Ca_{0.15}O_{1.85} + 1.70\ H_2. \qquad (4)$$

The products, however, was not a cubic zirconia solid solution but the mixture of m-ZrO_2 and $Ca(OH)_2$ after heating at 500° or 600°C under 100 MPa for 3 h. Small amounts of cubic zirconia, less than ~15%, and $CaZrO_3$ were produced at 700°C in addition to large amounts of monoclinic zirconia. Thus, the cubic zirconia and $CaZrO_3$ seems to be formed by a solid state reaction between monoclinic zirconia and $Ca(OH)_2$ at higher temperatures. From the expectation of the reaction between zirconium metal and solute Ca(aq.) in the solution, a more soluble calcium salt, calcium nitrate, was used as a stabilizer. The starting mixture corresponding to the equation as follows;

$$0.15\ Ca(NO_3)_2 + 0.85\ Zr + 0.95\ H_2O \rightarrow Zr_{0.85}Ca_{0.15}O_{1.85} + 0.3\ NH_3 + 0.5\ H_2, \qquad (5)$$

yielded cubic zirconia phase above 500°C. Excess amounts of zirconium, $Zr/O \geqslant 0.460$ derived from eq. (5), decomposed H_2O and nitrate completely to produce ZrN and ZrH_2 as seen in Fig. 8. The apparent yield fraction of cubic zirconia showed a maximum at around $Zr/O \approx 0.63$ at 700°C. The cubic zirconia has never been formed by the reaction between zirconium and pure H_2O as described before. The formation of cubic zirconia by the reaction between zirconium and $Ca(NO_3)_2$ solution, therefore, should be the effect of calcium and/or nitrogen. The cubic zirconia phase had a lattice parameter

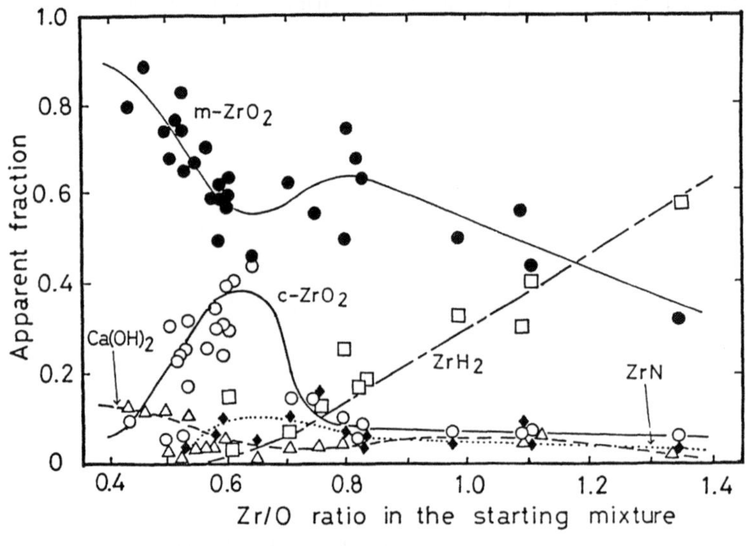

Fig. 8. Apparent fraction of each phase in the products between Zr metal and $Ca(NO_3)_2$ solution at 700°C under 100 MPa for 3 h.

ranged a = 0.507~0.510 nm, which is much smaller than a = 0.515 nm for 15% calcium stabilized zirconia, but is rather close to a = 0.509 nm of the cubic phase of pure zirconia. This result and the formation of free $Ca(OH)_2$ as a by-product in the products (see Fig. 8) suggested that the cubic phase might be stabilized both by calcium and nitrogen. This suggestion was confirmed by the facts that an energy dispersive X-ray spectroscopy associated with TEM revealed the existence of calcium in the aggregation of zirconia crystallites, and that a cubic zirconia could be formed by the reaction between zirconium and NH_4NO_3 solution under the similar hydrothermal conditions, 700°C, 100 MPa, 3 h with an excess zirconium as shown in Fig. 9. That is, nitrogen can also form a cubic zirconia phase. The cubic phase containing nitrogen had a lattice parameter of a = 0.508 nm, which is in good agreement with that of nitrogen stabilized cubic zirconia phase previously reported.[18]

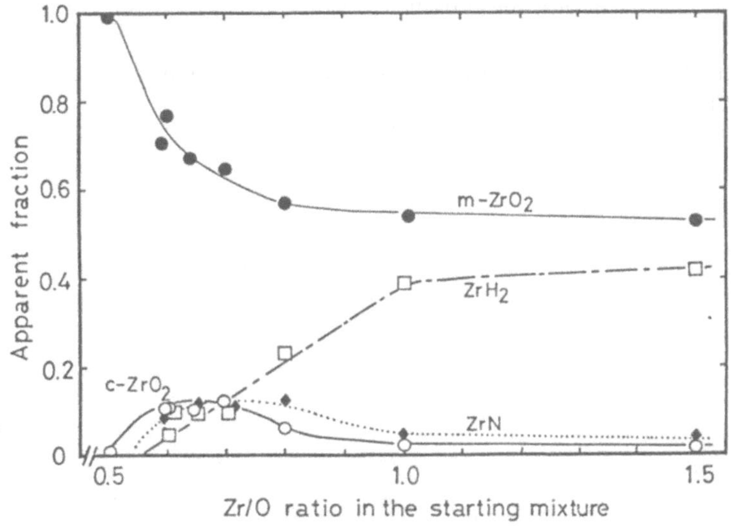

Fig. 9. Apparent fraction of each phase in the products between Zr metal and NH_4NO_3 solution at 700°C under 100 MPa for 3 h.

The average crystallite size was 18 nm and 25 nm for the cubic zirconia and monoclinic zirconia phases, respectively, formed by the reaction between zirconium and $Ca(NO_3)_2$ solution. Sometimes, large cubic zirconia particles ~250 nm in size (Fig. 10) were observed in the products. These large particles appeared to be grown in a high temperature-high pressure solution by a dissolution/precipitation mechanism in early stage of reaction. In intermediate and later stages of reaction, however, the solution (water) was consumed by the reaction with zirconium as indicated by the eqs. (4) and (5), so that the average crystallite size would remain small in these reactions.

Similarly, a cubic zirconia phase could be formed in the products between zirconium and a $Y(NO_3)_3$ solution under the same hydrothermal conditions, 700°C, 100 MPa, 3 h, as an expected reaction;

$$0.94\ Zr + 0.06\ Y(NO_3)_3 + 1.43\ H_2O \rightarrow Zr_{0.94}Y_{0.06}O_{1.97} + 0.09\ NH_3 + 1.295\ H_2. \quad (6)$$

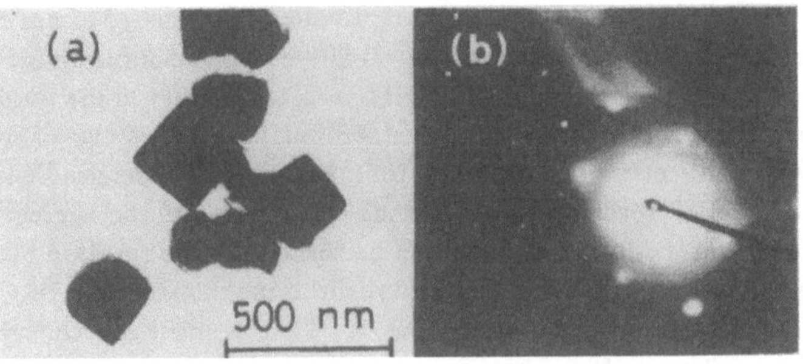

Fig. 10. Relatively large cubic ZrO_2 particles produced from Zr metal and $Ca(NO_3)_2$ solution at 700°C under 100 MPa for 3 h.
(a) Transmission electron micrograph
(b) Electron diffraction pattern

The products, however, contained a large amount of monoclinic zirconia, and ZrH_2, ZrN and Y_2O_3(YOOH), Fig. 11.

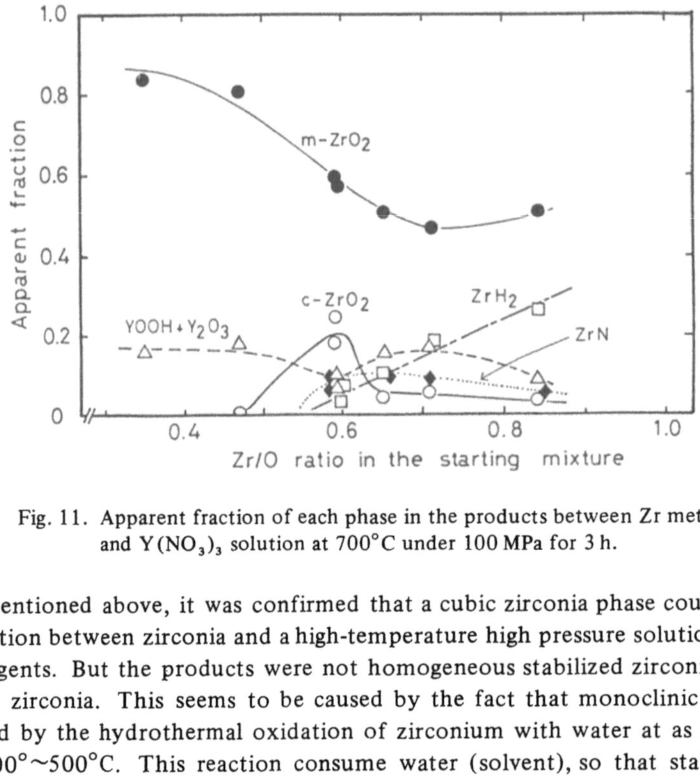

Fig. 11. Apparent fraction of each phase in the products between Zr metal and $Y(NO_3)_3$ solution at 700°C under 100 MPa for 3 h.

As mentioned above, it was confirmed that a cubic zirconia phase could be formed by the reaction between zirconia and a high-temperature high pressure solution containing stabilizer agents. But the products were not homogeneous stabilized zirconia but mainly monoclinic zirconia. This seems to be caused by the fact that monoclinic zirconia was first formed by the hydrothermal oxidation of zirconium with water at as low temperatures as 300°~500°C. This reaction consume water (solvent), so that stabilizer agents would be precipitated as solids.

In order to obtain homogeneous stabilized zirconia, therefore, it is desirable to keep the stabilizer agents in homogeneous solutions up to later stages of reaction. A large amount of dilute solutions containing stabilizer agents will be preferable for this purpose. Such works are now in progress.

Acknowledgement

This paper is a review including partly the theses of Dr. E. Tani and Mr. S. Kikugawa for Dr. Sci. in Eng. and M. Sci. in Eng., respectively. The authors are thankful for them.

References

1) C. R. Veale: "Fine Powders, Preparation, Properties and Uses", Halsted Press Div., Wiley & Sons, New York, 1972.
2) D. W. Johnson, Jr.: Nonconventional Powder Preparation Techniques, Am. Ceram. Soc. Bull., **60** [2] 221–224 & 243 (1981).
3) S. Sōmiya and M. Yoshimura: Ultrafine Zirconia Particles Part I & II (in Japanese), Kino Zairyo (Function & Materials), **3** [4] 51–58, [5] 56–62 (1983).
4) E. Tani, M. Yoshimura and S. Sōmiya: Hydrothermal Preparation of Ultrafine Monoclinic ZrO_2 Powders, J. Am. Ceram. Soc., **64** [12] C-181 (1981).
5) E. Tani, M. Yoshimura and S. Sōmiya: Formation of Ultrafine Tetragonal ZrO_2 Powder Under Hydrothermal Conditions, J. Am. Ceram. Soc., **66** [1] 11–14 (1983).
6) M. Yoshimura and S. Sōmiya: Fabrication of Dense Non-stabilized ZrO_2 Ceramics by Hydrothermal Reaction Sintering, Am. Ceram. Soc. Bull., **59** [2] 246 (1980).
7) M. Yoshimura and S. Sōmiya: Hydrothermal Reaction Sintering of Monoclinic Zirconia, pp. 453–63 in "Science and Technology of Zirconia, Advances in Ceramics, Vol. 3", A. H. Heuer and L. W. Hobbs, ed., The American Ceramic Society, Columbus, 1981.
8) M. Yoshimura and S. Sōmiya: Synthesis and Sintering of Zirconia Fine Powders by Hydrothermal Reactions from Zirconium Metal and High temperature-High pressure Solutions, pp. 417–22 in "Sintering-Theory and Practice", Materials Science Monographs, Vol. 14, D. Kolar, S. Pejovnik and M. M. Ristic, ed., Elsevier Pub. Co., Amsterdam, 1982.
9) M. Yoshimura, S. Kikugawa and S. Sōmiya: Preparation of Zirconia Fine Powders by Reactions between Zirconium Metal and High temperature-High pressure Solutions, pp. 793–96 in "High Pressure in Research and Industry, Proc. of 8th AIRAPT Conf. & 19th EHPRG Conf., Vol. 2", C-M. Backman, T. Johannison and L. Tegner, ed., Arkitektkopia ISBN, Uppsara, Sweden, 1982.
10) H. Toraya, M. Yoshimura and S. Sōmiya: Preparation of Fine Monoclinic Hafnia Powders by Hydrothermal Oxidation, J. Am. Ceram. Soc., **65** [5] C-72 (1982).
11) H. Toraya, M. Yoshimura and S. Sōmiya: Hydrothermal Oxidation of Hf Metal Chips in the Preparation of Monoclinic HfO_2 Powders, J. Am. Ceram. Soc., **66** [2] 148–50 (1983).
12) M. Yoshimura, S. Kikugawa and S. Sōmiya: Preparation of Alpha Alumina Fine Powders by Hydrothermal Oxidation Method (in Japanese), Funtai Oyobi Funmatsu Yakin, **30** [5] 35–38 (1983).
13) M. Yoshimura, S. Kikugawa and S. Sōmiya: Alumina-Zirconia Fine Powders Prepared by Hydrothermal Oxidation (in Japanese), Yogyo Kyokai Shi, **91** [4] 182–188 (1983).

14) H. Toraya, M. Yoshimura and S. Sōmiya: Reaction Kinetics in the Hydrothermal Oxidation of Hf, J. Am. Ceram. Soc., **66** [11] 818–22 (1983).
15) V. A. Kuznetsov: Crystallization of Titanium, Zirconium and Hafnium Oxides and Some Titanate and Zirconate Compounds under Hydrothermal Conditions, J. Cryst. Growth, 3/4, 405–10 (1968).
16) M. Yoshimura and S. Sōmiya: Preparation of Fine Zirconia Powders by Hydrothermal Techniques (in Japanese), pp. 45–60, Zirconia Ceramics 1, Ed. by S. Sōmiya, Uchida Rokakuho Pub. Co., Ltd., Tokyo, 1983.
17) R. C. Garvie: Occurence of Metastable Tetragonal Zirconia as a Crystallite Size Effect, J. Phys. Chem., **69** [4] 1238–43 (1965).
18) N. Claussen, R. Wagner, L. J. Gauckler and G. Petzow: Nitride-Stabilized Cubic Zirconia, J. Am. Ceram. Soc., **61** [7–8] 369–70 (1978).

Reprinted from *Z. anorg. allg. Chem.*, **540/541** (1986) 251–258.

Cubic Eu-doped Hafnia Ultrafine Particles Crystallized under Hydrothermal Conditions

SHIGEYUKI SŌMIYA*, MASAHIRO YOSHIMURA, HIDEO TORAYA, and YASUHITO FUSHII

Yokohama (Japan), Research Laboratory of Engineering Materials, & Dept. of Materials Science and Engineering Tokyo Institute of Technology

Dedicated to Professor Paul Hagenmuller on his 65th Birthday

Abstract. The crystallization of HfO_2—Eu_2O_3 amorphous co-precipitates was studied under hydrothermal conditions. The precipitates with the weight ratio of HfO_2:Eu_2O_3 = 80:20 and 70:30 co-precipitated by NH_4OH from the solution $HfOCl_2$ and $EuCl_3$ and dried at 90 °C for 100 h, yielded a single phase cubic solid solution $Hf_{1-x}Eu_xO_{2-x/2}$ (x = 0.230 and 0.339) with a fluorite lattice by hydrothermal treatment at 300 °C under 10 MPa for 3 h. These products consist of non-aggregated, ultrafine, narrow-sized and single-crystalline particles of about 10 nm and 8 nm in size for 20 wt% and 30 wt% Eu_2O_3 samples, respectively. It was confirmed by the TEM observation, X-ray diffraction and specific surface area measurement by BET method which gave almost the same particle size. The maximum compositional fluctuation in the products was less than 2 wt% by energy dispersive spectroscopy for each particle and by precise X-ray profile analysis of all the diffraction lines of the cubic phase using a Williamson-Hall plotting ($\beta \cos \Theta/\lambda$ vs $\sin \Theta/\lambda$).

Ultrafeine, unter hydrothermalen Bedingungen kristallisierte Partikel von kubischem, Eu-dotierten Hafniumdioxid

Inhaltsübersicht. Die Kristallisation von amorphen Niederschlägen von HfO_2—Eu_2O_3 wurde unter hydrothermalen Bedingungen untersucht. Die aus der Lösung von $HfOCl_2$ und $EuCl_3$ mit NH_4OH im Gewichtsverhältnis von HfO_2:Eu_2O_3 = 80:20 und 70:30 gefällten und bei 90°C in 100 h getrockneten Niederschläge, ergeben eine einphasige, kubische feste Lösung von $Hf_{1-x}Eu_xO_{2-x/2}$ (x = 0,230 und 0,339) mit einem Fluorit-Gitter nach hydrothermaler Behandlung bei 300 °C unter 10 MPa für 3 h. Diese Produkte bestehen aus nicht aggregierten, ultrafeinen und einkristallinen Partikeln von etwa 10 bzw. 8 nm für die 20 bzw. 30 Gew.-% Eu_2O_3-enthaltenden Proben. Dies wurde durch TEM-, Röntgen- und BET-Messungen bestätigt. Die maximale Schwankung in der Zusammensetzung der Produkte betrug weniger als 2 Gew.-% auf Grund von EDS-Messungen für jedes Partikel und durch genaue Röntgen-Profilanalysen aller Diffraktionslinien der kubischen Phase mittels einer Williamson-Hall-Darstellung ($\beta \cos \Theta/\lambda$ als Funktion von $\sin \Theta/\lambda$).

1. Introduction

HfO_2 has similar chemical properties as ZrO_2. Hafnia has become important in applications as engineering ceramics with high strength and toughness, electronic materials or solid electrolytes. The materials in the system HfO_2—Eu_2O_3 have also been expected as high temperature materials, optical materials or even as control rods of nuclear plant.

Few researches have been reported about the $HfO_2-Eu_2O_3$ system. SCHEIDECKER et al. [1] suggested a phase diagram of this system indicating no ordering phase. ZHIDOVINOVA et al. [2] and GUNDOVIN et al. [3], however, identified a pyrochlore phase by X-ray diffraction or by i.r. spectra. As for liquidus line, the results of SCHEVCHENKO et al. [4] are higher than that of SCHEIDECKER et al. [1] by 100 ~ 300 °C. Those are the investigations of phases by the solid-solid reaction or fusion-solidification reaction. But still fewer reports have been found about the preparation of fine powders which is very important for production of ceramic materials applicable for practical uses.

Hydrothermal crystallization, which was carried out in the present study is a method to prepare fine crystalline powders under hydrothermal conditions from amorphous or less crystalline precipitates as starting materials [5, 6]. With this method, TANI et al. [5, 6], NISHIZAWA et al. [7], MITSUHASHI et al. [8] obtained ultrafine particles of ZrO_2. The authors also reported about mixed powders of $Al_2O_3-TiO_2$ [9]. In the present paper, we describe the preparation of ultrafine particles of $HfO_2-Eu_2O_3$ solid solutions and the morphological and compositional analyses of these particles.

2. Experimental

Starting materials were $HfOCl_2 \cdot 8\,H_2O$ (containing 0.70 wt% Zr, Kojundo Kagaku Co. Ltd., Japan) and $EuCl_3$ (purity 99.9%, Shin-etsu Kagaku Industry Co. Ltd., Japan). The precipitates with the weight ratios of $HfO_2:Eu_2O_3 = 80:20$ and 70:30 were co-precipitated by NH_4OH from the mixed solutions of $HfOCl_2$ and $EuCl_3$. The co-precipitate was washed by distilled water until NH_4^+ and Cl^- ions could not be detected in the filtrate and then dried at 90 °C in air for 100 h.

Hydrothermal crystallization was performed at 300 °C, 10 MPa, for 3 h with re-distilled water in a gold-lined Glaylock type autoclave. The details of apparatus and experimental conditions were described in another paper [10].

Lattice parameters, crystallite sizes and lattice strains were measured by X-ray diffraction (XRD) with carbon monochromated CuKα using a diffractometer with a rotating anode UR-200 (Rigaku Electric Co. Ltd., Japan). Si Metal (Kinzoku Kagaku Co. Ltd. purity 99.99%) was used as an internal standard to measure lattice parameters. In order to determine crystallite sizes and compositional fluctuations of produced particles, precise X-ray profile analysis [11, 12] was carried out, where true diffraction profiles were deconvoluted using standard materials, and then crystallite sizes were calculated from their integrated breadths. The calcination of 30 wt% $Eu_2O_3-HfO_2$ co-precipitates at 1400 °C for 4 h provided cubic solid solution crystallized sufficiently as the standard material. All the profiles of diffraction lines required high accuracy achieved by counting for 200 second with every 0.05° scanning step.

Specific surface area of products were measured by means of a flowing BET method using mixture of He and N_2 gasses. The produced particles were observed by a transmission electron microscope (TEM, JEM-200CX, Jeol Co. Ltd., Japan) and analyzed by energy dispersive spectroscopy (EDS). Particle size was determined by Krummbein method, where maximum lengths of every particles are counted along a fixed direction on the TEM Microphotographs.

3. Results and Discussion

3.1. Produced particles

Co-precipitates dried at 90 °C for 100 h were amorphous. They yielded fluorite-type cubic hafnia solid solutions as shown in Fig. 1 after hydrothermal crystallization at 300 °C, under 10 MPa, for 3 h. The product are cubic fluorite solid solutions $Hf_{1-x}Eu_xO_{2-x/2}$, x = 0.230 and x = 0.339 for 20 wt% Eu_2O_3 and 30 wt% Eu_2O_3, respectively.

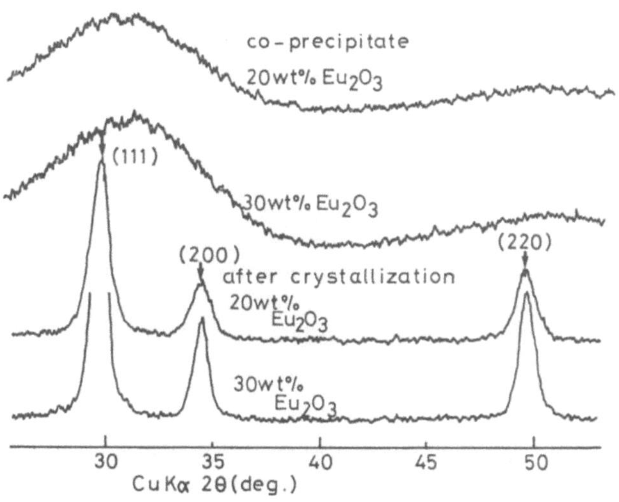

Fig. 1 X-ray diffraction patterns of amorphous coprecipitates and fluorite-type hafnia solid solutions crystallized hydrothermally at 300 °C, 10 MPa, for 3 h from HfO_2—Eu_2O_3 coprecipitates dried at 90 °C, for 100 h.

TEM photographs of the product indicate uniform and non-aggregated particles as seen in Fig. 2 for 20 wt% Eu_2O_3 and in Fig. 3 for 30 wt% Eu_2O_3. The particles have roundish or faceted polyhedral morphologies. Each particle is distinguishable from another. Particle size distributions by Krummbein method from TEM photographs are fairly narrow as shown in Fig. 4 (20 wt% Eu_2O_3) and Fig. 5 (30 wt% Eu_2O_3), where more than 400 particles were measured. As for 20 wt% Eu_2O_3 sample, 90 percent of particles belong between 4 and 12 nm in size, and the average size calculated from mean volume of particles in Fig. 4 is 9.2 nm. The products of 30 wt% Eu_2O_3 samples have the average size of 7.4 nm with about threefourths of whole particles between 4 and 8 nm. These results indicate that the hydrothermal crystallization produced ultrafine particles with uniform size and shape.

Crystallite size of 20 wt% and 30 wt% Eu_2O_3 samples, which were calculated from the Scherrer's equation with Scherrer's coefficient of 1.047 using the broadening of the X-ray diffraction line (111), were 10.6 nm and 8.4 nm, respectively. The specific surface area measured by BET method were 57.6 m^2/g for 20 wt% Eu_2O_3 and 76.2 m^2/g for 30 wt% Eu_2O_3 samples, respectively. These values indicate that the average diameter of particles were 10.4 nm for 20 wt% Eu_2O_3 and 8.1 nm for 30 wt% Eu_2O_3 samples based upon the assumptions that all the particles have same diameter and same theoretical density based on their lattice parameters.

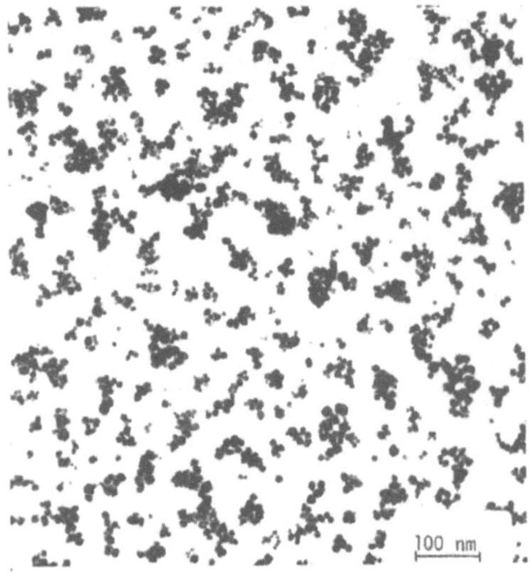

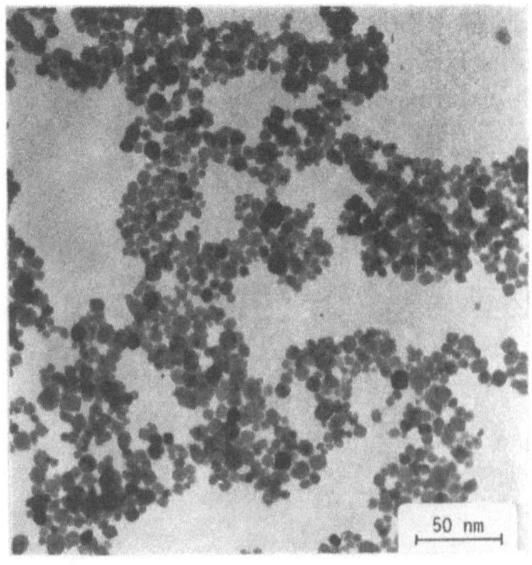

Fig. 2 Fig. 3

Fig. 2 Transmission electron micrograph of 20 wt% Eu_2O_3—HfO_2 cubic solid solution crystallized hydrothermally at 300°C, 10 MPa, for 3 h from the coprecipitate.

Fig. 3 Transmission electron micrograph of 30 wt% Eu_2O_3—HfO_2 cubic solid solution crystallized hydrothermally at 300°C, 10 MPa, for 3 h from the coprecipitate.

Particles sizes from TEM observation, crystallite sizes from XRD and diameter of particles from specific surface area are almost identical as shown in Table 1. This fact confirms that the products crystallized hydrothermally are composed of non-aggregated, uniform and single crystalline particles.

Table 1 Particle sizes from transmission electron micrograph, X-ray line profile and specific surface area by BET method of the Eu_2O_3—HfO_2 cubic solid solution crystallized hydrothermally at 300°C, 10 MPa, for 3 h from the coprecipitates.

HfO_2/Eu_2O_3	Surface Area (m^2/g)	Particle Size (nm) [BET]	Crystallite Size (nm) [X-ray]	Particle Size (nm) [TEM]
80/20	57.6	10.4	10.6	9.2
70/30	76.2	8.1	8.4	7.4

3.2. The composition of produced particles

Compositional distribution is very important as the evaluation of particles especially in multi-component systems. KLUG and ALEXANDER [13] suggested that fine crystallite size, mechanical lattice strains and stacking faults resulted in X-ray line broadening. In addition, compositional fluctuation also causes line

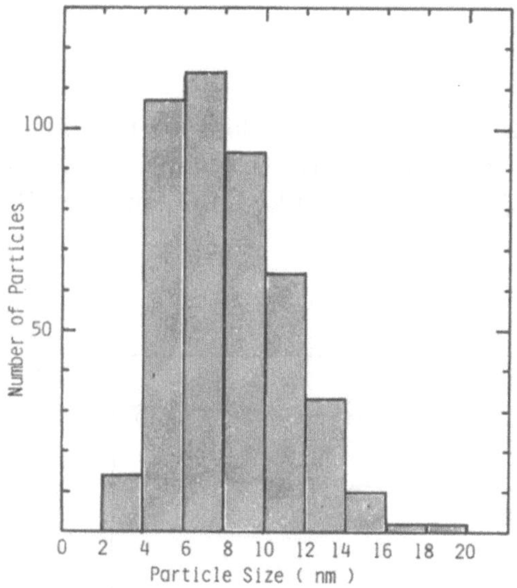

Fig. 4

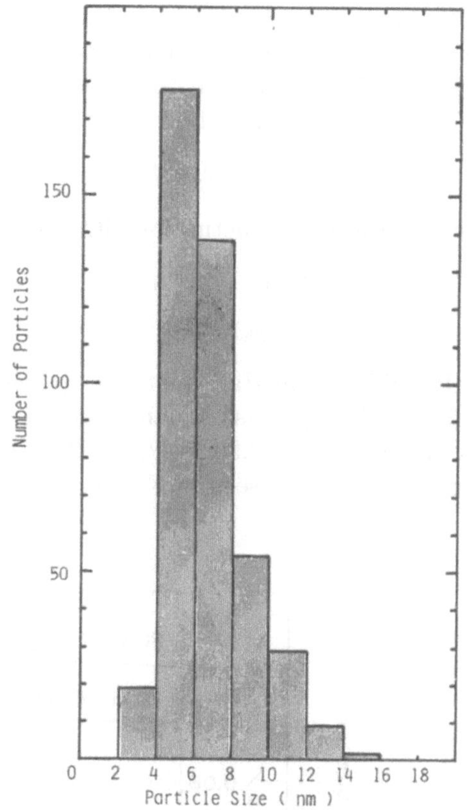

Fig. 5

Fig. 4 Particle size distribution by Krummbein method from transmission electron micrograph of 20 wt% Eu_2O_3—HfO_2 cubic solid solution crystallized hydrothermally at 300°C, 10 MPa, for 3 h from the coprecipitate.

Fig. 5 Particle size distribution by Krummbein method from transmission electron micrograph of 30 wt% Eu_2O_3—HfO_2 cubic solid solution crystallized hydrothermally at 300°C, 10 MPa, for 3 h from the coprecipitate.

broadening, because the lattice parameters varied with their compositions in the systems of solid solutions. HALL [14] proposed the separation of the effect of crystallite size in line broadening from that of lattice strain using the following equation,

$$\beta \cos \Theta/\lambda = 1/\varepsilon + \eta \sin \Theta/\lambda \tag{1}$$

where β is integral breadth, Θ is Bragg angle, λ is wave length, ε is crystallite size and η means lattice strain.

In the products by the hydrothermal crystallization, the effect of residual mechanical lattice strain and stacking faults resulting from twin are negligible, because they are non-aggregated and single crystalline particles. This indicates that the origin of η is only compositional fluctuation in the present case.

Using Williamson-Hall plotting [15] ($\beta \cos \Theta/\lambda$ vs $\eta \sin \Theta/\lambda$), the slope η is almost equal to the variation of interplanar distance $\Delta d/d$, that is, equal to the variation of lattice parameter $\Delta a_0/a_0$ in the cubic system. The lattice parameter a_0 varies with the composition (C:$EuO_{1.5}$ mol%) as the following equation [10].

$$a_0 = 0.5117 + 2.3987 \times 10^{-4}\,C\ (nm). \quad (2)$$

The compositional fluctuation ΔC, therefore, can be described in the equation (3).

$$\eta \doteqdot \Delta a_0/a_0 = (2.3987 \times 10^{-4}\,\Delta C)/a_0$$
$$\therefore \Delta C = (\eta a_0)/(2.3987 \times 10^{-4}). \quad (3)$$

In Williamson-Hall plotting as shown in Fig. 6, the slope of the line is 2.154×10^{-4} by least square method for the 30 wt% Eu_2O_3 sample. It means ΔC value to be ± 2.30 mol% (± 2.15 wt%). As for 20 wt% Eu_2O_3 sample, the slope is still less as seen in Fig. 6.

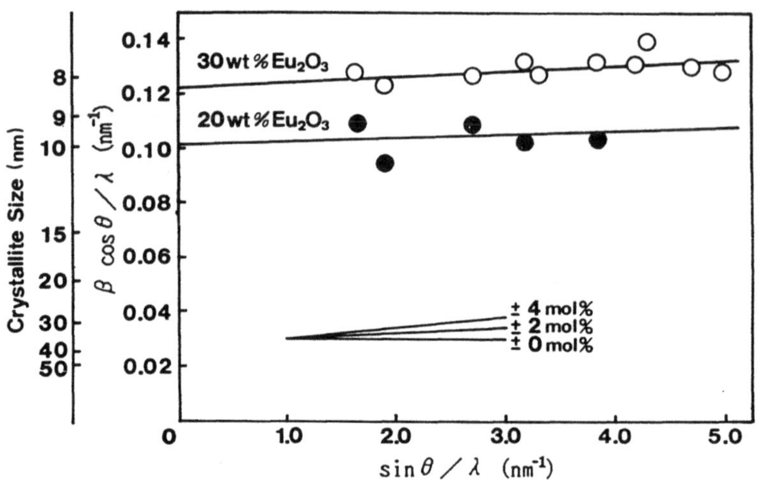

Fig. 6 Williamson-Hall plot for 30 wt% cubic solid solution crystallized hydrothermally at 300 °C, 10 MPa, for 3 h from the coprecipitate.

SCHEIDECKER et al. [1] reported another equation of lattice parameter for HfO_2—Eu_2O_3 solid solutions. That is,

$$a_0 = 0.5087 + 5.24 \times 10^{-4}x\ (nm)$$
$$x:\ Eu_2O_3\ mol\% \quad (4)$$

From this a_0 value, ΔC becomes to be ± 1.44 mol%. Both ΔC indicate little compositional fluctuation in the produced solid solutions.

These facts are in accordance with the result of EDS analysis. As shown in Table 2 and 3, compositional differences among particles found by EDS analysis are within experimental errors, which are estimated up to ~10% for such ultra-fine particles because a long time exposure for a small area (100 nm × 100 nm) was needed to avoid the radiation damage of a particle by a concentrated electron beam.

Table 2 Quantitative EDS analysis of 20 wt% Eu_2O_3—HfO_2 cubic solid solution particles crystallized hydrothermally at 300 °C, 10 MPa, for 3 h from the coprecipitate.

Particle No.	Composition (wt%) HfO_2	Eu_2O_3
1	78.8	21.2
2	79.9	20.1
3	88.6	11.4
4	80.0	20.0
5	81.1	18.9
6	79.7	20.3
7	80.5	19.5
8	78.5	21.5
9	80.1	19.1
Average	80.9	19.1
Standard Deviation		2.9

Table 3 Quantitative EDS analysis of 30 wt% Eu_2O_3—HfO_2 cubic solid solution particles crystallized hydrothermally at 300 °C, 10 MPa, for 3 h from the coprecipitate.

Particle No.	Composition (wt%) HfO_2	Eu_2O_3
1	68.1	31.9
2	69.7	30.3
3	69.7	30.3
4	71.3	28.7
5	70.6	29.4
6	71.1	28.9
Average	70.1	29.9
Standard Deviation		1.1

KAKEGAWA et al. [16, 17] investigated the compositional fluctuation of powders produced by reactions through solid-solid mixing or wet-dry method also using the Williamson-Hall plotting. Few reports, however, have been found on the compositional fluctuation about non-aggregated, single-crystalline and uniform particles. The present results based upon X-ray profile analysis, also confirmed by EDS analysis, revealed that the products by hydrothermal crystallization are uniform not only in size and morphology but also in composition.

In the preparation of powders by solid state reactions, the products are generally inhomogeneous [16—18] through aggregation or heterogeneous nucleous growth. YAMAGUCHI et al. [18] indicated that the average crystallite size derived from X-ray is in disagreement with that from BET measurements for ZnO powders by thermal decomposition.

4. Summary

The crystallization of HfO_2—Eu_2O_3 amorphous co-precipitates from the mixture of $HfOCl_2$ and $EuCl_3$ solution by NH_4OH, was studied under hydrothermal condition at 300 °C under 10 MPa, for 3 h.

1. The products were cubic hafnia solid solution $Hf_{1-x}Eu_xO_{2-x/2}$ of which particle size was about 10 nm and 8 nm for 20 wt% Eu_2O_3 and 30 wt% Eu_2O_3, respectively.
2. The products were composed of non-aggregated and single crystalline particles which have narrow size distribution and uniform morphology.
3. The maximum compositional fluctuation in the products was less than about 2 wt% by EDS for each particle and by precise X-ray profile analysis of all diffraction lines of the cubic phase using a Williamson-Hall plotting.

References

[1] Scheidecker, R. W.; Wilder, D. R.; Moeller, H.: J. Am. Ceram. Soc. **60** (1977) 501.
[2] Zhidovinova, S. V.; Strekalovskii, V. N.; Pal'guev, S. F.: Inorg. Mater. **11** (1976) 1551.
[3] Gundovin, N. V.; Spiridonov, F. M.; Komissarova, L. N.; Petrov, K. I.: Russ. J. Inorg. Chem. **20** (1975) 325.
[4] Schevchenko, A. V.; Lopato, L. M.; Stegnii, A. I.; Ruban, A. K.; Devernyakov, V. S.; Pasichnyi, V. V.: Inorg. Mater. **17** (1980) 741.
[5] Tani, E.; Yoshimura, M.; Sōmiya, S.: J. Am. Ceram. Soc. **64** (1981) C-181.
[6] Tani, E.; Yoshimura, M.; Sōmiya, S.: J. Am. Ceram. Soc. **66** (1983) 11.
[7] Nishizawa, H.; Yamasaki, N.; Matsuoka, K.; Mitsushio, H.: J. Am. Ceram. Soc. **65** (1982) 343.
[8] Mitsuhashi, T.; Ichihara, M.; Tatsuke, U.: J. Am. Ceram. Soc. **57** (1974) 97.
[9] Hattori, T.; Fushii, Y.; Mohri, J.; Yoshimura, M.; Somiya, S.: Yogyo Kyokai Shi **92** (1984) 699.
[10] Yoshimura, M.; Fushii, Y.; Toraya, H.; Tagawa, T.; Imai, H.; Sōmiya, S.: Yogyo Kyokai Shi **93** (1985) 548.
[11] Toraya, H.; Yoshimura, M.; Sōmiya, S.: J. Appl. Crystallogr. **16** (1983) 653.
[12] Toraya, H.; Marumo, F.: Report of the Research Laboratory of Engineering Materials, Tokyo Inst. Techn. **5** (1980) 55.
[13] Klug, H. P.; Alexander, L. E.: "X-ray Diffraction Procedures", John Wiley & Sons, New York 1974, pp. 618—708.
[14] Hall, W. H.: Proc. Phys. Soc. **62** (1949) 741.
[15] Louer, D.; Mesnier, M. T.; Niepce, J. C.: J. Mater. Sci. **19** (1984) 716.
[16] Shirasaki, S.; Takahashi, K.; Kakegawa, K.: J. Am. Ceram. Soc. **56** (1973) 430.
[17] Kakegawa, K.; Mohri, J.; Shirasaki, S.; Takahashi, K.: J. Am. Ceram. Soc. **67** (1984) C-2—C-3.
[18] Yamaguchi, T.; Hirokawa, S.; Watanabe, S.: Unpublished work, Proc. of Annual Meeting of the Ceramic Soc. Japan, May 23, 1985 Okayama, pp. 267—268.

Bei der Redaktion eingegangen am 13. Februar 1986.

Anschr. d. Verf.: Shigeyuki Sōmiya, Masahiro Yoshimura, Hideo Toraya, Yasuhito Fushii,
Research Laboratory of Engineering Materials, and Department of Materials Science and Engineering Institute of Technology Tokyo,
4259 Nagatsuta, Midori, Yokohama 227/Japan

Reprinted from *Advances in Ceramics*, Vol. 21, *Ceramic Powder Science* (American Ceramic Society, Inc., 1987) pp. 43–55.

Hydrothermal Processing of Ultrafine Single-Crystal Zirconia and Hafnia Powders with Homogeneous Dopants

SHIGEYUKI SŌMIYA AND MASAHIRO YOSHIMURA

Tokyo Institute of Technology
Research Laboratory of Engineering Materials
4259, Nagatsuta, Midori
Yokohama 227 Japan

ZENJIRO NAKAI, KAZUMITSU HISHINUMA, AND TAKAO KUMAKI

Chichibu Cement Co., Ltd.
R&D Division
2-1-1, Tsukimicho
Kumagaya 360 Japan

The crystallization of HfO_2-Eu_2O_3 amorphous coprecipitates from the mixture of $HfOCl_2$ and $EuCl_3$ solution by NH_4OH, was studied under hydrothermal conditions at 300°C under 10 MPa, for three hours. The products were cubic hafnia solid solution $Hf_{1-x}Eu_xO_{2-x/2}$ with particle sizes about 10 nm and 8 nm for 20 wt% Eu_2O_3 (x = 0.230) and 30 wt% Eu_2O_3 (x = 0.339), respectively. The products had a nonaggregated, uniform morphology. The maximum compositional fluctuation in the products was less than ≈2 wt% by EDS for each particle and by precise X-ray profile analysis of all diffraction lines of the cubic phase using a Williamson-Hall plotting. Y-PSZ (3 mol% Y_2O_3-doped ZrO_2) powders were homogeneously precipitated under hydrothermal conditions at 160° to 220°C under 2 to 7 MPa. The products were cubic/tetragonal fine powders about 12 nm in size at 220°C. They became tetragonal at about 16 nm by calcination at 800°C for one hour. The compositional fluctuation was less than ±0.75 mol% for 2.93 mol% Y_2O_3 in the products.

A number of techniques for preparing fine powders involve a liquid solution as an initial step. Multicomponent fine powders have also been prepared from the solutions.[1-4] The solutions, if well prepared, are homogeneous on an atomic scale. The most critical step in solution techniques is removal of the solvent. The degree of homogeneity of the residue, in general, depends on the solvent removal technique, but it is much less than that of the starting solution.

Coprecipitation techniques[1,2] are most widely used for the preparation of multicomponent fine powders. In these techniques, the homogeneous distribution of the components is not assured in the precipitates because segregation occurs during precipitation.[2,4] To improve the homogeneity and crystallization, the precipitates are generally heat-treated by calcination after filtration and drying.

Since these processes bring about the aggregation of the powders, a grinding or milling process is necessary to prepare dispersable powders.[3,4] This processing sequence is not so simple and is susceptible to contamination, particularly during grinding or milling.

The authors have used hydrothermal processing to prepare well-crystallized fine powders of ZrO_2 with a homogeneous shape and size distribution.[5-7] These crystalline fine powders are nonaggregated and easily dispersible in aqueous media. The homogeneity of the dopant concentration is also an important characteristic of these powders. In the present paper, we describe Eu-doped HfO_2 powders[8,9] using the hydrothermal crystallization method, including analysis of the homogeneity of dopant and Y-PSZ (3 mol% Y_2O_3-doped ZrO_2) powders by hydrothermal homogeneous precipitation.

Eu-Doped HfO_2 Powders

Experimental Procedure

Starting materials were $HfOCl_2 \cdot 8H_2O$* (containing 0.70 wt% Zr and 99.9% pure $EuCl_3$.† The precipitates with the weight ratios of $HfO_2:Eu_2O_3$ = 80:20 and 70:30 were coprecipitated by NH_4OH from the mixed solutions of $HfOCl_2$ and $EuCl_3$. The coprecipitate was washed using distilled water until NH_4^+ and Cl^- ions could not be detected in the filtrate; it was then dried at 90°C in air for 100 hours.

Hydrothermal crystallization was performed at 300°C, 10 MPa, for three hours with redistilled water in a gold-lined Glaylock-type autoclave. Details of the apparatus and experimental conditions are described elsewhere.[8]

Lattice parameters, crystallite sizes, and lattice strains were measured by X-ray diffraction (XRD) with carbon monochromated Cu$K\alpha$ using a diffractometer with a rotating anode UR-200.‡ Silicon metal§ (99.99% pure) was used as an internal standard to measure lattice parameters. To determine crystallite sizes and compositional fluctuations of produced particles, precise X-ray profile analysis[10,11] was carried out, where true diffraction profiles were deconvoluted using standard materials, and then crystallite sizes were calculated from their integrated breadths. Calcination of 30 wt% Eu_2O_3-HfO_2 coprecipitates occurred at 1400°C after four hours provided the cubic solid solution crystallized sufficiently as the standard material. All the profiles of the diffraction lines required high accuracy, achieved by counting for 200 seconds with every 0.05° scanning step.

The specific surface area of products was measured by means of a flowing BET method using a mixture of He and N_2 gases. The particles were observed by transmission electron microscopy¶ and analyzed by energy dispersive spectroscopy (EDS). Particle size was calculated by maximum lengths of every particle along a fixed direction on the TEM microphotographs.

Results and Discussion

Produced Particles: Coprecipitates dried at 90°C for 100 hours were amorphous. They yielded fluorite-type cubic hafnia solid solutions, as shown in Fig. 1 after hydrothermal crystallization at 300°C, under 10 MPa, for three hours. The products

---

*Kojundo Kagaku Co. Ltd., Tokyo, Japan.
†Shin-etsu Kagaku Industry Co., Ltd. Tokyo, Japan.
‡Rigaku Electric Co. Ltd., Tokyo, Japan.
§Kinzoku Kagaku Co. Ltd., Tokyo, Japan.
¶TEM, JEM-200CX, JEOL Co. Ltd., Tokyo, Japan.

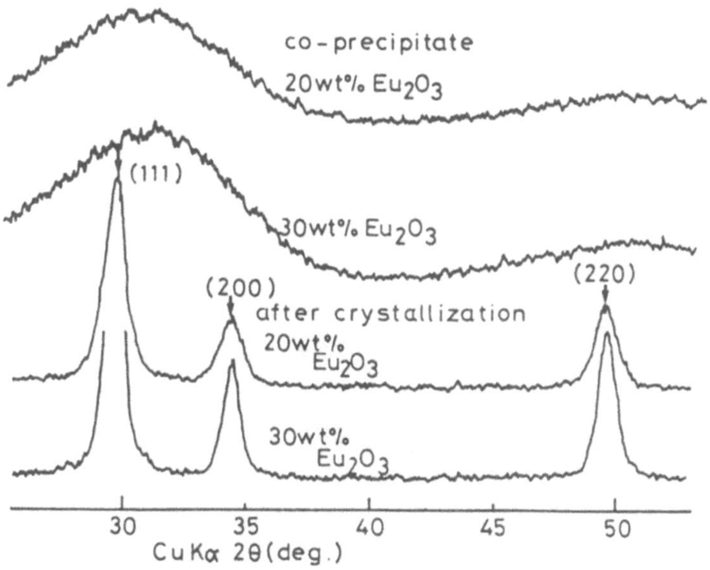

Fig. 1. X-ray diffraction patterns of amorphous coprecipitates and hafnia solid solutions crystallized hydrothermally at 300°C, 10 MPa, for 3 h from HfO_2-Eu_2O_3 coprecipitates dried at 90°C for 100 h.

are cubic fluorite solid solutions $Hf_{1-x}Eu_xO_{2-x/2}$, $x = 0.230$ and $x = 0.339$ for 20 wt% Eu_2O_3 and 30 wt% Eu_2O_3, respectively.

Transmission electron micrographs of the product indicate uniform and nonaggregated particles, as seen in Fig. 2 for 20 wt% Eu_2O_3 and in Fig. 3 for 30 wt% Eu_2O_3. The particles have faceted polyhedral morphologies. Each particle is distinguishable from another. Particle-size distributions by maximum diameters in TEM photographs are fairly narrow, as shown in Fig. 4 (20 wt% Eu_2O_3) and Fig. 5 (30 wt% Eu_2O_3), where more than 400 particles were measured. As for the 20 wt% Eu_2O_3 sample, 90% of the particles fall between 4 and 12 nm in size, and the average size calculated from the mean volume of particles in Fig. 4 is 9.2 nm. The products of 30 wt% Eu_2O_3 samples have an average size of 7.4 nm, with about three-fourths of the whole particles between 4 and 8 nm. These results indicate that the hydrothermal crystallization produced ultrafine particles with uniform size and shape.

The crystallite sizes of 20 wt% and 30 wt% Eu_2O_3 samples, calculated from Scherrer's equation with Scherrer's coefficient of 1.047, using the broadening of the X-ray diffraction line (111), were 10.6 and 8.4 nm, respectively. The specific surface areas measured by the BET method were 57.6 m^2/g for the 20 wt% Eu_2O_3 and 76.2 m^2/g for the 30 wt% Eu_2O_3 samples. These values indicate that the average particle diameters were 10.4 nm for 20 wt% Eu_2O_3 and 8.1 nm for 30 wt% Eu_2O_3 samples based on the assumptions that all the particles have the same diameter and the same theoretical density based on their lattice parameters.

Particle sizes from TEM observation, crystallite sizes from XRD, and particle diameters from specific surface area are almost identical, as shown in Table I. This fact confirms that the products crystallized hydrothermally are composed of nonaggregated, uniform, single-crystal particles.

Composition of the Produced Particles: Compositional distribution is very important for the evaluation of particles, especially in multicomponent systems.

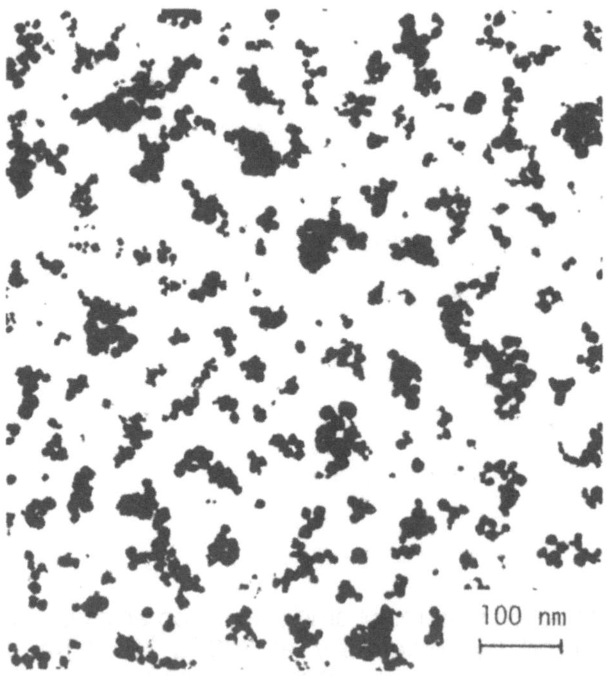

Fig. 2. Transmission electron micrograph of 20 wt% Eu_2O_3-HfO_2 cubic solid solution, crystallized hydrothermally at 300°C, 10 MPa, for 3 h from the coprecipitate.

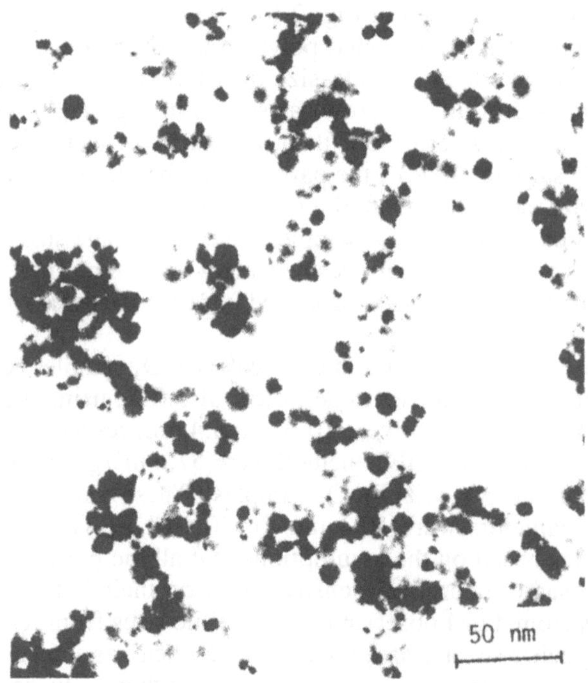

Fig. 3. Transmission electron micrograph of 30 wt% Eu_2O_3-HfO_2 cubic solid solution, crystallized hydrothermally at 300°C, 10 MPa, for 3 h from the coprecipitate.

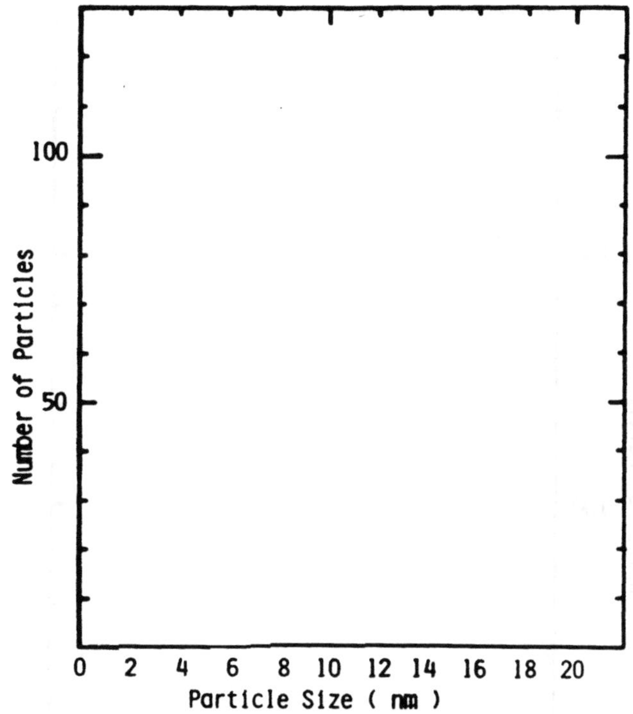

Fig. 4. Particle-size distribution by maximum diameter method from transmission electron micrograph of 20 wt% Eu_2O_3-HfO_2 cubic solid solution, crystallized hydrothermally at 300 °C, 10 MPa, for 3 h from the coprecipitate.

Klug and Alexander[12] suggested that fine crystallite size, mechanical lattice strains, and stacking faults resulted in X-ray line broadening. In addition, compositional fluctuation causes line broadening, because the lattice parameters varied with their compositions in the systems of solid solutions. Hall[13] proposed separating the effect of crystallite size in line broadening from that of lattice strain using the following equation,

$$\beta \cos\theta/\lambda = 1/\epsilon + \eta \sin\theta/\lambda \qquad (1)$$

where β is the integral breadth, θ the Bragg angle, λ the wavelength, ϵ the crystallite size, and η the lattice strain.

In the materials produced by hydrothermal crystallization, the effects of residual mechanical lattice strain and stacking faults resulting from twinning, etc. are negligible, because they are nonaggregate and single-crystal particles. This indicates that the origin of the line-broadening term by η is only a compositional fluctuation in the present case.

Using Williamson-Hall plotting[14] ($\beta \cos\theta/\lambda$ vs $\eta \sin\theta/\lambda$), the slope η is dependent on the variation of interplanar distance $\Delta d/d$, that is, equal to the variation of lattice parameter $\Delta a_0/a_0$ in the cubic system. The lattice parameter a_0 varies with the composition (C:$EuO_{1.5}$ mol%) as the following equation[8]:

$$a_0 = 0.5117 + 2.3987 \times 10^{-4} \cdot C \text{ (nm)} \qquad (2)$$

The compositional fluctuation ΔC, therefore, can be described in Eq. (3).

$$\eta \fallingdotseq \Delta a_0/a_0 = (2.3987 \times 10^{-4}\ \Delta C)/a_0 \qquad (3)$$

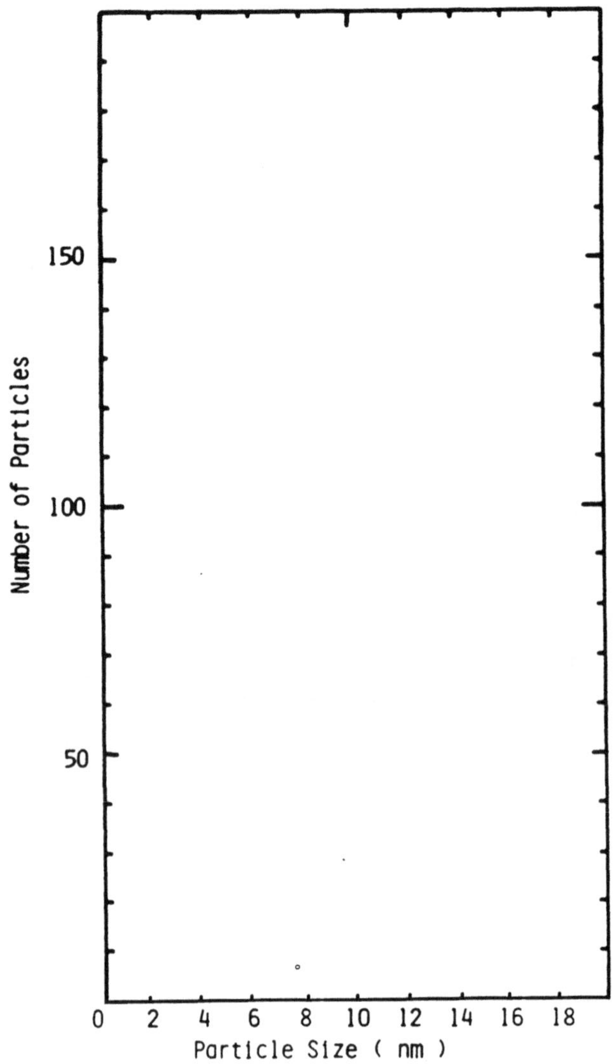

Fig. 5. Particle-size distribution by maximum diameter method from transmission electron micrograph of 30 wt% Eu_2O_3-HfO_2 cubic solid solution, crystallized hydrothermally at 300 °C, 10 MPa, for 3 h from the coprecipitate.

Table I. Particle Sizes of Eu_2O_3-HfO_2 Cubic Solid Solution

HfO_2/Eu_2O_3	Surface Area (m^2/g)	Particle Size (nm) [BET]	Crystallite Size (nm) [X-ray]	Particle Size (nm) [TEM]
80/20	57.6	10.4	10.6	9.2
70/30	76.2	8.1	8.4	7.4

From transmission electron micrograph, X-ray line profile, and specific surface area by BET method, crystallized hydrothermally at 300 °C, 10 MPa, for 3 h from the coprecipitates.

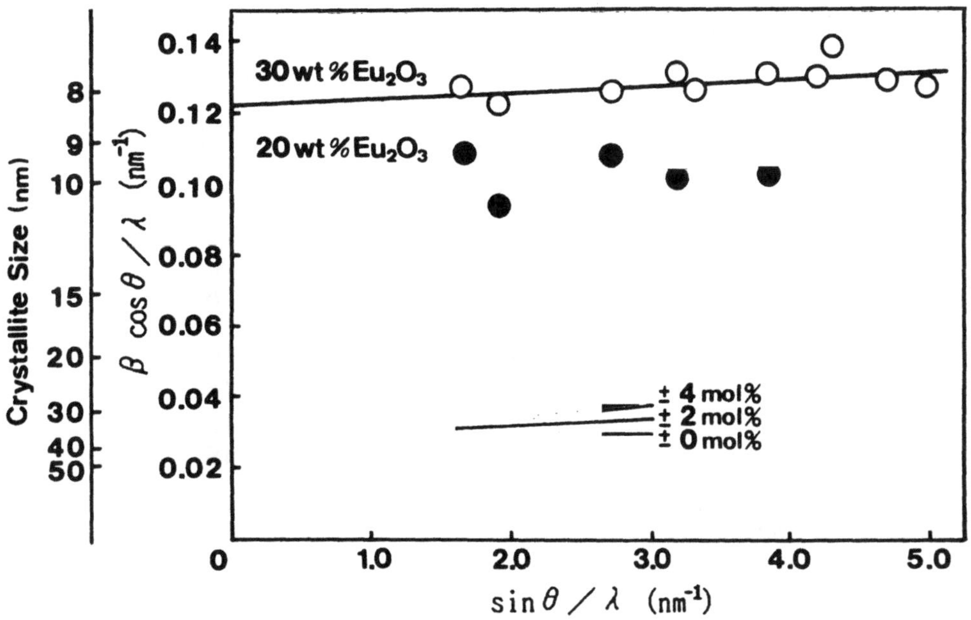

Fig. 6. Williamson-Hall plot for Eu-doped HfO_2 cubic solid solution, crystallized hydrothermally at 300°C, 10 MPa, for 3 h from the coprecipitate.

In Williamson-Hall plotting, as shown in Fig. 6, the slope of the line is 2.154×10^{-4} by the least-squares method for the 30 wt% Eu_2O_3 sample, which means that the ΔC value is ± 2.30 mol% (± 2.15 wt%). As for the 20 wt% Eu_2O_3 sample, the slope is still less, as seen in Fig. 6.

These facts are in accordance with the result of EDS analysis, as shown in Tables II and III. Compositional differences among particles found by EDS analysis are within experimental error, estimated up to ≈10% for such ultrafine particles, because a long exposure for a small area (100 by 100 nm) was needed to avoid the radiation damage of a particle by a concentrated electron beam.

Shirasaki et al.[15] and Kakegawa et al.[16] investigated the compositional fluctuation of powders produced by reaction through solid-solid mixing or a wet-dry method, also using the Williamson-Hall plotting. Few reports, however, have been found on the compositional fluctuation about nonaggregated, single-crystal, uniform particles. The present results based upon X-ray profile analysis, also confirmed by EDS analysis, revealed that the products by hydrothermal crystallization are uniform not only in their size and morphology, but also in their composition.

Y-PSZ Powders

Experimental Procedure

Starting materials were $ZrOCl_2 \cdot 8H_2O$ and $YCl_3 \cdot 6H_2O$. The mixed solution containing ZrO_2 + 3 mol% Y_2O_3 as 1 mol/L concentration was heated with urea: $(NH_2)_2CO$ of 120% for the equivalent quantity in a Zr-lined autoclave at 140° to 220°C under 2 to 7 MPa for 0 to 24 hours. The $(NH_2)_2CO$ decomposes into NH_3 and CO_2 through reaction with H_2O:

$$(NH_2)_2CO + H_2O \longrightarrow 2NH_3 + CO_2 \quad (4)$$

resulting in a change of the solution's pH to ≈8. A homogeneous precipitation

Table II. Quantitative EDS Analysis of 20 wt% Eu_2O_3-HfO_2 Cubic Solid Solution Particles

Particle No.	Composition (wt%)	
	HfO_2	Eu_2O_3
1	78.8	21.2
2	79.9	20.1
3	88.6	11.4
4	80.0	20.0
5	81.1	18.9
6	79.7	20.3
7	80.5	19.5
8	78.5	21.5
9	80.1	19.1
Average	80.9	19.1
	Standard deviation 2.9	

Crystallized hydrothermally at 300°C, 10 MPa, for 3 h from the coprecipitate.

Table III. Quantitative EDS Analysis of 30 wt% Eu_2O_3-HfO_2 Cubic Solid Solution Particles

Particle No.	Compositon (wt%)	
	HfO_2	Eu_2O_3
1	68.1	31.9
2	69.7	30.3
3	69.7	30.3
4	71.3	28.7
5	70.6	29.4
6	71.1	28.9
Average	70.1	29.9
	Standard deviation 1.1	

Crystallized hydrothermally at 300°C, 10 MPa, for 3 h from the coprecipitate.

condition has then been established under hydrothermal conditions. The precipitate of hydrous ZrO_2 with Y_2O_3 crystallizes in yttria-doped ZrO_2 in the hydrothermal conditions. The crystallized products were separated from the solution and washed with distilled water by decantation and centrifugal separation, then dried in air at 120°C for 12 hours. Ethanol washing was also applied before drying in some cases.

For a practical process to eliminate NH_4Cl completely, the products were calcined at 500° to 1100°C for one hour. After calcination and washing with water, the products were wet-milled with water in an alumina mill with PSZ balls for 12 hours, then dried at 120°C for 10 hours. Analysis of the produced powders is the same as that of Eu-doped HfO_2. The sinterability of these PSZ powders was also studied.

Results and Discussion

The hydrothermally treated products of homogeneous precipitation were gelatinous below 150°C but crystallized, mostly in the cubic/tetragonal phase, above 160°C. The optimum quantity of urea was 120% of the equivalent one, and

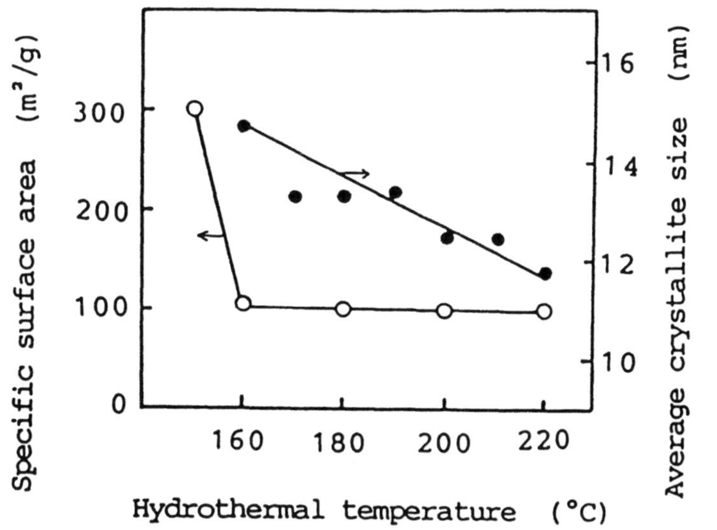

Fig. 7. Variation of surface area and crystallite size of PSZ (3 mol% Y_2O_3) powders with hydrothermal temperature using 1 mol/L solution, for 5 h.

crystallization was complete after five hours at 180°C. The cubic and tetragonal phases should be metastable, but were indistinguishable since the diffraction peaks of (004) and (400) were nonseparable in the products. Calcination of the products above 800°C brought about the apparent separation of (004) and (400), indicating the formation of tetragonal zirconia (t-ZrO_2). The monoclinic contents in the products decreased with hydrothermal temperature, to <10% above 200°C. The crystallite size of the cubic/tetragonal phase decreased from 15 to 11.6 nm with increasing temperature from 160° to 220°C, as shown in Fig. 7. The BET surface area, however, remained constant at ≈100 m²/g, regardless of the temperature. This surface area corresponds to ≈10 nm-sized spherical particles. Since this particle size is close to the crystallite size of the product at 220°C, the product appears to be well-dispersed, single-crystal particles with homogeneous size.

Calcination of the products brought about a slight increase in the crystallite size up to 900°C, but rapid increase above 950°C. For example, the crystallite of d_x = 11.6 nm with the products at 220°C increased to d_x = 15.8 nm by calcination at 800°C for one hour. This calcined product had a median diameter of 16.2 nm with the size distribution histogram of Fig. 8 from the TEM photograph (Fig. 9). These results demonstrate that the calcined powder also consists of well-crystallized, single-crystal particles with homogeneous size.

Energy dispersive X-ray spectroscopy with TEM revealed the compositional fluctuation in these fine particles to be within 25% for the ratio of Y/Z counts, corresponding to ±0.75 mol% fluctuation for 2.93 mol% Y_2O_3. The Williamson-Hall plotting (Fig. 10) for this product also indicated no measurable compositional fluctuation.

These results clearly indicate that Y-PSZ powders prepared by hydrothermal homogeneous precipitation have homogeneous shape and size distribution and dopant composition.

The product calcined at 800°C, composed of almost tetragonal ZrO_2, is shown in Fig. 11. This t-ZrO_2 was transformable by mechanical stress, because about 30% of the t-ZrO_2 was transformed into m-ZrO_2 by grinding for 30 minutes

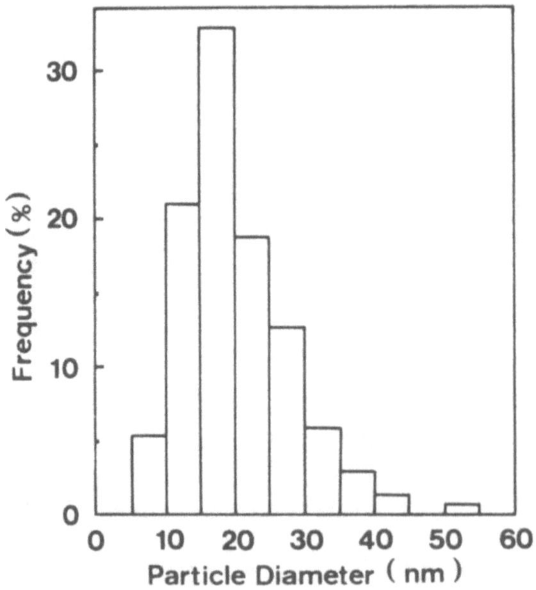

Fig. 8. Particle-size distribution by maximum diameter method from transmission electron micrograph of PSZ (3 mol% Y_2O_3) powders by hydrothermal homogeneous precipitation at 220 °C under 7 MPa for 5 h, after calcination at 800 °C for 1 h.

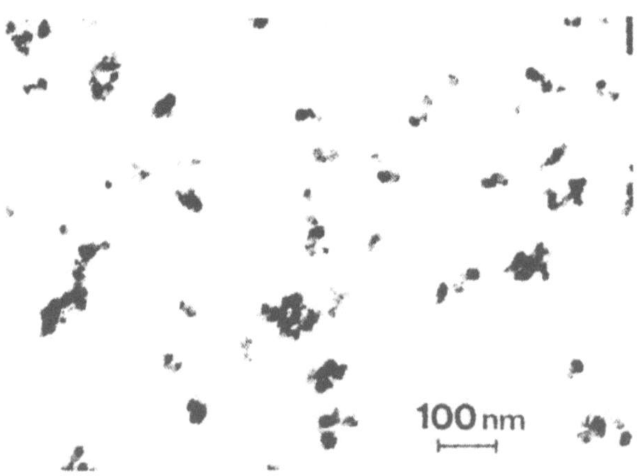

Fig. 9. Transmission electron micrograph of PSZ (3 mol% Y_2O_3) powders by hydrothermal, homogeneous precipitation at 220 °C under 7 MPa for 5 h, after calcination at 800 °C for 1 h.

in an agate mortar, as indicated in the figure. Similarly, ball-milling with water for 12 hours also yielded 30 to 50% m-ZrO_2 from calcined t-ZrO_2 powders. The transformed ratio of $t \longrightarrow m$ was dependent on the calcination temperature and also the ethanol washing before calcination (Fig. 12).

A typical chemical analysis of the calcined powder is given in Table IV. The hydrothermally produced Y-PSZ powders are sinterable to full density (>99%)

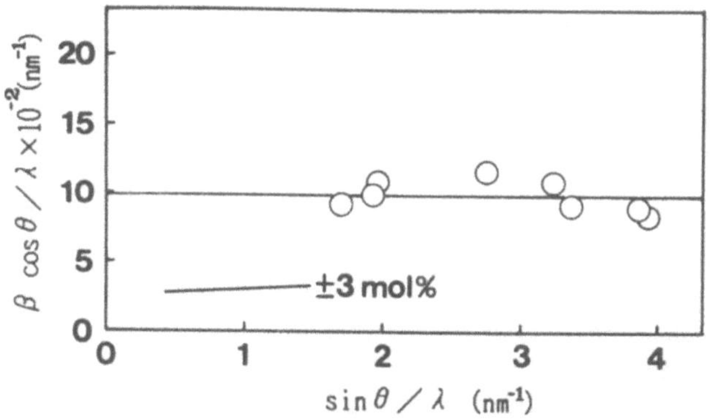

Fig. 10. Williamson-Hall plotting of the product of PSZ (3 mol% Y_2O_3) powders by hydrothermal homogeneous precipitation at 220 °C under 7 MPa for 5 h, after calcination at 800 °C for 1 h.

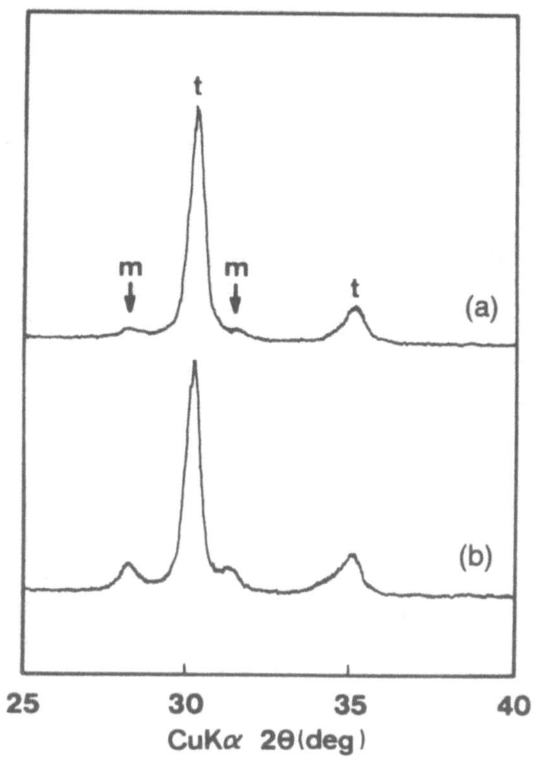

Fig. 11. X-ray diffraction pattern of product of (*a*) PSZ (3 mol% Y_2O_3) powders by hydrothermal homogeneous precipitation at 220 °C under 7 MPa for 5 h, after calcination at 800 °C for 1 h. (*b*) Ground for 30 min in agate mortar.

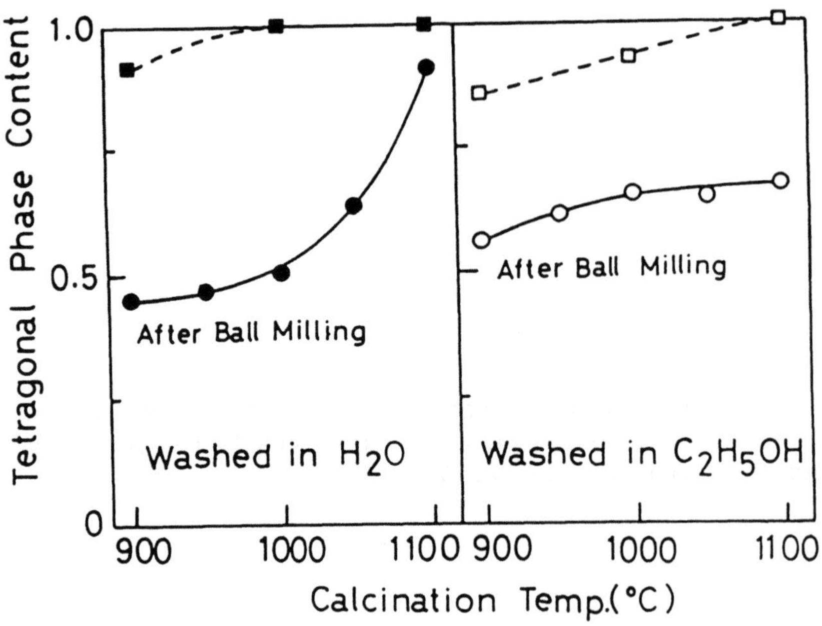

Fig. 12. Variation of tetragonal phase contents in the products of PSZ (3 mol% Y_2O_3) powders with ball-milling, calcined temperature, and washing media.

Table IV. Typical Chemical Analysis Data for PSZ (3 mol% Y_2O_3) Powders Prepared by Hydrothermal Homogeneous Precipitation

ZrO_2	Y_2O_3	Al_2O_3	SiO_2	Fe_2O_3	Na_2O
94.7	5.2* ‖‖ [2.93 mol%]	0.02	0.07	0.02	0.01 (wt%)

After calcination at 1000°C and ball-milling.
*By XMA, Cl <0.05 wt%.

Table V. Typical Properties of Y-TZP Ceramics Prepared from PSZ (3 mol% Y_2O_3) Powders by Hydrothermal Homogeneous Precipitation Method

Sintered density (1400°C, 2 h)	6.00 (g/cm^3)
Bending strength (Three-point bending test, 30 mm span)	1012 (MPa)
Compressive strength	>2.9 (GPa)
Elastic modulus (Ultrasonic pulse method)	200 (GPa)
Coefficient of linear expansion (20° ≈ 1100°C)	11 × 10^{-6} (1/K)
Fracture toughness (M.I. method)	6 ≈ 8 (MN · m$^{-3/2}$)
Content of tetragonal ZrO_2	≈100(%)

at 1400°C for two hours. Table V summarizes typical properties of the Y-TZP ceramics prepared from these powders. Details of the sinterability of these powders are reported elsewhere.[17]

Conclusion

Hydrothermal processing, the hydrothermal crystallization of the coprecipitate of Eu-Hf hydrous oxides or hydrothermal homogeneous precipitation of Y-Zr oxides, leads to the formation of single-crystal fine powders with homogeneous size and shape and dopant concentration. The crystallite size is controllable from <10 to >50 nm by various experimental parameters including temperature, concentration of the precipitant, time, washing media, calcination, etc.

Acknowledgment

The authors are thankful to Mr. Y. Fushii and K. Ioku for their experimental aid, to Dr. H. Toraya for X-ray profile analysis, and to Jeol Co., Ltd., for TEM use.

References

[1]C. R. Veale, Fine Powders, Preparation, Properties and Uses. Wiley & Sons, New York, 1972.

[2]D. W. Johnson, Jr., "Nonconventional Powder Preparation Techniques," *Am. Ceram. Soc. Bull.,* **60** [2] 221–24, 243 (1981).

[3]J. L. Pentecost, "Powder Preparation Processes"; pp. 1–13 in Treatise on Materials Science and Technology, Vol. 9, Ceramic Fabrication Processes. Edited by F. F. Y. Wang. Academic Press, New York, 1976.

[4]D. W. Johnson, Jr. and P. K. Gallagher, "Reactive Powders from Solution"; pp. 125–39 in Ceramic Processing before Firing. Edited by G. Y. Onoda, Jr. and L. L. Hench. Wiley & Sons, New York, 1978.

[5]M. Yoshimura and S. Sōmiya, "Preparation of Fine Zirconia Powders by Hydrothermal Techniques"; pp. 45–60 in Zirconia Ceramics 1 (in Japanese). Edited by S. Sōmiya. Uchida Rokakuho Publishing Co., Tokyo, 1983.

[6]E. Tani, M. Yoshimura, and S. Sōmiya, "Hydrothermal Preparation of Ultrafine Monoclinic ZrO_2 Powder," *J. Am. Ceram. Soc.,* **64** [12] C-181 (1981).

[7]E. Tani, M. Yoshimura, and S. Sōmiya, "Formation of Ultrafine Tetragonal ZrO_2 Powder under Hydrothermal Conditions," *J. Am. Ceram. Soc.,* **66** [1] 11–14 (1983).

[8]M. Yoshimura, Y. Fushii, H. Toraya, T. Tagawa, H. Imai, and S. Sōmiya, "Preparation of Cubic $Hf_{1-x}Eu_xO_{2-x/2}$ (x = 0.230, 0.339) Ultrafine Crystalline Particles by Hydrothermal Crystallization," *Yogyo-Kyokai-Shi,* **93** [9] 548–53 (1985).

[9]S. Sōmiya, M. Yoshimura, H. Toraya, and Y. Fushii, "Cubic Eu-doped Hafnia Ultrafine Particles Crystallized under Hydrothermal Conditions," *Z. Anorg. Allgem. Chem.,* in press.

[10]H. Toraya, M. Yoshimura, and S. Sōmiya, "A Computer Program for the Decomposition of X-ray Diffraction Profiles with the Composite of Pearson type VII Functions," *J. Appl. Crystallogr.,* **16** [6] 653–57 (1983).

[11]H. Toraya and F. Marumo, "Application of Total Pattern-Fitting to X-ray Powder Diffraction Data," Report of Research Laboratory of Engineering Materials, Tokyo Inst. Tech., **5,** 55–64 (1980).

[12]H. P. Klug and L. E. Alexander; pp. 618–708 in X-ray Diffraction Procedures. Wiley & Sons, New York, 1974.

[13]W. H. Hall, *Proc. Phys. Soc.,* **62,** 741–43 (1949).

[14]D. Louër, M. T. Mesnier, and J. C. Niepce, "X-ray Diffraction Line Broadening of Cadmium Oxide Produced by Cadmium Hydroxide Decomposition," *J. Mater. Sci.,* **19** [3] 716–22 (1984).

[15]S. Shirasaki, K. Takahashi, and K. Kakegawa, "Ferroelectric-Paraelectric Phase Transition in Lead Titanate Containing Lattice Defects," *J. Am. Ceram. Soc.,* **56** [8] 430–35 (1973).

[16]K. Kakegawa, J. Mohri, S. Shirasaki, and K. Takahashi, "Preparation of $Pb(Zr,Ti)O_3$ through the Use of Cupferron," *J. Am. Ceram. Soc.,* **67** [1] C-2–C-3 (1984).

[17]S. Sōmiya, M. Yoshimura, Z. Nakai, K. Hishinuma, and T. Kumaki, "Microstructure Development of Hydrothermal Powders and Ceramics," Ceramic Microstructures '86, Role of Interfaces, University Conference on Ceramic Science, University of California, Berkeley, July 28–31, 1986.

Part 5

Hydrothermal Synthesis

Reprinted from *J. Japan. Assoc. Min. Petr. Econ. Geol.*, **71** (1976) 193–200.

FLUID INCLUSIONS IN HYDROTHERMALLY SYNTHESIZED BRUCITE CRYSTALS

Kazuo Nakamura, Shin-ichi Hirano and Shigeyuki Sōmiya

The Research Laboratory of Engineering Materials, The Tokyo Institute of Technology, Tokyo

Brucite crystals were grown hydrothermally from $Mg(OH)_2$ powder in 7.5 and 15 mol% NaOH solutions. It is established from the observation of the surface microtopographs that the crystals grew by the spiral mechanism. The filling degree of numerous fluid inclusions in the crystals grown under various conditions of pressure, temperature and concentration of solution are measured. The P-T-F.D. relations of NaOH solution are discussed on the basis of these measurements.

Introduction

Determination of physicochemical parameters of aqueous solutions under high temperatures and pressures is essentially important in the investigations of mineral formation in lithosphere and for the development of hydrothermal crystal growth. Several papers have been published reporting the measurements of the solubility of NaCl (Sourirajan and Kennedy, 1966) and P-V-T relations of NaCl solution (Lemmlein and Klevtsov, 1961) under high temperatures and pressures. In spite of that NaOH solution is as important as NaCl solution in natural hydrothermal solutions, only limited data have been published as to these parameters of NaOH solution. Pressure-temperature-filling degree (P-T-F.D.) relations of 4N NaOH solution were studied up to 500°C and 1200 atm. by Liebertz (1969). Urusova (1974) also studied the vapour pressure of NaOH solution up to 550°C and 30 mol% concentration.

A vast amount of researches have been made in the field of economic geology to understand the conditions of mineral formation through the investigations of fluid inclusions. In this study the temperature conditions of mineral formation are deduced from the measurements of filling degree of fluid inclusions, filling temperatures, followed by pressure correction. However, uncertainty always remains in the case of fluid inclusion studies in natural crystals as to their origins, i.e., whether they are primary or secondary inclusions. This uncertainty is eliminated in the case of fluid inclusions in synthetic crystals.

Nakamura, Hirano and Sōmiya (1975) studied the hydrothermal growth of brucite crystals from MgO powder in alkali hydroxide solutions. Fluid inclusions were observed in the brucite crystals. However, due to the consumption of H_2O by the reaction of MgO with H_2O to form brucite, the concentration of the solution in these inclusions is not kept unchanged from the initial concentration. Therefore in order to study the P-T-F.D. relation using fluid inclusions, it is necessary to use starting materials which do not react with the solution. Thus, the present authors have

(Manuscript received March 22, 1976)

made hydrothermal growth of brucite crystals using $Mg(OH)_2$ powder as starting materials. The growth mechanism of the brucite crystals was studied by observing the surface microtopograph of crystal faces, and the P-T-F.D. relation of solution under high temperatures and pressures is calculated approximately from the study of fluid inclusions in the crystals.

EXPERIMENTAL

Test tube type cold seal vessels were employed and water was used as a pressure medium. A gold capsule, 5 mm in outer diameter, 4.7 mm in inner diameter, and 35 mm in length, including starting materials was subjected to the desired temperature and pressure conditions in the vessel. Pressure was measured with a Heise bourdon tube pressure gauge. Temperature was measured at the outside wall of the vessel using a platinel thermocouple calibrated against the melting point of gold (1064.4°C). The accuracies containing fluctuations during the run and error were ±20 kg/cm² and ±3°C.

Magnesium hydroxide powder used as starting material was synthesized hydrothermally from MgO powder and redistilled water at 500°C, 1000 kg/cm² and for 1 week. Magnesium oxide powder was prepared by calcining Mg $(NO_3)_2$ at 600°C for 24 hours which was obtained from Mg metal (99.95% pure, Furukawa Magnesium Co., Ltd., Japan) with reagent grade HNO_3 solution (Wako Pure Chemical Industry, Ltd., Japan). Solutions used were 7.5 mol% (15.3 wt%, 4.9 m) and 15 mol% (28.1 wt%, 9.8 m) NaOH solution.

The gold capsule was sealed by an electric arc. Many capsules were prepared simultaneously, because NaOH solution reacts with CO_2 in air if it is kept for a long time. The capsule was inserted into the vessel and treated for 1 week under a given temperature and pressure. Temperatures employed were 400 and 500°C and pressures levels of 500 to 2500 kg/cm². Temperature difference between the top and the bottom of the capsule was 15°C. The temperature at the center of the capsule was controlled at a desired temperature. After the treatment, the vessel was quenched into cold water. The capsule removed from the vessel was cut open and the crystals were washed with redistilled water and then observed with the optical microscope

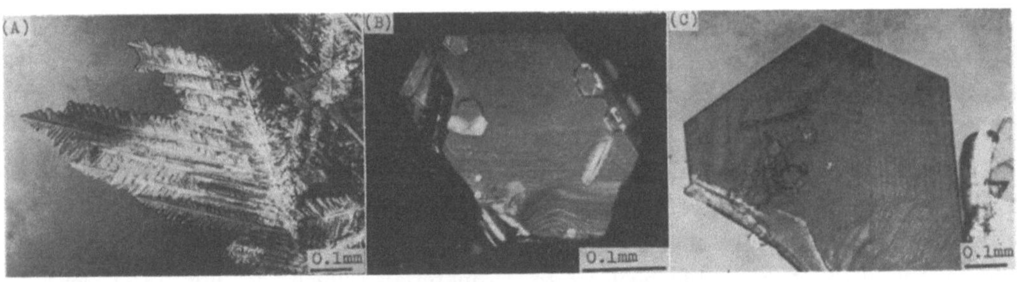

Fig. 1. Microphotographs of brucite crystals

(A). Dendritic crystals grown at 400°C and 1000 kg/cm² in 15 mol% NaOH solution.

(B). Spiral patterns originated from large surface defect on the (0001) face of brucite grown at 500°C and 500 kg/cm² in 15 mol% NaOH solution.

(C). Superimposed step patterns on the opposite (0001) surfaces of a thin crystal grown at 500°C and 500 kg/cm² in 7.5 mol% NaOH solution.

(Vanox, Olympus Optical Co., Ltd.). The filling degree is defined as the ratio of volume of the solution to the total volume of the fluid inclusion. In the present work, the areal ratio measured on thin section was approximated this value.

Results and Discussion

Crystal growth of brucite

The brucite crystals are generally transparent and hexagonal platy, bounded by the {0001} and $\{10\bar{1}1\}$ or {0001}, $\{10\bar{1}1\}$ and $\{10\bar{1}\bar{1}\}$. Dendritic crystals were found in the case of 15 mol% NaOH solution. Fig. 1 (A) shows dendritic crystals obtained at 400°C and 1000 kg/cm² in 15 mol% NaOH solution. They are considered to have been formed during the period of quenching due to supersaturation shoot up, and not during principal growth period.

Although growth rates were not measured quantitatively, it was observed that the crystals obtained at 500°C in 15 mol% NaOH solution were larger than those obtained at 400°C in 7.5 mol% NaOH solution, crystals as large as 3 mm having been obtained in the run at 500°C in 15 mol% NaOH solution for 1 week. Pressure effect on the growth rates was observed not to be detectable.

Surface microtopographs consisting of growth steps were observed under the differential interference microscope. Fig. 1 (B) shows the step patterns on the (0001) face of brucite grown at 500°C and 500 kg/cm² in 7.5 mol% NaOH solution. The growth steps originate from large defects on the surface, such as those created by the contact of other crystals, and are parallel to the edges of the crystal. Fig. 1 (C) also shows the step patterns on the (0001) face of brucite grown at 500°C and 500 kg/cm² in 7.5 mol% NaOH solution. The step patterns on both (0001) faces are seen superimposed in this figure. It is interesting to note that the step density of one surface is about 5 times higher than that of the other face. According to Burton, Cabrera and Frank's theory on crystal growth (1951), step separations of a spiral are inversely proportional to the natural logarithm of the saturation ratio, provided that the layer systems are originated from screw dislocations of the same Burgers vector. It is therefore conjectured that the saturation ratio on the denser surface was higher than that on the other surface. This difference is conjectured to be caused by the position of the thin, flaky crystal and the disturbance of the convection in the solution in the capsule.

From these observations, it is considered that the brucite crystals grow by the spiral mechanism.

Fluid inclusions and P-T-F.D. relations of NaOH solution

A large number of fluid inclusions flattened parallel to the (0001) and bounded by other crystallographic faces or curved faces are seen in the crystals. Fig. 2(A) is a representative fluid inclusion bounded by well developed faces, i.e. negative crystal. Fig. 2(B) shows those bounded by irregularly curved faces. In this photograph there are also fluid inclusions whose straight edges are parallel to the edges of the crystal. Fig. 2(C) shows a channel extending outwards and outcropping on the surface. It is not certain as to whether this channel is formed by dissolution after the growth or during growth. However, the channel seems to start necking down at a point. Many fluid inclusions seen in the crystals may have been formed in this

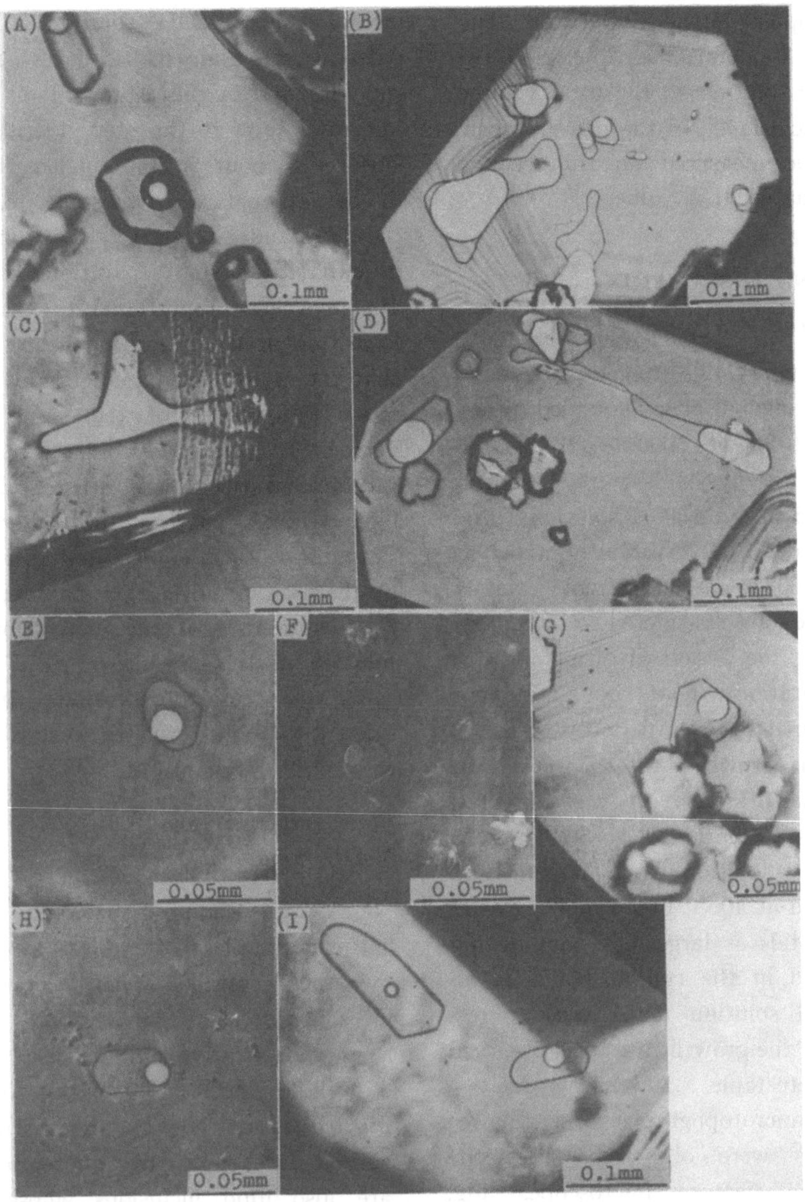

Fig. 2. Microphotographs of fluid inclusions in brucite crystals.
(A) Fluid inclusion of negative crystal shape in brucite crystal grown at 500°C and 2500 kg/cm² in 15 mol% NaOH solution. (B) Fluid inclusions with irregular shape and hexagonal shape in brucite crystal grown at 500°C and 500 kg/cm² in 7.5 mol% NaOH solution. (C) Channel extending outwards of crystal grown at 400°C and 2500 kg/cm² in 15 mol% NaOH solution. (D) Necking down phenomenon of fluid inclusion obtained at 500°C and 500 kg/cm² in 7.5 mol% NaOH solution. (E), (F), (G) and (H). Examples of fluid inclusions obtained at 500°C and (E) 1000 kg/cm², (F) 1500 kg/cm², (G) 200 kg/cm² and (H) 2500 kg/cm², respectively. (I). Primary and secondary fluid inclusions obtained at 400°C and 1500 kg/cm² in 15 mol% NaOH solution.

Table 1. Filling degree of fluid inclusions formed under different conditions. Data of pure water are calculated from Holser and Kennedy (1958, 1959), and Burnham, Holloway and Davis (1959).

	400℃			500℃		
Pressure (kg/cm^2)	pure water	7.5mol% NaOH	15mol% NaOH	pure water	7.5mol% NaOH	15mol% NaOH
500	0.571	0.739	0.810	0.247		0.726
1000	0.690	0.815	0.838	0.521	0.700	0.760
1500	0.749	0.846	0.842	0.625	0.764	
2000	0.788		0.870	0.685	0.795	0.814
2500	0.819	0.880	0.883	0.730	0.807	0.835

way. Initial irregular forms of fluid inclusions may be stabilized to polygonal ones, since less stable portion of the inclusions are dissolved and precipitation takes place on more stable crystallographic faces. An example of the so-called necking down phenomenon is clearly at the elongated inclusion seen in Fig. 2(D) in a more advanced state. The inclusion has just started to separate by necking down at the narrowest portion connecting the two portions which are in the different levels in the crystal.

The filling degree measured on crystals grown under different conditions are shown in Table 1. The data, whose error is less than $\pm 1.5\%$, are the average of measurements of about 10 fluid inclusionsf or each condition. Unreliable data due to experimental failure or small number of measurements are omitted from the table. The filling degree at 500°C and 500 kg/cm^2 in 7.5 mol% NaOH solution is also omitted, since the measured values vary in a wide spectrum from 0.373 to 0.693, probably due to the condition below critical point. This was consistent with the data by Urusova (1974). The filling degree of pure water is the reciprocal value of specific volume, about which Holser and Kennedy (1958, 1959) and Burnham, Holloway and Davis (1969) reported.

Figs. 2(E)$\sim$(H) show examples of fluid inclusions formed under different conditions. Figs. 2(E), (F), (G) and (H) are the fluid inclusions obtained in 7.5 mol% NaOH solution at 500°C and (E) 1000 kg/cm^2, (F) 1500 kg/cm^2, (G) 2000 kg/cm^2 and (H) 2500 kg/cm^2, respectively. It is clearly seen in these photographs that the filling degree becomes gradually higher as the increase of pressure.

There are however exceptions in the filling degrees in a few cases. Fig. 2(I) shows two fluid inclusions formed at 400°C and 1500 kg/cm^2 in 15 mol% NaOH solution. The filling degree of the inclusion on the right is 0.842, whereas that of the left 0.975. Under this condition, the average fluid inclusions show the value of 0.842. Another phenomenon was also observed, where the filling degree was much smaller than the other fluid inclusions. It is conjectured that these inclusions are secondary ones which are subjected to change after their formation.

The original fluid contains $Mg(OH)_2$ and NaOH as solutes. Therefore the data in Table 1 can not be regarded as showing true P-T-F.D. relations of 7.5 and 15 mol% NaOH solutions in a strict sense. Carlson, Peppler and Wells (1953) reported that the

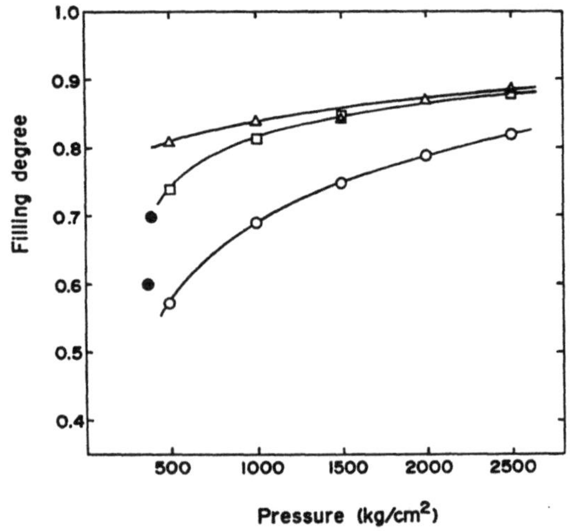

Fig. 3 (A). Filling degree of NaOH solution at 400°C versus pressure.
o pure water (Holser *et al.* & Burnham *et al.*)
• 6.8 mol% NaOH solution (Liebertz)
▫ 7.5 mol% NaOH solution
▵ 15 mol% NaOH solution

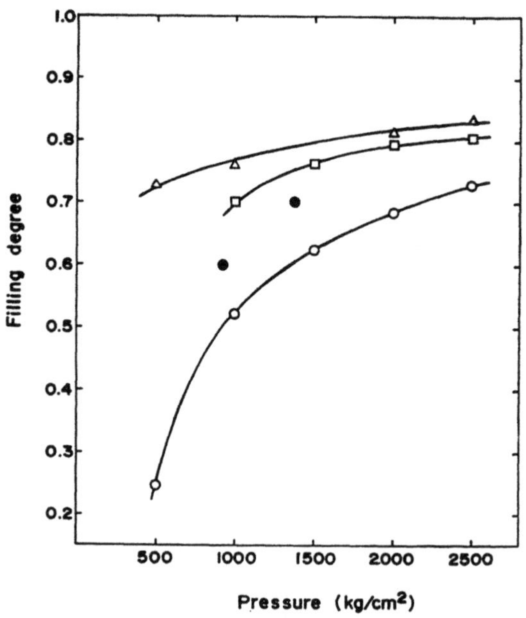

Fig. 3 (B). Filling degree of NaOH solution at 500°C versus pressure.
o pure water (Holser *et al.* & Burnham *et al.*)
• 6.8 mol% NaOH solution (Liebertz)
▫ 7.5 mol% NaOH solution
▵ 15 mol% NaOH solution

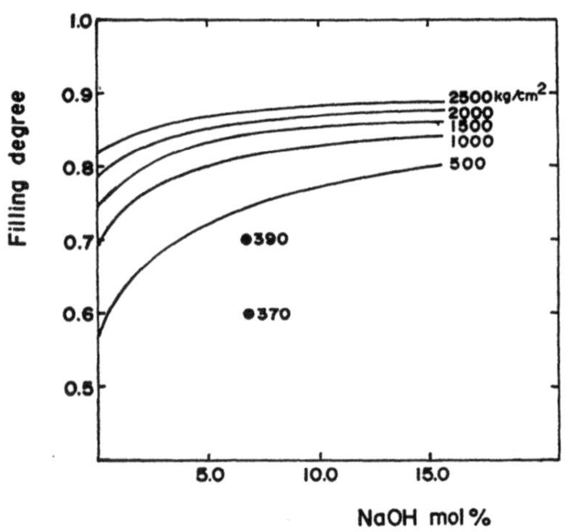

Fig. 4(A). Filling degree of NaOH solution at 400°C versus NaOH concentration.
—— Present work • Liebertz

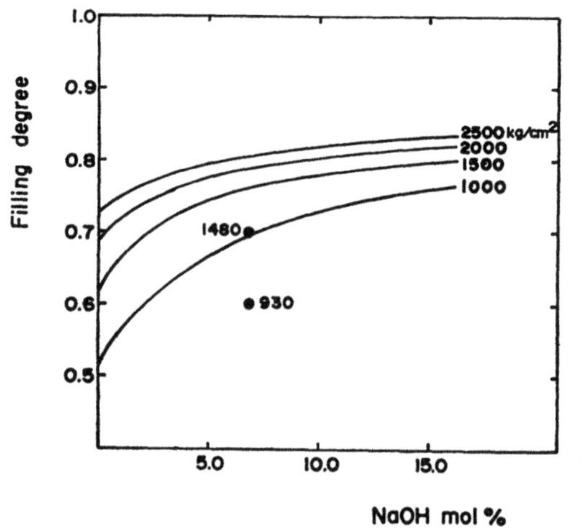

Fig. 4(B). Filling degree of NaOH solution at 500°C versus NaOH concentration.
—— Present work • Liebertz

solubility of $Mg(OH)_2$ at 250°C is 10^{-5} moles/liter. Although we have no exact data of the solubility of $Mg(OH)_2$ under the conditions employed in the present study, it may safely be said that the dominant factor affecting the change of filling degree is NaOH itself. The filling degrees of NaOH solution are plotted as a function of pressure in Fig. 3, (A) 400°C and (B) 500°C. Fig. 4 shows the filling degrees plotted against the concentration of NaOH, (A) 400°C and (B) 500°C. From these figures, it is seen that the filling degrees increase as the increase of NaOH mol%. Moreover the lower the concentration and the pressure, the greater are the slopes of increase of the filling degree.

The data of Liebertz are also plotted in Figs. 3 and 4. Taking his data at 400°C into consideration (Fig. 3 (A)), the pressure changes only negligibly small even if the filling degree changes from 0.6 to 0.7. In Fig. 3(A), the values of the present result are greater than his data. This difference may be responsible for the solubility of $Mg(OH)_2$. But it seems that the present result is reasonable since the discrepancy between the two data is not so large. If the fact that the accurate measurement of P-T-F.D. relations of solution is difficult and laborious is taken into consideration, the present simple method using fluid inclusion is considered to be very useful as an approximate method.

So far we have had not enough data of P-T-F.D. (P-T-V) relations of aqueous solutions at high temperatures and pressures except for NaCl solution. Therefore only the data of P-T-F.D. relations of pure water (Kennedy, 1950) have been used to estimate the pressure generated at high temperature, such as in the case of crystal growth using an autoclave. It is clear from the present results, however, that the pressure estimated from the data on pure water is much higher than the actual pressure. For example, in the case of pure water 1000 kg/cm² can be attained at 400°C and filling degree of 0.7, whereas in the case of 7.5 mol% NaOH solution the pressure will be less than 500 kg/cm² under the same conditions. In the other words such a behavior of solution may cause large erorr in the pressure correction for the filling temperature measurements in estimating mineral formation temperature.

It should be stressed that the studies of P-T-F.D. relations on various aqueous solutions are essentially important in the field of geology and hydrothermal crystal growth to know the real pressure, where minerals are formed, and also it is desirable to re-check the data on the natural fluid inclusion by the use of crystals grown under the defined growth conditions.

References

Burnham, C.W., Holloway, J.R. and Davis, N.F. (1969), The specific volume of water in the range 1000 to 8900 bars, 20 to 900°C. *Am. Jour. Sci.*, **267A**, 70–95.

Burton, W.K., Cabrera, N. and Frank, F.C. (1951), The growth of crystals and the equilibrium structure of their surfaces. *Phil. Trans.*, **243A**, 299–358.

Carlson, E.T., Peppler, R.B. and Wells, L.S. (1953), Studies in the system magnesia-silica-water at elevated temperatures and pressures. *Jour. Res. Nat. Bur. Stand.*, **51**, 179–184.

Holser, W.T. and Kennedy, G.C. (1958), Properties of water. Part IV. Pressure-volume-temperature relations of water in the range 100–400°C and 100–1400 bars. *Am. Jour. Sci.*, **256**, 744–753.

Holser, W.T. and Kennedy, G.C. (1959), Properties of water. Part V. Pressure-volume-temperature relations of water in the range 400–1000°C and 100–1400 bars. *Am. Jour. Sci.*, **257**, 71–77.

Kennedy, G.C. (1950), Pressure-volume-temperature relations in water at elevated temperatures and pressures. *Am. Jour. Sci.*, **248**, 540–564.

Lemmlein, G.G. and Klevtsov, P.V. (1961), Relations among the principal thermodynamic parameters in a part of the system H_2O-NaCl. *Geokhimiya*, **2**, 133–142.

Liebertz, J. (1969), P.T. f-Werte von 4n K_2CO_3-, Na_2CO_3- und NaOH- Losungen under hydrothermalen Bedingungen. *Chemie. Ing. Techn.*, **41**, 1231–1232.

Nakamura, K., Hirano, S. and Sōmiya, S. (1975), Hydrothermal growth of brucite crystals. *Jour. Am. Ceram. Soc.*, **58**, 349–350.

Sourirajan, S. and Kennedy, G.C. (1962), The system H_2O-NaCl at elevated temperatures and pressures. *Am. Jour. Sci.*, **260**, 115–141.

Urusova, M.A. (1974), Phase equilibria and thermodynamic characteristics of solutions in the systems NaCl-H_2O and NaOH-H_2O at 350–550°C. *Geokhimiya*, **9**, 1360–1366.

Reprinted from *Report of the Research Laboratory of Engineering Materials, Tokyo Institute of Technology*, No. 5 (1980) 153–161.

Hydrothermal Synthesis of Topaz Crystals

By
Shigeyuki SŌMIYA*, Shin-ichi HIRANO**,
Masahiro YOSHIMURA*** and Hironori SHIMA****

Abstract

Syntheses of topaz crystals under hydrothermal conditions were carried out at temperatures between 300°C and 800°C under pressures from 49 to 153 MPa. Topaz crystals max. 20 μm long and 1.5 μm thick, could be obtained. Single phase topaz was prepared from the mixture $2AlF_3 + 4Al(OH)_3 + 3SiO_2$ and pure water, at temperatures from 450°C to 650°C under 98 MPa. The addition of KF, KOH or K_2CO_3 as a mineralizer yielded the formation of K-containing crystals such as kalsilite, leucite or K_3AlF_6, etc. The addition of KCl, NaCl, HCl or H_2SO_4 appeared to have little effects on topaz synthesis but rather negative effects on crystal growth. The substitution of F^- ions by OH^- ions in synthesized topaz showed a tendency to increase at low temperatures and under high pressures.

Key Words: topaz, hydrothermal growth, crystal growth, synthesis of topaz, effects of mineralizer, substitution of F^- by OH^-.

1. Introduction

Topaz, $Al_2[SiO_4](OH, F)_2$, is one of the minerals occuring in acidic igneous rocks such as granites, granite pegmatites and rhyolite and is often found in veins or cavities of such rocks.[1] The synthesis of topaz was primarily reported by Schober and Thilo.[2] They obtained the topaz crystals max. 40 μm long, from a starting mixture of Al_2O_3, SiF_4 and SiO_2 at 750°C to 950°C. Coes[3] synthesized topaz under high pressures (1960 MPa, 900°C) from AlF_3, $Al(OH)_3$ and SiO_2. Recently, Rosenberg[4] prepared topaz by hydrothermal methods using AlF_3, Al_2O_3 and SiO_2 largely under 196 MPa below 800°C or 49 MPa above 800°C, and studied the compositional relationship of natural and synthetic topaz. Abdel Rehim[5] reported the synthesis of topaz by sintering of SiO_2 and AlF_3 in 1976. These reports, however, have little description of the synthetic conditions and of the crystals obtained, for example, no photographs have been given in the literatures.

* Professor, Dr. Sci. in Eng.
** Associate Professor when the initial part of this work had been done, Dr. Sci. in Eng., Present Address; Synthetic Crystal Research Laboratory, Faculty of Eng., Nagoya University, Nagoya
*** Associate Professor, Dr. Sci. in Eng.
**** Graduate Student, Present address: Takaoka, Toyama

In order to search for factors required to prepare a larger topaz crystal, we have studied the synthesis of topaz by hydrothermal reactions in detail. Here, the effects of starting materials, temperatures, pressures and solutions (mineralizer) on topaz synthesis are reported.

2. Experimental

On the basis of following assumed reactions,

$$2AlF_3 + 4Al(OH)_3 + 3SiO_2 \longrightarrow 3Al_2SiO_4F_2 + 6H_2O \quad (1)$$

$$2AlF_3 + 2Al_2O_3 + 3SiO_2 \longrightarrow 3Al_2SiO_4F_2 \quad (2)$$

$$2AlF_3 + 6Al(OH)_3 + 4Si \longrightarrow 3Al_2SiO_4F_2 + Al_2SiO_4(OH)_2 + 8H_2 \quad (3)$$

$$Al_4Si_4O_{10}(OH)_8 + 4AlF_3 \longrightarrow 4Al_2SiO_4F_2 + 4HF + 2H_2O \quad (4)$$

$$2AlF_3 + SiO_2 + 2H_2O \longrightarrow Al_2SiO_4F_2 + 4HF \quad (5)$$

the following starting mixtures were prepared by mixing reagent grade chemicals, AlF_3,* $Al(OH)_3$*, Al_2O_3☆ and SiO_2☆,

a) $2AlF_3 + 4Al(OH)_3 + 3SiO_2$

b) $2AlF_3 + 2Al_2O_3 + 3SiO_2$

c) $2AlF_3 + 6Al(OH)_3 + 4Si$

d) Georgia Kaolin; $Al_4Si_4O_{10}(OH)_8 + 4AlF_3$

e) $2AlF_3 + SiO_2$.

These mixtures were added to a gold capsule of 2.6 mm (or 4.6 mm in particular cases) in inner diameter, 0.2 mm in wall thickness and 50 mm in length, containing a known amount of re-distilled water prepared by a quartz distilator. The amounts of water and mixtures ranged typically 40–100 mg and 20–100 mg, respectively. The capsule was sealed and weighed. The capsule was loaded into a Roy and Tuttle[6] test tube type hydrothermal apparatus and the pressure was applied up to the half of the required pressure. Then the temperature was raised gradually up to the desired tempera-tube type hydrothermal apparatus and the pressure was applied up to half of the required pressure. Then the temperature was raised gradually up to the desired temperature and carefully releasing the excess pressure. Two platinel thermocouples capsule. The temperature was maintained within ±3°C by an electric controller (EC 53/2, Ohkura Electric Co. Ltd., Tokyo, Japan).

After the completion of the experimental run, the vessel was immersed into cold water for quenching. The capsule was weighed to check leakage. The products in the capsule were washed with water by the decantation method, then examined with an optical microscope (Vanox, Olympus Optical Co. Ltd., Tokyo, Japan) and a scanning electron microscope (JSM T-200, JEOL Co. Ltd., Tokyo, Japan) and also identified with an X-ray diffractometer (JDX-7F, JEOL Co. Ltd., Tokyo, Japan) using Ni-filtered CuKα radiation through a graphite monochrometer.

In order to study the effects of the mineralizer, the solutions of KF, KOH, K_2CO_3,

---

* Wako Pure Chemicals Co. Ltd., Tokyo, Japan.
☆ Kanto Chemicals Co. Ltd., Tokyo, Japan.

KCl, NaCl, HCl, H_2SO_4, NH_4F, NH_5F_2 or their mixtures were used instead of pure water. These solutions were prepared from reagent grade chemicals and had various concentrations.

3. Results and Discussions

3.1 Effects of Starting Mixtures on Topaz Synthesis

Table 1 summarizes the results of typical hydrothermal syntheses using five starting mixtures and pure water. Single phase topaz could be synthesized from the starting mixtures (a), (b) and (c) at 500°C or 600°C under 98 MPa. The mixtures (d) and (e), however, lead to the formation of AlF_3 with topaz at 500°C. These results may indicate that AlF_3 did not dissolve or was hydrolized completely in HF solution according to reaction (3) or (4) at lower temperatures. The other mixture, $2AlCl_3 + SiO_2 + 6KF$, did not result in the formation of topaz. The mixtures (a) or (b) were mainly used in this work.

Table 1. Effects of starting mixtures on the syntheses of topaz under hydrothermal conditions with pure water

Run No.	Starting Mixture†	Temp. (°C)	Press. (MPa)	Duration (days)	Products
197	a	500	105.8	7	Topaz*
196	a	600	101.4	7	Topaz*
33	b	500	100.9	7	Topaz
21	b	600	99.0	3	Topaz
141	c	600	97.0	7	Topaz☆
166	d	500	98.0	7	Topaz + AlF_3
189	d	550	105.4	7	Topaz* + AlF_3
190	d	650	100.9	7	Topaz*
58	e	500**	90.2	7	AlF_3 + tr. Topaz
144	e	600	97.0	7	Topaz

† Starting mixture a) $2AlF_3 + 4Al(OH)_3 + 3SiO_2$
b) $2AlF_3 + 2Al_2O_3 + 3SiO_2$
c) $2AlF_3 + 6Al(OH)_3 + 4Si$
d) $Al_4Si_4O_{10}(OH)_8 + 4AlF_3$
e) $2AlF_3 + SiO_2$

* Sharp X-ray diffraction pattern

☆ Broad X-ray diffraction pattern

** with temperature difference of 50°C between the top and the bottom of the capsule.

tr. Trace

3.2 Effects of Temperatures and Pressures on Topaz Syntheses

Table 2 summarizes the results of typical hydrothermal runs using pure water and the mixture (a) or (b). Single phase topaz was prepared at temperatures between 450°C and 650°C under 98 MPa pressure. At lower temperatures, un-identified phase(s) which seems to be hydrate(s) of alminium fluoride were found, while corundum crystals were observed in the products at high temperatures. The amount of corundum increased with the temperature. Even at 650°C or 600°C, the formation of corundum was observed when the capsule had leaked as the solution in the capsule was diluted by water in the pressure vessel. Temperatures affected the formation of topaz, pressures of 49.0 to 153.4 MPa, however, appeared to have less influence on the hydrothermal synthesis of topaz.

Table 2. Effects of temperatures and pressures on the syntheses of topaz under hydrothermal conditions with pure water

Run No.	Starting Mixture	Temp. (°C)	Pressure (MPa)	Duration (days)	Products
234	a	300	98.0	7	$AlF_{1.65}(OH)_{1.35} \cdot xH_2O$
236	a	400	98.0	7	Topaz + U_1☆
220	a	450	102.9	7	Topaz
197	a	500	105.8	7	Topaz
196	a	600	101.4	7	Topaz
186	a	700	125.0	1	Topaz + AlF_3
2	b	700	98.5	3	* Topaz + Corundum
187	a	700	99.0	1	* Topaz + Corundum
183	a	800	148.0	1	* Corundum + Topaz
21	b	600	99.0	7	Topaz
22	b	600	109.8	7	Topaz
5	b	700	97.0	3	Topaz + tr. Corundum
217	a	500	152.9	4	Topaz
210	a	550	49.0	7	Topaz
214	a	550	68.6	7	Topaz
193	a	600	150.9	12	Topaz

☆ Un-identified phase, probably a hydrate.
* With slight leakage. Gold capsules were corroded by solutions containing HF produced by the hydrolysis of AlF_3. All capsules had leaked in the runs at 800°C.

3.3 Effects of Solutions (mineralizers) on Topaz Synthesis

The effects of mineralizers were widely studied using various solutions in place of pure water. Table 3 shows typical results. In the solutions involving a large amount of KOH, KF or K_2CO_3, K-containing crystals such as kalsilite $KAlSiO_4$, leucite KSi_2AlO_6 or K_3AlF_6 were formed instead of topaz. It suggests that these crystals would have lower solubilities than topaz in those solutions. In the dilute potassium solution, a small amount of an un-identified phase U_1 was detected in addition to the topaz. Corundum was also produced in these solutions at higher temperatures. In the solutions of KCl,

Table 3. Effects of solutions (mineralizer) on the syntheses of topaz under hydrothermal conditions

Run No.	Starting Mixture	Solution (M)☆	Temp. (°C)	Press. (MPa)	Duration (days)	Products
110	a	KOH(2.5)	600	100.9	7	Kalsilite
113	a	KF(2.5)	600	98.0	7	Kalsilite
119	a	KF(2.5)+KCl(ex)	650	98.0	7	Leucite
121	a	KOH(2.5)+KCl(ex)	650	95.1	7	U_1+Corundum+Leucite
123	a	KF(0.5)+KCl(ex)	650	99.0	7	K_3AlF_6
69	a	K_2CO_3(1.0)	500*	92.1	7	Kalsilite
198	a	KOH(0.5)	600	112.7	7	Topaz+Corundum+U_1
17	b	KOH(0.5)	700	98.0	3	Topaz+Corundum
136	a	KF(0.5)	600	99.5	7	Topaz+U_1
11	b	KF(0.5)	700	100.5	3	Topaz+Corundum
82	a	KCl(ex)	625*	96.0	7	Topaz
133	a	KCl(ex)	550*	144.6	7	Topaz+Zunyite
81	a	KCl(3.6)	625*	98.0	7	Topaz
134	a	KCl(3.6)+NaCl(4.6)	600	97.0	7	Topaz
73	a	KCl(1.0)+HCl(1.0)	500*	102.9	7	Topaz+Zunyite
105	a	NaCl(4.6)	650	101.9	7	Topaz
107	a	NaCl(ex)	600	100.6	7	Topaz
83	a	HCl(1.0)	625*	117.6	7	Topaz
151	a	HCl(6.0)	500	102.4	7	Zunyite
46	a	H_2SO_4	500*	95.1	7	Topaz
176	d	NH_4F(4.7)	700	100.9	7	Topaz
203	a	NH_4F(4.7)+HCl	500	98.0	7	AlF_3+Topaz
207	a	NH_5F_2(6.0)	650	103.4	7	U_2+U_3

☆: Molarity of solutions
ex: Exess amounts of solutes added to the saturated solutions at room temperature.
*: With temperature difference of 50°C between the top and the bottom of the capsule.
U_1, U_2, U_3: un-identified phases

NaCl, HCl or H_2SO_4, single phase topaz was produced at temperatures above 600°C. However, topaz crystals were worse crystallized in NaCl or KCl solutions than pure water, because X-ray pattern showed a broad topaz pattern for the products in NaCl or KCl solutions. Topaz plus zunyite crystals were formed in the solutions of KCl or HCl below 600°C. Single phase zunyite was obtained at 500°C in HCl solutions. These results indicate that the zunyite crystals might contain a considerable amount of Cl^- ions. Un-identified phases U_2 and U_3 were produced in the solutions of $NH_4F \cdot HF$. These phases seemed to be soluble to water because they diminished during decantation with water.

The hydrothermal runs with a temperature gradient ($\Delta T = 50^{\circ}C$ between the top and the bottom of the capsule) did not show significant effects on syntheses and crystal growth of topaz crystals.

3.4 Prepared Topaz Crystals

Some typical topaz crystals are shown in Figs. 1–3. Topaz crystals had the habit of rods, typically 10–20μm long and 1μm thick. Multiple growths from a nucleation site are often seen as shown in Fig. 2. Fig. 4 shows large AlF_3 crystals and some aggregates of small topaz crystals in the product of the run at 700°C for 1 day. These large AlF_3 crystals might change into topaz with time since no AlF_3 crystals were detected in any products at 700°C for 3 days (run No. 3, 5, 6, 7, 8). The growth rate of topaz, however, seems to be very slow. The crystals at the top of the inner wall of the capsule grown for 3 days at 700°C were still 20μm in length as seen in Fig. 5.

Some other crystals synthesized in this work are shown in Fig. 6. The crystals of AlF_3 had the form of cubes or prisms, while kalsilite crystalized in the shape of hexagonal plates or polyhedrons.

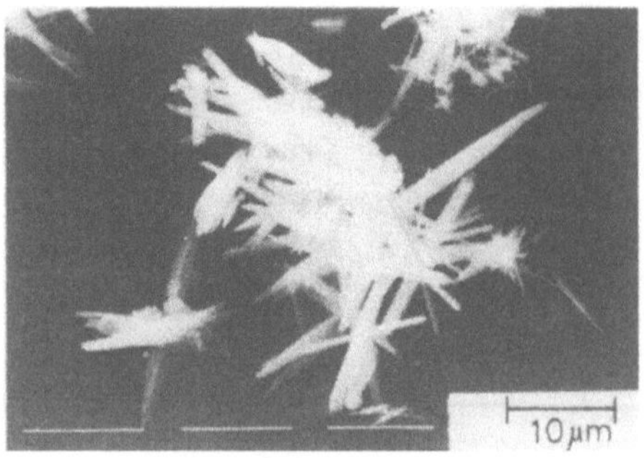

Fig. 1. SEM photograph of topaz crystals synthesized at 600°C, under 109.8 MPa for 7 days from $2AlF_3 + 2Al_2O_3 + 3SiO_2$ and water (run No. 22).

Fig. 2. SEM photograph of topaz showing multiple growth from a nucleation site, produced at 600°C under 109.8 MPa for 7 days (run No. 22).

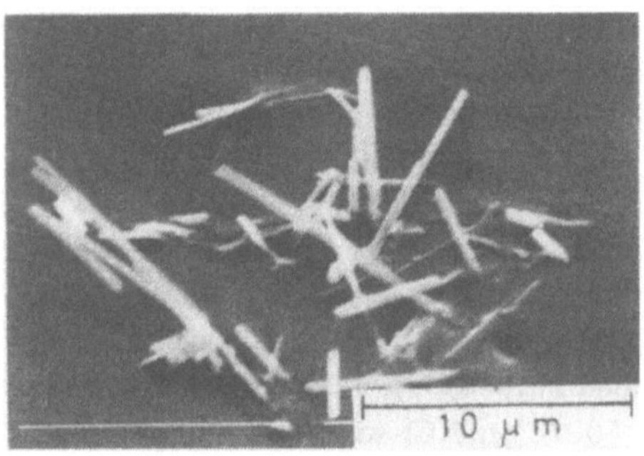

Fig. 3. SEM photograph of topaz synthesized at 625°C with ΔT=50°C under 117.6 MPa for 7 days from $2AlF_3$ + $4Al(OH)_3$ + $3SiO_2$ and lM HCl solution (run No. 83).

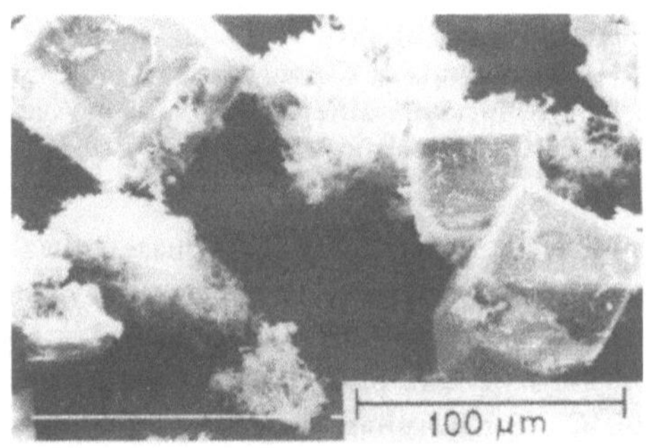

Fig. 4. SEM photograph of large AlF_3 crystals and the aggregate of small topaz crystals, produced at 700°C under 125.0 MPa for 1 day from $2AlF_3 + 4Al(OH)_3 + 3SiO_2$ and water (run No. 186).

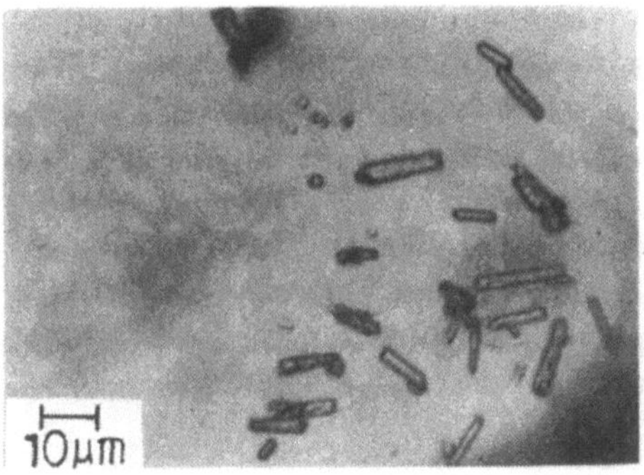

Fig. 5. Photomicrograph of topaz crystals grown at 700°C under 97.0 MPa for 3 days from $2AlF_3$ + $2Al_2O_3$ + $3SiO_2$ and water (run No. 5).

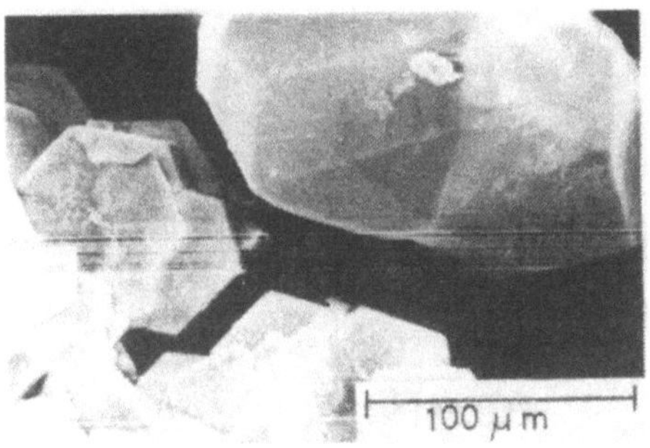

a) Kalsilite (run No. 113)

b) Corundum with topaz (run No. 2)

c) AlF_3 (run No. 186)

d) Leucite (run No. 119)

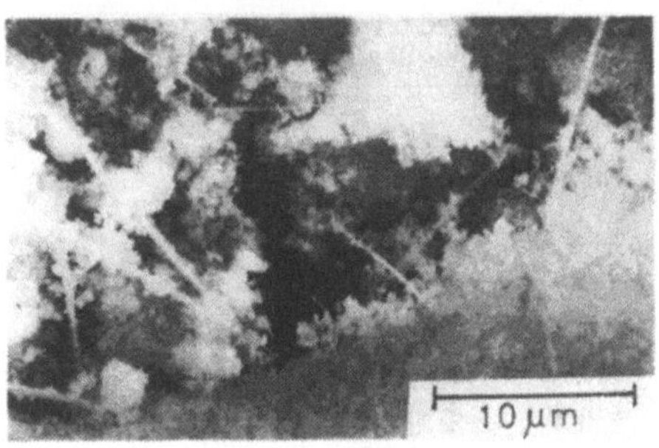

e) Zunyite (run No. 151)

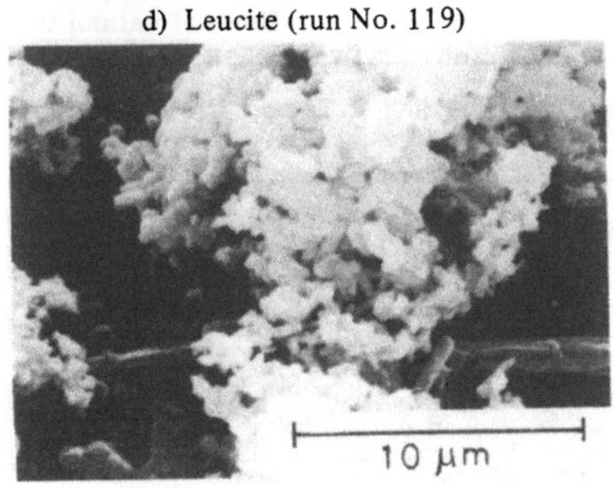

f) U_1 phase (run No. 136)

Fig. 6. SEM photographs of crystals prepared in present work.

3.5 Substitution of Fluorine Ions by Hydroxyl Ions in Synthetic Topaz

In a natural topaz, hydroxyl ions (OH^-) can substitute for fluorine ions (F^-) up to ~32%.[1] According to this substitution, the unit cell of topaz is expanded particularly along the b-axis[4] because F^- and/or OH^- ions are located in planes perpendicular to the b-axis of the unit cell. This phenomenon was also found in synthetic topaz as seen in Table 4. The unit cell parameter b_o of synthetic topaz is shown in Fig. 7 as a function of preparation temperatures and pressures. The tendency of larger b_o values in the topaz synthesized at lower temperatures and under high pressures might correspond to the tendency of OH to substitute in synthetic topaz. However, the

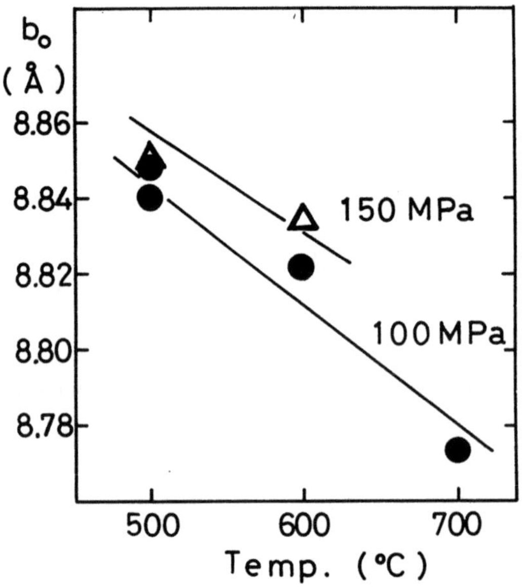

Fig. 7. Variation of b_o cell parameters of topaz with the preparation temperatures or pressures.

Table 4. Cell parameters of hydrothermally synthesized topaz. Those of natural topaz are taken from the reference (4).

Run No.	Starting Mixture	Temp. (°C)	Pressure (MPa)	Duration (day)	Cell parameters (Å)			
					a_o	b_o	c_o	V(Å³)
194	a	500	153.4	12	4.667	8.850	8.404	347.1
193	a	600	150.9	12	4.659	8.835	8.401	345.8
197	a	500	105.8	7	4.666	8.849	8.405	347.0
25	b	500	101.4	3	4.666	8.839	8.405	346.7
21	b	600	99.0	3	4.663	8.822	8.410	345.9
3	b	700	100.0	3	4.664	8.773	8.412	344.2
Natural topaz I (~2% OH)					4.652	8.789	8.394	343.2
Natural topaz II (~32% OH)					4.666	8.836	8.389	345.9

range of observed b_o values, from 8.773 Å to 8.850 Å, are considerably wider than that of natural topaz, from 8.789 Å for ~2%OH^- to 8.836 Å for ~32%OH^-. Further studies including chemical analyses are needed to clarify these discrepancies, but we would like to point out here that the topaz formation under the hydrothermal conditions can be written by the following equation (6) instead of equations (1) or (2) assumed previously,

$$2AlF_3 + 4Al(OH)_3 \ [\text{or } 2Al_2O_3] + 3SiO_2 \longrightarrow 3Al_2SiO_4F_{2-x}(OH)_x + xHF + (6-x)H_2O \quad \ldots\ldots\ (6)$$

The hydrothermal synthesis of topaz, thus, from the starting mixtures (a)–(e) and pure water has been carried out practically in HF solutions which change in the concentrations during the experimental run. This might be one of the reasons for slow growth rates of topaz crystals and/or the observed less or rather negative effects of mineralizers added upon the crystal growth. The most important factor to grow topaz crystals therefore seems to be the control of the fugacities (concentrations) of HF and H_2O in the solution during hydrothermal growth runs.

4. Summary

(1) Topaz crystals (max. 20μm long and 1.5μm thick) could be synthesized from AlF_3, $Al(OH)_3$ or Al_2O_3 and SiO_2 or Si under hydrothermal conditions, 400°C–800°C, 49–153 MPa.

(2) Single phase topaz is synthetized from $2AlF_3 + 4Al(OH)_3 + 3SiO_2$ and water at temperatures between 450°C and 650°C under 98.0 MPa.

(3) The addition of KF, KOH or K_2CO_3 as a mineralizer yields K-containing crystals such as Kalsilite, Leucite, K_3AlF_6, etc.

(4) The addition of KCl, NaCl, HCl or H_2SO_4 appeared to have little effects on topaz synthesis but rather negative effects on crystal growth.

(5) The substitution of F^- by OH^- in synthetic topaz showed a tendency to increase at low temperatures and under high pressures.

Acknowledgment

The authors thank to Mr. S. Itoh for the preparation of SEM photographs of crystals.

References

1) W. A. Deer, R. A. Howie and J. Zussman: "Rock Forming Minerals" vol. 1, pp. 145–150 Longmans, Green and Co. Ltd. London, 1962.
2) R. Schober and E. Thilo: Ber. deutsch. Chem. Ges., **73B**, 1219 (1940).
3) L. Coes: J. Am. Ceram. Soc., **38** (1), 298 (1955).
4) P. E. Rosenberg: Am. Mineral., **57**, 169 (1972).
5) A. M. Abdel Rehim: Proc. Eur. Sym. Therm. Anal., **1**, 325 (1976).
6) R. Roy and O. F. Tuttle: "Physics and Chemistry of the Earth 1", p. 138, Pergamon Press, London and New York, 1956.

Reprinted from Hiroshi Watanabe, Shuichi Iida and Mitsuo Sugimoto (eds), *Ferrites: Proceedings of the International Conference*, Japan, September–October 1980 (Center for Academic Publications, Japan, 1981) pp. 429–432.

Hydrothermal Synthesis and Sintering of $LaCrO_3$

Masahiro YOSHIMURA, Shen-tai SONG,* AND Shigeyuki SŌMIYA
Laboratory of Hydrothermal Syntheses, Research Laboratory of Engineering Materials, and Department of Materials Science and Engineering, Tokyo Institute of Technology, Midori-ku, Yokohama, Japan

Pure $LaCrO_3$ was synthesized from the mixture of La_2O_3 and Cr hydroxide under hydrothermal conditions of 100 MPa at temperatures as low as 700°C. Synthesized specimen of $LaCrO_3$ consisted of homogeneous fine grains with the size of 0.7 ± 0.2 μm. These grains were subsequently sintered by using hydrothermal reactions to remove H_2O as the form of H_2. A fine-grained and well-sintered body could be obtained under 100 MPa at 1000°C or above.

1. INTRODUCTION

The perovskite type compounds $LaMO_3$ (M=3d elements) are important materials in electromagnetic studies and applications. $LaCrO_3$ and its solid solutions are used for high temperature applications due to their high melting points >2400°C, good electronic conductivity and chemical stability against oxidation and corrosion at high temperatures. Therefore, there have been many attempts to fabricate pure and dense specimens.[1–7] Unfortunately, this material could not be prepared easily or sintered well by normal solid state reaction methods.[2,3] In the preparation of $LaCrO_3$, for example, firing of the mixture of La_2O_3 and Cr_2O_3 did not result in the single phase $LaCrO_3$ at 1400°C in air.[4] Intermediate products such as $La_2Cr^{6+}O_6$ are also formed at the temperature range from 600 to 900°C during heating,[5] then they react with excess Cr_2O_3 at higher temperatures. Pure $LaCrO_3$ could be prepared by evaporation and pyrolysis of the mixed particular organic solutions of lanthanum and chromium.[6,7] The required temperature for reaction is about 800°C.

Sintering of $LaCrO_3$ needs much higher temperatures as 1600°C or above. At such a high temperature, a difficulty appears to be caused from the volatilization of Cr oxide compounds in oxidizing atmospheres.[4,8–10] In order to reduce this volatility, the oxygen activity (P_{O_2}) in the atmosphere must be maintained in the order of 10^{-11} to 10^{-12} atm, which is the P_{O_2} near the equilibrium with Cr(s) and Cr_2O_3(s).

The major purposes of the present work were to study the synthesis and sintering of $LaCrO_3$ under hydrothermal conditions, where above difficulties could be greatly reduced.

* On leave from the Central Iron and Steel Research Institute, Peking, China.

2. EXPERIMENTAL

(1) *Hydrothermal Synthesis of $LaCrO_3$*

Chromium hydroxide was used as a starting material to improve the reactivity with La_2O_3. The amorphous hydrated chromium hydroxide was precipitated from the solution of chromic nitrate (guaranteed reagent) by mixing with an NH_4OH solution, filtered, washed, and then dried at 90°C for 24 h in air. The content of Cr_2O_3 in this hydroxide was analyzed by TG-DTA and the measurements of ignition losses. This hydroxide was mixed in an agate mortar with lanthanum oxide (99.9%), whose La_2O_3 content was also analyzed by TG-DTA and ignition loss measurements, in order to make the composition of La_2O_3: Cr_2O_3=1: 1 (mol) exact.

The mixutre of about 0.1 g was sealed by welding in a Pt capsule 2.7 mm inner diameter, 0.15 mm thick and 35 mm long, then heated in a test-tube type vessel under the pressue of 100 MPa H_2O at temperatures of 300 to 700°C for periods of 3 to 24 h. During heating, released water according to the following raction leads to a hydrothermal condition in the capsule,

$$La_2O_3+2[Cr(OH)_3\cdot nH_2O] \longrightarrow 2LaCrO_3+(3+2n)H_2O\,. \qquad (1)$$

After quenching at the completion of the run, all products were examined by X-ray powder diffraction method and scanning electron microscopy.

(2) *Hydrothermal Reaction Sintering of $LaCrO_3$*

The hydrothermal reaction sintering is a recently developed technique,[11–14] where metal powder and water or a dilute solution were generally used as starting materials. When they are heated in a hydrogen-permeable capsule under a pressure of ambient gas or liquid, they produce active fine powder of oxide and hydrogen gas. As the hydrogen gas can permeate out of the capsule, the oxide is compressed isostatically by the ambient gas pressure and then the sintering is allowed in the capsule. This method is, therefore, one of the HIP (hot isostatic pressing) method accompanied with hydrothermal reaction(s), and enabled us to fabricate dense (99%) and fine-grained (2–3 μm) ceramics of Cr_2O_3[11–13] and monoclinic ZrO_2[14] at 1000°C under 100 MPa. This technique has been applied to the sintering of pure $LaCrO_3$ in the present work by the following two methods;

Method (a): Excess Cr powder was added in the capsule containing La_2O_3 and chromium hydroxide described above in the hydrothermal synthesis. This Cr powder was expected to decompose released water according to Eq. (1) by the reaction,

$$2Cr+3H_2O \longrightarrow Cr_2O_3+3H_2 \quad (2)$$

thus, the produced $LaCrO_3$ and also Cr_2O_3+Cr would be subsequently compressed and then sintered into solid bodies.

Method (b): The mixture of La_2O_3 or $La(OH)_3$, Cr, and water were sealed in a Pt capsule expecting the direct synthesis and sintering of $LaCrO_3$ by the reaction,

$$La_2O_3+2Cr+3H_2O \longrightarrow 2LaCrO_3+3H_2. \quad (3)$$

These sintering experiments were carried out in a high-gas pressure apparatus at temperatures from 750 to 1400°C under 100 MPa for 3 h.

3. Results and Discussion

(1) *Hydrothermal Synthesis of $LaCrO_3$*

Table I summarizes the results of hydrothermal syntheses. The products of 300°C, $La(OH)_3$ and a small amount of $LaCO_3OH$ identified by X-ray powder diffraction, indicate that the chromium hydroxide might exist still in an amorphous state, and that the starting lanthanum oxide powder was contaminated by CO_2 gas probably absorbed from ambient atmosphere during preparation. The formation of $LaCrO_3$ could be observed at temperatures as low as 400°C under hydrothermal conditions. These temperatures are significantly lower than those above 1000°C in ordinal synthesis of $LaCrO_3$ by solid state reaction in air.[2,4] Thus it seemed that water accelerated the formation of $LaCrO_3$ as a mineralizer under hydrothermal conditions at 400°C under 100 MPa, where $La(OH)_3$ or CrOOH is stable in the binary system La_2O_3-H_2O[15] or Cr_2O_3-H_2O,[16] respectively.

Table I. X-ray diffraction analysis of the products prepared hydrothermally under 100 MPa.

Temperature (°C)	Period (h)	Identified phases
300	3	$La(OH)_3$, $LaCO_3OH$
400	3	$LaCrO_3$, $LaCO_3OH$
400	24	$LaCrO_3$, $LaCO_3OH$
500	3	$LaCrO_3$, $LaCO_3OH$
600	3	$LaCrO_3$, trace $LaCO_3OH$
700	3	$LaCrO_3$

Fig. 1. $LaCrO_3$ powders prepared hydrothermally at 700°C under 100 MPa for 3 h.

As reaction temperatures were elevated, the production of $LaCrO_3$ increased while the by-product $LaCO_3OH$ decreased. Pure $LaCrO_3$ was synthesized above 700°C. The $LaCrO_3$ crystals prepared hydrothermally at 700°C under 100 MPa for 3 h have homogeneous grains with the size of 0.7 ± 0.2 μm as seen in Fig. 1. The lattice parameters, a=5.477, b=5.511, and c=7.758 Å, are in good agreement with those for pure $LaCrO_3$ in the references.[2,3,6] These results revealed that fine-grained pure $LaCrO_3$ crystals could be prepared under hydrothermal conditions at 400 to 700°C.

(2) *Hydrothermal Reaction Sintering of $LaCrO_3$*

Fine-grained pure $LaCrO_3$ crystals prepared in the capsule were subsequently allowed to sinter under an isostatic gas pressure at higher temperatures of 700 to 1400°C. As Cr metal reacts with water at these temperatures[12,13] according to the reaction (2), released water by the reaction (1) can be decomposed into H_2 permeable out of the capsule. The outer ambient gas pressure, therefore, enabled to compress the inner products to sinter as illustrated in Fig. 2.

A sintered body of pure $LaCrO_3$ could be pre-

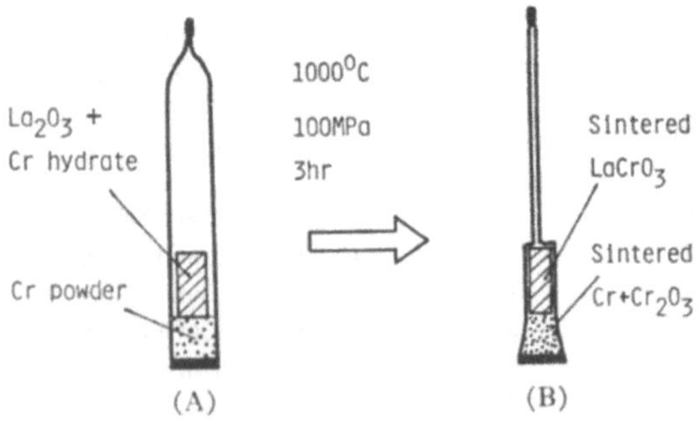

Fig. 2. Schematic illustration of hydrothermal reaction sintering of $LaCrO_3$ by method (a). (A) Before sintering, (B) after sintering.

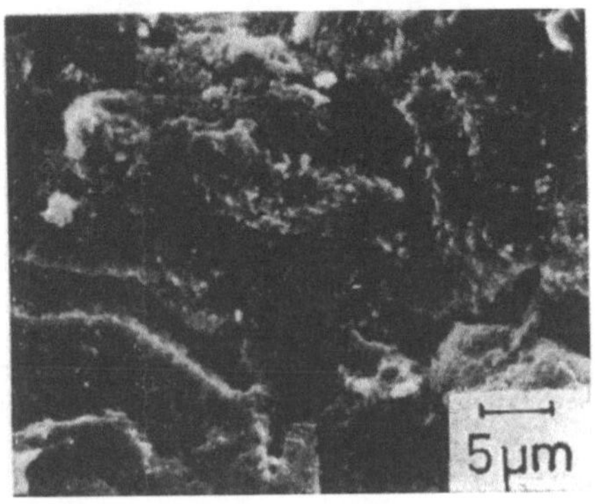

Fig. 3. Fractured surface of $LaCrO_3$ ceramic sintered hydrothermally at 1000°C under 100 MPa for 3 h.

pared by sintering at 1000°C at 100 MPa for 3 h. This ceramic consists of fine grains with less than 1 μm as seen in Fig. 3. The pores in the middle of this figure are indicating that residual gas(es) in the capsule prevented further densification of this ceramic. Such residual gases appear not to be H_2O because enough Cr powder was encapsuled to decompose H_2O, but to be the other gases absorbed during the preparation of the starting mixture. Because of the presence of the residual gases, the sintering of $LaCrO_3$ by this method has poor reproducibility. Well-sintered bodies could not be prepared even at higher temperatures of 1100 to 1400°C. Under an extreme condition where less compression had occurred, $LaCrO_3$ grains showed a typical step growth at the surfaces as seen in Fig. 4. These grains seem to have been grown almost freely in the pressured atmosphere. To improve more compression on $LaCrO_3$ grains by removing the residual gas(es) in the capsule, the capsule was punctured then sealed again by welding. The re-sealed capsule was treated again at temperatures between 1000 and 1400°C under 100 MPa, i.e., hot isostatic pressing. The fractured surface of a typical sample sintered by this method is shown in Fig. 5.

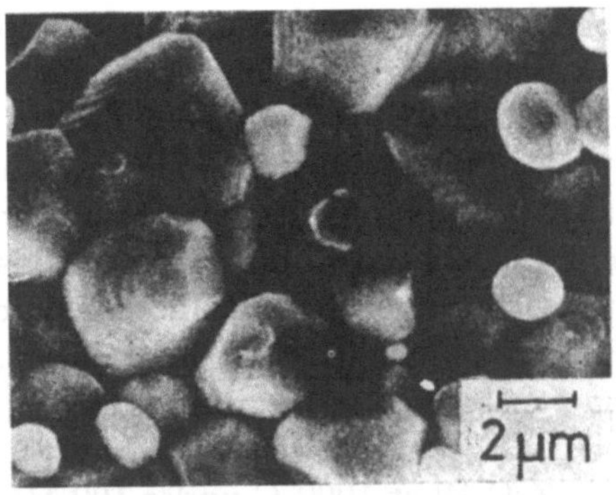

Fig. 4. Fractured surface of a $LaCrO_3$ body obtained at 1400°C under 100 MPa for 3 h. The step growths can be seen on the surfaces of grains.

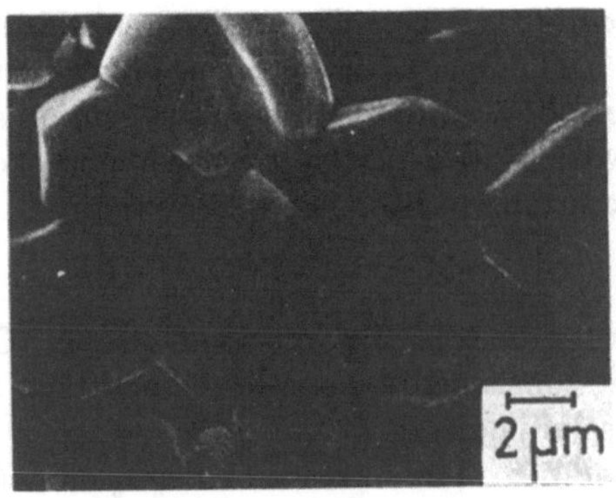

Fig. 5. Fractured surface of $LaCrO_3$ ceramic sintered hydrothermally at 1000°C then hot-isostatically pressed at 1400°C under 100 MPa for 3 h.

In the experiments by method (b), the products always consisted of the mixture of $LaCrO_3$, La_2O_3 or $La(OH)_3$, and Cr_2O_3 or/and Cr. $La(OH)_3$ and Cr were mainly identified in the product below 700°C. $LaCrO_3$ was formed above 1000°C, and the yield of $LaCrO_3$ increased with the increase of reaction temperatures. This method, however, resulted neither in the formation of pure $LaCrO_3$ nor in the well-sintered body. These results indicate that the formation of $LaCrO_3$ is shown by the series of the reactions,

$$La_2O_3+2Cr+3H_2O \longrightarrow 2La(OH)_3+2Cr \longrightarrow La_2O_3+Cr_2O_3+3H_2 \longrightarrow 2LaCrO_3+3H_2 \quad (4)$$

and that the reaction between La_2O_3 and Cr_2O_3 is much slower than that between La_2O_3 and Cr hydrate described before.

Hydrothermal reaction sintering of $LaCrO_3$ is still in progress to obtain denser bodies. This technique seems to be a unique method to prepare ultra-fine grained ceramics of materials, which are difficult to sinter by ordinary sintering methods.

ACKNOWLEDGMENTS

The authors are thankful to Grant-in-Aid for Scientific Research No. 443015 (1979–1980) and the Asahi Glass Foundation for Industrial Technology for a financial support.

REFERENCES

1) A. M. Anthony, G. Benezech, F. Cabannes, M. Faucher, M. Foex, V. Loc, and D. Yerouchalmi, *Proc. IUPAC 3rd Int. Symp. High Temperature Technology* (Butterworths, London, 1969) p. 213.
2) D. B. Meadowcroft, Energy Convers. **8** (1968) 185.
3) G. H. Jonker, Physica **20** (1954) 1118.
4) D. B. Meadowcroft, J. M. Wimmer, Am. Ceram. Soc. Bull. **58** (1979) 610.
5) R. Berjoan, G. Benezech, J.-P. Coutures, M. Foex, C. R. Acad. Sci. Paris **280C** (1975) 601.
6) J. Faber, Jr., M. H. Muller, W. L. Procarione, A. T. Aldred, H. W. Knott, and H. U. Anderson, *Proc. Conf. High Temperature Sciences Related to Open-Cycle Coal Fired MHD Systems, Argonne, Illinois*, 1977; Gov. Rep. Announce. (U. S.) **77** (1977) 11B.
7) L. Groupp, H. U. Anderson, J. Am. Ceram. Soc. **59** (1976) 449.
8) T. Sasamoto, T. Sata, Yogyo Kyokai Shi **74** (1971) 408 [in Japanese].
9) H. U. Anderson, R. Murphy, K. Humphrey, B. Rossing, A. Aldred, W. L. Procarione, R. J. Ackermann, and J. L. Bates, *The Rare Earths in Modern Science and Technology*, eds. G. J. McCarthy, J. J. Rhyne (Plenum Press, New York, 1978) p. 55.
10) H. C. Graham, H. H. Davis, J. Am. Ceram. Soc. **54** (1971) 89.
11) S. Hirano, S. Sōmiya, J. Am. Ceram. Soc. **59** (1976) 534.
12) S. Sōmiya, S. Hirano, M. Yoshimura, S. Itoh, and H. Kanai, *Proc. Int. Symp. Factors in Densification and Sintering of Oxide and Non-Oxide Ceramics*, eds. S. Sōmiya, S. Saito (Gakujutsu Bunken Fukyukai, Tokyo Institute of Technology, Tokyo, 1979) p. 267.
13) S. Sōmiya, M. Yoshimura, and H. Kanai, to be published in Zairyokagaku (J. Mater. Sci. Soc. Jpn).
14) M. Yoshimura, S. Sōmiya, Am. Ceram. Soc. Bull. **59** (1980) 246.
15) B.H.T. Chai, S. Mroczkowski, J. Cryst. Growth **44** (1978) 84.
16) M. W. Shafer, R. Roy, Z. Anorg. Allg. Chem. **276** (1954) 275.

Reprinted from *Yogyo-Kyokai-Shi*, **90**(2) (1982) 91–95.

Synthesis of $LaCrO_3$ Under Hydrothermal Conditions

Masahiro YOSHIMURA, Shen-tai SONG* and Shigeyuki SŌMIYA

(Laboratory for Hydrothermal Syntheses, Research Laboratory of Engineering Materials and Department of Materials Science and Engineering, Tokyo Institute of Technology 4259, Nagatsuta, Midori-ku, Yokohama-shi 227)

The synthesis of $LaCrO_3$ from the mixture of chromium hydroxide and La_2O_3 was studied under hydrothermal conditions and in air. $LaCrO_3$ was directly prepared under hydrothermal conditions of above 400°C under 100 MPa H_2O pressure in contrast to the formation of $LaCrO_3$ via La_2CrO_6 above 700°C in air. $LaCrO_3$ synthesized under hydrothermal conditions was well crystallized with homogeneous grain size of $0.7 \pm 0.2\,\mu m$ at temperatures of 700°∼1000°C, while that synthesized in air was an aggregate with irregular grains.

[Received April 28, 1981]

Key-words : $LaCrO_3$, Hydrothermal synthesis, Fine powder, La_2CrO_6

1. Introduction

Lanthanum chromite $LaCrO_3$ had been studied earlier from electromagnetic or crystallographic interest[1),2)]. In recent years, $LaCrO_3$ or doped $LaCrO_3$ have widely been studied as a refractory conducting material because of its electronic conductivity, high melting point (>2400°C) and oxidation or corrosion resistivity. Particularly, the doped $LaCrO_3$ materials with high electronic conductance have been interested as an electrode for MHD[3),4)], fuel cell[5)] or electric arc[6)], and a heating element[7)] or a conducting lead for ZrO_2-base heating elements[8)]. In these studies or applications, $LaCrO_3$ and doped $LaCrO_3$ have generally been synthesized by solid state reactions of oxide mixtures with heating at high temperatures above 1000°C in air. These procedures need repeat firing and grinding to complete the reaction due to slow diffusion processes in solids. Therefore we have studied the synthesis of $LaCrO_3$ under hydrothermal conditions in comparison with the usual firing under atmospheric pressure. Here the effects of the temperatures and atmospheres on the synthesis of $LaCrO_3$ is described.

2. Experimental

Chromium hydroxide was used as a starting material to improve the reactivity with La_2O_3. The chromium hydroxide was precipitated from 3 N chromic nitrate solution prepared from $Cr(NO_3)_3 \cdot 9H_2O$ (guaranteed reagent grade, Kanto Chem. Co., Ltd., Tokyo, Japan)

* On leave from the Central Iron and Steel Research Institute, Peking, the People's Republic of China

by adding into 8 N NH_4OH solution. The precipitate was filtered, washed by distilled water thoroughly, then dried at 90°C for 24 h in air. The contents of Cr_2O_3 in this hydroxide were analyzed by TG–DTA (Thermoflex TG-DTA, Rigaku Denki Co., Ltd., Tokyo, Japan) and ignition loss measurements. This chromium hydroxide was mixed with lanthanum oxide (>99.9%, Shinetsu Chem. Co., Ltd., Japan), of which La_2O_3 contents were also analyzed by TG–DTA and ignition loss measurements, to the exact composition of $La_2O_3:Cr_2O_3=1:1$ (mol) thoroughly in an agate mortar with aceton.

The mixture of about 0.1 g was sealed with an electric arc in a Pt capsule 2.7 mm ID, 0.15 mm thick and 35 mm long. The capsule was heated in a test-tube type pressure vessel of Stellite 25 under the pressure of 100 MPa H_2O at 300°~700°C for 3~24 h. The temperature was measured by calibrated Platinel thermocouples attached on the outside wall of the vessel.

In order to study the influences of atmospheres, additional experiments were carried out where ~0.06 g chromium powder (>99%, Wako Pure Chem. Co., Ltd., Tokyo) was added on the top of the mixture in the capsule, Chromium would react with H_2O to produce Cr_2O_3 and H_2 at high temperatures, so that a reducing atmosphere was created in the capsule.

The synthesis of $LaCrO_3$ from the starting mixture was also studied in air by the TG–DTA equipment. The mixture was heated at 600°~1200°C for 3 h in a Pt crucible then cooled rapidly.

All samples prepared were examined by X-ray powder diffraction using a diffractometer with a rotating Cu anode and a graphite monochromater (RU-7, Rigaku Denki Co., Ltd., Japan) and scanning electron microscopy using an equipment (JSM T-200, JEOL Co. Ltd., Japan).

A high purity silicon (>99.99%, Kinzoku Kagaku Lab., Tokyo) was added as an internal standard to determine the lattice parameters of $LaCrO_3$.

3. Results and discussion

3.1 Characterization of starting materials

The chromium hydroxide prepared was fine-grained aggregate as seen in **Fig. 1**, and amorphous on the basis of the X-ray diffraction. The composition of this material appeared to be $Cr_2O_3 \cdot 6.42\,H_2O$ from the ignition loss measurements. Based upon the X-ray diffraction, small amount of $La(OH)_3$ was detected in used La_2O_3 (**Fig. 2**). The water contents of this material was determined to be $La_2O_3 \cdot 0.95\,H_2O$ by ignition loss measurements and TG-DTA. **Table 1** summarizes the characterization of the starting materials.

3.2 Synthesis of $LaCrO_3$ under hydrothermal conditions

Table 2 shows the results of hydrothemal experiments. The product heated at 300°C was identified to be crystallized $La(OH)_3$ and small amount of $LaCO_3OH$[9] by X-ray diff-

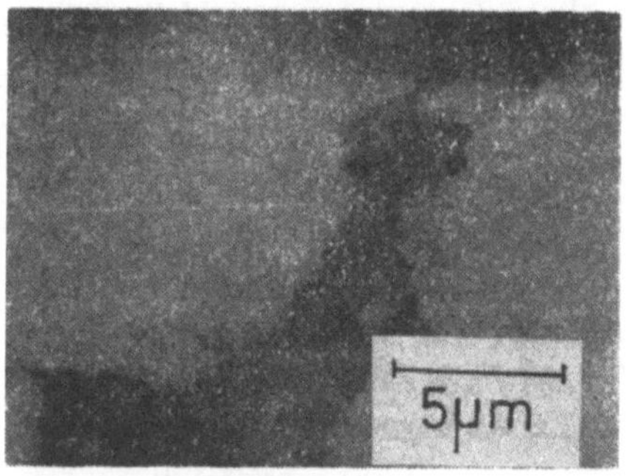

Fig. 1. SEM photograph of prepared chromium hydroxide.

Fig. 2. SEM photograph of lanthanum oxide.

Table 1. Characterization of starting materials.

Material	X-ray analysis	Water contents n values in $M_2O_3 \cdot nH_2O$
Chromium hydroxide	Amorphous	6.42
Lanthanum oxide	$La_2O_3 + La(OH)_3$*	0.95

* A small amount of CO_2 might be absorbed though no CO_2 containing phases could be detected.

raction (**Fig. 3**). Chromium hydroxide might exist still in amorphous state because no crystallized chromium compounds could be detected in this product. The formation of $LaCO_3OH$ indicates the contamination of starting La_2O_3 by CO_2 gas absorbed from atmosphere. Similar phenomena have often been observed in the studies in the system La_2O_3–H_2O under hydrothermal conditions[10),11)]. The crystal $LaCO_3OH$ is significantly stable under hydrothermal conditions particularly high pressures as 100 MPa or above[10)] in the presence of CO_2 components, even they were in a small amount, in the system. $LaCO_3OH$ remained at 500°C and traced at 600°C, but has not been detected above 700°C. The product of 400°C was almost well crystallized $LaCrO_3$ indicating the reaction between $La(OH)_3$ and chromium hydroxide with the release of H_2O. Scanning electron photomicrograph of this product (**Fig. 4**) revealed fine-grained $LaCrO_3$ of about 0.1 μm in diameter and hydrothermally grown $LaCO_3OH$ crystals of 1~3 μm in edge. As the reaction temperature elevated, the amount and the size of $LaCO_3OH$ crystals decreased while the production of $LaCrO_3$ increased. Only well crystallized $LaCrO_3$ could be observed in the product at 700°C. The $LaCrO_3$ crystals has homogeneous grain size of 0.7 ±0.2 μm (**Fig. 5**). The hydrothermally prepared $LaCrO_3$ has an orthorhombic symmetry (*Pbnm*) with the lattice parameters a=5.477 Å, b=5.511 Å and c= 7.758 Å, which are in good agreement with the previous data[2),12)~14)], a=5.476~5.479 Å, b=5.512~5.515 Å and c=7.752~7.758 Å for pure $LaCrO_3$ prepared in air at high temperatures above 1000°C. These results indicate that formation of $LaCrO_3$ has been completed in a short period of 3 h at temperatures as low as 700°C under hydrothermal conditions due to the action of supercritical water as a mineralizer.

Table 2. X-ray diffraction analysis of the products prepared hydrothermally under 100 MPa.

Temperature (°C)	Period (hr)	Identified phases
300	3	$La(OH)_3$ (vs), $LaCO_3OH$ (w)
400	3	$LaCrO_3$ (vs), $LaCO_3OH$ (w)
400	24	$LaCrO_3$ (vs), $LaCO_3OH$ (w)
500	3	$LaCrO_3$ (vs), $LaCO_3OH$ (vw)
600	3	$LaCrO_3$ (vs), $LaCO_3OH$ (tr)
700	3	$LaCrO_3$ (vs)

vs—very strong, w—weak, vw—very weak, tr—trace

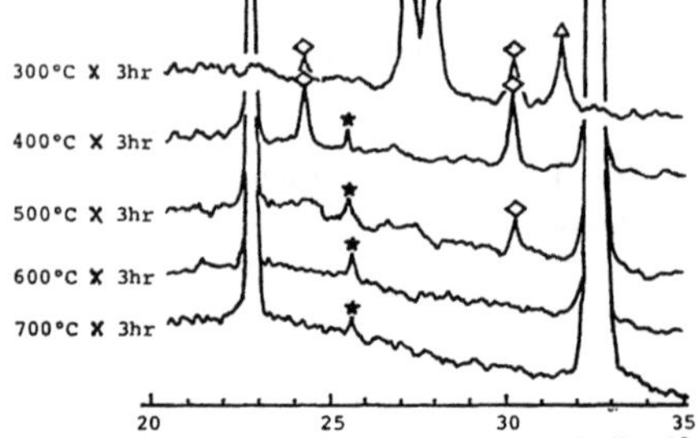

Fig. 3. X-ray powder pattern of the products prepared under hydrothermal conditions.

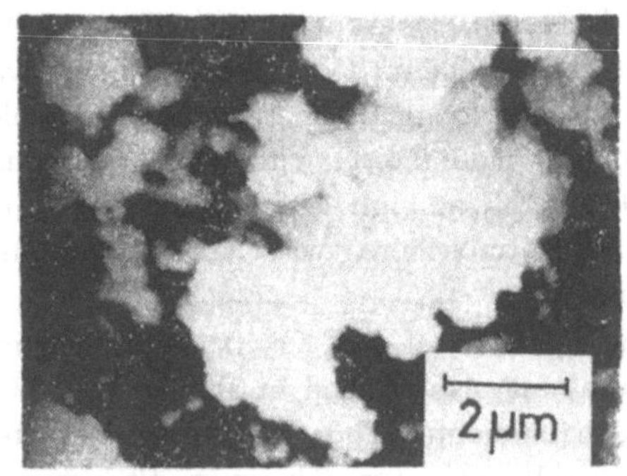

Fig. 4. SEM photograph of the product prepared under hydrothermal condition (at 400°C, 100 MPa, for 3 h).

Fig. 5. SEM photograph of the product prepared under hydrothermal condition (at 700°C, 100 MPa, for 3 h).

In the experiments with Cr powder, $LaCrO_3$ was also synthesized in the capsule at temperatures above 400°C. The products in the experiments with Cr powder were entirely identical with those in the experiments without Cr powder. A reducing atmosphere created by the hydrogen produced by the reaction

$$2\,Cr+3\,H_2O \rightarrow Cr_2O_3+3\,H_2$$

at temperatures above 600°C[15], had little effect on the formation of $LaCrO_3$. These results indicate that $LaCrO_3$ is stable under a reducing conditions, though oxygen deficient Ca-doped $LaCrO_3$ has been prepared[16] under a similar hydrothermal conditions employed in the present study. The water was decomposed into hydrogen which could permeate through the wall of the Pt capsule at high temperatures, so that a weight loss could be observed after hydrothermal treatments. The elimination of water contents, 12% or 18% at 600°C, corresponding to the observed weight losses did also show no significant effects on the formation and the crystal growth of $LaCrO_3$.

3.3 Synthesis of $LaCrO_3$ in an atmospheric pressure

When the same starting mixture of La_2O_3 and chromium hydroxide as that in hydrothermal experiments was heated in air with an atmospheric pressure, the formation of $LaCrO_3$ was observed at higher temperatures above 700°C as seen in **Table 3.** At lower temperatures as 600°C, no $LaCrO_3$ was detected but the mixture of La_2CrO_6 and small amounts of La_2O_3 and Cr_2O_3. According to the results by TG-DTA and X-ray diffraction for the samples heated at various temperatures, the reaction process to form $LaCrO_3$ in air appeared to be as follows,

Table 3. X-ray diffraction analysis of the products prepared in air (0.1 MPa).

Temperature (°C)	Period (hr)	Identified phases
300°C	3	La_2O_3 (vs), $La(OH)_3$ (s)
400°C	3	La_2O_3 (vs), LaOOH (s), Cr_2O_3 (vs)
500°C	3	La_2O_3 (vs), LaOOH (w), Cr_2O_3 (vs)
550°C	3	La_2O_3 (vs), Cr_2O_3 (vs)
600°C	3	La_2CrO_6 (vs), La_2O_3 (w), Cr_2O_3 (w)
700°C	3	$LaCrO_3$ (vs), La_2CrO_6 (w), La_2O_3 (vw), Cr_2O_3 (vw)
800°C	3	$LaCrO_3$ (vs), La_2CrO_6 (vw)
900°C	3	$LaCrO_3$ (vs)
1200°C	3	$LaCrO_3$ (vs)

vs—very strong, s—strong, w—weak, vw—very weak

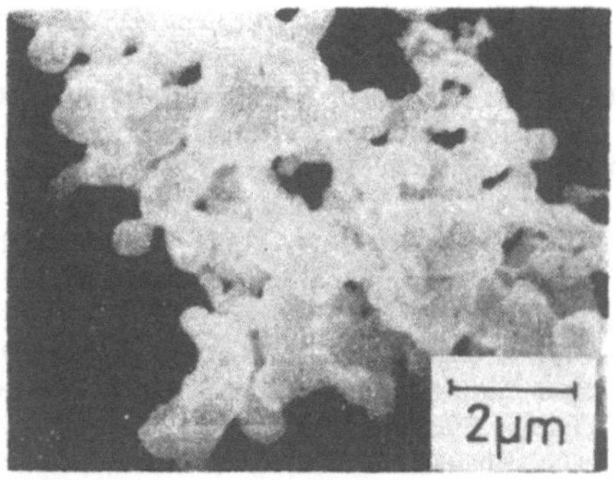

Fig. 6. SEM photograph of the product prepared in air (at 1200°C, 0.1 MPa, for 3 h).

$$La_2O_3 \text{ containing } La(OH)_3+\text{Cr hydroxide} \xrightarrow{\sim 440°C} La_2O_3 \text{ containing } LaOOH+Cr_2O_3 \xrightarrow{\sim 550°C} La_2O_3+Cr_2O_3 \xrightarrow{>600°C} La_2CrO_6+Cr_2O_3 \xrightarrow{>900°C} LaCrO_3$$

The formation of La_2CrO_6 with the oxidation of chromic oxide is characteristic[17] in the reaction between La_2O_3 and Cr_2O_3 in air. The powder of $LaCrO_3$ prepared in air at temperatures above 900°C aggregated irregularly as shown in **Fig. 6.**

4. Summary

Pure $LaCrO_3$ was synthesized at low temperatures as 700°C under hydrothermal conditions with 100 MPa pressure for 3 h from La_2O_3 and chromium hydroxide. $LaCrO_3$ was well crystallized with homogeneous grain size of $0.7\pm0.2\,\mu m$. The water seemed to accelerate the reaction between La_2O_3 or La hydroxide and Cr_2O_3 or Cr hydroxide as a mineralizer. The formation process under hydrothermal conditions appears to be simpler than that via the formation of La_2CrO_6 as an intermediate compound in air.

References

1) I. Náray-Szabó, *Muegyetemi Kozlemenyek*, **1**, 30 (1974).
2) S. Geller, *Acta Cryst.*, **10**, 243 (1957).
3) A.M. Anthony, G. Benezech, F. Cabannes, M. Faucher, M. Foex, Vutien Loc and D. Yerouchalmi, Proc. of the IUPAC 3rd Int. Symp. on High Temperature Technology, 213-33, Butterworths, London, 1969.
4) D.B. Meadowcroft, P.G. Meier and A.C. Warren, *Energy Conversion*, **12**, 145 (1972).
5) H. Kleinschmager and A. Reich, *Z. Naturforsch.*, **27**, 363 (1972).
6) S. Kose, *Kagaku Kogyo*, 72 (1974).
7) S. Kose, *Shinkinzoku Kogyo*, **1973**, 265-70.

8) S.T. Song, H.Y. Pan, J.R. Feng and Z. Wang, *J. Chinese Silicate Soc.*, **7**, 238-43 (1979).

9) B.H.T. Chai and S. Mroczkowski, *J. Cryst. Growth*, **44**, 84-97 (1978).

10) M.W. Shafer and R. Roy, *J. Am. Ceram. Soc.*, **42**, 564 (1959).

11) T.R. Narayanan Kutty, M.N. Viswanathiah and J.A.K. Tareen, *Proc. Indian Acad. Sci.*, **87 A**, 69-74 (1978).

12) C.E. Gooden and G.J. McCarthy, Powder Diffraction File 24-1016, Joint Committee on Powder Diffraction Standard USA (1974).

13) C.P. Khattak and D.E. Cox, *Mater. Res. Bull.*, **12**, 463 (1977).

14) J. Faber, Jr., M.H. Muller, W.L. Procarione, A.T. Aldred, H.W. Knott and H.U. Anderson, Proc. of Conf. on High Temperature Sciences Related to Open-Cyclet Coal Fired MHD Systems, Argonne, Illinois, 1977, CONF-770411-1, Government Reports Announcements, vol. 77, No. 26, 11 B.

15) S. Sōmiya, M. Yoshimura and H. Kanai, to be published in *Zairyokagaku* (*J. of the Mater. Sci. Soc. of Japan*).

16) S.T. Song, M. Yoshimura and S. Sōmiya, to be published in *Zairyokagaku* (*J. of the Mater. Sci. Soc. of Japan*).

17) R. Berjoan, G. Benezech, J-P. Coutures and M. Foex, *C.R. Acad. Sci. Paris*, **280 C**, 601-03 (1975).

Reprinted from *Proceedings of the First International Symposium on Hydrothermal Reactions*, Japan, 22–26 March 1982, pp. 509–518.

Effects of Starting Materials and Temperature on the Hydrothermal Synthesis of Aluminum Orthophosphate

Kiyoshi Hasegawa, Keiichi Minegishi and *Shigeyuki Sōmiya
Central Laboratory of Research and Development Chichibu Cement Co.,Ltd.
2-1-1 Tsukimi-cho, Kumagaya-shi, Saitama-ken 360, Japan
*Research Laboratory of Engineering Materials, Tokyo Institute of Technology
4259 Nagatsuda-cho, Midori-ku, Yokohama-shi, Kanagawa-ken 227, Japan

ABSTRACT

In order to understand and improve the hydrothermal techniques for the growth of $AlPO_4$, we have investigated the effects of the starting materials and the temperature on the hydrothermal synthesis of $AlPO_4$.
Four kinds of aluminate with phosphoric acid used as starting materials were enclosed in silver capsules.
They were heated over the temperature range of from 200°C to 500°C for 48 hrs in the Bridgeman type autoclave which was 70 % filled. The reaction products were studied by XRD, SEM and DTA.
In the case of sodium aluminate, needle-like $AlPO_4$ crystals of 100 ~ 500 μm in length were synthesized at 400°C by the hydrothermal reaction, and well-faced $AlPO_4$ crystals developed up to 1000 μm at 500°C.
With regard to aluminum hydroxide, needle-like crystals could be observed at 400°C by the addition of NaOH solution.
The results are discussed on the view of solubility of $AlPO_4$ and the adsorption of ions on the crystal surface.

Introduction

Aluminum orthophosphate, $AlPO_4$, has been suggested as one of the most attractive materials for surface acoustic wave devices because of its large piezoelectric coefficients and small frequency-temperature coefficients(1,2,3).

Therefore, the reproducible techniques for growth of good quality single crystals have been improved upon many laboratories. In recent years, Kolb and Laudise(4), Yoshimura and others(5) have succeeded in growing large high quality single crystals.

In order to understand and improve upon the hydrothermal techniques for the growth of $AlPO_4$, we have investigated the effects of the starting materials and the temperature on the hydrothermal synthesis of $AlPO_4$.

Experimental

The flow chart of the experimental procedures is shown in Fig. 1.

The starting materials for aluminates were reagent-grade $AlPO_4$, Al_2O_3, $Al(OH)_3$ and $NaAlO_2$.

Four kinds of 0.001 mol aluminates with 0.3 ml of 6.1 M phosphoric acid were enclosed in silver capsules and were settled in a Bridgeman-type autoclave.

The autoclave which was 70 % filled was heated over the temperature range of from 200 °C to 500 °C for 48 hrs, and was quenched in cold water to prevent redissolution of the products.

The reaction products were examined by X-ray powder diffractometer and scanning electron microscope.

Moreover, the crystallographic inversions of the products were investigated by DTA.

Results and Discussion

1. Hydrothermal reaction of $AlPO_4$ with phosporic acid

X-ray powder diffraction patterns of reagent grade $AlPO_4$ and its hydrothermal reaction products are presented in Fig. 2.

The reagent-grade $AlPO_4$ was almost amorphous, but it crystallized into berlinite due to the hydrothermal reaction.

Figure 3 shows the scanning electron micrographs of the products.

The crystal surfaces of products at 200 °C and 300 °C were relatively smooth, while those at 400 °C were rough.

These results are probably because of the negative temperature coefficient of $AlPO_4$ solubility in H_3PO_4, which was reported by J.M. Stanley(6) and E.D. Kolb(7).

That is to say, as the supersaturation of $AlPO_4$ in H_3PO_4 at 200 °C and 300 °C was low, it seemed that crystals grew slowly, so smooth surfaces were formed.

On the contrary, as the supersaturation was high at 400 °C, the rate of crystal growth was large, consequently rough crystal surfaces were obtained.

2. Hydrothermal reaction of Al_2O_3 with phosphoric acid

Figure 4 shows X-ray powder diffraction patterns of reagent-grade Al_2O_3 and its hydrothermal reaction products.

As shown in Fig. 4, reagent-grade Al_2O_3 was identified to be α-Al_2O_3 (corundum), and its products were a mixture of $AlPO_4$ as the main product and α-Al_2O_3 as the unreactive product.

Figure 5 represents the micrographs of the products.

As the reaction temperature increases, the crystal surfaces gradually become rough, which is simillar to the case of the reaction of $AlPO_4$ with H_3PO_4.

In the case of the reaction at 500 °C automorphic crystals could not be observed.

3. Hydrothermal reaction of $Al(OH)_3$ with phosphoric acid

Figure 6 shows X-ray powder diffraction patterns of reagent-grade $Al(OH)_3$ and its hydrothermal reaction products.

Reagent-grade $Al(OH)_3$ and its reaction with H_3PO_4 were identified to be gibbsite and $AlPO_4$, repectively.

The micrograph of the hydrothermal reaction products is shown in Fig. 7.

Trigonal crystals grown up to 0.8 mm in length were observed at 200 °C.

In the case of the reaction at 500 °C, fine crystals having rough surfaces were obtained.

4. Hydrothermal reaction of $NaAlO_2$ with phosphoric acid

X-ray diffraction patterns of sodium aluminate and those of its products are shown in Figs. 8 and 9, respectively.

Reagent-grade sodium aluminate was confirmed to be a mixture of $NaAlO_2$ and its hydrate($2NaAlO_2 \cdot 3H$ O).

The reaction products at 200 °C was a mixture of $AlPO_4$ and aluminum phoshate hydrates, but the products formed above 300 °C was monophase of $AlPO_4$.

In Fig. 10 which shows the micrographs of the products, needle-like $AlPO_4$ crystals of 100 ~ 500 µm in length could be observed at 400 °C, and well-faced $AlPO_4$ crystals developed up to 1000 µm at 500 °C.

The reasons for the needle-like crystals formed seems to be due to the large growth rate and anisotropic growth caused by adsorption of Na^+ ions on the certain surfaces of $AlPO_4$.

This will be discussed in later part of this paper.

5. The effect of Na^+ ion on the morphology of $AlPO_4$

In order to consider the effect of Na^+ ion on the morphology of $AlPO_4$, the morphologies of the products prepared at 400 °C from $Al(OH)_3$, H_3PO_4 and NaOH solutions as the starting materials were iuvestigated.

Figure 11 shows the micrographs of the products, which were identified to be berlinite.

Photo (A) represents the micrographs of $AlPO_4$ prepared from $Al(OH)_3$ and 6.1 M H_3PO_4 as the starting materials. No needle-like crystals were observed.

Photo (B) represents the micrographs of $AlPO_4$ prepared from $Al(OH)_3$, 9.15 M H_3PO_4 and 10 N NaOH as the starting materials.

Needle-like $AlPO_4$ crystals of 100 ~ 500 µm in length were synthesized.

These results suggest that with regard to the reaction of $Al(OH)_3$ with H_3PO_4, needle-like crystals can be obtained at 400 °C by the addition of NaOH.

Figure 12 shows DTA curves of the berlinites obtained by the hydrothermal reactions at 400 °C.

A sharp endothermic peak was observed at 574 °C for the berlinite prepared from $NaAlO_2$ and was observed at 582 °C for that prepared from amorphous $AlPO_4$.

These endothermic peaks are due to the $\alpha - \beta$ phase transition reported by W.R. Beck 8) and others.

The $\alpha - \beta$ phase transition peak of berlinite prepared from

$NaAlO_2$ shows a temperature about 10 °C lower than that of berlinite prepared from amorphous $AlPO_4$.

This seems to be responsible for Na contained in the berlinite.

Actually it was determined by the spectrographic analysis that the berlinite prepared from $NaAlO_2$ contained about 0.1 wt % of Na_2O as the impurity.

Conclusions

From the results of this study follwing conclusions may be drawn :

(1) Good quality $AlPO_4$ crystals, whose surfaces were relatively smooth, could be synthesized at 200 °C from amorphous $AlPO_4$ and $Al(OH)_3$.

(2) In the case of sodium aluminate, needle-like $AlPO_4$ crystals of 100 ~ 500 μm in length were synthesized at 400 °C.

(3) With regard to $Al(OH)_3$, needle-like crystals could to observed at 400 °C due to the addition of NaOH.

(4) The α - β phase transition peak of berlinite prepared from $NaAlO_2$ show a temperature about 10 °C lower than that of berlinite from amorphous $AlPO_4$.

References

1. Z.P. Chang and G.R. Barsch, "Elastic Constants and Thernal Expansion of Berlinite," IEEE Trans. on Sonics and Ultrasonics, SU-23, 127, 1976.

2. R.M. O'Connell and P.H. Carr, "High Piezoelectric Coupling, Temperature Compensated Cuts of Berlinite, $AlPO_4$, for SAW Application," IEEE Trans. on Sonics and Ultrasonics SU-24, 376, 1977.

3. D.G. Morency, W. Soluch, J.F. Vetelino, S.D. Mittleman, D. Harmon, S. Surek, J.C. Field, and G. Lehmann, "Experimental Measurement of the SAW Properties of Berlinite," Appl. Phys. Lett., 33, 117, 1978.

4. E.D. Kolb and R.A. Laudise, "Hydrothermal Synthesis of Aluminum Orthophosphate," J. Crystal Growth, 43, 313, 1978.

5. J. Yoshimura, S. Taki and S. Maruya, "Hydrothermal Synthesis of Berlinite Crystals." Yogyo-Kyokai-Shi, 89 8 , 433-439 (1981)

6. J.M. Stanley, "Hydrothermal Synthesis of Large Aluminum Phosphate Crystals," Ind. Eng. Chem.,46, 1684-1689 (1954)

7. E.D. Kolb, R.L. Barns, J.C. Grenier and R.A. Laubility, "Solubility, Crystal Growth and Perfection of Aluminum Orthophosphate," Proc. of 33rd Annual Symposium on Frequency Control, 88-97, (1979)

8. W.R. Beck, "Crystallographic Inversions of the Aluminum Orthophosphate Polymorphs and their Relation to those of Silica," J.Amer. Ceram. Soc.,32, 147-151 (1949)

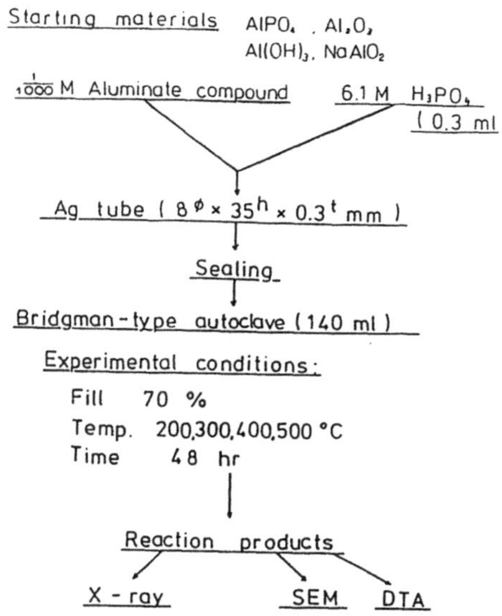

Fig.1 Experimental techniques

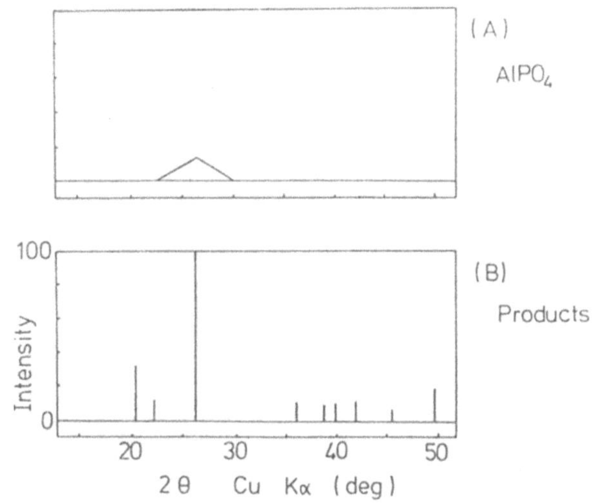

Fig.2 X-ray diffraction patterns of $AlPO_4$ and its products.

Fig.3 Scanning electron micrographs of products.
(Starting materiales : $AlPO_4$, 6.1 M H_3PO_4
Reaction time : 48hr)

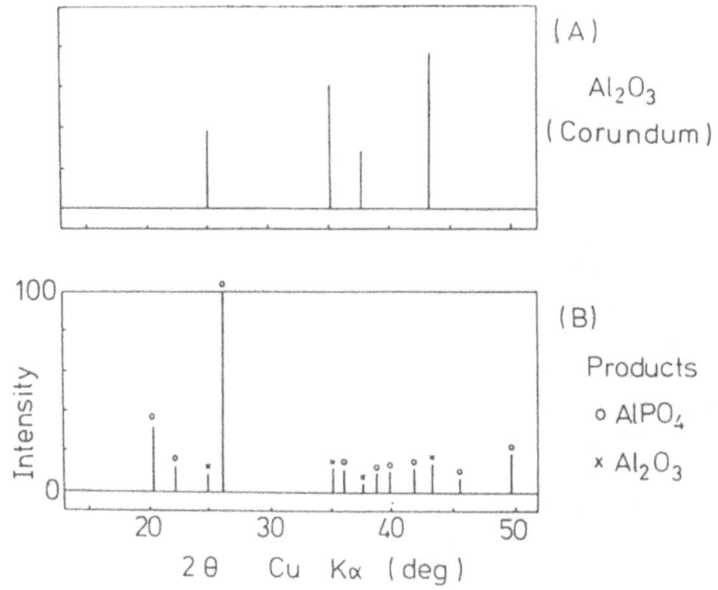

Fig.4 X-ray diffraction patterns of Al_2O_3 and its products.

Fig.5 Scanning electron micrographs of products.
(Starting materiales : Al_2O_3 , 6.1 M H_3PO_4
Reaction time : 48 hr)

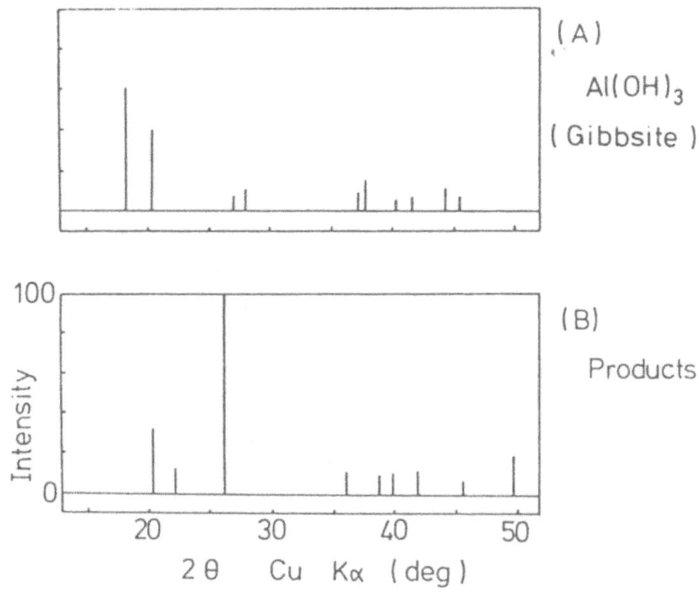

Fig.6 X-ray diffraction patterns of $Al(OH)_3$ and its products.

Fig.7 Scanning electron micrographs of products.
(Starting materiales : $Al(OH)_3$, 6.1 M H_3PO_4
Reaction time : 48 hr)

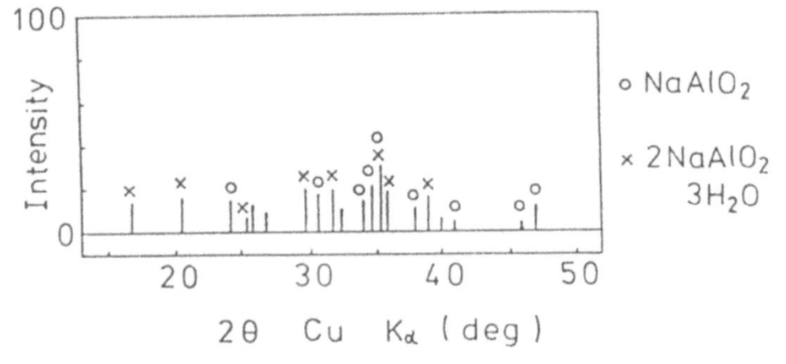

Fig.8 X-ray diffraction patterns of $NaAlO_2$ and its hydrate

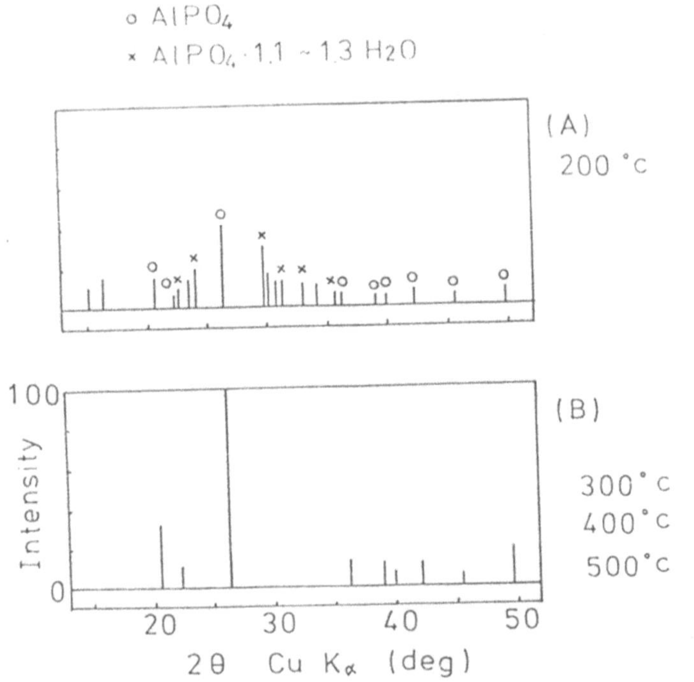

Fig.9 Variation of X-ray diffraction patterns of products.

Fig.10 Scanning electron micrographs of products.
(Starting materiales : $NaAlO_2$, 6.1 M H_3PO_4
Reaction time : 48 hr)

(A)
Starting materials :
$Al(OH)_3$ $\frac{1}{1000}$ mol
6.1 M H_3PO_4 0.3 ml

(B)
Starting materials :
$Al(OH)_3$ $\frac{1}{1000}$ mol
9.15 M H_3PO_4 0.2 ml
10 N NaOH 0.1 ml

Fig.11 Effect of sodium ion on morphology of product
(Reaction temperature : 400°C
Reaction time : 48 hr)

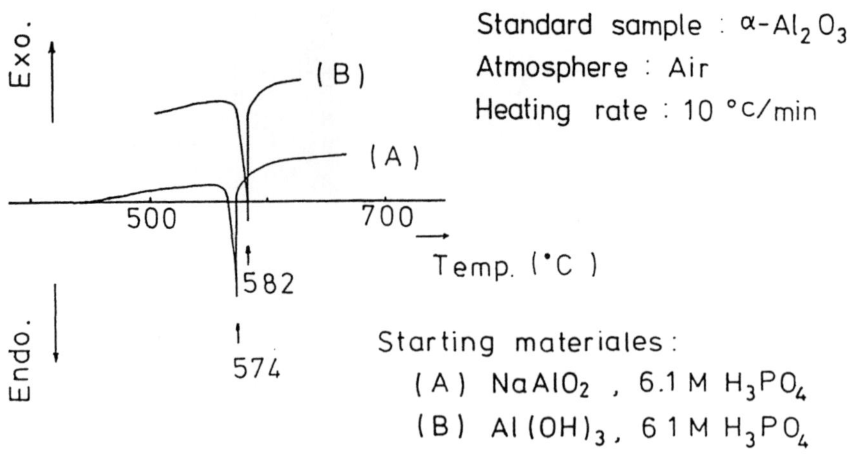

Fig.12 DTA curves of $AlPO_4$

Reprinted from *Proceedings of the First International Symposium on Hydrothermal Reactions*, Japan, 22–26 March 1982, pp. 649–657.

Hydrothermal Synthesis of Oxides of Manganese Single Crystals

J.A.K.TAREEN AND B.BASAVALINGU
Mineralogical Institute, University of Mysore
Manasa Gangotri,Mysore-570006,India

T.R.N.KUTTY
Materials Research Laboratory
Indian Institute of Science,Bangalore-560012

G.T.BHANDGE
Regional College of Education,Mysore-570006

M.YOSHIMURA AND S.SŌMIYA
Tokyo Institute of Technology,Yokohama,Japan

ABSTRACT

Hydrothermal runs were carried out in the systems, $MnCl_2$-NaOH-H_2O and $MnCl_2$-CuCl-NaCH-H_2O. In manganese copper system, single crystals of hausmanite(Mn_3O_4) were stable from 200 to 400°C, while manganosite(MnO) and pyrochroite $Mn(OH)_2$ stabilized at higher temperatures. In $MnCl_2$-NaOH-H_2O system single crystals of manganosite (MnO) and pyrochroite were synthesised.

Introduction:

Oxides of manganese occur in several polymorphic forms and owing to the variable valency of manganese it is difficult to synthesize them with controlled stoichiometry. Many investigators have attempted to hydrothermally synthesize crystals of manganese oxides in both strongly acid [1] and alkaline solutions[2-5]. The pyrolusite crystals β=MnO_2 with 1-2mm in size were grown in a solution of nitric acid by Fedosova[1]. Yamomoto and Endo[4] crystallised α-MnO_2 using KOH solutions but here potassium was found as an impurity which stabilizes the MnO_2 structure. Other oxides of manganese like Mn_5O_8, α-Mn_2O_3, Mn_3O_4 and β-MnO_2 were also reported by these authors to have been obtained in 0.1 N RbOH solution, where γ-MnOOH was hydrothermally treated at 300°C with addition of 0.1 N H_2SO_4 to manipulate the pH condition of the fluid phase, The MnO and Mn_2O_3 crystals were prepared [6] ranging in size from 1 to 2mm by heating 1 M K_2MnO_4 solution at temperature of 800°C at a pressure of 2.5 K.bars. They have reported phase diagram for K_2MnO_4-H_2O system wherein MnO is stable above 625°C at $<$ 1 K.bar. Apart from this report the single crystal synthesis of manganosite(MnO), hausmanite(Mn_3O_4) and pyrochroite $Mn(OH)_2$ has not been reported by hydrothermal means, The phase diagrams reported by Klingsberg and Roy[7] show the stability of pyrochroite,$Mn(OH)_2$ and manganosite, (MnO) in the MnO-H_2O system. The stability of manganite Mn_2O_3.H_2O and bixbyite,Mn_2O_3 as stable phases in the Mn_2O_3-H_2O system in a temperature range of 25-500°C and pressure range of 100-3000 bars have been demonstrated by these authors.

We have been investigating the hydrothermal stability of the oxides of manganese and the copper manganates using a combination of mineralisers. The present note reports some of the results of this investigation, wherein different oxides of manganese as well as manganese hydroxide single crystals could be synthesized.

Experimental:

This work essentially concerns the hydrothermal investigations of the systems, $MnCl_2$-NaOH-H_2O and $MnCl_2$-CuCl-NaOH-H_2O. The starting materials in all the hydrothermal runs were pure metal chloride salts and CO_2 free 20M NaOH solutions. In the case of copper manganese system, the atomic ratio of copper to manganese was maintained at 1:2(total weight of substrate 100 mg). Thoroughly mixed and ground chloride salts were taken in platinum ampaules with 3 to 4 drops of 20M NaOH solution and the ampoules were sealed by crimping the open end and welding with electric arc. In the case of pure manganese system, manganese (II) chloride (about 100 mg) was taken along with 3 to 4 drops of 20M NaOH. Experiments were conducted in Tuttle-type cold seal stellite reactor vessels using tempress unit HR-IB-4 and the temperature was measured with milivolt potentiometer using calibrated chromel-alumel thermocouples, where the hot end was inserted in a bore driven in the autoclave close to the platinum ampoule. The pressurising fluid in the autoclaves was distilled water and the pressure was read on the bourden guages. The duration of the experiments range between 70-80 hours. The vessels were quenched, after the experiment, to room temperature using air jets. The products were washed well with hot distilled water and examined under optical and scanning electron microscopes and identified with X-ray diffraction photographs.

Results and discussion:

The details of experiments conducted and the product stabilized are given in Table 1. In the pure manganese system, the stability of manganosite, MnO was found to extend as far as 700°C starting from 400°C. Pyrochroite, $Mn(OH)_2$ is also found stable along with manganosite at higher temperatures i.e. above 500°C.

In the system with copper and manganese two different oxides of manganese namely manganosite, MnO and hausmanite, Mn_3O_4 were found stable. It is surprising that copper has a relatively very high solubility compared to manganese and in none of the experiments which were conducted in the range of 300-800°C, copper appeared in any solid form. However, it has very well acted as a structure stabilizer for the growth of tetragonal Mn_3O_4 crystals, particularly at temperatures as low as 200°C, while at identical temperature, manganese chloridehydrate was stable in the $MnCl_2$-NaOH-H_2O system. The scanning electron micrographs are given for hausmanite, Mn_3O_4, fig.1, pyrochroite, $Mn(OH)_2$, fig.2, and manganosite, MnO crystals, fig.3a and 3b. Manganosite shows prominent hopper type step growth.

Manganosite, MnO is an antiferromagnetic insulator and hausmanite, Mn_3O_4 a ferrimagnetic material.

Since the order of pressures employed are relatively low, it will be possible to conduct the crystal growth of these compounds in large thick walled ampoules using larger autoclaves wherein the pressure within the ampoule and outside could be balanced by the appropriate percent fill. By giving sufficient time, larger crystals can be grown.

Acknowledgement:

The authors gratefully acknowledge the encouragement given by Prof.M.N.Viswanathiah during the course of this investigation.

Reference:

1. S.P. Fedosova, Pyrolusite crystal synthesis, Science. IX, 59 (1972).
2. I.Ismail-Zade, S.P.Fedosova and B.N.Litvin, Phase formation in the system $Na_2O-Mn_2O_3-O_2-H_2-H_2O$ Researches in the field of Inorganic Physical Chemistry. Baku, 341 (1971).
3. I.Ismail-Zade and B.N.Litvin, Hydrothermal synthesis of Na-Mn silicates and germanates single crystals Aserb.Chem.J.6, 135 (1969)
4. N.Yamomoto and T.Endo. Single crystal growth of α-MnO_2. Jap.J.Appl.Phys.13, 723 (1974).
5. N.Yamomoto, K.Masao and T.Toshio, A new preparation method of Mn_5O_{18}, Jap.J.Appl.Phys. 12, 1827-1828 (1973).
6. T.Endo, S.Kume, M.Shimada and M.Koizumi, Synthesis of Potassium manganse oxides under hydrothermal conditions. Miner.Mag. 39, 559-563 (1974).
7. Cyrusklingsberg and Rustum Roy, Stability and intercovertibility of phases in the system Mn-O-OH. Am.Mineral. 44 819-838(1959).

TABLE 1
DETAILS OF THE HYDROTHERMAL RUNS

Run.No.	Temp. °C	Pressure in PSI	Product	Remarks
(a) $MnCl_2$-NaOH-H_2O systems:				
MIN 55	200	500	$MnCl_2 4H_2O$	Powder
MIN 56	300	2000	$Mn(OH)_2$	Flaky
MIN 57	400	6500	MnO	Crystals
MIN 58	500	6500	MnO	Crystals
			$Mn(OH)_2$	Flaky
MIN 59	600	6500	MnO	Crystals
			$Mn(OH)_2$	Flaky
(b) $MnCl_2$-CuCl-NaOH-H_2O systems:				
MIN 52	200	1500	Mn_3O_4	Crystals
MIN 48	300	5500	Mn_3O_4	Crystals
MIN 49	400	3000	Mn_3O_4	Crystals
MIN 50	500	5500	MnO	Crystals
			$Mn(OH)_2$	Flaky
MIN 51	600	9500	MnO	Crystals
			$Mn(OH)_2$	Flaky
MIN 47	700	7500	MnO	Crystals
			$Mn(OH)_2$	Crystals and Flaky
MIN 53	800	10000	MnO	Crystals
			$Mn(OH)_2$	Flaky

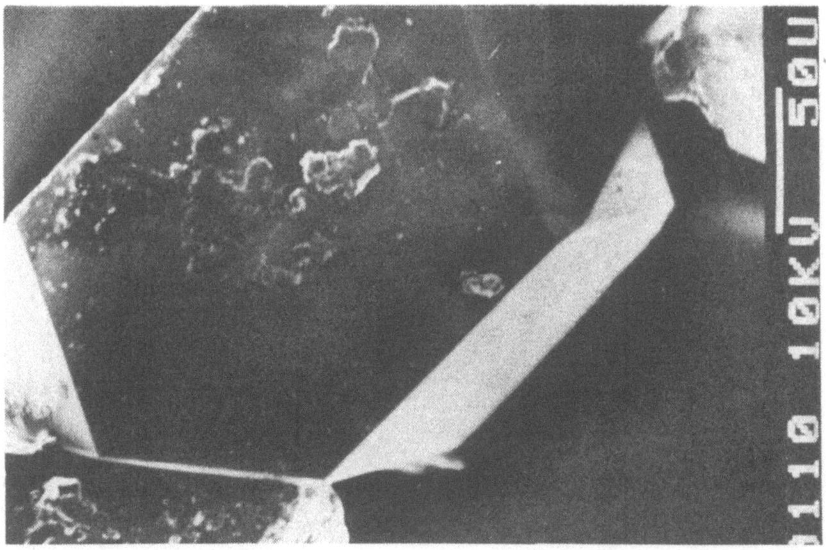

Fig.1 Hausmanite (Mn_3O_4) Single Crystal

Fig.2(a) Pyrochroite $Mn(OH)_2$

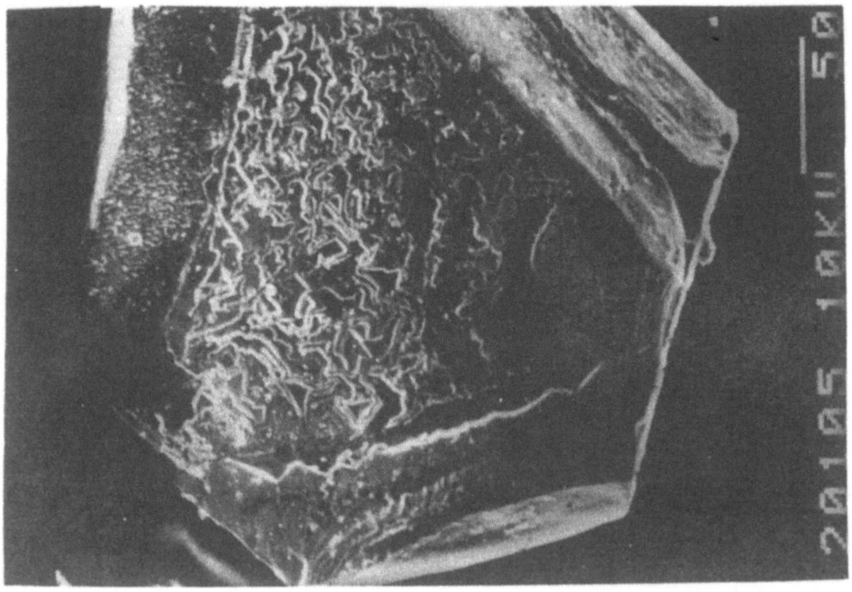

Fig.2(b) Pyrochroite Crystal $Mn(OH)_2$

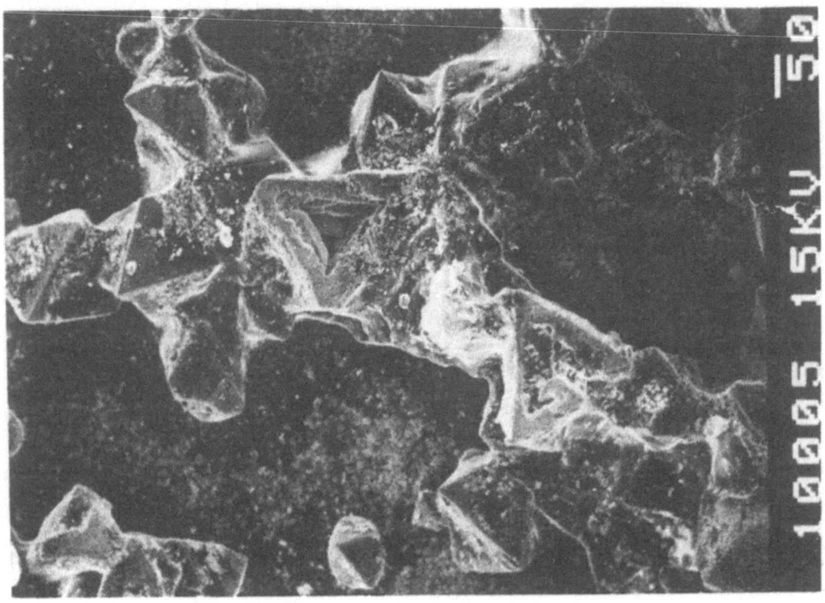

Fig.3(a) Manganosite MnO

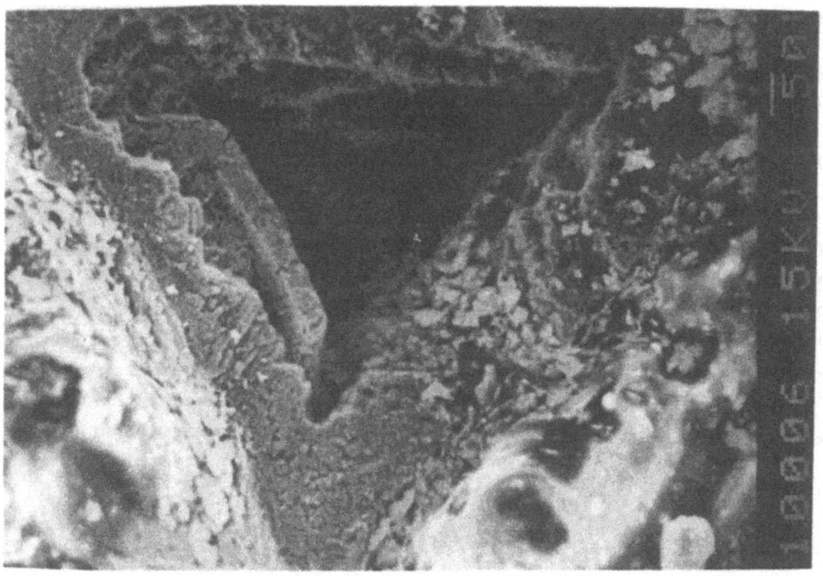

Fig.3(b) Manganosite (MnO)

Reprinted from *Proceedings of the International Meeting on Chemical Sensors* (Kodansha, Japan, 1983) pp. 198–202.

Hydrothermal Synthesis and Growth of $La_{1-x}Sr_xFeO_3$ ($0 \leq x \leq 0{\cdot}1$) Crystals

Masahiro YOSHIMURA, Kazuhito YAMASAWA, and Shigeyuki SŌMIYA
Research Laboratory of Engineering Materials and Department of Materials Science, Tokyo Institute of Technology,
4259 Nagatsuta, Midori-ku, Yokohama-shi 227, Japan

ABSTRACT: Synthesis and crystal growth of $La_{1-x}Sr_xFeO_3$ have been studied under hydrothermal conditions. The optimum conditions to synthesize $LaFeO_3$ were found to be 425-600°C with 50 wt% KOH solution under 100 MPa. $LaFeO_3$ single crystals have been grown in those conditions up to 0.8 mm in diameter with the temperature gradient of 5°C for 168 h. The perovskite formation regions from the mixture of lanthanum oxide, iron oxide and $SrCl_2$ solution of the composition $La_{1-x}Sr_xFeO_3$ (x=0.02, 0.10) were similar but a little narrower than those of $LaFeO_3$. The hydrothermal crystal growth brought about the crystals of maximum 0.3 mm at x=0.02 and 0.03 mm at x=0.10.

1. INTRODUCTION

Perovskite-type oxides have widely been studied in catalyst and sensor sciences and technologies due to their unique chemical and electromagnetic properties. Sr-doped lanthanum ferrite ($La_{1-x}Sr_xFeO_3$) have also widely been investigated electromagnetically[1-4], for a catalyst[5] and for a humidity sensor[6].

These studies have used only polycrystalline samples prepared by solid state reaction of component oxides. Polycrystalline samples are convenient and practical in preparation and in application, however, they generally include extrinsic nature due to the existence of surfaces and grain boundaries, which are the least caracterized part in the sample; sometimes the most active or the most contaminated by impurities, etc.. Single crystals are, therefore, evidently desirable to characterize intrinsic bulk properties and to evaluate these materials correctly.

The single crystals of $RFeO_3$ (R:rare earth) have been synthesized by flux[7], floating zone[8] and hydrothermal[9] methods for heavy rare earth ions but those for La have hardly been studied yet. Thus we intended to synthesize and grow the $LaFeO_3$ and Sr-doped $LaFeO_3$ crystals under hydrothermal conditions. Since hydrothermal crystal growth is a relatively low-temperature growth, the crystals have generally less thermal strain, less impurities and better homogeneity of doping ions. In this report, we describe the synthetic and growth conditions of $La_{1-x}Sr_xFeO_3$ crystals under hydrothermal conditions.

2. EXPERIMENTAL

The starting materials are La_2O_3 (99.99% Shin-etsu Chem. Co. Ltd.) and Fe_2O_3 (99.9% Kōjundo Chem. Ind. Co. Ltd.) and guaranteed reagent $SrCl_2 \cdot 6H_2O$. Approapriate amounts of them were sealed with a mineralizer solution in a Au or Pt capsule (2.7 mm ID, 30 mm long) by welding with a carbon arc. Among the solutions studied; KOH, KNO_3, KCl, NaOH, LiOH or H_2O, KOH was the most effective mineralizer to synthesize $LaFeO_3$ crystals, so that KOH solutions were used for crystal growth. The powders of La_2O_3, Fe_2O_3 and $SrCO_3$ (guaranteed reagent) were mixed in an alumina ball mill with water for 6 h, and ball-milled again without water for 6 h. The mixture was fired at 1200°C for 16 h in O_2.

The sealed capsule was heated in a test-tube type pressure vessel at 300—700°C under the pressure of 100 MPa for 24—168 h. The temperature was measured by two Platinel thermocouples attached on the outside wall of the vessel at the positions coresponding to the top and the bottom of the capsule. Temperature gradients (ΔT) were controlled to 0°C in the case of synthesis but to 5—20°C in the crystal growth. After these hydrothermal treatments, the vessel was quenched in a cold water. The products were washed repeatedly with distilled water, then examined by X-ray diffraction and scanning electron microscope.

3. RESULTS AND DISCUSSION

3.1. Hydrothermal synthesis of $La_{1-x}Sr_xFeO_3$

3.1.1. Perovskite formation region

The formation region of the Perovskite $LaFeO_3$ from the component oxides is shown in Fig. 1 as a function of temperatures and KOH concentrations under hydrothermal conditions of 100 MPa after 48 h. No perovskite was formed below 300°C, $La(OH)_3$ and α-Fe_2O_3 only were detected in the products. $LaFeO_3$ was formed at higher temperatures in concentrated KOH solutions. It became the sole product at ca. 600°C in 10—50 wt% KOH solutions. $LaFeO_3$, however, had the tendency of re-decomposition to $La(OH)_3$ and α-Fe_2O_3 at higher temperatures in highly concentrated KOH solutions after longer treatments. So that the single phase region of $LaFeO_3$ shifted to lower temperature, ca. 350—570°C, in 50 wt% KOH solution after 168 h.

Addition of $SrCl_2$ solutions for Sr doping to the perovskite brought about the decrease of single phase region of perovskite and the formation of magnetite (Fe_3O_4) instead of α-Fe_2O_3 at high temperatures as shown in Fig. 2 (x=0.02) and in Fig. 3 (x=0.10).

These results indicate that the perovskite formation became difficult by doping of Sr because probably the oxygen fugacity of the system in the capsule shifted toward reduction.

3.1.2. Morphology and size of crystals

The shape and size of $LaFeO_3$ crystals were influenced by the temperature, duration and KOH concentration. A dilute KOH solution (0.5 wt%) yielded only aggregated fine particle (< 1 μm) of perovskite at 500°C for 48 h, while a moderately concentrated solutions (> 5 wt%) yielded larger (> 1 μm) crystals with apparent (100) faces. The crystals synthesized in a concentrated (53 wt%) solution had a complete cubic habit with (100) faces with ca. 40 μm in size under the same conditions. These results indicate that the solubility of $LaFeO_3$ increased with the concentration of KOH as same as observed in $YFeO_3$[10].

When 53 wt% KOH solution was used, the size of crystals increased with increase in temperatures up to 500°C, but decreased above 600°C due to redissolution of crystals, of which corners and edges became round. Without the temperature gradient the $LaFeO_3$ crystals did not grow larger than 40 μm, therefore they had homogeneous size distribution as shown in Fig. 4 after a long duration (168 h) during which smaller crystals were diminished to give larger ones by Ostwald' ripening mechanism.

Similar results were observed for $La_{1-x}Sr_xFeO_3$ crystals. In conclusion, the optimum conditions to synthesize or grow these crystals to be at 450—600°C under 100 MPa in 53 wt% KOH solution.

3.2 Hydrothermal crystal growth of $La_{1-x}Sr_x FeO_3$

$La_{1-x}Sr_xFeO_3$ single crystals have been grown in the optimum conditions determined. The $LaFeO_3$ crystals have been grown up to 0.8 mm in diameter with ΔT=5°C at 500°C for 168 h (Fig. 5). They had the habit with apparent (100), (110) and (111) faces. The increase of ΔT brought about the increase in nucleation by enhanced super-saturation and transportation but rather the decrease in the size of each crystal.

Higher growing temperature than 600°C resulted in the formation of $La(OH)_3$ crystals on the $LaFeO_3$ crystals by redecomposition of $LaFeO_3$ as mentioned above.

In the case of the Sr-doped $LaFeO_3$ crystal growth, the size and the shape of the grown crystals were almost identical ,(Fig. 6) from $La_{1-x}Sr_xFeO_3$ polycrystal synthesized by the solid state reaction ,or (Fig. 7) from the component oxides and $SrCl_2$ solutions.

The crystal growth of perovskite seemed to become difficult with the Sr doping. The maximum size of crystals at x=0.02 was ca. 0.1 mm as seen in Figs. 6 and 7 at 500°C, ΔT=5°C, under 100 MPa using 50 wt% KOH solution for 168 h. (At 600°C with ΔT=5°C, crystals grew up to ca. 0.3 mm (Fig. 8) from component oxides and $SrCl_2$ solution with 50 wt% KOH solution.) Much smaller crystals of 0.03 mm (Fig. 9) could be grown at x=0.10 under the same conditions; 500°C, ΔT=5°C under 100 MPa using 50 wt% KOH solution for 168 h.

These results indicate that the Sr-doped $LaFeO_3$, where formation of Fe^{4+} ions and/or oxygen defects is needed to compensate the charge decrease by the doping of Sr^{2+} into $La^{3+}Fe^{3+}O_3$, is less stable than undoped $LaFeO_3$ under hydrothermal conditions studied. Fe^{4+} formation appears to be difficult by Sr^{2+} doping under hydrothermal conditions because Sr doping rather results in the shift of oxygen fugacity toward reduction. This is consistent with narrowing of perovskite formation region and Fe_3O_4 formation with the doping of Sr^{2+} in the system as indicated in Figs. 2 and 3. Less Sr-doped crystals, therefore, would be synthesized and/or grown than the starting composition under the hydrothermal conditions. In fact, the lattice parameters of crystals prepared hydrothermally are a little larger than those of samples prepared by a solid state reaction and a floating-zone method as seen in Table 1.

As described above, the hydrothermal growth of $La_{1-x}Sr_xFeO_3$ crystals is difficult. But larger crystals will be grown by the use of a larger container and seed crystal. Otherwise, floating zone method also will yield larger crystals of those materials. Such works are underway in our laboratory.

4. REFEREMCES

1) U. Shimony and J.M. Knudsen, Phys. Rev., 114, 361 (1966)
2) P.K. Gallagher and J.B. MacChesney, Symp. Farady Soc., 1, 40 (1967)
3) T. Sasamoto, J. Mizusaki, M. Yoshimura, W.R. Cannon and H.K. Bowen, Yogyo Kyokai Shi (J. Ceram. Soc. Japan), 90 [1], 32 (1982)
4) M. Takano and Y. Takeda, Funtai oyobi Funmatsu Yakin (J. Japan Soc. Powder and Powder Metal.), 28 [6], 203 (1981)
5) T. Arakawa, S. Tsuchi-ya and J. Shiokawa, Mat. Res. Bull., 16, 97 (1981)
6) W.B. Li, H. Yoneyama and H. Tamura, Nippon Kagaku Kaishi (J. Chem.Soc. Japan), 1982, 761 (1982)
7) H. Taguchi and Y. Takahashi, Jap. J. Appl. Phys., 20 [6], 1117 (1981)
8) J.P. Remeika, J. Am. Ceram. Soc., 78, 4259 (1956)
9) K. Kitamura and S. Kimura, J. Cryst. Growth, 48, 469 (1980)
10) L.N. Demianets, A.N. Lobachev and L.V. Usov, Inorg. Mater., 12, 430 (1976)
11) E.D. Kolb, D.L. Wood and R.A. Laudise, J. Appl. Phys., 39 [2], 1362 (1968)
12) M. Yoshimura, K. Yamasawa and S. Sōmiya, Yogyo Kyokai Shi (J. Ceram. Soc. Japan), 90 [9], 521 (1982)

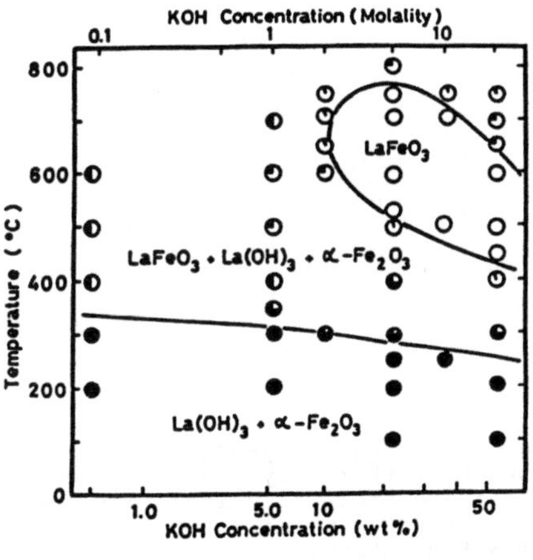

LaFeO$_3$ content (wt%): ●-0, ●-5, ◐-25, ◐-50, ◐-75, ◐-95, ○-100,

Fig. 1. x=0

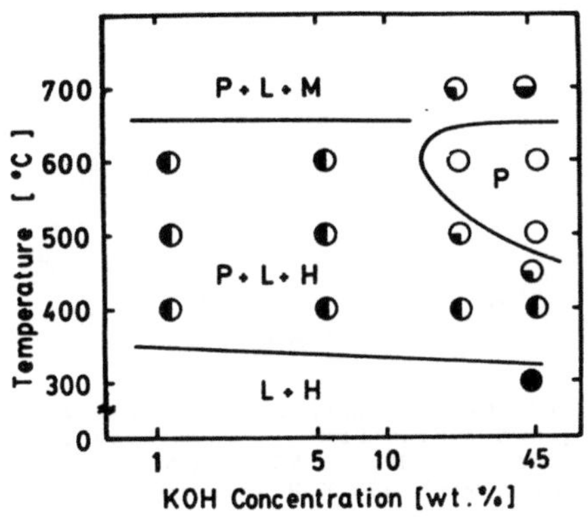

P : Perovskite , M : Fe_3O_4
H : α-Fe_2O_3 , L : $La(OH)_3$
Fig. 2. x=0.02

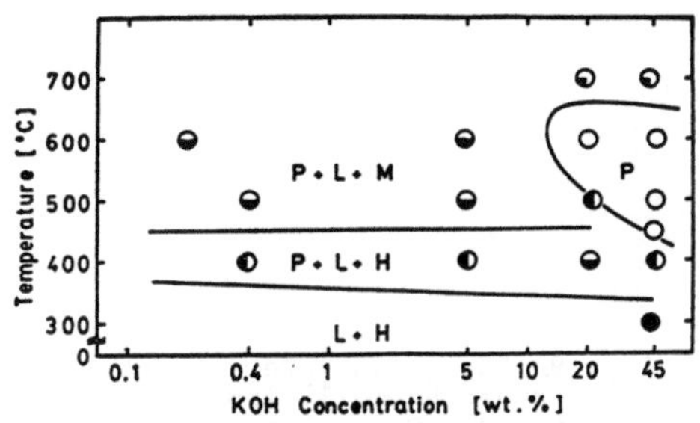

P : Perovskite , M : Fe_3O_4
H : α-Fe_2O_3 , L : $La(OH)_3$
Fig. 3. x=0.10

Fig. 1 — 3.
Effect of temperature and KOH concentration on the formation of Perovskite phase ($La_{1-x}Sr_xFeO_3$) under 100 MPa for 48 h.

Fig. 4. $LaFeO_3$ crystals synthesized hydrothermally under 100 MPa at 400°C for 168 h with 53% KOH solution.

Fig. 5. $LaFeO_3$ crystal grown hydrothermally under 100 MPa at 500°C, ΔT=5°C, for 168 h with 53% KOH solution.

Fig. 6. $La_{0.98}Sr_{0.02}FeO_3$ crystal grown hydrothermally from polycrystal under 100 MPa at 500°C (ΔT=5°C) for 168 h with 50% KOH solution.

Fig. 7. $La_{0.98}Sr_{0.02}FeO_3$ crystal grown hydrothermally from component oxides and $SrCl_2$ solution under 100 MPa at 500°C (ΔT=5°C) for 168 h with 50% KOH solution.

Fig. 8. $La_{0.98}Sr_{0.02}FeO_3$ crystal grown hydrothermally from component oxides and $SrCl_2$ solution under 100 MPa at 600°C (ΔT=5°C) for 168 h with 50% KOH solution.

Fig. 9. $La_{0.90}Sr_{0.10}FeO_3$ crystal grown hydrothermally from component oxides and $SrCl_2$ solution under 100 MPa at 500°C (ΔT=5°C) for 168 h with 50% KOH solution.

Table 1. Lattice Parameters of $La_{1-x}Sr_xFeO_3$ by Various Preparation Methods

Compound	Lattice Parameter (nm)				Preparation conditions
	a	b	c	pseudocubic	
$LaFeO_3$	0.5559	0.5565	0.7859	0.3931	Hy, 500°C, 48 h, 53% KOH
	0.5558	0.5559	0.7860	0.3930	Hy, 400°C, 168 h, 53% KOH
	0.5554	0.5565	0.7855	0.3930	SR, 1000°C, in O_2
$La_{0.98}Sr_{0.02}FeO_3$	0.5558	0.5560	0.7861	0.3931	Hy, 500°C, 48 h, 45% KOH
	0.5557	0.5555	0.7854	0.3928	SR, 1400°C, 14 h, in air
	0.5556	0.5556	0.7855	0.3928	Floating Zone, in O_2
$La_{0.90}Sr_{0.10}FeO_3$	0.5556	0.5558	0.7859	0.3929	Hy, 500°C, 48 h, 40% KOH
	0.5542	0.5548	0.7846	0.3922	SR, 1300°C, 12 h, in air
	0.5539	0.5546	0.7847	0.3921	Floating Zone, in O_2

Hy : Hydrothermal (100 MPa), SR : Solid state reaction

Reprinted from *Journal of Materials Science Letters*, 4 (1985) 19–21.

Synthesis of magnetite using ilmenite under hydrothermal conditions

M. G. M. U. ISMAIL,* M. YOSHIMURA, S. SŌMIYA
Laboratory for Hydrothermal Syntheses, Research Laboratory of Engineering Materials, Tokyo Institute of Technology, 4259 Nagatsuta, Midori, Yokohama 227, Japan

Ilmenite ($FeTiO_3$) is mainly used as a raw material for the production of titanium dioxide (TiO_2) pigment and titanium metal. The TiO_2 content in natural ilmenite varies from 42 to 54%, and the other main chemical constituents are the oxides of iron. The methods adopted [1] for the production of TiO_2 were mainly based on digestion of ilmenite with inorganic acids or treatment with hot gases (e.g. Cl_2, H_2S, I_2 etc.) at elevated temperatures. In these processes the iron is removed as the corresponding salts of acid or the gases. The iron salts formed in most of the methods do not have any commercial value and discarding of these iron compounds causes pollution problems.

The present study was mainly focused to decompose ilmenite under hydrothermal conditions and to get products consisting of titanium and iron compounds of commercial value.

Natural ilmenite obtained from the beach sand deposit at Pulmoddai in Sri Lanka was used in this investigation. The chemical analysis of the ilmenite showed that it contained 53.61% TiO_2, 20.67% FeO, 20.95% Fe_2O_3 and 0.95% MnO. Ilmenite was powdered to pass through a 200 mesh sieve and was used in the hydrothermal experiments. 10 M KOH, 10 M NaOH and 10 M LiOH solutions were used in this investigation. Known weights of ilmenite and alkali hydroxide solutions were introduced into the one-end sealed 2.7 mm inner diameter and 35 mm in length gold capsules and sealed hermetically. Both end-sealed capsules were placed in the test tube [2] type hydrothermal equipment and the pressure was raised to the desired value. Then the temperature was raised maintaining the pressure constant. At the end of the treatment the capsules were quenched in water. The phases present in the product formed were determined by X-ray diffraction (XRD) methods. (XRD equipment model RU 200, Rigaku Denki Ltd was used.) The lattice parameter of the magnetite formed by the decomposition of ilmenite was determined by using silicon (99.99% purity, Kinzoku Kagaku Kenkyujyo) as the internal standard. Scanning electron microscope (SEM) photographs were also taken (SEM T 200 JEOL, Japan). Ilmenite was completely decomposed [3] under hydrothermal conditions in 10 M KOH solution at 500° C, when the pressure was between 25 and 35 MPa for 63 h of treatment. Ilmenite was stable above a pressure of 35 MPa at 500° C in 10 M KOH solution. The products formed by complete decomposition of ilmenite were magnetite ($Fe_{3-x}O_4$) and $K_2O \cdot 4TiO_2$. The lattice parameter of the magnetite ($Fe_{3-x}O_4$) formed was $a = 0.8467$ nm which is larger than the stoichiometrically pure magnetite, where $a = 0.839\,6(7)$ nm [4]. The increase in lattice parameter is due to the substitution of Ti^{4+} in the magnetite lattice and forms the $Fe_{3-x}O_4$–Fe_2TiO_4 solid solution [5, 6]. On increasing the treatment temperature to 800° C, while maintaining the pressure at 30 MPa, ilmenite decomposed completely in 10 M KOH solution for 24 h of treatment. The lattice parameters of the magnetite formed at 600, 700 and 800° C and at 30 MPa were determined and found to decrease with increasing temperature. The lattice parameter of magnetite formed at 800° C, 30 MPa and 24 h by the decomposition of ilmenite in 10 M KOH solution was $a = 0.841\,3(8)$ nm.

The decrease in lattice parameter of magnetite with increase of temperature under hydrothermal conditions was due to the exsolution of Ti^{4+} from $Fe_{3-x}O_4$–Fe_2TiO_4 solid solution to form nearly stoichiometrically pure magnetite. Similar behaviour was reported by Vincent *et al.* [7] in

*Present address: Ceylon Institute of Scientific and Industrial Research, PO Box 787, Colombo 7, Sri Lanka.

Figure 1 **SEM photograph of magnetite formed by the complete decomposition of ilmenite in 10 M KOH solution at 800° C, 30 MPa and 24 h of treatment.**

Figure 2 **SEM photograph of the products formed by the complete decomposition of ilmenite in 10 M NaOH solution at 800° C, 30 MPa and 24 h of treatment.**

their heating experiments on titaniferrous magnetite.

The Curie points of the magnetite formed by the decomposition of ilmenite at 600, 700 and 800° C at 30 MPa and for 24 h using 10 M KOH solution were determined and found to increase with increase of temperature. The Curie point of the magnetite formed at 800° C, was 573° C, whereas for stoichiometric magnetite the Curie point is 580° C (Akimoto *et al.* [8]).

The SEM photograph of magnetite formed by the complete decomposition of ilmenite in 10 M KOH solution at 800° C, 30 MPa and 24 h is given in Fig. 1. The magnetite formed was of octahedral habit.

Ilmenite also completely decomposed in 10 M NaOH solution at 800° C, 30 MPa and 24 h of treatment. The products formed were magnetite ($Fe_{3-x}O_4$) and a nonstoichiometric sodium iron titanium oxide. The lattice parameter of the magnetite formed is given in Table I and it is larger than that for the magnetite formed in 10 M KOH and 10 M LiOH solutions under the same conditions.

The nonstoichiometric sodium iron titanium oxide formed was of orthorhombic structure [9] with $a = 0.908\,3(6)$ nm, $b = 0.295\,4(1)$ nm and $c = 1.071\,8(3)$ nm. The SEM photograph of the product formed in 10 M NaOH solution is given in Fig. 2, and the needle-like compound is the non-stoichiometric sodium iron titanium oxide.

Only a partial decomposition of ilmenite was observed in 10 M LiOH solution at 800° C, 30 MPa and 24 h of treatment. The lattice parameter of magnetite formed was determined and is given in Table I.

TABLE I Lattice parameter of magnetite formed by the decomposition of ilmenite in different solutions at 800° C, 30 MPa and 24 h of treatment

Solution used	Lattice parameter of magnetite (nm)
10 M KOH	0.841 3(8)
10 M NaOH	0.844 7(5)
10 M LiOH	0.841 5(3)
Stoichiometric magnetite (4)	0.839 6(7)

This investigation clearly showed that nearly stoichiometrically pure magnetite can be synthesized by the decomposition of ilmenite in 10 M KOH solution at 800° C, 30 MPa and in 24 h of treatment.

Acknowledgements
The authors are also indebted to Dr Takagi for carrying out the magnetic measurements of magnetite.

One of the authors (MGMUI) thanks the Japanese Association for the International Education, Japan, for the fellowship granted, during which this work was carried out at the Laboratory for Hydrothermal Syntheses at Tokyo Institute of Technology, Nagatsuta, Yokohama, Japan.

References
1. N. C. KOTHARI, *Int. J. Miner. Process.* (1974) 297.
2. R. ROY and O. F. TUTTLE, "Physics and Chemistry of the Earth", Vol. 1 (Pergamon Press, New York, 1956) pp. 138–80.
3. S. HIRANO, M. G. M. U. ISMAIL and S. SŌMIYA, *J. Amer. Ceram. Soc.* **58** (1976) 227.
4. JCPDS, Powder diffraction file 19-629.
5. E. Z. BASTA, *Mineral. Mag.* **31**, 237 (1975), 431.
6. D. H. LINDSLEY, Mineralogical Society of America short course notes, "Oxide Minerals", Vol. 3 (Mineralogical Society of America, 1976).
7. E. A. VINCENT, J. B. WRIGHT and R. CHEVALLIER, *Mineral. Mag.* **31** (1957) 625.
8. S. AKIMOTO, T. KATSURA and M. YOSHIDA, "Contribution from Division of Geomagnetism and Geoelectricity Geophysical Institute", Series II, No. 78 (Tokyo University, Japan, 1958).
9. T. ISHIGURO, K. TANAKA, F. MARUMO, M. G. M. U. ISMAIL, S. HIRANO and S. SŌMIYA, *Acta Crystallogr.* **B34** (1978) 3346.

Received 16 March and accepted 27 March 1984

Reprinted from *Journal of Solid State Chemistry*, **60** (1985) 182–187.

Synthesis of Vanadium Sulfides under High Pressure

MASAO YOKOYAMA, MASAHIRO YOSHIMURA,* MASATAKA WAKIHARA, SHIGEYUKI SŌMIYA,* AND MASAO TANIGUCHI†

*Department of Chemical Engineering, Tokyo Institute of Technology, Ookayama, Meguro-ku, Tokyo 152 and *Department of Material Science and Engineering, Tokyo Institute of Technology, 4259, Nagatsuta-cho, Midori-ku, Yokohama 227, Japan*

Received December 6, 1984; in revised form April 8, 1985

Vanadium sulfides were synthesized in a temperature range of 350–750°C and in a pressure range of 10–250 MPa (1 MPa = 9.87 atm), with an apparatus used for hydrothermal synthesis. The nonstoichiometric compositional range of the $V_{1+x}S_2$ phase, which cannot be prepared under atmospheric pressure, is $VS_{1.661}$–$VS_{1.732}$ ($0.155 < x < 0.204$). It was impossible to synthesize vanadium disulfide VS_2 under the present experimental conditions. Equilibrium phase diagrams for the V_5S_8–S system under 100 and 200 MPa have been proposed on the basis of the present results. Also, the phase relationship between V_5S_8 and $V_{1+x}S_2$ has been established and a pressure–temperature phase diagram for the V–S system has been drawn.

Introduction

In the vanadium–sulfur system, various nonstoichiometric compounds are present and have recently been investigated with respect to composition and physical properties. For example, monosulfide VS and disulfide VS_2 have NiAs- and CdI_2-type structures, respectively, and V_5S_8 (monoclinic structure) becomes an antiferromagnet below (approximately 30 K (*1, 2*). These interesting properties are due to the behavior of the *d* electrons of vanadium atoms in every second metal layer of vanadium sulfide.

Vanadium disulfide is considered as one of the important compounds used as cathode materials in high energy secondary batteries in which other transition metal disulfides have been used (*3, 4*). A comprehensive report, however, has not yet been published concerning the synthesis of the VS_2 using the dry process.

A few years ago some of the present authors (*5–7*) reported the phase equilibria of the V–S system as having a compositional range of VS–V_5S_8 under controlled sulfur partial pressure in the range from 10^{-14} to 1 atm. According to their investigations, the vanadium sulfide in equilibrium with sulfur vapor pressure of 1 atm is the V_5S_8 ($VS_{1.6}$); also, considerably higher pressure is indispensable in order to obtain VS_2 at high temperatures. Onoda *et al.* (*8*) attempted to synthesize VS_2 by thermal decomposition of VS_4 at 700–800°C under 20 kbar using a piston–cylinder apparatus. However, they were not able to synthesize VS_2 but ob-

† To whom all correspondence should be addressed.

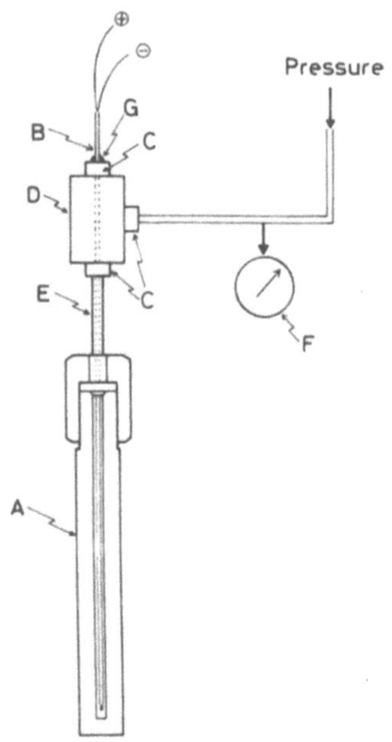

FIG. 1. Schematic diagram of the apparatus for temperature measurement under high pressure: (A) test-tube type vessel (stellite); (B) seath thermocouple; (C) ground; (D) tee; (E) nipple; (F) pressure gauge; (G) welding.

tained $VS_{1.8\pm0.1}$ instead (trigonal $V_{1+x}S_2$: $x = 0.11$). The trigonal $V_{1+x}S_2$ phase ($x = 0.2$) was also obtained by thermal decomposition of higher sulfides under reduced pressure by Jellinek (9), but the details of the composition range about the $V_{1+x}S_2$ phase have not been reported thus far. For this reason, we have studied the phase equilibria of V–S system under high pressure using an apparatus usually employed for hydrothermal synthesis.

Experimental

1. Apparatus

In the present work, we used an apparatus (Fig. 1) generally employed for high-pressure hydrothermal synthesis. A test-tube type vessel is made of stellite (an alloy of Co, Cr, W, etc.). A noble metal (Au, Pt, etc.) capsule which contains the sample was inserted into the vessel (o.d. 25.4 mm. i.d. 6.0 mm). It was heated and then subjected to a given pressure. The maximum temperature and pressure of this apparatus were set at 800°C (±5°C) and 400 MPa (±5 MPa), respectively, for safety reasons. A thermocouple was usually attached to the outside of the vessel and at the same position as that of the capsule during the reactions. As the inside temperature of the vessel actually corresponds to the reaction temperature, we checked the difference between the inside and the outside temperatures using a seath thermocouple (Fig. 1) before starting the experiment. The inner temperature was lower than the outside one by 20–35°C.

2. Experimental Procedure

Vanadium sulfides as starting materials were prepared by the following methods: (a) reaction of NH_4VO_3 (Wako pure chemical 99.9%) with H_2S gas at 400°C for 6 hr, and then at 1000°C for 20 hr to obtain a composition corresponding roughly to $VS_{1.5}$; and (b) synthesis of the VS_4 phase, in which both the sample prepared by (a) and pure sulfur powder (99.9%) were mixed in the S/V ratio of 4:5 and then sealed in a Pyrex tube followed by heating at 450°C for 3 days. The unreacted sulfur was transported to the cold zone of the tube by heating the sample zone at about 400°C. VS_4 phases with compositions in the range $VS_{3.2}$–$VS_{3.5}$ were always prepared by this method.

Next, we mixed the above starting material with sulfur powder in appropriate ratios. About 100 mg of the mixed sample was put into a gold capsule (o.d. 3.0 mm. i.d. 2.6 mm, and L 35 mm) and was sealed by arc-melting (a.c. 25–30 V). (The ratio "S/V" of the sealed material will be labeled as "sealed composition" VS_y in this paper.) Then the capsule was heated high pressure for 1–10 days (sulfur-rich samples were left for longer times). The temperature range

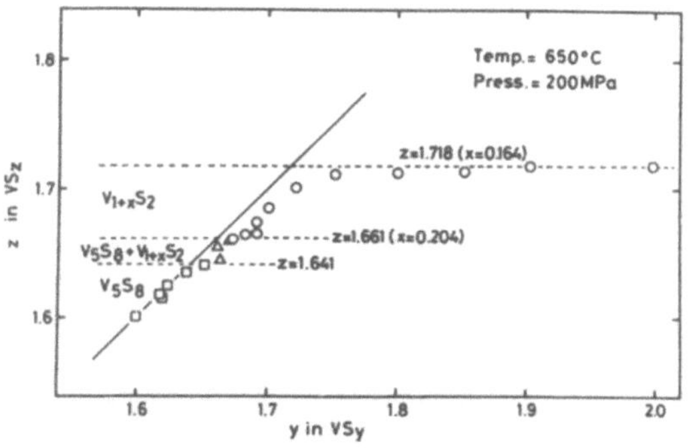

FIG. 2. Relation between the sulfur composition of the sealed materials (y) and that of the products (z) for the V–S system at 650°C under 200 MPa.

was 350–750°C and the pressure range was 10–25 MPa for the reactions. Finally, the capsule was quenched to room temperature before the pressure was released. The product contained unreacted sulfur which was removed by heating the products at 200°C in vacuum.

Phase identification of vanadium sulfide samples was made by X-ray powder diffractometry with CuK_α radiation (Rigaku Denki Co. Ltd.), and the compositions were determined by oxidation of the sulfide to V_2O_5 air at 600°C for 24 hr:

$$\underset{a(\mathrm{g})}{VS_z} \longrightarrow \underset{b(\mathrm{g})}{VO_{5/2}}$$

where $z = 2.837a/b - 1/0.589$.

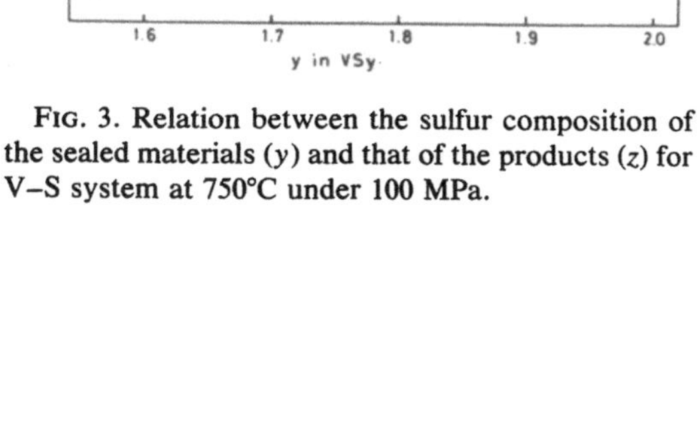

FIG. 3. Relation between the sulfur composition of the sealed materials (y) and that of the products (z) for V–S system at 750°C under 100 MPa.

After the reactions, it was observed that a slight amount of vanadium sulfide reacted with the gold capsule, and a very fine layer of a new Au–V–S compound was present only on the inner surface of the capsule. The compositional analysis showed the compound to be AuV_3S_5. Details concerning the properties of AuV_3S_5 will be published elsewhere.

Results and Discussion

1. Dependence of Sulfur Content (z) of the Products VS_z Composition (y) in VS_y

The relation between sulfur content (z) of synthesized samples and sealed composition (y), at 650°C under 200 MPa and at 750°C under 100 MPa, are shown in Figs. 2 and 3, respectively. According to Rau (10), the saturated vapor pressure of sulfur at 650 and 750°C are 9.030 and 22.63 atm, respectively. Therefore sulfur exists always in the liquid state under the present experimental conditions. When y is less than 1.7, z is nearly equal to y at 650°C under 200 MPa. The z value is, however, constant (1.718) when y is larger than approximately 1.72. Therefore, the maximum sulfur content in VS_z in equilibrium with liquid sulfur under high pressure in this condition is $VS_{1.718}$. The V_5S_8 phase is stable below $z = 1.641$, and the $V_{1+x}S_2$ phase is present in the compositional range of $1.661 < z < 1.178$ ($0.164 < x < 0.204$) at 650°C under 200 MPa, as shown in Fig. 2. The upper limit of the compositional range of the V_5S_8 phase ($z = 1.641$) agrees well with the results obtained by Saeki *et al.* (*11*). The nonstoichiometric compositional range of the $V_{1+x}S_2$ phase is found to be $VS_{1.670}$–$VS_{1.695}$ at 750°C under 100 MPa, as shown in Fig. 3.

2. Equilibrium Phase Diagram for V_5S_8–S System under High Pressure

Figure 4 shows the temperature dependence of the maximum sulfur content of the

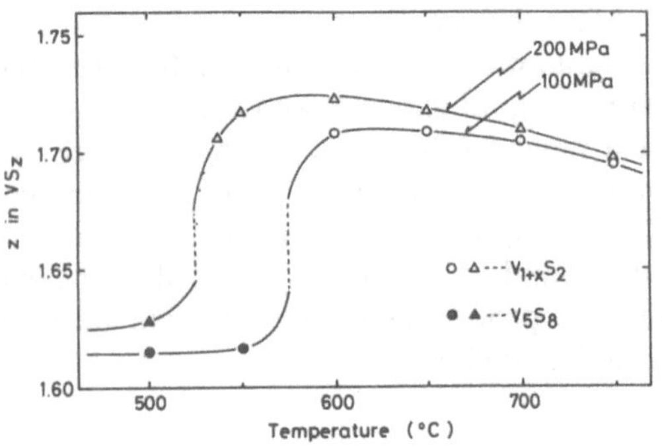

FIG. 4. Temperature dependence of maximum sulfur content in VS_z for products under pressure of 100 and 200 MPa.

products VS_z in equilibrium with the liquid sulfur under high pressure as mentioned in the previous section. The temperature dependence (350–750°C) of the maximum sulfur content in VS_z under 100 MPa is shown in Fig. 5. The $V_{1+x}S_2$ phase is stable at temperatures above 575°C, and the V_5S_8 phase is stable below this temperature. Under 200 MPa the $V_{1+x}S_2$ phase exists at temperatures above 525°C, and the V_5S_8 phase exists below this temperature (Fig. 4). It is seen from Fig. 5 that the VS_4 phase is stable at a temperature lower than 450°C under 100 MPa. Based on the present results, the equilibrium phase diagrams for the V_5S_8–S systems under 100 and 200 MPa are presented in Figs. 6 and 7, respectively. The stable single-phase region of $V_{1+x}S_2$ was not present when synthesized under atmospheric pressure. However, it was encountered under 100 and 200 MPa, as shown in these figures. The stable region under 200 MPa is slightly wider than that under 100 MPa. The following metatectic reaction may take place at points A and B (Fig. 6):

$$V_5S_8 + \text{Liq.} = V_{1+x}S_2$$

Also, the same reaction may take place at points A′ and B′ (Fig. 7) and a different

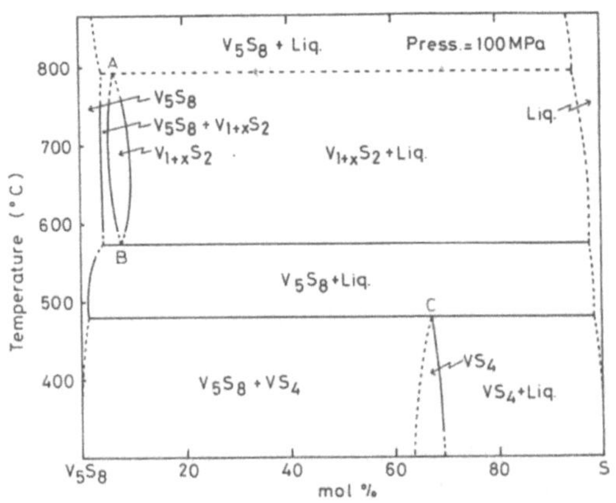

FIG. 6. Equilibrium phase diagram for V_5S_8–S system under a pressure of 100 MPa.

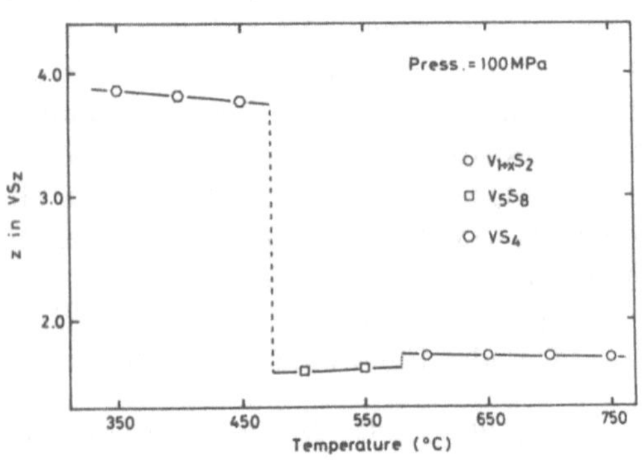

FIG. 5. Maximum sulfur content in VS_z of the product over the temperature range from 350 to 750°C under a pressure of 100 MPa.

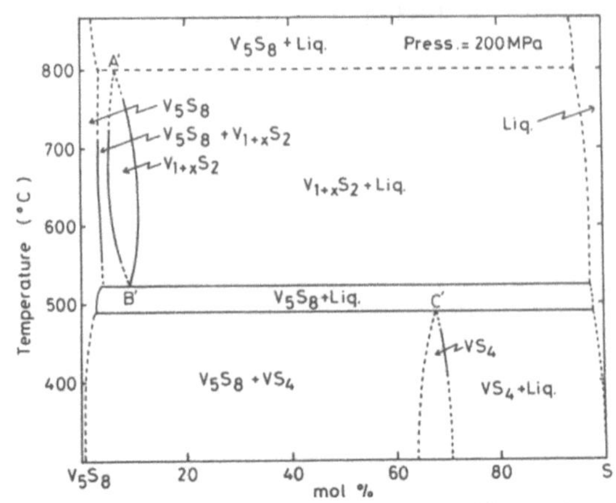

FIG. 7. Equilibrium phase diagram for V–S system under a pressure of 200 MPa.

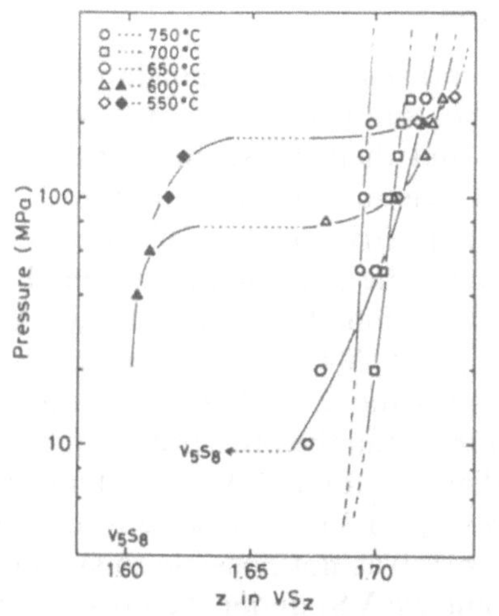

FIG. 8. Pressure dependence of the sulfur content of the products at constant temperature. Solid symbols represent the V_5S_8 phase; open circles represent $V_{1+x}S_2$.

metatectic reaction may take place at points C (Fig. 6) and C′ (Fig. 7):

$$VS_4 = V_5S_8 + \text{Liq.}$$

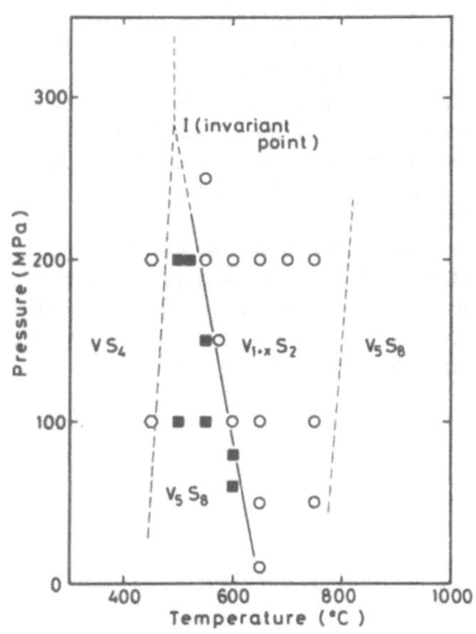

FIG. 9. Pressure–temperature phase diagram for the V–S system.

3. Pressure Dependence of the Maximum Sulfur Content of the Products

The pressure dependence of the maximum sulfur content of the products VS_z in equilibrium with the liquid sulfur at constant temperature is shown in Fig. 8. It is seen from the figure that the sulfur content (z) increases with increasing pressure at each temperature. Under the present experimental conditions (at 550°C under 250 MPa), the maximum sulfur-rich compound was found to be $VS_{1.732}$ ($V_{1+x}S_2$ phase). Accordingly, it is assumed that VS_2 may only be prepared under ultra-high pressures. As shown in Fig. 8, the phase transformation at V_5S_8 to $V_{1+x}S_2$ may occur at 550° and 600°C under 70 and 180 MPa, respectively, but the vanadium sulfides which belong to $V_{1+x}S_2$ could only be obtained above 650°C.

4. Pressure–Temperature Phase Diagram for the V–S System

The phase diagram of Fig. 9 shows the relationship between pressure and temperature for the transformation of the V_5S_8 phase to $V_{1+x}S_2$. It is found that the dP/dT line of the phase boundary has a negative slope (the solid line in Fig. 9). The VS_4 phase is stable below 450°C, and the extrapolated boundary line between the VS_4 and the V_5S_8 phases joins with the boundary line between the V_5S_8 phase and the $V_{1+x}S_2$ phase at point I in Fig. 9. At this point, four phases, i.e., VS_4, V_5S_8, $V_{1+x}S_2$, and liquid sulfur, coexist. Therefore, it represents an invariant point in the binary (V–S) system according to the phase rule ($f = c - p + 2$). The phase boundary between the V_5S_8 phase at high temperature and the $V_{1+x}S_2$ phase has not yet been determined due to our experimental limitations.

5. Compositional Dependence of Lattice Parameters of Vanadium Sulfides

The measured lattice parameters of the NiAs–CdI_2-type subcell (pseudo-orthohex-

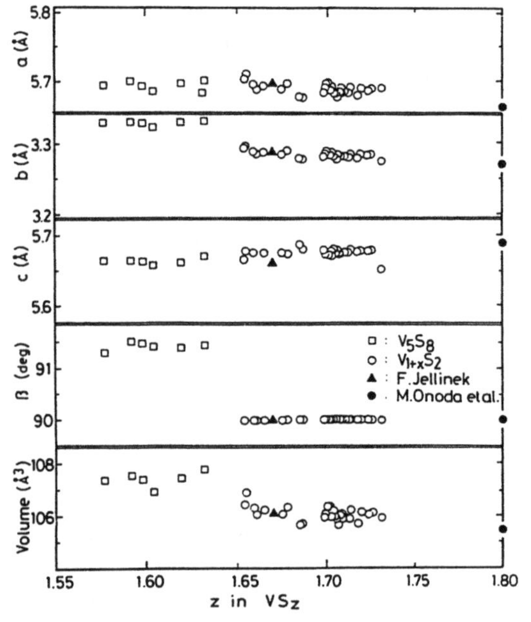

FIG. 10. Compositional dependence of lattice parameters in the range from $VS_{1.55}$ to $VS_{1.8}$.

agonal) of the $V_{1+x}S_2$ with the composition $VS_{1.58}$–$VS_{1.63}$ are shown in Fig. 10, along with those of $VS_{1.67}$ reported by Jellinek (*9*), and of $VS_{1.8}$, reported by Onoda *et al.* (*8*). It is seen from the Fig. 10 that the parameters suddenly change close to the sulfur composition (z in VS_z) of approximately 1.65, particularly for cases b with respect to volume. This significant change indicates the phase transformation of V_5S_8 to $V_{1+x}S_2$. The lattice parameters of the $V_{1+x}S_2$ phase obtained by us are in good agreement with those of Jellinek (*9*) and of Onoda *et al.* (*8*).

Conclusion

In the present work, vanadium sulfides with higher sulfur compositions were synthesized and their phase relationships under high pressure were determined. The nonstoichiometric range of the $V_{1+x}S_2$ phase under the present experimental conditions was found to be $VS_{1.661}$–$VS_{1.732}$ ($0.155 < x < 0.204$). The equilibrium phase diagrams for the V_5S_8–S system under 100 and 200 MPa and the pressure–temperature phase diagram for the V–S system under pressures up to 300 MPa were also drawn. However, it was not possible to synthesize stoichiometric VS_2 under the present set of conditions.

References

1. Y. OKA, K. KOSUGE, AND S. KACHI, *Phys. Lett. A*, **50,** 311 (1974).
2. A. B. DE VRIES AND C. HAAS, *J. Phys. Chem. Solids* **34,** 651 (1973).
3. D. W. MURPHY, F. J. DI SALVO, AND J. N. CARIDES, *J. Solid State Chem.* **29,** 339 (1979).
4. D. W. MURPHY, J. N. CRIDES, AND F. J. DI SALVO, *Mater. Res. Bull.* **12,** 825 (1977).
5. M. WAKIHARA, T. UCHIDA, AND M. TANIGUCHI, *Mater. Res. Bull.* **11,** (1976).
6. M. WAKIHARA, T. UCHIDA, AND M. TANIGUCHI, *Met. Trans. B* **9,** 29 (1978).
7. T. UCHIDA, H. HINODE, M. WAKIHARA, AND M. TANIGUCHI, *Met. Trans. B* **12,** 71 (1981).
8. M. N. Onoda, S. Yamaoka, K. Yukino, K. Kato, and I. Kawada, *J. Less-Common Met.* **24,** 341 (1976).
9. F. JELLINEK, *Ark. Kemi* **20,** 447 (1963).
10. H. RAU, T. R. N. KUTY, AND J. R. F. GUEDES DE CARVALHO, *J. Chem. Thermodyn.* **5,** 833 (1973).
11. M. SAEKI, M. ONODA, AND M. NAKAHIRA, *J. Crystal Growth* **24/25,** 154 (1974).

Reprinted from *Journal of Crystal Growth*, **79** (1986) 215–218.

HYDROTHERMAL SYNTHESIS OF GALLIUM ORTHOPHOSPHATE CRYSTALS

Shin-ichi HIRANO, Kazuo MIWA and Shigeharu NAKA

Department of Applied Chemistry, Faculty of Engineering, Nagoya University Furo-cho, Chikusa-ku, Nagoya 464, Japan

Some orthophosphates are known to have polymorphism similar to SiO_2. The low temperature form of gallium phosphate, $GapO_4$, is especially attractive due to its electrical character superior to α-quartz. In the present work, single crystals of $GaPO_4$ have been grown hydrothermally at temperatures below 400°C in phosphoric acid solution. Highly pure $GaPO_4$ nutrient has been prepared by the reaction of Ga_2O_3 with $NH_4H_2PO_4$ and subsequent hydrothermal treatment. The solubility increased with increase of pressure and decreased with temperature increase. The development of the prism $\{10\bar{1}0\}$ face was suppressed with the occurrence of {0001} (*Z*-plane) by growth under high pressure and temperature conditions.

1. Introduction

The orthophosphate crystals $M(III)PO_4$, with M = Al, Ga, Fe or Mn, have an isomorphous structure derived from α-quartz by substituting Si with M(III) and P(V). Among the crystals, aluminum orthophosphate, $AlPO_4$, has been of great interest because of its large piezoelectric coefficients and small temperature coefficients for resonators [1]. The room temperature modifications of $AlPO_4$ and $GaPO_4$ crystallize in the space group $P3_{2,1}21$. A detailed study on the structure has been performed only for $AlPO_4$ [2]. Compared with the extensive studies on $AlPO_4$, little has been done on $GaPO_4$ crystals for studying the polymorphic transformation and the substitution of the tetrahedral site [3–7]. No detailed data on the synthesis of low temperature modification of $GaPO_4$ (Ga-berlinite) have been available.

This work describes the solubility of $GaPO_4$ in H_3PO_4 hydrothermal solution and the hydrothermal synthesis of Ga-berlinite. The polymorphic transformation of α-quartz type to cristobalite-type was also examined using hydrothermally grown high quality single crystal of Ga-berlinite.

2. Experimental

The starting powders of $GaPO_4$ were prepared by the reaction of a stoichiometric mixture of Ga_2O_3 (99.99% purity) and $NH_4H_2PO_4$ (> 99.0% purity) at 800°C for 24 h, and by the subsequent hydrothermal treatment at 30 MPa and 300°C for 24 h in 8.25 molarity of H_3PO_4 solution [8]. The recrystallized product was confirmed to consist only of the low temperature form of $GaPO_4$.

Solubility measurements were carried out by a weight loss method in a welded gold or a platinum capsule of about 1 cm^3 in inner volume with a cold seal type pressure vessel. The pressure was measured with a calibrated Heise gauge with an accuracy of ±0.5% and the temperature was measured with a calibrated sheathed thermocouple set inside the pressure vessel and controlled within ±1°C. Following a run, the pressure vessel was quenched in cold water, the remnant grains were removed from the capsule, quickly rinsed on a filter, dried and weighed. Examinations for the phases present and capsule leaks were carefully made after each experimental run.

In order to remove the special problems associated with a negative temperature coefficient of solubility, a series of calibration runs were carried out to establish equilibrium and to ensure the weight loss method as in the solubility measurement of $AlPO_4$ by Kolb et al. [9]. Varying the duration that the crystalline grains were held at the desired pressure and temperature condition, the equilibrium was found to be established in less than two days. Therefore, the solubility was determined by experimental runs longer than three days.

Synthesis runs of Ga-berlinite were performed by the following two methods on the basis of the negative temperature coefficient (retrograde) of solubility; (1) the inversely placed method with the nutrient at lower temperature region; (2) the temperature increase method at a constant heating rate of 30°C/day up to 300°C.

The synthesized crystals were identified by X-ray diffraction at room temperature and at high temperature up to 1200°C, and analyzed by DTA-TG. The thermal expansion coefficient of a grown crystal was measured with a high temperature X-ray diffractometer. The morphology of the grown single crystals was observed with an optical microscope and a polarized light microscope.

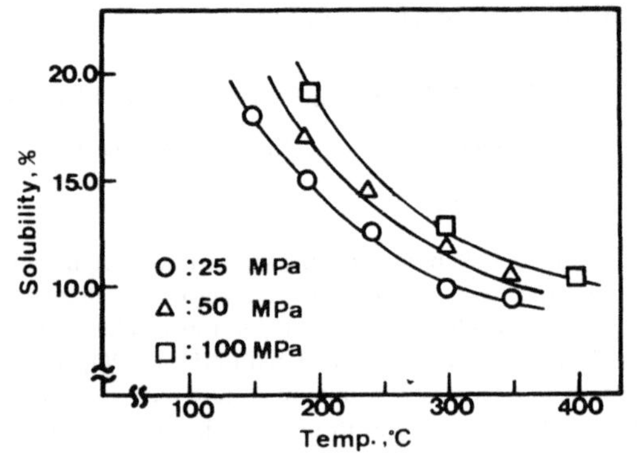

Fig. 1. Weight percent solubility of $GaPO_4$ as a function of temperature in 7.25 m H_3PO_4 solution at several pressures.

3. Results and discussion

3.1. Solubility

Solubility was measured as a function of temperature, pressure and concentration of solution over the range 150–400°C, 25–100 MPa and 1.00 m–8.25 m (molar) H_4PO_4. No foreign phase other than $GaPO_4$ was detected over the entire pressure and temperature conditions.

Fig. 1 shows the solubility of $GaPO_4$ as a function of temperature in 7.25 m H_3PO_4 at various pressures. These data reveal the important feature that the temperature coefficient of solubility $(\partial S/\partial T)_p$ is negative at constant pressure over the range 150–400°C. The Van 't Hoff equation is quite well obeyed with ΔH, the heat of reaction with the solvent, from the slopes of the linear relation of fig. 2 being -1.8 ± 0.2 kcal/mol.

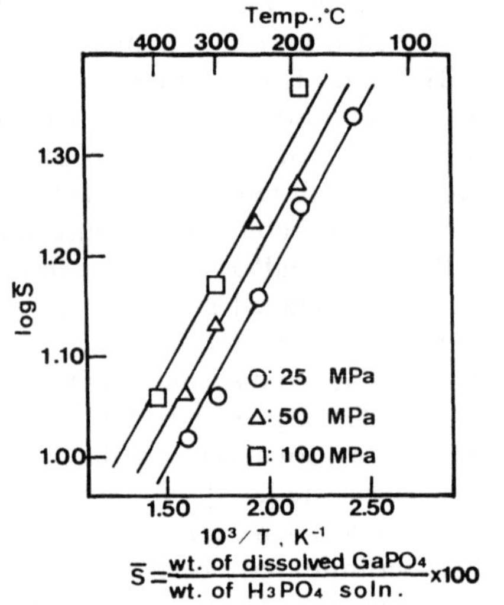

Fig. 2. Log solubility of $GaPO_4$ as a function of $1/T$ in 7.25 m H_3PO_4 solution at several pressures.

Fig. 3 shows the solubility dependence of $GaPO_4$ on the concentration of H_3PO_4 solution at 240°C and 25 MPa. The solubility of $GaPO_4$ increased with an increase in the concentration of H_3PO_4 solution at constant temperature and pressure conditions. The data in fig. 3 are replotted in fig. 4 as molarity of dissolved $GaPO_4$ against molarity of H_3PO_4 solvent. Except at low H_3PO_4 concentration, a good linear relationship can be observed with a slope of 5, as found in the dissolution of $AlPO_4$ [9]. The following possible reaction is considered to take part in the dissolution, if the

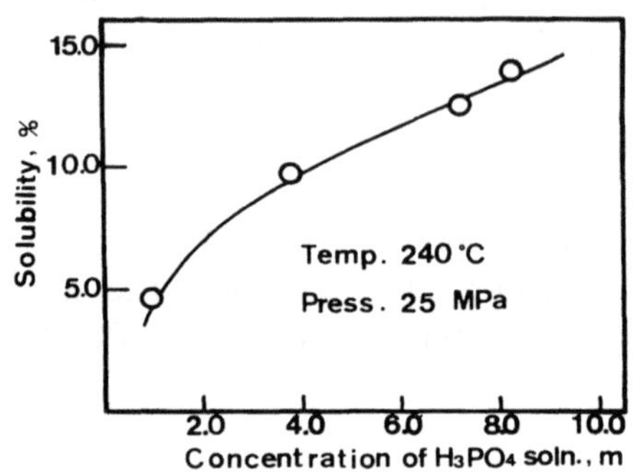

Fig. 3. Solubility of $GaPO_4$ as a function of concentration of H_3PO_4 solution at 240°C and 25 MPa.

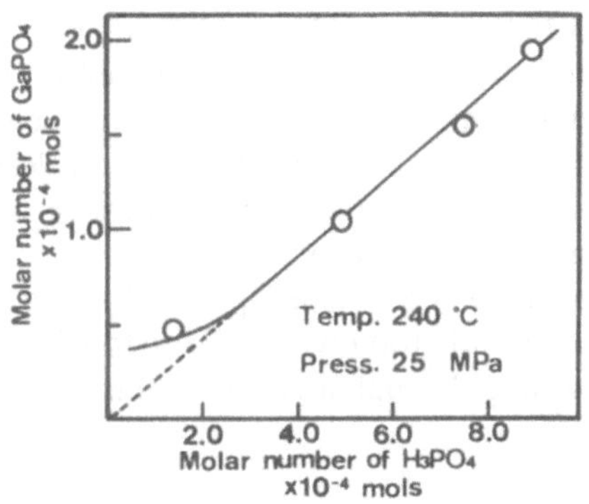

Fig. 4. Plot of molar number of solvent H_3PO_4 versus molar number of dissolved $GaPO_4$ (from the data of fig. 3).

reaction goes to completion at this condition:

$$GaPO_4 + 5\,H_4PO_4 \rightleftarrows Ga(H_2PO_4)_6^{3-} + 3H^+.$$

Fig. 5 shows the dependence of the solubility of $GaPO_4$ on pressure in 7.25 m H_3PO_4 solution at constant temperature. The result indicates that the solubility increases with pressure, suggesting a dissolving reaction in which the specific volume of the chemical species on the right in the above reaction scheme is smaller.

3.2. *Synthesis and properties of $GaPO_4$ crystals*

Coarse single crystals were synthesized by spontaneous nucleation in the high temperature zone of the lower part of the capsule and typical crystals are shown in fig. 6. Two kinds of morphologies were observed depending on the growth condition. With increases in the growth temperature and pressure, the development of M-plane $\{10\bar{1}0\}$ (prism faces) was found to be suppressed and associated with the occurrence of *Z*-plane $\{0001\}$ (basal faces). At lower temperature and pressure, the crystals tended to be bounded by prism $\{10\bar{1}0\}$ faces and terminated by small major rhombohedral $\{10\bar{1}1\}$ faces and minor rhombohedral $\{01\bar{1}1\}$ faces as shown in fig. 6a.

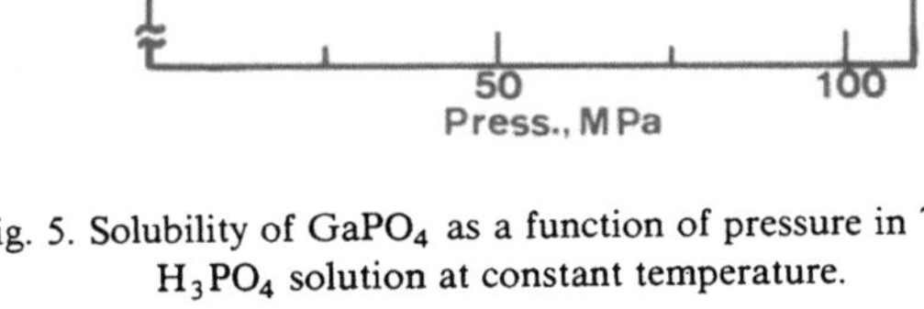

Fig. 5. Solubility of $GaPO_4$ as a function of pressure in 7.25 m H_3PO_4 solution at constant temperature.

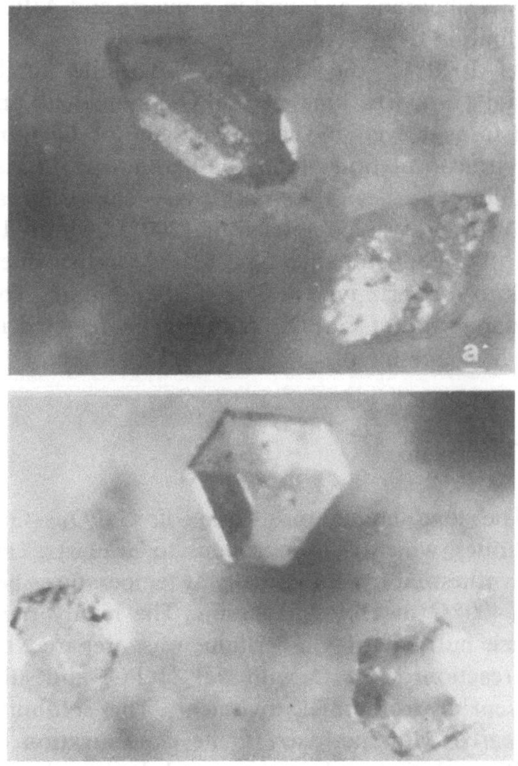

Fig. 6. Microscopic photographs of α-quartz type $GaPO_4$ single crystals grown hydrothermally in 7.25 m H_3PO_4 solutions at (a) 240 °C and 25 MPa, and (b) 400 °C and 100 MPa for 3 days (scale bar: 0.1 mm).

The thermal expansion coefficients on *a*- and *c*-axes are as follows: $\bar{\alpha}_a = 1.79 \times 10^{-5}\ K^{-1}$ and $\bar{\alpha}_c = 0.46 \times 10^{-5}\ K^{-1}$. These values are in good agreement with those calculated from the data of the high temperature X-ray diffraction of $GaPO_4$ powder [10]. The thermal expansion ($\bar{\alpha}_a$) along the normal to the screw axis is greater than that ($\bar{\alpha}_c$) along the screw axis of linkages of GaO_4 and

PO_4 tetrahedrons, as found in α-quartz and $AlPO_4$ (berlinite).

At 1000°C, the starting Ga-berlinite single crystal (α-quartz type of $GaPO_4$) remained untransformed even by heat-treatment for 60 min. The transformation of the Ga-berlinite to β-cristobalite type began to take place sluggishly at 1100°C and was completed at 1200°C within 10 min. The β-cristobalite type $GaPO_4$ transformed back to α-quartz type at 850°C after 30 min. The β-quartz type of $GaPO_4$ could not be detected in any heat treatment.

4. Conclusion

The low temperature form of $GaPO_4$ (Ga-berlinite), which is isomorphous to α-quartz, can be synthesized hydrothermally at temperatures below 400°C in H_3PO_4 solution. The highly pure coarse nutrient of Ga-berlinite was prepared by the reaction of Ga_2O_3 with $NH_4H_2PO_4$ and subsequent hydrothermal treatment. The solubility increased with increases in the concentration of H_3PO_4 solution and pressure, and decreased with temperature (retrograde solubility). The ratio of molarity of solvent H_3PO_4 to dissolved $GaPO_4$ was 5 at constant pressure and temperature.

The spontaneously grown single crystals at lower temperature and pressure were bounded by prism $\{10\bar{1}0\}$ faces and terminated by minor $\{01\bar{1}1\}$ and major $\{10\bar{1}1\}$ rhombohedral faces. The development of basal {0001} faces became more noticeable and associated with the less development of prism faces by increases in the growth temperature and pressure.

The α-quartz type $GaPO_4$ transformed sluggishly at around 1000°C to β-cristobalite type without passing through any intermediate phase of β-quartz type.

References

[1] Z.P. Chung and G.R. Barsch, IEEE Trans. Sonics Ultrasonics SU-23 (1976) 127.
[2] G. Schwarzenbach, Z. Krist. 123 (1966) 11.
[3] A. Goiffon, G. Bayle, R. Asters, J.C. Jumas, M. Maurin and E. Philipport, Rev. Chim. Minéral. 20 (1983) 338.
[4] U. Krau and G. Lehmann, Z. Naturforsch. 30a (1975) 28.
[5] A. Requardt and G. Lehmann, Phys. Status Solidi (b) 127 (1985) 695.
[6] A. Perloff, J. Am. Ceram. Soc. 39 (1956) 83.
[7] E.C. Shafer and R. Roy, J. Am. Ceram. Soc. 39 (1956) 330.
[8] S. Hirano, K. Shibata, K. Miwa and S. Naka, J. Japan. Assoc. Crystal Growth 12 (1985) 100.
[9] E.D. Kolb, R.L. Barns, J.C. Grenier and R.A. Laudise, in: Proc. 33th Annual Symp. on Frequency Control, 1979, p. 88;
E.D. Kolb, R.L. Barns, R.A. Laudise and J.C. Grenier, J. Crystal Growth 50 (1980) 404.
[10] K. Kosten and H. Arnold, Z. Krist. 152 (1980) 119.

Reprinted from *Journal of Crystal Growth*, 79 (1986) 219–222.

HYDROTHERMAL SYNTHESIS AND PROPERTIES OF $BaPb_{1-x}Bi_xO_3$

Shin-ichi HIRANO and Seiji TAKAHASHI

Department of Applied Chemistry, Faculty of Engineering, Nagoya University, Furo-cho, Chikusa-ku, Nagoya 464, Japan

Superconducting $BaPb_{1-x}Bi_xO_3$ (BPB) single crystals can be grown hydrothermally in KCl solution. In the present work, the solubility of BPB in KCl solution was measured in order to understand and improve the hydrothermal growth technique. It was found that the solubility increased with temperature, pressure and concentration of KCl solution, and decreased with increase in the composition x under the same hydrothermal conditions. The optimum solubility and solubility difference to grow large and euhedral crystals were determined to be about 3% and 0.7%, respectively. Hydrothermally grown single crystals of $BaPb_{0.75}Bi_{0.25}O_3$ showed a transition temperature T_c of 11.7 K for the superconducting state and a narrowest temperature width, T_c, of 1.8 K for the transition.

1. Introduction

It was found by Sleight et al. in 1975 that the solid solution system between $BaPbO_3$ (metallic electric conductor) and $BaBiO_3$ (semiconductor) shows superconductivity in the narrow composition range of $0.05 \leqslant x \leqslant 0.30$ for $BaPb_{1-x}Bi_xO_3$ (BPB) [1]. T_c, the onset to the superconducting state, has been reported to be around 13 K on the basis of electrical resistivity for a polycrystalline specimen with the composition $x = 0.25$. This T_c value is the highest among the superconductors without including a transition element, and is also very high for superconductive oxides, including M_xWO_3 and $Li_{1+x}Ti_{2-x}O_4$. Thanh et al. reported that in the superconductive composition range this system yielded semimetallic material with a carrier density of the order of 10^{21} cm^{-3}, which is exceptionally low for superconductors with a high T_c [2]. The Fermi energy E_F and the density of states $N(E_F)$ were calculated by Tani et al. [3] using Thanh's results. Their results revealed that the absolute value of $N(E_F)$ was about 10^{21} states/eV cm^3 which was one to two orders of magnitude smaller than that of the usual high T_c superconductors. It has been considered that high T_c superconductors were required to have both a strong electron–phonon interaction and a large $N(E_F)$ on the basis of the BCS theory and the strong-coupling theory, but these data suggest that the superconductivity of this system may be attributable to other mechanisms. Such results have been observed on polycrystalline samples, which show a broad transition from the normal to the superconducting state as measured by Khan et al. [4]. Itoh et al. showed that the specific anomaly at T_c could not be observed on these polycrystalline samples [5]. These problems have been understood to be caused by inhomogeneity or by the grain boundary effects in polycrystalline specimens.

Demands to grow homogeneous and well-qualified single crystals of BPB have been increasing greatly in order to characterize the intrinsic superconductive features in this oxide. BPB single crystals have been grown by flux methods above 900°C [6–10]. The superconducting transition of crystals grown by flux methods, however, was unfortunately broad. The results were considered to be caused by the structural transition and oxygen deficiency in BPB crystals grown at temperatures above 500°C. We reported that single crystals of BPB could be grown by a hydrothermal method below 500°C and the transition temperature width was the narrowest among the single crystals so far [11]. This paper describes the solubility of BPB in the KCl hydrothermal solution to determine the optimum growth condition and the property of BPB single crystal.

2. Experimental

On the basis of the growth experiment of BPB under hydrothermal conditions [11], the growth conditions were fixed as follows; temperature in the dissolution zone of a gold capsule, 400°C; pressure, 1000 kg/cm^2; 4.5m KCl solution as a solvent; temperature difference, 30°C. The solubility of BPB in this hydrothermal solution was measured by a weight-loss method. Weighed single crystals grown hydrothermally were sealed hermetically into a gold capsule, 3.0 mm in outer diameter, 0.15 mm thick and 66 mm in length with KCl solution as a solvent. The sealed capsule was placed in a test tube type pressure vessel and was subjected to hydrothermal treatment without a temperature gradient. The temperature was measured with a calibrated sheathed thermocouple set inside the pressure vessel. After an experimental run, the pressure vessel was quenched in cold water.

The dependence of the solubility on compositions of BPB was measured at hydrothermal temperatures of 350 to 450°C and 1000 kg/cm^2 in 4.5m KCl solution using the crystals with $x = 0$, 0.20, 0.25 of BPB. The dependence of the solubility on concentrations of solvents was determined at 450°C and 1000 kg/cm^2 in KCl solution with concentrations of 1.0m to 4.5m using a crystal of BPB with $x = 0.25$. The relation of the solubility to the hydrothermal pressures was measured at 450°C and pressures of 500 to 1250 kg/cm^2 in 4.5m KCl solution using crystals of BPB with $x = 0.25$. A duration of 3 days was found by the preliminary work to be long enough to reach the equilibrium in this system.

3. Results and discussions

3.1. Solubility of BPB

Solubility–temperature curves of BPB, with $x = 0$, 0.20, 0.25 under hydrothermal conditions, are shown in fig. 1. The solubility of BPB in hydrothermal solution increased with temperature of hydrothermal treatment and increased steeply from 400°C to 450°C. We reported [11] that

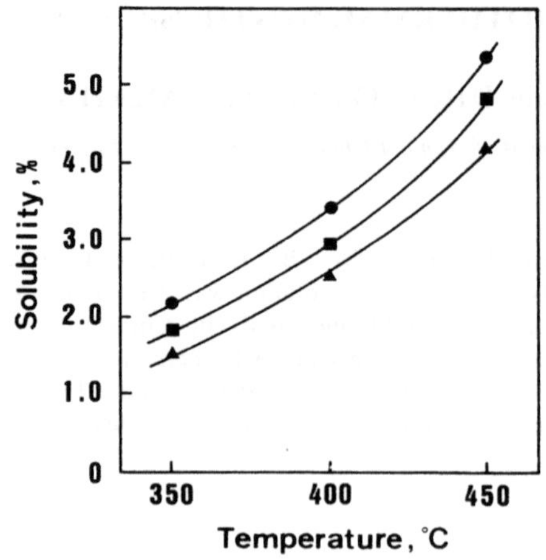

Fig. 1. Solubility–temperature curves for $BaPb_{1-x}Bi_xO_3$ with (●) $x = 0$, (■) $x = 0.20$ and (▲) $x = 0.25$, at 1000 kg/cm^2 in 4.5m KCl solution.

many small crystals were grown at 450°C in the growth zone of a gold capsule, while a few large crystals were grown at 400°C. The solubility data indicate that the degree of supersaturation at 450°C was much higher than that at 400°C with a temperature difference. Thus many small crystals can be grown spontaneously on the walls of gold capsules due to high supersaturation. The solubility was also found to be lower with an increase in x at a fixed temperature and pressure. The higher growth temperature was required to grow large and euhedral crystals with an increase in x, which is consistent with the decrease in solubility with increasing x shown in fig. 1.

Fig. 2 shows the Van 't Hoff relation of solubility. The linear relation between log S and $1/T$ demonstrates that BPB dissolves in the KCl hydrothermal solution by the same dissolution mechanism under hydrothermal conditions with a dissolution enthalpy of about 9.6 kcal/mol.

Fig. 3 shows the relationship between the solubility and concentration of KCl solutions and solubility increases with concentration. The dependency of the solubility on pressures is shown in fig. 4 and solubility increases with pressure, indicating a solubility increase in the higher density of solution. This fact suggests that the dissolution reaction involves decreasing the volume of the solute species.

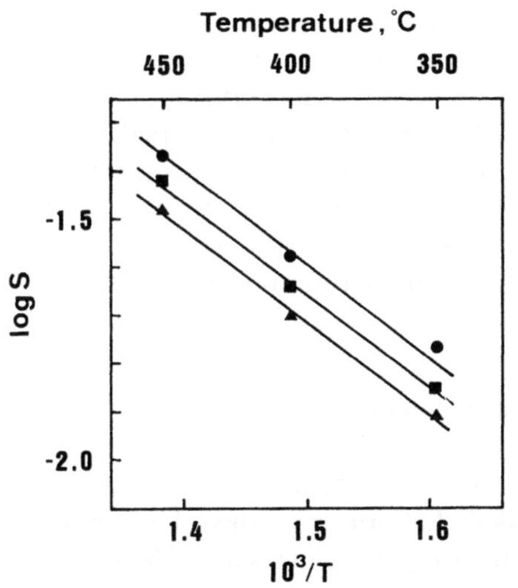

Fig. 2. The Van 't Hoff relation of solubility of $BaPb_{1-x}Bi_{x2}$ with (●) $x = 0$, (■) $x = 0.20$ and (▲) $x = 0.25$, at 1000 kg/cm^2 in 4.5m KCl solution.

3.2. *Superconducting property of grown single crystals*

Fig. 5 shows susceptibility–temperature curves for single crystals of BPB with $x = 0.20$ and 0.25. The data on a single crystal grown by a flux method are also shown. The transition temperature, T_c, and the transition width, ΔT_c, of single crystals of BPB with $x = 0.20$ were 6.8 and 1.8 K, and with $x = 0.25$ they were 11.7 and 1.8 K,

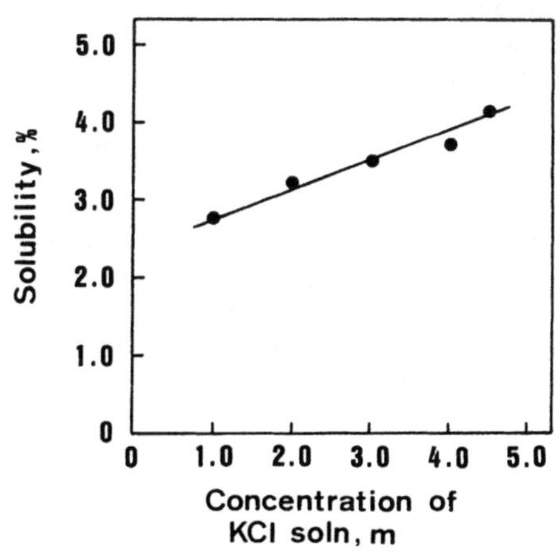

Fig. 3. Dependence of solubility on concentration of KCl solution at 450 °C and 1000 kg/cm^2.

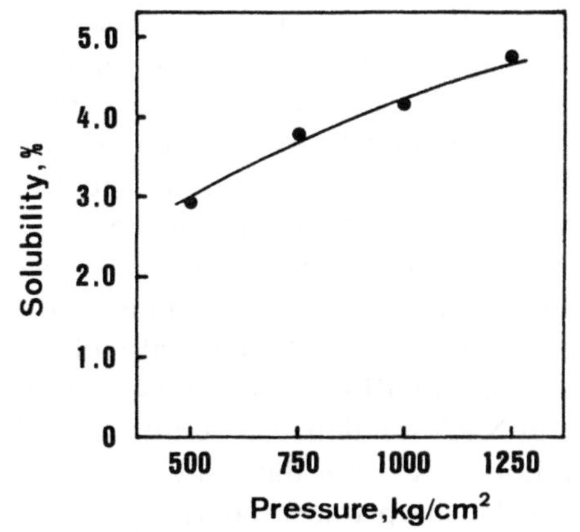

Fig. 4. Solubility–pressure curve for $BaPb_{0.75}Bi_{0.25}O_3$ at 450 °C in 4.5m KCl solution.

respectively. It was found that T_c depended strongly on x. Therefore, precise control of the composition of single crystals of BPB is required in order to grow single crystals of BPB with high T_c. Fortunately, the hydrothermal method has been found to be promising to grow homogeneous BPB single crystals with the desired composition [11].

The transition from the normal state to the superconducting state of single crystals grown by a flux method was very broad with a larger ΔT_c value, which has been attributed to inhomogeneity and oxygen deficiency in the grown crystals [12].

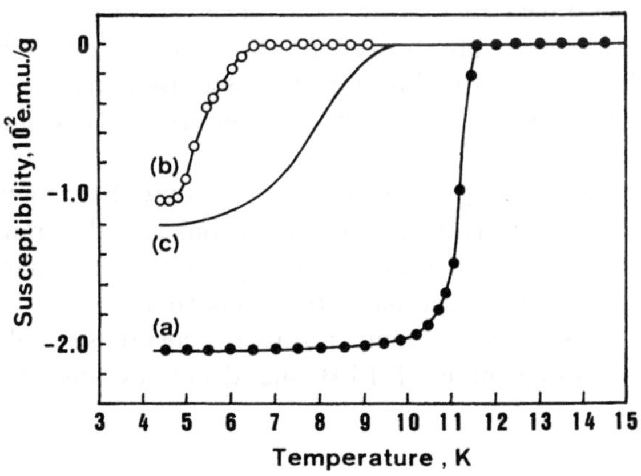

Fig. 5. Susceptibility–temperature curves for the single crystals of $BaPb_{1-x}Bi_xO_3$ with: (a) $x = 0.25$ grown hydrothermally; (b) $x = 0.20$ grown hydrothermally; (c) $x = 0.29$ grown by flux method [12].

In contrast, the transition is very sharp and the narrowest among the data so far published on single crystals by the hydrothermal method. The result indicates that hydrothermally grown single crystals of BPB are homogeneous and free from oxygen defects due to growth below 500 °C, above which the oxygen deficiency has been known to be inevitable, leading to the broad transition.

The same sharp transition was also observed on single crystals of BPB produced by hydrothermal growth in NaCl solution. The results show that the substitution of K^+ in a Ba^{2+} site is not the major effect on the sharp transition as reported by Chu et al. [13].

4. Conclusion

The solubility of BPB in the hydrothermal solution increased with temperature, pressure and concentration of KCl solution in hydrothermal treatment. The optimum conditions to grow large euhedral single crystals were as follows; growth temperature of 400 °C, ΔT of about 30 °C (0.7% of solubility difference between the dissolving and growth zones) and pressure of 1000 kg/cm^2 in 4.5m KCl solution. The solubility decreased with an increase of composition x in BPB. This indicates that a higher growth temperature is required in order to grow BPB crystals with higher x values.

A linear relationship was observed in the plot of log S (solubility) versus $1/T$, as expected from the Van 't Hoff law for the dissolution reaction, with ΔH of 9.6 kcal/mol for the heat of dissolution.

Large single crystals of BPB could be grown under such hydrothermal conditions and the transition width of these single crystals was the narrowest among the data so far reported.

The transition temperature T_c depends strongly on composition of BPB and decreases abruptly with slight deviations from $x = 0.25$. Precise control of the composition is required to reach a high T_c in homogeneous single crystals.

A hydrothermally grown single crystal of $BaPb_{0.75}Bi_{0.25}O_3$ showed a transition temperature T_c of 11.7 K and the narrowest ΔT_c of 1.8 K, whose superior properties were attained by the homogeneity and freedom from oxygen deficiency.

Acknowledgements

The authors gratefully acknowledge Professor S. Tanaka's Laboratory (Tokyo University, Tokyo, Japan) for support in measuring the superconducting properties of crystals.

References

[1] A.W. Sleight, J.L. Gillson and P.E. Bierstedt, Solid State Commun. 17 (1975) 27.
[2] T.D. Thanh, A. Koma and S. Tanaka, Appl. Phys. 22 (1980) 205.
[3] T. Tani, T. Itoh and S. Tanaka, J. Phys. Soc. Japan 49 Suppl. A (1980) 309.
[4] Y. Khan, K. Nahm, M. Rosenberg and H. Willer, Phys. Status Solidi (a) 39 (1977) 79.
[5] T. Itoh, K. Kitazawa and S. Tanaka, J. Phys. Soc. Japan 53 (1984) 2668.
[6] A. Katsui and M. Suzuki, Japan. J. Appl. Phys. 21 (1982) L157.
[7] A. Katsui, Japan. J. Appl. Phys. 21 (1982) L553.
[8] A. Katsui, Y. Hidaka and H. Takagi, J. Crystal Growth 66 (1984) 228.
[9] G.K. Wertheim, J.P. Remeika and D.N.E. Buchanan, Phys. Rev. B26 (1982) 2120.
[10] S.V. Zaitsev-Zotov, A.V. Kuznetsov, E.A. Protasov and V.N. Stepankin, Soviet Phys.-Solid State 26 (1984) 1928.
[11] S. Hirano and S. Takahashi, J. Crystal Growth 78 (1986) 408.
[12] H. Hasegawa, M. Naito, K. Kitazawa, S. Tanaka, S. Takahashi and S. Hirano, in: Abstracts 24th Basic Ceramic Soc. Meeting, Japan, 1986, p. 162.
[13] C.W. Chu, S. Huang and A.W. Sleight, Solid State Commun. 18 (1976) 977.

Part 6

Hydrothermal Reactions

Reprinted from *Proceedings of the Fourth International Conference on High Pressure*, Kyoto, April 1974, pp. 418–423.

GRAPHITIZATION OF CARBON IN PRESENCE OF CALCIUM COMPOUNDS UNDER HYDROTHERMAL CONDITION BY USE OF HIGH GAS PRESSURE APPARATUS

SHIN-ICHI HIRANO, KAZUO NAKAMURA, AND SHIGEYUKI SŌMIYA

Tokyo Institute of Technology, Res. Lab. of Eng. Mat.
Ookayama, Meguro-ku, Tokyo, Japan

On the basis of the geological background of natural graphite formation, the graphitization of carbon was studied under conditions of the coexistence of water and calcium carbonate, using a newly designed, internally heated high gas pressure apparatus or cold seal pressure vessels. Water was found to react with carbon to form low molecular gaseous compounds and also to proceed the graphitization under 10000kg/cm^2 at 900°C, while no graphitic component could be detected by the treatment under 10000kg/cm^2 at 1400°C without water or calcium compound. It was confirmed that the graphitization was remarkably accelerated by the coexistence of water and calcium carbonate, the graphitic component beginning to appear in the specimen treated under 1000kg/cm^2 at 600°C.

1. Introduction

Graphite is one of the chief representatives of elementary substances among rock-forming minerals. It is widely distributed geographically in the world. Geologically, it is mostly limited to areas where older rocks, such as Precambrian and lower Palezoic are exposed. This is particularly true for flaky graphite and for vein graphite. Disseminated flaky graphite is associated with older rocks subjected to a high degree of regional metamorphism. Occurrence is therefore mainly limited to regions in which such rocks as limestone, gneisses, schists, slate and quartz exist. The genesis theory of flaky graphite is not so well understood because the original form of the mother rock is conjectural, the original composition and structure having been oblisterated by shearing stresses and high temperatures during metamorphosis. There have been only some suggestions on the mineralogical background.[1)2)] Unfortunately on account of the lack of experimental data on the mineralogical formation of graphite, it appears very doubtful that geological data will cast much light on the mechanism of the formation of graphite. Only a few literatures are available to discuss and experiment about the mechanism of the formation of graphite.[3)- 8)] French and Rosenberg studied with externally heated cold seal pressure vessel and found the formation of black carbonaceous material, which was proved to be disordered carbon, under moderate conditions. Giardini et al. investigated the reaction of calcite with hydrogen in a cold seal pressure vessel over a range of pressures up to 800 atm and temperatures to 970°C. Their data, however, have been considered to be hot enough to obtain the well crystallized graphite and thus understand the effect of the coexistence of water or minerals on the graphitization of carbon under pressure.

The temperature and pressure of the graphite formation have been estimated geologically as the temperature lower than 1000°C and the pressure of several kilobars. Taking hints from the geological phenomena, one can expect to reproduce the occurrence of natural graphite on a laboratory scale. One of the authors(S. H.) studied the effect of coexisting minerals on the graphitization of carbon under pressure, using a piston cylinder type apparatus.[9)] The results indicated the remarkable acceleration effect on graphitization of the coexisting minerals, such as limestone, calcium hydroxide, calcium oxide, alumina, magnesia, silica, sodium carbonate, alkaline earth fluorides etc. and also water absorbed in calcium carbonate, under about 3 kb. The mechanism of the acceleration by coexisting minerals under pressure was considered to be through the chemical reaction to form intermediates as metals or carbides and gaseous compounds.

Graphite has also been recognized to occur in hydrothermal veins as in the deposit at Ticonderoga, New York. On the geological view points, water has generally assumed to act the part of the reaction species or a kind of mineralizer associated with the coexisting minerals on the natural graphite formation. In the present work, an internally heated, high gas pressure apparatus was newly constructed to carry out the experiments under the isostatic pressure up to 10000kg/cm^2 so as to confirm the formation of well crystallized graphite. The effects of the coexistence of water or of both water and calcium compounds on the graphite formation were studied under pressures up to 10000kg/cm^2 and temperatures up to 1400°C.

2. Development of Internally heated, high gas pressure apparatus

Graphitization studied under the pressure lower than 2000kg/cm^2 were carried out with a cone-in-cone, cold seal test tube type pressure vessel which was heated externally. The experiments under pressures higher than 2000kg/cm^2 and at temperatures higher than 900°C were performed in a platinum capsule with a newly designed, internally heated high gas pressure apparatus illustrated later. The apparatus consists mainly of a pressure intensifying system, a pressure vessel and a heating system, associated with the pressure and temperature controlling system, connections of which are shown in Fig.1 and the specification of the apparatus is illustrated in Table 1.

Table 1. Specification of High gas pressure apparatus

Intensifier	Capacity cc/stroke	First stage	567
		Second stage	49
	Pressure kg/cm²	First stage	577
		Second stage	7000
Pressure vessel	Pressure	max. 10000 kg/cm²	
	Temperature	max. 1500°C	
	Space of sample chamber	13 mm ID 50mm long	

2.1 Pressure Intensifier

The pressure intensifier consists of the two stage moving pistons and cylinders, both pistons being uniaxially connected each other to make them move back and forth at the same time. The first stage of the intensifier receives pure argon of $50kg/cm^2$ pressure from a commercial cylinder supply and compresses and then inserts it hydraulically into the second stage as the pressure of $577kg/cm^2$.

The intensifier was designed in such way that once the head of the piston in the first stage chamber of the intensifier reached to the certain point of the cylinder, it stoped moving further and returned automatically, that is the piston in the second stage started to compress the gas up to $7000kg/cm^2$ at room temperature. Phot. 1 shows the appearance of the intensifier. The total length of the intensifier is about 200cm, and the width is about 45cm. Safety devices of the rupture cylinder type are set in the line of the intensifying system, one is the cylinder as $680kg/cm^2$ in

Photo 1. Appearance of the intensifier

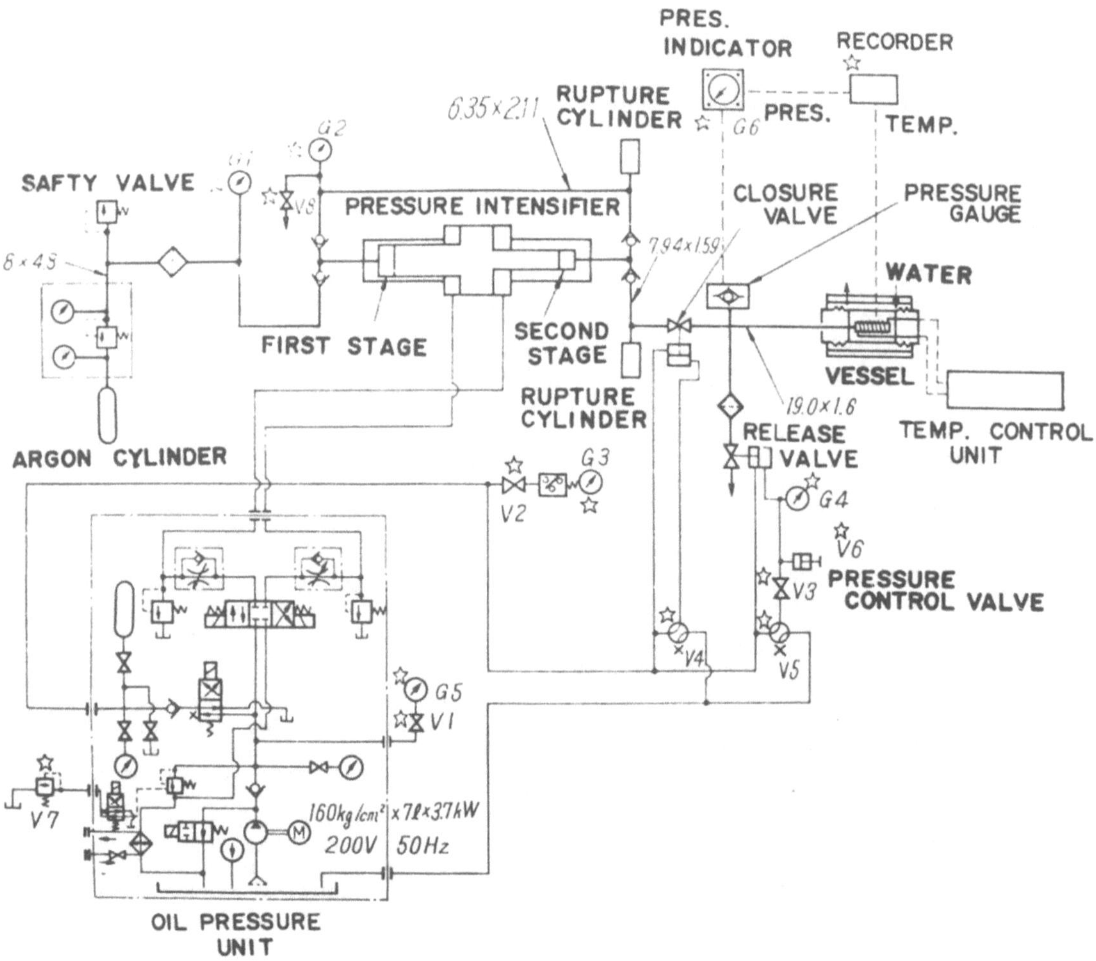

Fig. 1 Schematic diagram of the apparatus

the first stage and the other as 8400kg/cm^2 in the second stage. In the intensifier, special cares were also paid for the design of the piston in order to avoid the pinching-off effect of the piston head and the selection of the packing materials making it sure to be free from oil and mist in the pressure system and then to enhance the operating efficiency. Fig. 2 shows the detail of the structure of the piston used.

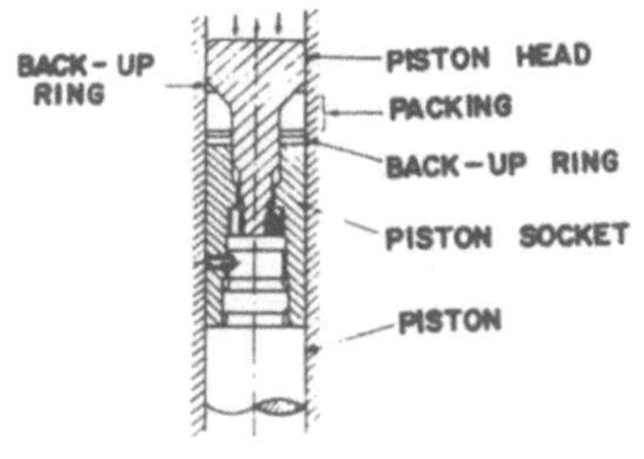

Fig. 2 Scheme of the piston used

2. 2 Pressure vessel

The outline of the pressure vessel is as shown in Fig. 3. The pressure vessel is used ordinally as vertical type and made of the shrunk double wall cylinder. The inner cylinder was fabricated from the modified heat treated steel (KNCM 11, modified SAE 4340, Brinell hardness about 360) by the autofrettage teqnique under 12500kg/cm^2.

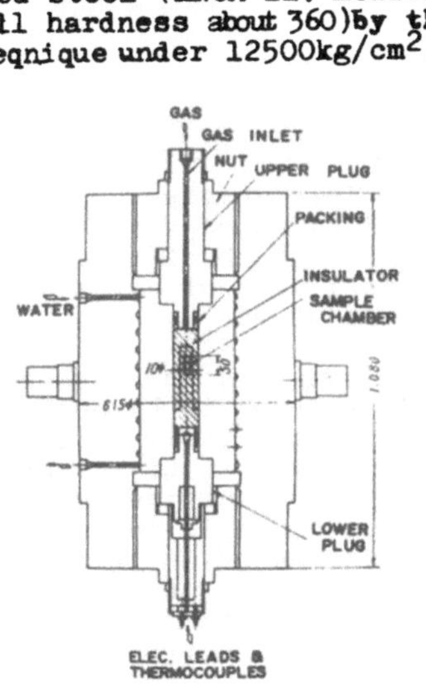

Fig.3 Outline of the pressure vessel

The inner surface of the second cylinder has helical grooves as cooling water passage. Sealing of the high pressure gas was ensured by using the modified Bridgman type self sealing method, taking care for the use of the composite packing sequences.

Electric leads, thermocouples and the electric furnace are fixed to the lower plug of the vessel in order to remove the upper plug only to set samples into the vessel. The pressure vessel can be turned around the pins attached to its center as shown in Fig. 3 for the convenience of preparing the interior of the vessel.

The details of the furnace and the parts of the insulator are shown in Fig.4. The single stage or the two stage heating element can be chosen, where the molybdenum wire of 1.5mm in diameter was worked upon in the wavy shape with change in height so as to obtain the desired heating zone. The input electric power can be supplied independently in the two stage heating system consisting of two separate molybdenum heating elements.

The most important part of the furnace was in the structure of the thermal insulater to get the most effective heat insulation for the convection of argon gas and to reduce the thickness of the insulating wall compared with the space of the sample chamber of 13mm in diameter. The typical configuration of the furnace and the insulator consisting of the end-sealed metal sheaths, refractory alumina disks and ceramic fibers is shown in Fig.4, with which the temperature at the inner wall of the first cylinder was controlled lower than 75°C even on the experiment at 1400°C under 10000kg/cm^2.

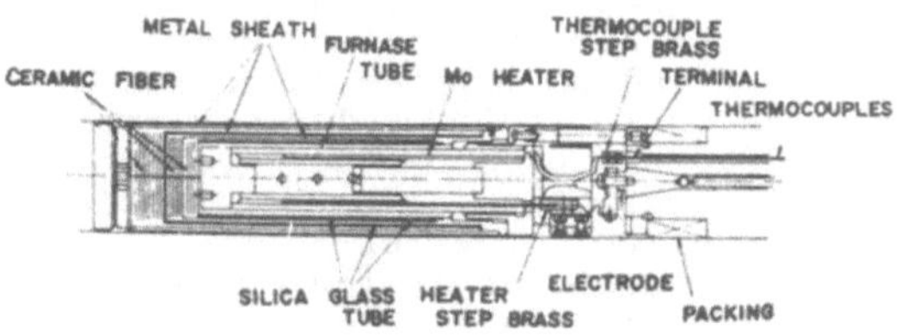

Fig.4 The typical configuration of the furnace and the insulator

The temperature can be controlled and measured with an automatic temperature control equipment through a programming unit connected with a Pt 6%Rh-Pt 30%Rh thermocouple by controlling the input power with a SCR power control unit. The pressure in the vessel can be measured with a quadrupole strain detector calibrated by means of a free piston type pressure calibrator. All valves in the pressure system are designed to enable the remote control with an oil pressure unit.

3. Experimental

A starting carbon sample used was a coke prepared by carbonization of polyvinylchloride and consequently heated up to 680°C in the flow of nitrogen gas. The carbon sample sized in 0.1 to 0.4mm was hermatically sealed into gold or platinum capsule, with pure water and/or calcium carbonate. The capsules were set in the vessel and heat-treated under pressures up to 10000kg/cm^2 and at temperatures up to 1400°C for various period to 3 days. The cold-seal type pressure vessel of 6.5mm in inner diameter was used for the experiment up to 2000kg/cm^2 or at temperatures lower than 900°C. The high gas pressure apparatus mentioned above was used for the experiments under pressures above 2000kg/cm^2 or temperatures higher than 900°C.

The carbon specimens obtained were examined by means of a X-ray powder diffractometer with Ni-filtered Cu Kα radiation (JDX 7F, JEOL Co.,Japan) and an electron microscope (JEM100C, JEOL Co.,Japan) The existance of the graphitic component in the carbon specimen obtained was observed on the X-ray diffraction line profiles and also checked by the selected area electron diffraction method. The amount of the graphitized component in the treated carbon specimen was calculated from the ratio of the area of the profile corresponding to the graphite to that of the turbostratic carbon part on the composite (002)X-ray diffraction line profiles.

The pressure was measured with a Heise gause or a kind of the quadrupole strain gause calibrated by means of the free piston type pressure calibrator. The temperature was measured with a platinel thermocouple calibrated at the melting points of pure gold and zinc or with a Pt 6%Rh-Pt 30%Rh thermocouple at temperatures higher than 900°C. The accuracies of temperature and pressure were within ±3°C and ±20kg/cm^2 under pressure lower than 2000kg/cm^2 and then ±50kg/cm^2 above 2000kg/cm^2, respectively.

4.Results and Discussions

4.1 Effect of the presence of water

It was observed that the starting carbon sample reacted with water above 400°C to form low molecular gaseous substances under pressure. No graphitic component could be detected in the carbon specimen heat-treated without water and/or calcium carbonate at temperatures up to 1400°C under any pressure up to 10,000kg/cm^2 even by the electron microscopic observation. The remarkable acceleration effect of water on the graphitization was recognized in the comparison of two kinds of experiments under 10,000kg/cm^2 and 900°C for 4 hours; one of the experiments was carried out without the presence of water and the other with the coexistence of water. The graphite formation was found to be accelerated by the treatment of carbon under hydrothermal condition with the coexistence of water as shown in electron microscopic photographs of Photos 2a and 2b.

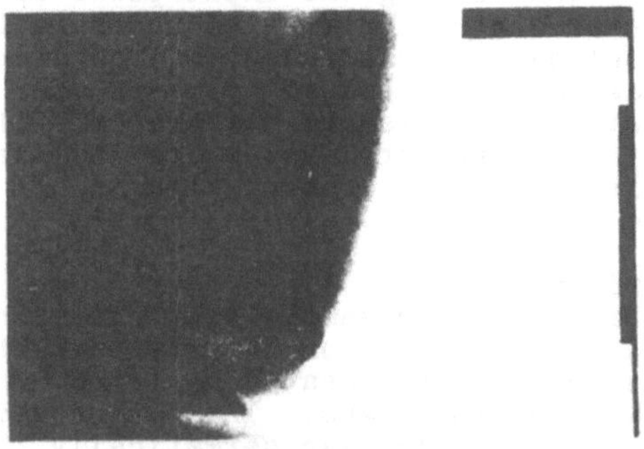

Photo 2a; Transmission electron micrograph of the graphitized component in the specimen treated under 10,000kg/cm^2 at 900°C with presence of water.

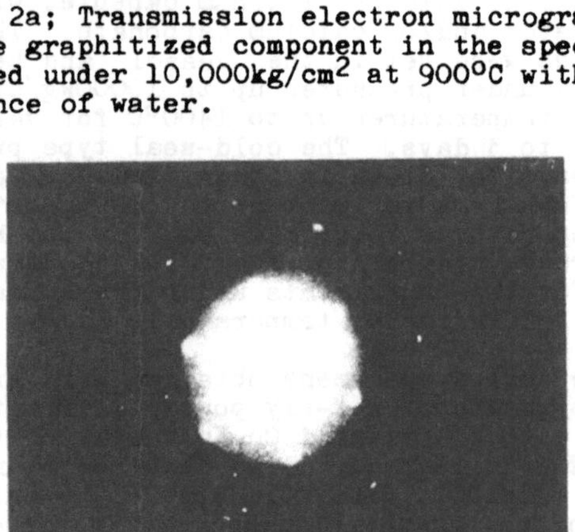

Photo 2b; Selected area electron diffraction micrograph of the graphitized component found in the specimen treated under 10000 kg/cm^2 at 900°C with the presence of water.

The six-fold electron diffraction spots in Photo 2b show the presence of the three dimensional structure of the ordered graphite.

On the other hand, the carbon specimen treated without the presence of water under the same condition showed only the diffraction rings corresponding to the disordered carbon as shown in Photos 3a and 3b.

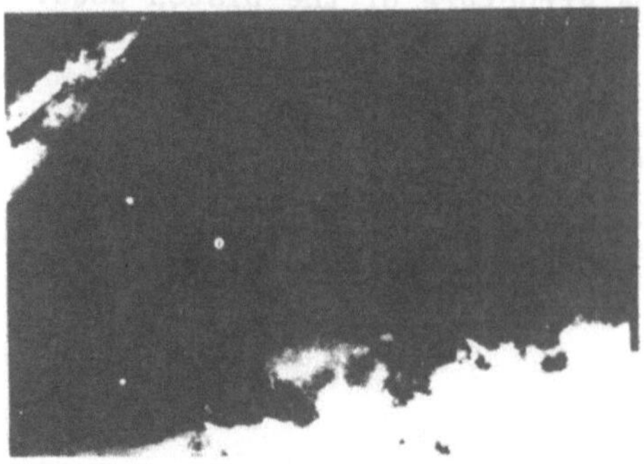

Photo 3a

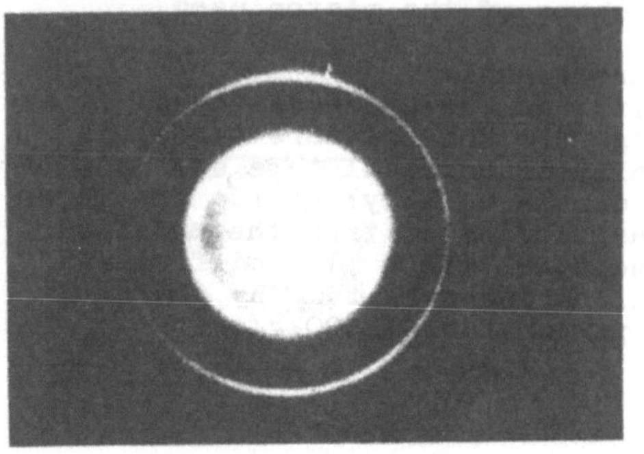

Photo 3b

Photos 3a and 3b;Transmission electron micrograph and selected area electron diffraction micrograph of the carbon specimen treated without water under 10,000kg/cm^2 at 900°C.

The amount of water to the carbon sample in the capsule was found to be optimum for 1 to 4 in weight with the fill of about 60%, considering the results of the graphite formation and the effectiveness of the treatment of specimen in the pressure vessel. The degree of the graphitization, however, was so low on the treatment only with the presence of water that no (002) X-ray diffraction line profile characteristic to the graphitic component could be observed even on the carbon specimen treated at 1400°C under 10000kg/cm^2. The facts seem to indicate that the presence of water shows the acceleration effect on the graphitization of carbon and plays the role of lowering the temperature to form the graphitic component, whereas the effect is not enough for considering the occurrence of the natural graphite.

4.2 Effect of the coexistence of water and calcium carbonate

The accelerating effect of calcium carbonate on the graphitization found with the piston-cylinder type pressure vessel[9] was confirmed by the experiment under 3000kg/cm^2 with the presence of calcium carbonate using the internally heated high gas pressure apparatus. The graphite component began to appear in the carbon specimen treated under 3000kg/cm^2 at 1000°C for 24 hours.

The coexistence of water with calcium

carbonate was found to further promote the graphitization drastically such that the formation of the three dimensional graphite crystal took place by the treatment at the temperature as low as 600°C under 1000kg/cm^2 for 4 hours as shown in Photos 4a and 4b.

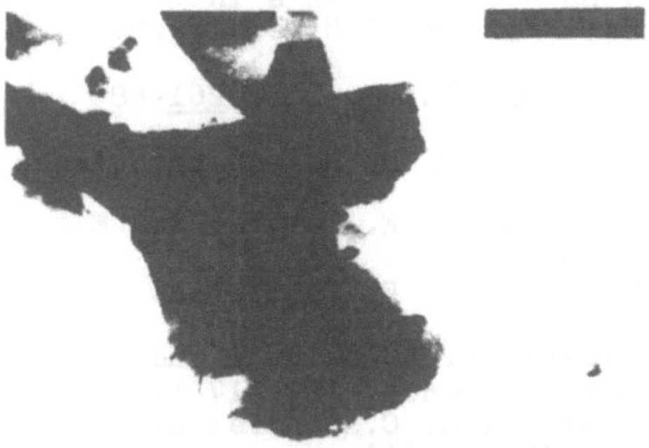

Photo 4a

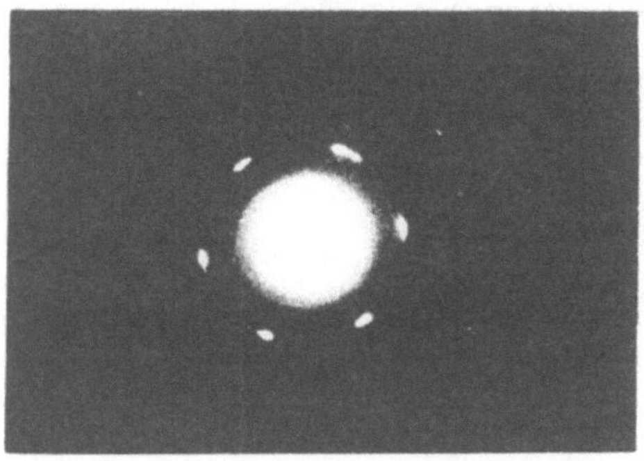

Photo 4b

Photos 4a and 4b; Transmission electron micrograph and selected area electron diffraction micrograph of the prticle found in the carbon specimen treated with the coexistence of water and calcium carbonate under 1000kg/cm^2 at 600°C for 4 hours.

Fig. 5 shows changes in (002) X-ray diffraction line profiles of the carbon specimens treated under 3000kg/cm^2 at 1000°C for 24hours with the coexistence of water and calcium carbonate and only with calcium carbonate. In the figure, the (002) X-ray diffraction line profile of the carbon specimen treated without both water and calcium carbonate under the same condition is shown as a reference. The change in the profile reflects the difference in the amount of the graphitized component in the carbon matrix. The sharp profile located at around 26.5° in 2θ corresponds to the graphitic component, compered to the broad profile at around 25.5° corresponding to the disordered (turbostratic structure) carbon. In the case of the result shown in Fig.5, the presence of water with calcium carbonate is evaluated to accelerate the graphite formation by about 15% in weight ratio, calculated from the ratio of the area corresponding to the profile of the graphitic component to that of the rest of the profile after the correction due to the preferred orientation of the graphitic component. The results also indicated that the graphitization was favoured by the increase of the temperature rather than pressure.

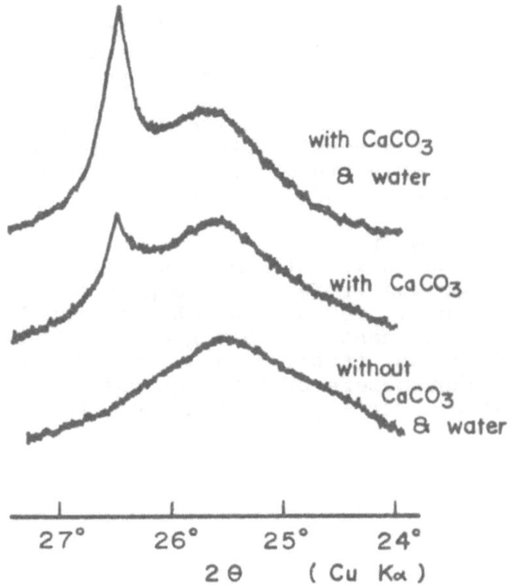

Fig.5; (002) X-ray diffraction line profiles of the carbon specimens treated under 3000kg/cm^2 at 1000°C for 24hours with the coexistence of both water and calcium carbonate, only with calcium carbonate and without any coexisting substance.

The calcium carbonate powder contacted with the carbon sample was crystallized into calcite crystal by the treatment with water to be obtained in the disk. The flaky graphite crystals were observed often on the surface of the disk of calcite crystal associated with carbon sample. The condition where the graphitization took place was considered to follow the enhancement of the reactivity of calcium carbonate and its melting phenomenon with the addition of water, according to Wyllie and Tuttle.[6)]

The effect of water in the presence of calcium carbonate on the graphitization of carbon seems to suggest one possibility of the formation of natural graphite under moderate conditions in the bed of limestone; the metasomatism or mineralization effect with magma or pneumatolysis could be considered as the mechanism for the genesis of natural graphite.

5. Conclusion

The internally heated, high gas pressure apparatus capable of the treatment of sample under 10,000kg/cm^2 at 1500°C was designed and constructed under considerations of the efficient operation with safety and then specific designs of the piston in the high pressure intensifier and of the structure of the furnace and insulator. The apparatus was insured to be able to operate usually under pressures up to 10,000kg/cm^2 and at temperatures up to 1400°C.

The studies on the graphitization of carbon were carried out under hydrothermal condition with or without calcium carbonate. It was found that the graphitization was accelerated by the presence of water and further by the coexistence of both water and calcium carbonate. The graphitization took place at 900°C under 10,000kg/cm^2 with the presence of water, while no graphitic component could be detected even on the carbon specimen treated under 10,000kg/cm^2 at 1400°C. The coexistence of water with calcium carbonate lowered the graphitization condition drastically to the temperature as low as 600°C and the pressure of 1000 kg/cm^2. The present results were considered to suggest the possibility of the formation of natural graphite under moderate conditions in the bed of limestone.

Acknowledgement
The authors wish to thank Dr. Tokiti Noda, Emeritus Professor of Nagoya Univ. and Mie Univ., for his valuable discussion and encouragement. We are grateful to the Kobe Steel Co. (Kobe, Japan) for supports of the manufacture of the high gas pressure apparatus. Our appreciation are also extended to Mr. E. Watanabe, JEOL Co. (Akishima, Japan) for his aid to the electron microscopic observation.

References

1) For an example, S. B. Seeley; "Encyclopedia of Chemical Technology", 4, 319 (1964), John Willey & Son's Inc.
2) J. D. Dana; "System of Mineralogy", 1 152 (1837)
3) P. E. Rosenberg; Am. J. Sci. 261 683 (1963)
4) B. M. French and P. E. Rosenberg; Science, 147 1283 (1965)
5) P. J. Wyllie and O. F. Tuttle; Am. Min. 44 453 (1959)
6) P. J. Wyllie and O. F. Tuttle; J. Petr. 1 1 (1960)
7) D. C. Giardini, C. A. Salotti and J. F. Lakner; Science, 159 317 (1968)
8) R. F. Müller and K. C. Condie; J. Geology, 70 400 (1964)
9) S. Hirano; Dissertation for Dr.Eng., Nagoya Univ.(1970)

Reprinted from *Report of the Research Laboratory of Engineering Materials, Tokyo Institute of Technology*, No. 7 (1982) 43–53.

Hydration of Calcium Aluminates (3:1, 12:7, 1:1, 1:2 and 1:6) in Phosphoric Acid Solutions at Various Temperatures

By

Kengo YAMAZAKI*, Toshihiko NAKAZAWA**, Masahiro YOSHIMURA*** and Shigeyuki SŌMIYA****

Abstract

Hydration of the compounds of the system CaO-Al_2O_3 (i.e. CA_n; C_3A, $C_{12}A_7$, CA, CA_2 and CA_6) in phosphoric acid solutions was studied at 35°C under a saturated pressure and at 100°C and 200°C by hydrothermal techniques under the pressure of 10 MPa. At 35°C, C_3A hydrated to form C_3AH_6 within 1 h in water, while $C_{12}A_7$, CA and CA_2 yielded C_2AH_8 after 1 day, which changed gradually into C_3AH_6 in water.

The addition of phosphoric acid into water retarded both the hydration of CA_n and the conversion of C_2AH_8 to C_3AH_6. A more addition of phosphoric acid stopped the hydration of CA_n even for 30 days. It seems that phosphoric acid forms an insoluble layer of amorphous aluminium phosphates on the surface of each CA_n grain. The hydration region, where C_3AH_6 and AH_3 or AH produced seemed to extend to higher H_3PO_4 content sides with higher temperatures and with longer duration in hydrothermal treatments, but CA_6 and α-A could hydrate neither at 100°C nor at 200°C.

The crystalline phosphate phases produced were only $CaHPO_4 \cdot 2H_2O$ at 35°C, $CaHPO_4$, $Ca_{10}(PO_4)_6(OH)_2$ and three types of $AlPO_4 \cdot 2H_2O$ at 100°C, and $CaHPO_4$, $Ca_{10}(PO_4)_6(OH)_2$ and three types of $AlPO_4$ at 200°C, respectively. Their formation depended upon the contents of H_3PO_4 and CA_n, and also upon the duration.

Key Words: influence of phosphoric acid, hydration, Al phosphates, hydrothermal treatment, $CaO \cdot nAl_2O_3$, Ca phosphates, CaO-Al_2O_3-H_3PO_4 system.

1. Introduction

The compounds present in the CaO-Al_2O_3 system are generally given as C_3A, $C_{12}A_7$, CA, CA_2 and CA_6*****. Among these compounds, C_3A, a normal constituent of portland cements, and $C_{12}A_7$, CA and CA_2, normal constituents of high-alumina cements, hydrate in the presence of water, while CA_6 is believed not to hydrate in water.[1)] The hydration products of CA_n at room temperature consist of CAH_{10}, C_2AH_8,

* Research Student M. Sci. in Eng., Present Address: Tōwa Refractory Engineering Co. Ltd., 4-7-9, Ginza, Chuo, Tokyo 104, Japan

** Research Student

*** Associate Professor, Dr. Sci. in Eng.

**** Professor, Dr. Sci. in Eng.

***** The standard cement nomenclature is used throughout CaO=C, Al_2O_3=A, H_2O=H, compounds of the system CaO-Al_2O_3=CA_n.

C_3AH_6 and AH_3 (gel or crystalline), of which relative proportions depend on the hydration conditions and curing periods.[2] In the case of the hydration of CA which is the main cementitious compound in high-alumina cement, CAH_{10}, C_2AH_8 and AH_3 gel are preferentially formed at lower temperatures. They are converted to C_3AH_6 and AH_3 (gibbsite) at high temperatures, according to the scheme given in Fig. 1.[3] C_3AH_6, however, is unstable above 225°C and converts into $C_4A_3H_3$. AH_3 is stable below 150°C, then changes to AH (boehmite) above 150°C under saturated steam pressures.[4]

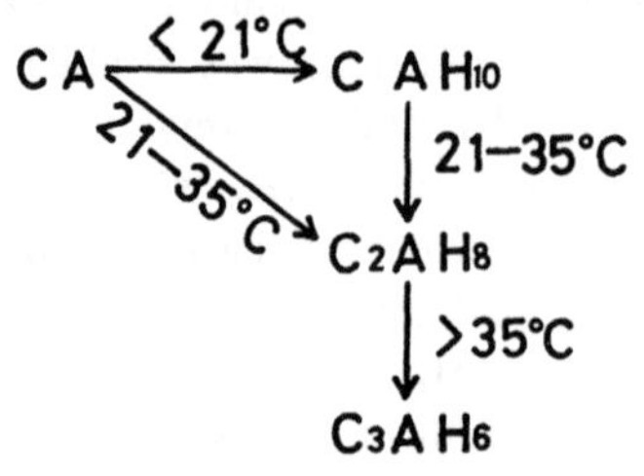

Fig. 1 Hydration of C A

Various phosphates are now widely used to reduce water requirements and prolong setting time for the alumina cements. Many acids and acidic compounds have a severe retarding action on cements.[5] However, little is known regarding the manner in which the individual retarder reacts with the components in cements.

We have systematically studied the behaviour of phosphoric acid on the compounds of the system $CaO-Al_2O_3$ and found the phosphoric acid to be a considerable retarder or inhibiter of CA_n to hydration in low concentrations[6–8] and to form phosphate bond in high concentrations.[9,10]

The purpose of the present work is to obtain informations on the products in the system $CA_n-H_3PO_4(H_2O)$ and to study systematically the influence of phosphoric acid on the hydration of CA_n at various temperatures. A curing temperature of 35°C and hydrothermal conditions of 100°C and 200°C under 10 MPa was chosen to emphasize the significant effects of phosphoric acid solutions on the mechanism of hydration and on the reaction products.

2. Experimental

2.1 Starting Materials

Five kinds of the compounds of the system $CaO-Al_2O_3$ were prepared from mixtures of guaranteed reagent grade $CaCO_3$ (Wakō-Pure Chem. Ind.) and α-Al_2O_3 (Iwatani-Sangyo Co. Ltd.). The mixtures to yield the products having CaO to Al_2O_3 molar ratios of 3, 12/7, 1, 1/2 and 1/6 were heated at 1360 to 1480°C for 4 hours in a crucible composed of a high purity tabular alumina and a high-alumina cement using a SiC furnace. The resulting sintered bodies were ground in an alumina ball-mill to pass a 150-mesh sieve. These procedures were repeated until pure respective phases could be produced by powder X-ray analysis.

2.2 Preparation of Hydrated Specimens

(1) Curing at 35°C

The CA_n powders were mixed with a phosphoric acid solution with a concentration of phosphoric acid (0, 0.2, 1, 5, 10 or 20 wt%) to the weight ratio of 0.5, 1 or 3 for solution/CA_n. These mixtures were poured in cylindrical plastic capsules (10 mmϕ x 50 mm), enclosed with a cap to maintain 100% relative humidity and kept in a thermostat at 35°C for 1 hour, 1 day, 10 days or 30 days. After the required curing time, the capsules were broken and the specimens were washed with acetone and filtered by suction.

(2) Hydrothermal treatment

The CA_n powders were mixed with phosphoric acid solutions with various concentrations, to have a weight ratio of 5 of solution/CA_n in teflon tubes (6 mm ID, 50 mm long). Both ends of the tubes were sealed with pylex glass plugs. Hydrothermal conditions were at 100°C and at 200°C under the pressure of 10 MPa. The pressure vessel used was an autoclave of 76 mm ID, 111 mm OD, 306 mm long.

2.3 Analysis of the Products

Specimens were examined by an X-ray diffractometer (Rotaflex RU-200, Rigaku Denki, Co. Ltd., Tokyo) with CuKα radiation, by a scanning electron microscope (JSM-T200, JEOL Co. Ltd., Tokyo) and by TG-DTA (Thermoflex, Rigaku-Denki Co. Ltd., Tokyo).

Table 1. X-ray powder diffraction analysis of hydrated CA_n pastes in phosphoric acid solutions as a function of various curing time at 35°C

Starting compounds		1 day				10 days				30 days			
		C_2AH_8	C_3AH_6	AH_3	C*	C_2AH_8	C_3AH_6	AH_3	C*	C_2AH_8	C_3AH_6	AH_3	C*
C_3A	P_0=1		vs		w		vs				vs		
	P_1=1	vw	vs		w	vw	vs		vw	w	vs		
	P_0=3		vs		w	w	vs			w	vs		
$C_{12}A_7$	P_0=1	m		vw	w		vs	w	w		vs	w	w
	P_1=1	vs		vw	w	w	vs	w	vw		vs	w	
	P_0=3	vs					vs	w			vs	w	
CA	P_0=1	vs		w	vw		vs	m			vs	m	
	P_1=1	vs		w	w	vs	w	w			vs	m	
	P_0=3	vs		w		m	vs	w			vs	m	
CA_2	P_0=1	vs		vs			vs	vs			vs	vs	
	$P_{.2}$=1	m		w	m		vs	vs			vs	vs	
	P_0=3	vs		vs	vw	w	vs	vs			vs	vs	

C*: C_3A, $C_{12}A_7$, CA, CA_2

vs: very strong, s: strong, m: moderate, w: weak, vw: very weak

P_0 =1: liquid/solid (wt. ratio) = 1, H_3PO_4 0%

P_1 =1: liquid/solid (wt. ratio) = 1, H_3PO_4 1%

$P_{.2}$=1: liquid/solid (wt. ratio) = 1, H_3PO_4 0.2%

P_0 =3: liquid/solid (wt. ratio) = 3, H_3PO_4 0%

3. Results and Discussion

3.1 Hydration of CA_n Cured at 35°C

C_3A, $C_{12}A_7$, CA and CA_2 hydrated in pure water or in low phosphoric acid solutions but CA_6 did not hydrate. Table 1 summarized the products of the hydration of CA_n in phosphoric acid solutions as a function of curing time at 35°C. C_3A hydrated in pure water quickly to form C_3AH_6 within 1 hour, but the addition of phosphoric acid retarded this hydration. The main hydrate of $C_{12}A_7$, CA and CA_2 was C_2AH_8 after 1 day, but C_2AH_8 gradually converted to C_3AH_6 in pure water or in dilute phosphoric acid solutions. The conversion from C_2AH_8 to C_3AH_6 was retarded in the mixtures with large ratio of H_2O/CA_n. The addition of phosphoric acid also retarded the conversion from C_2AH_8 to C_3AH_6. The hydration reactivity was dependent on the ratio CaO/Al_2O_3, i.e., the larger CaO/Al_2O_3 ratio of CA_n the faster rate of hydration.

Fig. 2 shows the reaction products in the system CA_n-H_3PO_4 (H_2O) cured at 35°C for 30 days. In this figure, (1) shows the hydration region (C_3AH_6 or $C_2AH_8 + AH_3$), (2) –a shows the nonhydration region with $CaHPO_4 \cdot 2H_2O$ phase and (2) –b shows the nonhydration region without any crystalline phase.

3.2 Products by Hydrothermal Treatment at 100°C and 200°C

Table 2 shows the products of CA_n in dilute phosphoric acid solutions by hydrothermal treatment at 100°C and at 200°C under the pressure of 10 MPa. Hydration products in this system were C_3AH_6 and AH_3 (gibbsite) at 100°C, and C_3AH_6 and AH (boehmite) at 200°C. The formation of various phosphate phases were strongly dependent on the pH of solutions. $Ca_{10}(PO_4)_6(OH)_2$ was produced in alkaline or neutral

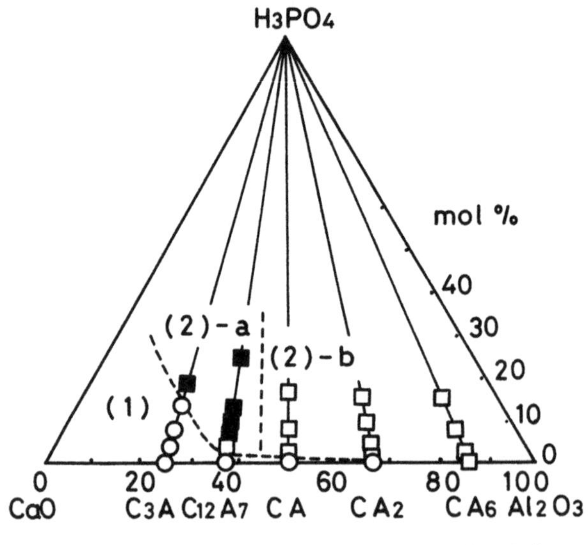

Fig. 2 Reaction products in the system $CaO-Al_2O_3-H_3PO_4(H_2O)$ at 35°C for 30 days. ○: Hydrates, ■: $CaHPO_4 \cdot 2H_2O + CA_n$, □: CA_n only.
(1) : Hydration region
(2) : Nonhydration region
–a : $CaHPO_4 \cdot 2H_2O$ phase
–b : amorphous phase

Table 2. The compounds produced in the system CA_n-dilute phosphoric acid solutions at 35°C, 100°C and 200°C

Temperature (°C)	Formula	Name
35, 100, 200	$Ca_3Al_2(OH)_{12}$	Tricalcium aluminium hexahydrate
35, 100	$Al(OH)_3$	Gibbsite
200	$AlOOH$	Boehmite
35	$CaHPO_4 \cdot 2H_2O$	Brushite
100, 200	$CaHPO_4$	Monetite
100, 200	$Ca_{10}(PO_4)_6(OH)_2$	Hydroxylapatite
100	$AlPO_4 \cdot 2H_2O$	Variscite
200	$AlPO_4$	Aluminium phosphate

Table 3. Free energy change (ΔG) for the reactions in the system CaO-Al_2O_3-H_3PO_4 (H_2O)

Reaction	ΔG_{298} (kcal/mol H_2O)
$3CAO \cdot Al_2O_3 + 6H_2O = 3CaO \cdot Al_2O_3 \cdot 6H_2O$	−6.65
$12CaO \cdot 7Al_2O_3 + 33H_2O = 4(3CaO \cdot Al_2O_3 \cdot 6H_2O) + 3(Al_2O_3 \cdot 3H_2O)$	−3.58
$3(CaO \cdot Al_2O_3) + 12H_2O = 3CaO \cdot Al_2O_3 \cdot 6H_2O + 2(Al_2O_3 \cdot 3H_2O)$	−1.82
$3(CaO \cdot 2Al_2O_3) + 21H_2O = 3CaO \cdot Al_2O_3 \cdot 6H_2O + 5(Al_2O_3 \cdot 3H_2O)$	−0.89
$Al_2O_3 + 3H_2O = Al_2O_3 \cdot 3H_2O$	−0.04
$3CaO \cdot Al_2O_3 \cdot 6H_2O + 3H_3PO_4 = 3CaHPO_4 + Al_2O_3 \cdot 3H_2O + 6H_2O$	−28*
$10(3CaO \cdot Al_2O_3 \cdot 6H_2O) + 18H_3PO_4 = 3Ca_{10}(PO_4)_6(OH)_2 + 10Al_2O_3 \cdot 3H_2O + 54H_2O$	−46*
$Al_2O_3 \cdot 3H_2O + 2H_3PO_4 = 2AlPO_4 + 6H_2O$	−5*

* (kcal/mol H_3PO_4)

solutions, $CaHPO_4$ in neutral and acidic solutions. In addition Lucin-type $AlPO_4 \cdot 2H_2O$ was formed in neutral solutions and Messbach-type $AlPO_4 \cdot 2H_2O$ in acidic solutions at 100°C, and Tridymite form $AlPO_4$ in neutral solutions and Cristobalite form $AlPO_4$ in acidic solutions at 200°C respectively.

3.3 Thermodynamic Aspects of the Products

To predict or confirm stable products, thermodynamic calculations were made for the products in the system CA_n-dilute phosphoric acid solutions. Table 3 shows that the hydration reactivity of CA_n decreased with the decrease of its CaO/Al_2O_3 ratio. While a CA_n reacts with phosphoric acid, the CA_n is difficult to react with H_2O. AH_3 slightly reacts with H_3PO_4 to form $AlPO_4$. Nevertheless, it is likely that $AlPO_4$ and AH_3 coexist in neutral solutions since the solubility product of AH_3 is much lower than that of $AlPO_4$.

3.4 Relation between Hydrates and Phosphates

Fig. 3 shows the changes of hydration region of CA_n in dilute phosphoric acid solutions at 35°C, at 100°C and at 200°C after 1 day, 10 days or 30 days, respectively. CA_6 and α-A did not form any hydrates even after hydrothermal treatments at 200°C. The hydration reactivity of CA_n decreased with the decrease of its CaO/Al_2O_3 ratio, so

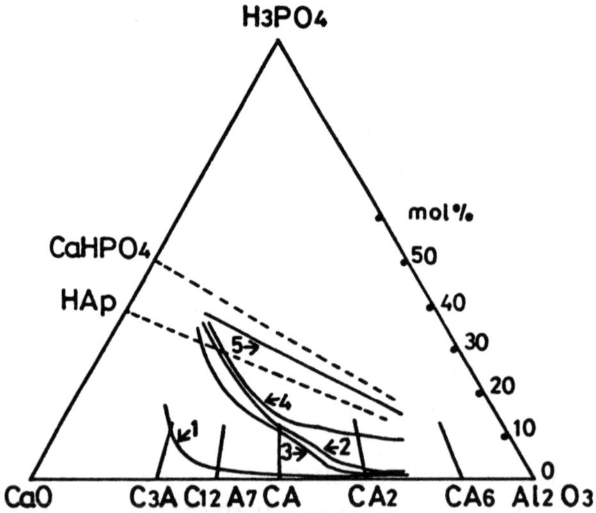

Fig. 3 The region where phosphoric acid inhibits the hydration reaction of CA_n.
1. curing 30 days at 35°C
2. curing 30 days at 100°C
3. curing 1 day at 200°C
4. curing 10 days at 200°C
5. curing 30 days at 200°C

the hydration of CA and CA_2 at 100°C could not completely proceed even after 30 days by the addition of a phosphoric acid.

Fig. 4 and Fig. 5 shows the changes of the products of Ca-phosphates and Al-phosphates respectively by the hydrothermal treatments at 200°C 10 MPa. The results at 100°C and at 200°C showed a similar tendency with a few exceptions, i.e., the formation of $AlPO_4 \cdot 2H_2O$ or AH_3 at 100°C and $AlPO_4$ or AH at 200°C. These figures show that the formation lines of $CaHPO_4$ and $AlPO_4$ tend to shift to a high-H_3PO_4 side in CA and CA_2 composition with the increase of hydrothermal-treating durations. In the region where the reaction with H_3PO_4 occured rather than hydration, $CaHPO_4$ or AH was tended to form instead of HAp or $AlPO_4$. These results at 200°C were summarized as follows;

(1) hydration in low-H_3PO_4 (in alkaline solutions)

$$CA_n \xrightarrow{H_3PO_4,\ H_2O} C_3AH_6 + HAp + AH$$

(2) in the case that hydration of CA_n was inhibited in high-H_3PO_4 (in neutral or weak

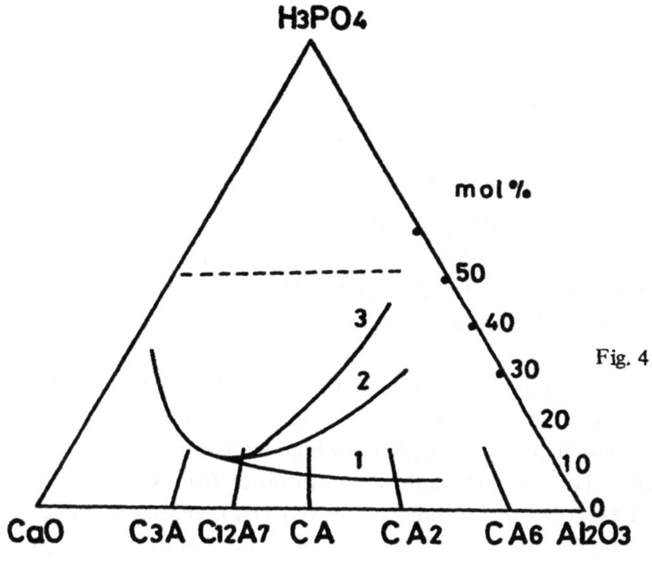

Fig. 4 Changes of the products of Ca–phosphate with hydrothermal treating durations at 200°C 10MPa.
line 1,2 and 3: the boundary between HAp and HAp+$CaHPO_4$, after 1 day, 10 days and 30 days respectively.

(- - - -) : the boundary between $CaHPO_4$ and $CaHPO_4$ + HAp.

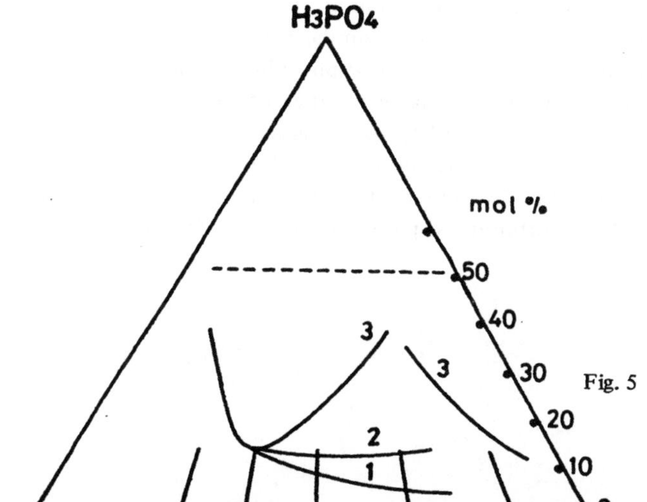

Fig. 5 Changes of the products of Al–phosphate with hydrothermal treating durations at 200°C 10MPa.
line 1,2 and 3: the formation boundary of tridymite form $AlPO_4$, after 1 day, 10 days and 30 days, respectively.

(- - - -) : the boundary between berlinite and berlinite + tridymite form $AlPO_4$ (+ cristobalite form $AlPO_4$).

acidic solutions)

$$CA_n \xrightarrow{H_3PO_4,\ H_2O} CA_n + CaHPO_4 + AlPO_4$$

(3) in the case that CA_n hydrated in high-H_3PO_4 (in weak alkaline or neutral solutions)

$$CA_n \xrightarrow{H_3PO_4,\ H_2O} C_3AH_6 + AH + HAp + CaHPO_4\ (+ AlPO_4)$$

(4) in high-H_3PO_4, amount of CaO < amount of H_3PO_4,

$$CA_n \xrightarrow{H_3PO_4,\ H_2O} AH + HAp + CaHPO_4 + AlPO_4.$$

These results and thermodynamical data suggested that the formation region of C_3AH_6 extended to the line from $CaHPO_4$ to Al_2O_3 and that since C_3AH_6 did not exist in higher H_3PO_4 side of this line, there coexisted AH_3 and $AlPO_4 \cdot 2H_2O$ or AH and $AlPO_4$.

3.5 SEM Observations

SEM photographs of the typical products in the system CA_n-H_3PO_4 (H_2O) at 35°C, 100°C and 200°C are shown in Fig. 6. In the hydration at 35°C, the main hydrate in the early stage consisted of hexagonal plates of C_2AH_8, which converted to rounded grains of C_3AH_6 in the later stages. (a), a micrograph of the product from $C_{12}A_7$ cured for 10 days at 35°C, shows the hexagonal or irregular plates (C_2AH_8), gel-like materials (AH_3) and polyhedra (C_3AH_6). The C_3A cured in low-phosphoric acid solutions at 35°C formed a hardened body since gel-like materials probably interlock many round grains of C_3AH_6 (b). The gel-like materials were identified with AH_3 (gibbsite) by TG-DTA but no Ca-phosphate produced by such a reaction, $C_3A \xrightarrow{H_3PO_4,\ H_2O} C_3AH_6 + AH_3 +$ Ca-phosphate. The hydration of CA_n was inhibited by a certain amount of phosphoric acid. Plate (c) shows the C_3A remaining unhydrated and the large needles of $CaHPO_4 \cdot 2H_2O$. Plate (d) shows $CaHPO_4 \cdot 2H_2O$ crystals. On the hydration at 100°C for C_3A (e) and at 200°C for CA_2 (f), the rounded grains of C_3AH_6 were similarly formed in pure water. Micrograph (g) of the product from $C_{12}A_7$ treated in low phosphoric acid solutions at 100°C, shows gel-like materials (AH_3 or AH), a large number of needles (HAp) and polyhedra (C_3AH_6). A very fine powder of $CaHPO_4$ was produced in the high-H_3PO_4 region, (h), while a large crystal of HAp was produced in the low-H_3PO_4 region, (i). The micrograph (j) of the product from CA_2 cured at 35°C in high-phosphoric acid solution, shows a hardened body by amorphous Al-phosphate bonds. The micrographs

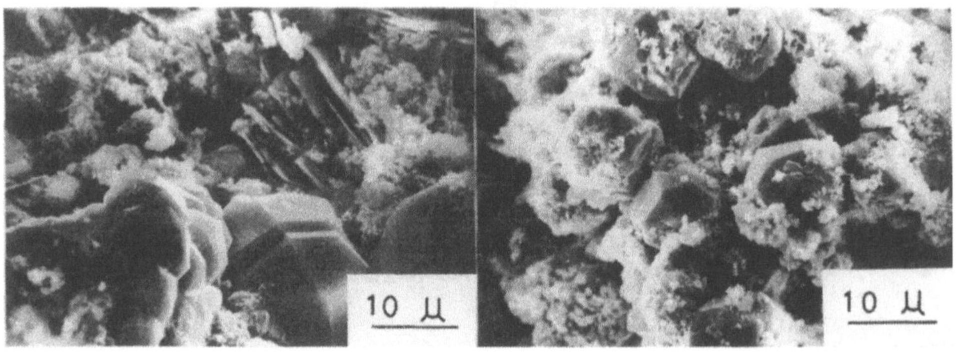

(a) (b)

Photo 1 SEM photographs of some typical products indilute phosphoric acid solutions at 35°C, 100°C and 200°C from various starting compounds.

(c) (d)

(e) (f)

(g) (h)

Photo 1 (cont'd)

(k) or (l) and (m) of the specimens treated hydrothermally in high phosphoric acid solutions at 100°C or at 200°C, show a continuous hardened body or aggregation of small crystallines weakly bonded by $AlPO_4 \cdot 2H_2O$ or $AlPO_4$.

3.5 Behavior of Phosphoric Acid for Hydration

At 35°C, low concentration of phosphoric acid retarded both the hydration of CA_n

(i)

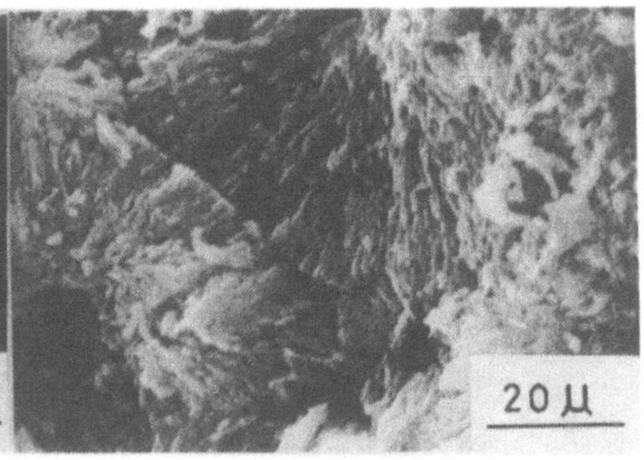

(j)

(k)

(l)

(m)

Photo. 1 (Contd.)

and the transformation of C_2AH_8 into C_3AH_6, whereas the addition of a still more amount of phosphoric acid inhibited hydration completely even after 30 days at 35°C and at 100°C. A specific critical ratio of H_3PO_4/CA_n, above which the hydration of CA_n could not proceed, existed for each CA_n compound. The reason of this retardation or inhibitation is still unknown.

Murakami et al.[11] suggested that the retarding action of the phosphoric acid was caused by the formation of protective film of insoluble calcium phosphates resulted from the reaction between phosphoric acid and the high-lime surface of cement grains. However, SEM micrographs showed that Ca-phosphates had a tendency to crystallize separately while Al-phosphate had a tendency to form an amorphous hardened body or to crystallize as an aggregated monolithic body. It suggests, therefore, that phosphoric acid forms $AlPO_4$-like materials on the surface of each CA_n and these insoluble layers retard or inhibit the hydration of CA_n. When these insoluble layers of Al-phosphates were insufficient to cover completely the surface of the CA_n grains, the hydration would proceed. It can interpret the fact that Ca-rich CA_n has the tendency of the hydration. Calcium ion might gradually break Al-O-P linkage as shown by Zachariasen rules[12] so that less Ca-rich CA_n would also hydrate at higher temperatures or after longer durations. At higher temperatures, the amorphous Al-phosphate layers crystallized in the form of Al-phosphates such as $AlPO_4 \cdot 2H_2O$ or $AlPO_4$. Thus, the hydration proceeds completely.

The phase assemblages observed in the present work at 200°C under 10 MPa have been different from those in the recent work in the system $CaO\text{-}Al_2O_3\text{-}P_2O_5\text{-}H_2O$ at 200°C under $\simeq 1.71$ MPa by Yuh and Rockett.[13] This might be caused from the differences in the pressure and the starting materials.

References

1) W. Eitel: "Silicate Science" V-Ceramics and Hydraulic Binders, (1966) p. 273–319.
2) "Yogyo-Kogaku handbook", Yogyo Kyokai, (1973) p. 1639–62.
3) A. C. C. Tseung and K. G. Carruthers: Trans. Brit. Ceram. Soc., **62**, 305–21 (1963).
4) H. F. W. Taylor: "The Chemistry of Cements", (1964) p. 256–58.
5) T. D. Robson: "High Alumina Cements and Concretes", (1962) p. 125–42.
6) K. Yamazaki, Y. Yamaguchi, M. Yoshimura and S. Sōmiya: Yogyo-Kyokai-Shi, **89**, 31–40 (1981).
7) K. Yamazaki, Y. Yamaguchi, T. Nakazawa, M. Yoshimura and S. Sōmiya: ibid., **89**, 661–68 (1981).
8) K. Yamazaki, Y. Yamaguchi, T. Nakazawa, M. Yoshimura and S. Sōmiya: ibid., **90**, 56–63 (1982).
9) K. Yamazaki, Y. Yamaguchi, M. Yoshimura and S. Sōmiya: ibid., **89**, 444–451 (1981).
10) K. Yamazaki, Y. Yamaguchi, T. Nakazawa, M. Yoshimura and S. Sōmiya: Nippon-Kagaku-Kaishi, 1422–28 (1981).
11) K. Murakami, H. Tanaka, K. Sato, K. Hashimoto and K. Ichimura: Yogyo-Kyokai-Shi, **78**, 253–63 (1968).
12) for example, W. D. Kingery: "Introduction to Ceramics" John Wiley & Sons Inc., (1975) p. 91–124.
13) Shang Jung Yuh and T. J. Rockett: J. Am. Ceram. Soc., **64**, 491–93 (1981).

Reprinted from *Journal of Nuclear Science and Technology*, **21**(1) (1984) 71–73.

Immobilization of Radioactive Wastes in Hydrothermal Synthetic Rock

Lithification of Silica Powder

Nakamichi YAMASAKI,
Kazumichi YANAGISAWA,
Satoshi KANAHARA, Mamoru NISHIOKA,
Kiyoshi MATSUOKA and Jumei YAMAZAKI

*Research Laboratory of Hydrothermal Chemistry, Faculty of Science, Kochi University**

Received April 11, 1983
Revised September 22, 1983

KEYWORDS: hydrothermal hot-pressing, hydrothermal synthetic rock, lithification, immobilization, radioactive wastes, silica powder, mechanical strength

As the use of the nuclear power is increased, the safe disposal of highly radioactive wastes has been required. It needs to immobilize the wastes in stable and dense solid(1)(2). The most popular procedure being incorporate the wastes into borosilicate glass is thermodynamically unstable relative to crystalline phases. Immobilization of the wastes in crystalline phases has been researched on supercalcine(3)(4) and synthetic rock(2)(4). However, the high temperature and the long reaction time are necessary to immobilize the wastes by these methods, because of the slow reaction in viscous melt or solid state of silica, titania, zirconia and other oxides.

Silica and silicates are naturally yielded and are most popular minerals. In this study, silica was chosen to be lithified under hydrothermal conditions. Silica is the main constituent of additives to immobilize the radioactive wastes in stable solids.

The possibility of applying the hydrothermal method to the immobilization of the radioactive wastes is of main interest in this note. It is expected that the wastes react with additives at much lower temperature under hydrothermal conditions and form hardened crystalline matrix by the hydrothermal hot-pressing technique(5). Moreover, the hydrothermal reaction is in closed system, which prevents evaporation of volatile materials.

1. Experimental

Starting materials were silicastone (from Fukushima Prefecture, Japan) and siliceous sinter (from Beppu City, Japan). Silicastone was low-quartz, and siliceous sinter was composed of noncrystalline silica (80%) and low-cristobalite (20%). They were powdered to pass through 300 mesh and mixed well in a ball mill. The simulated radioactive waste used was JW-D (Japan Atomic Energy Research Institute). The waste (28 %) was added to the silica mixture containing 70% silicastone.

The apparatus for the hydrothermal hot-pressing was shown in **Fig. 1**. The autoclave (Fig. 1(a)) was made of Fe with a liner of Hastelloy C and could be rapidly heated by an induction heater. The hot-press cell (Fig. 1(b)) made of Hastelloy C was composed of the inner and outer cases. The inner case 1.4 cm in inner diameter was divided into three parts in order to remove the content of the cell without its destruction. The content of the cell was allowed to be compressed through the gland packing by the push rod.

Sodium hydroxide solution was added to the mixture of the starting materials in the ratio of 1 cm^3 solution to 4 g mixture. The mixture was kneaded well, stuffed into the hot-press cell and pressed vertically at 45 kg/cm^2. The cell was put into the autoclave, in which distilled water or NaOH solution had been poured. When the content of the cell was not compressed during the hydrothermal treatment, 15 cm^3 of NaOH solution was poured to confirm the wetting of silica grains. As a result, the cell was in liquid phase. Under continuously compressed conditions, the addition of a large amount of a solution resulted in the flowout of the silica paste into the solution. Under these conditions, 1 cm^3 of water was added not to dry

* *Akebono-cho, Kochi-shi 780.*

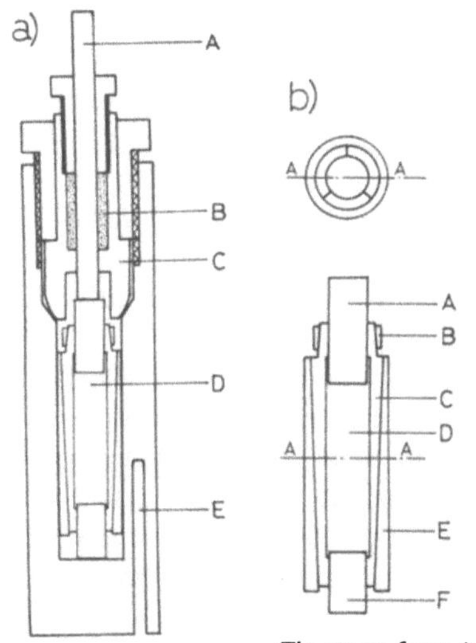

The upper figure is the cross section of AA.

A: Push rod
B: Gland packing
C: Cone
D: Hot press cell
E: Thermocouple well

A: Upper rod
B: Ring
C: Inner case
D: Raw material
E: Outer case
F: Lower rod

(a) Hot-press autoclave (b) Hot-press cell

Fig. 1 Apparatus for hydrothermal hot-pressing

the surface of the silica paste. The cell was in vapor phase.

After the autoclave was sealed, it was heated to a proposed temperature at the rate of 30°C/min. The temperature was kept for 30 min. During the hydrothermal treatment, the content of the cell was compressed at 0 or 160 kg/cm^2.

The autoclave was cooled to room temperature by an electric fan. The lithified silica was taken out from the cell, soaked in boiling water for 30 min, and dried at 120°C. It was identified by X-ray powder diffraction method and its surface was observed by an optical microscope. The compressive strength of the lithified silica (1.6 cm long and 1.4 cm diameter) was measured. The bending strength was measured by three point bending.

2. Results and Discussion

(1) Uncompressed Conditions

The effect of the concentration of NaOH solution on the compressive strength was shown in **Fig. 2**. As the concentration was increased from 3 to 5 N, the strengh was increased remarkably. The strength had a tendency to increase with the treatment temperature. It was, however, reduced when silica was lithified at high temperature above 300°C in the case of the concentrated solution above 10 N. Under these conditions, a large amount of silica was dissolved from the cell into the solution and recrystallized at the bottom of the autoclave. These results suggested that the solubility of silica into the concentrated NaOH solution at high temperature might be too increased for recombination of silica grains.

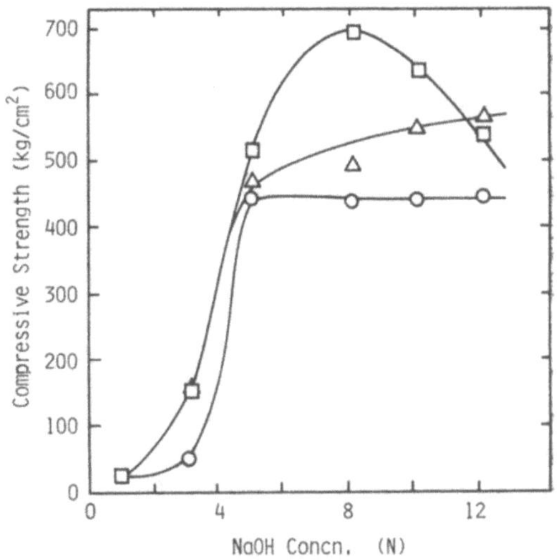

□: 300°C, △: 250°C, ○: 200°C

Fig. 2 Effect of NaOH concentration on compressive strength of lithified silica (50% silicastone)

(2) Hydrothermally Hot-pressed Conditions

The mechanical strength of the hydrothermally hot-pressed silica was increased with the treatment temperature (**Fig. 3**). The hot-pressed products had the higher strength than those produced under uncompressed conditions. It was considered that continuous compressing reduced pores and kept silica grains in contact with each other.

The microstructure of the lithified silica **Photo. 1** showed that the polished surface was dotted with large grains of silicastone. X-ray powder diffraction patterns of the hydrothermally lithified silica were the same to those of starting materials. The sample

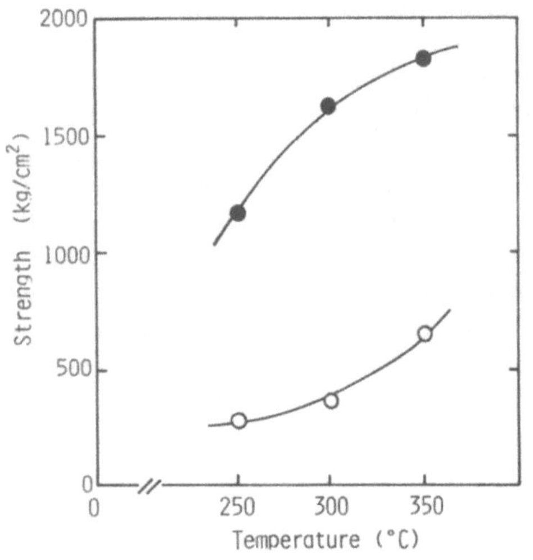

Fig. 3 Compressive strength (●) and bending strength (○) of silica (50% silicastone) hot-pressed by use of 5 N NaOH

composed of 50% silicastone had higher strength than the end members, silicastone and siliceous sinter. These results suggested that the hydrothermal lithification of silica was explained by the dissolution-deposition of silica in NaOH solutions. Siliceous sinter seemed to be more soluble than silicastone, because it was much finer powder[6]. It was concluded that siliceous sinter contributed to linking of the silicastone particles.

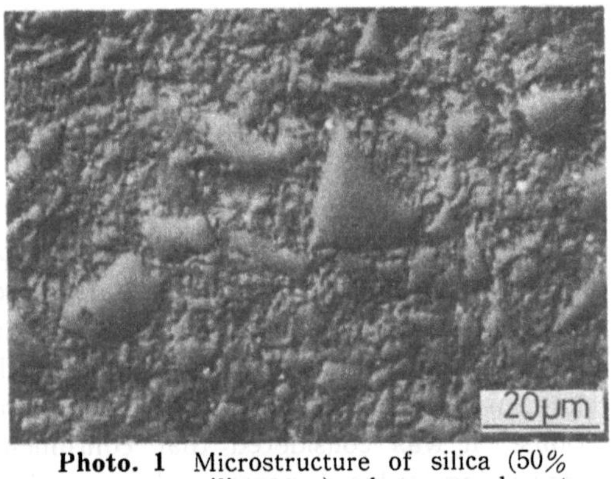

Photo. 1 Microstructure of silica (50% silicastone) hot-pressed at 300°C by use of 5 N NaOH

(3) Immobilization of Simulated Highly Radioactive Waste

The starting material (28% waste) kneaded with 5 N NaOH solution was hot-pressed 250°C for 30 min. The compressive strengh of the product was 890 kg/cm^2, which was higher than that of borosilicate glass or supercalcine[4]. Its X-ray powder diffraction patterns showed that the waste reacted with silica to produce new compounds. The phase identified was only low-quartz, at present. It is considered that new phases, probably silicates produced by the reaction of the waste with silica, contributed to linking of silicastone grains.

Strontium, which has the long half life, is one of the most dangerous elements in the radioactive waste, and its behavior is noticeable in the waste form. After the hydrothermal hot-pressing, no Sr was detected in the water added into autoclave. In order to confirm the immobilization of Sr, the hydrothermal reaction of strontium hydroxide and silica (silicastone) was carried out in NaOH solution. Sodium hydroxide played a role as a mineralizer and produced no crystalline compounds. When the mixture of 40 mol% Sr was heated above 250°C under alkaline hydrothermal conditions, strontium silicates ($SrSiO_3$ and Sr_2SiO_4) were produced and no Sr was detected in the solution. These results indicated that Sr in the radioactive waste could be immobilized by the reaction with silica.

It has been shown that the hydrothermal hot-pressing technique is available for the lithification of silica containing the simulated radioactive waste. Farther experiments on the optimum conditions to lithify the silica containing the waste, the immobilization mechanism of the waste into the silica matrix including the behavior of Cs, and its leaching resistance are now in progress.

The authors wish to thank Dr. S. Muraoka (Japan Atomic Energy Research Institute) for offering the simulated waste (JW-D).

—REFERENCES—

(1) TERAI, R.: *Ceramics Jpn.*, 11, 226 (1976).
(2) RINGWOOD, A. E.: *"Safe Disposal of High Level Nuclear Reactor Wastes"*, (1978), Australian Nat. Univ. Press.
(3) TSUNODA, N.: *Ceramics Jpn.*, 12, 913 (1977).
(4) TERAI, R.: *ibid.*, 16, 916 (1981).
(5) MATSUOKA, K., *et al.*: *Rep. Res. Lab. Hydrothermal Chem.*, 3, 8 (1979).
(6) OSTWALD, W.: *Z. Phys. Chem.*, 34, 503 (1900).

Reprinted from *Journal of Nuclear Science and Technology*, **21**(7) (1984) 558–560.

Immobilization of Radioactive Wastes in Hydrothermal Synthetic Rock, (II)

Hydrothermal Synthesis of Pollucite

Kazumichi YANAGISAWA, Satoshi KANAHARA, Mamoru NISHIOKA and Nakamichi YAMASAKI

*Research Laboratory of Hydrothermal Chemistry, Faculty of Science, Kochi University**

Received February 13, 1984

KEYWORDS: pollucite, hydrothermal synthesis, cesium, immobilization, siliceous sinter, radioactive wastes

The last note[1] revealed the hydrothermal hot-pressing technique was capable for the immobilization of radioactive wastes. By this technique, the simulated high-level radioactive waste reacted with silica mixture of silicastone and siliceous sinter, and simultaneously formed a hardened body by hot-pressing under alkaline hydrothermal conditions. It is difficult to understand clearly the reaction of the waste with silica matrix, because the waste has many components. It is one of the meaningful methods for understanding it to clarify the stable products by the reaction of each element in the waste with silica matrix under alkaline hydrothermal conditions. It was confirmed that Sr, one of the most dangerous elements in the waste, reacted with silica under hydrothermal conditions to produce strontium silicate ($SrSiO_3$)[1]. This note describes the hydrothermal reaction of Cs, another dangerous element, with silica matrix in NaOH solution. Pollucite, a ceramic host structure for Cs, was produced by the reaction of Cs with siliceous sinter which was amorphous silica containing Al.

1. Experimental

Siliceous sinter from Kagoshima Prefecture was used as a starting material. It was amorphous, according to X-ray diffraction (Fig. 1). The result of its chemical analysis was shown in **Table 1**.

Table 1 Chemical analysis of siliceous sinter (%)

SiO_2	Al_2O_3	Fe_2O_3	CaO	MgO	Other oxides	Ignition loss
71.7	12.9	2.3	2.4	0.4	5.7	4.6

Powders of siliceous sinter and $CsNO_3$ (99%, Mitsuwa's Pure Chem.) were put into a microautoclave, together with NaOH solution (4 cm^3). The autoclave[2] was lined by Hastelloy-C and its inner volume was 10 cm^3. It was heated by an induction heater to a desired temperature, and the temperature was kept constant for 10 min. The heater[2] was rocked back and forth 40 times per minute, in order to stir the content in the autoclave.

After the hydrothermal treatment, the content in the autoclave was washed well with warm water, filtered and dried in air. It was observed by an optical and a scanning electron microscope with energy dispersive X-ray attachment interfaced to computer, and identified by X-ray powder diffraction method.

2. Results and Discussion

When siliceous sinter and $CsNO_3$ reacted in 5-N NaOH solution under the hydrothermal conditions above 200°C, reaction products were white powder and translucent lump. The lump was water glass, because it was dissolved in warm water. The X-ray diffraction patterns of the white powder (**Fig.** 1) were well agreed with the published data of pollucite[3]. The photograph (**Photo.** 1) of the white powder produced at 300°C from siliceous sinter 1 g and $CsNO_3$ 0.4 g in 5-N NaOH solution, showed that the powder was composed of bulky and needle crystals. The results of their energy dispersive X-ray analysis (Fig. 2) revealed that bulky crystals consisted mainly of Na, Cs, Al and Si, and

* *Akebono-cho, Kochi-shi 780.*

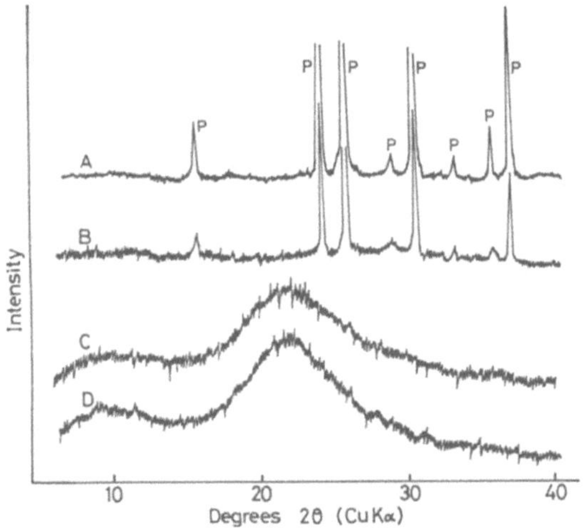

Peaks marked with P are agreed with the published data of pollucite.

Fig. 1 X-ray diffraction patterns of white powder produced from siliceous sinter 1 g and $CsNO_3$ 0.2 g in 5-N NaOH soln. for 10 min at 300°C (A), 200°C (B) and 100°C (C); Starting siliceous sinter is (D).

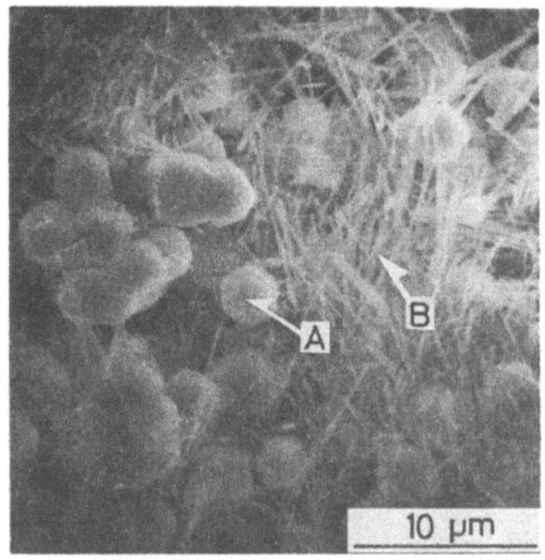

A bulky crystal (A) and needle crystals (B) were analyzed by energy dispersive X-ray analysis.

Photo. 1 SEM photograph of white powder produced from siliceous sinter 1 g and $CsNO_3$ 0.4 g in 5-N NaOH soln. at 300°C for 10 min

needle crystals of Na, Ca and Si. Needle crystals are not suitable for the analysis, because another substance may exist behind them. A little amount of Cs and Al detected in needle crystals might come from bulky crystals behind them. Bulky crystals must be pollucite and needle crystals might be sodium calcium silicate. Needle crystals were produced at 300°C, but not below 200°C. No X-ray diffraction from needle crystals was observed, because their amount was very small in comparison with bulky crystals. The bulky crystal in **Fig. 2** included 38 a% Na against Cs. It was considered to be a solid solution between pollucite $CsAlSi_2O_6$ and analcime $NaAlSi_2O_6 \cdot H_2O$, though no evidence was shown by X-ray diffraction patterns. It had been reported that a solid solution of pollucite with analcime was indistinguishable from pollucite by X-ray diffraction[4].

Pollucite was produced above 200°C, but not at 100°C (see Fig. 1). In dilute NaOH solution (1-N), pollucite was not produced even at 300°C, but siliceous sinter remained.

These results suggested that the hydrothermal reaction to produce pollucite was controlled by the dissolution of siliceous sinter.

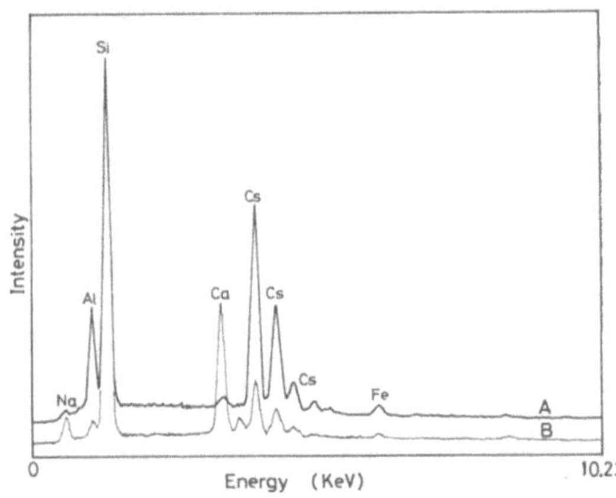

Fig. 2 Energy dispersive X-ray analysis of a bulky crystal (A) and needle crystals (B) in Photo. 1

Pollucite produced at 300°C from siliceous sinter 1 g and $CsNO_3$ 0.2 g in 5-N NaOH solution, included 55 a% Na against Cs. The amount of Cs in the pollucite was reduced to half in that produced from $CsNO_3$ 0.4 g. The following relation was observed; Cs/Na in a starting solution: Cs/Na in pollucite= 1:16. It means that Cs is concentrated in the solid phase from a mixed solution of Cs and Na.

Cesium aluminosilicates such as $CsAlSi_2O_6$ (pollucite), $CsAlSi_5O_{12}$ and $CsAlSiO_4$ are possible hosts for immobilization of Cs in radioactive wastes[(4)~(10)]. It has been pointed out that pollucite is a better host than the others for Cs immobilization based on its high resistance to alteration under hydrothermal conditions[(9)(10)]. Pollucite is crystallized, when aluminosilicates reacted with Cs[(11)(12)], or when ceramic nuclear waste forms were altered under hydrothermal conditions[(13)]. Besides, pollucite is the naturally occurring mineral. It is concluded that pollucite is one of the most stable phases under depository conditions and is preferable phase for Cs immobilization.

This study revealed that pollucite was produced by the hydrothermal reaction of Cs with siliceous sinter in NaOH solution. Siliceous sinter and NaOH solution were used as silica matrix and sclerotic, respectively, in the hydrothermal hot-pressing technique for the immobilization of radioactive wastes. The technique may be able to immobilize Cs in radioactive wastes into pollucite structure.

The authors wish to express their thanks to Análysis Center in Horiba Co. for energy dispersive X-ray analysis.

—References—

(1) Yamasaki, N., Yanagisawa, K., Kanahara, S., Nishioka, M., Matsuoka, K., Yamazaki, J.: *J. Nucl. Sci. Technol.*, 21[1], 71 (1984).

(2) Yamasaki, N., Yasui, T., Mastuoka, K.: *Environ. Sci. Technol.*, 14, 550 (1980).

(3) *ASTM* 29-407.

(4) Komarneni, S., White, W.B.: *"Scientific Basis for Nuclear Waste Management"*, Vol. 3, (Moore, J.G., ed.), p. 387~96 (1981), Plenum, New York.

(5) Gallargher, S.A., McCarthy, G.J., Smith, D.K.: *Mater. Res. Bull.*, 12, 1183 (1977).

(6) McCarthy, G.J.: *Nucl. Technol.*, 32, 92 (1977).

(7) Strachan, D.M., Shulz, W.W.: *Am. Ceram. Soc. Bull.*, 58, 865 (1979).

(8) Adl, T., Vance, E.R.: *J. Mater. Sci.*, 17, 849 (1982).

(9) Komarneni, S., Roy, R.: *J. Am. Ceram. Soc.*, 66, 471 (1983).

(10) Komarneni, S., White, W.B.: *Nucl. Chem. Waste Manage.*, 3, 169 (1982).

(11) McCarthy, G.J., Komarneni, S., Scheetz, B.E., White, W.B.: Ref. (4), Vol. 1, (McCarthy, G.J. ed.), p. 329~39 (1979).

(12) Komarneni, S., White, W.B.: *Clays Clay Miner.*, 29, 299 (1981).

(13) Sasaki, S., Komarneni, S., Roy, R.: *Am Ceram. Soc. Bull.*, 62, 649 (1982).

Reprinted from *Journal of Nuclear Science and Technology*, **23**(6) (1986) 550–558.

Immobilization of Radioactive Wastes in Hydrothermal Synthetic Rock, (III)

Properties of Waste Form Containing Simulated High-Level Radioactive Waste

Kazumichi YANAGISAWA, Mamoru NISHIOKA and Nakamichi YAMASAKI

*Research Laboratory of Hydrothermal Chemistry, Faculty of Science, Kochi University**

Received November 8, 1985

The high-level radioactive waste form (21.8% waste, 47.7% α-quartz, 20.5% amorphous aluminosilicate, 10% $Al(OH)_3$) was produced with the addition of 10-N NaOH solution under hydrothermal hot-pressing conditions at 350°C and 66 MPa for 6 h. In order to characterize the waste form, following tests and measurements were performed: Soxhlet flow leach tests (97°C), static leach tests in deionized water at low temperatures (40, 90°C) and under hydrothermal conditions (100∼350°C), crystalline phase determination, microstructure observation, compressive strength measurement, weight loss measurement by heat treatment and thermal conductivity measurement.

The waste form was mainly composed of α-quartz and had porous structure. As for leach rates determined by static low temperature leach tests, the leach rates of the waste form were much lower than a concrete form but slightly higher than a glass form. The waste form was stable under hydrothermal conditions in comparison with a glass form. It had high compressive strength and thermal resistance. Its thermal conductivity was higher at room temperature than that of a glass form.

KEYWORDS: ***chemical properties, comparative evaluations, high-level radioactive wastes, hydrothermal hot-pressing, immobilization, leachability, physical properties, quartz, solidification, thermal conductivity***

I. INTRODUCTION

The most generally accepted concept for disposal of high-level radioactive waste is to incorporate it into a solid waste form that is placed into a deep geologic repository[1]. The waste form is a first barrier in multiple barrier system.

Many kinds of the waste forms have been proposed. Immobilization of high-level radioactive waste into borosilicate glass is considered to be the most realizable process. It is, however, obvious that glass is thermodynamically unstable relative to an assemblage of crystalline phases[2]. Ceramic waste forms, for example, supercalcine[3][4], SYNROC[2][5] and high-alumina tailored ceramics[6][7], have been researched to immobilize the waste into crystalline phases. Production of glass and ceramic waste forms requires high temperatures above 1,000°C, which may cause evaporation of volatile radioactive isotopes and corrosion of equipments. Concrete waste form is appealing because of its ease of processing, though it has rather high leach rates of Cs and Sr[8][9].

The last note[10] revealed that a simulated high-level radioactive waste could be solidified into a silica matrix by hydrothermal hot-pressing technique. This technique has

* *Akebono-cho, Kochi-shi 780.*

following advantage: (1) simplicity of an apparatus, (2) ease of operation, (3) low reaction temperatures up to 350°C, (4) closed system reaction (solidification in an autoclave), which prevents evaporation of volatile materials, (5) immobilization into stable crystalline phases under hydrothermal conditions (for instance, Sr→$SrSiO_3$, Cs→$CsAlSi_2O_6$)(10)(12).

In order to produce a waste form with high mechanical strength and low leachability, the composition of the silica matrix and the optimum conditions for hydrothermal hot-pressing have been determined(11). In this paper, properties of the waste form produced under the optimum conditions, especially leachability, are studied.

II. EXPERIMENT

1. Production of Waste Form

A silica matrix was a mixture of silicastone powder passed through 300 mesh (α-quartz, 99.5% SiO_2, from Fukushima Prefecture), siliceous sinter (amorphous and friable rock, 72% SiO_2 -13% Al_2O_3(12), from Kagoshima Prefecture) and $Al(OH)_3$ (Wako Pure Chem. Ind. Ltd., Osaka, Japan). The composition of a simulated high-level radioactive waste from reprocessing of nuclear fuels is shown in **Table 1**. It was estimated by Banba *et al.* of Japan Atomic Energy Research Institute (JAERI)(13). In the waste, Na was not included because NaOH solution was added as mineralizer for hydrothermal solidification. The waste was produced by mixing the appropriate chemicals in a ball mill.

Table 1 Composition of simulated high-level radioactive waste (%)

Fe_2O_3	39.05	SrO	1.69[a]
MoO_3	8.66	Sm_2O_3	1.64
Nd_2O_3	8.24	MnO_2	1.29
ZrO_2	8.21	TeO_2	1.12[d]
CeO_2	6.77	CoO	1.00
NiO	4.92	Y_2O_3	1.00
Cs_2O	4.88[a]	Rb_2O	0.61[a]
BaO	3.12[c]	Gd_2O_3	0.19
La_2O_3	2.54	CdO	0.16
Pr_6O_{11}	2.47	Ag_2O	0.15[a]
Cr_2O_3	2.26	Sb_2O_3	0.13[b]

Added as [a] nitrate, [b] chloride, [c] carbonate, and [d] metal.

The silica matrix (15.64 g) and the waste (4.53 g) was mixed well to prepare a starting dry powder (47.7% silicastone, 20.5% siliceous sinter, 10% $Al(OH)_3$, 21.8% waste as oxides). The dry powder was kneaded with 10-N NaOH solution (2.5 cm^3) to prepare a starting wet powder. The apparatus for hydrothermal hot-pressing was previously reported(10). The wet powder was transferred into the hot-pressing cell 1.4 cm I.D., and the cell was put into the autoclave. The autoclave was heated to 350°C at the rate of 30°C/min by an induction heater, and the temperature was kept constant for 6 h. During the hydrothermal treatment, the powder in the cell was continuously compressed at 66 MPa. After the treatment, the autoclave was cooled down to room temperature by an electric fan, and a solidified body was taken out of the cell.

2. Leach Test

Three types of leach tests were performed: (1) flow leach test by Soxhlet type apparatus at 97°C, (2) static leach test under hydrothermal condition up to 350°C, (3) modified MCC-1 low temperature leach test at 40 and 90°C.

(1) Flow leach test: Soxhlet type apparatus was constructed of fused silica glass. Cylindrical specimens for the test (1.4 cm in diameter, about 0.8 cm in height) were fabricated by diamond-sawing and dry-abrading. The test was performed for 29 d.

(2) Static leach test under hydrothermal conditions: The autoclave for the test, 27.7 cm^3 in inner volume, was made of Fe and lined by Ti. The leachant was deionized water

($15\,cm^3$). Cubic specimens ($5\times5\times5\,mm^3$) were fabricated by diamond-sawing and dry-abrading, giving a specimen surface area/leachant volume ratio $1/10\,cm^{-1}$. The test was performed under hydrothermal conditions from 100 to 350°C for 1 d. Monolithic glass specimens of JW-A glass (21.8% waste loading, prepared by JAERI) were also used for the test.

(3) MCC-1 static leach test[14]: The leach container, $130\,cm^3$ in inner volume, was made of Teflon. The leachant was deionized water ($40\,cm^3$) and cubic specimens ($8.2\times8.2\times8.2\,mm^3$), fabricated by diamond-sawing and dry-abrading, had surface area of $4\,cm^2$. The test was performed at 40 and 90°C up to 28 d. The leachant was changed to fresh deionized water when weight loss of a specimen was measured, that is, 3, 7, 14 d in leaching time in this modified MCC-1 leach test.

All leach tests were conducted in duplicate. After the leach tests, each leachant was analyzed by atomic absorption spectrometry. Weight loss was calculated by weight change of a specimen dried at 110°C before and after the leach tests, and it was normalized by dividing by geometric surface area of the starting specimen. Elemental mass loss appeared in figures was also normalized by dividing by only geometric surface area of the specimen. Weight loss and elemental mass loss in figures indicated the sum up to a leaching time.

3. Characterization

(1) Phase Determination and Microstructure Observation

Crystalline phases of the waste form produced by hydrothermal hot-pressing were identified by X-ray powder diffraction. A polished section was examined by a scanning electron microscope. The specific surface area was measured by BET method using N_2 gas as an absorbent.

(2) Compressive Strength Measurement

The solidified waste form (1.4 cm in diameter) was cut by diamond-sawing to 1.6 cm long and dried at 110°C to prepare specimens for compressive strength measurement. Uniaxial compression tests were performed at room temperature. The compression axis was perpendicular to the base of the cylindrical specimens. The crosshead travel rate of a mechanical machine was 0.02 cm/min.

(3) Thermal Properties

Thermal diffusivity α and specific heat C_p were measured at 25 to 900°C in vacuum (10^{-3} mmHg) by the laser flash method, and thermal conductivity k was calculated by the equation;

$$k=C_p\,\alpha\,\rho \qquad (\rho\text{: apparent density}).$$

The waste forms were heated from 300 to 900°C for 6 h, and weight loss by evaporation was measured. Soxhlet leach tests of the heated specimens were also performed for 1 d. The waste form heated at 500°C was quenched into water and then its compressive strength was measured.

III. Results and Discussion

1. Leachability

Leach rate and weight loss of the waste form by Soxhlet leach tests are shown in **Fig. 1**. The leach rate was remarkably reduced with the increase in leaching time, because the weight loss up to 6 h was very large. The same phenomenon was observed when FUETAP concretes (concretes formed under elevated temperatures and pressures) were leached[8]. After the leach tests for 29 d, the leach rate reached $5.3\,g/m^2{\cdot}d$, which was higher than

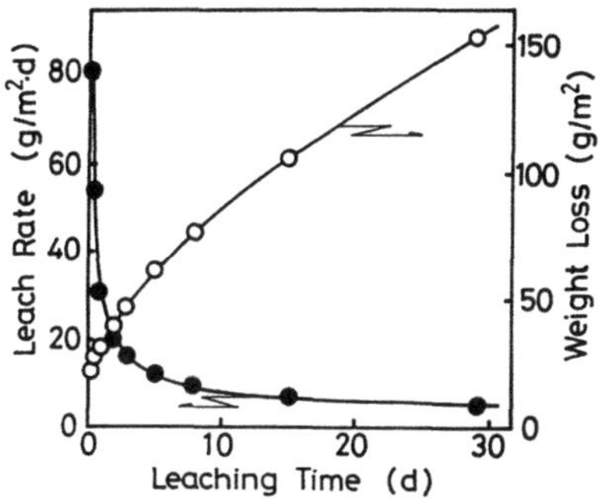

Fig. 1 Leach rate and weight loss by Soxhlet leach tests

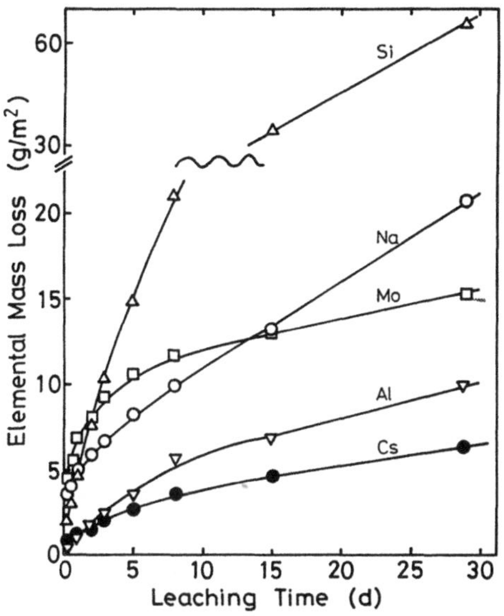

Fig. 2 Elemental mass loss by Soxhlet leach tests

that of a glass form (2 g/m²·d for total weight loss by Soxhlet leach tests)[15]. It may be reduced by further long-time leach tests.

The high leach rate by short-time leach tests resulted from porous structure of the waste form. During hydrothermal treatment, pores of the waste form had been filled with the mineralizer, which might include Na, Si, Al and soluble elements of the waste components. These elements were not immobilized enough only by drying at 110°C before leach tests, and were easily released by an initial washout through porous structure of the waste form.

Elemental mass loss by Soxhlet leach tests is shown in **Fig. 2.** In the leachants, Sr was not detected. The amount of each element released from the waste form by the tests up to 6 h was very large. The release rate of each element was reduced up to 8 d and was constant over 8 d.

Results of MCC-1 leach tests are shown in **Fig. 3.** The weight loss by 3 d leach tests was very large. According to the elemental mass loss, Na and Mo were easily released by the short-time leach tests. In the waste components, MoO_3 might change to sodium molybdate under alkaline hydrothermal conditions.

In comparison with the results of the different leach tests, the weight loss by Soxhlet leach tests was largest, because in the tests leaching temperature was higher than the others and distilled water was continuously refluxed. The weight loss by MCC-1 leach tests at 90°C was higher than that at 40°C. According to the elemental mass loss, release of Si from the waste form contributed greatly to the increase of the weight loss by soxhlet leach tests.

Normalized elemental mass loss is usually given by the following equation[14]:

$$NLi = Mi/Fi/SA,$$

where NL*i*: Normalized elemental mass loss (g/m²)
M*i*: Mass of element "*i*" in leachant (g)
F*i*: Fraction of element "*i*" in unleached waste form (unitless)
SA: Specimen surface area (m²).

It was difficult to estimate F*i* in the waste form produced by hydrothermal hot-pressing, because the waste form included water which evaporated at high temperatures. Assuming that a batch of the starting powder gives a waste form of 20 g (21.8% waste), leach rate of each element (NL*i*/leaching time) are shown in **Table 2.** The leach rate of Na and Mo was large in all leach tests. The leach rate of Cs, Al and especially Si was increased by

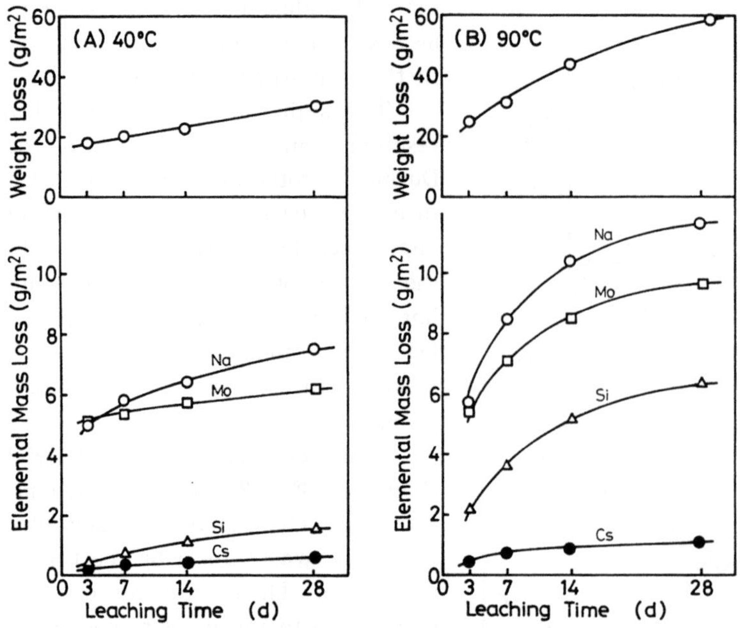

Fig. 3 Weight loss and elemental mass loss by modified MCC-1 leach tests in deionized water at 40°C (A) and 90°C (B)

Table 2 Normalized elemental leach rate ($g/m^2 \cdot d$)

Element	Soxhlet†	MCC-1†† 40°C	MCC-1†† 90°C
Na	24.57	9.37	14.64
Cs	21.47	2.21	3.98
Al	7.11	0.57	1.67
Si	7.76	0.19	0.77
Mo	38.87	17.56	27.42

† Soxhlet leach tests at 97°C for 29 d
†† Modified MCC-1 static leach tests for 28 d in deionized water

Table 3 Comparative leach data ($g/m^2 \cdot d$; MCC-1 leach tests at 90°C in deionized water for 28 d)

Element	HHP	BSG[9]	SYN[9]	TC[9]	FUE[9]
Cs	3.98	1.12	0.75	0.45	48
Sr	<0.09	<0.001	0.33	0.0011	0.27

HHP: waste form produced by hydrothermal hot-pressing
BSG: borosilicate glass
SYN: Synroc
TC: tailored ceramic
FUE: FUETAP concrete

Soxhlet leach tests more than that of Na and Mo.

The leach rate of the waste form produced by hydrothermal hot-pressing was compared with that of other forms (**Table 3**). The release rate of Cs from the waste form was much lower than that of FUETAP concrete but was higher than that of borosilicate glass. The waste form had low release rate of Sr. Yamasaki *et al.*[16] described that a solidified body, which was produced by hydrothermal hot-pressing technique from a powder of borosilicate glass containing a simulated high-level radioactive waste, had lower Sr release rate than the original borosilicate glass. It is expected that hydrothermal reaction of Sr with glass or SiO_2 may immobilize Sr in more leach resistant structure than glass.

Weight loss of the waste form due to leaching under hydrothermal conditions in deionized water for 1 d is shown in **Fig. 4**, together with that of JW-A glass form. The waste form produced by hydrothermal hot-pressing had smaller weight loss at high leaching temperature above 250°C than that of the glass form, but the weight loss of the former was higher below 200°C. Elemental mass loss of each element (**Fig. 5**) showed

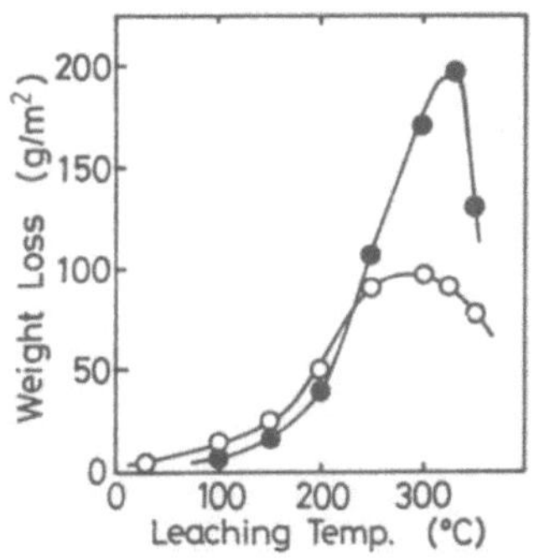

Fig. 4 Weight loss of waste form (○) produced by hydrothermal hot-pressing, and of JW-A glass form (●), by leach tests under hydrothermal conditions in deionized water for 1 d

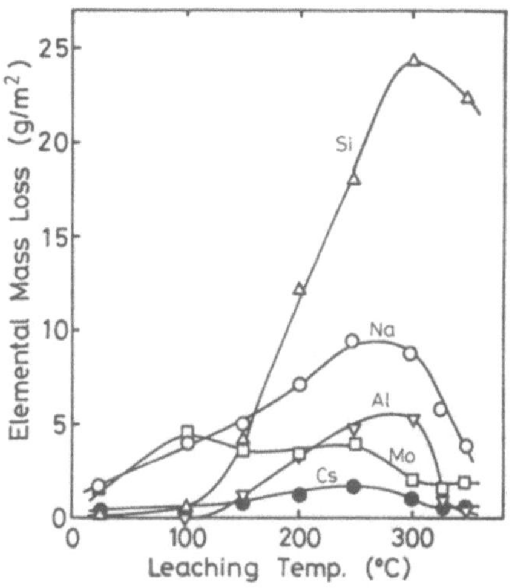

Fig. 5 Elemental mass loss from waste form by leach tests under hydrothermal conditions for 1 d

that release of Na and Mo was large at low leaching temperature and release of Si, Na and Al caused the large weight loss at high temperature.

When the glass form was leached above 150°C, a reaction layer was observed on its surface. The thickness of the layer was increased with the increase in leaching temperatures. The glass form leached above 325°C was swelled and surrounded by a yellow reaction layer, about 1 mm in thickness (**Photo. 1**(a)). On the other hand, the waste form produced by hydrothermal hot-pressing had no reaction layer and maintained the shape before the leach tests (Photo. 1(b)).

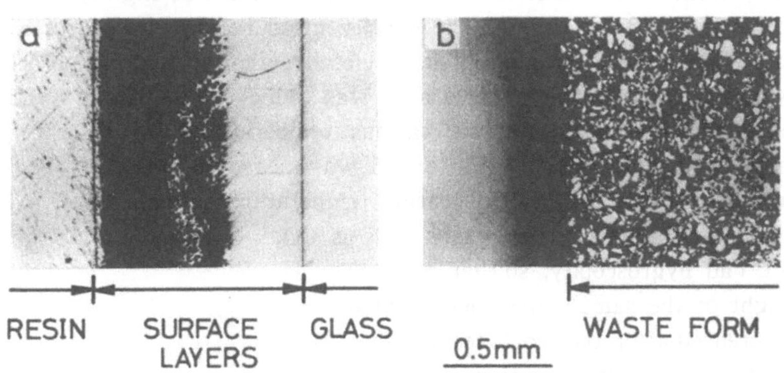

(a) JW-A glass form (b) Waste form produced by hydrothermal hot-pressing

Photo. 1 Cross section of sample specimens after leached in deionized water at 325°C for 1 d

2. Other Properties

According to X-ray powder diffraction patterns (**Fig. 6**), the waste form was mainly composed of α-quartz, the main constituent of the matrix. The diffraction patterns showed that MoO_3 and Nd_2O_3 in the waste were disappeared by hydrothermal hot-pressing and peaks due to α-Fe_2O_3, CeO_2 and ZrO_2 diffraction were observed with low intensity. When

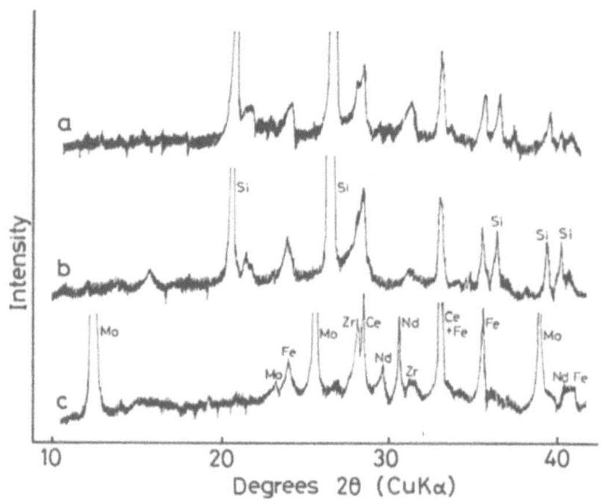

Mo: MoO_3, Fe: α-Fe_2O_3, Zr: m-ZrO_2
Nd: Nd_2O_3, Si: α-quartz, Ce: CeO_2

Fig. 6 X-ray powder diffraction patterns of simulated waste (c), waste form produced by hydrothermal hot-pressing (b), and waste form after heated at 900°C for 6 h (a)

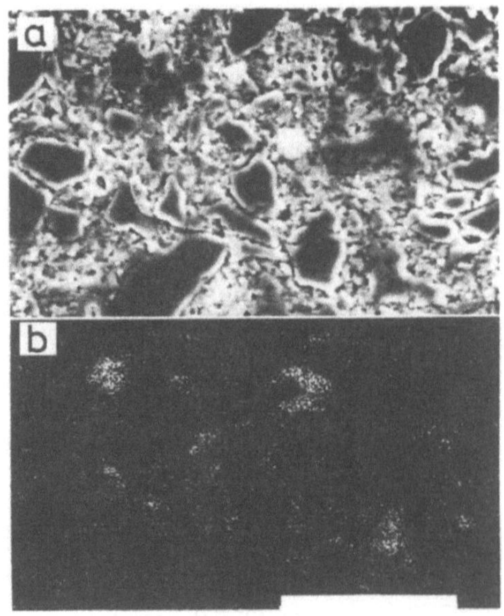

Photo. 2 Polished section of waste form (a) and its Fe Kα X-ray image (b) (White bar : 60 μm)

Al$(OH)_3$ was not added to the starting powder, ZrO_2 was disappeared. A few new peaks with low intensity, probably due to hydrothermal reaction products, were also observed, though they were not identified yet.

A polished section of the waste form is shown in **Photo.** 2(a). Large grains were α-quartz of the matrix. The Fe Kα X-ray image (Photo. 2(b)) showed that Fe existed between α-quartz grains. The waste form had many pores and was slightly water permeable. Its apparent density was 2.3 g/cm^3 and the density measured by the use of a pycnometer was 2.73 g/cm^3. The waste form might have closed pores, so that its porocity was estimated to be over 20%. Its specific surface area was found to be 10 m^2/g by BET method.

After the waste form was heated at various temperatures for 6 h, weight loss by evaporation and by Soxhlet leach tests for 1 d was measured (**Fig.** 7). The evaporation loss was increased with the increase in heating temperature, probably due to evaporation of hydrated water and anions in the waste such as CO_3^{-2}, NO_3^- and Cl^-. The waste form heated at 500°C had hygroscopity, so that in some cases the weight of the sample specimens heated at 500°C was increased after the leach tests. This result showed that reversible hydration and dehydration reaction occurred in the waste form. The weight loss by the leach tests was decreased by heating the waste form above 300°C (Fig. 7). Elemental mass loss of each element was decreased by the heat treatment, but it was not changed so much relative to the total weight loss. Release of Mo was especially decreased by heating the waste form above 700°C.

The waste form was not melted and had no

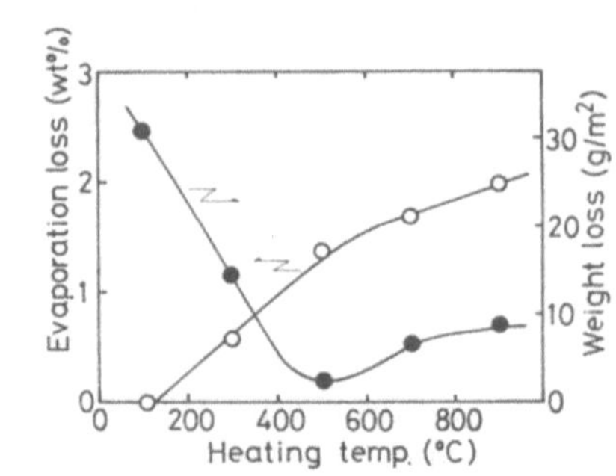

Fig. 7 Effect of heat treatment on evaporation loss and weight loss by Soxhlet leach tests for 1 d

cracks even by heating at 900°C. The heat treatment at 900°C gave no large changes in X-ray powder diffraction patterns (Fig. 6(a)).

Compressive strength of the waste form was about 200 MPa, which was higher than that of FUETAP concrete (40~100 MPa)[8], and similar to that of a glass form (270 MPa)[17]. After quenched from 500°C, the waste form had no cracks and its compressive strength was about 120 MPa. On the other hand, many cracks were observed in quenched JW-A glass form.

Thermal conductivity, specific heat and thermal diffusivity of the waste form are shown in **Fig.** 8 against temperatures, together with those of a glass form[17]. The thermal conductivity of the waste form was reduced with the increase in temperature up to 500°C and then increased above 700°C. This change in the thermal conductivity might result from the fact that the waste form was porous and included volatile materials. The thermal conductivity of the waste form was larger at room temperature than that of the glass form, but smaller at about 500°C.

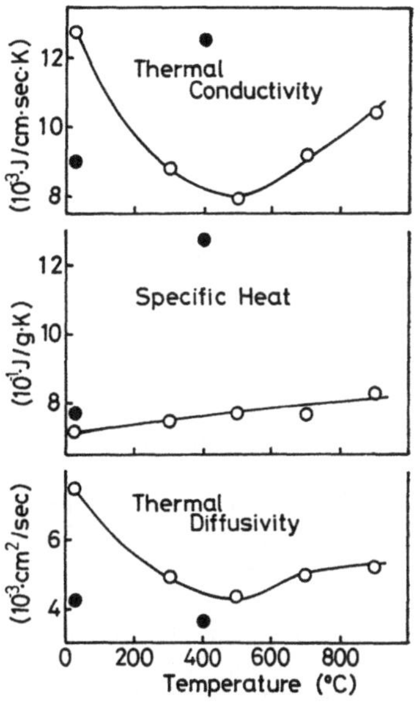

Fig. 8 Thermal properties of waste form (○) and of glass form (●)[17]

IV. Conclusion

The simulated high-level radioactive waste was solidified by hydrothermal hot-pressing technique. The waste form (21.8% waste loading), produced under the conditions at 350°C and 66 MPa for 6 h by the use of 10-N NaOH solution, was characterized as following:

(1) Silica rich matrix: The waste form was mainly composed of α-quartz, the main constituent of the silica matrix, and included the waste components such as α-Fe_2O_3, CeO_2 and ZrO_2, as crystalline phases.

(2) Porous structure: It contained over 20% porosity and its specific surface area was 10 m^2/g.

(3) High stability at high temperatures and under hydrothermal conditions: It was not melted even at 900°C, though weight loss, mainly due to evaporation of hydrated water, was observed by heat treatment. It had high thermal shock resistance and its stability under hydrothermal conditions was extremely high relative to a glass form.

(4) Medium thermal conductivity: Its thermal conductivity was higher at room temperature but lower at 500°C than a glass form.

(5) High leach rate by short-time leach tests: Weight loss of the waste form by short-time leach tests was large, due to release of Na and Mo. These elements, which had been included in the mineralizer after hydrothermal hot-pressing, were easily released by an initial washout through the porous structure of the waste form.

(6) Low leach rate of Cs and Sr: The amount of weight loss by the different leach tests in deionized water was in the order, Soxhlet flow leach tests at 97°C>modified MCC-1 static low temperature leach tests at 90°C>modified MCC-1 at 40°C. According to leach rates determined by MCC-1 leach tests at 90°C, release rate of Cs and Sr from the waste form was much lower than a concrete form, but slightly higher

relative to a glass form.

ACKNOWLEDGMENT

The authors wish to express their thanks to Drs. S. Muraoka and T. Murakami, Japan Atomic Energy Research Institute for their offering JW-A glass form, to Analysis center of Horiba Co. for help in microstructure observation, and to Rigaku Co. for help in thermal conductivity measurement.

—REFERENCES—

(1) McVAY, G. L., BUCKWALTER, C. Q.: Effect of iron on waste-glass leaching, *J. Am. Ceram. Soc.*, **66**, 170~174 (1983).
(2) RINGWOOD, A. E.: *"Safe Disposal of High-level Nuclear Reactor Waste: A new strategy"*, p. 1~64 (1978), Aust. Nat. Univ. Press.
(3) McCARTHY, G. J., DAVIDSON, M. T.: Ceramic nuclear waste form: (I), Crystal chemistry and phase formation, *Am. Ceram. Soc. Bull.*, **54**, 782~786 (1975); *ibid*: (II), A ceramic waste composite prepared by hot pressing, *ibid.*, **55**, 190~194 (1976).
(4) McCARTHY, G. J.: High level waste ceramics; Materials consideration, process simulation, and product characterization, *Nucl. Technol.*, **32**, 92~105 (1977).
(5) RINGWOOD, A. E., *et al.*: Immobilization of high level nuclear reactor wastes in SYNROC, *Nature*, **278**, 219~223 (1979).
(6) HARKER, A. B., JANTZEN, C. M., MORGAN, P. E. D., CLARKE, D. R.: Tailored ceramic nuclear waste forms: preparation and characterization, *"Scientific Basis for Nuclear Waste Management"*, (edited by MOORE, J. G.), p. 139~146 (1981), Plenum, New York.
(7) MORGAN, P. E. D., CLARK, D. R., JANTZEN, C. M., HARKER, A. B.: High-alumina tailored nuclear waste ceramics, *J. Am. Ceram. Soc.*, **64**, 249~258 (1981).
(8) DOLE, L. R., *et al.*: Cementitious radioactive waste hosts formed under elevated temperatures and pressures (FUETAP Concrete), *"Scientific Basis for Nuclear Waste Management"*, (edited by TOPP, S. V.), p. 585~593 (1982), North-Holland, New York.
(9) BERNADZIKOWSKI, T. A., *et al.*: High-level nuclear waste form performance evaluation, *Am. Ceram. Soc. Bull.*, **62**, 1364~1368; 1390 (1983).
(10) YAMASAKI, N., *et al.*: Immobilization of radioactive waste in hydrothermal synthetic rock; Lithification of silica powder, *J. Nucl. Sci. Technol.*, **21**, 71~73 (1984).
(11) YANAGISAWA, K., NISHIOKA, M., YAMASAKI, N.: Immobilization of radioactive wastes by hydrothermal hot-pressing, *Am. Ceram. Soc. Bull.*, **64**, 1563~1567 (1985).
(12) YANAGISAWA, K., KANAHARA, S., NISHIOKA, M., YAMASAKI, N.: Immobilization of radioactive wastes in hydrothermal synthetic rock, (II), Hydrothermal synthesis of pollucite, *J. Nucl. Sci. Technol.*, **21** 558~560 (1984).
(13) BANBA, T., *et al.*: Study of actual and simulated HLW compositions, *JAERI-memo* 8982, (1980).
(14) STRACHAN, D. M., BARNES, B. O., TURCOTTE, R. P.: Standard leach tests for nuclear waste materials, Ref. (6), p. 347~354.
(15) ISHIGURO, K., *et al.*: Growth of surface layer during the leaching of the simulated waste glass and its barrier effects on the leaching, *"Scientific Basis for Nuclear Waste Managements"*, (edited by BROOKINS, D. G.), p. 135~142 (1983), Plenum, New York.
(16) YAMASAKI, N., YANAGISAWA, K., NISHIOKA, M.: Hydrothermal Immobilization of glass powder containing simulated high level radioactive waste, *J. At. Energy Soc. Jpn.*, (in Japanese), **28** 266~273 (1986).
(17) KAMEI, F., IGARASHI, H., SASAKI, N., NAGAKI, H.: *Preprint 1983 Annu. Mtg. AESJ*, (in Japanese), F34, (1983).

Part 7

Hydrothermal Decomposition and Corrosion

Reprinted from *Proceedings of the Fourth International Conference on High Pressure*, Kyoto, April 1974, pp. 190–193.

DECOMPOSITION OF ILMENITE UNDER HYDROTHERMAL CONDITION (PRELIMINARY WORK)

M.G.M.U. ISMAIL, SHIN-ICHI HIRANO, AND SHIGEYUKI SŌMIYA

Tokyo Institute of Technology
Res. Lab. Eng. Mat.; Ookayama, Meguro-ku, Tokyo 152, Japan

Complete decomposition of ilmenite was observed by treating with hydrothermal solution of 10MKOH under certain optimum conditions. The decomposition products were magnetite and $K_2O(TiO_2)_n$ type of compounds. The stiochiometry of magnetite and n of the $K_2O(TiO_2)_n$ compound formed depend on the conditions employed for the decomposition. At temperatures below 800°C and pressure of 300 kg/cm^2 magnetite formed by decomposition existed as magnetite-ulvospinel binary spinel solid solution. Formation of potassium hexa titanate [$K_2O(TiO_2)_6$] was favoured at temperatures above 550°C.

Ilmenite is the main raw material used for the production of titanium and its compounds. The methods so far adopted to decompose ilmenite were mainly based on treating with inorganic acids or under gaseous atmospheres at high temperatures. Kozyrin and Nikolskii (1968) in their experimental decomposition of ilmenite under near hydrothermal condition in the presence of acids and alkalies; brookite, scholarmite and a small amount of sphene were observed as the alteration products. The present study was mainly focussed on the possibilities of decomposing ilmenite by chemical reaction with hydrothermal solution.

Experimental,

A sample of magnetically separated ilmenite from beach sand was used as the starting material and the X-ray studies of this sample showed the presence of impurities. To remove these impurities the sample was subjected to magnetic separation and then treated with HCl acid at 100°C temperature for 20 minutes. After washing the treated sample with hot water it was dried at 120°C in an oven. This was ground to particle size of 200-300 mesh and used in the hydrothermal studies. The chemical analysis of the purified sample is given in the table 1.

Table 1.
Analysis of purified Ilmenite

Component	% by wt
TiO_2	55.33
FeO	19.72
Fe_2O_3	23.66
MnO	0.91
MgO	0.25
CaO	None
H_2O	None

The mineralizer solution was introduced into the one end sealed 35mm in length and 3mm in diameter gold capsule using a micro syringe. Then the required quantity of the ilmenite sample was introduced into the capsule and sealed with an electric arc. The capsule was placed in a test tube type cold seal hydrothermal equipment and the pressure was raised gradually to the desired value and care was taken to prevent over shooting either temperature or pressure in the initial heating period. At the end of each run the vessel was quenched in cold water. The temperature and pressure were maintained within ± 5°C and 10kg/cm^2 respectively of the desired value. The capsules were cut opened and the run products were examined with X-ray diffraction method using Fe filtered Co Kα radiation for the decomposed products.

Results and Discussion,

The effect of mineralizer KOH, temperature pressure, ilmenite to water ratio, period of treatment, particle size and the degree of fill on the decomposition of ilmenite under hydrothermal condition were studied at first.

1) The effect of concentration of mineralizer KOH, was studied using 2M, 5M, 10M and 20M solutions. The rate of decomposition of ilmenite was accelerated by increasing the concentration of KOH solution. (Illustrated in figure 1.)

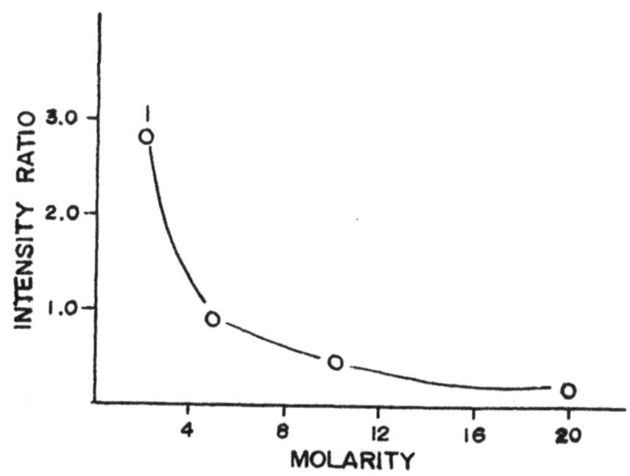

Figure 1; Effect of concentration of mineralizer KOH on the decomposition. At 500°C, 500kg/cm^2 and 72 hours of treatment time.
Intensity ratio—I_{Il}/I_{uk}
I_{Il}--Intensity of Ilmenite peak (104)
I_{uk}--Intensity of maximum intense unknown compound peak.

2) Experiments which were carried out at 400°C, 500°C, 600°C, 700°C and 900°C temperatures showed that the rate of decomposition increases with the increase of temperature. (Refer figure 2)

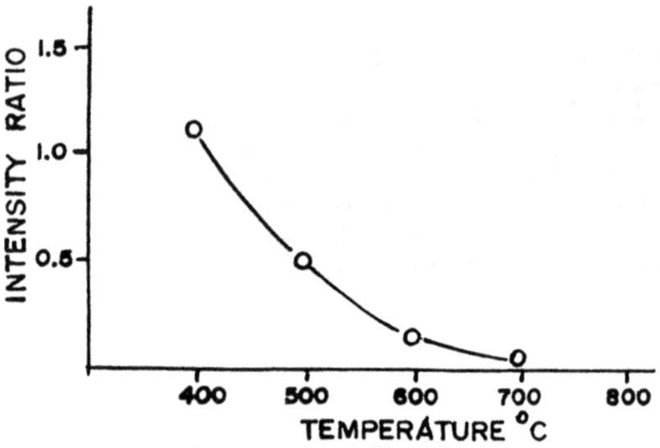

Figure 2; Effect of temperature on the decomposition. Pressure 500kg/cm^2, 10M KOH Mineralizer, 72 hours of treatment time.
Intensity ratio--I_{Il}/I_{uk}
I_{Il}--Intensity of Ilmenite peak (104)
I_{uk}--Intensity of maximum intense unknown compound peak

3) The variation of limenite to water ratio, affected the rate of decomposition in such a way that the more the water content the faster the rate of reaction. (Illustrated in figure 3)

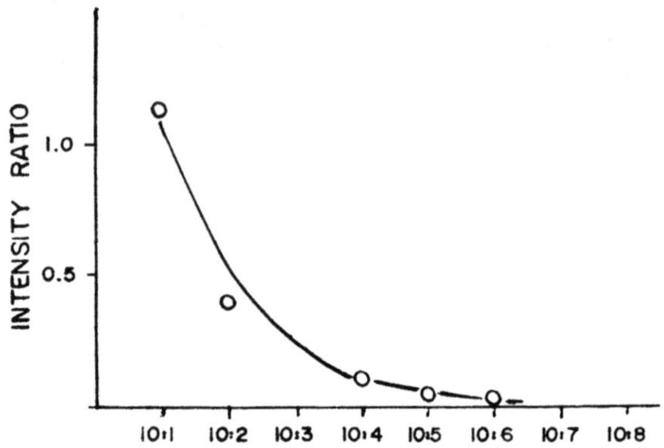

Figure 3; effect of water on the decomposition at 500°C, 500kg/cm^2 pressure, 10M KOH Mineralizer and 72 hours of treatment time.
Intensity ratio--I_{Il}/I_{uk}
I_{Il}-Intensity of ilmenite peak (104)
I_{uk}-Intensity of maximum intense unknown compound peak

4) The dependence of pressure on the rate of decomposition of ilmenite showed that ilmenite became unstable to decompose with increasing pressure and stable over 400kg/cm^2 and that the optimum pressure range to decompose ilmenite completely was concluded between 250-350kg/cm^2. (Illustrated in figure 4)

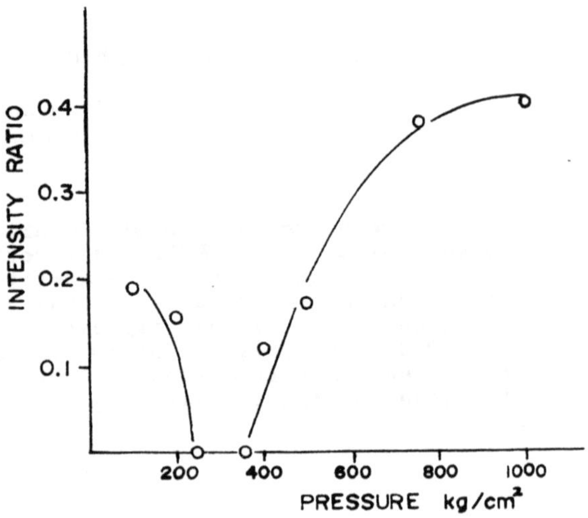

Figure 4; Effect of pressure on the decomposition. At 500°C, 10M KOH Mineralizer, 5:3 Ilmenite to water ratio, 72 hours of treatment time.
Intensity ratio--I_{Il}/I_{uk}
I_{Il}--Intensity of ilmenite peak (104)
I_{uk}--Intensity of maximum intense unknown compound peak

5) Runs which were carried out to study about the minimum period of treatment necessary to decompose ilmenite completely at a temperature of 500°C, 300kg/cm^2 pressure, 5:3 ilmenite to water ratio, 10M KOH mineralizer and about 70% fill, showed complete decomposition can be obtained in 63 hours of treatment.

6) Studies on increase of degree of fill and decrease of particle size of ilmenite on the hydrothermal decomposition, showed the same type of behaviour.
From the basic study already carried out on the effect of above variables on the hydrothermal decomposition of ilmenite, it is clear that ilmenite can be decomposed completely only when treated under certain optimum conditions. Now we can write the decomposition reaction of ilmenite as

$$3FeTiO_3 + H_2O \longrightarrow Fe_3O_4 + 3TiO_2 + H_2$$

Also the detection of formation $K_2O(TiO_2)_n$ compound revealed that the TiO_2 formed by above reaction react with KOH to form $K_2O(TiO_2)_n$ compounds; n of these compounds depend on the conditions employed for decomposition.

Further investigation on the effect of water on the decomposition revealed that the addition of excess water made ilmenite to decompose completely in a shorter period of treatment, when treated at 500°C, 300kg/cm^2 pressure, 10M KOH mineralizer solution and about 70% fill. Minimum duration of treatment required for complete decomposition and the unit cell dimension of Fe_3O_4 formed are listed in Table 2.

Table 2 Effect of water on Magnetite

Ilmenite:water ratio	Minimum period required for complete decomposition. hrs.	Lattice parameter of Magnetite Å
5 : 3	63	8.467
5 : 4	39	8.458
1 : 1	24	8.448

1 : 2	24	8.446

For the determination of lattice parameter of magnetite, the decomposed product was initially washed with very dilute HNO_3 acid taking precautions not to effect the magnetite and $K_2O(TiO_2)_n$ compounds, and followed by hot water until the filtrate was free of acid. To the washed sample excess water was added and magnetite was separated using a magnet. After drying the sample in an oven at 100°C, the lattice parameter was determined using Si as an internal standard with Fe filtered Co Kα radiation. Instrumental conditions on the JEOL Co.(JDX-7F) X-ray diffractometer were scanning speed ¼°/minute, time constant 1 second, chart speed 20mm/minute, scattering slit 1°, divergence slit 1°, receiving slit 0.1°, Voltage 35kV and current 10mA. Refering to data listed in Table 2, it is clear that the minimum duration of treatment necessary for complete decomposition decreases with increase of water content and also this has an effect on the purity of magnetite formed. Different authors have given various values for the unit cell dimension of magnetite and in this study we take 8.396Å as lattice parameter of stiochiometric magnetite as determined by Basta (1957).

The samples which were subjected to longer duration of treatment than required for complete decomposition, maintaining the other conditions same, we observed the same phenomena and the effect of longer period of treatment on the lattice parameter of magnetite is illustrated in the figure 5. Formation of $K_2O(TiO_2)_6$ compound favoured when treated for a longer duration.

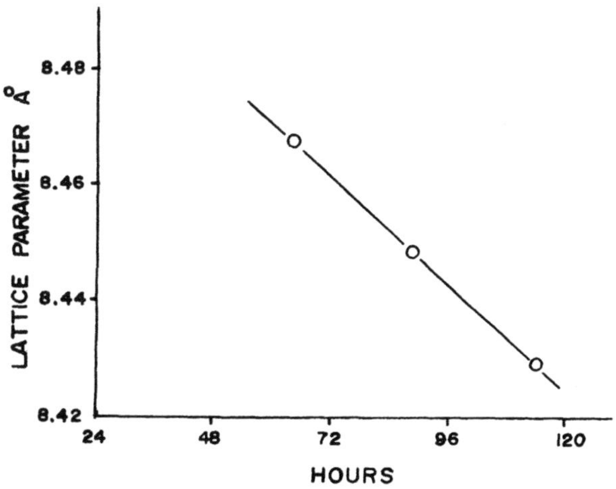

Figure 5, Effect of longer duration of treatment on the magnetite formed. Fixed conditions: Pressure 300kg/cm^2, 10M KOH mineralizer, 5 : 3 ilmenite to water ratio, temperature 500°C about 70% fill.

Runs which were carried out at 550°C, 600°C 700°C temperatures and ilmenite to water ratios 5 : 3, 5 : 4 and maintaining the other conditions fixed, showed ilmenite completely decomposed in 24 hours, when ilmenite to water ratio was 5 : 4 and in 48 hours when ilmenite to water ratio was 5 : 3 at temperatures equal to or over 550°C. Figure 6 illustrate the variation of lattice parameter of magnetite formed with temperature.

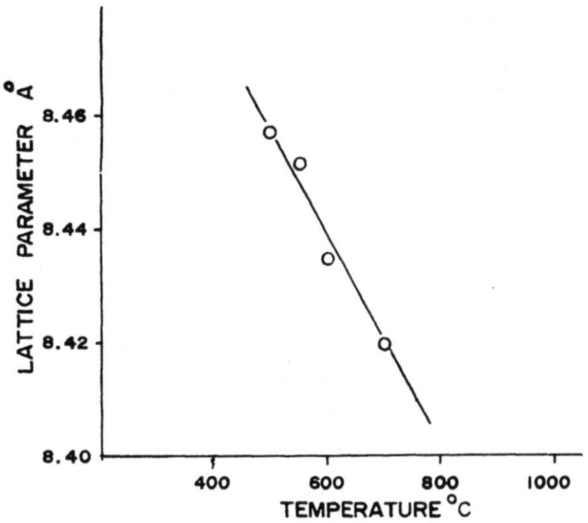

Figure 6; Effect of temperature on the lattice parameter of magnetite formed. Fixed conditions 300kg/cm^2 pressure, 10M KOH mineralizer, 24 hours of treatment time, 5 : 4 ilmenite to water ratio and about 70% fill.

This study revealed that the stiochiometry of magnetite formed by decomposition depend on the amount of water employed, duration of treatment and temperature. In all three cases discussed above we observed a decrease of lattice parameter of magnetite with respect to increase of respective variables.

Vincent, Wright and Chevallier (1957) in their heating experiments on natural titaniferous magnetite observed the same phenomena and they explained the higher value obtained for lattice parameter of magnetite was due to existence of Fe_3O_4-Fe_2TiO_4 binary inverse spinel solid solution. On heating, this solid solution, Fe_2TiO_4 starts separating and hence the lattice parameter decreases. The value for lattice parameter of Fe_2TiO_4 given by them was 8.491Å. Making use of this explanation, we can say in the initial case the decrease of lattice parameter of magnetite was due to increase of fugacity of water, which made part of Fe_2TiO_4 in Fe_3O_4-Fe_2TiO_4 solid solution to decompose to magnetite and TiO_2. But the lattice parameter of magnetite formed by using ilmenite to water ratio 1 : 2 is much larger than that of stiochiometric Fe_3O_4. This showed the fugacity of water has only a slight influence on formation of stiochiometric magnetite.

In the second case the decrease of lattice parameter of magnetite was much sharper compared to the first case. This can be explained due to separation of Fe_2TiO_4 from Fe_3O_4-Fe_2TiO_4 solid solution.

In the third case the decrease of lattice parameter of magnetite is very sharp and by extrapolating the graph it can be seen that the stiochiometric magnetite can be prepared above 800°C temperature when treated under given fixed conditions. Formation of only $K_2O(TiO_2)_6$ compound was observed with increase of temperature.

Acknowledgment
The authors are pleased to thank Prof. J Ossaka and Mr. J. Hirabayashi for the

analysis of the ilmenite sample.

References

(1). N.A.Kozyrin and A.L.Nikoll'ski "Experimental decomposition of sphene and ilmenite under near hydrothermal condition", Izv. Vyssh. Ucheb. Zared. Geol. Razed. 11 (1) 145-8 (1968) [Chemical Abstract, vol.68, 1968, 97402X)

(2). B.Z.Basta, "Accurate determination of the cell dimensions of magnetite" The Mineralogical Mag. 431-442 (1957)

(3).E.A.Vincent,J.B.Wright and R.Chevallier "Heating experiments on some natural titaniferous magnetites", Mineralogical Magazine 31, 624-655 (1957)

****

Questions and Answers

Q : S.Kume —— The starting material contains a fairy amount of Fe_2O_3. The existence of Fe^{3+} ions should influence the decomposition reaction of ilmenite under hydrothermal conditions. An experiment using pure ilmenite is desirable.

A : M.G.M.U. Ismail —— You are right. This experiment was carried out in order to know the decomposition of natural ilmenite sand at first. We are proceeding the experiment using pure synthesized ilmenite to study the effect of the presence of Fe^{3+} ion.

Reprinted from *Proceedings of the First International Symposium on Hydrothermal Reactions*, Japan, 22–26 March 1982, pp. 669–680.

Decomposition of Ilmenite under Hydrothermal Condition

M.G.M.U. ISMAIL, SHIGEYUKI SŌMIYA*
Ceylon Institute of Scientific and Industrial Research,
Colombo, Sri Lanka.

*Laboratory for Hydrothermal Synthesis, Research Laboratory for Engineering Materials, Tokyo Institute of Technology, Nagatsuda, Yokohama.

1. Introduction

A knowledge of the stability of iron-titanium oxides under hydrothermal conditions is important in the interpretative petrology, which provides informations on the conditions under which economic depositis of iron-titanium oxide minerals may have formed. Ilmenite is one of the important mineral existing in the iron-titanium oxide system, which has a vast industrial potential. In nature ilmenite exists as sand and rocks, but as ilmenite-magnetite or ilmenite-hemattite variety. The present study was carried out in order to study the stability of ilmenite in different solutions; alkali hydroxides, alkali carbonates, alkali chlorides and alkali bicarbonates under hydrothermal conditions. This type of research will provide certain amount of important data for geochemists in predicting the formation of ilmenite and other related minerals in the lithosphere. On the other hand the results gained from these experiments will open a new era, which will provide better method of decomposing ilmenite commercially to produce iron or titanium compounds, compared to the already existing industrial processes.

2. Experimental Procedure

2.1 Starting material for hydrothermal studies,

Natural ilmenite sand separated from beach sand in Sri Lanka was used as the starting material. This was treated with 3N HCl at 80°C for 20 minutes for further purification. The treated ilmenite was washed with hot water several times until the filtrate was free of acids and dried in an oven at 100°C for few hours. The purified ilmenite was ground to pass 200 mesh sieve and used for hydrothermal studies. The chemical analysis of the purified ilmenite is given in Table I, and these results showed a higher concentration of Fe_2O_3 in the purified ilmenite.

2.2 Capsule preparation for hydrothermal studies

Gold capsules of 2.7 mm in inner diameter and 35 mm in length were used for the hydrothermal treatment of the samples. The mineraliser solution was introduced into the capsules using a microsyringe and followed by the addition of the required quantity of purified ilmenite. In all the runs the quantity of ilmenite added was maintained within 50 mgs. After sealing the capsule it was introduced into the Roy and Tuttle[1] type test tube hydrothermal apparatus and pressure was raised to the desired value. Then the temperature was increased gradually by carefully maintaining the pressure constant. The temperature was measured using a platinel thermocouple calibrated at the melting point of gold (1064. 4°C). The pressure was measured using a 12" Heise gauge. The temperature was maintained with in ± 3°C and 10kg/cm^2 respectively of the desired value. After treating for the required period the test tube was quenched into cold water. The capsule was cut opened and the products were examined with Fe filtered Co Kα and Ni filtered Cu Kα radiations and by observation under an optical microscope. The magnetite formed by the decomposition of ilmenite was seperated using a magnet and used for the determination of lattice constant. Pure Si (99.999%) was used as an internal standard in this investigation. The powder pattern of the new phase formed was obtained by using single crystal. Single crystal Weissenberg technique was employed for the determination of the structure and the cell constants of the new phase. The chemical analysis of the new phase was carried out using EPMA method.

3. Results and Discussion

3.1 Ilmenite-Alkali Hydroxides - H_2O System

In this investigation KOH, NaOH and LiOH solutions were employed and the effect of these solutions on the stability of ilmenite under hydrothermal conditions were studied by carrying out a series of experiments.

3.1.1 Effect of concentration of alkali hydroxide solution on the stability of ilmenite

The runs carried out by using 2m, 5m, 8m and 10m KOH and NaOH solutions and 500°C, 500 kg/cm^2, ilmenite to water ratio of 5:2 and for the duration of 72 hours, showed the accelaration of the rate of decomposition of ilmenite with the increase of concentration of the mineralizer solutions. The results are illustrated in the figure 1. The products formed by the decomposition of ilmenite in KOH solution were magnetite and $K_2O\ (TiO_2)_4$. In the case of NaOH solution the products were magnetite and a new phase, which was identified as a Sodium Iron Titanium bronze compound.

On the other hand LiOH solution showed an entirely different type of behaviour; the recrystallisation of ilmenite was favoured in 1m and 2m LiOH solutions. Only a slight decomposition of ilmenite was observed in 5m LiOH solution.

3.1.2 Increase of temperature on the stability of ilmenite under hydrothermal conditions,

The pressure, duration of treatment, ilmenite to water ratio and the concentration of KOH and NaOH solutions were maintained constant as in the previous experiment at 500 kg/cm^2, 72 hours, 5:2 and 10m respectively. The runs carried out at 400°C, 500°C, 600°C and 700°C, showed the increase of rate of decomposition of ilmenite with the increase of temperature. The results are graphically illustrated in figure 2. The products formed by the decomposition of ilmenite were magnetite and $K_2O\ (TiO_2)_n$ type of compounds, where n=4 or 6. At temperatures lower than 600 °C, a compound of n=4 was formed and at temperatures above 600 °C, fibrous n=6 compound was formed. In the case of NaOH solution the products formed were magnetite and a sodium iron titanium bronze compound. The runs performed using 5m LiOH solution at different temperatures, did not increase the rate of decomposition of ilmenite as in KOH or NaOH solutions.

3.1.3 Variation of ilmenite to water ratio on the stability of ilmenite

Ilmenite to water ratio was varied between 10:1 and 10:8 and the runs were carried out by keeping the other conditions constant. Solutions of 10m KOH and 10m NaOH were used in this investigation. Increase of quantity of water favoured a higher decomposition rate for ilmenite in both solutions. But in the case of 5m LiOH solution the increase of quantity of water did not enhance the rate of decomposition of ilmenite remarkably.

3.1.4 Effect of pressure on the stability of ilmenite under hydrothermal conditions

The experiments carried out between 100 and 2000 kg/cm^2 pressures, maintaining the temperature at 500°C, duration of treatment 72 hours, ilmenite to water ratio 5:3 and using 10m KOH and 10m NaOH solutions showed the complete unstability of ilmenite at pressures between 250 kg/cm^2 and 350 kg/cm^2. The results obtained for 10m KOH solution was illustrated in figure 3. Increase of pressure from 100 kg/cm^2 to 250 kg/cm^2, the ilmenite became unstable and the complete decomposition of ilmenite was observed between 250 kg/cm^{2}and 350 kg/cm^2. stability of ilmenite was observed to increase above 350 kg/cm^2. The products formed by the complete decomposition of ilmenite in KOH solution were magnetite and $K_2O\ (TiO_2)_4$, and in NaOH solution were magnetite and a sodium iron titanium iron bronze.In the case of 5m LiOH solution the stability of ilmenite was observed between 250 kg/cm^2 and 350 kg/cm^2 pressures. Increase of pressure beyond 350 kg/cm^2 also did not show any considerable change in stability of ilmenite.

3.1.5. Minimum duration required for complete decomposition

Using 10m KOH and 10m NaOH solutions at 500°C, 300 kg/cm^2 pressure and ilmenite to water ratio of 5:3, minimum duration required for complete decomposition was investigated. These runs showed the complete decomposition of ilmenite could be obtained in 63 hours of duration. The products formed were magnetite and $K_2O\ (TiO_2)_4$ in the case of KOH solution and magnetite and a sodium iron titanium bronze in the case of NaOH solution. The increase of ilmenite to water ratio to 5:4 favoured to decompose ilmenite completely in a shorter duration of 39 hours under same temperature and pressure. The products formed were same as in the previous experiment. Considering the results obtained from these experiments, the decomposition reaction of ilmenite under hydrothermal conditions can be written as follows :

in KOH solution,

$$FeTiO_3 + H_2O \longrightarrow Fe_{3-x}O_4 + TiO_2 + H_2$$

$$nTiO_2 + 2KOH \longrightarrow K_2O\ (TiO_2)_n + H_2O$$

where n=4 or 6

in the case of NaOH solution,

$$FeTiO_3 + NaOH + H_2O \longrightarrow Fe_{3-x}O_4 + (Na_p Fe^{2+}_q).[Ti_r Fe^{2+}_s Fe^{3+}_t) O_{2-\delta}]_n + H_2$$

The lattice constant of magnetite formed by the complete decomposition of ilmenite in 10m KOH solution at 500°C and 300 kg/cm^2 and ilmenite to water ratio of 5:3 was larger than that reported on stoichiometric magnetite by Basta[2]. This is an indication of the substitution of Ti into the magnetite lattice, which causes an increase in cell dimension. Experiments were carried out to find out the conditions necessary to prepare stoichiometric magnetite by the decomposition of ilmenite under hydrothermal conditions. In this study the increased temperature and the longer duration of treatment played a positive role in forming the stoichiometric magnetite. It was concluded that the conditions necessary to prepare stoichiometric magnetite were 800°C, 300 kg/cm^2 and 5:4 ilmenite to water ratio for 24 hours of duration. The formation of fibrous $K_2O\ (TiO_2)_6$ was observed, when the decomposition reaction was carried out above 600°C.

3.1.6. Investigation on the new phase formed by the decomposition of ilmenite in NaOH solution

An experiment, which was carried out at 500°C and 300 kg/cm^2 pressure using a mixture of TiO_2 (rutile) and Fe powder in the ratio 5:1 (weight basis) as the starting materials in 10m NaOH solution, resulted in the formation of the same compound as that formed by the decomposition of ilmenite in NaOH solution under hydrothermal conditions. Single crystals of this compound were grown at 650°C and 1000 kg/cm^2 in 10m NaOH solution, starting with TiO_2(rutile) and Fe powder in the ratio 5:1 (weight basis) for durations of one week. The weissenberg studies of the single crystals of this compound showed the structure as orthorhombic. Crystallographical data of this new compound is given table 2. The EPMA analysis of this new compound showed the presence of a higher percentage of Na. The possible chemical formula for this compound was derived on the basis of the Mössbauer and EPMA studies carried out on one of the phase formed under increased oxygen partial pressure. More detail description of the derivation of the chemical formula of this non stoichiometric compound is givenby the authors[2]. Mössbauer studies at lower temperatures and the structural analysis of this new compound are underway for more precise determination of the chemical formula. The powder pattern of this compound is given in Table 3.

3.1.7 Influence of Fe_2O_3 and other impurities presence in natural ilmenite on the decomposition

The chemical analysis of the purified natural ilmenite sand showed the presence of Fe_2O_3. In order to study the effect of Fe^{3+} on the decomposition of ilmenite, pure ilmenite was synthesized under hydrothermal conditions starting with an equimolar mixture of metallic Fe and Ti powders by Hirano[4] et al. The complete decomposition of synthesized ilmenite was observed at 500°C and 300 kg/cm^2. This shows clearly that there is no detectable effect of Fe^{3+} ion present in natural ilmenite, on the decomposition of ilmenite under hydrothermal conditions.

3.2 Ilmenite - Alkali Carbonate - H_2O System

Using 10m K_2CO_3 and 10m Na_2CO_3 solutions runs were carried out to study the stability of ilmenite under hydrothermal conditions. The experiments carried out at 500°C and 300 kg/cm^2 for 72 hours of duration, resulted in a partial decomposition of ilmenite. The temperature was increased to 600°C maintaining the pressure at 300 kg/cm^2 and the stability of ilmenite was studied in 10m K_2CO_3 and 10m Na_2CO_3 solutions for 72 hours of duration. In this case also only a partial decomposition of ilmenite was observed. Furthermore the investigation on the effect of increase of pressure on the stability of ilmenite showed the same results. In the case of K_2CO_3 solution products formed by the partial decomposition of ilmenite were magnetite and $K_2O\ (TiO_2)_4$ or $K_2O\ (TiO_2)_6$. The 10m Na_2CO_3 solution served to form magnetite and freaudenbergite ($Na_2O.\ Fe_2O_3.\ 7TiO_2$) from the partial decomposition of ilmenite.

3.3 Ilmenite - Alkali Chloride - H_2O System

Solutions of 10m KCl and 10m NaCl were used in this investigation. The experiments carried out at 500°C and 300 kg/cm^2 for 72 hours of duration of treatment resulted in a partial decomposition of ilmenite in both solutions. The effects of increase of pressure and temperature on the decomposition of ilmenite were studied separately by maintaing the other conditions constant. These runs also did not show a considerable effect on the rate of decomposition. But in all the cases the partial decomposition of ilmenite was observed and the products were $FeCl_3$ and TiO_2.

3.4 Ilmenite - $KHCO_3$ - H_2O System

Same series of investigation was followed for this system also, using 10m $KHCO_3$ solution. Only a partial decomposition of ilmenite was observed at 500°C, 300 kg/cm2 and for 72 hours of duration of treatment. Increases of temperature and pressure hardly effected the decomposition of ilmenite in 10m $KHCO_3$ solution. The products formed by the partial decompositionwere magnetite and $K_2O\ (TiO_2)_4$.

4. Conclusions

The results of these experiments predict the stability of ilmenite under higher pressure than 300 kg/cm^2 and at least upto 2000 kg/cm^2. This is an indication that the exsolution of ilmenite in nature would have occured under such higher pressures.

From the consideration of the industrial aspect of the results obtained from these experiments, it is clear that 10m KOH solution is the most suitable solution for the decomposition of ilmenite to produce the fibrous potassium hexa-titanate and magnetite directly. This process has some advantages compared with the processes used for the decomposition of ilmenite at present. To obtain the oxide of titanium it's necessary to retreat the initial product in the processes existing today where inorganic acid or Cl_2 gas are employed for the decomposition of ilmenite. In the case of the process described in the present paper, titanium dioxide or titanate were formed directly by the decomposition of ilmenite under hydrothermal conditions. This process was found to have a great advantage to avoid any pollution as encountered in the inorganic acid treatment process or Cl_2 gas process.

Acknowledgements

This work was carried out at the Hydrothermal Synthesis Laboratory of the Tokyo Institute of Technology, Tokyo, Japan. The author express his thanks to Professor Shigeyuki Sömiya and Dr. Shin-Ichi Hirano for the direction gain in carrying out this work.

The chemical analysis of the ilmenite was carried out at the Ishihara Sangyo Co. Ltd.

References

1. Roy, R. and Tuttle, O.F., 1956, Investigations under Hydrothermal Conditions: Physics and Chemistry of the earth, V. 1, P. 138 - 180.

2. Basta, E.Z. 1957, Accurate Determination of the call dimensions of Magnetite: Mineralogical Magazine, V. 37, P. 431-442.

3. Hirano, S., Ismail, M.G.M.U., and Sómiya, S., Crystal Growth of Sodium Iron Titanium Bronze Compounds under Hydrothermal Conditions: Materials Research Bulletin Vol. 11, 1976

4. Hirano, S. and Sõmiya, S., 1975, Crystal Growth of Ilmenite under Hydrothermal Conditions: Abstract of the 20th Jinko Kobutsu Toronkai, P. 92 .

Table 1. Chemical analysis of purified ilmenite

Component	% by weight
FeO	17.07
Fe_2O_3	23.62
TiO_2	54.41
MnO	0.93
MgO	1.03
Al_2O_3	0.64
SiO_2	0.92

Table 2. Crystal data of the new phase

Structure	-	Orthorhombic
Space group	-	Pbnm
a	-	9.13 Å
b	-	10.82 Å
c	-	2.93 Å
Density	-	3.4 g/cm³
Chemical formula	-	$(Na_{0.6}Fe^{2+}_{0.1})\left[(Ti_{1.55}Fe^{2+}_{0.3}Fe^{3+}_{0.13})\ O_{4-x}\right]$

Table 3 , Powder pattern of the new phase

H K L	I	$Sin^2\theta_{obs.}$	$Sin^2\theta_{cal.}$
0 2 0	w	0.0203	0.0203
1 2 0	vs	0.0274	0.0274
2 0 0	vs	0.0286	0.0285
2 2 0	w	0.0488	0.0488
1 0 1	w	0.0765	0.0764
1 1 1	vs	0.0816	0.0815
3 2 0	vs	0.0848	0.0848
1 2 1	s	0.0968	0.0967
3 3 0	vw	0.1100	0.1099
3 0 1	vw	0.1334	0.1334
4 2 0	w	0.1343	0.1344
2 4 1	w	0.1789	0.1790
2 6 0	w	0.2110	0.2112
5 0 1	vw	0.2477	0.2475

vs - very strong , s - strong , w - weak
vw - very weak

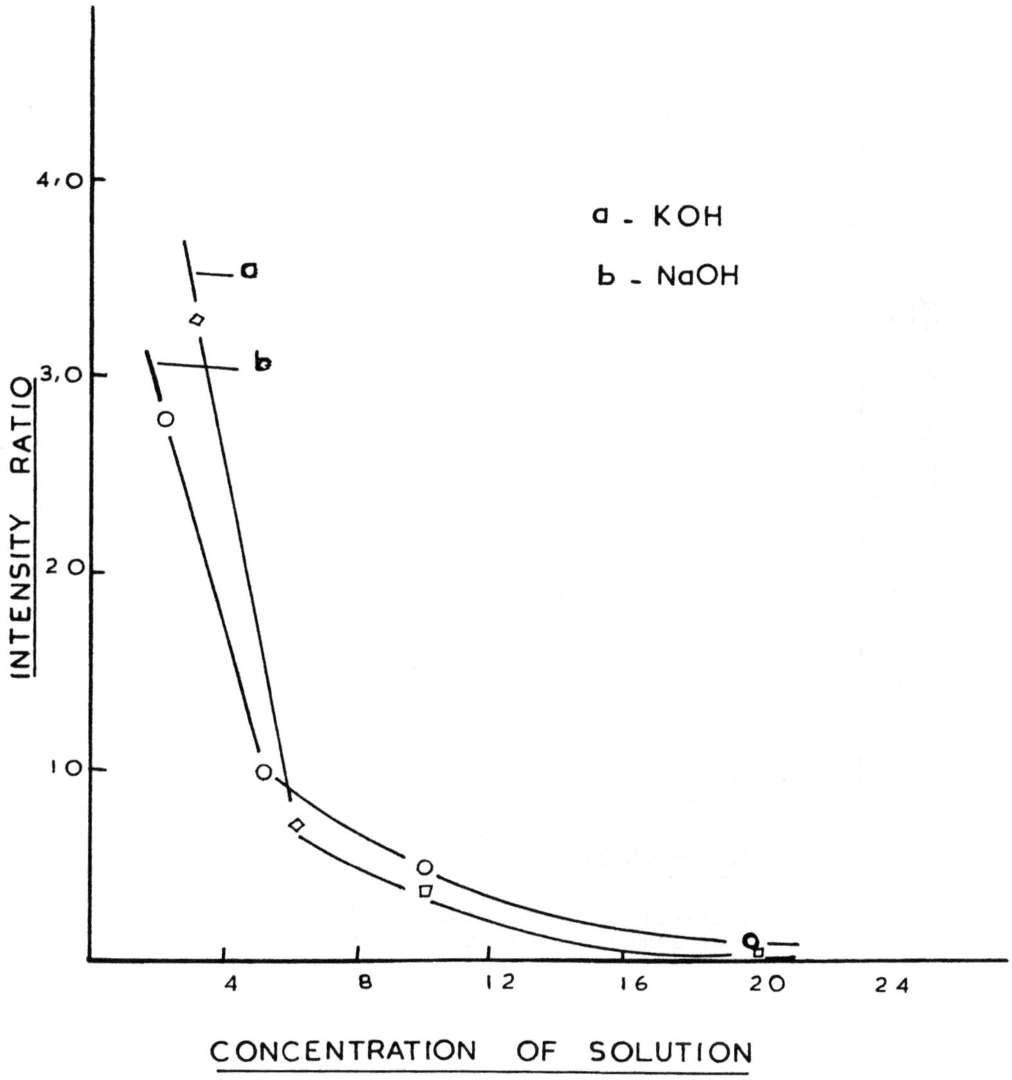

Figure 1, Effect of concentration of KOH and NaOH solutions on the decomposition of ilmenite

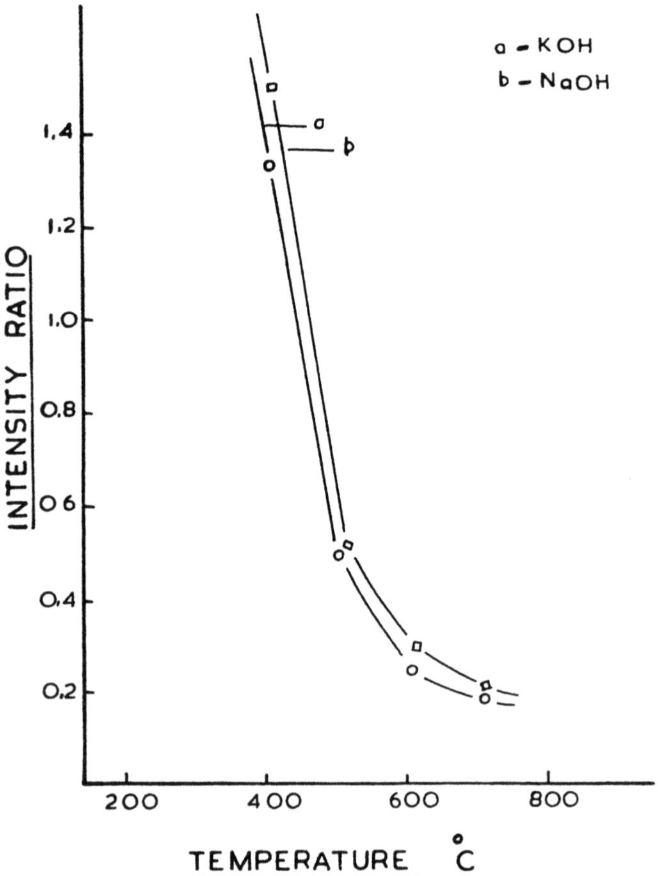

Figure 2, Effect of temperature on the decomposition of ilmenite for KOH solution

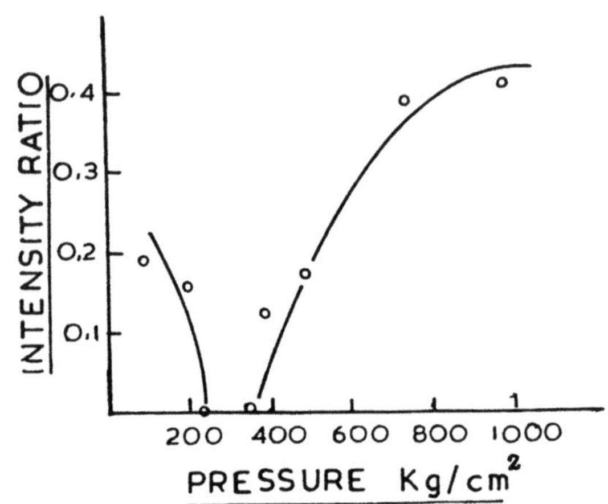

Figure 3, Effect of pressure on the decomposition of ilmenite

Reprinted from *Report of the Research Laboratory of Engineering Materials, Tokyo Institute of Technology*, No. 4 (1979) pp. 117–122.

Reaction Products of Si_3N_4 in Ammonia Solution at High Temperature Under High Pressure

By

Shigeyuki SŌMIYA*, Giichi SUDOH**,
Mitsuo TANAKA*** and Yasubumi FUKAYA***

Abstract

Si_3N_4 was decomposed and oxidized at high temperature under high pressure in NH_3aq and changed to amorphous SiO_2 or low form cristobalite.

Key Words: Si_3N_4, hydrothermal reaction, NH_3aq at high temperature and pressure, cristobalite.

1. Introduction

Si_3N_4 is one of the most important materials to be used as a high temperature structural material and has been studied by many researchers on its physical and chemical properties.[1] The oxidation behavior of Si_3N_4 at high temperature is characterized by the formation of the amorphous SiO_2 at early stage of the reaction and its recrystalization into low cristobalite in succession.[2,3] Moreover, the oxidation behavior of Si_3N_4 is affected by additives, impurities and other conditions, e.g. with greater oxidation rate in wet O_2 than in dry.[4] This paper describes on the reaction prodacts of Si_3N_4 in ammonia solution at high temperature under high pressure. There is a report that the amorphous SiO_2 was formed from Si_3N_4 at 700 to 750°C under the normal pressure in air.[2] We found that the oxidation and the decomposition of Si_3N_4 was greatly accelerated in ammonia solution under hydrothermal conditions, from 200 to 500°C with corresponding pressures from 500 to 1500 atm.

2. Experiment and Results

In Table 1, Si_3N_4 powder*, ammonia solution** and SnO powder*** used in this investigation are shown.

* 99.9%, Obtained from Soekawa Rikagaku K.K.

** Reagent Grade, Shuzui Hikotaro, Shoten

*** Shuzui Hikotaro, Shoten

An autoclave used in this investigation is shown in Figs. 1a and 1b. The apparatus 1a and 1b were used for heating at 200°C and 300 to 500°C, respectively. Si_3N_4 powder

---

* Professor, Director of Laboratory for Hydrothermal Syntheses, Dr. Sci. in Eng.

** Head of Res. Dept., Central Lab. of Res. & Develop., Chichibu Cement Co. Ltd. Dr. Sci.

*** Senior and Res. Chem., Central Lab. of Res. & Develop., Chichibu Cement Co. Ltd.

Table 1. Materials and mixing conditions

Material system	Treating conditions	Si_3N_4	NH_3aq.	SnO	Autoclaving apparatus
Si_3N_4-NH_3aq.	200°C 10,30 days	500 mg	40 ml (5 mol%)	—	I a (58 ml)
Si_3N_4-NH_3aq.	300°,500 °C, 500~1500 atm. 4 days	200 mg	0.50 ml (15 mol%)	—	I b (2.5 ml)
Si_3N_4-SnO -NH_3aq.	300°,500 °C, 500~1500 atom. 4 days	100 mg	0.50 ml (15 mol%)	100 mg	I b (2.5 ml)
	—	99.9 % purity (α, β mix)	99 % purity (JIS-GR)	99 % purity	—

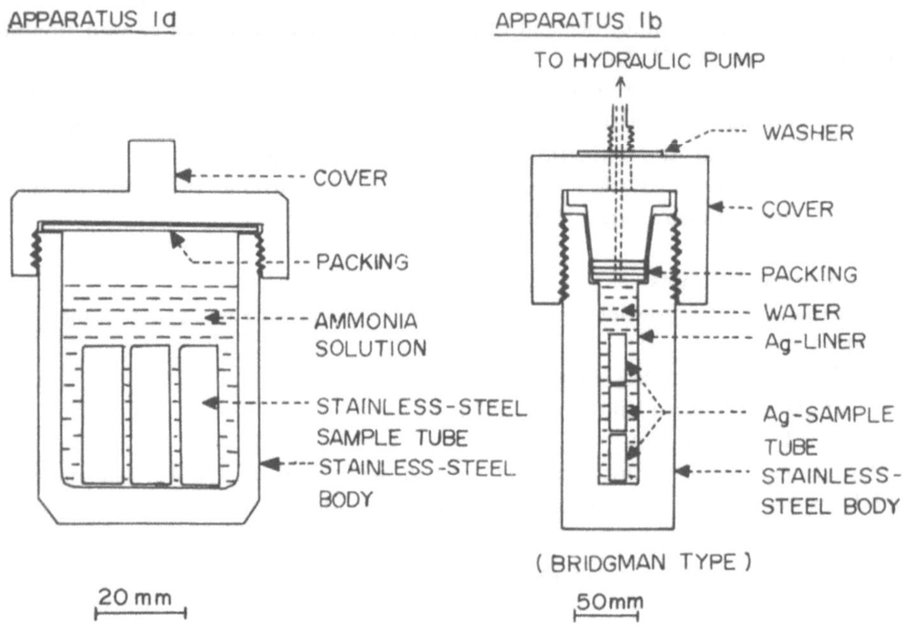

Fig. 1. Autoclaving apparatus

was put into the stainless steel tubes. They were closed with pinchers and put into the autoclave with 5 mol% ammonia solution of which volume was 70% of the apparatus. The capped appratus was put into an electric oven for heating. In the case of the apparatus 1b, Ag-tubes were filled with Si_3N_4 powder and ammonia solution and were weld. Then, these sealed tubes were put into the autoclave which was filled with water and joined to a hydraulic pump and a pressure gage. The temperature was held by the electric furnace with the controller.

The samples treated in this apparatus under various conditions were quenched and washed by pure water. X-ray diffraction analyses were carried out on the powder samples with Cu Kα radiation.

Fig. 2 and Table 2 show X-ray diffraction patters of the samples which were treated at 200°C for 10 and 30 days in ammonia solution. Peak intensities of the Si_3N_4 diffraction lines were decreased as the treating time increased. On the other hand, a new diffraction peak appeared broadly at Cu Kα 2θ 21.4° near the most strongest peak of low cristobalite 22.0° 2θ angle, and shifted to higher angle as the treating time increased together with the appearance of a new-peak at 35.7° 2θ angle.

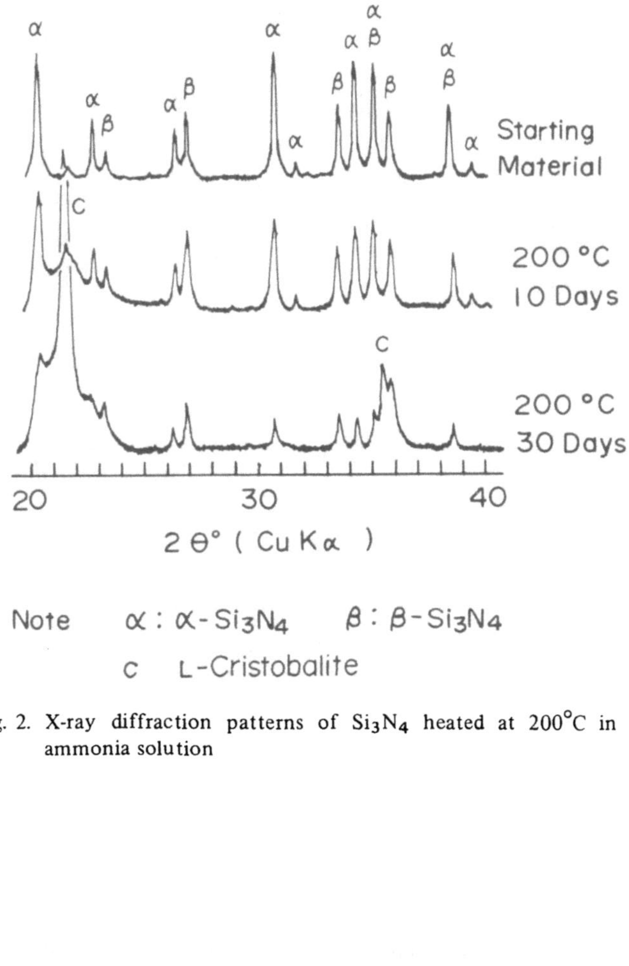

Fig. 2. X-ray diffraction patterns of Si_3N_4 heated at 200°C in ammonia solution

Table 2. X-ray diffraction peaks of Si_3N_4 which were treated under various conditions

hkl	Si_3N_4-NH_3 sol. (5 mol %)		Si_3N_4-NH_3 sol. (15 mol %)					Si_3N_4-SnO-NH_3 sol. (15 mol %)					L-Cristobalite
	200°C 10days	200°C 30days	300°C, 4days			500°C 4days		300°C, 4days			500°C 4days		(ASTM 4-0379)
	—	—	500 atm	1000 atm	1500 atm	500 atm	1000 atm	500 atm	1000 atm	1500 atm	500 atm	1000 atm	
101	21.4°	21.5°	N.F.	21.4° broad	21.5° broad	21.6° broad	21.6° broad	N.F.	21.5° broad	21.5° broad	21.8°	21.8°	22.0° (4.04 Å)
200	N.F.	35.7°	N.F.	N.F.	N.F.	N.F.	N.F.	N.F.	N.F.	N.F.	36.0°	36.0°	36.05° (2.489 Å)

Note : (β)Cristobalite (ASTM 4-0359)

h k l	2θ°(Cu Kα)	d (Å)
1 0 1	21.4	4.15
2 2 0	35.45	2.53

N.F. : Not found

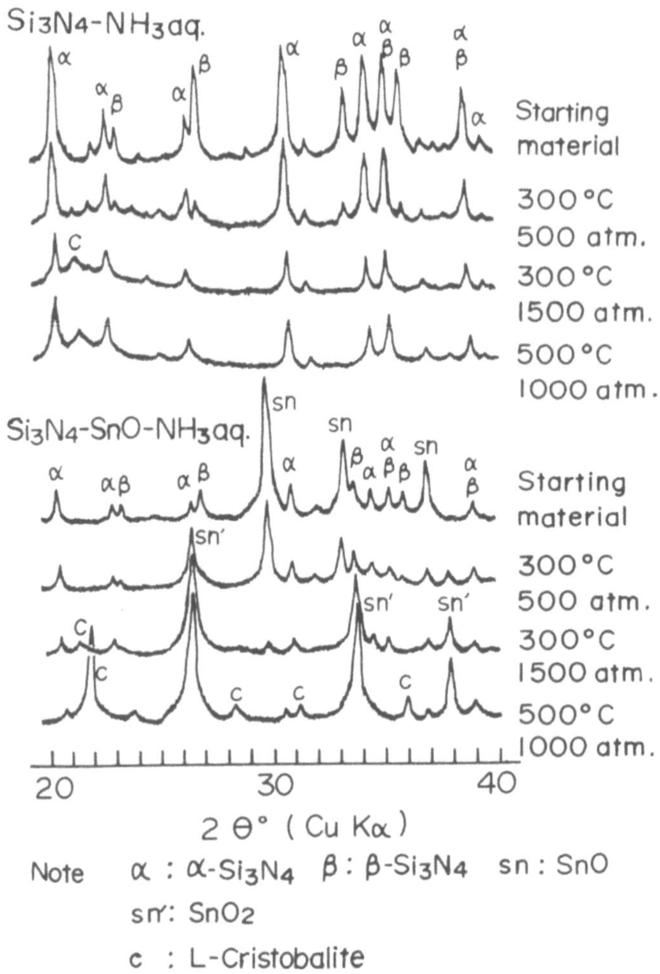

Fig. 3. X-ray diffraction patterns of Si_3N_4 heated at 300°C and 500°C in ammonia soultion

X-ray diffraction patterns of the sample treated at 300 and 500°C are shown in Fig. 3 and Table 2. In the system Si_3N_4-NH_3aq, the peak intensity of Si_3N_4 decreased and shifted to higher angle without sharpness as the pressure increased from 500 to 1500 atm. A new broad peak observed in the sample treated at 500°C appeared at 21.6° 2θ angle as the same of one at 300°C. The reaction product mentioned above in the system Si_3N_4-NH_3aq was assigned to be an amouphous SiO_2 or low-cristobalite. X-ray diffraction peaks of the sample treated at 300°C in the system Si_3N_4-SnO-NH_3aq, appeared more clearly at 21.5° 2θ angle above 1000 atm. On the other hand, the sample which was treated at 500°C showed the sharp peak at 21.5° 2θ angle with other peaks of low-cristobalite. The peak at 21.8° 2θ angle didn't shift as the testing pressure increased, but it shifted by increasing temperature from 300 to 500°C.

It is known that the oxidation of Si_3N_4 in air began at 750°C under the normal air pressure. The higher pressure of water vapor may allow the oxidation temperature to be down. It might be assumed that the water vapor played a role as an effective catalizer or a solvent in the oxidation of Si_3N_4.[3] It seems that at early stage of the oxidation of Si_3N_4 the amorphous SiO_2 is formed and successively transforms to the crystaline low-cristobalite.

In our experiment, probably, the amorphous SiO_2 was formed at the comparatively low temperature and pressure. Krejci and Ott[5] and Warren[6] reported that the artificial silica-gel showed the diffused X-ray diffraction peaks of low-cristobalite which differ from them of low-quarz and tridymite.

It may be assumed that the SiO_2 structure increases the periodicity by itself and shifts to the low-cristobalite structure with rising temperature and pressure. When O_2 oxidizes the surface of Si_3N_4 particle, the water vapor acts as a catalizer and helps the decomposition of Si_3N_4.

In the case of the system Si_3N_4-SnO-NH_3aq low-cristobalite might be formed more easily as the polymerization of the decomposed SiO_2 was accelerated by H^+ ion from the oxidation of SnO.

3. Summary

(1) The oxidezing decomposition of Si_3N_4 was easy when Si_3N_4 was in the ammonia solution from 300 to 500°C and from 500 to 1500 atm, and the oxidation began above 200°C under very low pressure namely 30 kg/cm^2 with the formation of the amorphous SiO_2 and/or cristobalite. The X-ray diffraction peak of low-cristobalite shifted to higher angle as the temperature and pressure were increased.

(2) When SnO was added to the system Si_3N_4-NH_3aq, the reaction product of Si_3N_4 was well crystalized into low-cristobalite. It may be a reason of the well-crystalized low-cristobalite that H^+ ion in the atmosphere generated by the oxidation of SnO affected the polymerization rate of the decomposed SiO_2.

References

1) D. R. Messier and M. M. Murphy, Silicon Nitride for Structural Applications. Annotated Bibliography, Army Materials & Mechanics Research Center, Watertown, Massachusetts 02172, U.S.A. (1975).
2) A. J. Kiehle, L. K. Hevng, P. J. Glelisse and T. J. Rockett, J. Am. Ceram. Soc., **58**, (1/2) 17 (1925).
3) L. Milton, J. F. Joseph, Am. Ceram. Soc. Bull., **57**, (4) 457 (1978).
4) S. C. Singhal, J. Am. Ceram. Soc. **59**, (1/2), 81 (1976).
5) L. Krejci, E. Ott, J. Phys. Chem., **35**, 2061 (1931).
6) B. E. Warren, J. Appl. Phys., **8**, 645 (1937).

Reprinted from *Environmental Science & Technology*, **14** (1980) 550–552.

Hydrothermal Decomposition of Polychlorinated Biphenyls

Nakamichi Yamasaki*

The Research Laboratory of Hydrothermal Chemistry, Faculty of Science, Kochi University, Akebono-cho, Kochi 780, Japan

Takaji Yasui and Kiyoshi Matsuoka

Department of Chemistry, Faculty of Science, Kochi University, Akebono-cho, Kochi 780, Japan

■ Decomposition of polychlorinated biphenyls (PCBs) by dechlorination was studied with a microautoclave. PCBs were decomposed completely in the presence of a methanol and sodium hydroxide solution under hydrothermal conditions of 300–320 °C and 180 kg/cm^2 pressure. A similar dechlorinating decomposition process using 4-chlorobiphenyl as a simple model compound was tested. A possibility was suggested for the industrial treatment of PCBs with a continuous pipeline system. The decomposed mixture was easily treated with activated bacteria.

Environmental contamination by polychlorinated biphenyls (PCBs), which are stable compounds, is a widespread social problem. Some methods (*1–3*) have been investigated for the decomposition of PCBs, including open systems using radiant rays or an oxygen flame of temperature >2000 °C.

Authors have confirmed that the chlorine atoms of PCBs are substituted by hydroxyl groups under the hydrothermal condition of a sodium hydroxide solution. A closed system with an autoclave for the hydrothermal dechlorination of PCBs was studied. The treatment proposed here is a safe, simple, and rapid method for the decomposition of PCBs. The released chloride from PCBs forms a chloride ion or sodium chloride, and the dechlorinated organic compounds are safely burned or are treated easily by activated sludge processes. A pipeline system such as the Zimmermann process (*4*) may be suitable rather than a large-volume autoclave for the treatment of PCBs on an industrial scale. In this study, various effects were examined to find out the most suitable conditions for the hydrothermal dechlorination of PCBs, and a possibility of industrial treatment was suggested with a pipeline system.

Experimental

Apparatus. Figure 1 shows the microautoclave that was used in this study. The stirring ball in the reaction chamber can be rolled up and down with the rocking motion of the heater. The induction heating method as shown in Figure 2 was used. The heating system is convenient for rapid temperature increasing (100 °C/min).

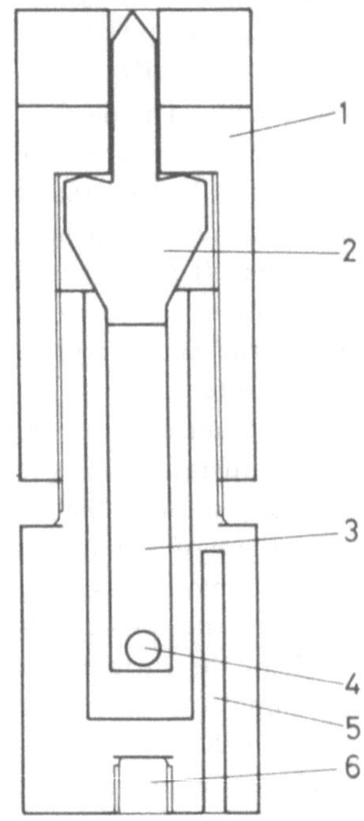

Figure 1. Microautoclave: (1) lid; (2) cone packing; (3) reaction chamber (60 mL); (4) ball for stirring; (5) thermocouple well; (6) socket of steel rod for taking out the autoclave

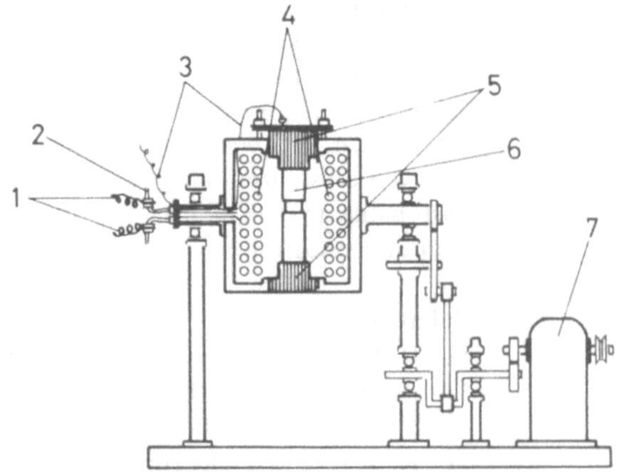

Figure 2. Induction heater: (1) lead wire for alternating current; (2) cooling water; (3) lead wire of thermocouple; (4) copper tube (induction coil); (5) steel blades of ferrosilicon; (6) microautoclave; (7) reduction gear

Test Sample. The sample of PCBs used in this study was KC-400, which was produced at Kanegafuchi Chemical Co. It was confirmed using a modified Carius method (*5*) that the PCB sample contained four chlorine atoms per molecule of PCB on the average.

Procedure. PCB (1–15 g) and 15–30 mL of solvent (a sodium hydroxide solution and dispersing agents) were poured into the microautoclave (inner volume: about 60 mL). The total volume of the mixture in the autoclave was adjusted to 30 mL (half the volume of the reaction chamber in Figure 1). The mixture in the autoclave was heated with an induction heater, and then cooled with a fan. The mixture was run off with water and hexane in a separating funnel. The aqueous and organic layers were separated. The aqueous solution was washed with hexane, and the organic layer was washed with a 0.25 M sodium hydroxide solution repeatedly. All aqueous solutions were mixed and decolored with 1 g of activated charcoal. The amount of chloride ion in the aqueous solution was determined with Mohr's titrimetric method. On the other hand, all hexane solutions were mixed and dried with 20 g of anhydrous calcium chloride and then evaporated to 10 mL. The amount of undecomposed PCBs in the hexane solution was determined using gas chromatography and mass spectrometry.

Results and Discussion

Dechlorination Conditions. The pressure in the reaction chamber was measured with a specially made autoclave equipped with a pressure gauge. The pressure decreased as the concentration of the sodium hydroxide solution increased. In this study, the pressure was ~170–200 kg/cm^2 at 300 °C.

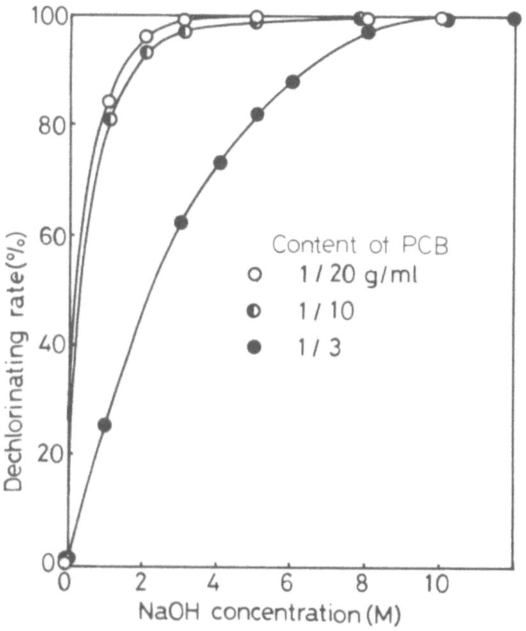

Figure 3. Effects of NaOH concentration on the dechlorination: total volume of PCB and NaOH solution, 30 mL; reaction time, 20 min; temperature, 350 °C

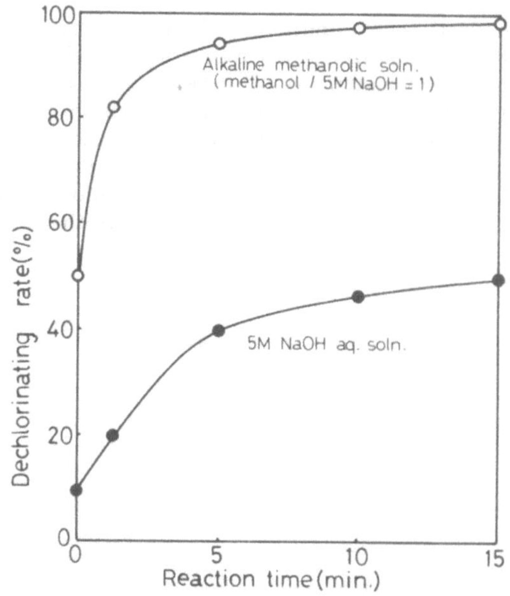

Figure 5. Effect of methanol: total volume of PCB (3 g) and 5 M NaOH solution or alkaline methanolic solution (equal volume mixture of methanol and 5 M NaOH solution), 30 mL; reaction temperature, 250 °C

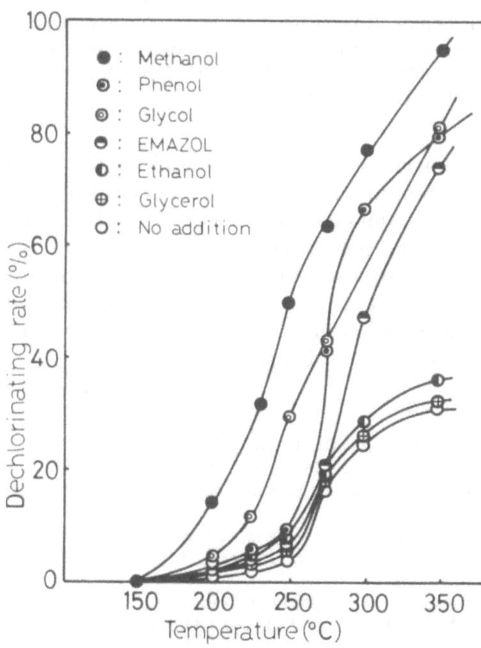

Figure 4. Effects of dispersing agents: total volume of PCB (3 g) and 5 M NaOH solution, 15 mL; dispersing agent, 15 mL; reaction time, 0 min

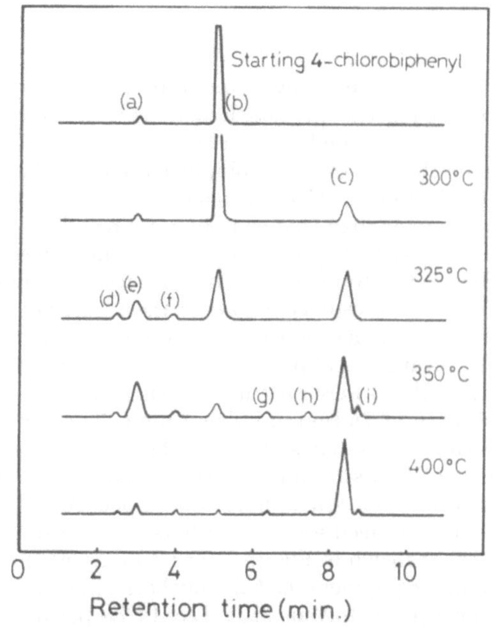

Figure 6. Gas chromatograms of hexane extracts from the 4-chlorobiphenyl solutions treated at various temperatures: reaction chamber volume of autoclave, 10 mL; reaction time, 3 min; 4-chlorobiphenyl, 0.2 g; 5 M NaOH solution, 5 mL

Figure 3 shows the effects of the concentration of the sodium hydroxide solution on the dechlorinating decomposition of PCBs in the absence of dispersing agents such as alcohol or glycol. A large amount of sodium hydroxide was needed for the complete dechlorination of PCBs even at 350 °C.

PCBs are generally insoluble in water, but their solubility is expected to increase under hydrothermal conditions. High temperatures are found to be more conducive to PCB decomposition than low temperatures under any conditions. A homogeneous condition may also be effective for the hydrolysis reaction. The effects of the following dispersing agents were tested: methanol, ethanol, ethylene glycol, glycerol, phenol, and Emazol (produced by KAO ATRAS Co.). Of these, methanol was the best agent for the PCB decomposition. The results are summarized in Figures 4 and 5. The recommended conditions for the dechlorination of PCBs with the microautoclave are: temperature, 300–320 °C; PCB, 1 g; methanol, 12.5 mL; and 5 M sodium hydroxide solution, 12.5 mL. PCBs were completely decomposed under these conditions within a few minutes. The decomposition of PCBs was confirmed using gas chromatography and mass spectrometry.

Mechanism of Dechlorination. The hydrolysis mechanism of chlorobenzene has been studied by Lüttringhaus and Ambros (*6*). Bottini and Roberts (*7*) have proposed a mechanism for the hydrolysis of halotoluene.

4-Chlorobiphenyl was used as a simple model compound for PCB dechlorination, treated in the same way as PCBs under hydrothermal conditions. The results are shown in Figure 6. The peaks in Figure 6 were assigned using infrared and mass spectrometry: (a) 2-chlorobiphenyl; (b) 4-chlorobiphenyl; (c) biphenyl-4-ol; (d) 4-phenoxybiphenyl; (e) 2-phenoxybiphenyl; (f) biphenyl-2-ol; (g, h, and i) di(4-biphenyl), di(2-biphenyl), or 2-biphenyl 4-biphenyl ether. The following mechanism is proposed for the dechlorinating decomposition of 4-chlorobiphenyl under hydrothermal conditions:

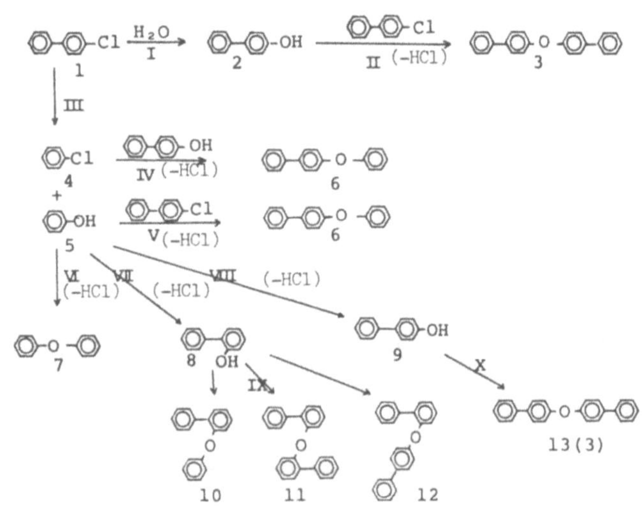

The main reaction at low temperature was suggested to be stage I in the reaction process. Reactions II–X, as well as reaction I, were presumed to occur at 325–400 °C. The decomposition mechanism of 4-chlorobiphenyl is a useful model for the dechlorination of PCBs, though the reaction of PCBs may be more complex than that of 4-chlorobiphenyl.

Bottini and Roberts proposed a benzyne mechanism for the hydrolysis of halotoluene (*7*). The formation of chlorinated dibenzofurans is presumed. However, these were not detected in the decomposition products of 4-chlorobiphenyl. In the case of PCBs, the analysis of products was not performed because the starting PCBs were mixtures of many isomers.

Possibility of Industrial Treatment. The continuous-pipeline capillary system seems to be the most suitable for the treatment of PCBs on an industrial scale. PCBs were decomposed within a few minutes above 300 °C, and a long pipeline may be needless. The authors assembled a "mini"-test plant, as shown in Figure 7. The Zimmermann process, using similar equipment, had already been applied to a waste disposal process (*4*). About 20 mL of PCBs per minute was able to be decomposed successively by the test plant.

Biological treatment was studied as a method for the decomposition of the PCB solution. A PCB solution dechlorinated by the proposed method was diluted with water to 2350 ppm of the biochemical oxygen demand (BOD) value. The solution was treated with a mixture of six Enzobac species (produced by the American Enzyme Co.). The BOD value was decreased by about 98% after 9 days. These results support the view that highly toxic materials such as chlorinated dibenzofuran do not exist. Furthermore, the waste solutions containing dechlorinated PCBs can be easily treated with activated sludge without hazard.

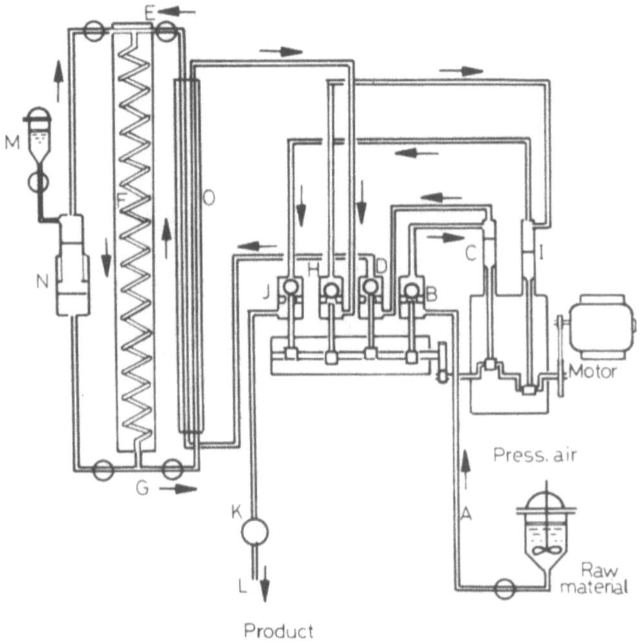

Figure 7. Flow diagram of a continuous pipeline system for industrial treatment. N and M are the injection systems for PCB treatment only, and the raw material tank is filled with an alkaline methanolic solution only. In the treatment of industrial waste solutions containing PCBs, N and M are not necessary and the raw material tank is filled with waste solution: (A) inlet pipe; (B) ball valve; (C) injection pump; (D) ball valve; (E) reactor column inlet; (F) reactor column; (G) reactor column outlet; (H) ball valve; (I) back pressure pump; (J) ball valve; (K) pressure control valve; (L) outlet; (O) heat exchanger

Literature Cited

(1) Nasu, H., *Sangyo Kogai*, **11**(3), 221 (1973).
(2) Imamura, A., *Kagaku*, **28**(7), 581 (1973).
(3) Komamiya, K., Morisaki, S., *Environ. Sci. Technol.*, **12**(10), 1205 (1978).
(4) Zimmermann, F. J., *Chem. Eng.*, **25**, 117 (1958).
(5) Carius, G. L., *Ann. Chem. Pharm.*, **136**, 129 (1865); *Z. Anal. Chem.*, **4**, 451 (1865).
(6) Lüttringhaus, A., Ambros, D., *Chem. Ber.*, **89**, 463 (1956).
(7) Bottini, A. T., Roberts, J. D., *J. Am. Chem. Soc.*, **79**, 4458 (1957).

Received for review January 29, 1979. Accepted January 29, 1980.

Reprinted from *Journal of the American Ceramic Society*, **58**(5–6) (1976) 277.

Decomposition of Ilmenite in KOH-H_2O Solutions Under High-Pressure Conditions

SHIN-ICHI HIRANO, M. G. M. U. ISMAIL, and SHIGEYUKI SŌMIYA*

OF ALL the naturally occurring minerals containing Ti as a major constituent only two, i.e. ilmenite and rutile, have been seriously considered as a suitable feedstock for either the metal-producing or pigment industries. Current methods for decomposing ilmenite[1] are based mainly on the treatment of ilmenite with sulfuric acid or chlorination at high temperatures. The object of the present work was to study the effects of KOH solution on the decomposition and stability of ilmenite under hydrothermal conditions.

Natural mineral ilmenite sand separated from beach sand was used as the starting material. This mineral sand was magnetically purified and treated with 3*N* HCl at 80°C for 20 min. Samples ground to 200 mesh were studied hydrothermally. Required amounts of the solution and the ilmenite sand (i.e. 60 mg of 10*M* KOH and 50 mg of ilmenite sand) were sealed hermetically in an Au capsule 35 mm long, 3 mm OD, and 0.15 mm thick. The capsule was placed in hydrothermal equipment described in Ref. 2 and the pressure and temperature were raised to the desired value. After the sample was quenched, the products were identified by X-ray diffraction and optical microscopy.

At 500°C and 500 kg/cm² pressure, experiments were performed in order to study the effects of the concentration of the KOH solution on the rate of decomposition of ilmenite. As shown in Fig. 1, the rate of decomposition increased with increasing molarity of the KOH solution.

The effects of temperature were studied using a 10*M* KOH solution under 500 kg/cm² pressure. The rate of decomposition of ilmenite increased with increasing temperature, as shown in Fig. 1. Increasing the quantity of water also caused a higher decomposition rate; as shown in Fig. 2, under conditions of 10*M* KOH, a 5:3

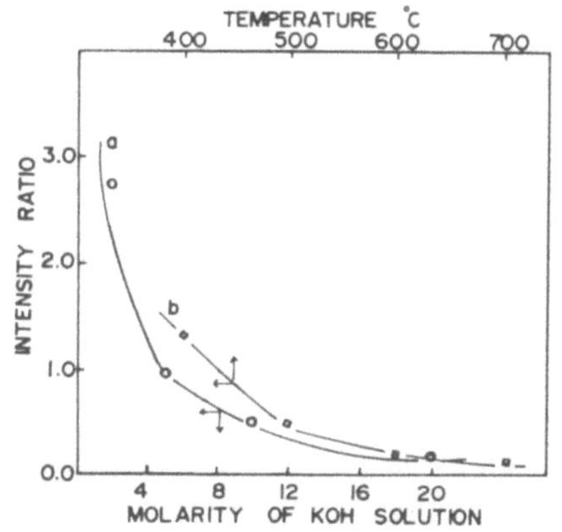

Fig. 1. Effect of concentration of KOH solution (curve *a*) and temperature (curve *b*) on hydrothermal decomposition of ilmenite. Intensity ratio = (intensity of ilmenite line (104) / (intensity of $K_2O(TiO_2)_4$ line ($d = 2.94$ Å)) for curve *a* and curve *b* for $T \leq 600$°C; intensity ratio = (intensity of ilmenite line (104)) / (intensity of $K_2O(TiO_2)_4$ line (311)) for $T > 600$°C.

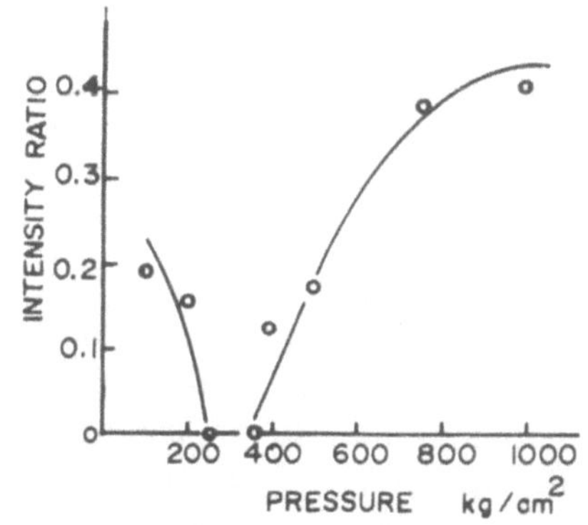

Fig. 2. Effect of pressure on decomposition of ilmenite under hydrothermal conditions. Intensity ratio = (intensity of ilmenite line (104)) / (intensity of $K_2O(TiO_2)_4$ line ($d = 2.94$ Å)).

ilmenite-to-water ratio at 500°C, ilmenite decomposed completely only at pressures between 250 and 350 kg/cm². The time required for complete decomposition of ilmenite at 500°C, 300 kg/cm² using a 10*M* KOH solution and maintaining a 5:3 ilmenite-to-water ratio was 63 h. The products obtained were magnetite and $K_2O(TiO_2)_4$. Under these same experimental conditions, increasing the duration of the treatment to 84 h resulted in the formation of $K_2O(TiO_2)_6$ (not fibrous) with magnetite.

The possibility of preparing stoichiometric magnetite by the decomposition of ilmenite under hydrothermal conditions was investigated by varying the quantity of water and increasing both the duration of treatment and temperature. The results indicate that increasing the water content has little effect on the preparation of stoichiometric magnetite via the decomposition of ilmenite. However, increasing the temperature and duration of treatment resulted in the formation of stoichiometric magnetite. For example, at 800°C, 300 kg/cm² pressure, a 5:4 ilmenite-to-water ratio, and a 10*M* KOH solution, stoichiometric magnetite and fibrous $K_2O(TiO_2)_6$ resulted after a 24-h treatment.

This type of behavior of magnetite resulted from the formation of an Fe_2TiO_4-$Fe_{3-x}O_4$ solid solution. High temperatures favored the separation of Fe_2TiO_4 from the solid solution to form stoichiometric magnetite. A fibrous type of $K_2O(TiO_2)_6$ was formed at temperatures >600°C, 300 kg/cm² pressure, and using 10*M* KOH solution as the result of a 24-h treatment.

Chemical analysis of the starting material showed the presence of Fe_2O_3. Whether or not Fe_2O_3 affects decomposition was studied in a series of experiments using pure ilmenite synthesized under hydrothermal conditions.[3] The results agreed well with those of experiments using natural ilmenite.

The present results indicate that ilmenite can be decomposed completely to magnetite and $K_2O(TiO_2)_4$ under hydrothermal conditions using a 10*M* KOH mineralizer at 500°C under 300 kg/cm² pressure in 63 h and to magnetite and fibrous $K_2O(TiO_2)_6$ at >600°C under 300 kg/cm² pressure for 24 h. It is, therefore, possible to obtain pure titanate by the magnetic separation of magnetite from the decomposed products.

Acknowledgments: The writers thank the Ceylon Institute of Scientific and Industrial Research of Sri Lanka for supplying the sample of natural ilmenite and the Ishihara Sangwo Co. Ltd. for the analysis of the ilmenite sample. M.G.M.U. Ismail is indebted to the scholarship program of the Ministry of Education of Japan for providing him with the opportunity to take part in the present work.

Received December 23, 1975; revised copy received February 11, 1976.
The writers are with the Research Laboratory of Engineering Materials, Tokyo Institute of Technology, Ookayama, Meguro-ku, Tokyo 152, Japan.
*Member, the American Ceramic Society.

[1] William Volk and H. H. Stotler, "Hydrogen Reduction of Ilmenite Ores in a Fluid Bed," *J. Metals*, **22** [1] 50–53 (1970).
[2] R. Roy and O. F. Tuttle; pp. 138–80 in Physics and Chemistry of the Earth, Vol. 1. Edited by L. H. Ahrens, Kalervo Rankama, Frank Press, and S. K. Runcorn. Pergamon Press, New York, 1956.
[3] Shin-ichi Hirano and Shigeyuki Somiya; "Crystal Growth of Ilmenite Under Hydrothermal Conditions"; abstract of the 20th Jinko Kobutsu Toronkai, p. 92 (1975).

Reprinted from *Journal of Materials Research*, **1** (1986) 100–103.

Oxidation of SiC powder by high-temperature, high-pressure H_2O

Masahiro Yoshimura, Jun-ichiro Kase, and Shigeyuki Sōmiya
Research Laboratory of Engineering Materials and Department of Materials Science and Engineering, Tokyo Institute of Technology, 4259 Nagatsuta, Midori, Yokohama, 227 Japan

(Received 10 October 1985; accepted 6 January 1986)

The reaction between SiC powder and H_2O has been studied at 400°–800 °C under 10 and 100 MPa. Silicon carbide reacted with H_2O to yield amorphous SiO_2 and CH_4 by the reaction $SiC + 2H_2O \rightarrow SiO_2 + CH_4$ above 500 °C. Cristobalite and tridymite crystallized from amorphous silica after the almost complete oxidation of SiC above 700 °C. The oxidation rate, as calculated from the weight gain, increased with temperature and pressure. The Arrhenius plotting of the reaction rate based on a Jander-type model gave apparent activation energies of 167–194 kJ/mol. Contrasted with oxidation in oxidative atmosphere, oxidation in H_2O is characterized by the diffusion of H_2O and CH_4 in an amorphous silica layer where the diffusion seemed to be rate determining. Present results suggest that the oxidation of SiC includes the diffusion process of H_2O in silica layers when atmospheres contain water vapor.

I. INTRODUCTION

Silicon carbide, as well as silicon nitride, has attracted great attention in recent years because of its potential application as one of high-temperature structural ceramics. Oxidation resistance is one of the most important properties in these applications. As for silicon carbide powders, the oxidation behavior, investigated for many samples in various oxidizing atmospheres, indicates a parabolic kinetics.[1–7] The reaction is believed to be as follows:

$$SiC + 2O_2 \rightarrow SiO_2 + CO_2 \,. \qquad (1)$$

Suzuki[3] and Jorgensen *et al.*[7] studied the effects of water vapor on the oxidation of silicon carbide powders. Singhal[8] also examined the effects of water vapor on the oxidation of hot-pressed silicon carbide. They argued that the oxidation rate was accelerated when water vapor was added to oxygen. However, the reaction of SiC by only H_2O has not been studied yet. We have studied the reactions between SiC powder and H_2O under hydrothermal conditions where almost only high-temperature, high-pressure H_2O exists. Furthermore, we will discuss the effect of H_2O on the oxidation of SiC in comparison with a similar study on Si_3N_4 reported previously.[9]

II. EXPERIMENTAL PROCEDURE

The starting material was SiC powder (Starck A10, average grain size 0.65 μm, 94% α form). The sample powder was placed in an Au tube (2.7 mm i.d., 0.15 mm thick, and 35 mm long) of which the top was flattened and bent by pressing to prevent the escape of solid materials. The tube was heated in a test-tube–type pressure vessel under pressures of 10 or 100 MPa of redistilled water. In the treatment the temperature was varied from 400° to 800 °C, and time for 0 to 72 h. After the treatment the pressure vessel was quenched, then the sample tube was dried and weighed to calculate the oxidation ratio of SiC. Solid products were examined by x-ray diffractometry (XRD) and scanning electron microscopy (SEM). The oxidation ratio was also checked by XRD through the decrease of the SiC phase. Gaseous products in the vessel were analyzed by a mass spectrometer.

III. RESULTS

Under 100 MPa H_2O, the oxidation of SiC was not observed at 400 °C but proceeded above 500 °C as a function of time (Fig. 1). Weight gain and appearance of a halo in x-ray diffraction indicated that amorphous silica was formed by the oxidation of SiC as the first step at every temperature above 500 °C. Cristobalite and tridymite crystallized above 700 °C after 8 h where SiC was almost completely changed into amorphous silica. Scanning electron micrographs showed that the surfaces of the SiC particles were covered with amorphous

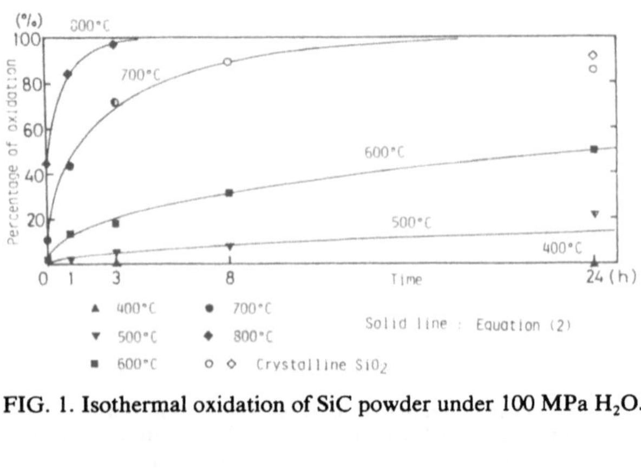

FIG. 1. Isothermal oxidation of SiC powder under 100 MPa H_2O.

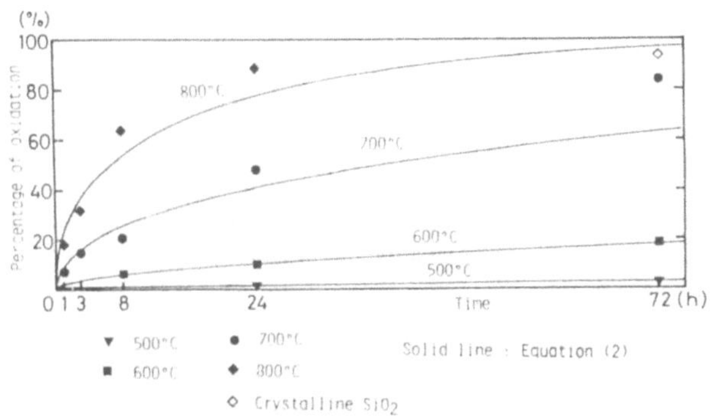

FIG. 2. Isothermal oxidation of SiC powder under 10 MPa H_2O.

silica by the oxidation and these silica grains grew and sintered during further hydrothermal treatments.

Under 10 MPa H_2O, the oxidation behavior was similar to that under 100 MPa H_2O but the oxidation rate was slower, as shown in Fig. 2. Amorphous silica was produced at every temperature and then crystallized to form cristobalite at 800 °C after 72 h.

The oxidation ratio was approximately represented by a Jander-type equation,[9] as illustrated with solid lines in Figs. 1 and 2:

$$[1-(1-\alpha)^{1/3}]^2-[1-(1-\alpha_0)^{1/3}]^2=Kt\,, \qquad (2)$$

where t is the time, α is the oxidation ratio at t h, α_0 is the oxidation ratio at 0 h, and K is the rate constant. Using this equation we can calculate the rate constants at every temperature. Arrhenius plots of the rate constants under both pressures (Fig. 3) suggest that H_2O pressure does not change the reaction mechanism but accelerates the reaction rate. Calculated apparent activation energies are 167 kJ/mol under 100 MPa H_2O and 194 kJ/mol under 10 MPa H_2O for the oxidation of SiC powder. These values are somewhat smaller than the values reported for the oxidation of SiC powders in dry air or oxygen (210–330 kJ/mol[1,4,5]). It should be also noted that the oxidation of SiC powder proceeds in H_2O at temperatures as low as 500 °C in contrast to oxidation above 900 °C in air or oxygen.

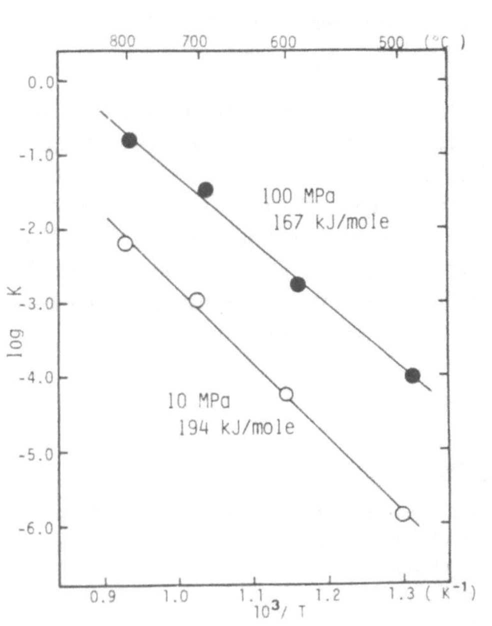

FIG. 3. Arrhenius plot of the hydrothermal oxidation isotherms of SiC powder. (Jander-type model.)

These differences suggest differing diffusion species in the silica layer between the oxidation in air or oxygen and in pressurized H_2O. Collected gases from the vessel contained a large amount of CH_4 and a small amount of CO_2. The formation of CH_4 indicates that the oxidation reaction is

$$SiC + 2H_2O \rightarrow SiO_2 + CH_4\,. \qquad (3)$$

This reaction is completely different from Eq. (1) where oxygen reacts with SiC to produce CO_2 as a product. Singhal[8] had expected the reaction to proceed as in Eq. (3) in water vapor, but he failed to detect the production of CH_4. Another possible reaction in pressurized H_2O is

$$SiC + 4H_2O \rightarrow SiO_2 + CO_2 + 4H_2\,. \qquad (4)$$

The Gibbs free energies of reactions (3) and (4) are calculated assuming the ideal gas laws to apply for CH_4, CO_2, and H_2 as follows:

$$\Delta G(3) = G^0_{SiO_2} + G^0_{CH_4} + RT\ln(P/0.1) - \left(G^0_{SiC} + 2G^0_{H_2O} + 2\int_{G^0}^{G} dG_{H_2O}\right), \qquad (5)$$

$$\Delta G(4) = G^0_{SiO_2} + G^0_{CO_2} + 4G^0_{H_2} + 5RT\ln(P/0.1) - \left(G^0_{SiC} + 4G^0_{H_2O} + 4\int_{G^0}^{G} dG_{H_2O}\right), \qquad (6)$$

where T is the requisite temperature (K), P is the requisite pressure (MPa), R is the gas constant, G^0_i is the Gibbs free energy of i under 0.1 MPa at T (Ref. 10), and $\int_{G^0}^{G} dG_{H_2O}$ is the difference of Gibbs free energy of H_2O between 0.1 MPa and pressure P MPa at temperature T (calculated from Ref. 11). Figure 4 gives the tempera-

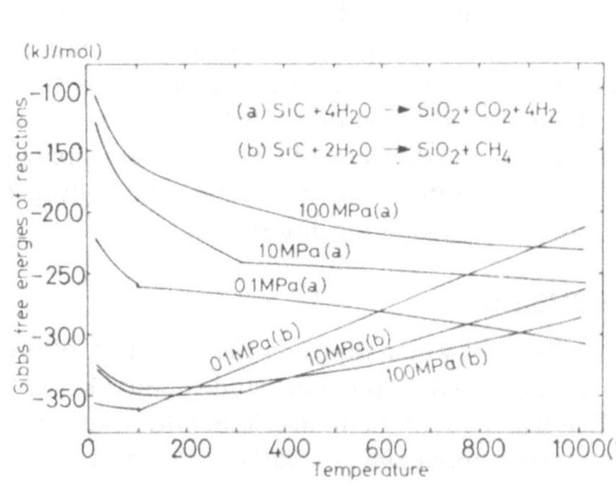

FIG. 4. Gibbs free energies of the reactions to produce CH_4 or CO_2 in high-temperature, high-pressure H_2O.

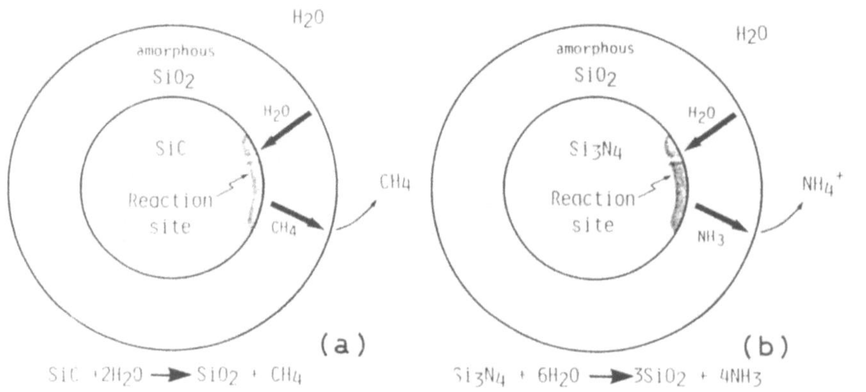

FIG. 5. Reaction model of hydrothermal oxidation of SiC and Si_3N_4 powder.

ture dependence of ΔG for various pressures and shows that SiC is a sufficiently strong reducing agent to make CH_4 rather than CO_2 as the stable carbon species in the hydrothermal capsule. Under 0.1 MPa the reaction which makes CO_2 is more stable than CH_4 above 600 °C. These calculations are completely compatible with Singhal's result.

We propose here a reaction model for SiC oxidation in H_2O [Fig. 5(a)]. After the amorphous silica layer is formed on the surface of SiC particles, H_2O and CH_4 diffuse from outside and from inside of the silica layer, respectively. A similar model, Fig. 5(b), was reported[9] for the oxidation of Si_3N_4 powder in pressurized H_2O where the apparent activation energies were 70–80 kJ/mol (under 100 MPa at 200°–400 °C and under 10 MPa at 200°–300 °C) and 130 kJ/mol (under 10 MPa at 700°–800 °C). The former value, which is similar to the activation energy of H_2O diffusion in amorphous silica [60–105 kJ/mol (Ref. 12,13)] would indicate that the oxidation of Si_3N_4 was controlled by H_2O diffusion rather than by O_2 diffusion in the silica layers.[9] The H_2O diffusion in the silica layer should be the same in the oxidation both of SiC and of Si_3N_4. The difference between the oxidation of SiC and Si_3N_4 in pressurized H_2O, therefore, would be due to the difference of produced gases CH_4 and NH_3. The former is a nonpolar gas while the latter is a polar gas. It is probable that nonpolar species are difficult to diffuse in amorphous silica phases. Thus the oxidation rate of SiC in H_2O seems to be controlled by the diffusion of CH_4 in the amorphous silica layer or separate diffusion of carbon and hydrogen if CH_4 has dissociated in the amorphous silica.

IV. DISCUSSION

The effects of water vapor on the oxidation of SiC have been discussed in previous studies. Suzuki[3] proposed that water vapor might accelerate the diffusion of oxygen, carbon monoxide, or carbon dioxide through the silica layer. Jorgensen *et al.*[7] considered that the diffusing species would be the same in the presence of water vapor and in the presence of oxygen, but water vapor might promote the transition of amorphous silica to cristobalite, which might change diffusion paths from a bulk diffusion to a grain-boundary diffusion.

Calculated Gibbs free energies of reaction by oxygen (1) are − 1175 and − 1016 kJ/mol under 0.1 MPa of oxygen pressure at 25° and 1000 °C, respectively. Therefore even if water vapor had been added to oxygen, oxygen must be a stronger oxidizing agent than water vapor. According to the present investigation it is probable that water vapor added to oxygen may cause the following two-step reactions in the oxidation of SiC:

$$SiC + 2H_2O \rightarrow SiO_2 + CH_4 , \qquad (3')$$

$$CH_4 + 2O_2 \rightarrow CO_2 + 2H_2O , \qquad (7)$$

or

$$SiC + 4H_2O \rightarrow SiO_2 + CO_2 + 4H_2 , \qquad (4')$$

$$4H_2 + 2O_2 \rightarrow 4H_2O , \qquad (8)$$

rather than the one-step reaction in Eq. (1):

$$SiC + 2O_2 \rightarrow SiO_2 + CO_2 .$$

Although reaction (4) is more stable than (3) at high temperature under atmospheric pressure, there is a possibility that Eq. (3) occurs as the one of the processes of the reactions. Since H_2O diffuses faster than oxygen in the silica layer, the overall reaction of (3) plus (7) or (4) plus (8) would proceed faster than the reaction (1). This is one of the reasons for acceleration of the oxidation of SiC in atmospheres containing H_2O.

REFERENCES

[1]T. Nakatogawa, J. Chem. Soc. Jpn. Ind. Chem. Sect. **57**, 348 (1954).
[2]H. Suzuki, Yogyo Kyokai Shi **65**, 88 (1957).
[3]H. Suzuki, Yogyo Kyokai Shi **67**, 157 (1959).
[4]G. Ervin, J. Am. Ceram. Soc. **41**, 347 (1958).
[5]R. F. Adamsky, J. Phys. Chem. **63**, 305 (1959).
[6]P. J. Jorgensen, M. E. Wadsworth, and I. B. Cutler, J. Am. Ceram. Soc. **42**, 613 (1959).

[7]P. J. Jorgensen, M. E. Wadsworth, and I. B. Cutler, J. Am. Ceram. Soc. **44**, 258 (1961).
[8]S. C. Singhal, J. Am. Ceram. Soc. **59**, 81 (1976).
[9]M. Yoshimura, J. Kase, and S. Sōmiya, Yogyo Kyokai Shi **94**, 129 (1986).
[10]D. R. Stull, and H. Prophet, JANAF *Thermochemical Tables*, NSRDS-NBS37 (National Bureau of Standards, Washington, D.C., 1971), 2nd ed.
[11]J. H. Keenan, F. G. Keyes, P. G. Hill, and J. G. Moore, *Steam Tables: Thermodynamic Properties of Water Including Vapor, Liquid, and Solid Phases* (International system of units-SI) (Wiley, New York, 1978).
[12]I. Burn and J. P. Roberts, Phys. Chem. Glasses **11**, 106 (1970).
[13]S. White, Nature (London) Phys. Sci. **230**, 192 (1971).

Reprinted from *Journal of the American Ceramic Society*, **69**(7) (1986) 583–584.

Dissolution and Reaction of Yttria-Stabilized Zirconia Single Crystals in Hydrothermal Solutions

MASAHIRO YOSHIMURA,* TAKEHIRO HIUGA, and SHIGEYUKI SŌMIYA*

Research Laboratory of Engineering Materials and Department of Materials Science, Tokyo Institute of Technology, 4259 Nagatsuta, Midori, Yokohama 227, Japan

Dissolution and reaction of yttria-stabilized zirconia (YSZ) single crystals were investigated in various solutions at 600° to 780°C under 100 MPa. YSZ crystals were not corroded in pure H_2O and neutral solutions such a LiF, LiCl, $NaNO_3$, KCl, KBr, K_2SO_4, and Na_2SO_4 even under severe conditions at 600°C, 100 MPa. They were, however, dissolved and reprecipitated in basic solutions such as NaF, K_2CO_3, KOH, NaOH, and LiOH with partial decompositon (destabilization) in the last three solutions. YSZ crystals were completely decomposed into *m*-ZrO_2 in acidic solutions of Li_2SO_4, H_2SO_4, and HCl, whereas they reacted with the solution and formed other compounds in KF, NH_4F, and H_3PO_4 solutions.

I. Introduction

YTTRIA-STABILIZED zirconia (YSZ) has been used widely in solid electrolytes for its ionic conductivity, special refractories for its thermal and chemical stability, and in toughening ceramics via its phase transformation. The effects of water on YSZ at high temperatures are important aspects of these applications. Recently, it has been recognized that water accelerates the degradation of Y-PSZ at ≃200°C,[1-3] where the transformation of metastable tetragonal zirconia into monoclinic zirconia has been observed. Some authors[4,5] have suggested that the reaction of yttria with water is responsible for the phase transformation. Others[6,7] have proposed that chemical adsorption of water on zirconia can cause the transformation. Such interfacial phenomena, interactions between water vapor and ceramic surfaces, particularly containing metastable tetragonal phases, are difficult to study. As the first step, therefore, we have studied the dissolution behavior of YSZ single crystals in various hydrothermal solutions.

II. Experimental Procedure

The specimens were commercial YSZ single crystals which were prepared by a skull-melting technique, and then cut and polished in pseudo-brilliant-type shapes. They were fully-stabilized cubic zirconia, with Y_2O_3 contents determined by energy-dispersive spectroscopic (EDS) analyses and lattice parameter measurements to be 14 mol%. A specimen was sealed in a Pt or Au capsule, 2.7 mm ID and 30 mm long, with one of the various aqueous solutions: H_2O; 15 and 30 wt% LiOH, NaOH, KF, LiF, NaF, and NH_4F; 30 wt% KOH, K_2CO_3, $NaNO_3$, KCl, LiCl, KBr, HCl, K_2SO_4, Na_2SO_4, and H_2SO_4; 25.7 wt% Li_2SO_4; concentrated H_3PO_4.

The capsule was heated in a test-tube-type pressure vessel with water at 100 MPa and temperatures up to 600° to 780°C for 0 to 168 h. After this treatment, the vessel was quenched in cold water. The specimen was washed, dried, and examined by scanning electron microscope, EDS, and X-ray diffraction. The solubility of the YSZ crystal was calculated from the weight loss.

Received July 12, 1985; revised copy received November 6, 1985; approved December 11, 1985.
*Member, the American Ceramic Society.

III. Results and Discussion

The behavior of YSZ crystals in the 30 wt% solutions at 600°C and 100 MPa for 24 h are classified into the following five categories:[8] (1) no detectable dissolution; (2) YSZ dissolution and YSZ precipitation; (3) partial decomposition, YSZ dissolved, and *m*-ZrO_2 precipitated; (4) complete decompositon into *m*-ZrO_2; (5) formation of compounds other than ZrO_2. These behaviors seemed to be dependent mainly on the basicity-acidity of solutions, as shown in Fig. 1 (the vertical column is degree of degradation; the horizontal column is only qualitative, because quantitative measures (like pH) are not available in hydrothermal solutions). Even under severe conditions (600°C under 100 MPa), YSZ crystals are not corroded by pure H_2O and many solutions of neutral salts, such as LiF, $NaNO_3$, KCl, LiCl, KBr, K_2SO_4, and Na_2SO_4. The weight loss of the specimen was within experimental error (<0.1 wt%) in these solutions except for LiCl (0.18 wt%) and $NaNO_3$ (0.1 wt%) (Fig. 2). No detectable change could be observed by SEM on the surfaces in these cases. In the 15 wt% KOH solution, on the other hand, small YSZ crystals, in which EDS showed yttrium contents, were reprecipitated along the surface polishing flaws (Fig. 3(*A*)). The weight loss of the specimen was 0.6 wt% in this case. In other alkaline (basic) solutions, larger weight losses were observed (Fig. 2). The sample surfaces were much more corroded and precipitates of *m*-ZrO_2 crystals were arranged in particular directions (Fig. 3(*B*)). The crystals did not contain yttrium components and showed typical *m*-ZrO_2 morphologies[9] when grown hydrothermally. The solubility of yttrium components appeared to be larger than zirconium components in these solutions. The latter, therefore, precipitated as monoclinic zirconia crystals. If an excess amount of yttria is added to the system, the alkaline solution can act as a good mineralizer for crystal growth[10] of cubic YSZ.

In acidic solutions, the selective dissolution of yttrium components is accelerated because yttrium ion is more basic than zirconium ion. This leaves a polycrystalline *m*-ZrO_2 layer on the surface of the specimen. H_2SO_4 solution brought about severe corrosion (weight loss) even in a short period of time (within 1 h), as seen in Fig. 2. Figure 4 shows the surface of the specimen covered by *m*-ZrO_2 whiskers after heating to 600°C at a rate of

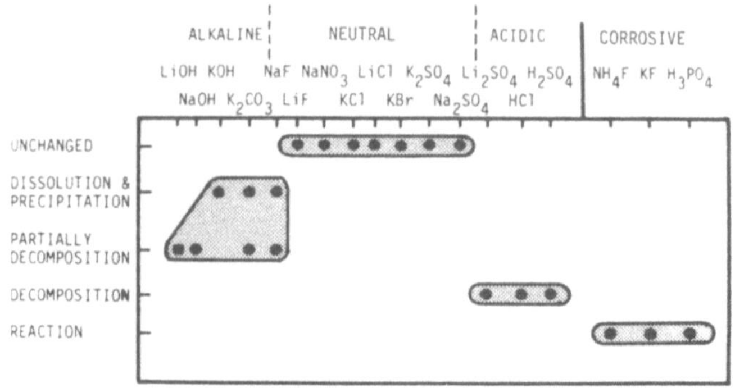

Fig. 1. Behavior of YSZ crystals in various hydrothermal solutions at 600°C, 100 MPa, for 24 h.

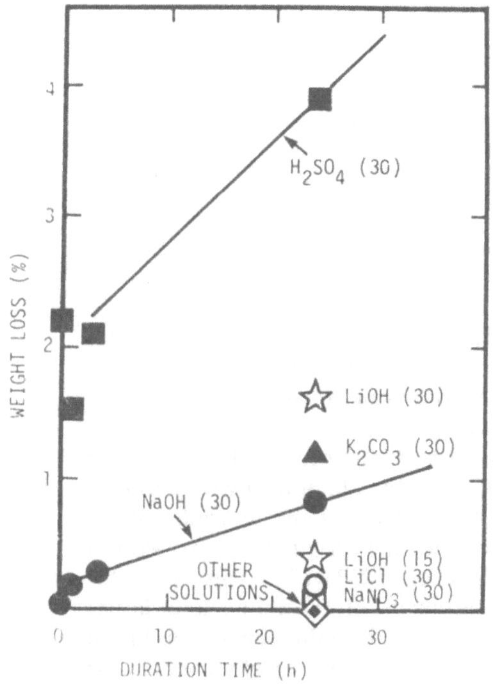

Fig. 2. Corrosion rate weight loss (%) of YSZ crystals in various high-temperature high-pressure solutions at 600°C, 100 MPa.

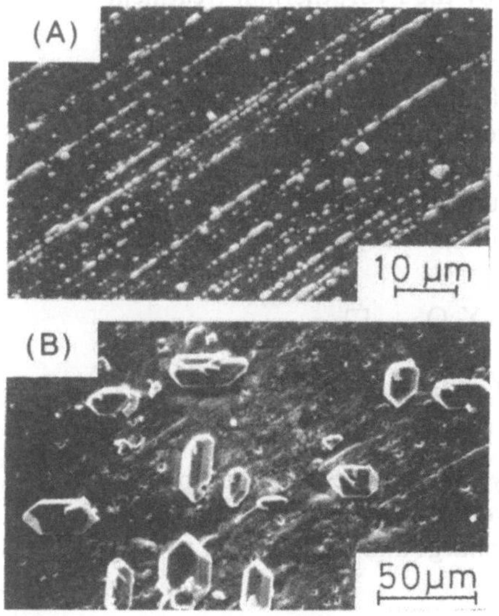

Fig. 3. Scanning electron micrographs of the surface of YSZ crystal after treatments at 600°C, 100 MPa, for 24 h: (*A*) in 15 wt% KOH solution, showing cubic zirconia grains which precipitated along the polishing flaws; (*B*) in 15 wt% LiOH solution, showing monoclinic zirconia crystals which were grown.

10°C/min and then quenching immediately (Fig. 4(*A*)) and with a 3-h growth period (Fig. 4(*B*)).

H_3PO_4 soluton similarly dissolves yttrium components selectively and yielded ZrP_2O_7 crystals on the surface. In solutions of NH_4F or KF, the surfaces of the specimen were signficantly corroded and changed to *m*-ZrO_2; the formation of YF_3 or an unidentified compound containing potassium and yttrium has also

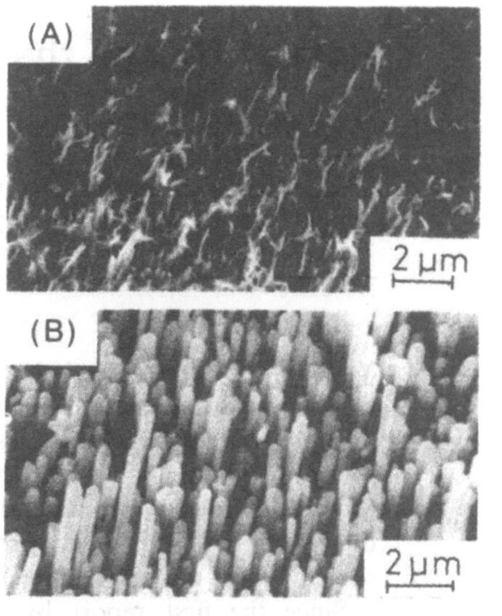

Fig. 4. Scanning electron micrographs of *m*-ZrO_2 whiskers on the surface of the YSZ crystal after treatment at 600°C, 100 MPa for (*A*) 0 and (*B*) 3 h, in 30 wt% H_2SO_4 solution.

been observed in these solutions, respectively.

The above results clearly indicate that the dissolution of yttrium components from YSZ is minimal in H_2O or neutral solutions even under severe hydrothermal conditions. If the reactivity of yttrium in YSZ is similar in Y-PSZ, the destabilization[5] of Y-PSZ by yttrium depletion or hydroxide formation is unlikely to occur in wet atmospheres and at low temperatures (200° to 300°C). The main effect of water on the tetragonal-to-monoclinic transformation would be chemisorption of H_2O to Zr–O–Zr bonds as suggested by Sato and Shimada[11] (Zr–O–Y bonds, also). This chemisorption brings about neucleating defects[12] for transformation through the formation of stress concentration sites,[7] rather than the release of the strain which acts to stabilize the tetragonal phase.[11] A study of the degradation of Y-PSZ in high-temperature high-pressure water is now in progress.

References

[1]K. Kobayashi, H. Kuwajima, and T. Masaki, "Phase Change and Mechanical Properties of ZrO_2-Y_2O_3 Solid Electrolyte after Aging," *Solid State Ionics*, **3/4**, 489–93 (1981).

[2]K. Nakajima, K. Kobayashi, and Y. Murata, "Phase Stability of Y-PSZ in Aqueous Solutions"; pp. 399–407 in Advances in Ceramics, Vol. 12, Science and Technology of Zirconia II. Edited by N. Claussen, M. Rühle, and A. H. Heuer. American Ceramic Society, Columbus, OH, 1984.

[3]K. Tsukuma, Y. Kubota, and T. Tsukidate, "Thermal and Mechanical Properties of Y_2O_3-Partially Stabilized Zirconia," *Yogyo Kyokaishi*, **92** [5] 233–41 (1984).

[4]H. Schubert, N. Claussen, and M. Rühle, "Surface Stabilization of Y-TZP," *Proc. Br. Ceram. Soc.*, **34**, 157–60 (1984).

[5]F. F. Lange, G. L. Dunlop, and B. I. Davis, "Degradation During Aging of Transformation-Toughened ZrO_2-Y_2O_3 Materials at 250°C," *J. Am. Ceram. Soc.*, **69** [3] 237–40 (1986).

[6]T. Sato, S. Ohtaki, and M. Shimada, "Transformation of Yttria Partially Stabilized Zirconia by Low Temperature Annealing in Air," *J. Mater. Sci.*, **20** [4] 1466–70 (1985).

[7]M. Yoshimura and S. Sōmiya; unpublished data presented at the International Workshop on Transformation Toughening, April 15–18, 1985, Lorne, Australia.

[8]M. Yoshimura, T. Hiuga, and S. Sōmiya, "Hydrothermal corrosion of Yttria Stabilized Zirconia (YSZ) Single Crystals"; pp. 137–47 in Zirconia Ceramics 2 (in Jpn.). Edited by S. Sōmiya and M. Yoshimura. Uchida Rokakuho Publishing Co., Tokyo, 1984.

[9]Y. Fujiki, T. Mitsuhashi, and Y. Suziki, "Hydrothermal Growth and Etching of ZrO_2 Crystals," *J. Am. Ceram. Soc.*, **55** [4] 223–24 (1972).

[10]M. Yoshimura, T. Hiuga, and S. Sōmiya, "Crystal Growth of Yttria Stabilized Zirconia under Hydrothermal Conditions," *J. Cryst. Growth*, **71** [1] 277–79 (1985).

[11]T. Sato and M. Shimada, "Transformation of Yttria-Doped Tetragonal ZrO_2 Polycrystals by Annealing in Water," *J. Am. Ceram. Soc.*, **68** [6] 356–59 (1985).

[12]I-Wei Chen and Y.-H Chiao, "Martensitic Transformations in ZrO_2 and HfO_2—An Assessment of Small Particle Experiments with Metal and Ceramic Matrices"; pp. 33–45 in Advances in Ceramics, Vol. 12, Science and Technology of Zirconia II. Edited by N. Claussen, M. Rühle, and A. H. Heuer. American Ceramic Society, Columbus, OH, 1984. □

Reprinted (with corrections) from *Journal of Materials Science Letters*, **6** (1987) 465–467.

Role of H_2O on the degradation process of Y–TZP

MASAHIRO YOSHIMURA, TATSUO NOMA, KATSUICHI KAWABATA, SHIGEYUKI SŌMIYA
Research Laboratory of Engineering Materials, and Department of Materials Science and Engineering, Tokyo Institute of Technology, 4259, Nagatsuta, Midori, Yokohama 227 Japan

Yttrium-doped tetragonal zirconia (Y–TZP) is expected to be a good structural ceramic material because of its high strength, high toughness and wear resistance. However, the instability (degradation) of Y–TZP around 150 to 400° C in air and above 100° C in water during ageing is a severe problem which must be solved before the direct application of this material.

Since the first report by Kobayashi *et al.* [1], many works [2–16] have been devoted to the study of degradation of Y–TZP. Their phenomological observations are summarized as follows:

1. The degradation proceeds most rapidly at temperatures of 200 to 300° C in (humid) air.
2. The degradation is caused by the t–m transformation accompanying micro- and macro-cracking.
3. The transformation progresses from the surface to the interior of the specimen.
4. Water or water vapour enhances the transformation.
5. The decrease in grain size and the increase in stabilizer content retard the transformation.

In order to explain these observations, some mechanical and chemical degradation mechanisms have been proposed. In the latter, the role of water has been suggested to form OH^- by the chemisorption at the surface of Y–TZP grains [9, 10], or to form $Y(OH)_3$ leading to the depletion of yttrium [12, 13]. In the present study we have examined the effects of H_2O on Y–TZP under hydrothermal conditions to emphasize the influence of water.

The Y–TZP specimen used has similar characteristics to those reported previously [5]; Toray Industries Inc., 2.66 mol % Y_2O_3, sintered 1450° C for 2 h then HIPed at 1400° C under 200 MPa for 1.5 h, grain size $\sim 0.5\,\mu$m, bending strength ~ 140 MPa, $3 \times 4 \times 36\,\text{mm}^3$ in size. The specimen was hydrothermally treated in re-distilled water under 5 and 100 MPa for < 48 h in a test-tube type vessel. After the treatment, the vessel was quenched in cold water, and then dried and weighed. The phases present and the lattice parameters were examined by X-ray diffraction using silicon and SiO_2 as the standard materials. Infrared spectroscopy and weight change were also studied for the hydrothermally treated and thereafter reheated samples.

The constituent phases of the starting sintered specimen were 90% tetragonal and 10% cubic. The tetragonal phase was transformed to monoclinic when well ground for X-ray diffraction. Its lattice parameters are given in the first column of the Table I. Hydrothermal treatment brought about the formation of monoclinic phase from the surface to the interior of the specimen. This phase had an expanded lattice compared to the starting lattice regardless of the temperature of the hydrothermal treatment, whereas the cubic phase changed neither in content nor lattice parameter. Subsequent re-heating, at 400° C in vacuum for 6 h or 1300° C in air for 1 h, of the hydrothermally treated samples brought about the returning of the lattice parameters to the original values with a decrease in weight of 0.233 wt %. This reversible expansion and shrinkage of the monoclinic lattice seems to be caused by the introduction and exclusion of OH^- in the lattice. If the depletion of yttrium by the formation of $Y(OH)_3$ or by the dissolution of Y^{3+} had occurred on hydrothermal treatment, such a reversible recovery of the lattice parameters could not be expected in a short period and/or at a lower temperature.

Infrared spectroscopy (Fig. 1) clearly showed the introduction of OH^- by the hydrothermal treatment and the exclusion of OH^- by the subsequent re-heating. Raman spectroscopy also confirmed the existence of OH^- in the sample treated hydrothermally. The above results undoubtedly indicate the introduction of OH^- into the zirconia lattice at least under the hydrothermal conditions studied. According to the weight change, the compositional change caused by H_2O is shown by

$$\mathrm{Zr}_{1-x}\mathrm{Y}_x\mathrm{O}_{2-x/2}\square_{x/2}$$

$$\xrightleftharpoons{\partial \mathrm{H_2O}} (\mathrm{Zr}_{1-x}\mathrm{Y}_x)\mathrm{O}_{2-2/x-\delta}(\mathrm{OH})_{2\delta}\square_{x/2-\delta}$$

$$x = 0.0518, \quad \delta = 0.0159$$

that is, the ≈ 60% oxygen vacancies were occupied by OH^-. These OH^- would migrate faster than O^{2-}, because OH^- has less charge and a size similar to O^{2-}.

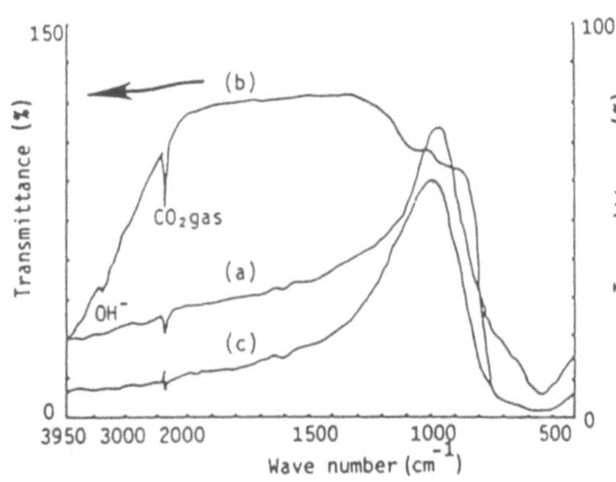

Figure 1 Infrared absorption spectra of the Y–TZP samples, (a) before hydrothermal treatment, (b) hydrothermally treated at 250° C under 100 MPa for 6 h, (c) reheated at 1300° C in air for 1 h.

TABLE I Latice parameters of monoclinic zirconia phases transformed from tetragonal phase after various treatment

Treatment	Conditions			Lattice parameters				
	Temp. (°C)	Pressure (MPa)	Time (h)	*a* (nm)	*b* (nm)	*c* (nm)	β (deg)	*V* (nm^3)
Before hydrothermal treatment				0.5166	0.5196	0.5319	98.61	0.1412
Hydrothermal treatment	250	5	6	0.5172	0.5217	0.5316	99.14	0.1416
	250	100	6	0.5175	0.5209	0.5319	99.12	0.1416
	400	100	6	0.5169	0.5217	0.5313	99.15	0.1415
	600	100	6	0.5174	0.5210	0.5319	99.13	0.1416
	800	100	6	0.5173	0.5212	0.5317	99.10	0.1416
	800	5	6	0.5173	0.5212	0.5314	99.12	0.1415
Re-heating after hydrothermal treatment (250° C, 100 MPa)	400	vac	6	0.5171	0.5203	0.5318	98.98	0.1413
	400	vac	10	0.5173	0.5201	0.5320	99.04	0.1413
	1300	air	1	0.5170	0.5193	0.5322	98.60	0.1413

Recently, Lange *et al.* [12] suggested that water vapour drew yttrium out of the solid solution to form $Y(OH)_3$ crystallite of 20 to 50 nm and that the yttrium-depleted volume element can act as monoclinic nuclei. Schubert [13] excluded the hydroxide formation by the diffusion of Y^{3+} because of its extreme slowness, but adopted the formation of $Y(OH)_3$ by faster diffusion of H_2O in the lattice. The existence of H_2O in the zirconia lattices has not yet been confirmed. In addition to this, even if the diffusion of H_2O were fast enough, the formation of $Y(OH)_3$ crystallite must be controlled by the slow diffusion of Y^{3+}. Supposing the $(D_{Y^{3+}})_{250^\circ C} \approx 1 \times 10^{-43}\,cm^2\,sec^{-1}$ extrapolated from the diffusion coefficient of Y^{3+} in 16 mol % Y–FSZ [17], we can estimate 10^{29} sec ($= 10^{21}$ year) for the time required for 1 nm diffusion according to the relation, $x \approx (Dt)^{1/2}$. The formation of $Y(OH)_3$ crystallite of 20 to 50 nm from the zirconia solid solutions, therefore, cannot be considered as an essential mechanism of the nucleation of the t–m transformation. The dissolution of Y^{3+} in the water is also disproved by analysis of the solvent carried out by Sato and Shimada [9] and our results on the hydrothermal dissolution of YSZ [18].

Based upon the above discussion where anion diffusion (not cation diffusion) should be responsible for the t–m transformation, and the observation that OH^- was formed at the surface and in the lattice of zirconia solid solutions, we propose the following degradation mechanism of Y–TZP by water (Fig. 2):

1st step: the chemical adsorption of H_2O at the surface.

2nd step: the formation of Zr–OH and/or Y–OH at the surface at which points stressed sites are created.

3rd step: the accumulation of the stresses by the migration of OH^- at the surface and in the lattice, to prepare nucleating defects [19].

4th step: the nucleation of monoclinic phase in the tetragonal grains; then the t–m transformation yields micro- and macro-cracking.

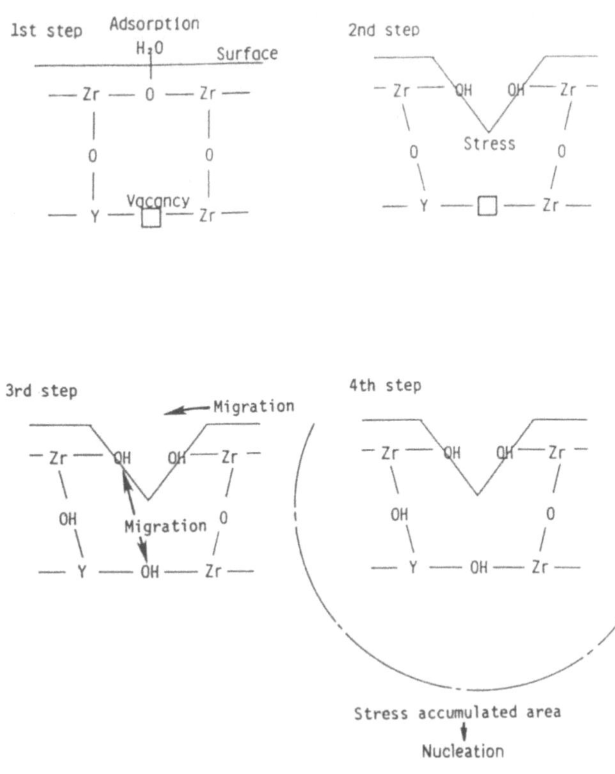

Figure 2 Proposed degradation process of Y–TZP by H_2O.

Saka *et al.* [15] and Nakanishi *et al.* [16] recently supposed the O^{2-} diffusion in the lattice to be responsible for the t–m transformation. But we suggest that OH^- rather than O^{2-} is the key, based upon the present experiments and faster diffusion of OH^- than O^{2-}. The formation of Zr–OH(Y–OH) at the surface of zirconias seems to be completely reasonable, because even hydrogen adsorbed on the surface of ZrO_2 calcined at 700° C can form Zr–OH on heating at 200° C according to the infrared study of Ohnishi *et al.* [15]. Sato and Shimada [9] also suggested the formation of Zr–OH at the surface, but they imagined that the formation of OH^- causes the release of strain which acts to stabilize the tetragonal phase. On the contrary, we think that the formation of Zr–OH brings about the stressed site.

Not only H_2O but also some polar gases such as NH_3 and non-aqueous solvents [9] of the molecules containing a lone-pair electron orbital opposite a proton donor site would also be adsorbed at the surface, and would then cut the Zr–O–Zr(Y) bonds to prepare stressed sites at the surface. Thus they also can enhance the t–m transformation through the accumulation of the stresses. Similarly to these chemical stresses, the mechanical stresses [13, 14] by mismatching the thermal expansion between the matrix and the dispersed grains, and heterogeneous thermal stresses also play a role in triggering the t–m transformation when the accumulated stresses overcome the nucleation barrier. The size, shape and location of t-grains as well as their composition may effect the accumulation of stresses in preparing nucleating defects [19].

As described above, the proposed mechanism seems to be consistent with other phenomenological observations concerning the degradation. A detailed study including macroscopic degradation of Y–TZP under hydrothermal conditions has been published elsewhere [20].

Acknowledgements

The authors thank Toray Industries Inc. for preparing the specimens, and Professor H. Imai and Dr T. Tagawa at our Institute for the infrared study.

References

1. K. KOBAYASHI, H. KUWAJIMA and T. MASAKI, *Solid State Ionics* **3/4** (1981) 489.
2. M. MATSUI, T. SOMA and I. ODA, in "Science and Technology of Zirconia II, Advances in Ceramics 12", edited by N. Claussen, M. Rühle and A. H. Heuer (American Ceramic Society, Columbus, Ohio 1984) pp. 371–81.
3. K. TSUKUMA, Y. KUBOTA and T. TSUKIDATE, *ibid.*, pp. 382–90.
4. M. WATANABE, S. IIO and I. FUKUURA, *ibid.*, pp. 391–98.
5. K. NAKAJIMA, K. KOBAYASHI and Y. MURATA, *ibid.*, pp. 399–407.
6. M. RÜHLE, N. CLAUSSEN and A. H. HEUER, *ibid.*, pp. 352–370.
7. T. MIYAZAWA, H. MITSUDA, H. TORII and H. OKINAKA, *Funtai oyobi Funmatsu-Yakin* **32** (4) (1985) 141.
8. T. SATO and M. SHIMADA, *J. Am. Ceram. Soc.* **67** (1984) C212.
9. *Idem, ibid.* **68** (1985) 356.
10. M. YOSHIMURA and S. SŌMIYA, unpublished work presented at the International Workshop on Transformation Toughening, 15–18 April 1985, Lorne, Australia.
11. T. SATO, S. OHTAKI and M. SHIMADA, *J. Mater. Sci.* **20** (1985) 1466.
12. F. F. LANGE, B. I. DAVIS and G. L. DUNLOP, *J. Am. Ceram. Soc.* **69** (1986) 237.
13. H. SCHUBERT, in "Zirconia Ceramics 7", edited by S. Sōmiya and M. Yoshimura (Uchida Rokakuho, Tokyo, 1986) pp. 65–81.
14. S. SCHMAUDER and H. SCHUBERT, *J. Am. Ceram. Soc.* **69** (1986) 534.
15. H. SAKA, K. KURODA, S. IIO, M. WATANABE and T. IMURA, in "Zirconia Ceramics 8", edited by S. Sōmiya and M. Yoshimura (Uchida Rokakuho, Tokyo, 1986) pp. 63–70.
16. N. NAKANISHI, T. SHIGEMATSU, T. SUGIMURA and H. OKINAKA, *ibid.*, pp. 71–85.
17. Y. OISHI, K. ANDO and Y. SAKKA, in "Advances in Ceramics", Vol. 7, edited by M. F. Yan and A. H. Heuer (American Ceramic Society, Columbus, Ohio, 1983) pp. 208–19.
18. M. YOSHIMURA, T. HIUGA and S. SŌMIYA, *J. Am. Ceram. Soc.* **69** (1986) 583.
19. I-WEI CHEN and Y-H. CHIAO, in "Science and Technology of Zirconia II, Advances in Ceramics 12", edited by N. Claussen, M. Rühle and A. H. Heuer (American Ceramic Society, Columbus, Ohio, 1984) pp. 33–45.
20. S. SŌMIYA, M. YOSHIMURA, T. NOMA and K. KAWABATA, Presented at the 3rd International Conference on Zirconia, 9–11 September, 1986, Tokyo.

T. OHNISHI, H. ABE, K. MARUYA and K. DOMEN, *J. Chem. Soc., Chem. Commun.*, 617 (1985).

Received 23 September
and accepted 10 November 1986

Reprinted from *Proceedings of Expermat '87*, Bordeaux, 24–27 November 1987.

Oxidation of Non-oxide Ceramics by High-Temperature High-Pressure Water

Masahiro YOSHIMURA, Jun-ichrō KASE, Makoto HAYAKAWA and Shigeyuki SŌMIYA

Research Laboratory of Engineering Materials, Tokyo Institute of Technology, Nagatsuta, Midori, Yokohawa 227 Japan

1. Introduction

Non-oxide ceramics, as silicon nitride, silicon carbide, zirconium nitride and zirconium carbide, are hard, strong and refractory materials which show considerable promise as structural materials in high temperature environments. They are chemically stable in inert conditions but under oxidizing conditions, particularly at high temperatures, some reactions do take place. Thus the oxidation/corrosion is the most serious failure in non-oxide ceramics in their applications. This problem has been extensively studied in various oxidative atmospheres. In air or oxygen, the oxidation is believed to proceed as follows;

$Si_3N_4 + 3O_2 \longrightarrow 3SiO_2 + 2N_2$ ----------------[1],

$SiC + 2O_2 \longrightarrow SiO_2 + CO_2$ ----------------[2],

$2ZrN + 2O_2 \longrightarrow 2ZrO_2 + N_2$ ----------------[3], and

$ZrC + 2O_2 \longrightarrow ZrO_2 + CO_2$ ----------------[4].

In the cases of Si_3N_4 and SiC, as the oxidation is proceeding, amorphous SiO_2 covers Si_3N_4 or SiC and forms a protective layer (1-7). Therefore, the oxidation is controlled by diffusion processes, and it proceeds very slowly. This mechanism is called as passive oxidation. In SiC, when oxygen pressures are low enough, the oxidation proceeds much faster, because gaseous SiO is produced as follows;

$SiC + O_2 \longrightarrow SiO + CO$ ----------------[5].

This mechanism is called as active oxidation (8). In the case of ZrN and ZrC, it has been reported that the reaction is controlled by phase boundary reactions below 1000℃ (9-15), and controlled by diffusion process above 1000℃ (13).

A certain number of reports (9,16-18) argued that the oxidation rate was accelerated when water vapor coexisted with oxygen, but the reaction of those non-oxide ceramics by H_2O only has hardly been studied yet. We have studied the oxidation of Si_3N_4,SiC (19-22), ZrN and ZrC (23) by H_2O under hydrothermal conditions where almost only high-temperature high-pressure water exists. Observed reactions are;

$Si_3N_4 + 6H_2O \longrightarrow 3SiO_2 + 4NH_3$ ----------------[6],

$SiC + 2H_2O \longrightarrow SiO_2 + CH_4$ ----------------[7],

$$2ZrN + 4H_2O \longrightarrow 2ZrO_2 + 2NH_3 + H_2 \quad \text{----------------[8], and}$$

$$ZrC + 2H_2O \longrightarrow ZrO_2 + CH_4 \quad \text{----------------[9].}$$

These reactions are quite different from those in oxidative atmospheres without H_2O. We discuss in the present paper the effects of H_2O on high-temperature degradation of Si_3N_4, SiC, ZrN and ZrC powders, and of Si_3N_4 and SiC ceramics based on the experimental results.

2. Experimental Procedure

The starting materials were Si_3N_4 powder(*1), SiC powder(*2), ZrN powder(*3), ZrC powder(*4), hot pressed Si_3N_4(*5) and reaction sintered SiC(*6) ceramics. Each sample weighed 60-100 mg was placed in an Au capsule of which top was narrowed by pressing to prevent the escape of solid materials but to allow the passing of fluids. The capsule was heated in a test-tube-type pressure vessel (volume 5.5ml) under the pressure of 10 or 100 MPa of redistilled water at temperatures of between 100° and 800℃ for 0 to 72 h. Most of the samples were heated first under a certain temperature then introduced the pressurized H_2O. After the treatment the sample in the pressure vessel was quenched by immersing into cold water, then the sample capsule was dried and weighed. Ceramic samples were directly weighed to measure the weight change. Solid products were examined by X-ray diffractometry(XRD), transmission electron microscopy(TEM), scanning electron microscopy(SEM) and X-ray energy dispersive spectroscopy(EDS). The oxidation ratio was calculated from the weight change or the integrated peak intensity of X-ray diffraction for powder samples but not for ceramic samples. Because ceramic samples behavéd both in weight gain by the oxidation and in weight loss by the dissolution of produced oxides. Gaseous products in the vessel were analyzed by gas chromatography, a mass spectrometer and Nessler's reagent.

(*1) Starck LC12, average grain size 0.5μm, >90% α-form

(*2) Starck A10, average grain size 0.65μm, α-form

(*3) Starck, -325mesh, Zr:N= 1:0.91(component ratio)

(*4) Starck, -325mesh, average grain size 3.2μm, Zr:C= 1:0.98(component ratio)

(*5) Toshiba Ceramics Co., Y_2O_3 5wt%, Al_2O_3 2wt%, Si_3N_4 93wt%, 4mmϕ x2mmt

(*6) Shin-Etsu Chemical Co., almost pure, 4mmϕ x2mmt

3. Results and Discussion

3-1 The oxidation of Si_3N_4 powder

Under 100 MPa H_2O, the oxidation of Si_3N_4 proceeded above 200℃ as a function of time (Fig.1). Weight gain and appearance of a halo around 2θ~16° -27° in X-ray diffraction (CuKα) indicated the formation of amorphous silica by the

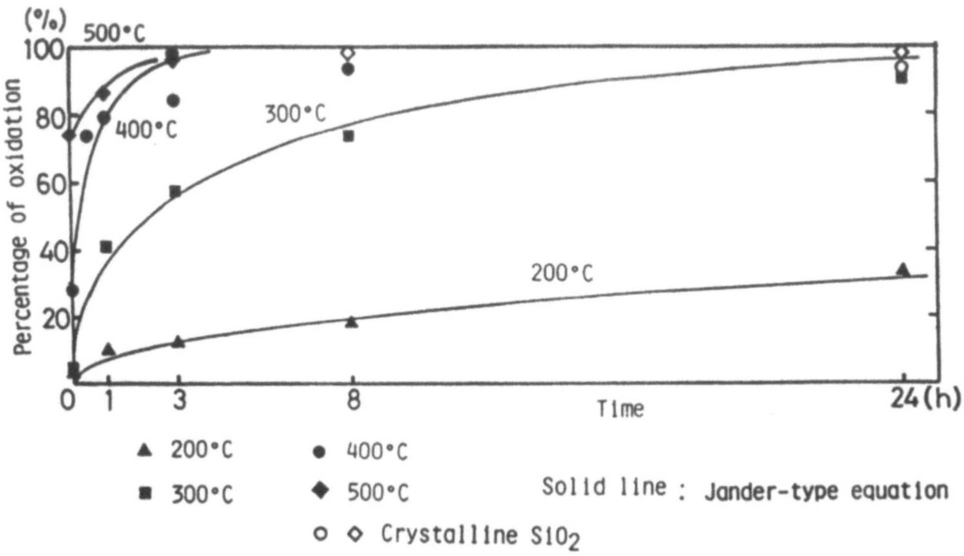

Fig.1 Isothermal oxidation of Si_3N_4 powder under 100 MPa H_2O.

oxidation of Si_3N_4 as the first step at every temperature. Cristobalite and Keatite crystallized above 400℃ after 8h when Si_3N_4 was almost completely changed into amorphous silica. Keatite which is one of the high-pressure phases of silica is reported to appear at 400° to 800℃ under 80 to 130 MPa (24). Scanning electron micrographs showed that amorphous silica grains covered the surface of Si_3N_4 particles and then these silica grains grew and sintered during further hydrothermal treatment. The production of NH_3 was confirmed with the Nessler's reagent. These results indicate the oxidation of Si_3N_4 by H_2O to be given by the equation[6].

Under 10 MPa H_2O, the oxidation behaviors were similar to those under 100 MPa H_2O at 200° -300℃ but quite different above 400℃ (Fig.2). The oxidation rates

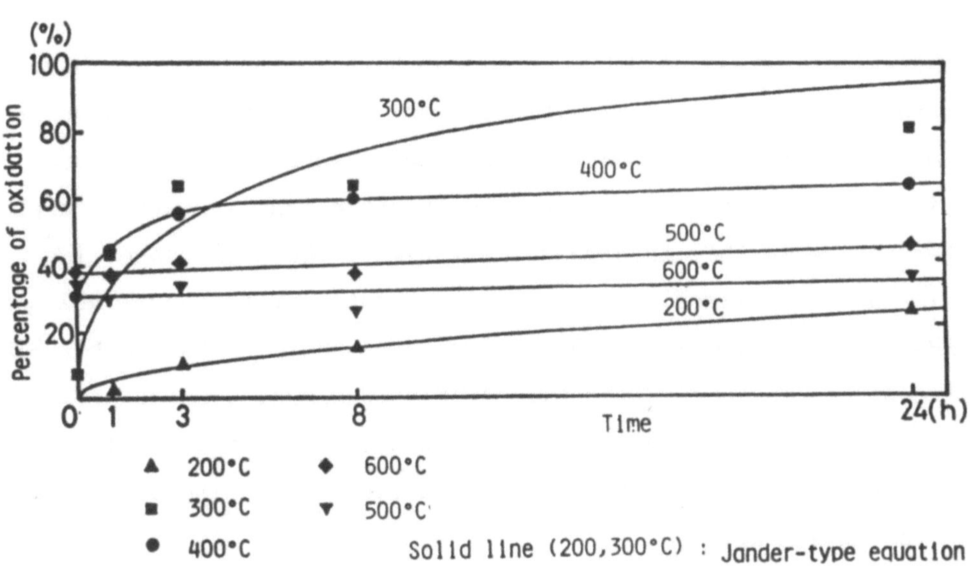

Fig.2 Isothermal oxidation of Si_3N_4 powder under 10MPa H_2O. (I), Heated after pressurized.

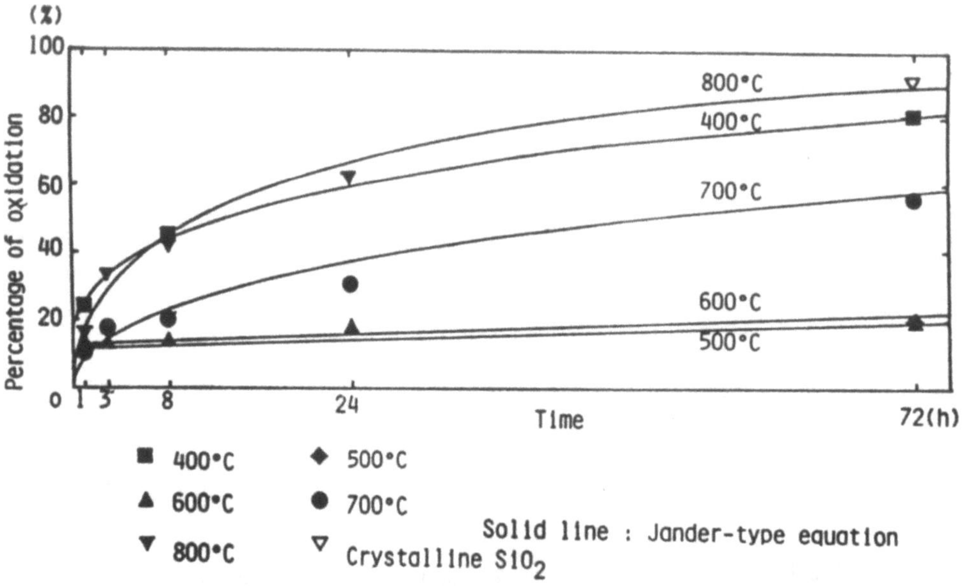

Fig.3 Isothermal oxidation of Si_3N_4 powder under 10MPa H_2O. (II), Pressurized after heated.

were extremely slower than under 100 MPa. This would be caused by the phase transformation of H_2O from liquid to vapor, which occurs about 310℃ under 10 MPa pressure, and by the decomposition of NH_3 into N_2 and H_2, which occurs above 430℃ under 10 MPa pressure. The other experiments where the 10 MPa water vapor was introduced into a vaccuum vessel kept at 400° to 800℃ indicated very slow oxidation rates at 500° -600℃ and increased rates at higher temperatures (Fig.3). Amorphous silica was also produced at every temperature, and then crystobalite was formed after 72h at 800℃.

3-2 The oxidation of SiC powder

The hydrothermal oxidation of SiC occurred at higher temperatures above 500℃ to yield amorphous silica, too (Fig4,5). The oxidation behaviors under 10 MPa H_2O resembled to those under 100 MPa but proceeded more slowly. Produced amorphous silica covered the SiC grains and then the grains started to grow and and to sinter during further treatment similarly to the Si_3N_4 case. The amorphous silica crystallized in cristobalite and tridymite above 700℃ after 8h under 100 MPa and cristobalite at 800℃ after 72h under 10 MPa, where, SiC had been almost completely changed into silica in both cases. Gaseous products in the vessel were proved to contain a large amount of CH_4 and a small amount of CO_2 by a mass spectrometric analysis. The formation of CH_4 indicates the reaction as the equation[7],

$$SiC + 2H_2O \longrightarrow SiO_2 + CH_4 \quad \text{------------------[7]}$$

rather than the reaction[10] to produce CO_2,

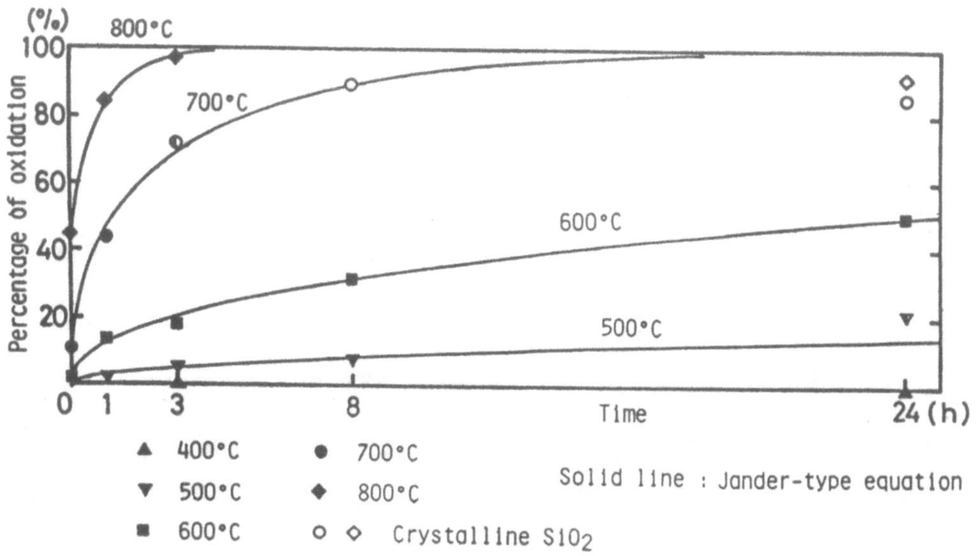

Fig.4 Isothermal oxidation of SiC powder under 100 MPa H_2O.

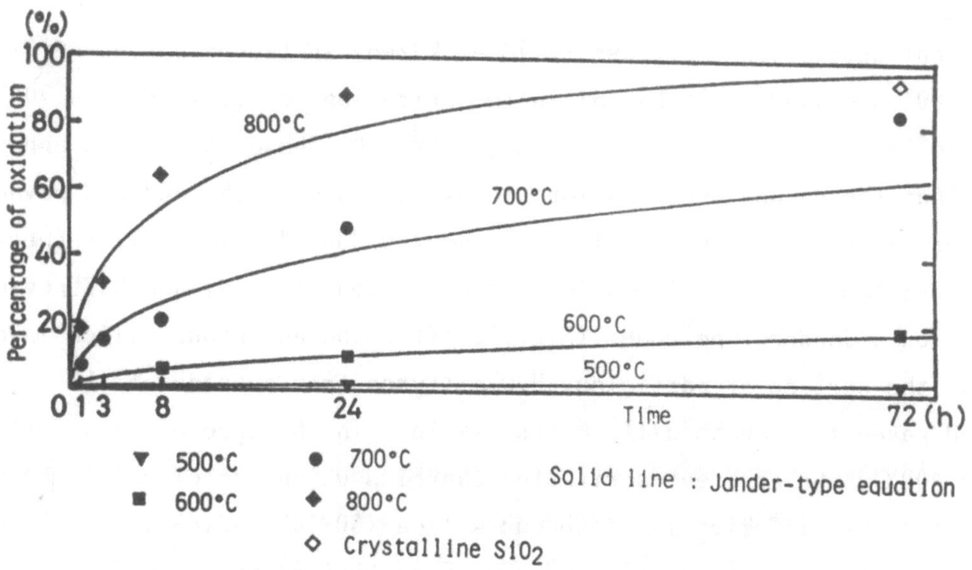

Fig.5 Isothermal oxidation of SiC powder under 10 MPa H_2O.

$SiC + 4H_2O \longrightarrow SiO_2 + CO_2 + 4H_2$ ------------------[10].

Thermodynamic calculations (20) support the experimental results and demonstrate that SiC is a sufficiently strong reducing agent to make CH_4 rather than CO_2 as the stable carbon species under the hydrothermal conditions studied.

3-3 Kinetics of the oxidation of Si_3N_4 and SiC powders

The oxidation ratio was approximately represented by a Jander-type equation(19), as illustrated with solid lines in figure 1-5;

$[1-(1-\alpha)^{1/3}]^2-[1-(1-\alpha_0)^{1/3}]^2 = (2k/r^2)t$ ------------[11],

where t is the time, α is the oxidation ratio at t h, α_0 is the oxidation ratio at 0 h, r is the radius of particle, and k is the rate constant. Figure 6 shows the Arrhenius plots of the rate constants calculated from the equation[11].

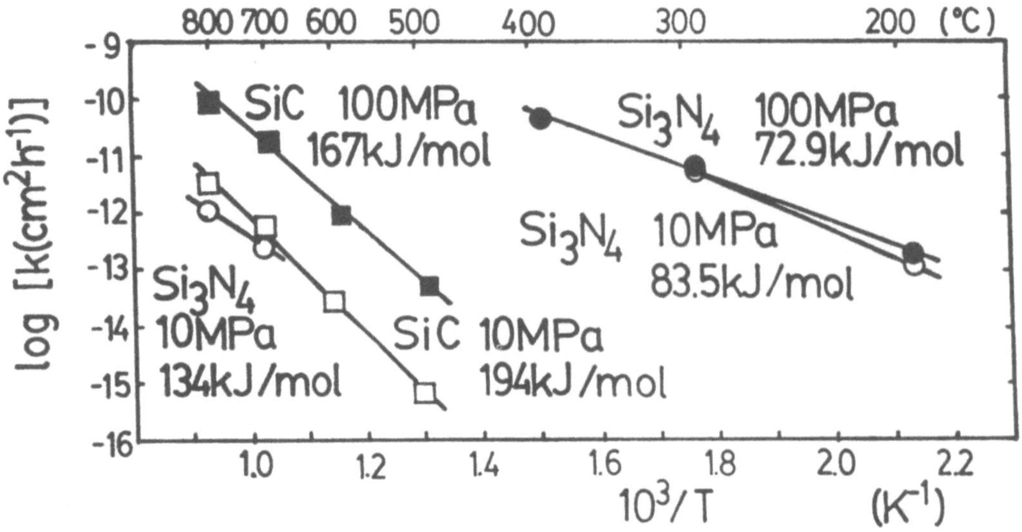

Fig.6 Arrhenius plot of parabolic rate constant for the hydrothermal oxidation of Si_3N_4 and SiC powders

The apparent activation energies are 73-84 kJ/mol in the oxidation of Si_3N_4 between 200° and 400℃, 134 kJ/mol in the oxidation of Si_3N_4 between 700° and 800℃, and 167-194 kJ/mol in the oxidation of SiC. These values are apparently smaller than the values reported for the oxidation of Si_3N_4 and SiC powders in dry air or oxygen (150-290 kJ/mol (1-4) and 210-330 kJ/mol (5-7), respectively). We have proposed reaction models for the oxidation of Si_3N_4 and SiC powders in H_2O based on a Jander-type model (Fig.7). After the amorphous silica layer is formed on the surface of particles, H_2O diffuses from outside of the silica layer and gaseous products diffuse from inside. In the case of Si_3N_4 between 200° and 400℃, the apparent activation energies which are close to the values reported for H_2O diffusion in amorphous silica (60-105 kJ/mol (25,26)) would indicate that the oxidation of Si_3N_4 was characterised by H_2O diffusion rather

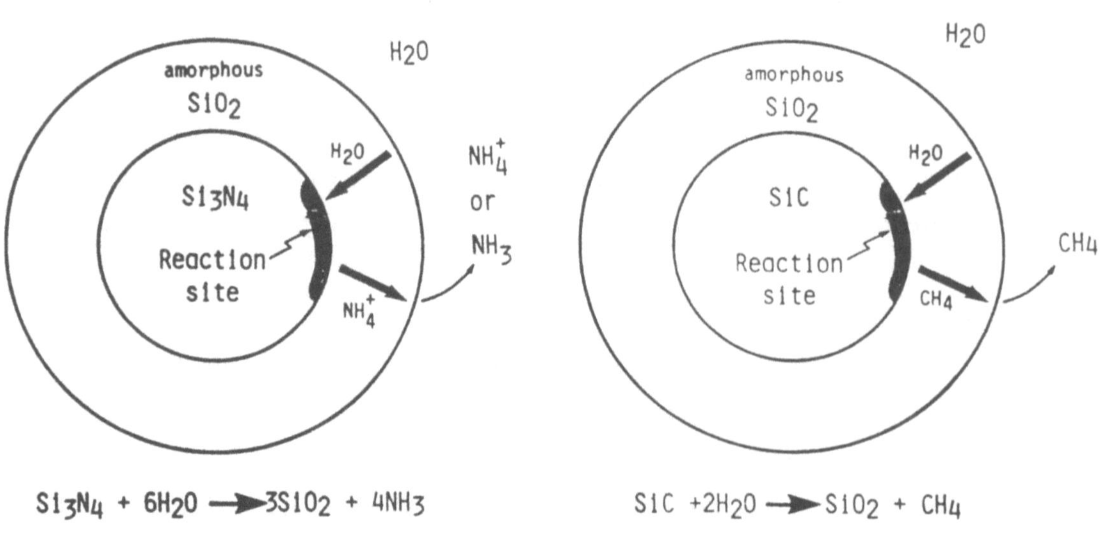

Fig.7 Reaction model of hydrothermal oxidation of Si_3N_4 and SiC powder.

than O_2 diffusion in the silica layers (120-300 kJ/mol (27-29)).

The difference in the activation energy for oxidation of Si_3N_4 between 200° and 400℃ and between 700° and 800℃ might be due to the difference of H_2O contents or microstructures in the amorphous silica and the difference of diffusional species diffusing from inside; probably between 200° and 400℃ the species is NH_3 (probably NH_4^+) and between 700° and 800℃ the species are N_2 and H_2. Since the H_2O diffusion in the silica layer should be similar both for the oxidation of Si_3N_4 and SiC, the different activation energies between them, would be due to the difference of produced gases; NH_3 or CH_4. NH_3 is a polar gas and easy to be ionized in liquid H_2O or amorphous silica, while CH_4 is a nonpolar gas in any conditions studied. Thus, NH_3 (probably NH_4^+) might diffuse as fast as H_2O (probably OH^-) in amorphous silica. However, it is necessary to make clear the states of gaseous products in the silica layer to prove these mechanisms.

In contrasting the oxidation in dry air or oxygen, the oxidation in pressurized H_2O proceeds very fast because of the faster diffusion of H_2O than oxygen in the silica layer. The effect of water vapor on the oxidation of SiC have been interpreted as the acceleration (16) of the diffusion of oxygen, CO and CO_2, or the promotion of the crystallization (17) of silica into cristobalite, which might change diffusion paths from a bulk diffusion to a grain-boundary diffusion. The present investigation, however, suggests that water vapor causes the reaction including diffusion process of H_2O. This seems to be one of the reasons for the acceleration of oxidation of SiC or Si_3N_4 in atmospheres containing H_2O.

3-4 The oxidation of ZrC powder

The oxidation of ZrC occurred above 210℃ to yield monoclinic zirconia under 100 MPa H_2O (Fig.8). With elevating temperature, more ZrC was converted to monoclinic ZrO_2. Above 340℃ the reaction was proceeded completely. Transmission electron micrographs for the sample treated at 350℃ for 3h under 100 MPa H_2O showed ultrafine ZrO_2 particles of about 10 nm forming clusters of particles. Gaseous products in the vessel were proved to contain a large amount of CH_4 and a small amount of H_2 but not to contain CO and CO_2 by a gas chromatography analysis. The formation of CH_4 indicates the reaction as the equation[9];

$$ZrC + 2H_2O \longrightarrow ZrO_2 + CH_4 \qquad [9].$$

The production of a small amount of H_2 must be due to the nonstoichiometry of zirconium carbide as the equation[12]

$$ZrC_{1-x} + 2H_2O \longrightarrow ZrO_2 + (1-x)CH_4 + 2xH_2 \qquad [12].$$

3-5 The oxidation of ZrN powder

The oxidation of ZrN occurred above 190℃ to yield monoclinic zirconia under

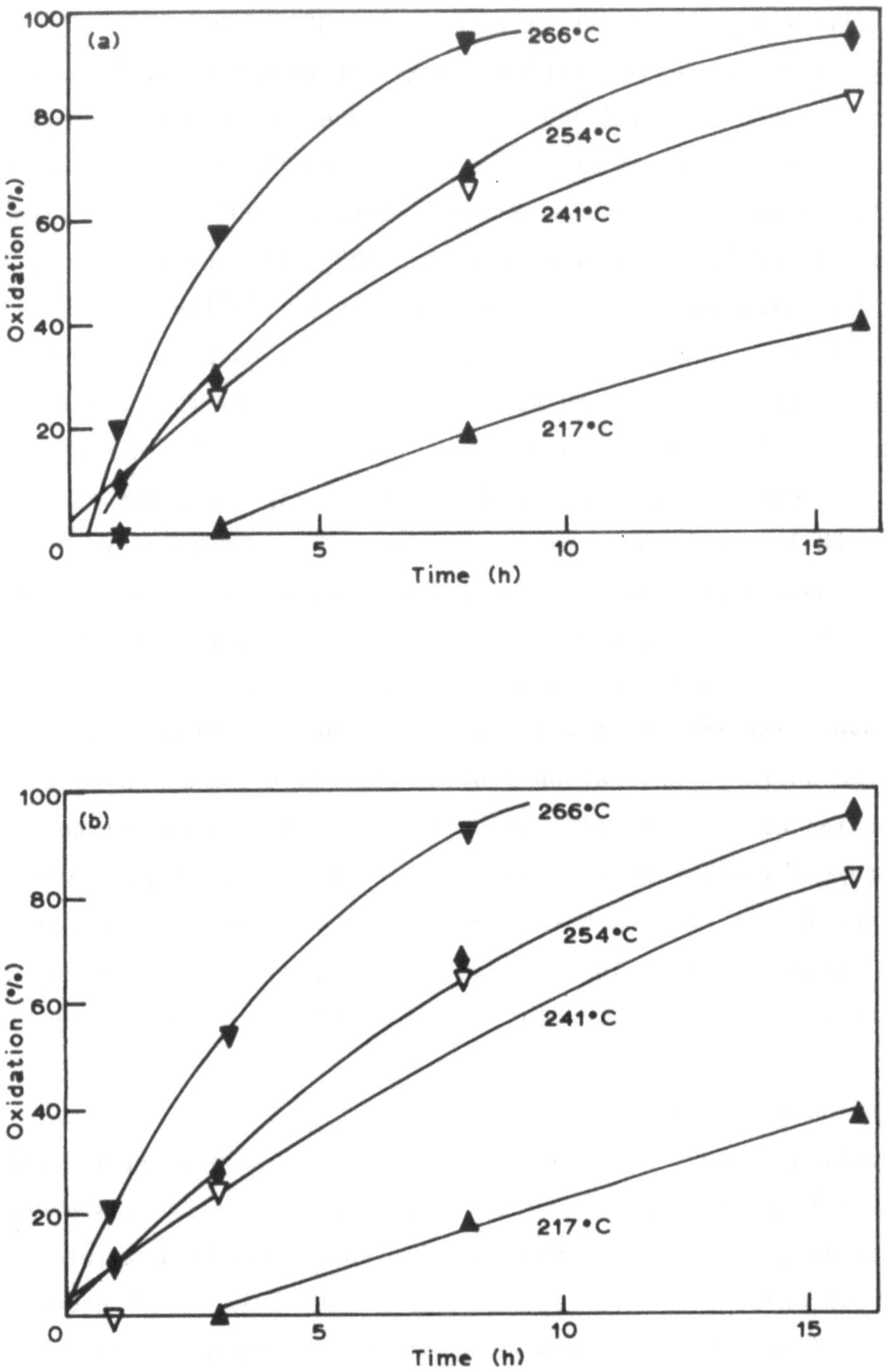

Fig.8 Isothermal oxidation of ZrC powders under 100 MPa H_2O. Solid lines are drawn by;

(a)Nucleation-controlled model

(b)Core-shrinking model

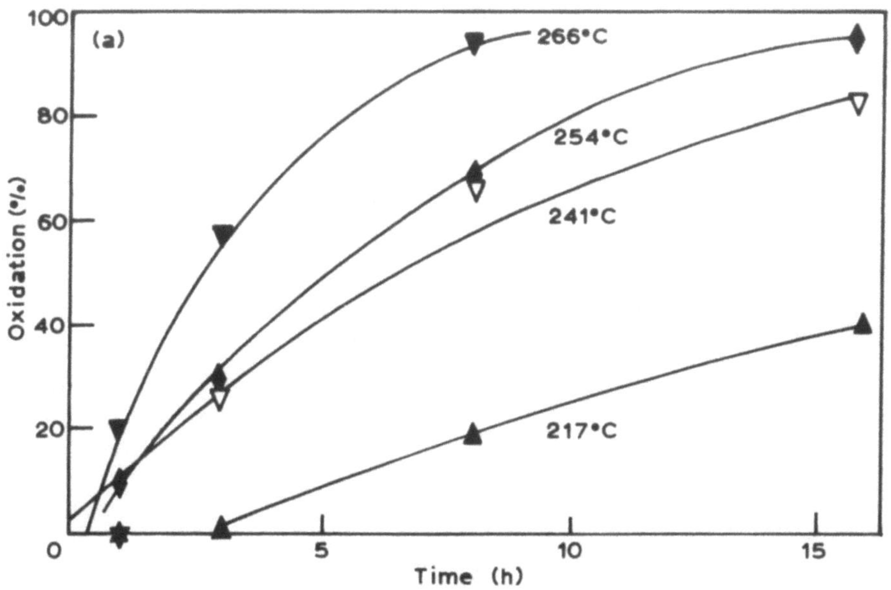

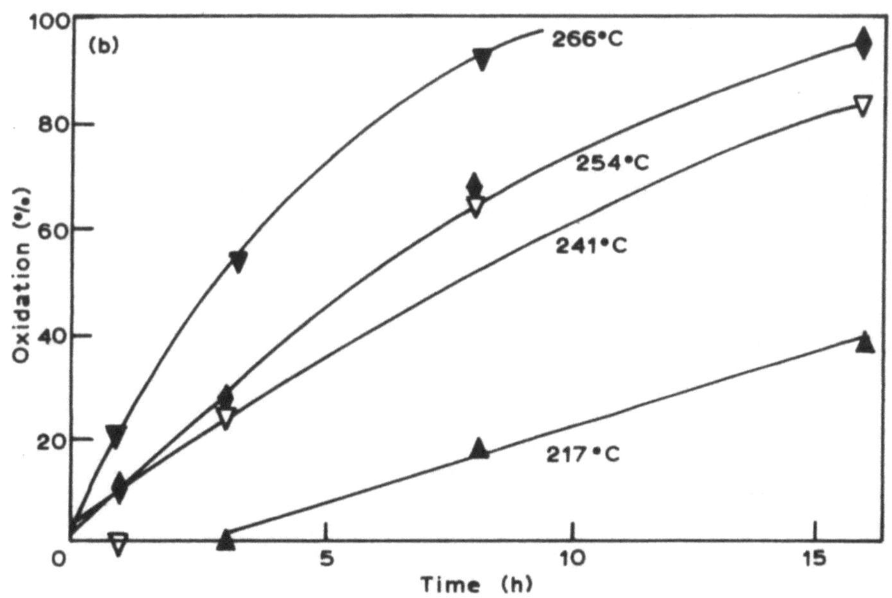

(c)Jander-type equation

(d)Prout-Tompkins equation

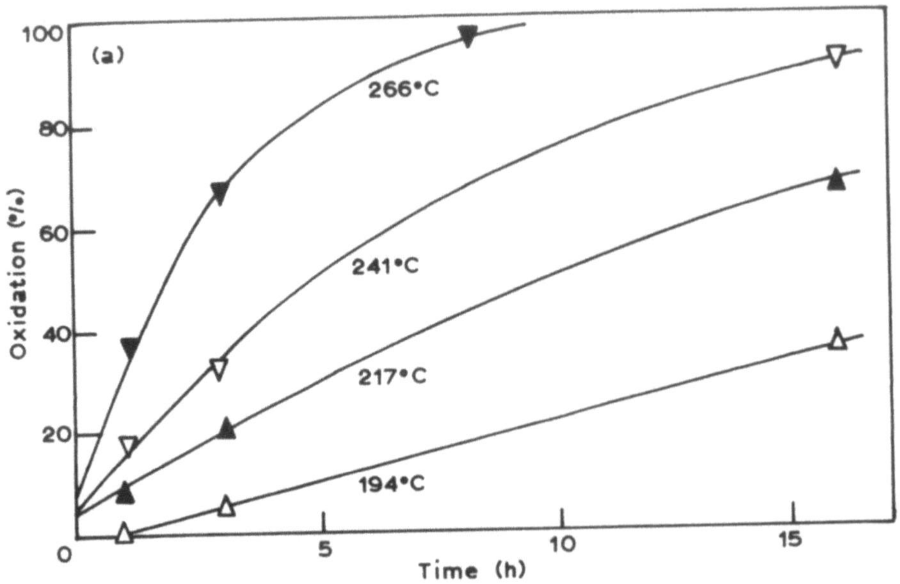

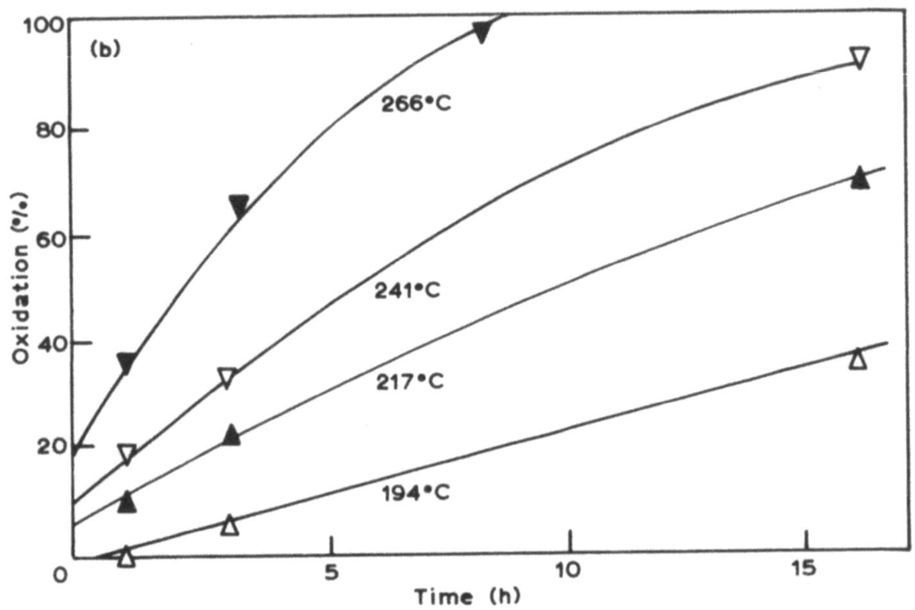

Fig.9 Isothermal oxidation of ZrN powders under 100 MPa H_2O. Solid lines are drawn by;

(a)Nucleation-controlled model

(b)Core-Shrinking model

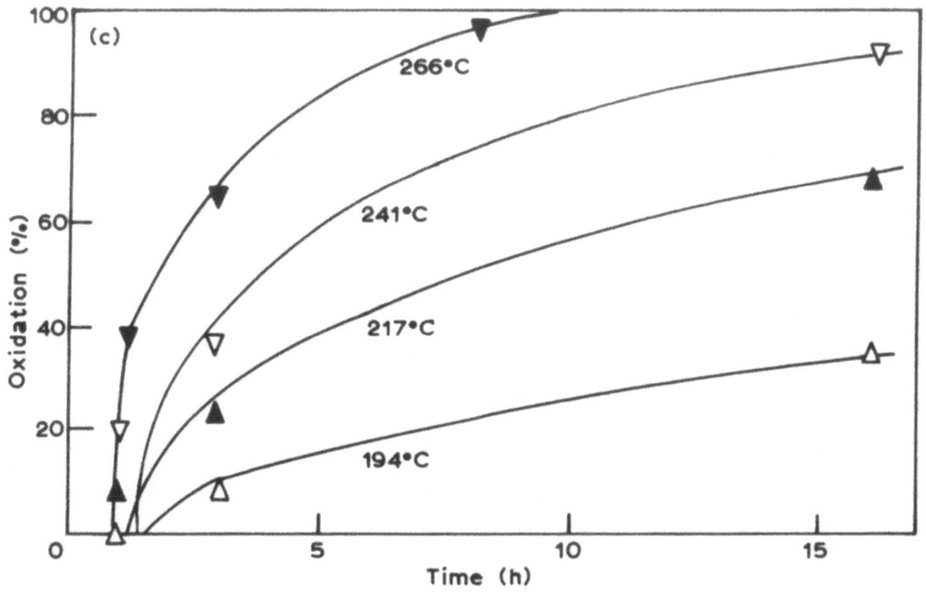

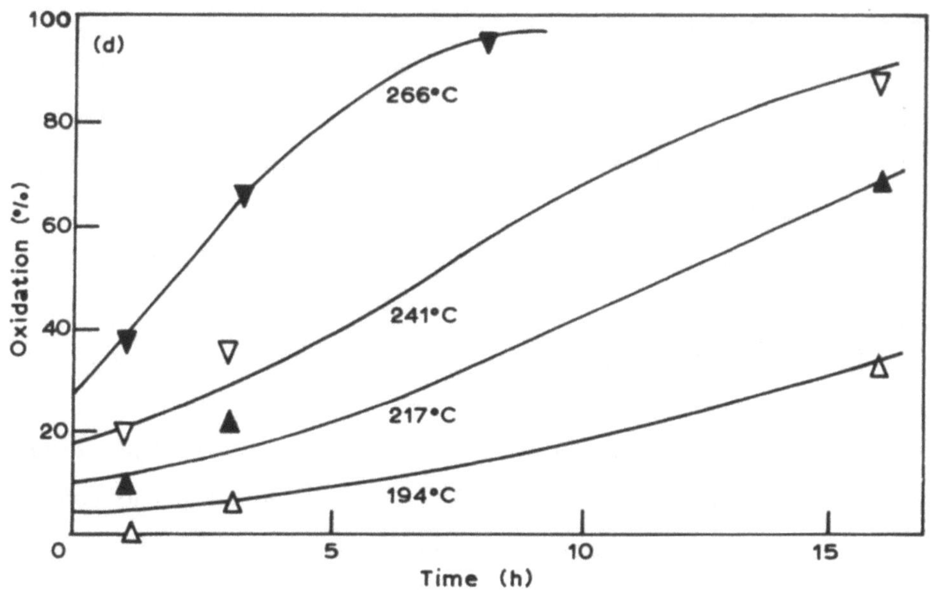

(c)Jander-type equation

(d)Prout-Tompkins equation

100 MPa H_2O (Fig.9). Higher temperatures yielded less ZrN and more monoclinic ZrO_2. The starting sample, ZrN, disappeared completely above 340℃. So far, it was similar to ZrC. But in the case of ZrN there appeared a very small peak of about 1% (the ratio of the integrated peak intensity) around 2θ~30° in X-ray diffraction (CuKα) patterns of the products of between 190° and 390℃. This peak can be identified as cubic zirconia or $ZrH_{0.6}N$. We cannot decide which phase appeared, but think it as a nitrogen containing phase according to the fact that the peak did not appear in the oxidation of ZrC by H_2O. From transmission electron micrographs for the samples treated at 350℃ for 3h under 100 MPa H_2O, the products consisted of clusters of ultrafine ZrO_2 particles. The production of NH_3 was confirmed with the Nessler's reagent. Other gaseous products in the vessel were proved to contain a large amount of H_2 but not to contain N_2 by a gas chromatography. The formation of NH_3 and H_2 indicates the reaction as the equation[8],

$$2ZrN + 4H_2O \longrightarrow 2ZrO_2 + 2NH_3 + H_2 \qquad [8].$$

3-6 Kinetics of the oxidation of ZrC and ZrN powders

The oxidation ratio was approximately represented by a nucleation-controlled model (eq.[13]) or by a core-shrinking model (eq.[14]), but it was not fitted by a Jander-type equation (eq.[11]) or a Prout-Tompkins equation (eq.[15]) (Fig.8,9). In figures 8 and 9, solid lines are;

$$-\ln(1-\alpha) + \ln(1-\alpha_0) = kt \qquad [13],$$

$$[1-(1-\alpha)^{1/3}] - [1-(1-\alpha_0)^{1/3}] = kt \qquad [14],$$

$$[1-(1-\alpha)^{1/3}]^2 - [1-(1-\alpha_0)^{1/3}]^2 = (2k/r^2)t \qquad [11], \text{ and}$$

$$\ln(\alpha/(1-\alpha)) - \ln(\alpha_0/(1-\alpha_0)) = kt \qquad [15],$$

where t is the time, α is the oxidation ratio at t h, α_0 is the oxidation ratio at 0 h, and k is the rate constant. They suggested that the reaction was not controlled by diffusion processes but controlled by the phase boundary reactions from the surface to the center of ZrC or ZrN grains. An Arrhenius plot of the rate constants gives apparent activation energies of 99 kJ/mol and 78 kJ/mol for the oxidation of ZrC and for ZrN by H_2O, respectively, on the basis of nucleation-controlled models, while on the basis of core-shrinking models an Arrhenius plot of the rate constants gives apparent activation energies of 82 kJ/mol and 63 kJ/mol for the oxidation of ZrC and for ZrN respectively.

We propose here reaction models for the oxidation of ZrC and ZrN powders in H_2O based on the model fittings and transmission electron micrographs (Fig.10). When ZrN reacted with H_2O, ultrafine ZrO_2 particles of about 10 nm were nucleated on the surface of ZrN grains with about 1μm or ZrC grains with about 3μm. These

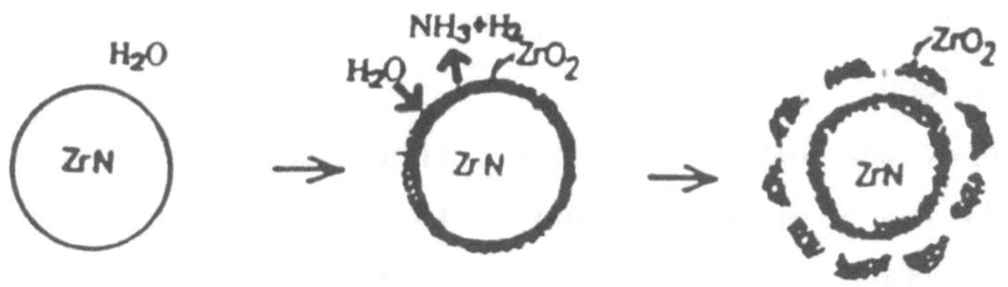

Fig.10 Reaction model of the hydrothermal oxidation of ZrN powder.

ZrO_2 fine particles did not form a protective layer for ZrN or ZrC particles, but formed clusters of particles because of large volume expansions of about 40% during oxidation. As the reaction proceeds, the surface of ZrN or ZrC particles went inner to the center of the particles.

The difference in the activation energies for the oxidation of ZrC and ZrN by H_2O might be due to the difference in the reaction processes. Here we assume that Zr-O bonding and C-H bonding formed simultaneously (Fig.11). This

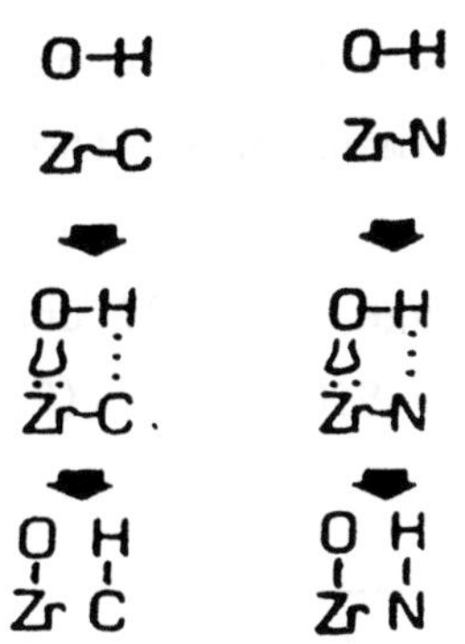

Fig.11 Elementary reaction step in the hydrothermal oxidation of ZrC and ZrN.

corresponded to the assumption that the formation of ZrO_2 nucleous and the formation of CH_4 gas occurred simultaneously. The bonding energies of O-H, C-H and N-H bondings are 424 kJ/mol, 335 kJ/mol and 356 kJ/mol respectively (30). Therefore cutting O-H bonding and forming C-H bonding simultaneously need the energy of 89 kJ/mol, whereas cutting O-H bonding and forming N-H bonding simultaneously need the energy of 68 kJ/mol. Thus the formation of ZrO_2 nucleous from ZrC needs about 20 kJ/mol more than the formation of ZrO_2 nucleous from ZrN. This difference, about 20 kJ/mol, corresponds to the difference in the activation energies for the oxidation of ZrC and ZrN. Therefore the difference of about 20 kJ/mol in the activation energies for the oxidation of ZrC and ZrN by H_2O might be caused by the difference between the formation of C-H bonding and N-H bonding. Those results are completely different from the oxidations of SiC and Si_3N_4 by H_2O where the kinetics were controlled by diffusion processes mentioned above. Furthermore the oxidation in pressurized H_2O proceeds at much lower temperatures than the oxidation in dry air or oxygen.

3-7 The oxidation of hot-pressed Si_3N_4 (HP-Si_3N_4) and reaction sintered SiC (RS-SiC) specimens

Under 100 MPa H_2O, the weight of ceramic specimens changed strongly depending both on the oxidation of specimens and on the dissolution of oxide layers produced on the surface. HP-Si_3N_4 reacted with H_2O and gave a little weight loss even at 200℃ (Fig.12). When the temperature was raised up to 800℃, the isotherm showed a weight gain by intense oxidations and almost whole samples changed into oxides, mainly silica after 72 h. At 400° and 600℃, we suppose following processes from the isotherms. The pressurized H_2O starts to dissolve the oxides as soon as they are produced on the surface of the specimen. As H_2O has been saturated with the oxides, the oxide layers began to thicken on the surface of specimen and give weight increases. Since the thickness of the oxide layers controls the oxidation rate by diffusion processes in the oxide layer, the oxidation and dissolution rates can be balanced after a certain period of the reaction to show a steady weight decrease. Yoshio et al. (31) also reported a steady weight loss as illustrated with a broken line in Fig.12 for pressureless-sintered Si_3N_4 specimens under the conditions of 300℃ and 8.6 MPa H_2O.

RS-SiC showed little weight change at 400℃ under 100 MPa H_2O, as had been expected from the examination for SiC powder (Fig.13). Weight loss or weight

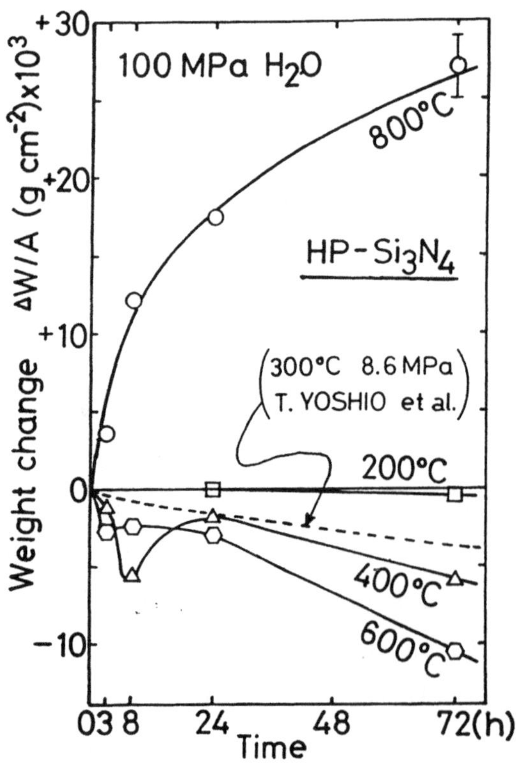

Fig.12 Weight change of HP-Si_3N_4 by hydrothermal treatment under 100 MPa H_2O

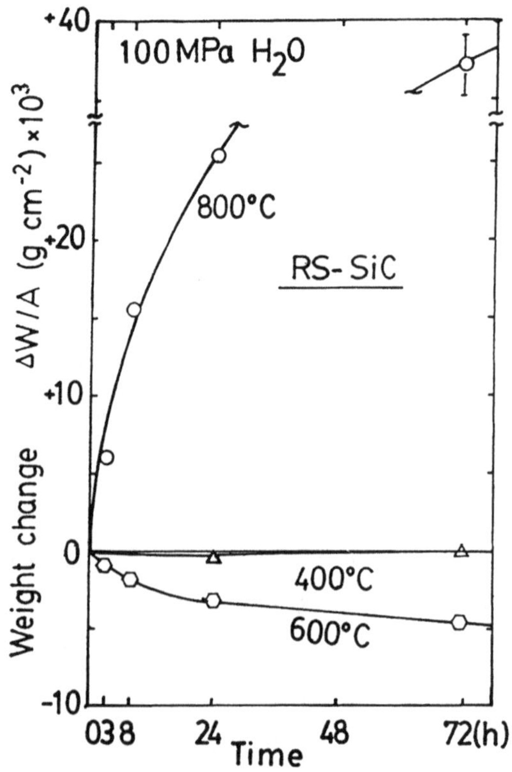

Fig.13 Weight change of RS-SiC by hydrothermal treatment under 100 MPa H_2O

gain was observed at 600° or 800℃, respectively, in the same manner as HP-Si_3N_4. Therefore, the dissolution behavior would be almost the same for RS-SiC or HP-Si_3N_4 ceramics regardless with or without the sintering additives and the produced gases of CH_4 or NH_3.

Under 10 MPa, the weight changes of the specimen were smaller than those under 100 MPa both for RS-SiC and HP-Si_3N_4. It is probable that vapor-like H_2O is difficult to dissolve the oxides formed on the surface of the specimen.

X-ray diffuraction and scanning electron micrographs indicated that amorphous silica and/or silicate were formed by the oxidation of SiC or Si_3N_4, respectively , as the first stage of the reactions at every temperature. Then those amorphous phases crystallized in the order of cristobalite, tridymite and quartz. This order seems to indicate the ease in crystallization of silica phases under the hydrothermal conditions. X-ray diffraction showed also the formation of α-$Y_2Si_2O_7$ on the surface of the specimen in the oxidation of HP-Si_3N_4 because of the reaction between silica and Y_2O_3 of a sintering additive.

RS-SiC was changed to silica keeping the original morphology of the surface under 10 MPa H_2O, while larger silica crystals were grown on the surface of RS-SiC in 100 MPa by successional dissolution/deposition processes. The oxidation products of HP-Si_3N_4 were amorphous silicates under 10 MPa H_2O, and three crystalline phases under 100 MPa H_2O. They are quartz crystals of about 10μm in size, and needle-like and minute crystals both of which contained large amounts of Y and Al.

4. Summary

Si_3N_4 and SiC powders are oxidized by H_2O even as low temperature as 200° and 500℃, respectively, to yield amorphous silica and NH_3 and CH_4, respectively. The oxidation of Si_3N_4 and SiC can be accelerated when atmospheres contain water vapor because the diffusion of H_2O would be faster than that of oxygen in the amorphous silica layer.

ZrN and ZrC powders are oxidized by H_2O also at low temperatures as 190° and 210℃, respectively, to yield monoclinic zirconia, H_2 and NH_3 and CH_4, respectively. The oxidation of ZrN and ZrC can be accelerated when atmospheres contain H_2O because the mechanism of the phase boundary reaction changed.

The degradation of HP-Si_3N_4 and RS-SiC were accelerated by the dissolution of the oxide layer when H_2O exists as a liquid-like phase.

5.References

1) R.M. Horton, J.Am.Ceram.Soc.,52,121-24 (1969).
2) D.Tétard, P.Lorthlary, P.Goursat and M.Billy, Rev.Int.Hautes.Temp.et Refract.,

10,153-59 (1973).
3) M.Mitomo and J.Sharp, Yogyo Kyokai Shi,84,33-36 (1976).
4) T.Sata and K.Fujii, ibid,90,110-118 (1982).
5) T.Nakatogawa, J.Chem.Soc.Jpn.Ind.Chem.Sect.,57,348-50 (1957).
6) G.Ervin, J.Am.Ceram.Soc.,41,347-52 (1958).
7) R.F.Adamsky, J.Phys.Chem.,63,305-07 (1959).
8) "Muki Kagaku Zensho XII-2 Silicon",pp.299, Edited by G.Iwasaki, Maruzen,1986.
9) R.W.Bartlett, M.E.Wadsworth and I.B.Cutler, Trans.Met.Soc.AIME,227(4), 467-73 (1963).
10) V.A.Terekhova, V.N.Kruglov and N.N.Bragin, Inorg.Mater.,18(3),330-35 (1982).
11) A.K.Kuriakose and J.L.Margrave, J.Electrochem.Soc.,111(7),827-31 (1964).
12) L.C.Dufuor, J.Simon and P.Barret, C.R.Acad.Sc.Paris,265(3),171-74 (1967).
13) Yu.D.Afonin, V.N.Shalaginov and A.R.Beketov, J.Appl.Chem.of USSR,56(10), 2030-33 (1983).
14) J.Desmaison, M.Billy and W.Smeltzer,"Reactivity of Solids",pp.107-12,Edited by J.wood, O.lindquvist C.Helgesson and N.G.Vannerberg, Plenum Press, New York and London, 1977.
15) M.Caillet, H.F.Ayedi, and J.Besson, J.Less Common Met.,51,305-322 (1977).
16) H.Suzuki, Yogyo Kyokai Shi,67,157-64 (1957).
17) P.J.Jorgensen, M.E.Wadsworth and I.B.Cutler, J.Am.Ceram.Soc.,44,258-61(1961).
18) S.C.Singhal, J.Am.Ceram.Soc.,59,81-82 (1976).
19) M.Yoshimura, J.Kase and S.Sōmiya, Yogyo Kyokai Shi,94,129-34 (1986).
20) M.Yoshimura, J.Kase and S.Sōmiya, J.Mater.Res,1(1),100-03 (1986).
21) M.Yoshimura, J.Kase and S.Sōmiya, "Chikka Keiso (Shin Sozai Series)",pp121-34, Edited by S.Sōmiya, M.Yoshimura and M.Mitomo, Uchida Rokakuho, Tokyo, 1987.
22) M.Yoshimura, J.Kase and S.Sōmiya, Proc.of 2nd international Symposium "Ceramic Materials and Components for Engine",pp.529-36, Edited by W.Bunk and H.Hausner (1986).
23) M.Yoshimura, M.Hayakawa and S.Sōmiya, Yogyo Kyokai Shi, in press.
24) R.K.Iler, "The Chemistry of Silica",pp.16, John Willey & Sons, New York,1979.
25) I.Burn and J.P.Roberts, Phys.Chem.Glasses,11,106-14 (1970).
26) S.White, Nature Phys.Sci.,230,192 (1971).
27) V.R.Haul and G.Dümbgen, Z.Elektrochem.,66,636-41 (1962).
28) E.W.Sucov, J.Am.Ceram.Soc.,46,14-20 (1963).
29) E.L.Williams, ibid,48,190-94 (1965).
30) "Lange's Handbook of Chemistry",pp.3.128-31, 13th ed., Edited by J.A.Dean, McGrow-Hill,1985.
31) T.Yoshio, K.Oda and K.O-oka, Yogyo Kyokai Shi,94,116-22 (1986).

Part 8

Hydrothermal Hot Pressing

Reprinted from *Yogyo-Kyokai-Shi*, **94** (1986) 1119–1124.

Solidification of Glass Powder by A Hydrothermal Hot-Pressing Technique

Mamoru NISHIOKA, Kazumichi YANAGISAWA and Nakamichi YAMASAKI

(Research Laboratory of Hydrothermal Chemistry, Faculty of Science, Kochi University 2-5-1, Akebono-cho, Kochi-shi 780)

A borosilicate glass powder was solidified with a NaOH solution or distilled water by the hydrothermal hot-pressing technique. The effect of hydrothermal conditions on the compressive strength was investigated, and the densification mechanism of the glass powder during the hydrothermal hot-pressing was discussed in terms of isothermal shrinkage. The glass powder was successfully solidified by hydrothermal hot-pressing at a reaction temperature lower than that of an ordinary pressure sintering. The solidified body obtained in the presence of distilled water (10 wt%) at 280°C had a high compressive strength of about 2300 kg/cm². The analysis of initial densification process of the glass powder in hydrothermal hot-pressing with Murray's rate equation revealed that the densification proceeds both by viscous flow and by rearrangement process. Analysis of the shrinkage rates with a rate equation of hydrothermal reaction suggested that the dissolution of particles into solution controlled the initial densification of the glass powder, and that the alkaline metal acted as a catalyst.

[*Received February 3, 1985*]

Key-words : *Hydrothermal hot-pressing, Shrinkage, Glass powder, Activation energy, Solidification*

1. Introduction

The solidification under hydrothermal conditions is used to produce thermal insulations and calcium silicate like structural boards. On mechanism of hydrothermal reaction, Kondo et al.[1),2)] discussed kinetics of lime-quartz and cement-quartz system. Kuczynski[3)] presented experimental evidence that sintering of glass powder occurred by viscous flow. Kinoshita et al.[4)] also reported that densification of glass powder proceeded by viscous flow during hot-pressing at high temperature.

In the previous papers[5),6)], the authors showed that the solidified body (compressive strength : about 2000 kg/cm^2) was formed from mixture of silica stone and siliceous sinter by hot-pressing under alkaline hydrothermal conditions, and revealed that its solidification was owing to linkage between the particles.

In the last paper[7)], it was reported powder of simulated nuclear waste glass was easily solidified by addition of distilled water with a rapid shrinkage at 280°-300°C, and the solidified body had low leachability and high compressive strength similar to that of the original glass. The authors expected that this method be acceptable to final geologic disposal of radioactive waste. Solidified bodies produced by hydrothermal hot-pressing technique are similar to ceramic compacts by sintering rather than those of cement or calcium silicate. However, the densification behavior of glass powder under the hydrothermal conditions was not understood. Few papers have been presented on hydrothermal hot-pressing technique, and the solidification mechanism was not clear.

In the present work, glass powder was solidified by hydrothermal hot-pressing technique. The effect of hydrothermal conditions on compressive strength was investigated. The solidification mechanism during hydrothermal hot-pressing was discussed by means of measurement of isothermal shrinkage.

The final aim of the present work was to obtain fundamental knowledge for immobilization of radioactive waste into glass matrix by hydrothermal hot-pressing technique.

2. Experimental procedure

2.1 Raw material

Composition of glass was shown in **Table 1.** This borosilicate glass contains 60.2 wt% SiO_2,

18.8 wt% B_2O_3, 4.59 wt% Al_2O_3, 4.02 wt% Li_2O. Particle distribution, spesific gravity, softening and transition point were shown in **Table 2.** The average particle size under 10 μm was 1.83 μm in diameter. SEM photograph of glass powder in **Fig. 1** showed that these particles are spheres in shape rather than angular.

Table 1. Composition of starting glass* (wt%).

SiO_2	B_2O_3	Li_2O	CaO	ZnO	BaO	Al_2O_3	Fe_2O_3
60.2	18.8	4.02	3.87	4.01	3.90	4.59	0.01

*Glass frit No. PF 800 P, offered by Power Reactor and Fuel Development Corporation.

Table 2. Characteristics of starting glass.

particle distribution	-10μm	73.7%
	10μm-40μm	25.4%
	40μm-	0.9%
softening point	820°C	
transition point	530°C	
specific gravity	2.45	

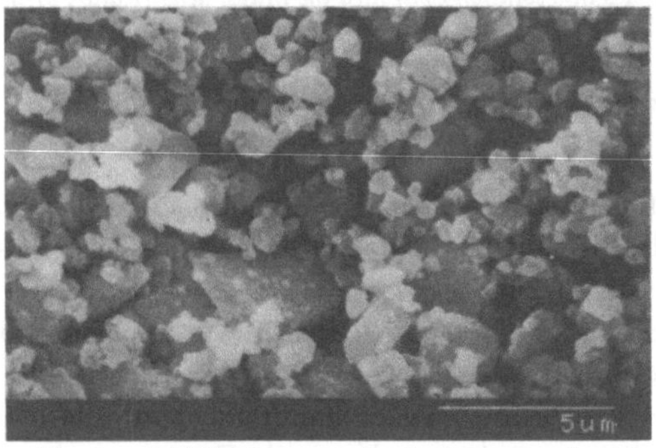

Fig. 1. SEM photograph of glass powder.

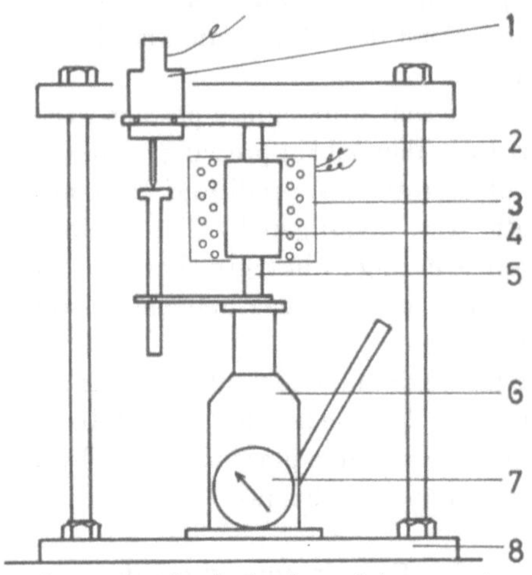

Fig. 2. Apparatus for hydrothermal hot-pressing.
1 : Dilatometer, 2 : Top piston, 3 : Induction heater, 4 : Autoclave, 5 : Bottom piston, 6 : Oil jack, 7 : Pressure guage, 8 : Rack

2.2 Apparatus

The apparatus for the hydrothermal hot-pressing was shown in **Fig. 2.** An autoclave for hydrothermal hot-pressing was previously reported[5]. It was cylindrical shape, and made of SCM steel. A solidified body is easily taken out without destruction because of tapering in diameter. The shrinkage of the starting material was determined by measuring the distance between top and bottom pistons during hydrothermal hot-pressing.

2.3 Solidification method

Glass powder (15 g) was kneaded well with NaOH solution or distilled water (1.5 ml) in the mortar to prepare the starting material. The material was put in the autoclave and cold pressed at 500 kg/cm^2. During the hydrothermal treatment, the starting material was continuously compressed at 260 kg/cm^2 by an oil jack and heated to a desired temperature by an induction heater. The heater was able to heat the autoclave from room temperature to 300°C in 4 min. The pressure and temperature were kept constant for 5 min-8 h.

2.4 Measurement of shrinkage

High rate of heating (75°C/min) was adapted for neglect of shrinkage during the raising temperature. The shrinkage occurred after the temperature reached the desired one. Measurement of shrinkage during holding period at the desired temperature, allowed to discuss kinetics of shrinkage.

After the treatment, the autoclave was cooled to room temperature and then solidified body was taken out. After drying at 110°C, the solidified body produced by hydrothermal hot-pressing technique was used to measure compressive strength and bulk density, and to observe fractured surface.

3. Results and discussion

3.1 Effect of hydrothermal hot-pressing conditions on density and compressive strength of solidified body

The powder of acidic mineral was easily solidified by addition of NaOH solution under the hydrothermal hot-pressing conditions and the solidified body had high compressive strength[5].

In this study, the starting glass powder contains 60.2 wt% silica. The effect of concentration of NaOH solution on the compressive strength and the density of the solidified body is shown in

Fig. 3. The density was increased with the increase in NaOH concentration. In the case of the reaction time for 5 min, the concentration of NaOH solution affected significantly the compressive strength of the solidified body. In the case of the reaction for 15 min, the concentration of NaOH solution had little effect on the compressive strength.

Effect of the reaction temperature and the reaction time on the compressive strength and the density of the solidified body by the use of distilled water was shown in **Fig. 4** and **Fig. 5**, respectively.

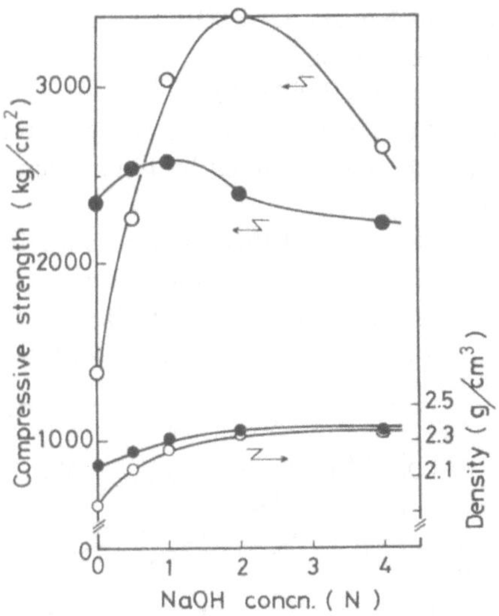

Fig. 3. Effects of NaOH concentration on compressive strength and bulk density of hardened bodies.
Hydrothermal hot-pressing conditions :
Reaction temperature : 280°C, Pressure : 260 kg/cm^2,
Reaction time : 5 min (○), 15 min (●)

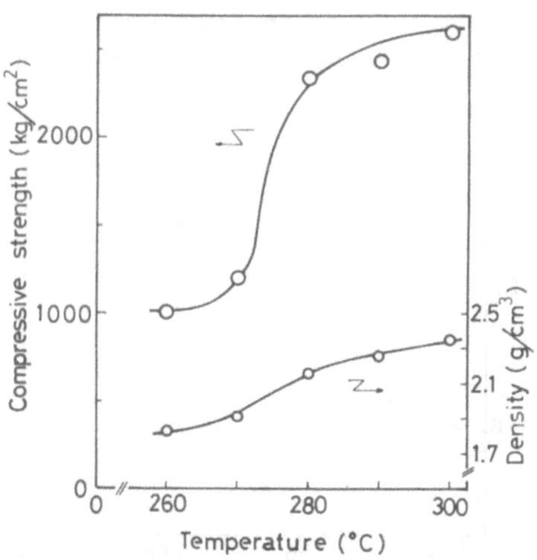

Fig. 4. Effects of reaction temperature on compressive strength and bulk density of solidified bodies.
Hydrothermal hot-pressing conditions :
Pressure : 260 kg/cm^2, Reaction time : 15 min

The density was linearly increased with the increase in the reaction temperature and the compressive strength was remarkably increased at 280°C (Fig. 4). This remarkable increase may occur by chemical reaction, since this temperature is much lower than that for ordinary sintering and vitrification of glasses (above 500°C).

When the reaction time changed from 10 min to 15 min, the compressive strength and the density were rapidly increased (Fig. 5). This increase may be affected by progress of linkage between the particles.

These results suggested that the glass powder was easily solidified without the aid of alkali at the temperatures much lower than that for ordinary sintering.

3.2 Initial densification process

The densification process of the glass powder

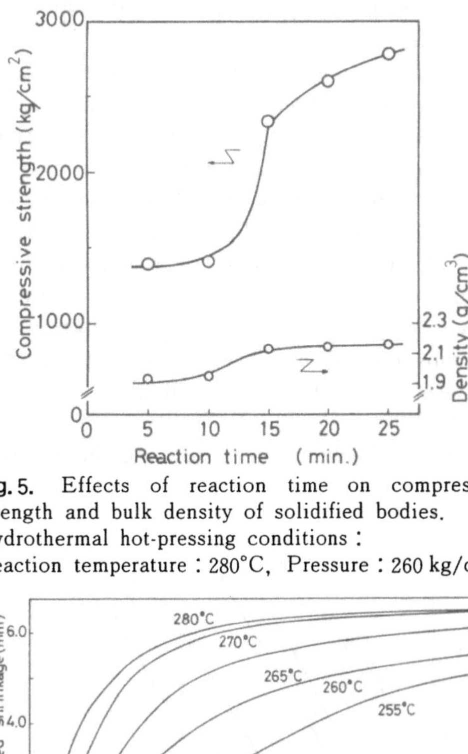

Fig. 5. Effects of reaction time on compressive strength and bulk density of solidified bodies.
Hydrothermal hot-pressing conditions :
Reaction temperature : 280°C, Pressure : 260 kg/cm^2

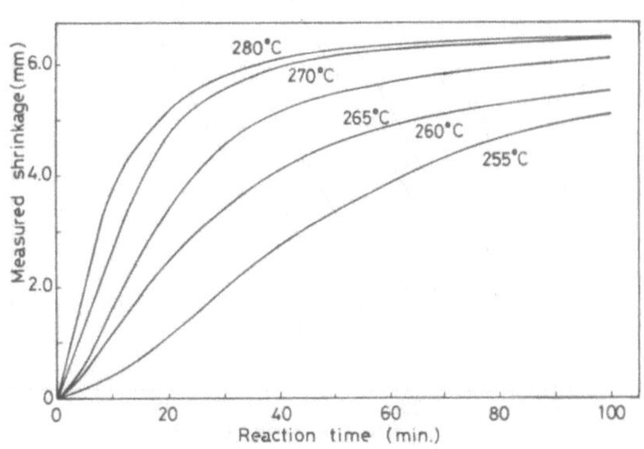

Fig. 6. Measured shrinkage during the hydrothermal hot-pressing.

by the use of distilled water was discussed by the measurement of the shrinkage during the hydrothermal hot-pressing. **Figure** 6 shows the relation between the uniaxial shrinkage and the reaction time at constant temperature. The rapid shrinkage was observed during the initial hydrothermal hot-pressing.

Several workers[8)-10)] presumed a mechanism of sintering from a slope of log shrinkage ratio vs. log reaction time.

Kingery et al.[11),12)] revealed that log shrinkage ratio vs. log time plot is represented by a slope of 1.3-1.4 when the densification takes place by rearrangement process during sintering in the presence of a liquid phase. **Figure** 7 shows log shrinkage ratio vs. log time for our data. The slope of initial portion of these curves are 1.5 for 0.5-5 min reaction. This value agrees approximately with that for rearrangement process. The initial densification of glass powder may occur by rearrangement of particles at 260°-290°C for 0.5-5 min.

On the other hand, Murray et al.[13)] shows a simple rate equation for pressure sintering by introducing the effect of pressure on sintering.

$$\ln(1-\rho)=-(3\sigma/4\eta)t+c$$

where ρ : relative density, σ : pressure, η : viscosity and t is time. This equation corresponds with plastic flow or viscous flow mechanism. Mangsen et al.[14)] and Vasilos[15)] obtained the value of viscosity of ceramic compacts during the pressure sintering by this equation. In this study, the viscosity of the glass is calculated from this equation.

Relation between ln $(1-\rho)$ and the reaction time, and the result of this calculation are shown in **Fig. 8** and **Fig. 9**.

The value of viscosity were calculated at 7.4×10^{11}-4.4×10^{12} poise. Temperature dependence of the calculated viscosity corresponds to the activation energy in 67 kcal/mol. The value agrees approximately with that for viscous flow during sintering of glass powder[3)]. The initial densification of the glass powder during the hydrothemal hot-pressing may proceed by viscous flow with rearrangement process.

Kondo et al.[1),2)] discussed the hydrothermal reaction mechanism in $Ca(OH)_2$-quartz system by following equation.

$$\{1-(1-\alpha)^{1/3}\}^N=kt$$

where t : the reaction time, α : the reaction ratio, and N is determined by the mechanism of reaction.

$N=2$ the usual diffusion rate-contrlling
$N=1$ the surface reaction rate-controlling
$N>2$ the diffusion rate-controlling : the reaction layer spread densely with the reac-

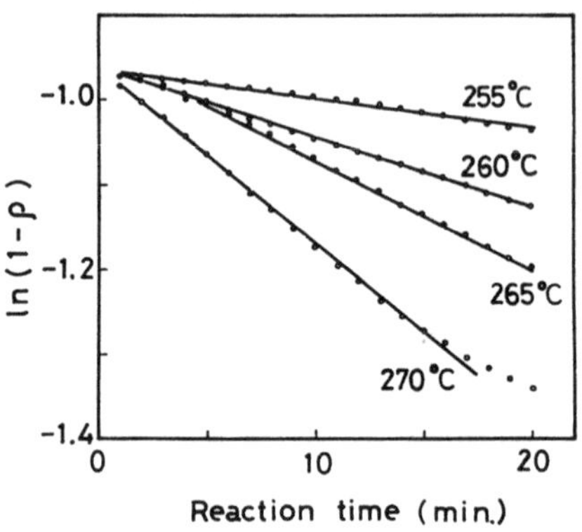

Fig. 8. Relation between ln $(1-\rho)$ and reaction time.

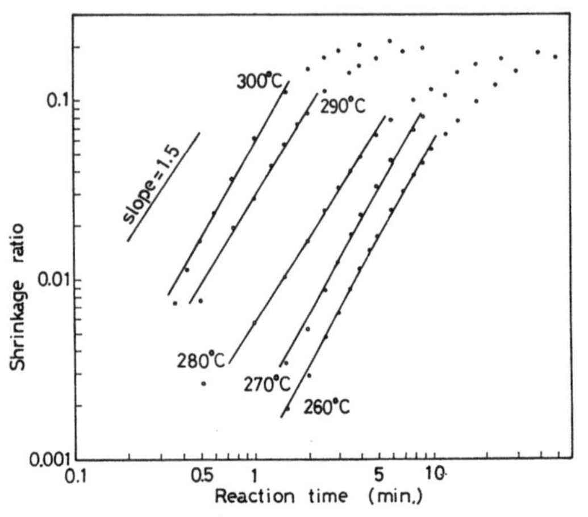

Fig. 7. Relation between log shrinkage ratio and log reaction time.

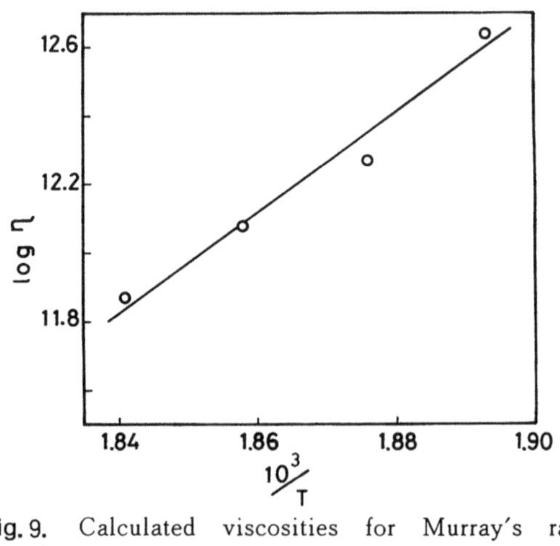

Fig. 9. Calculated viscosities for Murray's rate equation.

tion time

$N<2$ the diffusion rate-controlling : the reaction layer spread coarsely with the reaction time

When the linkage among the particles controlles the densification of powder, in the same behavior as sintering, its reaction rate may be approximately shrinkage rate. The densification of powder during the hydrothermal hot-pressing occurs by the linkage between the particles[5].

In this study, the densification process was analyzed based on the shrinkage rate. That is, the shrinkage ratio was used in place of the reaction ratio in Kondo's equation.

The shrinkage ratio was calculated by following equation.

$$x=\Delta L/L$$

where x : shrinkage rate, ΔL : the measured shrinkage on each time, L is the total shrinkage after the treatment. The relation between the shrinkage ratio and the reaction time is shown in **Fig. 10.**

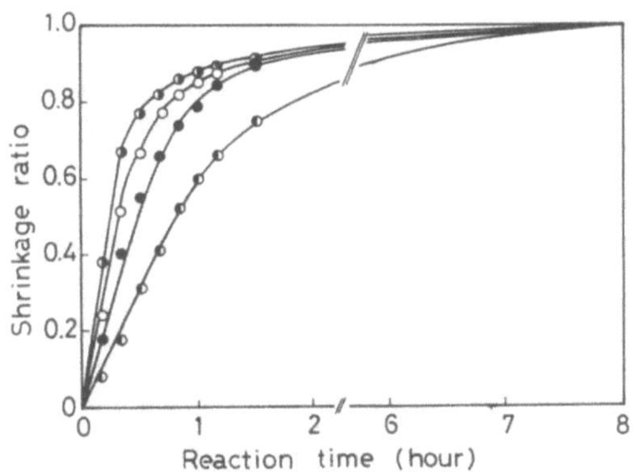

Fig. 10. Relation between shrinkage ratio and reaction time.
Reaction time is 255°C(◐), 260°(●), 265°C(○), and 270°C(◑).

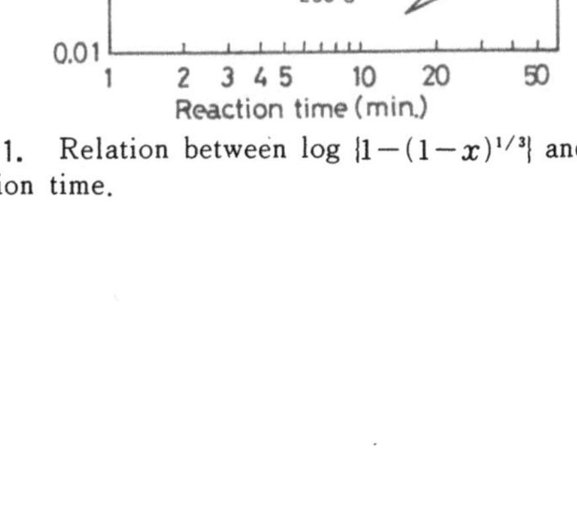

Fig. 11. Relation between log $\{1-(1-x)^{1/3}\}$ and log reaction time.

The rate equation $\{1-(1-x)^{1/3}\}^N=kt$ is determined by obtaining the value of N. **Fig. 11** shows log $\{1-(1-x)^{1/3}\}$ vs. log t. The value of N is closer to unity from 5 to 20 min reaction and this reaction may depends on the surface reaction rate-controlling. Arrhenius plot is approximately on a straight line and the calculated apparent activation energy is 60 kcal/mol (Fig. 12). This activation energy is high comparing with that for a glass containing various alkaline metals[7].

The glass powder was formed to the solidified body by the use of 0.5 N NaOH or 1.0 N NaOH solution, because of examination of the effect of alkaline metal on the densification. These activation energy are calculated from Kondo's equation as shown in **Fig. 12.** The activation energy was reduced with the increase in NaOH concentration. The results show that the alkaline metal acts as catalyst and the dissolution of particles in solution controls the initial densification of the glass powder.

3.3 Observation of the fractured surface

SEM photographs of the solidified bodies are shown in **Fig. 13.** The solidified body for reaction time 10 min has large void spaces among the particles. These void spaces are reduced by linkage among the particles with reaction time. The densification of the solidified body proceeds sufficiently under the conditions at 280°C and 90 min reaction. In the case of the ordinary pressure sintering, initial densification depends on rearrangement process by sliding and rotation of particles, and in an ultimate stage, pores are reduced and dissapear by the self-diffusion[16].

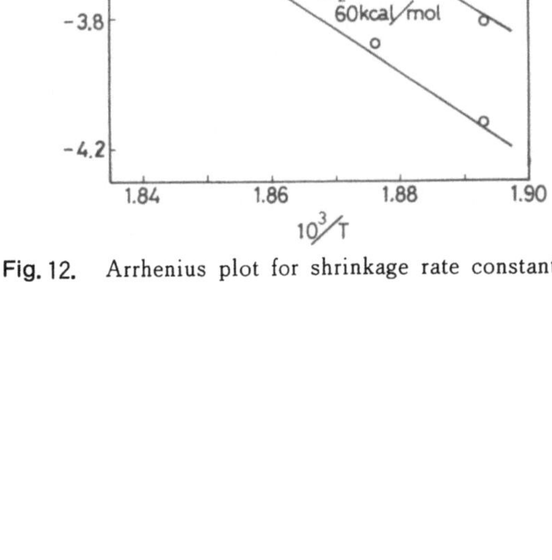

Fig. 12. Arrhenius plot for shrinkage rate constant.

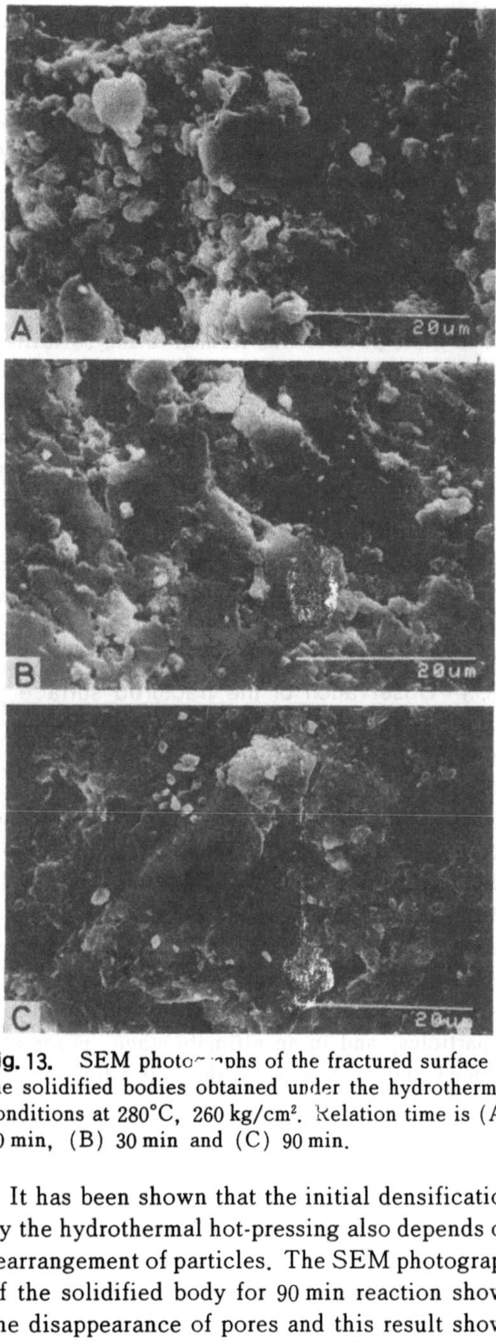

Fig. 13. SEM photographs of the fractured surface of the solidified bodies obtained under the hydrothermal conditions at 280°C, 260 kg/cm². Relation time is (A) 10 min, (B) 30 min and (C) 90 min.

It has been shown that the initial densification by the hydrothermal hot-pressing also depends on rearrangement of particles. The SEM photograph of the solidified body for 90 min reaction shows the disappearance of pores and this result shows that the densification of the solidified body has progressed at this ultimate stage.

4. Conclusion

Borosilicate glass powder was hot-pressed under the hydrothermal conditions and the isothermal shrinkage of powder compact was measured during the hydrothermal hot-pressing. The compressive strength of the formed solidified bodies was measured. The densification mechanism of the glass powder was discussed and the following results were obtained.

(1) The glass powder was easily solidified at 280°C even by the use of distilled water instead of alkaline solution, and the solidified body had high compressive strength (about 2300 kg/cm²).

(2) The initial densification of the glass powder analyzed with Murray's rate equation. The calulated activation energy agreed with that for viscous flow during sintering of the glass powder, and its densification proceeded with rearrangement process.

(3) The densification was analyzed based on the shrinkage rate. The dissolution of particles in solution controlled the initial densification of the glass powder, and the alkaline metal acted as catalyst.

(4) The glass powder was successfully solidified by the hydrothermal hot-pressing technique at much lower temperature comparing with the ordinary pressure sintering.

Acknowledgements The authors thanks Power Reactor and Fuel Developement Corporation for offering starting glass materials.

References

1) R. Kondo, K. Lee and M. Daimon, *Yogyo-Kyokai-Shi*, 84, 573-78 (1976).
2) M. Harada, Y. Hagiwara, V. H. Nguyen, K. Asaga and M. Daimon, ibid., 90, 123-29 (1982).
3) G. C. Kuczynski, *J. Appl. Phys.*, 20, 1160-63 (1949).
4) M. Kinoshita and H. Wakabayashi, *Osaka-Kogyo-Gizyutsushikenzyo-Kiho*, 24, 208-13 (1973).
5) N. Yamasaki, M. Nishioka, K. Yanagisawa and S. Kanahara, *Yogyo-Kyokai-Shi*, 92, 150-52 (1984).
6) N. Yamasaki, K. Yanagisawa, S. Kanahara, M. Nishioka, K. Matsuoka and J. Yamazaki, *J. Nucl. Sci. Technol.*, 21, 71-73 (1984).
7) N. Yamasaki, K. Yanagisawa and M. Nishioka, *Nihon-Genshiryoku-Gakkai-shi*, in press.
8) S. S. Cole, Jr., *J. Am. Ceram. Soc.*, 55, 296-99 (1972).
9) H. V. Anderson, *J. Am. Ceram. Soc.*, 50, 235-38 (1967).
10) J. M. Neve and R. L. Coble, *J. Am. Ceram. Soc.*, 57, 274-75 (1974).
11) W. D. Kingery, *J. Appl. Phys.*, 30, 301-06(1959).
12) W. D. Kingery and M. D. Narasimhan, *J. Appl. Phys.*, 30, 307-10 (1959).
13) P. Murray, E. P. Rodgers and A. E. Williams, *Trans. Brit. Ceram. Soc.*, 53, 474-510 (1954).
14) T. Vasilos, *J. Am. Ceram. Soc.*, 43, 517-19 (1960).
15) G. E. Mangsen, W. A. Lambertson and B. Best, *J. Am. Ceram. Soc.*, 43, 55-59 (1960).
16) W. Komatsu and T. Shimohira in "Kagaku Sosetsu No. 9", Tokyo Daigaku Syuppan Kai (1975) p. 256.

Reprinted from *Journal of Materials Science Letters*, **5** (1986) 355–356.

A hydrothermal hot-pressing method: apparatus and application

N. YAMASAKI, K. YANAGISAWA, M. NISHIOKA, S. KANAHARA
Research Laboratory of Hydrothermal Chemistry, Faculty of Science, Kochi University, 2-5-1 Akebono-cho, Kochi-shi 780, Japan

Loose particles occurring in sediment are transformed into sedimentary rock by a process termed lithification, which is a chemical process that reduces the original porosity by compaction and cementation [1]. A hydrothermal hot-pressing method is intended for artificial lithification. In this study, silica powder including mineralizer solution is compressed under hydrothermal conditions from outside an autoclave for hydrothermal hot-pressing and converted to a solid body.

Few studies have been carried out on hot-pressing under hydrothermal conditions. Roy *et al.* [2, 3] prepared very high strength materials by hot-pressing cement paste under conditions up to 250° C and 345 MPa. A small amount of water (0.093 g water per 1 g cement) was added to anhydrous cement and the degree of hydration was around 30% for hardened materials [3]. In this case, vapour pressure was very small, because water was consumed for hydration of cement. The hydrothermal reaction sintering method [4–6] has been used to produce sintered oxide materials from metals. In this method, water reacts with metals to prepare oxides and is completely consumed.

The hydrothermal hot-pressing method has two characteristics: continuous compression of samples under hydrothermal conditions and space for water retreat. Compression accelerates compaction of starting powder and prevents development of cracks by shrinkage. Compaction depends on the rate at which water can be expelled from the starting powder. The hydrothermal hot-pressing method requires the space for water retreat, into which water included in the starting powder is released. Without the space, water exists in pore space of the starting powder and hinders compaction of the powder.

Alkaline solution, used as a mineralizer, increases solubility of silica and is easily saturated under hydrothermal conditions. Under such conditions, cementation of silica is accelerated by means of dissolution and deposition. Furthermore, solubility of many inorganic compounds increases under hydrothermal conditions and the compounds react with each other to produce new compounds with low solubility. The reaction products may contribute to cementation.

Two types of autoclaves for hydrothermal hot-pressing are shown in Figs 1 and 2. The autoclave shown in Fig. 1b requires a hot-pressing cell (Fig. 1a). The cell, made of Hastelloy C, is composed of an inner and an outer case. The inner case, 1.4 cm i.d., is divided into three parts in order to remove a solidified body from the cell without its destruction. The outer case is used to keep the inner case in cylindrical shape.

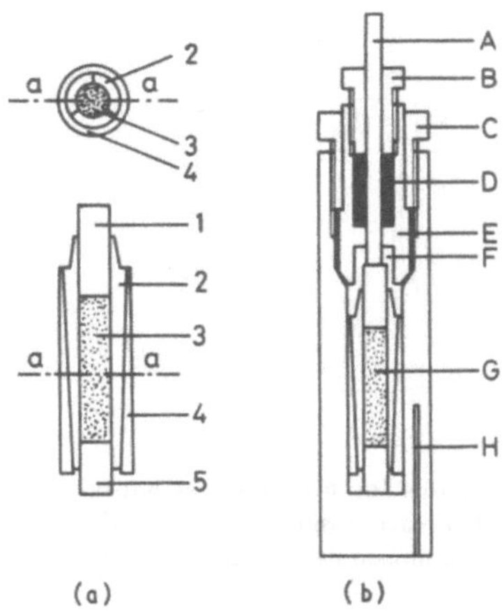

Figure 1 (a) Cell and (b) autoclave for hydrothermal hot-pressing. (a) 1, 5, piston; 2, inner case; 3, sample; 4, outer case. The upper figure is a cross-section of aa. (b) A, Push rod; B, nut for pushing gland packing; C, nut for pushing cone; D, gland packing (copper and cotton yarn); E, cone; F, space for water retreat; G, sample; H, well for thermocouple.

The autoclave (Fig. 1b) is made of iron with a liner of Hastelloy C. It has the space for water retreat (about 25 cm^3) unoccupied by the cell. The starting powder in the cell is uniaxially compressed during hydrothermal treatment, via the push rod of the autoclave with gland packing to prevent leakage.

The autoclave shown in Fig. 2 is made of iron and is a cylinder with a cylindrical chamber. The starting powder is transferred directly into the chamber. One end of the chamber is larger than the other, in order to remove a solidified body with ease. The starting

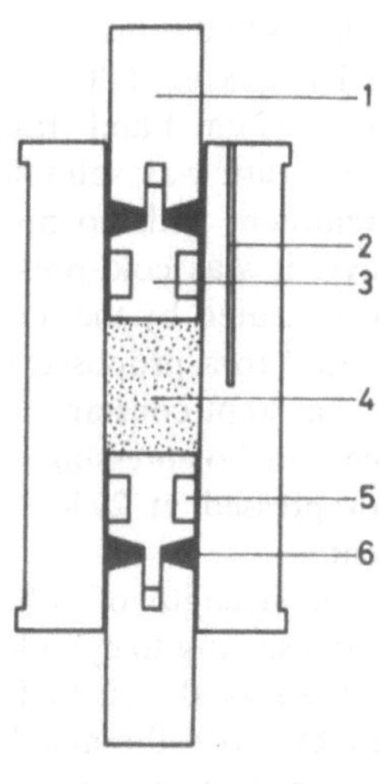

Figure 2 Autoclave for hydrothermal hot-pressing: 1, push rod; 2, well for thermocouple; 3, piston; 4, sample; 5, space for water retreat; 6, gland packing (Teflon containing carbon).

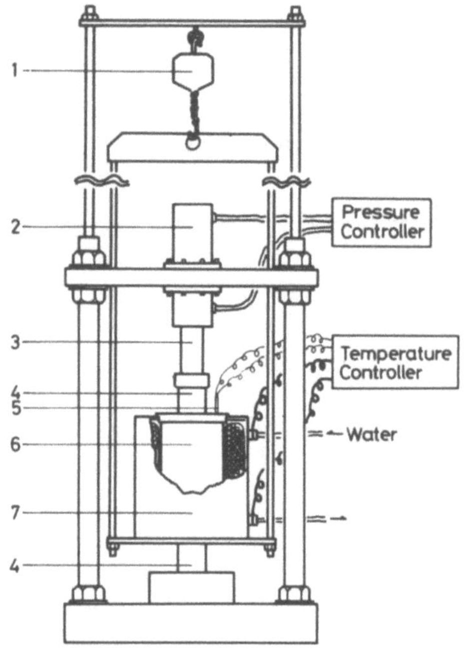

Figure 3 Apparatus for hydrothermal hot-pressing: 1, hoist for adjustment of furnace position; 2, pump; 3, ram; 4, push rod of autoclave; 5, thermocouple; 6, autoclave; 7, induction furnace.

powder is compressed by pistons from above and below. The pistons have the space for water retreat of varying volume. Gland packing between piston and push rod prevents leakage. It is easy to produce a large-sized autoclave of this type, because of its simple cylindrical shape with no screw. A large autoclave 6 cm i.d. has been produced.

A schematic general view of the apparatus for hydrothermal hot-pressing is shown in Fig. 3. The apparatus is composed of autoclave, furnace, pump and frame. An induction heater, produced from copper tube coil, is used to heat a large autoclave quickly. Insulation of the coil is assured by passing water through the tube.

An equivalent mixture of low-quartz (powder of quartzite, passed through 200 mesh) and amorphous silica (powder of siliceous sinter, porous and friable rock deposited by waters of hot springs) was solidified under hydrothermal conditions, using the autoclave shown in Fig. 1. The mixture (18 g) was kneaded with 5 N NaOH solution (2 cm^3) and transferred into the cell. Before the mixture was solidified by the usual hydrothermal treatment with no pressure from outside the autoclave, it was cold-pressed at 180 MPa. The autoclave was heated by the induction heater at the rate of $30° min^{-1}$ to a proposed temperature and the temperature was kept constant for 30 min. In the case of hydrothermal hot-pressing, the mixture was continuously compressed at 27 MPa during hydrothermal treatment.

The compressive strength of solidified specimens shaped by diamond sawing to cylinders, 1.4 cm diameter and 1.6 cm high, is shown in Fig. 4. Hardened bodies produced by hydrothermal hot-pressing had higher strength than those produced by usual hydrothermal treatment. During hydrothermal treatment, the push rod came into the autoclave because of

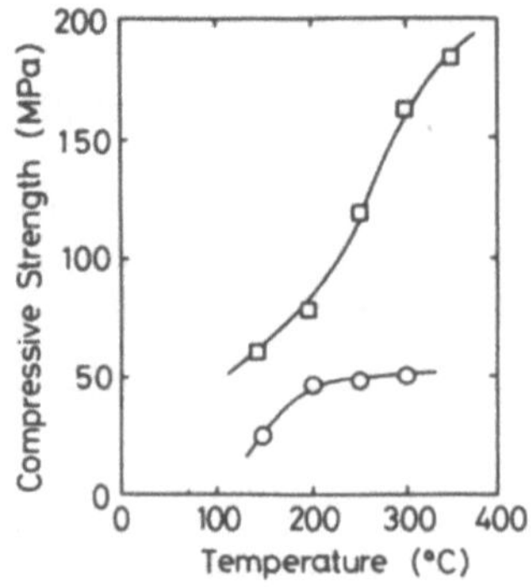

Figure 4 Compressive strength of silica specimens solidified by hydrothermal hot-pressing (□) and by usual hydrothermal treatment (○).

shrinkage of the starting powder. The highest compressive strength and density of the solidified specimens were achieved by hydrothermal hot-pressing at 350° C.

Powder of the following substances, among others, can be solidified by the hydrothermal hot-pressing method under conditions up to 350° C and 140 MPa: crystalline and amorphous silica, silicates, mixtures containing silica ($SiO_2 + Fe_2O_3$, ZrO_2, $Ca(OH)_2$, $Al(OH)_3$, plastic, etc.), cement, calcium carbonate [7], zirconia [8], barium titanate [9]. A translucent body of cubic zirconia is obtained from co-precipitated gel by hydrothermal hot-pressing at 350° C and subsequent sintering at 1350° C [8]. A solidified body of barium titanate is prepared from a mixture of TiO_2 by this method [9]. Cement specimens solidified by the method have high compressive strength over 150 MPa. The method is also useful for solidifying industrial wastes such as fly ash, coal ash, radioactive wastes from nuclear power plants, etc. Finally, high-level [10] and low-level radioactive wastes [11] have been successfully immobilized into a silica matrix by hydrothermal hot-pressing.

References

1. G. M. FRIEDMAN and J. E. SANDERS, "Principles of Sedimentology" (Wiley, New York, 1978) p. 144.
2. D. M. ROY, G. R. GOUDA and A. BOBROWSKY, *Cement Concrete Res.* **2** (1972) 349.
3. D. M. ROY and G. R. GOUDA, *J. Am. Ceram. Soc.* **56** (1973) 549.
4. S. HIRANO and S. SOMIYA, *ibid.* **59** (1976) 534.
5. M. YOSHIMURA and S. SOMIYA, *Am. Ceram. Soc. Bull.* **63** (1980) 246.
6. H. TORAYA, M. YOSHIMURA and S. SOMIYA, *J. Am. Ceram. Soc.* **63** (1982) C-159.
7. K. MATSUOKA, Y. KUWABARA, N. YAMASAKI, M. MITSUSHIO and J. YAMAZAKI, *Rep. Res. Lab. Hydrothermal Chem.* **3** (1979) 8.
8. N. NISHIZAWA, H. TEBIKA and N. YAMASAKI, *Yogyo-Kyokai-Shi* **92** (1984) 420.
9. Unpublished work.
10. K. YANAGISAWA, M. NISHIOKA and N. YAMASAKI, *Am. Ceram. Soc. Bull.* to be published.
11. Y. NISHIHARA, T. KASHIWAI and N. YAMASAKI, presented at Waste Management 85, Arizona, March 1985.

Received 12 August
and accepted 25 September 1985

Reprinted from *American Ceramic Society Bulletin*, **64**(12) (1985) 1563–1567.

Immobilization of Radioactive Wastes by Hydrothermal Hot Pressing

KAZUMICHI YANAGISAWA, MAMORU NISHIOKA, and NAKAMICHI YAMASAKI
Kochi University, Akebono-cho, Kochi-shi 780, Japan

Simulated high-level radioactive waste was solidified into a silica-rich matrix containing α-quartz (70 wt%) and amorphous aluminosilicate (30 wt%) by a hydrothermal hot-pressing method. The waste and the matrix were mixed with an NaOH solution and then treated hydrothermally at ≤350°C. During the treatment, the starting powder was compressed at ≤66 MPa from the outside of the autoclave. The high reaction temperature and pressure gave the waste form with large compressive strength and low leachability. The highest compressive strength was achieved by using 10*N* NaOH solution. The long reaction time of 6 h reduced the weight loss due to leaching. The addition of 10 wt% $Al(OH)_3$ to the starting powder also reduced the leachability by immobilizing Na into aluminosilicates. The leach rate of the waste form (21.8 wt% waste, 10 wt% $Al(OH)_3$, 68.2 wt% matrix) produced under the optimum conditions (10*N* NaOH solution, 350°C, 66 MPa, 6 h) was 7 g/m²·d as determined by 14-d Soxhlet leach tests.

The immobilization of high-level radioactive wastes into borosilicate glass is the most popular process to produce waste forms. The process is appealing because of its relative simplicity and utilization of conventional glass-making technology. In a recent quantitative evaluation of candidate waste forms, the high process rating for borosilicate glass and its intermediate product performance score resulted in the overall top-ranking position.[1] It is, however, obvious that glass is thermodynamically unstable relative to a chemically equivalent assemblage of crystalline phases.[2] Under some circumstances, it may devitrify to form the stable crystalline assemblage.

The immobilization of the wastes into crystalline phases has been researched for supercalcine,[3,4,5] synroc,[2,6,7] high-alumina tailored ceramics,[8,9,10] and so on. The equipment for these processes is rather complicated, though the waste forms produced by the processes have excellent properties in comparison with the glass forms.

The processes mentioned above require temperatures above 1000°C, which may cause evaporation of volatile radioactive isotopes and corrosion of process equipment.

Concretes formed under elevated temperatures and pressures (FUETAP concretes) are cost-effective hosts for the wastes, because of its ease of processing.[11] The FUETAP concretes, however, have high leach rates for cesium and strontium in comparison with other waste forms.[1,11]

The hydrothermal hot-pressing technique has been used to produce dense bodies from inorganic powder such as silica,[12,13] aluminosilicate and calcium carbonate.[14] Two recent papers[12,15] revealed that the technique was useful for solidification of silica containing simulated high-level radioactive waste. The hydrothermal hot-pressing technique has the following advantages: The apparatus is simple and its operation is easy. The reaction temperature is less than 350°C, much lower than that used to produce borosilicate glass or ceramic waste forms. The hydrothermal reaction occurs in a closed system, which prevents the evaporation of volatile materials. Waste forms produced by this technique are expected to be stable under repository conditions because they are produced under hydrothermal conditions. Elements in the wastes may react with the silica matrix to produce stable crystalline phases such as silicates and aluminosilicates. It has been confirmed by separate experiments that strontium reacted with α-quartz to produce strontium silicates,[12] and cesium reacted with amorphous aluminosilicate to produce pollucite ($CsAlSi_2O_6$) in a sodium hydroxide solution under hydrothermal conditions.[16]

In this paper, simulated high-level radioactive waste was solidified using a silica matrix composed of α-quartz and amorphous aluminosilicate by the hydrothermal hot-pressing technique. The effect of hot-pressing conditions on compressive strength and leachability of the waste forms was investigated.

Experimental Procedure

Solidification

Starting materials for silica matrix were α-quartz (quartzite from Fukushima Prefecture, Japan) and amorphous aluminosilicate (siliceous sinter, porous, and friable rock deposited by the waters of hot springs, from Kagoshima Prefecture, Japan). Their chemical analyses are shown in Table I. Quartzite was ground to pass through 200 mesh and mixed with siliceous sinter in a ball mill to get a mixture of 70 wt% quartzite and 30 wt% siliceous sinter.

The composition of a simulated high-level radioactive waste from reprocessing of nuclear fuels is shown in Table II; it was estimated by Japan Atomic Energy Research Institute (JAERI), and Tc, Pm, Ru, Rh, Pd, and actinides were simulated by Mn, Nd, Fe, Co, Ni, and Ce, respectively.[17] In the waste, Na was not included because NaOH solution was added as mineralizer for hydrothermal solidification. The waste was produced by mixing the appropriate chemicals in a ball mill.

Table I. Chemical Analyses of Starting Materials

	Amount of constituent (wt%)						
Material	SiO_2	Al_2O_3	Fe_2O_3	CaO	MgO	Other oxides	Ignition loss
Quartzite	99.5	0.2	0.02	0.02	0.01	0.25	0
Siliceous sinter	71.7	12.9	2.3	2.4	0.4	5.7	4.6

Received January 16, 1985; revised copy received June 3, 1985; approved July 19, 1985.

Table II. Composition of Simulated High-Level Radioactive Waste

Oxide	Amount (wt%)	Oxide	Amount (wt%)
Fe_2O_3	39.05	SrO	1.69‡
MoO_3	8.66	Sm_2O_3	1.64
Nd_2O_3	8.24	MnO_2	1.29
ZrO_2	8.21	TeO_2	1.12§
CeO_2	6.77	CoO	1.00
NiO	4.92	Y_2O_3	1.00
Cs_2O	4.88*	Rb_2O	0.61*
BaO	3.12‡	Gd_2O_3	0.19
La_2O_3	2.54	CdO	0.16
Pr_6O_{11}	2.47	Ag_2O	0.15*
Cr_2O_3	2.26	Sb_2O_3	0.13†

*Added as nitrate. †Added as chloride. ‡Added as carbonate. §Added as metal.

The silica matrix (70 wt% α-quartz and 30 wt% amorphous aluminosilicate) was mixed well with the waste (21.8 wt% waste loading). The mixture (20 g) was kneaded with NaOH solution (2.5 cm^3) to prepare the starting wet powder. The powder had no plasticity because of its low water content. It was transferred into the hot-pressing die (Fig. 1(B)). The die, made of Hastelloy C, was composed of inner and outer cases. The inner case, 1.4 cm in inner diameter, was divided into three parts in order to take out a solidified body from the die without its destruction. The outer case was used to keep the inner case in cylindrical shape. The die was put into the autoclave (Fig. 1(A)), and heated to temperatures from 100° to 350°C at the rate of 30°C/min by an induction heater. The autoclave was made of Fe with a liner of Hastelloy C, and had the open space (about 25 cm^3) unoccupied by the die. The open space was required for solidification by hydrothermal hot-pressing in order to expel the water from the wet powder and to compact the particles of the powder. The powder in the die was uniaxially compressed at 17 to 66 MPa during hydrothermal treatment, via piston of the autoclave with gland packing to prevent leakage. The temperature and the pressure were kept constant for 0 to 6 h. After the hydrothermal treatment, the autoclave was cooled down to room temperature by an electric fan.

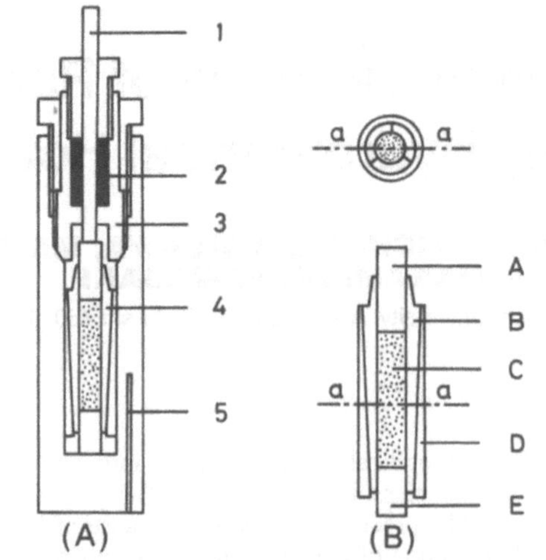

Fig. 1 Apparatus for hydrothermal hot-pressing. (A) Autoclave, showing (1) piston, (2) gland packing, (3) cone, (4) hot-pressing die, and (5) thermocouple well. (B) Hot-pressing die (cross section at *a–a* is shown at top), including (A) upper rod, (B) ring, (C) inner case, (D) wet starting powder, (E) outer case, and (F) lower rod.

Characterization of Waste Forms

The solidified waste forms were cut with a diamond saw to 1.6-cm-long sections and dried at 110°C to prepare specimens for strength and density measurement. Uniaxial compression tests were performed at room temperature to measure compressive strength. The compression axis was perpendicular to the base of the cylindrical specimens. The crosshead travel rate of a mechanical machine was 0.02 cm/min. Density of the waste form was measured from its weight and apparent volume. The crystalline phases were identified by X-ray powder diffraction, and a polished section was examined by a scanning electron microscope.

Leach Tests

Static leach tests were performed under hydrothermal conditions up to 350°C. The autoclave for the tests, 27.7 cm^3 in inner volume, was made of Fe and lined by Ti. The leachant was 15 cm^3 of deionized water. The cubic specimens (5 by 5 by 5 mm^3) were fabricated by diamond-sawing and dry-abrading the waste forms, giving a specimen surface area/leachant volume ratio of 1/10 cm^{-1}. The time for the tests was selected to be one day because the initial release rate under hydrothermal conditions was large enough to estimate the effect of hydrothermal hot-pressing conditions on the leachability of the waste forms. Monolithic glass specimens of JW-A (21.8 wt% waste loading, prepared by JAERI) were also used for the tests, in order to compare their release rate with that of the waste forms produced by hydrothermal hot-pressing.

Flow leach tests were performed at 97°C in Soxhlet type apparatus which was constructed of fused silica glass. Cylindrical specimens for the tests (1.4 cm in diameter, 0.8 cm in height) were fabricated by diamond-sawing and dry-abrading. The tests were performed for 14 d.

After the leach tests, each leachant was analyzed by atomic absorption spectrometry. Weight loss was calculated by weight change of a specimen dried at 110°C before and after the leach tests, and it was normalized by dividing by geometric surface area of the starting specimen.

Results and Discussion

Effect of Hydrothermal Hot-pressing Conditions on Properties of Waste Forms

The weight loss by leach tests at 200°C for 1 d, the density and the compressive strength of the waste forms are shown in Fig. 2, vs the hydrothermal hot-pressing conditions; NaOH concentration, reaction temperature, pressure, and time.

When the concentration of NaOH solution was increased from 3 to 10*N*, the compressive strength and the density were increased (Fig. 2(A)). Above 12*N*, the strength was reduced and the density was kept constant at 2.7 g/cm^3. The highest compressive strength was achieved by the use of 10*N* NaOH solution. The weight loss was exponentially increased with the increase in NaOH concentration.

The density was linearly increased along with the increase in the reaction temperature (Fig. 2(B)). The compressive strength was also increased above 150°C and a peak appeared at 100°C, at which amorphous aluminosilicate began to dissolve in a NaOH solution.[13] The waste form with high compressive strength and low weight loss was produced at 350°C.

The high reaction pressure increased the compressive strength and the density, but reduced the weight loss (Fig. 2(C)). The reaction time up to 30 minutes increased the strength and the density (Fig. 2(D)). The long reaction time (6 h) reduced the weight loss.

These results suggested that a waste form with high compressive strength and resistance to leaching could be produced at high temperature (350°C) and high pressure (66 MPa) for a long reaction time (6 h) by the use of 10*N* NaOH solution.

Crystalline Phases and Microstructures

According to X-ray powder diffraction, the waste forms were mainly composed of α-quartz, the main constituent of the silica matrix. The diffraction patterns showed that MoO_3, Nd_2O_3, and ZrO_2 disappeared by hydrothermal hot pressing and peaks due to only α-Fe_2O_3 and CeO_2 in the waste components were observed with low intensity. A few new peaks with low intensity, probably due to hydrothermal reaction products, were also observed, though they were not identified.

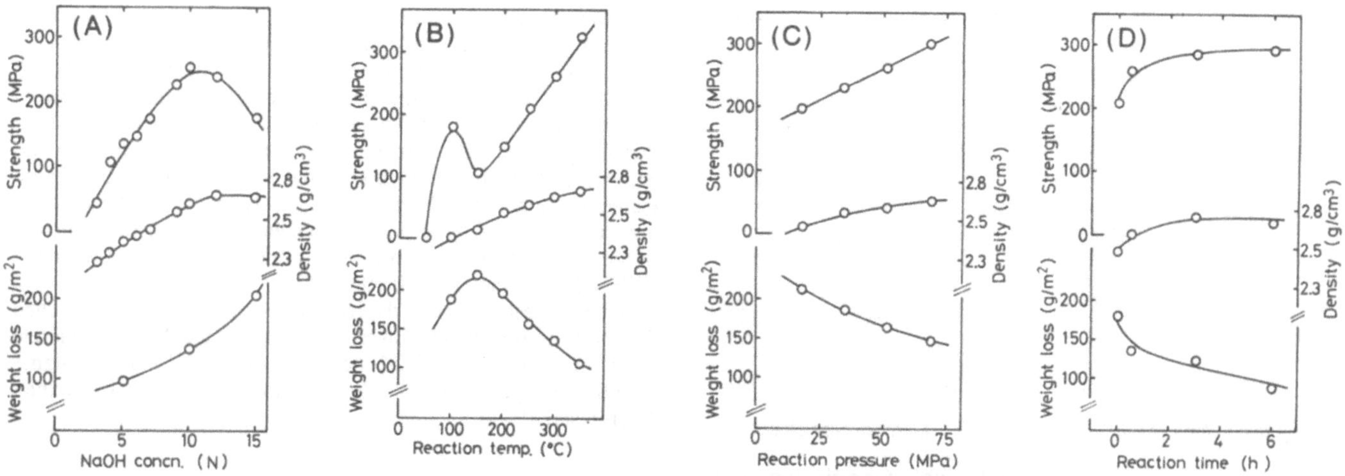

Fig. 2. Compressive strength, density, and weight loss of waste forms vs (A) NaOH concentration, (B) reaction temperature, (C) reaction pressure, and (D) reaction time. Each hydrothermal hot-pressing condition was varied from standard conditions, i.e., 10*N* NaOH solution, 300°C, 49 MPa, 30 min.

A polished section of the waste form (21.8 wt% waste, 10 wt% $Al(OH)_3$, 68.2 wt% matrix) produced at 350°C and 66 MPa for 6 h by the use of 10*N* NaOH solution, is shown in Fig. 3. The Si*Kα* X-ray image showed that large grains were α-quartz of the matrix. Elements of the waste components existed between α-quartz grains. As seen in the photograph, the waste form had many pores and contained as much as 25% porosity. Its specific surface area was found to be 10 m^2/g by BET method using N_2 gas as an absorbent.

Weight Loss Dependency on Leaching Temperatures

The waste forms produced under the following two kinds of hydrothermal hot-pressing conditions were leached at various temperatures in deionized water for one day. The leach tests for JW-A glass form were also performed.

Condition (*A*): 10*N* NaOH solution, 300°C, 49 MPa, 30 min
Condition (*B*): 10*N* NaOH solution, 350°C, 66MPa, 6 h

The waste form produced under the condition (*A*) had higher weight loss at any leaching temperatures than others (Fig. 4(A)). The weight loss above 150°C was greatly reduced when the condition (*B*)was used. Its weight loss curve against the leaching temperature was similar to that of the glass form, though it was slightly higher than the latter at leaching temperatures below 200°C.

When the glass form was leached above 150°C, a reaction layer was produced on its surface. The layer produced below 300°C was yellowish translucent thin plate and was spontaneously stripped off when the glass form was taken out of the leachant. The glass form leached above 325°C was swelled and surrounded by yellow reaction layer, about 1 mm in thickness. On the other hand, the waste forms produced by hydrothermal hot pressing had no reaction layer and maintained the shape before the leach tests at any temperatures.

Elemental mass loss of Na, Si, Mo, and Cs was increased with the increase in leaching temperature up to 325°C (Fig. 4(B)), and reduced at 350°C. Sr was not detected. A large amount of Na was constantly released below 200°C. The release of Na caused the large weight loss.

Addition of $Al(OH)_3$

In order to immobilize Na, $Al(OH)_3$ was added to the starting material. The addition of $Al(OH)_3$ up to 10 wt% reduced the weight loss, as well as the compressive strength (Fig. 5); the addition of over 10 wt% had no effect on either. The density had the same tendency.

Sodium aluminosilicates such as analcite ($NaAlSi_2O_6 \cdot H_2O$) were easily produced from α-quartz and $Al(OH)_3$ within 10 min in 5*N* NaOH solution under hydrothermal conditions at 300°C, though the amorphous aluminosilicate of the matrix component gave no aluminosilicates by the same hydrothermal treatment.[16] Thus, Na might be immobilized in aluminosilicates by the reaction of $Al(OH)_3$ and α-quartz during hydrothermal hot pressing.

The weight loss of the waste form including 10 wt% $Al(OH)_3$ is plotted against the leaching temperatures in Fig. 6. The weight loss below 150° and above 300°C was reduced in comparison with that

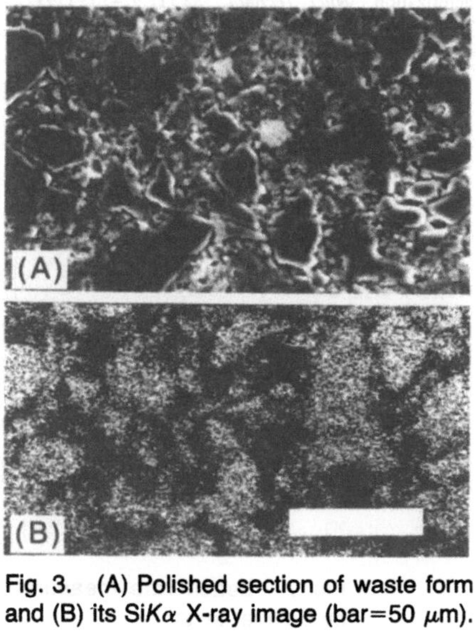

Fig. 3. (A) Polished section of waste form and (B) its Si*Kα* X-ray image (bar=50 μm).

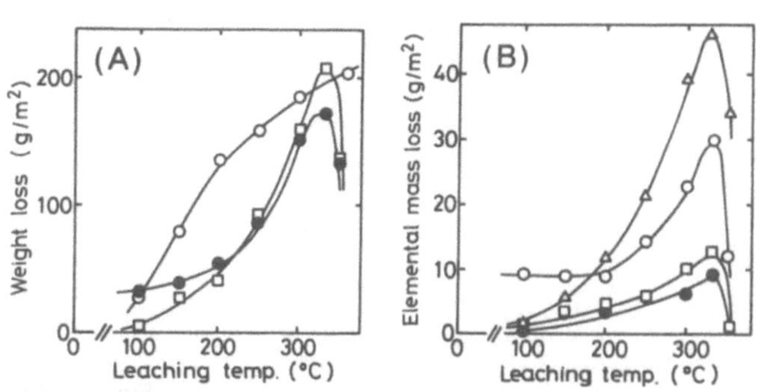

Fig. 4. (A) Effect of leaching temperature on weight loss of waste forms produced hydrothermally under conditions (○) (*A*) and (●) (*B*) and of (□) JW-A glass form. (B) Elemental mass loss of waste form produced under condition (*B*); (○) Na, (△) Si, (□) Mo, (●) Cs.

Kazumichi Yanagisawa

Mamoru Nishioka

Nakamichi Yamasaki

Kazumichi Yanagisawa is a research associate at the Research Laboratory of Hydrothermal Chemistry, Faculty of Science, Kochi University, Kochi, Japan. He graduated from the Faculty of Engineering at Shinshu University in 1976 and earned an M.E. in materials science from Tokyo Institute of Technology in 1979. Mr. Yanagisawa served as a visiting researcher at Oxford University during 1977–78.

Mamoru Nishioka is a technical officer at the Research of Hydrothermal Chemistry. He is a graduate of Anan Technical College.

Nakamichi Yamasaki is an associate professor and director of the Research Laboratory of Hydrothermal Chemistry. He earned a B.S. in chemistry at Kochi University. His current research involves hydrothermal alteration and synthesis of minerals, hydrothermal treatment of radioactive wastes, and hydrothermal decomposition of stable organic compounds.

of the waste form including no $Al(OH)_3$. Elemental mass loss of Na, Si, Mo, and Cs was also reduced by the addition of $Al(OH)_3$.

Leach Rate by Soxhlet Leach Tests

Soxhlet leach tests were done for the waste form (waste 21.8 wt%, $Al(OH)_3$ 10 wt%, matrix 68.2 wt%) produced under the condition (*B*) (10*N* NaOH, 350°C, 66 MPa, 6 h). The leach rate was remarkably reduced with the increase in leaching time (Fig. 7), because the weight loss for a short time up to 6 h was very large. The high leach rate by short-time leach tests was explained as follows: The waste forms produced by hydrothermal hot pressing have many pores, which have been filled with the mineralizer during hydrothermal treatment. The mineralizer may include Na, Si, Al, and soluble elements of the waste components. These elements are not immobilized enough only by drying at 110°C and are easily released by an initial washout through porous structure of the waste forms. The same phenomenon was observed when the FUETAP concretes were leached.[11]

After the tests for 14 d, the leach rate of the waste form reached 7 g/m^2d. It was much lower than that of the Fuetap concrete (48 g/m^2d for Cs by MCC-1 leach tests at 90°C),[11] but was higher than that of a glass form (2 g/m^2d for total weight loss by Soxhlet leach tests).[18]

Conclusion

The simulated high-level radioactive waste was immobilized in the silica matrix by hydrothermal hot-pressing. The high compressive strength of the waste forms was achieved by the use of 10*N* NaOH solution as mineralizer. The high reaction temperature and pressure increased the strength and decreased the leach weight loss. The consolidation of the waste forms was achieved in 30 min; however, longer reaction times reduced the weight loss. The addition of 10 wt% $Al(OH)_3$ to the starting material improved the leachability of the waste forms but also reduced the compressive strength. The optimum conditions to produce waste forms with the high compressive strength and the resistance to leaching were as follows within the conditions used in this study: addition of 10 wt% $Al(OH)_3$, 10*N* NaOH solution, 350°C, 66 MPa, and 6 h.

The leach rate of the waste form (21.8 wt% waste loading) produced under the optimum conditions was found to be 7 g/m^2d by Soxhlet leach tests for 14 d. The leach rate by short-time leach test up to 6 h was very large, but it was reduced remarkably with the increase in the leach time.

Acknowledgment

The authors wish to express their thanks to Drs. S. Muraoka and T. Murakami (Japan Atomic Energy Research Institute) for their offering JW-A glass form.

References

[1] T. B. Bernadzikowski et al., "High-level Nuclear Waste Form Performance Evaluation," *Am. Ceram. Soc. Bull.*, **62** [12] 1364–1368, 1390 (1983).
[2] A. E. Ringwood, "Safe Disposal of High Level Nuclear Reactor Wastes: A New Strategy," Australian Nat. Univ. Press, Canberra, Australia, 1978.
[3] G. J. McCarthy and M. T. Davidson, "Ceramic Nuclear Waste Form: I, Crystal Chemistry and Phase Formation," *Am. Ceram. Soc. Bull.*, **54** [9] 782–786 (1975).
[4] G. J. McCarthy and M. T. Davidson, "Ceramic Nuclear Waste Form: II, A Ceramic Waste Composite Prepared by Hot Pressing," *ibid.*, **55** [2] 190–194 (1976).
[5] G. J. McCarthy, "High Level Waste Ceramics: Materials Considerations, Process Simulation, and Product Characterization," *Nucl. Technol.*, **32** [1] 92–105 (1977).
[6] A. E. Ringwood et al., "Immobilization of High Level Nuclear Reactor Wastes in Synroc," *Nature*, **278**, 219–233 (1979).

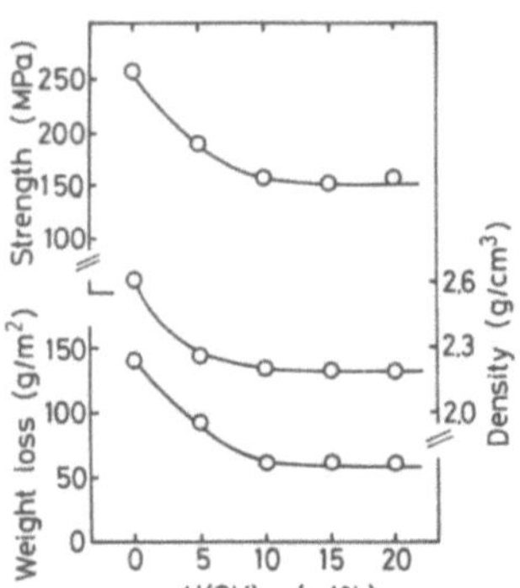

Fig. 5. Effect of $Al(OH)_3$ addition on properties of waste forms produced under condition (*A*).

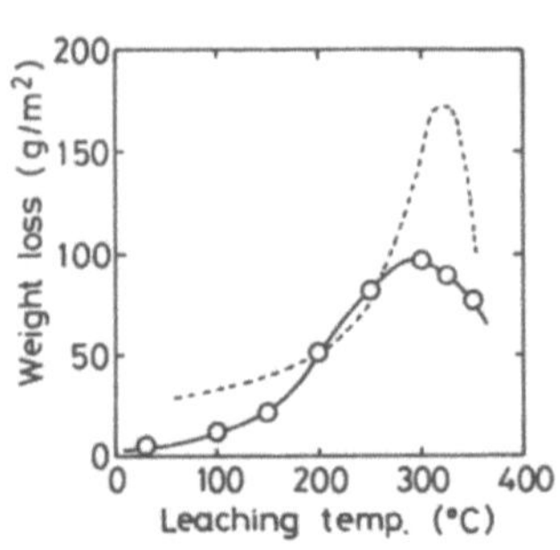

Fig. 6. Weight loss of waste forms with 10 wt% $Al(OH)_3$ (solid line) and no $Al(OH)_3$ (dashed line) produced under condition (*B*); leach time was 1 d.

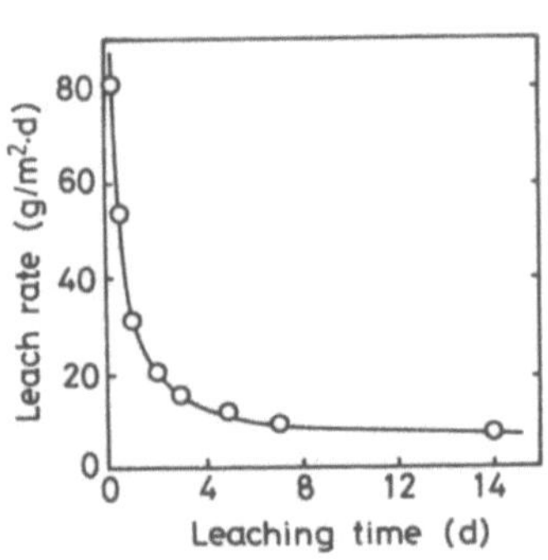

Fig. 7. Leach rate measured by Soxhlet tests.

[7]J. L. Woolfrey and D. M. Levins, "Effects of Water Chemistry on the Leach Resistance of Synroc C," pp. 663–70 in Scientific Basis for Nuclear Waste Management VII; Edited by G. L. McVay. North-Holland, New York, 1984.
[8]A. B. Harker et al., "Tailored Ceramic Nuclear Waste Forms: Preparation and Characterization," pp. 139–46 in Scientific Basis for Nuclear Waste Management, Vol. 3; Edited by J. G. Moore, Plenum, New York, 1981.
[9]P.E.D. Morgan et al., "High-Alumina Tailored Nuclear Waste Ceramics," *J. Am. Ceram. Soc.*, **64** [5] 249–258 (1981).
[10]C. M. Jantzen et al., "Leaching of Polyphase Nuclear Waste Ceramics," *J. Am. Ceram. Soc.*, **65** [6] 292–300 (1982).
[11]L. R. Dole et al., "Cementitious Radioactive Waste Hosts Formed under Elevated Temperatures and Pressures (Fuetap Concretes)," pp. 585–93 in Scientific Basis for Nuclear Waste Management; Edited by S. V. Topp, North-Holland, New York, 1982.
[12]N. Yamasaki et al., "Immobilization of Radioactive Wastes in Hydrothermal Synthetic Rock: Lithification of Silica Powder," *J. Nucl. Sci. Technol.*, **21** [1] 71–73 (1984).
[13]N. Yamasaki et al., "Aggregate Formation of Silica under Hydrothermal Conditions," *Yogyo-Kyokai-shi*, **92** [3] 150–52 (1984).
[14]K. Matsuoka et al., "Hot-Pressing of Calcium Carbonate under Hydrothermal Conditions," *Rep. Res. Lab. Hydrothermal Chem.*, **3** [3] 8–11 (1979).
[15]N. Yamasaki et al., "Shrinkage Process and Compressive Strength of High-Level Radioactive Waste Forms by Hydrothermal Hot-Pressing Method," *Yogyo-Kyokai-Shi*, **93** [3] 151–53 (1985).
[16]K. Yanagisawa et al., "Immobilization of Radioactive Wastes in Hydrothermal Synthetic Rock: Hydrothermal Synthesis of Pollucite," *J. Nucl. Sci. Technol.*, **21** [7] 558–60 (1984).
[17]T. Banba et al., "Study of Actual and Simulated HLW Compositions," JAERI-memo 8982 (1980).
[18]K. Ishiguro et al., "Growth of Surface Layer During the Leaching of the Simulated Waste Glass and its Barrier Effects on the Leaching," pp. 135–42 in Scientific Basis for Nuclear Waste Management VI; Edited by D. G. Brookings, North-Holland, New York, 1983. □

Part 9

Hydrothermal Hot Isostatic Pressing

Reprinted from *Journal of Materials Science Letters*, **3** (1984) 57–59.

Fabrication of phosphate-bonded mica ceramics by hot isostatic processing

T. HATTORI, A. YAHATA, J. MOHRI
Department of Industrial Chemistry, Faculty of Engineering, Chiba University, 1-33 Yayoi, Chiba 260, Japan

M. YOSHIMURA, S. SŌMIYA
Research Laboratory of Engineering Materials, Tokyo Institute of Technology, 4259 Nagatsuta, Midori, Yokohama 227, Japan

In the previous paper [1], we reported on hot isostatic pressing of nonadditive mica ceramics using a Roy–Tuttle-type pressure vessel. Powder compacts of nonadditive fluorphlogopite encapsulated in a borosilicate glass tube were isostatically hot-pressed at a selected temperature between 600 and 800° C under 49 to 98 MPa of water. Consequently synthetic mica without any additives was able to be sintered to a density of 2.60 g cm^{-3} (91.2% of theoretical density) with no grain growth under 98 MPa at 800° C for 45 min.

Many types of mica ceramics such as glass-bonded [2], phosphate-bonded [3], fused-cast [4], and hot-pressed [4] types have been obtained in addition to nonadditive mica ceramics. Hirao [3] reported on phosphate-bonded mica ceramics in which H_3PO_4 was added in an amount of 2.5 to 10 wt% to various mixing ratio of fluorphlogopite to fluortetrasilicic mica followed by pressing and firing at a selected temperature between 1000 and 1200° C in air. Minimum total porosity obtained was 5.7% using 1 : 1 mixture at the temperature of 1200° C for 30 min, of which temperature was nearly melting point of fluortetrasilicic mica.

In the present paper we attempted to fabricate phosphate-bonded mica ceramics at as low a temperature as possible by hot isostatic processing.

Starting materials used are commercial fluorphlogopite*, nominally 80 to 325 mesh, and guaranteed reagent grade 85% ortho phosphoric acid†. Mica was dried at 120° C for 24 h and then mixed with 5 and 10 wt% H_3PO_4 for 90 min, respectively, using an agate mortar in a glove box with dry nitrogen gas. The mixtures were hand-moulded in a bakelite mould 5 mm in diameter and 45 mm long, and isostatically pressed at 200 MPa. These pellets had a green density between 2.2 and 2.3 g cm^{-3}. And these were found, using thermogravimetric analysis, to be lost in weight by about 3.9% and 8.7% in the samples of 5 and 10 wt% H_3PO_4, respectively, up to about 380° C and to be almost unchanged above 400° C in both samples. So, after degassing at 400° C for 1 h under 1.33×10^{-3} Pa (1×10^{-5} torr) using oil diffusion pump, one pellet embedded in BN powder was sealed in a borosilicate glass tube 6 mm in outer diameter (OD) and 20 mm long.

The encapsulated specimen was isostatically hot-pressed for 0.5 to 3 h at a selected temperature between 500 and 700° C under the pressure of 50 to 150 MPa with water at high temperature and

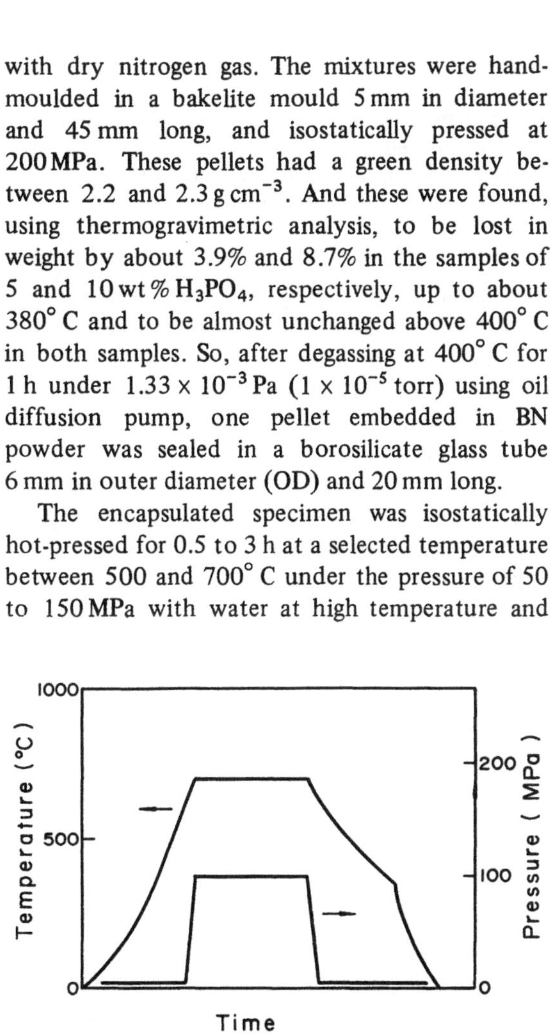

Figure 1 Schematical time schedule of hot isostatic processing used in the experiment.

*PDM-7 Mica, Topy Industries, Ltd., Tokyo, Japan.
†Wako Pure Chemical Industries, Ltd., Tokyo, Japan.

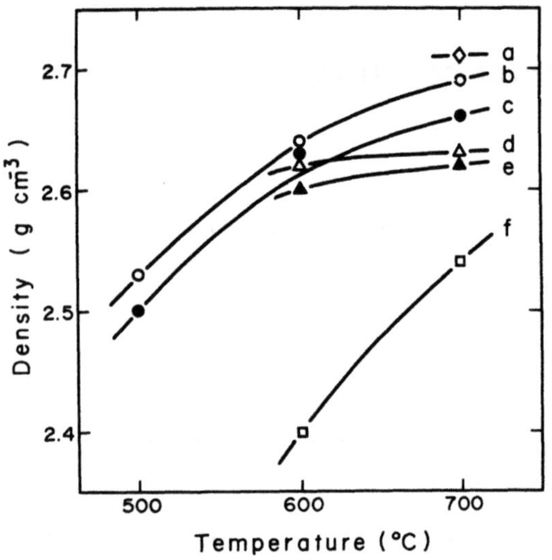

Figure 2 HIPing results of mica with 10 wt% (a–c), 5 wt% (d, e), and 0 wt% (f) H_3PO_4. (a) 150 MPa, 3 h, (b) 100 MPa, 1.5 h, (c) 50 MPa, 1.5 h, (d) 100 MPa, 1.5 h, (e) 50 MPa, 1.5 h, (f) 147 MPa, 1.5 h.

pressure in a test-tube-type pressure vessel [5]. Time schedule of the process is schematically shown in Fig. 1, in which temperature was firstly raised in order to soften the glass container prior to pressing. After HIPing the pellet was removed from its capsule and the density was measured by the Archimedes technique using ethyl alcohol.

The HIPing results are shown with a part of those of nonadditive mica for comparison in Fig. 2 which indicates that the pellets of mica with H_3PO_4 were easily sintered to higher density than the other. Especially, when mica with 10 wt% H_3PO_4 was HIPed under 150 MPa at 700° C for 3 h, it was densified to the density of 2.71 g cm^{-3}, corresponding to above 95% of theoretical density of fluorphlogopite. Fig. 3 shows a microstructure of a fracture surface of this specimen. Platelike mica crystals are observed to be highly interlocked.

X-ray diffraction analysis of mica before and after HIPing was conducted using Ni-filtered Cu*Kα* radiation. As shown in Fig. 4a, magnesium fluoride was found to be included as an impurity in the starting material. Unmarked diffraction patterns in the same figure are almost belonged to fluorphlogopite. After HIPing at 600° C (Fig. 4b) and 700° C (Fig. 4c) under 100 MPa for 1.5 h, new diffraction lines are found between 20 and 35°. These lines would perhaps be a part of those of either magnesium phosphate or wagnerite. But many diffraction lines of both compounds overlap with those of fluorphlogopite, respectively. Therefore, it is not clearly decidable from these profiles which compound is produced. In either case, such a compound would be useful as a binder in the sintering of fluorphlogopite.

Acknowledgement
Authors would like to thank Topy Industries Ltd. for supplying the sample.

References
1. T. HATTORI, A. SHIGEMORI, J. MOHRI, M. YOSHIMURA and S. SŌMIYA, *J. Amer. Ceram. Soc.* 65 (1982) C-142.
2. D. G. GROSSMAN, *ibid.* 55 (1972) 446.
3. M. HIRAO, *Amer. Ceram. Soc. Bull.* 55 (1976) 788.

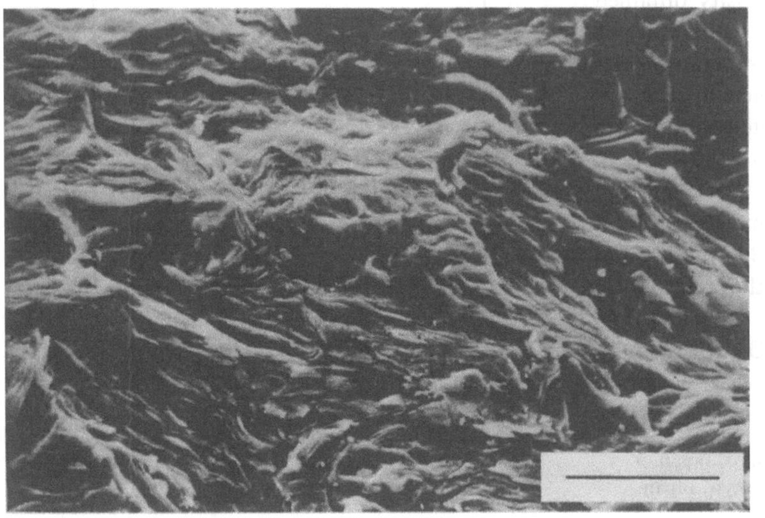

Figure 3 Scanning electron micrograph of a fracture surface of the specimen isostatically hot-pressed under 150 MPa at 700° C for 3 h (bar = 50 μm).

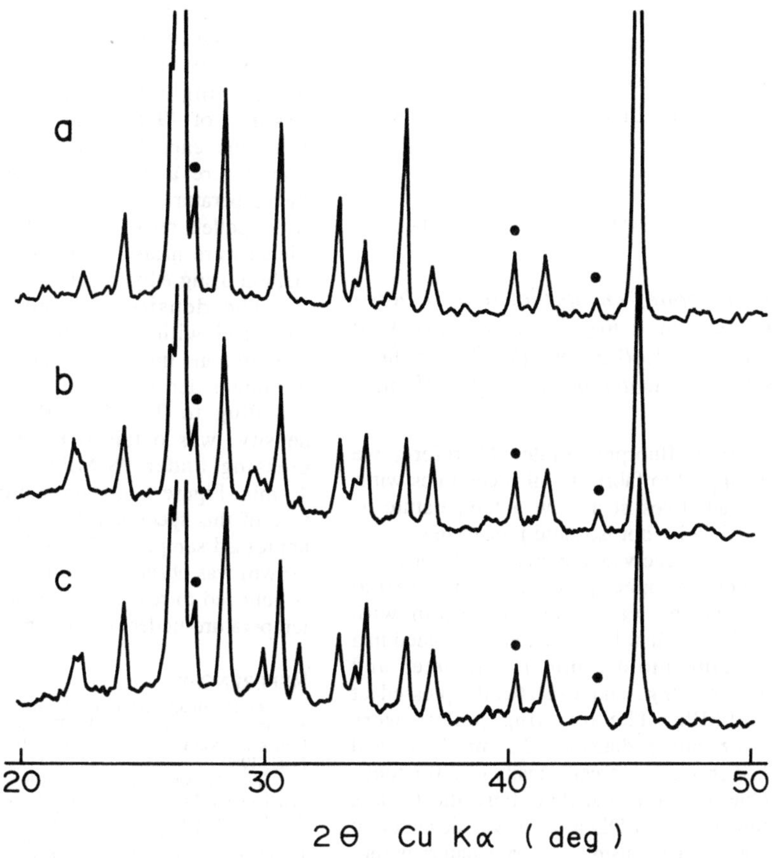

Figure 4 X-ray diffraction patterns of starting material (a) and mica with 10 wt % H_3PO_4 after HIPing under 100 MPa for 1.5 h at 600° C (b) and 700° C (c), respectively.

4. R. A. HATCH, R. A. HUMPHREY, W. EITEL and J. E. COMEFORO, US Bur. Mines Rept. Invest., No. 5337 (1957) 67.
5. R. ROY and O. F. TUTTLE, "Physics and Chemistry of the Earth", Vol. 1 (Pergamon, Oxford, 1956) p. 138.

Received 6 June and accepted 30 June 1983

Reprinted from *Journal of the American Ceramic Society*, **65**(9) (1982) C-142.

Fabrication of Nonadditive Mica Ceramics by Hot Isostatic Processing

TAKEO HATTORI, ATSUSHI SHIGEMORI, AND JUN-ICHI MOHRI

Department of Industrial Chemistry, Faculty of Engineering, Chiba University, 1–33 Yayoi, Chiba 260, Japan

MASAHIRO YOSHIMURA* AND SHIGEYUKI SŌMIYA*

Research Laboratory of Engineering Materials, Tokyo Institute of Technology, 4259 Nagatsuta, Midori, Yokohama 227, Japan

Powder compacts of synthetic mica (fluorphlogopite) encapsulated in a borosilicate glass tube were isostatically hot-pressed in a Roy-Tuttle-type pressure vessel. Synthetic mica was sintered to a density of 2.60 g/cm³ (91.2% of theoretical density) without additives under 98 MPa of water at 800°C for 45 min.

MUCH attention has been focused recently on mica-containing glass-ceramics.[1] These ceramics can be machined easily to precision tolerances with ordinary machine tools. Their machinability results from their unique microstructure, which consists of a highly interlocked array of platelike mica crystals which are randomly dispersed in a glassy matrix.

So far, glass-bonded,[2,3] phosphate-bonded,[4–6] fused-cast,[6] and hot-pressed[6,7] mica ceramics have been obtained. Hatch *et al.*[6] reported on hot-pressed synthetic mica-ceramics in which fluorphlogopite was densified to 2.80 g/cm^3 (98.2% of theoretical density) by treatment at 1200°C and 13.7 MPa (2000 psi) for 60 min. This hot-pressing temperature was considered to be too high compared with the melting point of fluorphlogopite. Therefore, we attempted to fabricate mica ceramics without additives at as low a temperature as possible by hot isostatic processing.

As received commercial fluormica* (fluorphlogopite) powder, nominally 80 to 325 mesh, was mixed for 5 min with ≈10 wt% distilled water, hand-molded in a bakelite mold 5 mm in diameter and 45 mm long, and isostatically pressed at 200 MPa. The resulting pellets were ≈4.8 mm in diameter, 2.2 mm high, and had an average green density of 2.0 g/cm^3. One pellet was sealed in a borosilicate glass tube 6 mm in OD after it was degassed at 400°C for 1 h using a rotary vacuum pump. A scanning electron micrograph of a fracture surface of the degassed pellet indicated (Fig. 1 (*A*)) that the microstructure consists of aggregates of relatively uniform size which contain many small flakes of mica.

The encapsulated specimen was isostatically hot-pressed at a selected temperature between 600° and 800°C with water at high temperature and pressure in a test-tube-type pressure vessel.[8] The heating and cooling rates were ≈15°C/min. Pressures of 49 to 98 MPa were applied when the capsule reached the selected temperature and were unloaded before the temperature was reduced. The densities of the pellets removed from the glass containers were measured by the Archimedes method using ethyl alcohol.

The densities of the specimens increased with increasing temperature, pressure, and time of treatment (Fig. 2). A maximum density of 2.60 g/cm^3, corresponding to 91.2% of the theoretical density, was obtained by isostatic hot-pressing under 98 MPa at 800°C for 45 min. Figure 1 (*B*) shows a fracture surface of this specimen. Compared with the unpressed sample of Fig. 1 (*A*), less grain growth was observed in this specimen; this is believed to result from treatment at low temperature under isostatic pressure.

CONTRIBUTING EDITOR — R. M. SPRIGGS

Received April 12, 1982; revised copy received May 11, 1982. Approved May 11, 1982.

*Member, the American Ceramic Society.

*PDM-6 Mica, Topy Industries, Ltd., Tokyo, Japan.

REFERENCES

[1] G. H. Beall and R. C. Doman; pp. 25–36 in Science of Ceramics, Vol. 10. Edited by H. Hausner. Deutsche Keramische Gesellschaft, Bad Honnef, FRG, 1980.

[2] P. S. Hessinger and T. W. Weber, "Development of a Synthetic Mica Ceramic Suitable for Use at 750°C," *Am. Ceram. Soc. Bull.*, **39** [1] 10–13 (1960).

[3] D. G. Grossman, "Machinable Glass-Ceramics Based on Tetrasilicic Mica," *J. Am. Ceram. Soc.*, **55** [9] 446–49 (1972).

[4] J. E. Comeforo, "Synthetic Mica Investigations: V, A Low-Shrinkage Machinable Ceramic of Phosphate-Bonded Synthetic Mica," *J. Am. Ceram. Soc.*, **37** [9] 427–32 (1954).

[5] M. Hirao, "Phosphate-Bonded Synthetic Mica Ceramics," *Am. Ceram. Soc. Bull.*, **55** [9] 788–91 (1976).

[6] R. A. Hatch, R. A. Humphrey, W. Eitel, and J. E. Comeforo, "Synthetic Mica Investigations: IX, Review of Progress from 1947 to 1955," *U.S. Bur. Mines Rept. Invest.*, **1957**, No. 5337; pp. 67–76.

[7] J. E. Comeforo, R. A. Hatch, R. A. Humphrey, and W. Eitel, "Synthetic Mica Investigations: I, A Hot-Pressed Machinable Ceramic Dielectric," *J. Am. Ceram. Soc.*, **36** [9] 286–94 (1953).

[8] R. Roy and O. F. Tuttle; pp. 138–80 in Physics and Chemistry of the Earth, Vol. 1. Edited by L. H. Ahrens, K. Rankama, and S. K. Runcorn. Pergamon, Oxford, 1956. □

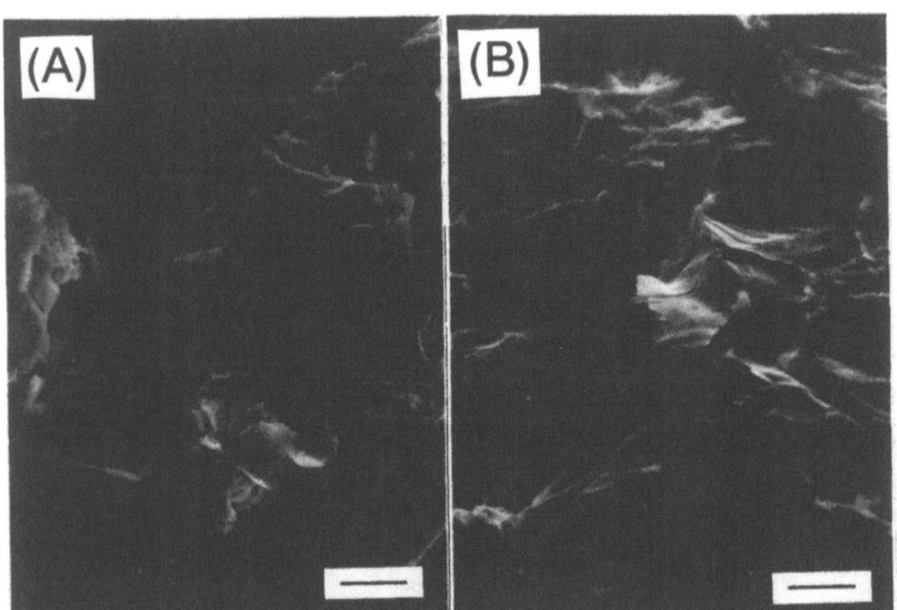

Fig. 1. Scanning electron micrographs of fracture surfaces of fluorphlogopite (*A*) before and (*B*) after isostatic hot-pressing at 800°C for 0.75 h under 98 MPa (bar=20 μm).

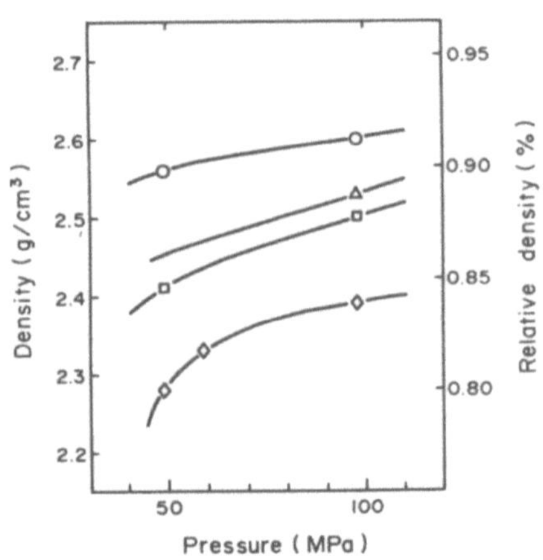

Fig. 2 Effects of isostatic hot-pressing conditions on density of fluorphlogopite: (○) 800°C, 0.75 h; (△) 700°C, 3 h; (□) 700°C, 1.5 h; and (◇) 600°C, 3 h.

Reprinted from *Powder Metallurgy International*, **16** (1984) 131–132.

Sintering of Synthetic Mica by a Hydrothermal Equipment

T. HATTORI*, J. MOHRI*, M. YOSHIMURA**, S. SŌMIYA**

Sintering of Synthetic Mica by a Hydrothermal Equipment
Green compacts of fluorphlogopite with 0, 5, and 10 wt% H_3PO_4 were respectively evacuated in a borosilicate glass tube at 400°C for 1 h and then sealed off. Each capsule was isostatically hot-pressed using a hydrothermal equipment with a test-tube-type pressure vessel. Synthetic mica with 10 wt% H_3PO_4 was sintered to a bulk density of 2,71 Mg/m^3 under 150 MPa of water at 700°C for 3 h.
powder metall. int. **16** (1984) [3] 131

Sintern von synthetischem Glimmer durch hydrothermale Behandlung
Grünlinge aus Fluorphlogopit mit Gehalten von 0,5 und 10 Gew.% H_3PO_4 werden getrennt in Rohren aus Borosilikatglas bei 400°C für 1 h unter Vakuum ausgelagert und dann in Ampullen eingeschmolzen. Jede Ampulle wurde in einem rohrähnlichen Druckgefäß mit Wasser als Wirkmedium heißisostatisch gepreßt. Synthetischer Glimmer mit 10 Gew.% H_3PO_4 wurde so bis zu einer Dichte von 2,71 Mg/m^3 unter einem Wasserdruck von 150 MPa bei 700°C für 3 h gesintert.

Frittage de mica synthétique dans une installation hydrothermale
Des comprimés crus de phlogopite fluorée avec respectivement 0,5 et 10% poids de H_3PO_4 ont été mis sous vide dans un tube de verre borosilicaté à 400°C pour une heure et scellés ensuite. Chaque capsule a été pressée isostatiquement à chaud en utilisant une installation hydrothermale avec un récipiant sous pression du type tube à essai. Du mica synthétique avec 10% poids de H_3PO_4 a été fritté sous 150 MPa d'eau à 700°C pendant 3 heures en atteignant une densité volumique de 2,71 Mg/m^3.

Introduction

Mica is known to be chemically and thermally stable and also to have a good electrical insulation and mechanical strength. Moreover mica is a famous machinable ceramic material which can be machined to precision tolerances using conventional tools. The machinability results from the fact that fracture propagation is very difficult across the vasal planes of mica. For this reason individual plate-like mica crystals are preferred to array randomly for machinable mica ceramics.

Mica ceramics have been produced mainly by hot-pressing technique except for the use of a fused cast process. This is usually a good technique to make a dense ceramic material with little or no grain growth. But even in this technique, excessive higher temperatures result in larger grain size. Moreover their microstructures tend to orient preferentially for anisotropic oxides [1], particularly for plate- or needle-like crystals.

In order to reduce this preferred orientation, pressurizing at a sintering temperature is more desirable to be isostatically rather than uniaxially. Hot isostatic pressing (HIP) may perhaps be the most adequate technique for this purpose.

On the other hand, many inorganic materials have been synthesized by a hydrothermal synthesis technique. This technique is conducted in a closed system usually using an autoclave or a test-tube-type pressure vessel [2].

We attempted a hot isostatic pressing of synthetic mica using this test-tube-type pressure vessel with water at high temperature and pressure. This pressure vessel is made of stellite 25 alloy which can be used for a considerably long time under the conditions of 800°C and 200 MPa.

* Department of Industrial Chemistry, Faculty of Engineering, Chiba University, Chiba, 260 Japan
** Research Laboratory of Engineering Materials, Tokyo Institute of Technology, Midori, Yokohama, 221 Japan

Experiments

Starting raw materials are commercial fluorphlogopite* and 85% phosphoric acid**. Fluorphlogopite dried at 120°C for 24 h was mixed with 0, 5, and 10 wt% H_3PO_4 in an agate mortar for 90 min in a glove box with dry N_2 gas and then hand-molded in a bakelite mold 5 mm in diameter and 45 mm long followed by isostatically pressing at 200 MPa. Green densities of these pellets were about 2,0 Mg/m^3 and 2,2 to 2,3 Mg/m^3 for non-additive and H_3PO_4-additive micas, respectively. As the green pellet contains connecting porosity, it is necessary to envelop it by a fluid impermeable container and also to evacuate before sealing it off prior to the HIPing. One pellet embedded in BN powder (Denka BN, Denki Kagaku Industries Ltd., Tokyo, Japan) was sealed in a borosilicate glass tube 6 mm in outer diameter and 20 mm long after degassing at 400°C for 1 h under $1{,}33 \cdot 10^{-3}$ Pa.

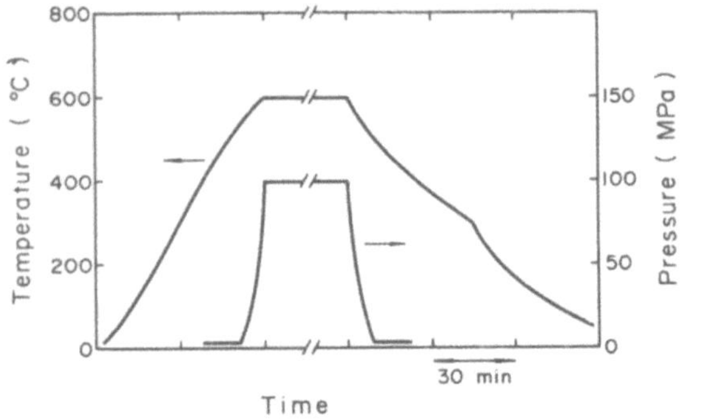

Fig. 1 A schematic drawing of the HIP schedule.

The encapsulated specimen was isostatically hot-pressed at temperatures ranging from 500°C to 800°C and at pressures varying from 49 to 150 MPa. It was firstly heated to the softening temperature of the glass and then pressure was applied. A schematic drawing of the schedule was shown in Figure 1.

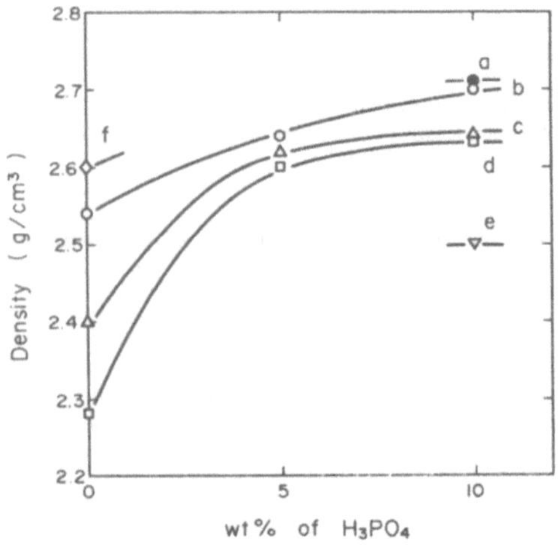

HIPing conditions:
(a) 700°C, 150 MPa, 3 h
(b) 700°C, 150 MPa, 1,5 h
(c) 600°C, 150 MPa, 1,5 h
(d) 600°C, 50 MPa, 3 h
(e) 500°C, 50 MPa, 1,5 h
(f) 800°C, 98 MPa, 0,75 h.

Fig. 2 HIPing results of synthetic mica
Pressures for non-additive samples were 147 and 49 MPa in (b, c) and (d), respectively.

* Topy Industries, Ltd., Tokyo, Japan
** guaranteed reagent grade, Wako Pure Chemical Industries, Ltd., Tokyo, Japan

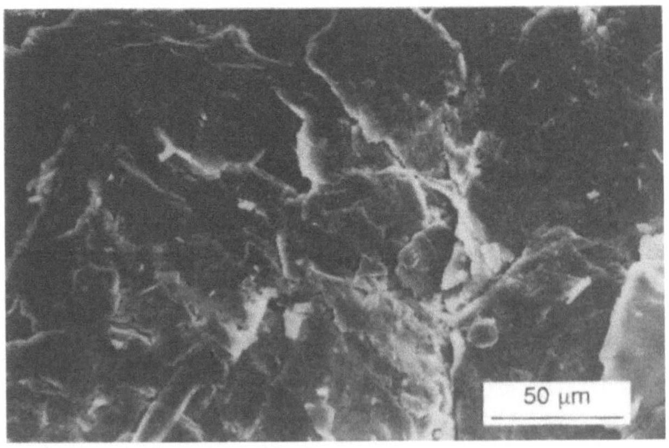

Fig. 3 A scanning electron micrograph of non-additive mica ceramics isostatically hot-pressed at 700°C for 1,5 h under 147 MPa by a hydrothermal equipment (bar ≙ 50 µm).

Density of the HIPed pellet was measured by the Archimedes replacement method using ethyl alcohol after removing from the glass container.

Results and Discussion

HIPing results are shown in Figure 2. The densities of HIPed specimens increased with increasing temperature, pressure, and time of HIP treatment and amounts of additive. Particularly the addition of H_3PO_4 affected the sintering of synthetic mica effectively.
Figure 3 shows a scanning electron micrograph of a fracture surface of non-additive mica isostatically hot-pressed at 700°C for 1,5 h under 147 MPa. Figure 4 shows that of mica with 10 wt% H_3PO_4 HIPed at 700°C for 0 h under 100 MPa. Densities of the specimens indicated in Figures 3 and 4 were 2,54 and 2,65 Mg/m^3, corresponding to 89% and 93% of theoretical density of pure fluorphlogopite, respectively. As shown particularly in Figure 4, isostatically hot-pressed mica ceramic material shows a complex microstructure; namely a highly interlocked array of plate-like crystals.

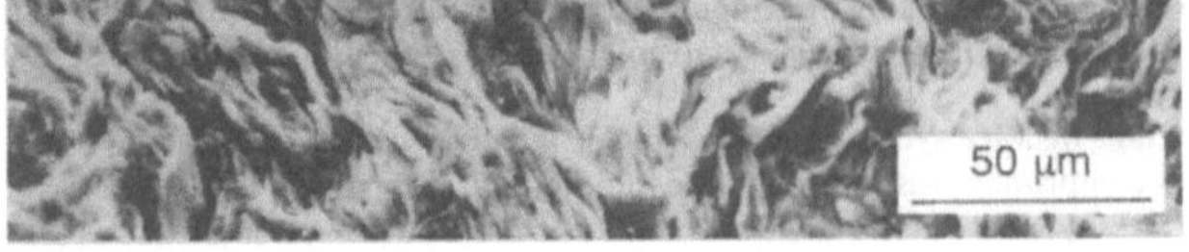

Fig. 4 A scanning electron micrograph of synthetic mica with 10 wt% H_3PO_4 isostatically hot-pressed at 700°C for 0 h under 100 MPa by a hydrothermal equipment (bar ≙ 50 µm).

References
1. Gardner, W. J., McClelland, J. D. and Richardson, J. H.: Modern Ceramics. Ed. by Hove, J. E. and Riley, W. C. John Wiley & Sons, Inc., New York (1965) pp. 215–234.
2. Roy, R. and Tuttle, O. F.: Physics and Chemistry of the Earth Vol. 1, Ed. by Ahrens, L. H., Rankama, K. and Runcorn, S. K. Pergamon, Oxford (1956) pp. 138–180.

Reprinted from G. S. Upadhyaya (ed.), *Sintered Metal-Ceramic Composites* (Elsevier Science Publishers BV, Amsterdam, 1984) pp. 153–155.

Sintering of Synthetic Mica by a Hydrothermal Hot Isostatic Process

TAKEO HATTORI, JUN-ICHI MOHRI[1], MASAHIRO YOSHIMURA and SHIGEYUKI SŌMIYA[2]
[1]Dept. of Industrial Chemistry, Faculty of Engineering, Chiba University, 1-33, Yayoi, Chiba 260, Japan
[2]Research Laboratory of Engineering Materials, Tokyo Institute of Technology, 4259, Nagatsuta, Midori, Yokohama 227, Japan

1. INTRODUCTION

Mica is well known as one of machinable ceramics.[1,2] Mica ceramics which are made of synthetic mica, generally fluorphlogopite, can be fabricated by a sintering of its powder compact or melting-nucleation-crystallization such as glass-ceramics.

A new technique of hot isostatic processing is currently used sintering for nonoxide materials such as silicon nitride.[3,4,5] This process offers several significant advantages; one of them is the application of the isostatic pressure to the body during high temperature, which provides higher density of the body with a controlled microstructure at relative lower temperature and pressure.

Therefore, two types of mica ceramics, nonadditive and phosphate-bonded types, were attempted to fabricate as low temperature as possible by hot isostatic processing.

2. EXPERIMENTAL PROCEDURE

An equipment used is a test-tube-type pressure vessel[6)], which is made of stellite 25 alloy. This alloy enables one to used safely at 800°C under 200 MPa for a long time. The vessel is 25 mm in diameter and 250 mm in length. The pressure transmitting fluid used was pressurized water by a simple handy-type pump.

Starting raw materials used were commercial fluorphlogopite* and 85% phosphoric acid**.

(1) Nonadditive-type Fluormica powder, nominally 80 to 325 mesh, was mixed for 5 min with about 10 wt% distilled water, hand-molded in a bakelite mold 5 mm in

---

*Topy Industries, Ltd., Tokyo, Japan
**Wako Pure Chemical Industries, Ltd., Tokyo, Japan

diameter and 45 mm in height, and isostatically pressed at 200 MPa for 5 min. The resulting pellets were about 4.8 mm in diameter, 2.2 mm in height and had an average green density of 2.0 g/cm^3.

(2) Phosphate-bonded type Fluormica dried at 120°C for 24 h was mixed with 5 and 10 wt% H_3PO_4 in an agate mortar for 90 min in a glove box with dry nitrogen gas and then pressed as mentioned above. Green densities of these pellets were 2.2-2.3 g/cm^3.

The one pellet embeded in BN powder* was sealed by welding in a borosilicate glass tube 6 mm in OD and 20 mm in length after degassing at 400°C for 1 h under $1.33x10^{-3}$ Pa ($1x10^{-5}$ torr).

The encapsulated specimen was hotpressed isostatically under the pressure of 49 to 150 MPa at temperatures between 500° and 700°C with fluid (water at high temperature and high pressure) in a test-tube-type pressure vessel for 0 to 3 h. Time schedule of the HIPing process is schematically drawn in Fig. 1.

After HIPing the specimen was removed from its capsule and its density was measured by the Archimedes replacement technique using ethyl alcohol.

3. RESULTS AND DISCUSSIONS

The HIPing results are shown in Figs. 2,3 and 4. Fig. 2 indicates the result of nonadditive-type fluormica. Fig. 3 and 4 show the results of 5 wt% and 10 wt% H_3PO_4-bonded-type. In any case the densities of the HIPed specimens increased with increasing temperature, pressure, and time of the HIP treatment and amounts of additive. Particularly the addition of H_3PO_4 affected to the sintering of synthetic mica effectively. For example, mica without additive was HIPed to the density 2.60 g/cm^3 which corresponds to 91.6% of theoretical density at 800°C under 98 MPa for 0.75 h and mica with 10 wt% H_3PO_4 was HIPed to the density of 2.70 g/cm^3 under the conditions of 700°C, 150 MPa, and 1.5 h which corresponds to above 95% of theoretical density of fluorphlogopite.

ACKNOWLEDGEMENTS

Parts of experimental materials were supplied by Topy Industries, Ltd. and Denki Kagaku Kogyo Ltd.

REFERENCES

1 R.A. Hatch, R.A. Humphrey, W. Eitel and J.E. Comeforo, U.S. Bur. Mines Rep. Invest., 1957 No. 5337: pp 67-76.

2 G.H. Beall and R.C. Doman, Science of Ceramics 10, ED. by H. Hausner, Deutsche Keramische Gessellschaft (1980) pp 25-36.

*Denki Kagaku Kogyo Ltd., Tokyo, Japan

3 H. Larker, in High Pressure Science and Technology, Vol. 2, Ed. by K.D. Timmerhaus and M.S. Barber (Plenum): (1979) pp 651-655.
4 Hun C. Yeh and Paul F. Sikora, Am. Ceram. Soc. Bull., 58 [4] (1979) pp 444-447
5 Shigeyuki Sōmiya, Masahiro Yoshimura, Tomonobu Suzuki and Hiroshi Nishimura, Taikabutsu, 31 [7] (1979) pp 332-339
6 Rustum Roy and O.F. Tuttle, in Physics and Chemistry of the Earth, Vol. 1. Ed. by L.H. Ahrens, K. Rankama and S.K. Runcorn (Pergamon) (1956) pp 138-180.

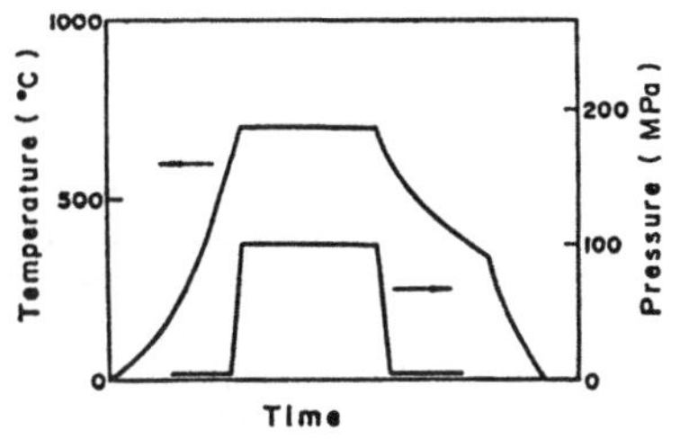

Fig. 1 Schematical time schedule of the HIPing process

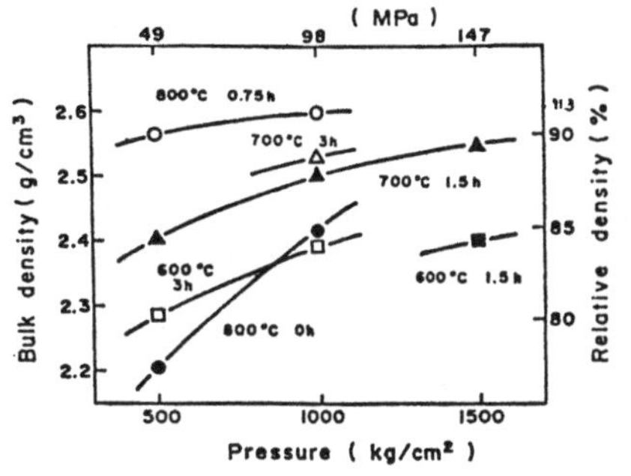

Fig. 2 Effects of HIPing conditions on density of nonadditive mica

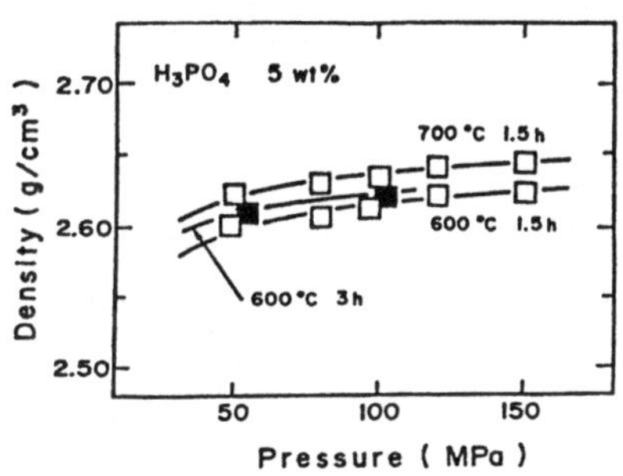

Fig. 3 Effects of pressures on density of mica with 5 wt% H_3PO_4

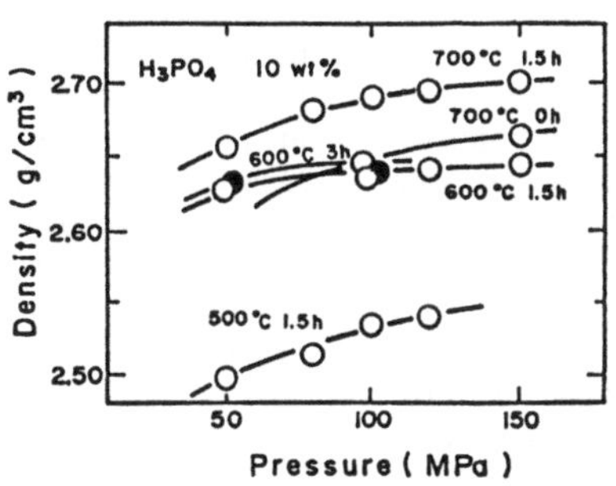

Fig. 4 Effects of pressures on density of mica with 10 wt% H_3PO_4

Reprinted from *Journal of the American Ceramic Society*, **69**(8) (1986) C-182–C-183.

High-Pressure Hot Isostatic Pressing of Synthetic Mica

TAKEO HATTORI

Department of Synthetic Chemistry, Faculty of Engineering, University of Chiba, 1-33 Yayoi, Chiba 260, Japan

MASAHIRO YOSHIMURA* AND SHIGEYUKI SŌMIYA*

Research Laboratory of Engineering Materials, Tokyo Institute of Technology, 4259 Nagatsuta, Midori, Yokohama 227, Japan

Powder compacts of synthetic mica (fluorphlogopite) encapsulated in a borosilicate glass tube were isostatically hot-pressed using high-pressure–high-temperature hot isostatic pressing equipment with Ar gas and a Mo heater. Synthetic mica was isostatically hot-pressed, while retaining the composition and structure of the starting material, to 96.8% of theoretical density without additives at 500 MPa and 700°C for 3 h.

HOT ISOSTATIC pressing[1] has recently been reviewed and described as a promising process in producing fully dense ceramics at temperatures significantly below those normally required for conventional sintering or hot-pressing. This process has been applied to the densification of not only some structural ceramics[2–9] but also other ceramics.[10–16]

Previously we reported the fabrication of nonadditive[11] and phosphate-bonded[14] mica ceramics by hot isostatic pressing with water at high temperatures and high pressures in a Roy-Tuttle-type pressure vessel.[17]

Nonadditive mica ceramics were isostatically hot-pressed at 98 MPa and 800°C for 3 h, only to a density of 2.60 g·cm^{-3}, corresponding to ≈91.2% of the theoretical density of fluorphlogopite, though mica with 10 wt% H_3PO_4 was isostatically hot-pressed to 2.71 g·cm^{-3} (above 95% of theoretical density) at 150 MPa and 700°C for 3 h. The pressure vessel, made of cobalt-base, wear-resistant alloy,* limits hot isostatic pressing conditions to 200 MPa at 800°C.[18]

So we tried to fabricate denser mica ceramics without additives at as low a temperature as possible under high pressure using high-pressure–high-temperature hot isostatic pressing equipment[4,10] with Ar gas and a Mo heater which enables hot isostatic pressing at 980 MPa and 1450°C.

The mica powder used was commercial fluorphlogopite.[†] The X-ray diffraction pattern was found to be identical with that of standard fluorphlogopite (ASTM, Powder Diffraction File No. 16-344) except for a few additional weak peaks presumably due to impurities present in small quantities. The powder was mixed for 5 min with a small amount of distilled water, hand-molded in a plastic[‡] mold 5 mm in diameter and 45 mm long, and isostatically pressed at 200 MPa. These pellets had an average green density of 2.0 g·cm^{-3}. One pellet embedded in BN powder was sealed in a borosilicate glass tube 6 mm in outer diameter and 20 mm long after degassing for 1 h at 400°C and 1.33×10^{-3} Pa (1×10^{-5} torr) using an oil diffusion pump.

Table I. Hot Isostatic Pressing Density of the Nonadditive Mica Ceramics

No.	Conditions	Density (%TD)*
1	800°C, 300 MPa, 3 h	96.1
2	700°C, 500 MPa, 3 h	95.8
3	700°C, 500 MPa, 3 h	96.8

*%TD=percent of theoretical density.

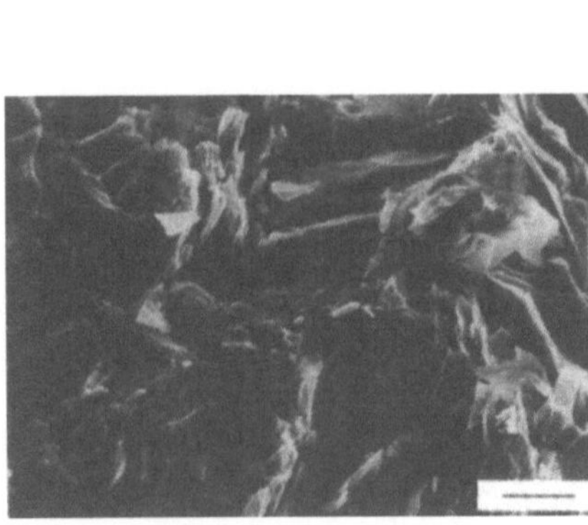

Fig. 1. Scanning electron micrograph of a fracture surface of the synthetic mica isostatically hot-pressed at 500 MPa and 700°C for 3 h (bar=10 μm).

The encapsulated specimen was isostatically hot-pressed for 3 h under the conditions of 700°C–500 MPa and 800°C–300 MPa in the hot isostatic pressing equipment using Ar gas as mentioned above. The time schedule of hot isostatic pressing was similar to that described previously;[11,14] namely, the temperature was first raised in order to soften the glass container before pressing. After hot isostatic pressing, the pellet was removed from its capsule and the density was measured by the Archimedes technique using ethyl alcohol.

The densities of the isostatically hot-pressed specimens are shown in Table I. All samples were successfully isostatically hot-pressed to a satisfactory high density while retaining the composition and structure of the starting material as revealed by X-ray analysis. As shown in the table, samples 2 and 3, whose temperature was 100°C lower than for sample 1, could be isostatically hot-pressed to a relative density nearly equal to or more than that of sample 1. It was found that the densification of synthetic mica was largely affected by high pressure applied at the sintering temperature.

Figure 1 shows a scanning electron micrograph of a fracture surface of the synthetic mica (sample 3) isostatically hot-pressed at 500 MPa and 700°C for 3 h, in which platelike mica crystals are observed to be randomly orientated and highly interlocked.

CONTRIBUTING EDITOR — V. J. TENNERY

Received January 27, 1986; revised copy received April 28, 1986; approved May 12, 1986.

*Member, the American Ceramic Society.
*Stellite 25, Union Carbide, New York, NY.
†PDM-7 mica, Topy Industries, Ltd., Tokyo, Japan.
‡Bakelite, Union Carbide.

REFERENCES

[1] H. D. Hanes, "Hot Isostatic Processing"; pp. 633–50 in High-Pressure Science and Technology, Vol. 2. Edited by K. D. Timmerhaus and M. S. Barber. Plenum Press, New York, 1979.

[2] H. Larker, "Hot Isostatic Pressing of Shaped Silicon Nitride Parts"; pp. 651–55 in High-Pressure Science and Technology, Vol. 2. Edited by K. D. Timmerhaus and M. S. Barber. Plenum Press, New York, 1979.

[3] T. Yamada, M. Shimada, and M. Koizumi, "Fabrication of Si_3N_4 by Hot Isostatic Pressing," *Am. Ceram. Soc. Bull.*, **60** [11] 1225–28 (1981).

[4] T. Hattori, J. Mohri, M. Yoshimura, and S. Sōmiya, "Hot Isostatic Processing of Ultrafine Silicon Nitride," *Yogyo Kyokaishi (J. Ceram. Soc. Jpn.)*, **91** [9] 423–25 (1983).

[5] D. W. Richerson and J. M. Wimmer, "Properties of Isostatically Hot-Pressed Silicon Nitride," *J. Am. Ceram. Soc.*, **66** [9] C-173–C-176 (1983).

[6] S. Sōmiya, M. Yoshimura, S. Fujiwara, K. Kondo, A. Sawaoka, T. Hattori, J. Mohri, and M. Araki, "Hot Isostatic Processing of Shocked Si_3N_4 Powder," *J. Am. Ceram. Soc.*, **67** [3] C-51–C-52 (1984).

[7] G. K. Watson, T. J. Moore, and M. L. Millard "Effect of Hot Isostatic Pressing on Reaction-Bonded Silicon Nitride," *J. Am. Ceram. Soc.*, **67** [10] C-208–C-210 (1984).

[8] D. M. Mieskowski and W. A. Sanders, "Microstructure of Reaction-Bonded Silicon Nitride Consolidated by Isostatic Hot-Pressing," *J. Am. Ceram. Soc.*, **68** [8] C-217–C-218 (1985).

[9] G. K. Watson, T. J. Moore, and M. L. Millard, "Effect of Hot Isostatic Pressing on the Properties of Sintered Alpha Silicon Carbide," *Am. Ceram. Soc. Bull.*, **64** [9] 1253–56 (1985).

[10] T. Hattori, F. Goseki, J. Mohri, M. Yoshimura, and S. Sōmiya, "Hot Isostatic Processing of Spinel Ceramics Prepared from the Powder by Freeze-Dried Method," *Yogyo Kyokaishi (J. Ceram. Soc. Jpn.)*, **90** [2] 82–86 (1982).

[11] T. Hattori, A. Shigemori, J. Mohri, M. Yoshimura, and S. Sōmiya, "Fabrication of Nonadditive Mica Ceramics by Hot Isostatic Processing," *J. Am. Ceram. Soc.*, **65** [9] C-142 (1982).

[12] M. Furukawa, "Alumina Ceramic Tools," *Am.*

Ceram. Soc. Bull., **62** [12] 1384–87 (1983).

[13]C. L. Hoenig and H. T. Larker, "Large Scale Densification of a Nuclear Waste Ceramic by Hot Isostatic Pressing," *Am. Ceram. Soc. Bull.*, **62** [12] 1389–90 (1983).

[14]T. Hattori, A. Yahata, J. Mohri, M. Yoshimura, and S. Sōmiya, "Fabrication of Phosphate-Bonded Mica Ceramics by Hot Isostatic Processing," *J. Mater. Sci. Lett.*, **3** [1] 57–59 (1984).

[15]A. B. Harker, P. E. D. Morgan, and J. F. Flintoff, "Hot Isostatic Pressing of Nuclear Waste Glasses," *J. Am. Ceram. Soc.*, **67** [2] C-26–C-28 (1984).

[16]O. Kwon and G. L. Messing, "Microstructural Changes in Hot Isostatically Pressed Alumina-Glass Composites," *J. Am. Ceram. Soc.*, **67** [3] C-43–C-45 (1984).

[17]R. Roy and O. F. Tuttle, "Investigations under Hydrothermal Conditions"; pp. 138–80 in Physics and Chemistry of the Earth, Vol. 1. Edited by L. H. Ahrens, K. Rankama, and S. K. Runcorn. Pergamon, Oxford, 1956.

[18]S. Sōmiya, "Electronic and Magnetic Materials by Hydrothermal Synthesis (1)," *J. Mater. Sci. Soc. Jpn.*, **13** [1, 2] 55–62 (1976). □

Part 10

Nitridation

Reprinted from *Journal of Materials Science Letters*, **4** (1985) 91–93.

Reactions of titanium metal powders with high-temperature high-pressure ammonia

S. SŌMIYA, M. YOSHIMURA, H. TORAYA
Research Laboratory of Engineering Materials, Tokyo Institute of Technology, Nagatsuta, Midori, Yokohama 227, Japan

Nitride powders as ceramic materials have been synthesized mainly by two methods: one employing the reduction of oxide powders and the succeeding nitridation with N_2 and/or NH_3 gases at high temperatures, and the other by the direct nitridation of metals or their hydrides. The latter method has a great affinity with the hydrothermal oxidation method, a technique for preparing fine oxide powders (20 to 35 nm) from metals, such as of zirconium [1], hafnium [2], titanium [3] or aluminium [4], by the reaction with high-temperature high-pressure water. The hydrothermal oxidation mechanism of hafnium metals has been studied in detail [5], and elucidated to proceed through two reaction paths: one is the direct oxidation of hafnium with high-temperature high-pressure water to form HfO_2 and H_2, and the other the formation of hafnium hydrides by the reaction of hafnium metals with released H_2 and the succeeding oxidation of hydrides. The parallelism is that, if metals react with high-temperature high-pressure ammonia to form nitrides and to release H_2, the H_2 will react with metals to form hydrides and these hydrides will be nitridated further with NH_3 as oxidized with H_2O in the hydrothermal oxidation. One of the advantages of hydrothermal oxidation is relatively lower temperatures (500 to 700° C) for the formation of fine oxide powders. The low reaction temperatures will be due to a high reactivity of pressurized water. Therefore, the formation of fine nitride powders can be expected from the reaction of some kinds of metals with pressurized ammonia at relatively lower temperatures compared to the conventional techniques.

The present study has been based on the parallelism mentioned above. An apparatus was first required to be developed for the study of the reactions of metals with ammonia in the pressure range of 10 to 200 MPa and the temperature range up to 800° C. Then, it was designed from the view point of security and simplicity in using virulent and combustible ammonia. This communication describes the apparatus developed and the study on the reactions of titanium metal powders with high-temperature high-pressure ammonia, which was conducted partly as a test of the apparatus.

Fig. 1 shows a schematic of the apparatus. Two sets of furnaces, pressure vessels and various sorts of valves are arranged symmetrically, where only one counterpart is presented in the diagram. The system should be evacuated by a rotary pump (C) before the injection of ammonia. The liquid ammonia tank (A) is set upside down [6] to drive the liquid through the pressure line by an ammonia vapour pressure of about 1 MPa. The ammonia can be compressed into a test-tube-type pressure vessel (D) made of stellite by a hand-operated diaphragm-type compressor (B) with a maximum working pressure of 70 MPa. All parts were connected with stainless steel tubings (1.5 mm inner diameter and 2.3 mm thickness). The inner pressure of the vessel can be read from a Bourdon-spring pressure gauge (F). The ammonia can be further compressed by heating the vessel. During heating, the valves (G and H) are closed, isolating the system (shaded part) from the supply. When the pressure rises above a prescribed pressure level, it can be decreased by discharging excess ammonia from an outlet (I) (through valve H) into water, which is connected to an aspirator.

The pressure vessel can be heated in an electric furnace with a platinum resistance heater (E). The quenching of the vessel can be made by lowering the furnace. Temperatures are measured by two platinel thermocouples attached to the wall of the pressure vessel, which can be kept within ± 5° C of a desired temperature. A part of the apparatus (enclosed by the broken line of the diagram) is set in a steel box to protect an operator from an

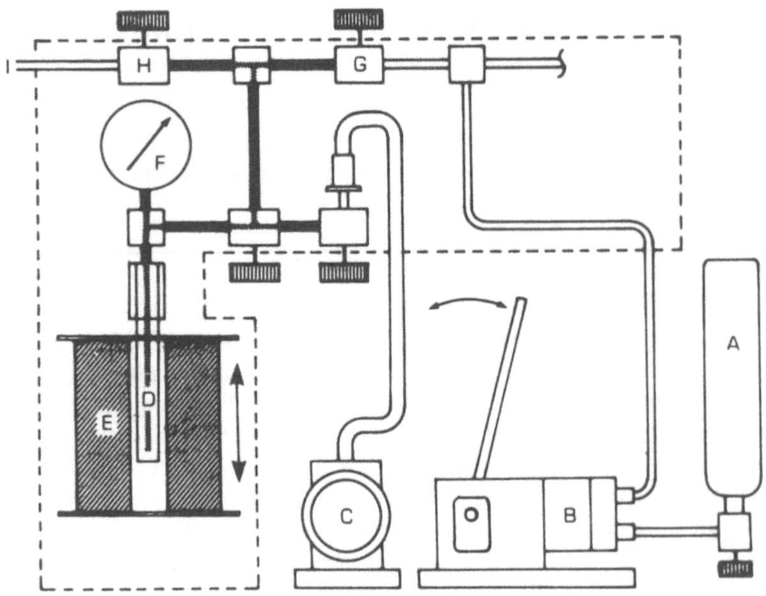

Figure 1 A scheme of the apparatus. (A) liquid ammonia tank, (B) hand-operated diaphragm-type-compressor, (C) rotary pump, (D) test-tube-type pressure gauge, (G and H) valves, (I) outlet connected to an aspirator.

accidental burst or a jet of ammonia gas. Three ventilators were placed behind the apparatus.

The titanium metal was selected as an object of the present study, because it easily reacts with H_2 to form titanium hydrides and is a good indicator of the decomposition of ammonia by $2NH_3 \rightarrow N_2 + 3H_2$. Furthermore, TiN powders are important industrial materials, being used where thermal and mechanical toughness is required. Titanium metal powder was packed into a 5 mm ID gold capsule (0.2 mm thick and 35 mm long). The capsule with an opening at its top was set at the bottom of the pressure vessel. The vessel was then filled with ammonia by a prescribed procedure. The capsules were heated at 400, 450, and 500° C under 30, 35 and 100 MPa for 5 to 30 days. Heated specimens were then examined by the X-ray powder diffraction technique.

The samples thus obtained exhibited various colours, from brown to green, blue and purple. The TiN powder is well known to exhibit a gold

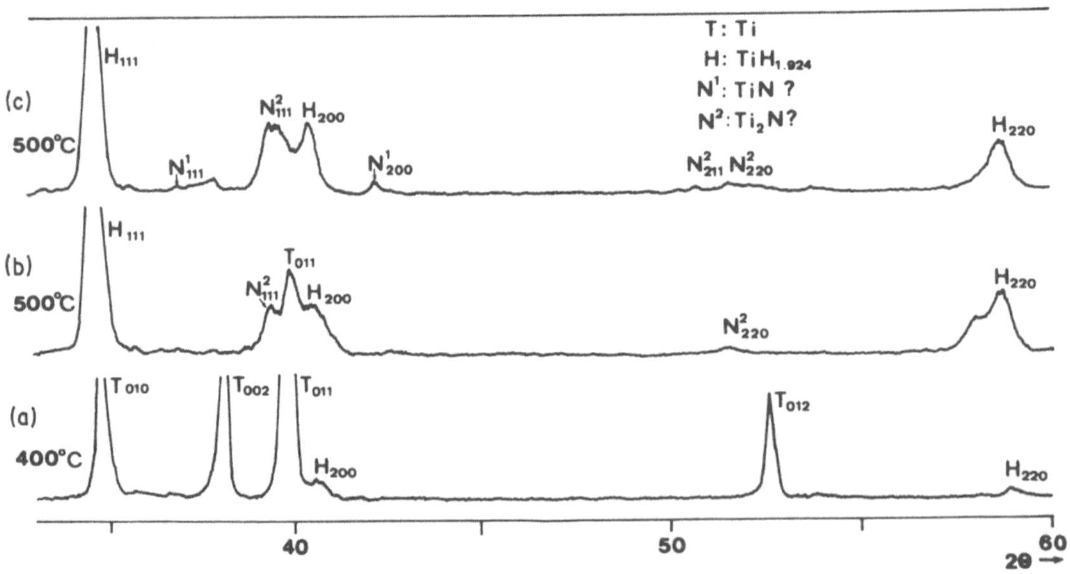

Figure 2 X-ray powder diffraction patterns of the products obtained during the runs at (a) 400° C (7 days), (b) 500° C (5 days), and (c) 500° C (7 days) under 30 MPa.

colour. Some metal particles with gold colouration seemed to be sprinkled among the specimens heated above 450° C. However, no product with a wholly gold colour could be obtained. Fig. 2 shows X-ray powder diffraction patterns of the specimens heated at 400 and 500° C. The specimen heated at 400° C always consisted of titanium metal and a small amount of titanium hydride, and those at 450 and 500° C formed a small amount of titanium, titanium hydride and a third phase, which was considered to be Ti_2N or TiN. However, the reflections, which were considered to be those of Ti_2N and TiN, were generally very weak in intensity and the most intense peak of Ti_2N (1 1 1) is partially overlapping with the reflection of titanium (0 1 1) or of $TiH_{1.92}$ (2 0 0). Thus the formation of titanium nitrides could not definitely be confirmed. The formation of titanium hydride can mostly occur above 450° C. The temperature for forming titanium hydride agreed well with the temperature of hydride formation in the hydrothermal oxidation of titanium [3]. As can be seen from Fig. 2, X-ray diffraction lines of titanium metal were broadened after the run even at 400° C, indicating that the cracking in the titanium metal grains occurred with hydriding and permeation of hydrogen into the metals. Concerning the formation of TiN powders, experiments at higher temperatures are now in progress.

References

1. M. YOSHIMURA, S. KIKUGAWA and S. SŌMIYA, Proceedings of the 9th AIRAPT and 19th EHPHG Conference (1981).
2. H. TORAYA, M. YOSHIMURA and S. SŌMIYA, *J. Amer. Ceram. Soc.* 66 (1983) 148.
3. M. YOSHIMURA, H. ŌHIRA and S. SŌMIYA, *Yogyo-Kyokai-Shi* to be published.
4. M. YOSHIMURA, S. KIKUGAWA and S. SŌMIYA, *J. Jpn. Soc. Powder Powder Metall.* 30 (1983) 207.
5. H. TORAYA, M. YOSHIMURA and S. SŌMIYA, *J. Amer. Ceram. Soc.* 66 (1983) 818.
6. J. F. BALASCIO, W. B. WHITE and R. ROY, *Mater. Res. Bull.* 2 (1967) 913.

Received 9 May
and accepted 26 June 1984

Reprinted from W. Bunk and H. Hausner (eds). *Proceedings of the Second International Symposium 'Ceramic Materials and Components for Engine'*, Germany, October 1986, pp. 529–536.

OXIDATION OF Si_3N_4 AND SiC BY HIGH TEMPERATURE – HIGH PRESSURE WATER VAPOR

M. Yoshimura, J. Kase and S. Sōmiya

1. Introduction

Silicon nitride and silicon carbide have attracted great attention in recent years because of their potential application as high-temperature structural ceramics. Oxidation resistance is one of the most important properties in these applications. In air or oxygen, the oxidation believed to proceed as follows ;

$Si_3N_4 + 3O_2 \longrightarrow 3SiO_2 + 2N_2$ ---[1], and

$SiC + 2O_2 \longrightarrow SiO_2 + CO_2$ ---[2].

A certain numbers of reports(1-3) argued that the oxidation rate was accelerated when water vapor coexisted with oxygen. However, the reaction of Si_3N_4 or SiC by H_2O only has not been studied yet. We have studied the oxidation of Si_3N_4(4) and SiC(5) powders by H_2O under hydrothermal conditions where almost only high-temperature high-pressure H_2O exist. Observed reactions are ;

$Si_3N_4 + 6H_2O \longrightarrow 3SiO_2 + 4NH_3$ ---[3], and

$SiC + 2H_2O \longrightarrow SiO_2 + CH_4$ ---[4].

These reactions are quite different from the reactions in oxidative atmosphere without H_2O. We discuss in the present paper the effects of H_2O on high-temperature degradation of Si_3N_4 and SiC powders and ceramics based on the experimental results.

2. Experimental Procedure

The starting materials were Si_3N_4 powder(*1), SiC powder(*2), hot pressed Si_3N_4(*3) and reaction sintered SiC(*4). Each samples weighed 60-90mg were placed in Au capsule of which top was narrowed by pressing to prevent the escape of solid materials. The capsule was heated in a test-tube-type pressure vessel(volume 5.5ml) under pressure of 10 or 100 MPa of re-distilled water at temperatures of between 200° and 800°C for 0 to 72 h. Most of the samples were heated under a requisite temperature then introduced the pressurized H_2O. After the treatment the sample in the pressure vessel was quenched by immersing into cold water, then the sample capsule was dried and weighed. Ceramic samples were directly

(*1) Starck LC12, average grain size 0.5μm, >90% α-form
(*2) Starck A10, average grain size 0.65μm, α-form
(*3) Toshiba Ceramics Co., Y_2O_3 5wt%, Al_2O_3 2wt%, Si_3N_4 93wt%, 4φx2t
(*4) Shin-Etsu Chemical Co., almost pure, 4mmφ x 2mmt

Research Laboratory of Engineering Materials, Tokyo Institute of Technology, Nagatsuta, Midori, Yokohama 227 JAPAN weighed to measure the weight change. The oxidation ratio was calculated from the weight change for powder samples but not for ceramic samples. Because ceramic samples behaved both in weight gain by the oxidation and in weight loss by the dissolution of produced oxides. Solid products were examined by X-ray diffractometry(XRD), scanning electron microscopy(SEM), and X-ray energy dispersive spectroscopy(EDS).

3. Results and Discussion

3-1 The oxidation of Si_3N_4 powder

Under 100 MPa H_2O, the oxidation of Si_3N_4 proceeded above 200°C as a function of time(Fig.1), Weight gain and appearence of a halo around 2θ≈16°-27° in X-ray diffraction(CuKα) indicated the formation of amorphous silica by the oxidation of Si_3N_4 as the first step at every temperature. Cristobalite and Keatite crystallized above 400°C after 8 h where Si_3N_4 was almost completely changed into amorphous silica. Keatite is one of the high-pressure phase of silica which is reported to appear at 400° to 800°C under 80 to 130 MPa(6), Scanning electron micrographs showed that amorphous silica grains covered the surfaces of the Si_3N_4 particles and then these silica grains grew and sintered during further hydrothermal treatment. The production of NH_3 was confirmed with the Nessler's reagent. These results indicates the oxidation of Si_3N_4 by H_2O to be given by the equation[1].

Under 10 MPa H_2O, the oxidation bahavior was similar to those under 100 MPa H_2O at 200°-300°C but quite different above 400°C(Fig.2). The oxidation rates were extremely slower than under 100 MPa. This would be caused by the phase transformation of H_2O from liquid to vapor, which occurs about 310°C under 10 MPa pressure. The other experiments where the 10 MPa water vapor was introduced into a vacuum vessel kept at 400°C to 800°C indicated very slow oxidation rates at 500°-600°C and increased rates at higher temperatures(Fig.3). Amorphous silica was also produced at every temperature under 10 MPa Cristobalite was formed after 72 h at 800°C.

Figure 4 shows the hydrothermal oxidation as functions of temperature, pressure and time. The oxidation rate increased with temperature at 200° to 300°C under 100 MPa as well as 10 MPa, but became lower at 500° to 600°C under 10 MPa, then increased again above 600°C. These results suggest the oxidation at 200°-300°C was caused by H_2O liquid and that at above 600°C under 10 MPa should be by H_2O vapor.

3-2 The oxidation of SiC powder

The hydrothermal oxidation of SiC occurred at higher temperatures above 500°C to yield amorphous silica, too(Fig.5,6). The oxidation behavior under 10 MPa H_2O resembled to that under 100 MPa but proceeded more slowly. Produced amorphous silica covered the SiC grains and then the grains started to grow and sinter during further treatment similarly to Si_3N_4. The amorphous silica crystallized in cristobalite and tridymite above 700°C after 8 h under 100 MPa H_2O and cristobalite at 800°C after 72 h under 10 MPa H_2O, where, SiC had been almost completely reacted with H_2O in both cases. Gaseous products in the vessel were proved to contain a large amount of CH_4 and a small amount of CO_2 by a mass spectrometric analysis. The formation of CH_4 indicates the reaction as the equation[4],

$$SiC + 2H_2O \longrightarrow SiO_2 + CH_4 \quad \text{---[4]}$$

rather that the reaction[5] to produce CO_2,

$$SiC + 4H_2O \longrightarrow SiO_2 + CO_2 + 4H_2 \quad \text{---[5]}.$$

Thermodynamic calculation(5) supports the experimental results and demonstrates that SiC is a sufficiently strong reducing agent to make CH_4 rather than CO_2 as the stable carbon species under the hydrothermal conditions studied.

3-3 Kinetics of powder oxidation

The oxidation ratio was approximately represented by a Jander-type equation(4), as illustrated with solid lines in figure 1-3,5 and 6 ;

$$[1-(1-\alpha)^{1/3}]^2 - [1-(1-\alpha_0)^{1/3}]^2 = (2k/r^2)t \quad \text{-----} \quad [6],$$

where t is the time, α is the oxidation ratio at t h, α_0 is the oxidation ratio at 0 h, r is the radius of particle, and k is the rate constant. Figure 7 showes the Arrhenius plots of the rate constants calculated from the equation[6]. The apparent activation energies are 73-84 kJ/mol in the oxidation of Si_3N_4 by liquid-like H_2O, 134kJ/mol in the oxidation of Si_3N_4 by vapor-like H_2O, and 167-194kJ/mol in the oxidation of SiC. These values are apparently smaller than the values reported for the oxidation of Si_3N_4 and SiC powderes in dry air or oxygen(150-290kJ/mol (7-10) and 210-330kJ/mol (11-13), respectively).

We propose here reaction models for the oxidation of Si_3N_4 and SiC powders in H_2O based on Jander-type model(Fig.8). After the amorphous silica layer is formed on the surface of particles, H_2O diffuse from outside of the silica layer and NH_3 or CH_4 diffuse from inside. In the case of Si_3N_4 by liquid-like H_2O, the apparent activation energies which are close to the values reported for H_2O diffusion in amorphous silica(60-105kJ/mol (14,15))would indicate that the oxidation of Si_3N_4 was characterised by H_2O diffusion rather than O_2 diffusion in the silica layers(120-300kJ/mol (16-18)).

The difference in the activation energy for oxidation of Si_3N_4 by liquid-like H_2O and vapor-like H_2O might be due to the difference in H_2O contents or microstructures in the amorphous silica. Since the H_2O diffusion in the silica layer should be similar both for the oxidation of Si_3N_4 and SiC, the different activation energies between them, would be due to the difference of produced gases ; NH_3 or CH_4. The former is a polar gas and easy to be ionized in liquid H_2O or amorphous silica, while the latter is a nonpolar gas in any conditions studied. Thus, NH_3(probably NH_4^+) might diffuse as fast as H_2O(probably OH^-) in amorphous silica layer, but CH_4 molecule would diffuse very slowly in amorphous silica. However, it is necessary to make clear the states of gaseous products in the silica layer to prove these mechanisms.

In contrasting to the oxidation in dry air or oxygen, the

oxidation in pressurized H_2O proceeds very fast because of the faster diffusion of H_2O than oxygen in the silica layer. The effects of water vapor on the oxidation of SiC have been interpreted as the acceleration(1) of the diffusion of oxygen, CO or CO_2, or the promotion of the crystallization(2) of silica into cristobalite, which might change diffusion paths from a balk diffusion to a grain-boundary diffusion. The present investigation, however, suggests that water vapor causes the reaction including diffusion process of H_2O. This seems to be one of the reasons for the acceleration of oxidation of SiC or Si_3N_4 in atmospheres containing H_2O.

3-4 The oxidation of hot-pressed Si_3N_4(HP-Si_3N_4) and reaction sintered SiC(RS-SiC) specimens

Under 100 MPa H_2O, the weight of ceramic specimens changed complicately influenced by both the oxidation and the dissolution of oxide layers produced on the surface. HP-Si_3N_4 reacted with the H_2O and gave a little weight loss even at 200°C(Fig.9). When the temperature was raised up to 800°C, the isotherm showed the weight gain from intense oxidation and almost all specimen changed into oxides, mainly silica after 72 h. At 400° and 600°C, we suppose following process from the isotherms. The pressurized H_2O starts to dissolve the oxides as soon as they are produced on the surface of the specimen. As H_2O has been saturated with the oxides, the oxide layers began to thicken on the surface of specimen and to give weight increases. Since the thickness of the oxide layer controles the oxidation rate by diffusion process in the oxide layer. The oxidation and disolution rates can be balanced after a certain period of the reaction to show a steady weight decrease. Yoshio et al.(19) also reported a steady weight loss as illustrated with a broken line in Fig.9 for pressureless sintered Si_3N_4 specimens under the conditions of 300°C 8.6 MPa H_2O.

RS-SiC showed little weight change at 400°C under 100 MPa H_2O, as had been expected from the examination for SiC powder (Fig.10). Weight loss or weight gain were observed at 600°C or 800°C, respectively, in the same manner as HP-Si_3N_4. Therefore, the dissolution behavior would be almost the same for RS-SiC or HP-Si_3N_4 ceramics regardless without or with the sintering aids and the produced gasses CH_4 or NH_3.

Under 10 MPa H_2O, the weight change of the specimen was smaller than under 100 MPa both for RS-SiC and HP-Si_3N_4. It is probable that vapor-like H_2O is difficult to disolve the oxides formed on the surface of the specimen.

X-ray diffraction and scanning electron micrographs indicated that amorphous silica or silicate was formed by the oxidation of SiC or Si_3N_4, respectively, as the first stage of the reactions at every temperature. Then those amorphous phases crystallized in the order of cristobalite, tridymite, and quartz. This order seems to indicate the ease in crystallization of phases under the hydrothermal conditions. X-ray diffraction showed also the formation of α-$Y_2Si_2O_7$ on the surface of the specimen in the oxidation of HP-Si_3N_4.

RS-SiC was changed to silica keeping the original morphology of the surface under 10 MPa H_2O(Fig.11-b), while larger silica crystals grew on the surface of RS-SiC in 100 MPa H_2O by successional dissolution/deposition processes(Fig.11-c). The oxidation products of HP-Si_3N_4 were amorphous silicates under 10 MPa H_2O(Fig.12-b), and three crystalline phases under 100 MPa H_2O. They are quartz crystals of about 10 μm in size(Fig.12-c), and needle-like and minute crystals both of which contained large amounts of Y and Al.

4. Summary

Si_3N_4 and SiC powder are oxidized by H_2O even as low temperature as 200° and 500°C, respectively, to yield amorphous silica and NH_3 and CH_4, respectively. The oxidation of Si_3N_4 and SiC can be accelerated when atmospheres contain water vapor bacause the diffusion of H_2O would be faster than that of oxygen in the amorphous silica layer. The degradation of these ceramic samples were accelerated by the dissolution of the oxide layer when H_2O exists as liquid-like phase.

5. References

1) H.Suzuki,Yogyo Kyokai Shi,67,157-64(1957).
2) P.J.Jorgensen,M.E.Wadsworth, and I.B.Cutler, J.Am.Ceram.Soc.,44,258-61(1961)
3) S.C.Singhal, J.Am.Ceram.Soc.,59,81-82(1976).
4) M.Yoshimura,J.Kase,S.Sōmiya,Yogyo Kyokai Shi,94,129-34 (1986)
5) M.Yoshimura,J.Kase,S.Sōmiya,J.Mater.Research, in press (1986)
6) R.K.Iler,"The Chemistry of Silica", John Wiley & Sons, New York(1979)p.16.
7) R.M.Horton,J.Am.Ceram.Soc.,52,121-24(1969).
8) D.Tétard,P.Lortholary,P.Goursat, and M.Billy, Rev.Int.Hautes.Temp.et Réfract.,10,153-59)1973).
9) M.Mitomo and J.H.Sharp,Yogyo Kyokai Shi,84,33-36(1976).
10) T.Sata and K.Fujii,ibid,90,110-18(1982).
11) T.Nakatogawa,J.Chem.Soc.Jpn.Ind.Chem.Sect.,57,348-50 (1957).
12) G.Ervin,J.Am.Ceram.Soc.,41,347-52(1958).
13) R.F.Adamsky,J.Phys.Chem.,63,305-07(1959).
14) I.Burn and J.P.Roberts,Phys.Chem.Glasses.,11,106-14(1970).
15) S.White,Nature Phys.Sci.,230,192(1971).
16) V.R.Haul,G.Dümbgen,Z.Elektrochem.,66,636-41(1962).
17) E.W.Sucov,J.Am.Ceram.Soc.,46,14-20(1963).
18) E.L.Williams,ibid.,48,190-94(1965).
19) T.Yoshio,K.Oda,K.O-oka,and K.Oda,Yogyo Kyokai Shi,94, 116-22(1986).

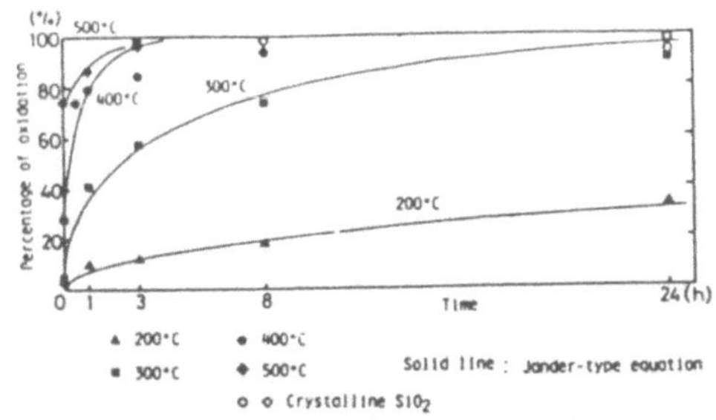

Isothermal oxidation of Si_3N_4 powder under 100 MPa H_2O.

Fig. 1

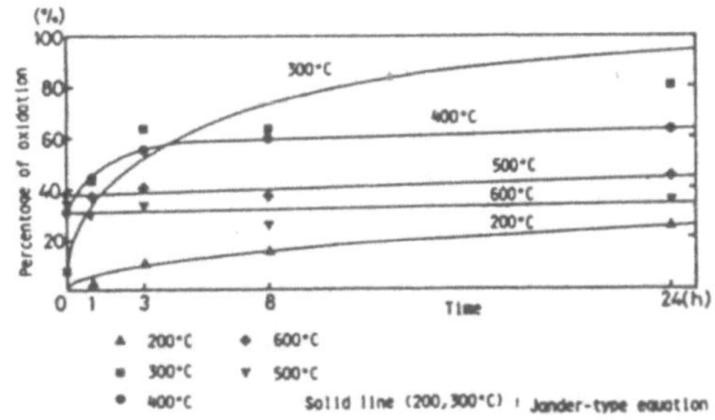

Isothermal oxidation of Si_3N_4 powder under 10MPa H_2O. (I), Heated after pressurized.

Fig. 2

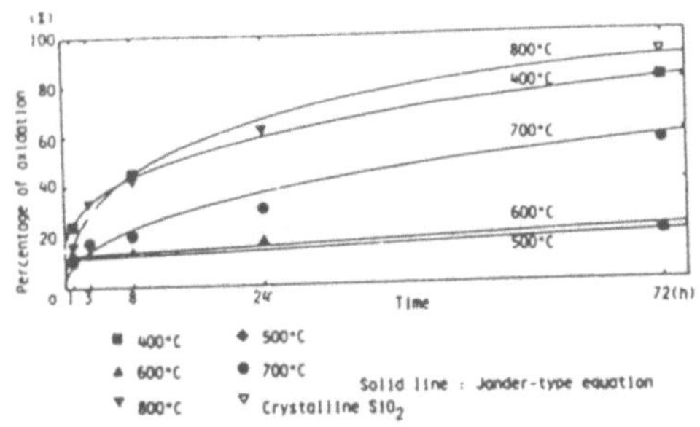

Isothermal oxidation of Si_3N_4 powder under 10MPa H_2O. (II), Pressurized after heated.

Fig. 3

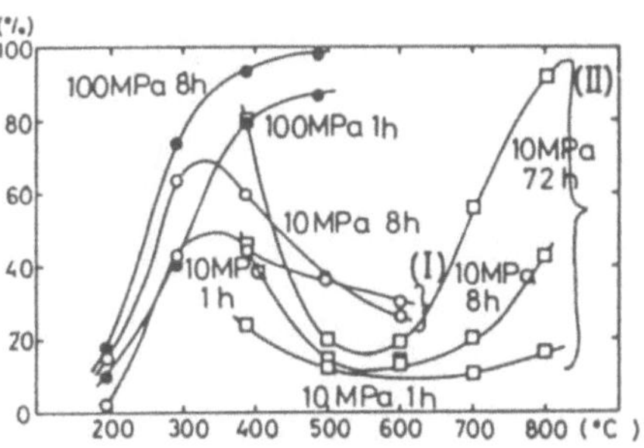

Variation of oxidation rate with temperature by the hydrothermal oxidation of Si_3N_4 powder.

Fig. 4

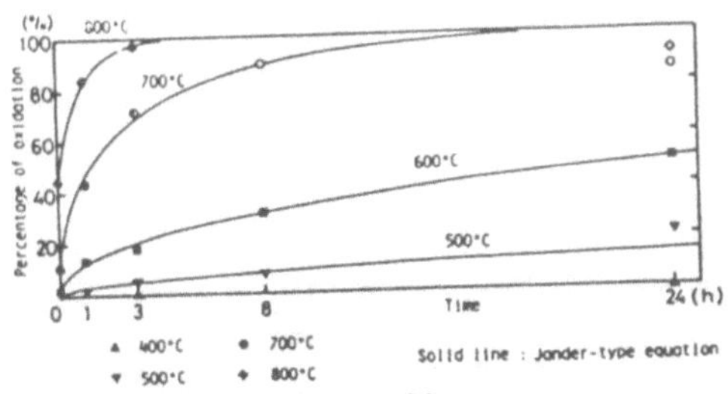

Isothermal oxidation of SiC powder under 100 MPa H_2O.

Fig. 5

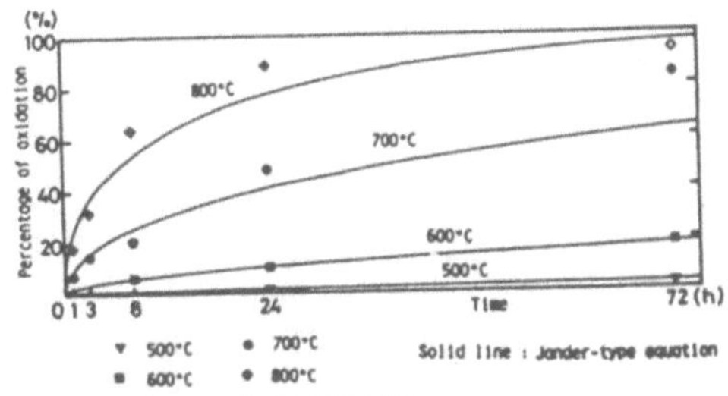

Isothermal oxidation of SiC powder under 10 MPa H_2O.

Fig. 6

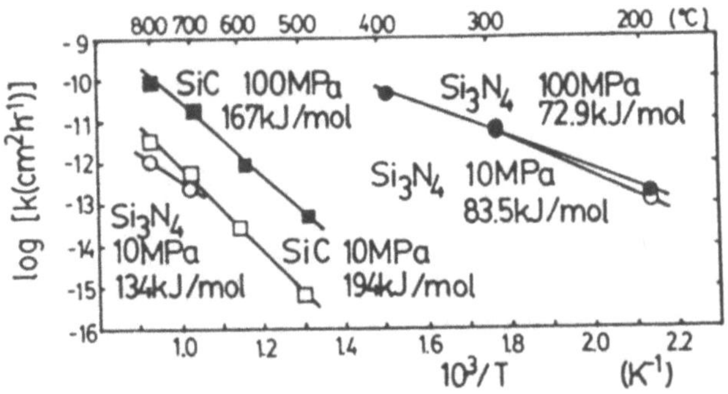

Fig. 7 Arrhenius plot of parabolic rate constant for the hydrothermal oxidation of Si_3N_4 and SiC powders

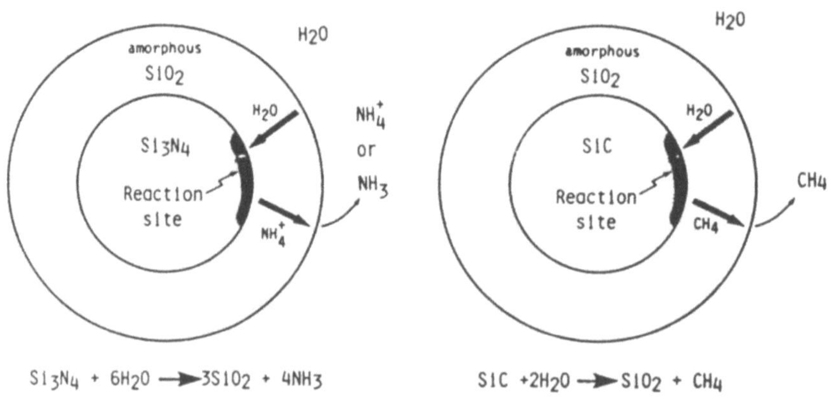

Fig. 8 Reaction model of hydrothermal oxidation of Si_3N_4 and SiC powder.

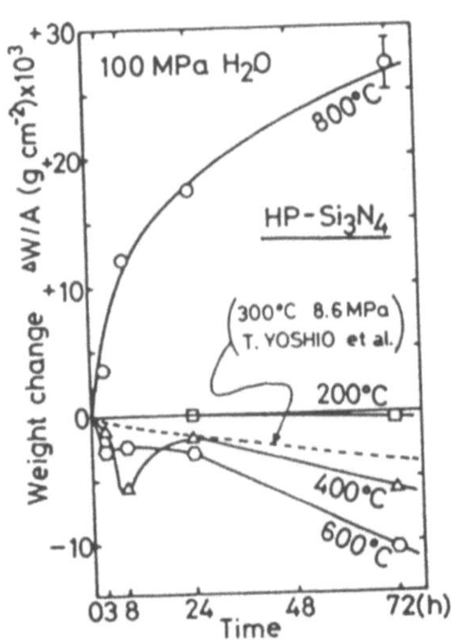

Weight change of HP-Si_3N_4 by hydrothermal treatment under 100 MPa H_2O

Fig. 9

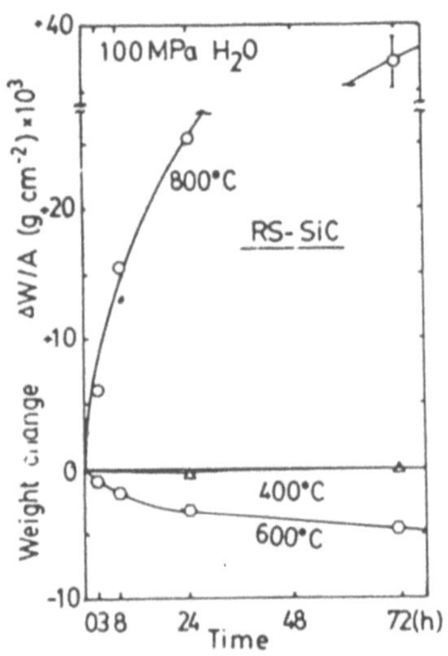

Weight change of RS-SiC by hydrothermal treatment under 100 MPa H_2O

Fig. 10

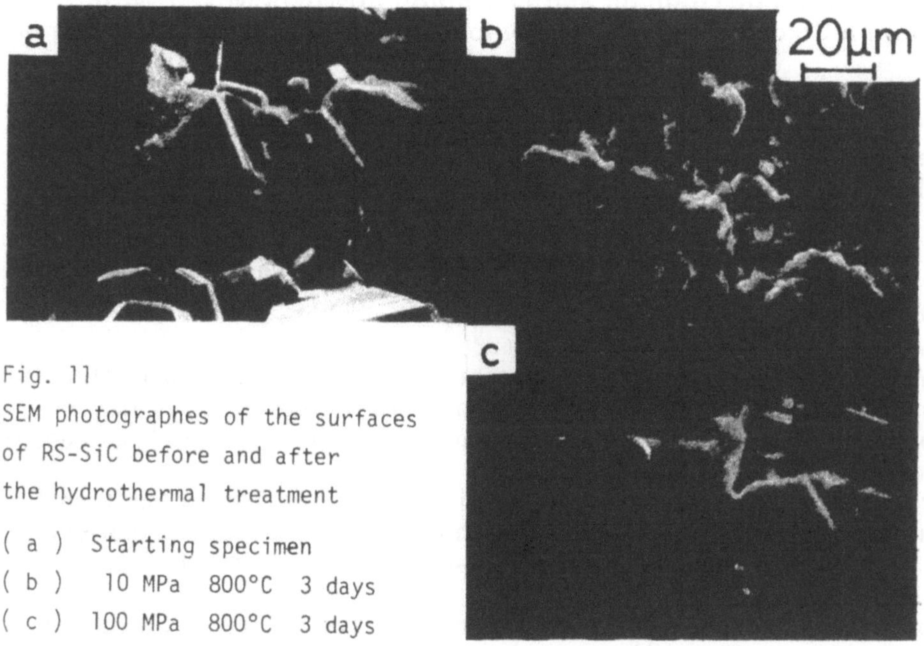

Fig. 11
SEM photographes of the surfaces of RS-SiC before and after the hydrothermal treatment

(a) Starting specimen
(b) 10 MPa 800°C 3 days
(c) 100 MPa 800°C 3 days

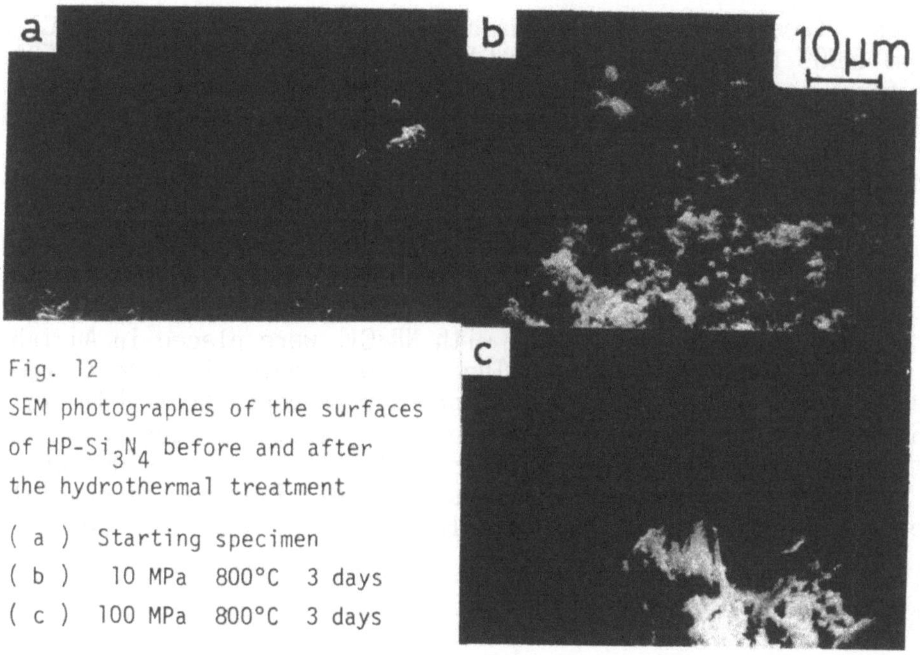

Fig. 12
SEM photographes of the surfaces of HP-Si_3N_4 before and after the hydrothermal treatment

(a) Starting specimen
(b) 10 MPa 800°C 3 days
(c) 100 MPa 800°C 3 days

Reprinted from Kaysser and Huppman (eds), *Proceedings of the Powder Metallurgy Conference*, Germany, 7–11 July 1986, pp. 179–182.

Nitridation of Titanium and Zirconium Powders Mixed with Ammonium Chloride under High Temperature–High Pressure Ammonia

Shigeyuki Sōmiya, Kazuo Suzuki and Masahiro Yoshimura

Research Laboratory of Engineering Materials, Tokyo Institute of Technology, 4259 Nagatsuta, Midori, Yokohama 227, Japan

I. INTRODUCTION

Titanium and zirconium nitrides are expected as high temperature structural materials, super-hard materials, surface protection materials, etc.,[1] because these nitrides have high melting point, chemical stability and high thermal-shock resistance.
These nitrides have been synthesized by direct nitridation, reduction & nitridation, thermal decomposition, or CVD methods. These methods, however, need generally high temperatures above 1000℃ and low pressures of nitridation gases below one atomosphere.
The present study used high-pressure ammonia to synthesize titanium and zirconium nitrides at lower temperature than previous works. According to our preliminary work[2] on the direct reaction of Ti with NH_3, the produced phase was not TiN but $TiH_{1.924}$ in bulk even under 30~100MPa NH_3 at 400°~500℃ for 5~30 days.
In the studies between Zr and NH_4Cl, Ohashi et al.[3] reported that β-ZrNCl was produced from 500° to 700℃. The β-ZrNCl decomposes into ZrN above 750℃ in vacuo. Therefore, we attempt here the nitride synthesis from metal with ammonium chloride at lower temperature in high pressure ammonia.

II. EXPERIMENTAL PROCEDURE

The starting materials are titanium powder(purity 99.1%up, Osaka Titanium,Co.), zirconium powder (purity 99.9%, Kojundo Chemical Laboratory,Co.) and ammonium chloride (a reagent grade, Wako Pure Chemical Co.Ltd.).
The sample powders, metal only or metal with NH_4Cl, were placed in Au tube 2.7mm i.d.,0.15mm thick and 50mm long. The tube was heated in a test-tube-type pressure vessel under 0.1 or 1MPa(initial pressure)ammonia from 300° to 700℃ for 3 to 168h. After the treatment, the pressure vessel was quenched in air.
The products were examined by X-ray diffraction(XRD), scanning electron microscopy(SEM) and transmission electron microscopy(TEM). The average crystallite size of the produced phases was also calculated by XRD using the Sherrer's formula.

III. RESULTS AND DISCUSSION

III—1 Reaction of Ti

The Ti powder reacted with 1MPa NH_3 at 700℃ for 3h to produce $TiH_{1.924}$ and Ti, but titanium nitride was not detected by XRD as the previous work.[2]
Figure 1 showed the amounts of every produced phases with temperature in the reaction of Ti with 10wt% (9.2 mol%) NH_4Cl under 1MPa NH_3 for 3h. Ti and NH_4Cl did not react at 300℃, but formed $TiH_{1.924}$ at 400℃, and the $TiH_{1.924}$ changed into TiN above 500℃. This result demonstrates the possible synthesis of TiN at low temperature as 500℃. The X-ray relative intensity of NH_4Cl phase was approximately constant at 300°~700℃. The samples of Ti-10wt% NH_4Cl under 0.1MPa, Ti-50wt% NH_4Cl under 0.1 and 1MPa NH_3 showed similar results.
When the sample of Ti-10wt% NH_4Cl was heated under 0.4MPa Ar gas not ammonia

gas, the product was only $TiH_{1.924}$ at 400 ℃ or above, and no TiN was formed (Fig.2). The amounts of $TiH_{1.924}$ showed maximum at 500℃ because of the dissociation of $TiH_{1.924}$ above 500℃. Above results clearly indicate that the production of TiN needs the coexistence of both NH_4Cl and NH_3 with Ti metal.
When the Ti-10wt%NH_4Cl mixture was reacted in NH_3 for 3h, TiN produced at maximum at 500℃, but $TiH_{1.924}$ was still remained with TiN. The single phase of TiN was yielded at 500℃ after 8h or longer as shown in Fig.3.
The reaction of Ti mixed with NH_4Cl in NH_3, therefore, would be

$$Ti \xrightarrow[400^\circ C]{NH_4Cl} TiH_{1.924} \xrightarrow[>500^\circ C]{NH_4Cl,\ NH_3} TiN$$

to yield TiN.
Figure 4 showed the SEM photograph of the products. Ti powder as starting material has the particle size of the tens of micrometer or less. The surface of the powder little changed at 300℃, but covered by small particles (probably $TiH_{1.924}$) at 400℃. The surface morphologies between 400℃ and above 500℃ did not change apparently even though the formation of TiN occured above 500℃ from $TiH_{1.924}$.
The produced TiN particles have less than ten nanometers as shown in Fig.5. The average crystallite size was 8~9 nm by the Scherrer's formula for TiN. Since the starting Ti powder had the crystallite size of >>150 nm and $TiH_{1.924}$ of about 20 nm, it is evident that the particles were broken into smaller one during hydridation and nitridation. Similar size reduction phenomena have been observed in the hydrothermal oxidation of metals.[4,5]
The produced TiN has the lattice parameter of 0.4232~0.4236nm corresponding to the N/Ti=0.80~1.05 according to Ehrlich.[6]

Ⅲ—2 Reaction of Zr

When the Zr powders were reacted with 1MPa NH_3 at 700℃ for 3h, the product was ZrH_2 not ZrN.
Figure 6 showed the amount of every produced phases with temperature by the reaction of Zr with 10mol%(6wt%)NH_4Cl under 1MPa NH_3 for 3h. ZrN was formed above 600℃ through the formation of ZrH and ZrH_2 above 400℃. A small amount of β-ZrNCl was also produced at 700℃ under the same conditions. The NH_4Cl phase was not detected by XRD in the products. The Zr-10mol% NH_4Cl mixture under 0.1MPa NH_3 gave similar results. ZrN produced above 600℃ in 1MPa NH_3 for 3h, but β-ZrNCl co-produced at 700℃. The single phase of ZrN has not been yielded even after 168h as shown in Fig.7.
The reaction of Zr mixed with NH_4Cl in NH_3, therefore, would be

$$Zr \xrightarrow[400\sim500^\circ C]{NH_3,\ NH_4Cl} ZrH,\ ZrH_2 \xrightarrow[\sim600^\circ C]{NH_3,\ NH_4Cl} ZrH_2 \xrightarrow[>600^\circ C]{NH_3,\ NH_4Cl} ZrN,\ \beta\text{-}ZrNCl$$

The scanning electron microscopy showed similar morphology changes of Zr powders during reactions to those of Ti powders as shown in Fig.4.
The average crystallite size was about 12nm by the Scherrer's formula for ZrN. Since the starting Zr powder had the crystallite size of >>150nm and ZrH_2 of about 20nm, it is evident that the size reduction occured during hydridation and nitridation similarly to the case of Ti.
The produced ZrN has the lattice parameters of 0.4559 ~0.4566nm corresponding to the N/Zr=1.25~1.30nm according to Ref.7.

IV. CONCLUSION

We studied nitride synthesis using high pressure ammonia.

(1) Titanium and Zirconium powders did not yield nitrides effectively by the treatment under 1MPa NH_3 at 700℃ for 3h.

(2) The addition of NH_4Cl was effective for the nitridation of Ti and Zr. TiN produced above 500℃, and ZrN produced above 600℃.

(3) The average crystallite size of TiN was 8 ~9 nm and of ZrN was about 12nm because of the breakdown of crystallite during hydridation and nitridation.

REFERENCES

/1/ L.E.Toth, "Transition Metal Carbide and Nitride" P.1-4, Academic Press, New York, N,Y (1971).

/2/ S.Sōmiya, M.Yoshimura and H.Toraya, J. Mater. Sci. Lett., 4, 91-93 (1985).

/3/ M.Ohashi, S.Yamanaka, T.Ishii and S.Hattori, Proceedings of the 23th Symposium for Basic Ceramics, Japan, P157 (1985).

/4/ M.Yoshimura, H.Ohira and S.Sōmiya, Yogyo-kyokai-shi, 93 〔7〕, 357-63 (1985).

/5/ H.Toraya, M.Yoshimura and S.Sōmiya, J. Am. Ceram. Soc., 66 〔2〕, 148-50 (1983).

/6/ P.Ehrlich, Z. anorg. allgem. Chem. 259 〔1-4〕, 1-41 (1949).

/7/ A.Yajima, Y.Segawa, K.Matsuzaki and Y.Saeki, Bull. Chem. Soc. Jpn., 56 〔9〕 2638-42 (1983).

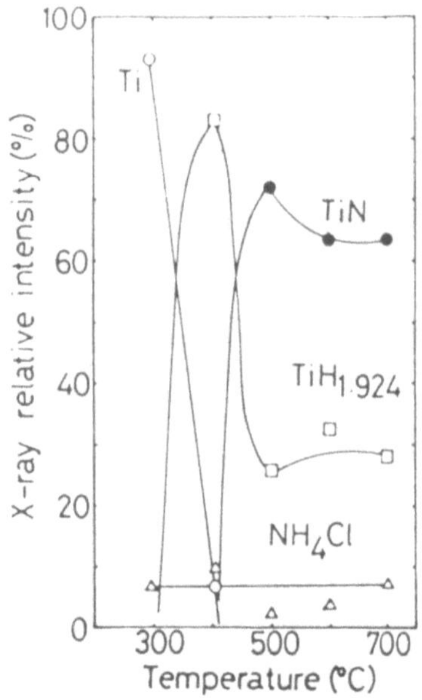

Fig.1
Phase contents with temperature in the reaction of Ti mixed with 10wt% NH_4Cl under 1MPa NH_3 for 3h.

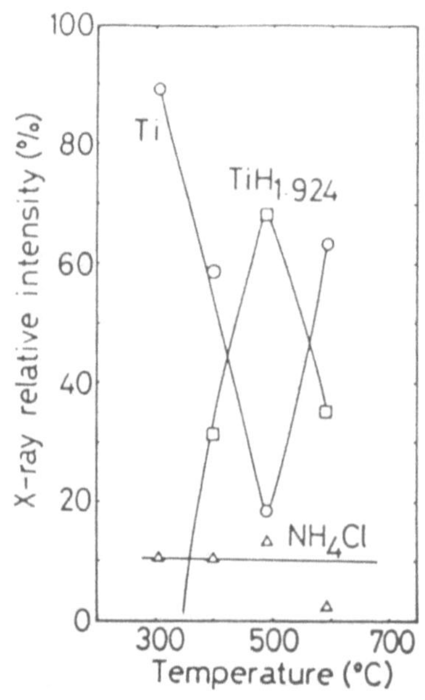

Fig.2
Phase contents with temperature in the reaction of Ti mixed with 10wt% NH_4Cl under 0.4MPa Ar for 3h.

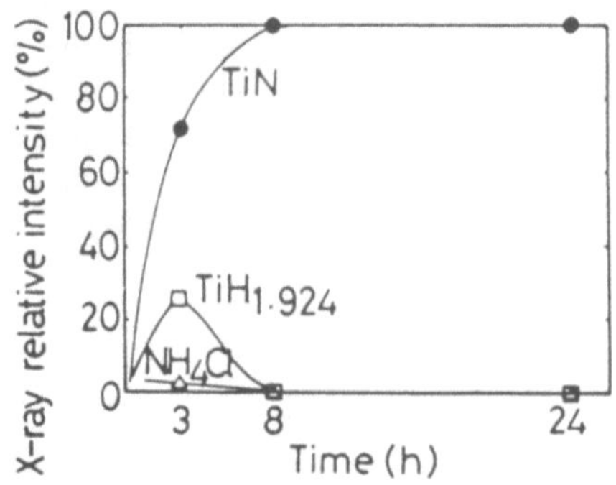

Fig.3
Phase contents with time in the reaction of Ti mixed with 10wt% NH_4Cl under 1MPa NH_3 at 500℃.

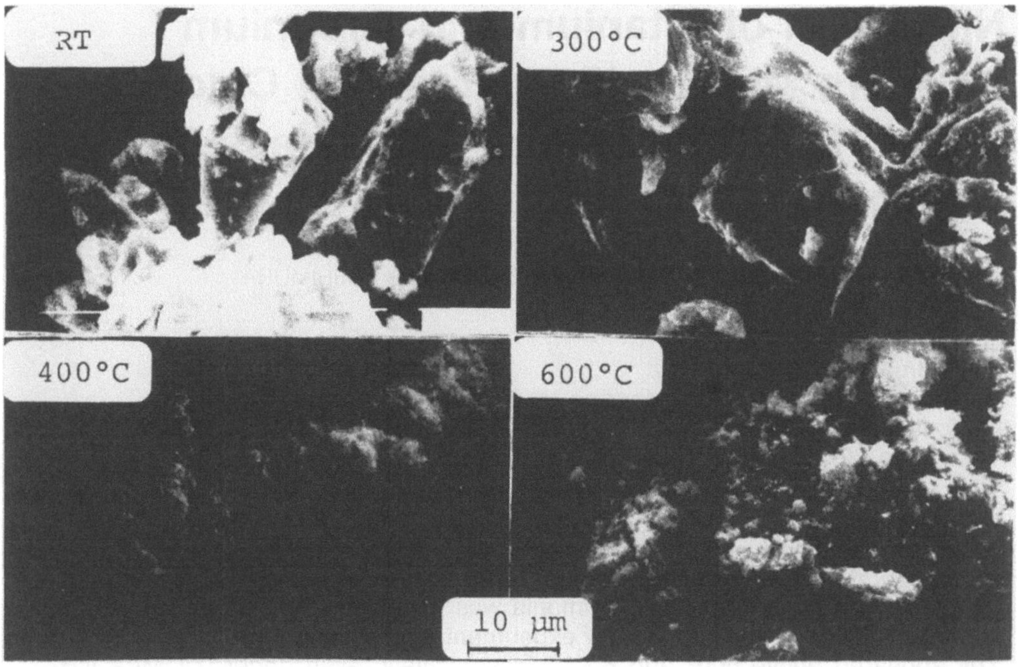

Fig.4
SEM photographs of products in the reaction of Ti mixed with 10wt% NH_4Cl under 1MPa NH_3 for 3h at various temperature.

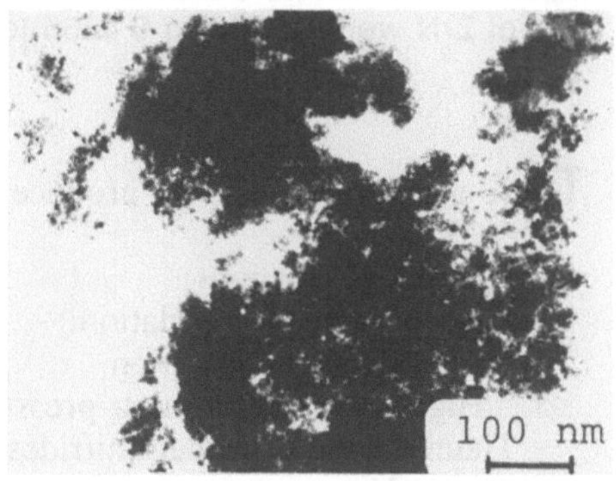

Fig.5 TEM photograph of TiN powder by the reaction of Ti mixed with 10wt% NH_4Cl under 1MPa NH_3 at 500℃ for 8h.

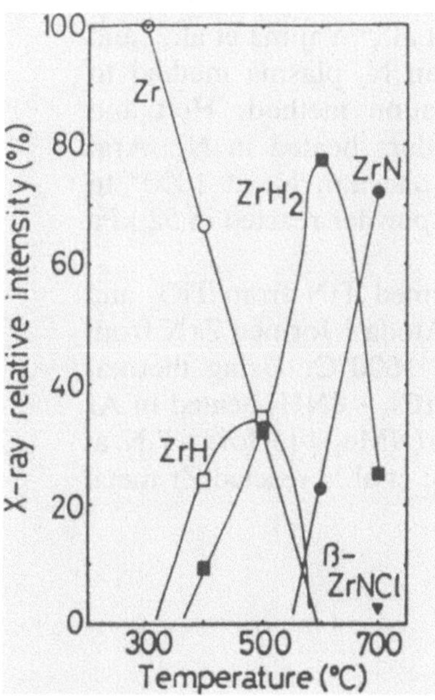

Fig.6
Phase contents with temperature in the reaction of Zr mixed with 10wt% NH_4Cl under 1MPa NH_3 for 3h.

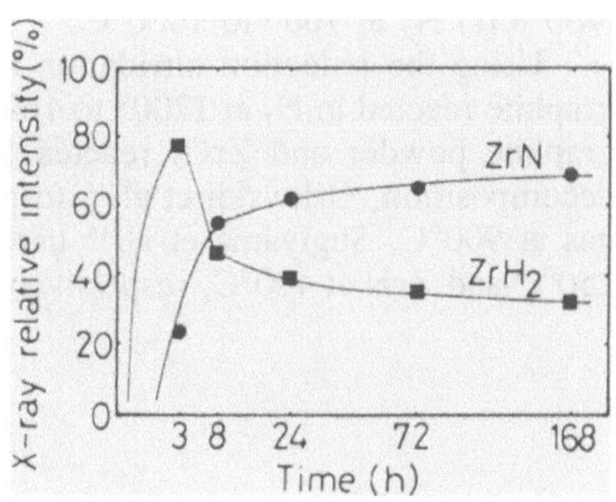

Fig.7
Phase contents with time in the reaction of Zr mixed with 10wt% NH_4Cl under 1MPa NH_3 for 3h.

Reprinted from *Advances in Ceramics*, Vol. 21, *Ceramic Powder Science* (American Ceramic Society, Inc., 1987) pp. 279–287.

Nitridation of Titanium and Zirconium Powders Mixed with Ammonium Chloride under High Temperature–High Pressure Ammonia

SHIGEYUKI SŌMIYA, KAZUO SUZUKI, AND MASAHIRO YOSHIMURA

Tokyo Institute of Technology
Research Laboratory of Engineering Materials
4259, Nagatsuta, Midori
Yokohama 227 Japan

Nitridation of Ti and Zr metal powders mixed with or without NH_4Cl under high-temperature and high-pressure ammonia was investigated. Without NH_4Cl, only hydrides of Ti or Zr appeared under experimental conditions; with NH_4Cl, TiN and ZrN appeared above 500° and 600°C, respectively, even under 0.1 MPa NH_3. Crystallite sizes for TiN and ZrN were 8 to 9 nm and 12 nm by X ray. Lattice parameters of TiN were 0.4232 to 0.4236 nm (corresponding to N/Ti = 0.80 to 1.05), and of ZrN were 0.4559 to 0.4566 (corresponding to N/Zr = 1.25 to 1.30).

There are several ways to produce nitrides of Ti and Zr;

1. CVD.
2. Direct nitridation.
3. Reduction and nitridation.
4. Thermal decomposition.
5. High-temperature–high-pressure ammonia methods.

Titanium and zirconium nitrides have been synthesized with CVD processes in NH_3-H_2-N_2 atmospheres using $TiCl_4$ and $ZrCl_4$ as the starting materials by Saeki et al.,[1] Kato et al.,[2] Hojo and Kato,[3] Kamiyama et al.,[4] Yajima et al.,[5] and Okabe et al.[6] Matsumoto[7] and Hayashi et al.[8] used an N_2 plasma method to synthesize Ti or Zr nitride. Using the direct nitridation method, Holt and Kingman[9] formed Ti nitride from Ti and NaN_3 powders heated in N_2. Arai et al.[10] formed Ti nitride from Ti powder heated in one atm N_2 at 1000° to 1400°C. Rosa and Smeltzer[11] formed Zr nitride from Zr powder reacted in 52 kPa (400 torr) N_2 at 700° to 1000°C.

Using the reduction-nitridation method, Arai[12] formed TiN from TiO_2 and graphite reacted in N_2 at 1200° to 1600°C. Okada and Atoda[13] formed ZrN from graphite powder and ZrO_2 reacted in N_2 at 1350° to 1600°C. Using thermal decomposition, Orlovskii et al.[14] formed Zr_3N_4 from $ZrCl_4 \cdot 8NH_3$ heated in Ar gas at 900°C. Sugiyama et al.[15] used $Ti(NMe_2)_4$ or $Zr(NMe_2)_4$ to form TiN at 250°C and ZrN at 400°C, respectively. Recently, Ohashi et al.[16] reacted Zr metal

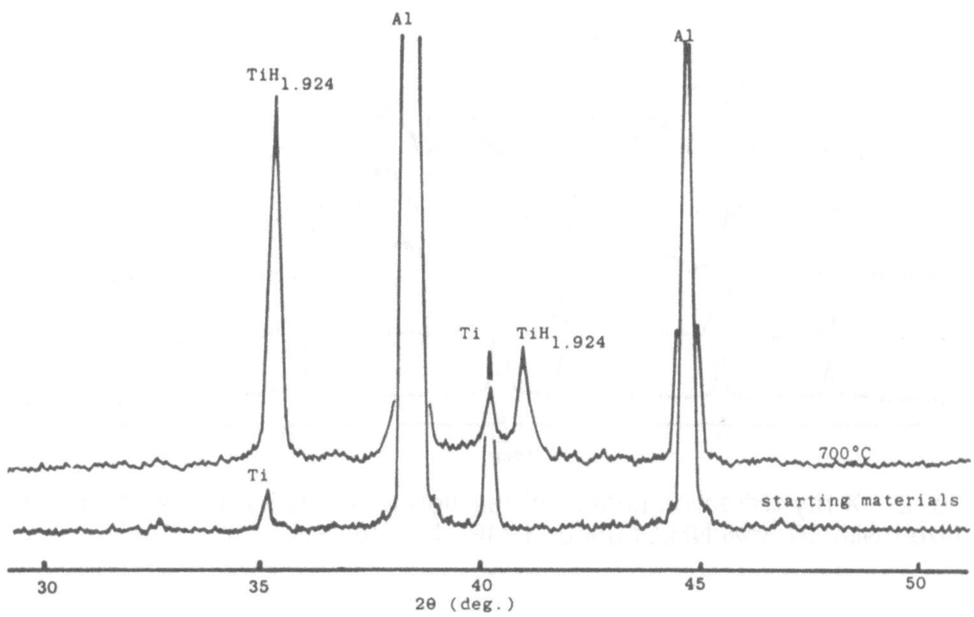

Fig. 1. X-ray diffraction pattern of reaction products between Ti and NH_3 under 1 MPa for 3 h.

powder and NH_4Cl in N_2 gas to form β-ZrNCl between 500° and 700°C and ZrN above 750°C. Sōmiya et al.[17] attempted to form Ti nitride under high pressures of ammonia. The phase produced was not TiN but $TiH_{1.924}$ in bulk, even under 30 to 100 MPa NH_3 at 400° to 500°C for 5 to 30 days. A report by Rabenau[18] summarizes several examples of nitride formation under high-pressure ammonia conditions.

Experimental Procedure

The starting materials are titanium powder* (purity ≥ 99.1%), zirconium powder† (purity 99.9%), and reagent-grade ammonium chloride.‡ The sample powders, metal only or metal with NH_4Cl, were placed in a gold tube (2.7 mm in diameter, 0.15 mm thick, and 50 mm long). The tube was heated in a test tube-type pressure vessel of Stellite 25§ under 0.1 or 1.0 MPa (initial pressure) ammonia from 300° to 700°C for 3 to 168 hours. After the treatment, the pressure vessel was quenched in air. The products were examined by X-ray diffraction (XRD), scanning electron microscopy (SEM), and transmission electron microscopy (TEM). The average crystallite size of the produced phases was also calculated by XRD using Scherrer's formula.

Results and Discussion

Ti Nitride

Reaction products of Ti with or without NH_4Cl under 1 MPa ammonia for 3 hours are shown in Figs. 1 and 2, respectively. An aluminum plate was used

---

*Osaka Titanium Co., Osaka, Japan,.
†Kojundo Chemical Laboratory Co., Osaka, Japan.
‡Wako Pure Chemical Co., Ltd., Osaka, Japan.
§Union Carbide, New York.

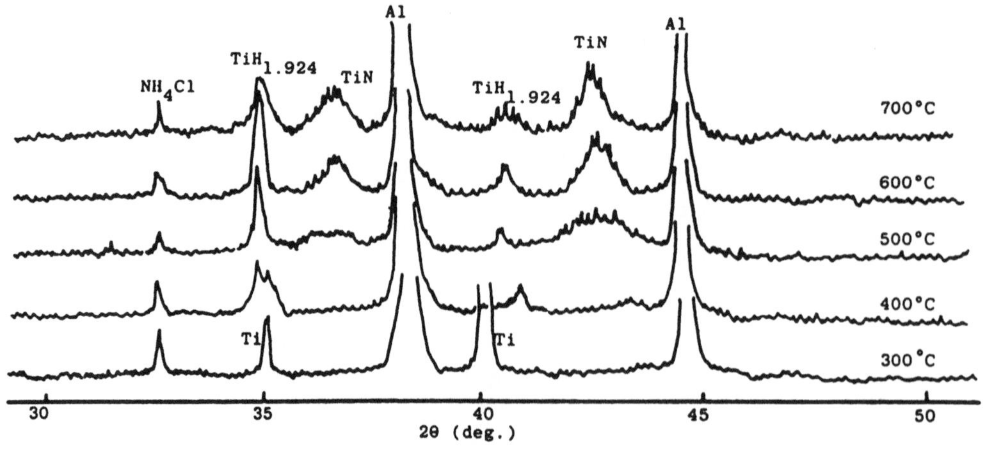

Fig. 2. X-ray diffraction pattern of reaction products by the reaction of Ti mixed with 10 wt% NH_4Cl under 1 MPa NH_3 for 3 h.

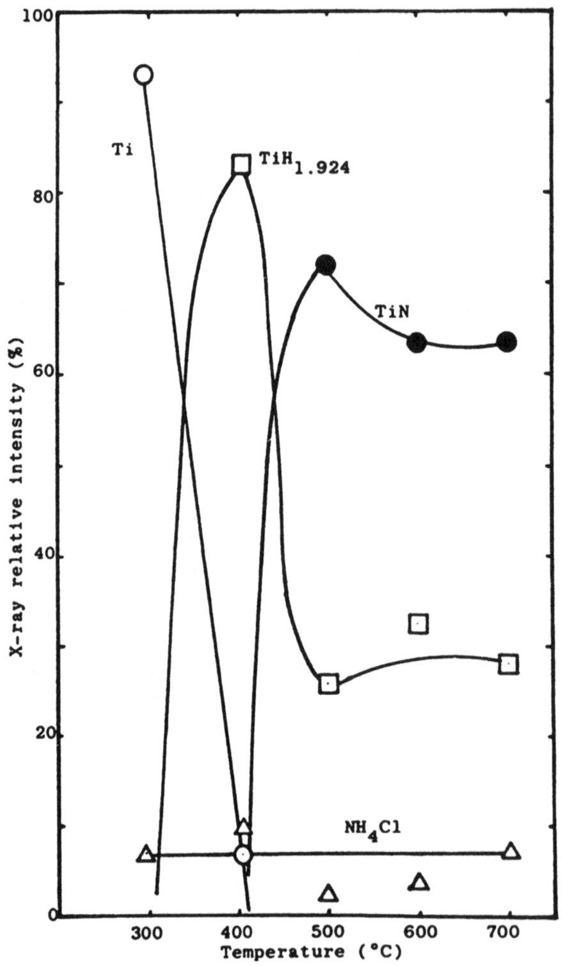

Fig. 3. Variation of amount of products with temperature by the reaction of Ti mixed with 10 wt% NH_4Cl under 1 MPa NH_3 for 3 h.

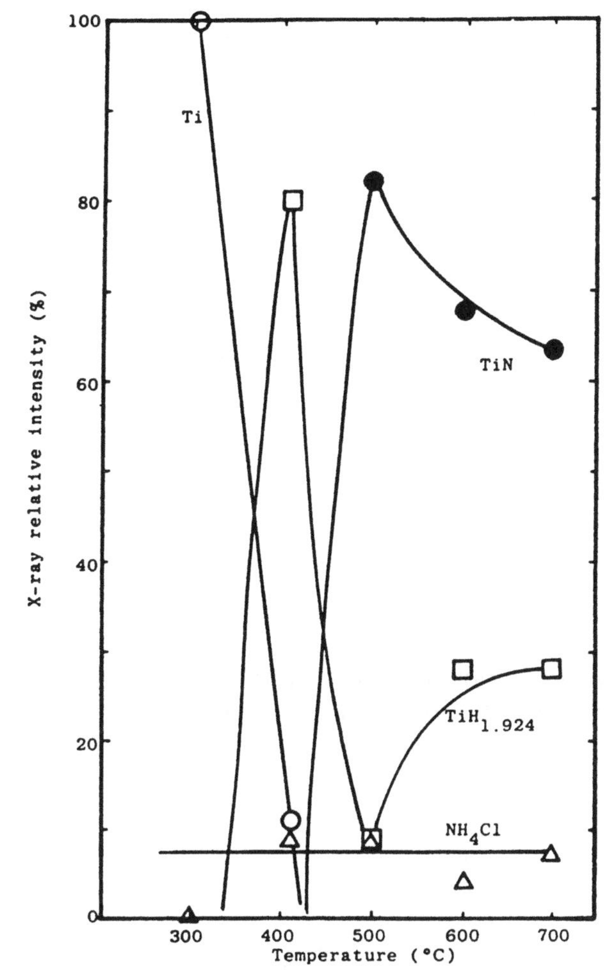

Fig. 4. Variation of amount of products with temperature by the reaction of Ti mixed with 10 wt% NH_4Cl under 0.1 MPa NH_3 for 3 h.

for powder X-ray diffraction. Therefore, Al peaks appeared in the figures. Without NH_4Cl, even at 700°C, $TiH_{1.924}$ appeared, but not TiN, as a reaction product. With NH_4Cl, TiN appeared above 500°C. The reaction products, by X-ray powder diffraction patterns at various temperatures for 3 hours and at different pressures of ammonia (1 and 0.1 MPa), are shown in Figs. 3 and 4, respectively. The effects of pressure on the reaction products are almost the same. Figure 5 shows the effect of time on amounts of reaction products at 500°C under 1 MPa ammonia for the mixture of Ti and 10 wt% NH_4Cl.

After 3 hours, the amount of $TiH_{1.924}$ decreased, and the amount of TiN increased. A TEM photograph of the TiN powder is shown in Fig. 6. The average crystallite size was determined to be 8 to 9 nm by X-ray diffraction. The lattice parameter of the synthesized TiN was 0.4232 to 0.4236 nm, corresponding to N/Ti = 0.80 to 1.05, according to Ehrlich.[19]

Zr Nitride

The reaction between Zr metal and high-temperature–high-pressure ammonia by the X-ray diffraction method is shown in Fig. 7. Even at 700°C under 1 MPa

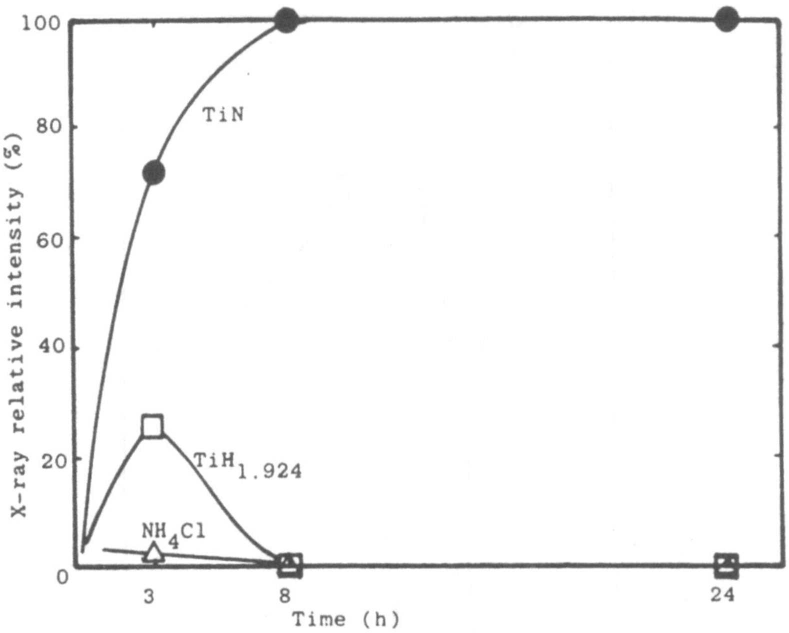

Fig. 5. Variation of amount of products with time by the reaction of Ti mixed with 10 wt% NH_4Cl under 1 MPa NH_3 at 500 °C.

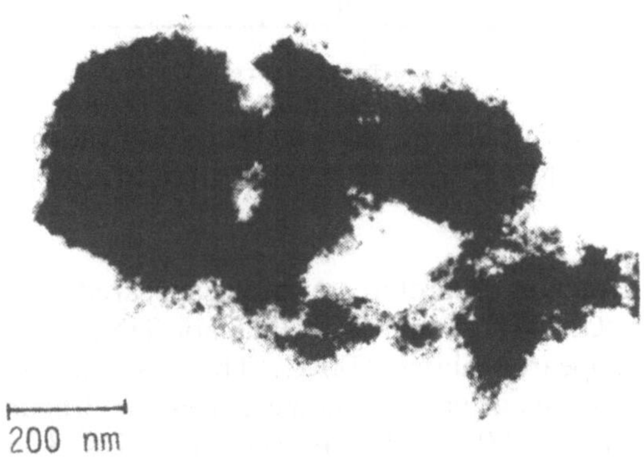

Fig. 6. Transmission electron micrograph of the product by the reaction of Ti mixed with 10 wt% NH_4Cl (1 MPa NH_3, 500 °C, 8 h).

ammonia for 3 hours, ZrH_2 appeared, but not the nitride. Reactions of Zr metal with 10 mol% NH_4Cl under 1 MPa ammonia for 3 hours showed ZrN above 600°C (Figs. 8 and 9). An increase in the temperature (from 300° to 600°C) caused the amount of Zr to decrease and hydrides to appear. A single phase of ZrN was not formed, even after 168 hours, as shown in Fig. 10. The average crystallite size of ZrN, determined by X-ray diffraction, was 12 nm. The lattice parameter

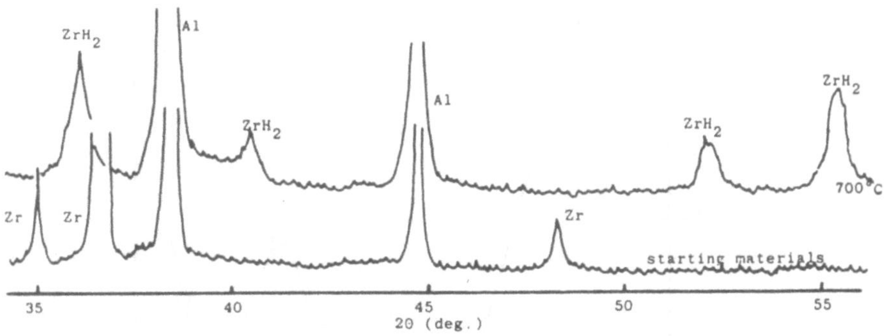

Fig. 7. X-ray diffraction pattern of reaction products between Zr and NH_3 under 1 MPa for 3 h.

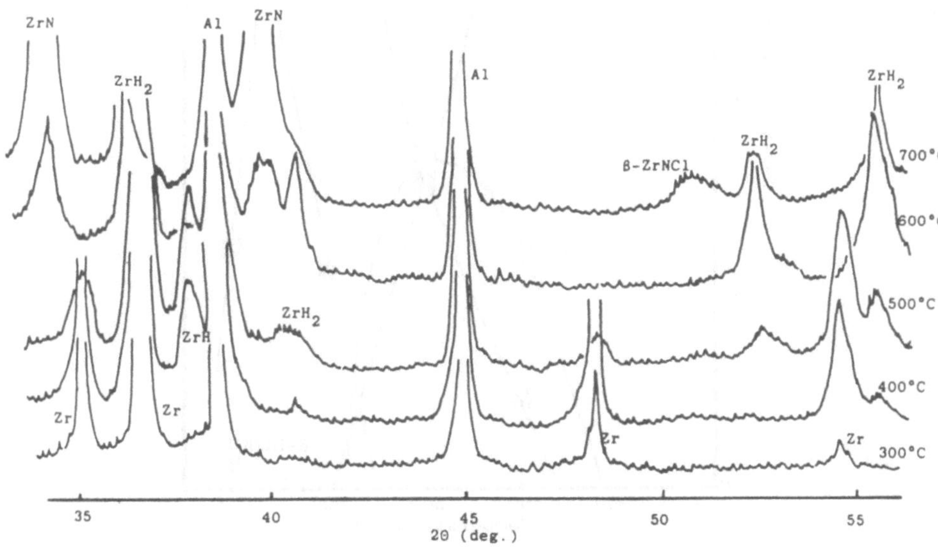

Fig. 8. X-ray diffraction pattern of reaction products by the reaction of Zr mixed with 10 mol% NH_4Cl under 1 MPa NH_3 for 3 h.

of the synthesized ZrN was 0.4559 to 0.4566 nm, corresponding to N/Zr = 1.25 to 1.30, according to Yajima et al.[5] A TEM picture of the synthesized ZrN is shown in Fig. 11.

Summary

The nitrides of Ti and Zr were synthesized under high-temperature–high-pressure ammonia.

1. The addition of NH_4Cl was effective for the nitridation of Ti and Zr.
2. The average crystallite size of TiN was 8 to 9 nm, and of ZrN about 12 nm due to breakdown of the crystallites during hydration and nitridation.
3. Titanium and zirconium powders did not yield nitrides effectively by the treatment under 1 MPa NH_3 at 700°C for 3 hours without NH_4Cl.

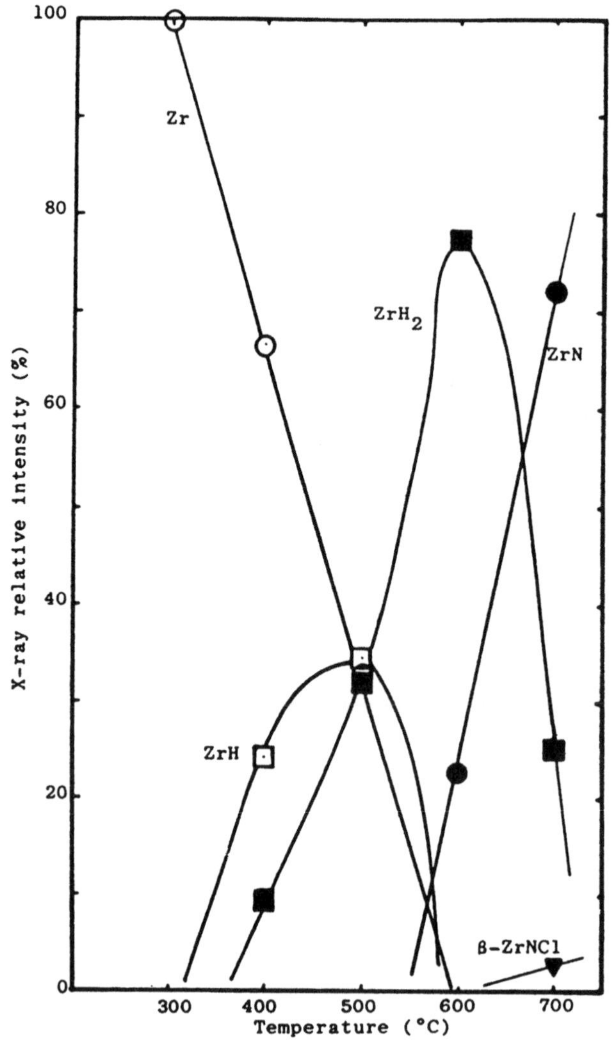

Fig. 9. Variation of amount of products with temperature by the reaction of Zr mixed with 10 mol% NH_4Cl under 1 MPa NH_3 for 3 h.

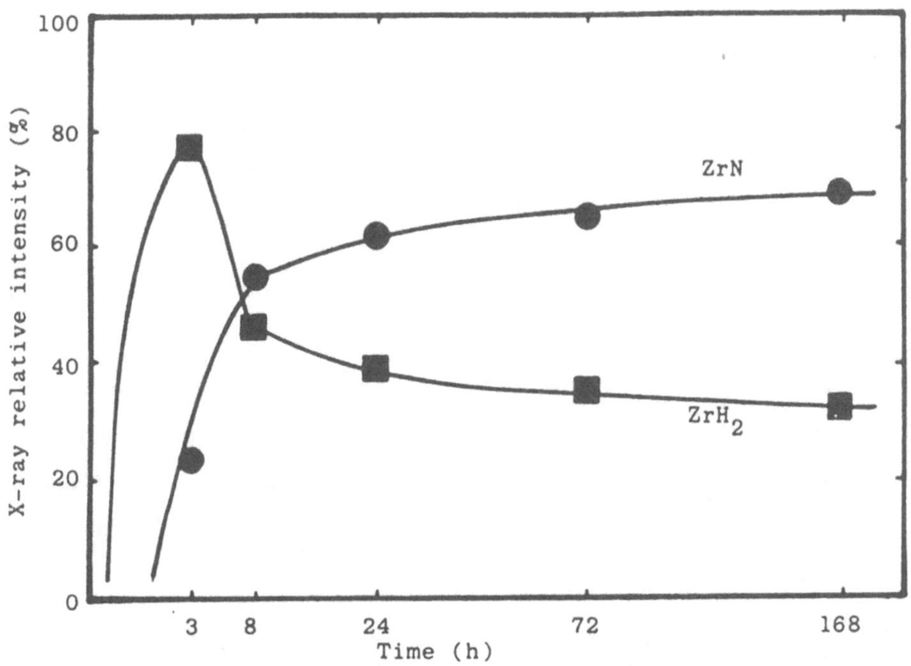

Fig. 10. Variation of amount of products with time by the reaction of Zr mixed with 10 mol% NH_4Cl under 1 MPa NH_3 at 600°C.

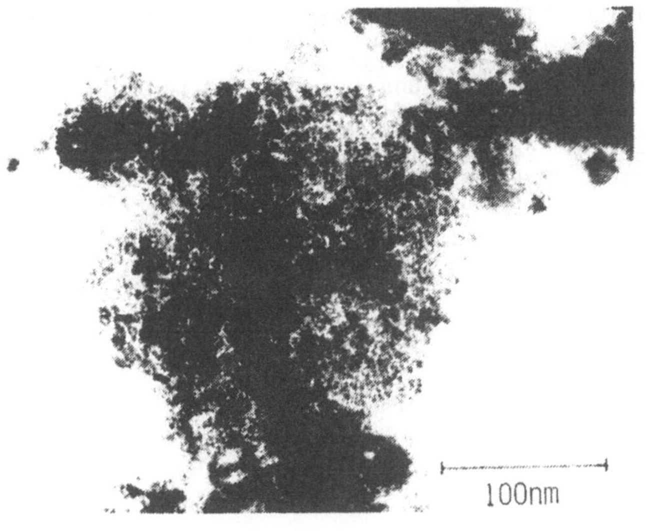

Fig. 11. Transmission electron micrographs of ZrN (with ZrH_2) from Zr mixed with 10 mol% NH_4Cl under 0.1 MPa NH_3 at 600°C for 8 h.

References

[1]Y. Saeki, R. Matsuzaki, A. Yajima, and M. Akiyama, "Reaction Process of Titanium Tetrachloride with Ammonia in the Vapor Phase and Properties of the Titanium Nitride Formed,"*Bull. Chem. Soc. Jpn,* **55** [10] 3193–96 (1982).

[2]A. Kato, J. I. Hojo, and T. Watari, "Some Common Aspects of the Formation of Nonoxide Powder by the Vapor Reaction Method," *Mater. Sci. Res.,* **17,** 123–35 (1984).

[3]J. I. Hojo and A. Kato, "Direct Observation of the Formation Process of TiN Particles by CVD Method," *Yogyo-Kyokai-Shi,* **89** [5] 277–79 (1981).

[4]Y. Kamiyama, T. Arakawa, T. Mori, T. Iida, and T. Mitamura, "Preparation of Zirconium Nitride Powder from Zirconium Chloride (IV)," *Yogyo-Kyokai-Shi,* **92** [1] 29–34 (1984).

[5]A. Yajima, Y. Segawa, R. Matsuzaki, and Y. Saeki, "Reaction Process of Zirconium Tetrachloride with Ammonia in the Vapor Phase and Properties of the Zirconium Nitride Formed," *Bull. Chem. Soc. Jpn.,* **56** [9] 2638–42 (1983).

[6]Y. Okabe, J. I. Hojo, and A. Kato, "Zirconium-Nitride Powders by the Vapor Phase Reaction of $ZrCl_4$-NH_3-H_2-N_2 System," *Yogyo-Kyokai-Shi,* **85** [4] 173–80 (1977).

[7]O. Matsumoto, "Formation of Titanium Nitride by Use of a Transferred Plasma Torch," *Denki Kagaku,* **35** [9] 625–30 (1967).

[8]Y. Hayashi, K. Okada, and S. Koide, "Synthesis of Nitriding Using Plasma Flame," *Yogyo-Kyokai-Shi,* **78** [9] 307–12 (1968).

[9]J. B. Holt and D. D. Kingman, "Combustion Synthesis of Transition Metal Nitrides," *Mater. Sci. Res.,* **17,** 167–75 (1984).

[10]Z. Arai, H. Hayashi, S. I. Nakamura, and N. Higashi, "Synthesis of Titanium Nitride from Titanium Powder and Nitrogen or Ammonia," *Rept. Government Industrial Res. Inst., Nagoya,* **11** [2] 119–25 (1962).

[11]C. J. Rosa and W. W. Smeltzer, "The Nitriding Kinetics of Zirconium in the Temperature Range 750–1000°C," *Electrochem. Technol.,* **4** [3–4] 149–53 (1966).

[12]Z. Arai, "Synthesis of Titanium Nitride [I]: Synthesis of TiN from TiO_2, Graphite and Nitrogen," *Rept. Government Industrial Res. Inst., Nagoya,* **5** [9] 454–59 (1956).

[13]S. Okada and T. Atoda, "Synthesis of Zirconium Nitride," *Yogyo-Kyokai-Shi,* **88** [5] 242–47 (1980).

[14]V. P. Orlovskii, N. V. Rudenko, and B. N. Ivanov-Emin, "Thermal Stability of Amine Complexed of Zirconium Tetrachloride," *Russ. J. Inorg. Chem.,* **12** [9] 1217–19 (1967).

[15]K. Sugiyama, S. Pac, Y. Takahashi, and S. Motojima, "Low Temperature Deposition of Metal Nitrides by Thermal Decomposition of Organometallic Compounds," *J. Electrochem. Soc.,* **122** [1] 1545–49 (1975).

[16]M. Ohashi, T. Ishihara, S. Yamanaka, and S. Hattori, "Synthesis and Reaction Mechanism of Layered Crystal β-ZrNCl from ZrH_2 and NH_4Cl"; Abstract for the 23rd Symposium on Basic Ceramics, p. 157, 1985.

[17]S. Sōmiya, M. Yoshimura, and H. Toraya, "Reaction of Titanium Metal Powders with High-Temperature High-Pressure Ammonia," *J. Mater. Sci. Lett.,* **4** [1] 91-93 (1985).

[18]A. Rabenau, "The Role of Hydrothermal Synthesis in Preparative Chemistry," *Angew. Chem. Int. Ed. Engl.,* **24,** 1026–40 (1985).

[19]P. Ehrlich, "Über die binaren Systeme des Titans mit den Elementen Stickstoff, Kohlenstoff, Bor und Beryllium," *Z. Anorg. Allgem. Chem.,* **259** [1-4] 1–41 (1949).

Part 11

Phase Diagrams

Reprinted from National Bureau of Standards SP-496, *Applications of Phase Diagrams in Metallurgy and Ceramics*, Proceedings of Workshop held at NBS, Gaithersburg, Md., 10–12 January 1977, pp. 508–515.

Phase Equilibria in the System MgO–RCl (R = Li, Na and K) Solution under Hydrothermal Conditions by Means of a Capsule Bursting Method

Shigeyuki Sōmiya, Kazuo Nakamura, Shin-ichi Hirano and Shinroku Saito

Research Laboratory of Engineering Materials
Tokyo Institute of Technology
Ookayama, Meguroku, Tokyo, 152, Japan

Introduction

Brucite $Mg(OH)_2$, is one of the raw materials to produce basic bricks. Brucite is obtained from sea water by precipitation or as natural rock which occurs in hydrothermal veins. The researches on the phase equilibrium in the system MgO-$Mg(OH)_2$ in pure water have been carried out by many researchers by means of various methods including the thermodynamic calculation. The phase equilibrium curves obtained, however, dispersed in wide temperature range because of the difficulty of quench of periclase, MgO.

It is well known that the hydrothermal solutions often contain alkali chlorides. But the literature that can be found on the phase equilibria in the system MgO-$Mg(OH)_2$ in alkali chloride solution, was scarce.

The purpose of the present work is to clear up the effect of the kind of the alkali chloride such as LiCℓ, NaCℓ and KCℓ on the phase equilibrium curve of the system MgO-$Mg(OH)_2$ up to 1,500 kg/cm^2 by means of a capsule bursting method.

Experimental

In this investigation test-tube-type hydrothermal vessels were employed and water was used as a pressure medium. Pressure was measured with a 12 inch Heise Bourdon tube pressure gauge. Temperature was measured with a platinel thermocouple calibrated against the melting points of gold (1,064.4°C) and zinc (419.5°C). Temperature was controlled with an automatic electronic controller combined with a SCR system (Ohkura Electric Co. Ltd., Tokyo, Japan) within ±3°C. The accuracy of pressure during a run was ±30 kg/cm^2.

The starting material of MgO was prepared by calcination at 600°C in Pt crucible for 24 hrs from magnesium nitrate, which had been obtained by the reaction of reagend grade nitric acid (Wako Pure Chemical Industry, Osaka, Japan) and magnesium metal (99.94% Furukawa Magnesium, Tokyo Japan). Five molar percent solutions were prepared from reagend grade alkali chloride (Wako Pure Chemical Industry) and redistilled water.

A gold capsule, 3.0 mm in OD, 2.7 mm in ID and 35 mm in length, including the desired quantity of MgO powder and 5 mol% RCℓ solution was sealed by an electric arc. The quantity of H_2O in the capsule was from 0.6 to 0.8 in molar ratio necessary to convert MgO into $Mg(OH)_2$.

The process of hydrothermal treatment are the same as in the study of the system MgO-$Mg(OH)_2$ in pure water(1); that is, pressure was initially applied to the desired value and then the temperature was increased to the desired value at the constant pressure. Samples were subjected to the constant temperature and pressure for a given duration. After the run pressure was released quickly to the atomospheric pressure and subsequently the vessel was quenched in cold water. This process is called "Capsule Bursting Method".

If the set point of temperature and pressure was on the higher temperature side of the equilibrium curve, the capsule would be burst by unreacted solution. In contrast, if it was on the lower temperature side, water inside the capsule was completely consumed up to form $Mg(OH)_2$. Therefore the capsule would not be burst but squeezed. After the run, the sample taken out of the capsule was examined by the Xray powder diffraction method using copper K_α radiation. The result was accepted when the results of Xray powder diffraction coincided with those of the capsule bursting method of two set of the capsules in one vessel.

Results and Discussion

The outside feature of the capsule showed the stable phase which existed in the capsule at each temperature and pressure in the system containing alkali chloride solution as in the system with pure water.

The phase equilibrium curves in LiCℓ solution, NaCℓ solution and KCℓ solution obtained are shown in Figs. 1, 2 and 3, respectively. Fig. 4 shows these curves including one in the system of MgO- pure H_2O which was studied by authors with the same method.[1] The equilibrium temperatures of each system at 500, 1,000 and 1,500 kg/cm^2 are shown in table 1. For example, equilibrium temperatures at 1,000 kg/cm^2 are 517°C in 5 mol% LiCℓ solution, 559°C in 5 mol% NaCℓ solution and 537°C in 5 mol% KCℓ solution. On the other hand it is 595°C in pure water.

All the equilibrium curves in RCℓ solution are in lower temperature side than that in pure water except for the region below about 200 kg/cm^2 in NaCℓ solution. The three equilibrium curves in RCℓ solution are nearly parallel and steeper than that in pure water. The effect of LiCℓ on lowering the equilibrium temperature is the greatest among the three kinds of solutions. The order of the effect, however, does not obey the order in the periodic table. It is known that such the effect is raised by the decrease of the fugacity of water due to the interaction of solute with water.[2]

From these results, the capsule bursting method is recognized as one of the acceptable methods for the study of phase equilibria under hydrothermal conditions whenever the reaction rate of hydration is rapid.

Reference

1) S. Hirano, K. Nakamura and S. Sōmiya
Am. J. Sci, submitted

2) H. L. Barnes and W. G. Ernst, Am. J. Sci, 261 129 (1963)

Table 1. Phase equilibrium temperatures of the system $Mg(OH)_2$-MgO in pure water, 5 mol% LiCℓ solution, 5 mol% NaCℓ solution at 500, 1,000 and 1,500 kg/cm^2.

	Pressure (kg/cm^2)		
	500	1,000	1,500
System of $Mg(OH)_2$-MgO			
pure water[1)]	567	595	614
5 mol% LiCℓ solution	505	517	524
5 mol% NaCℓ solution	553	559	562
5 mol% KCℓ solution	530	537	544

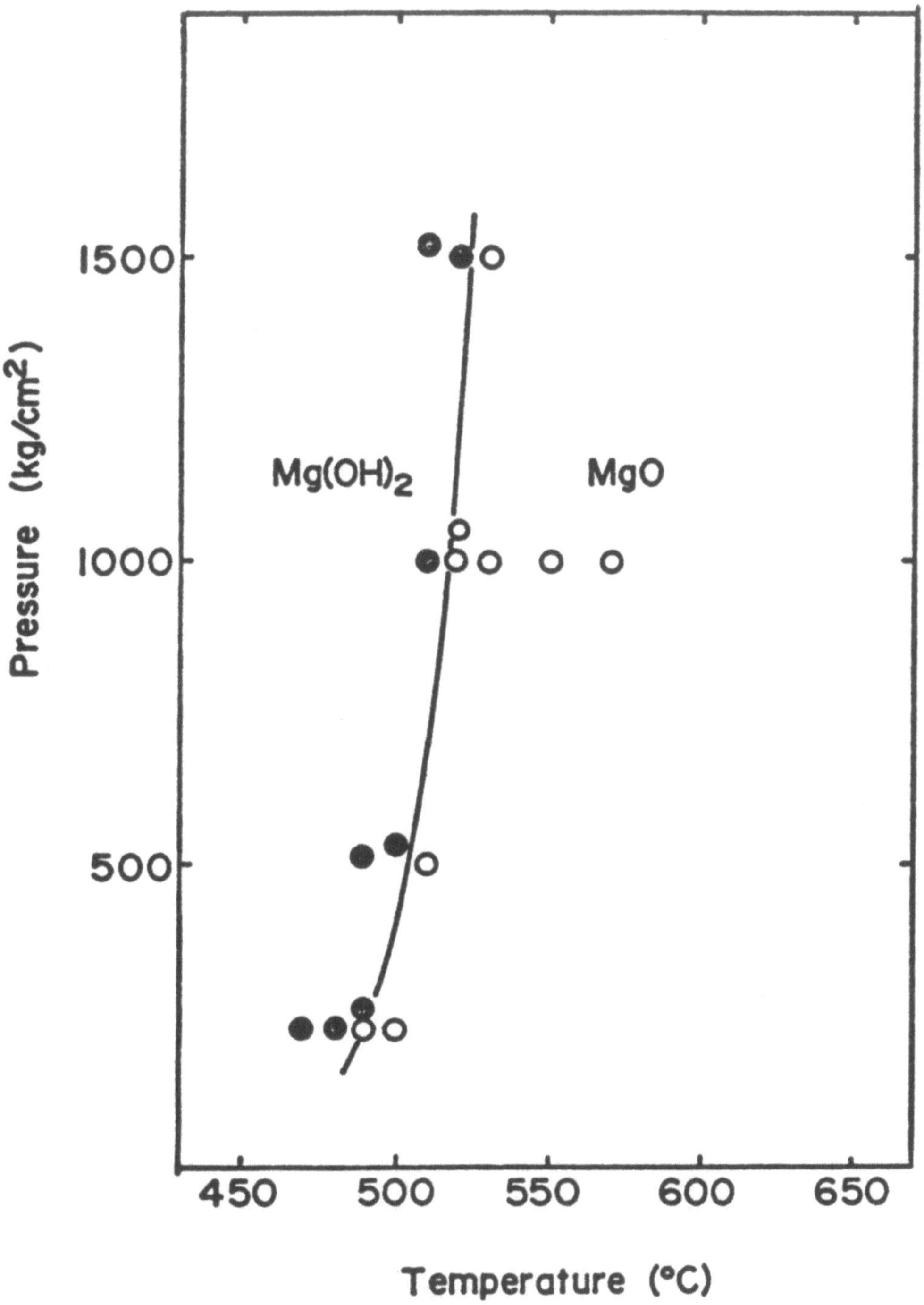

Fig.1 P-T diagram of the system $MgO-Mg(OH)_2$ in 5 mol% LiCl solution.

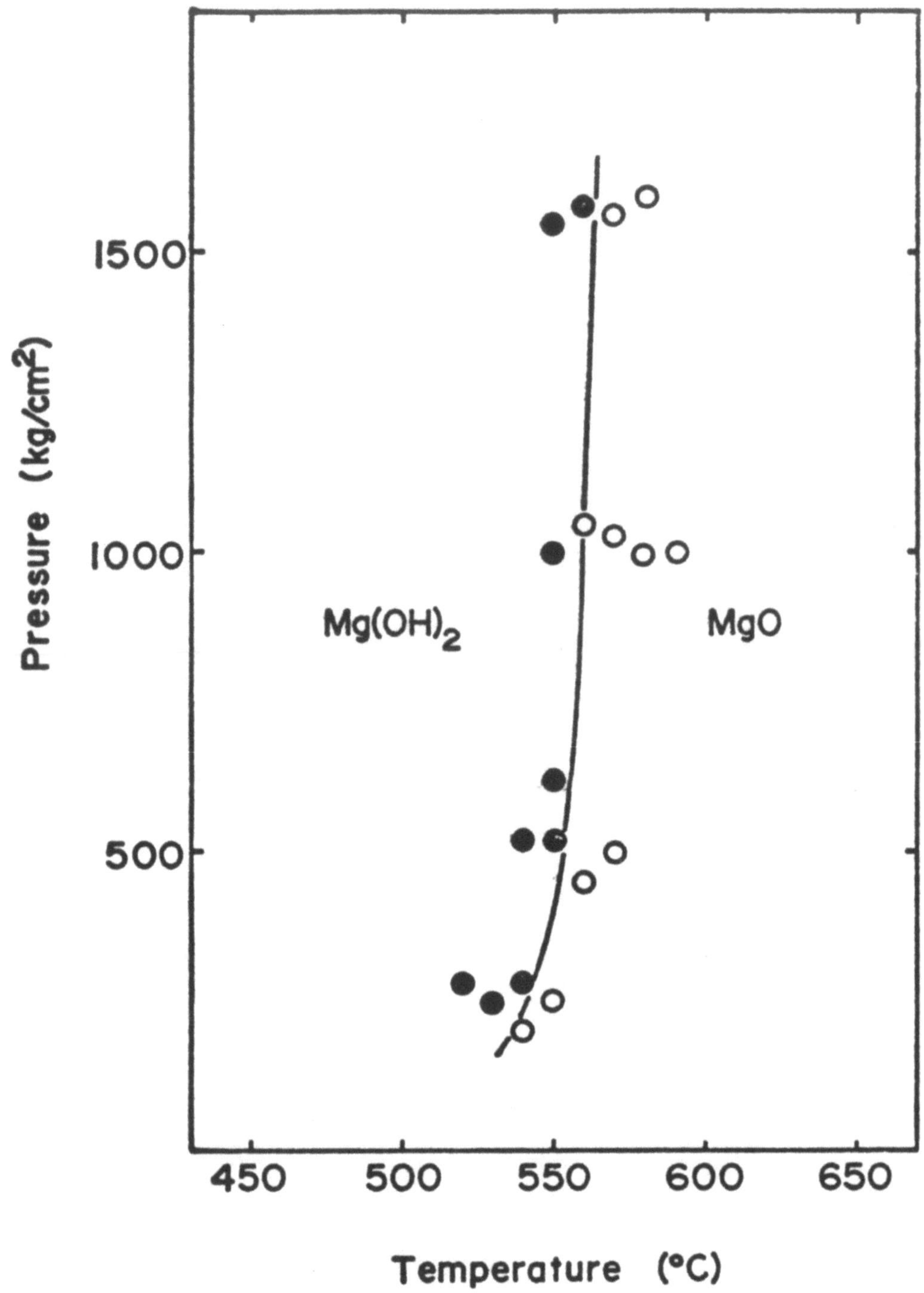

Fig.2 P-T diagram of the system MgO-Mg(OH) in 5 mol% NaCl solution.

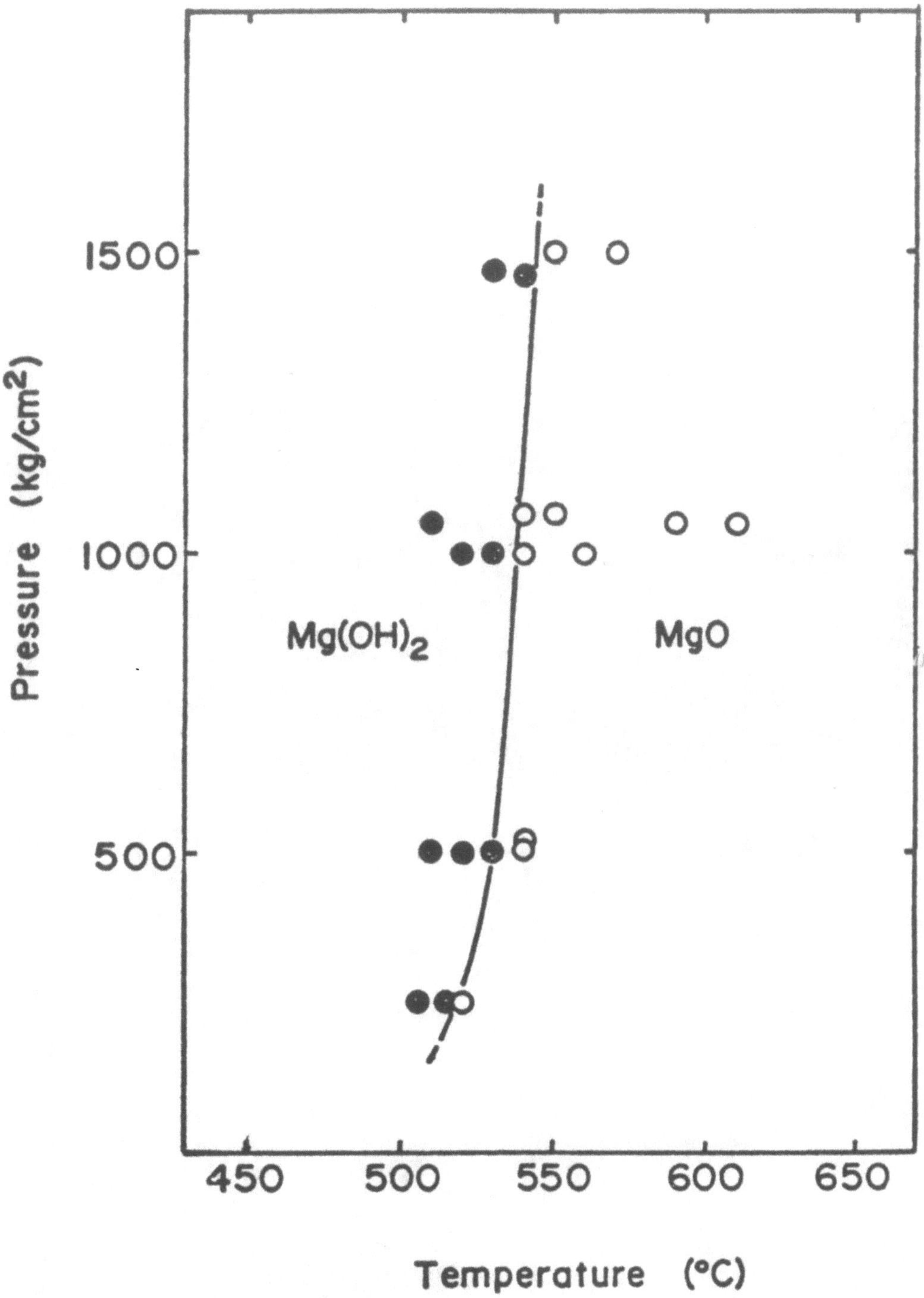

Fig.3 P-T diagram of the system $MgO-Mg(OH)_2$ in 5 mol% KCl solution.

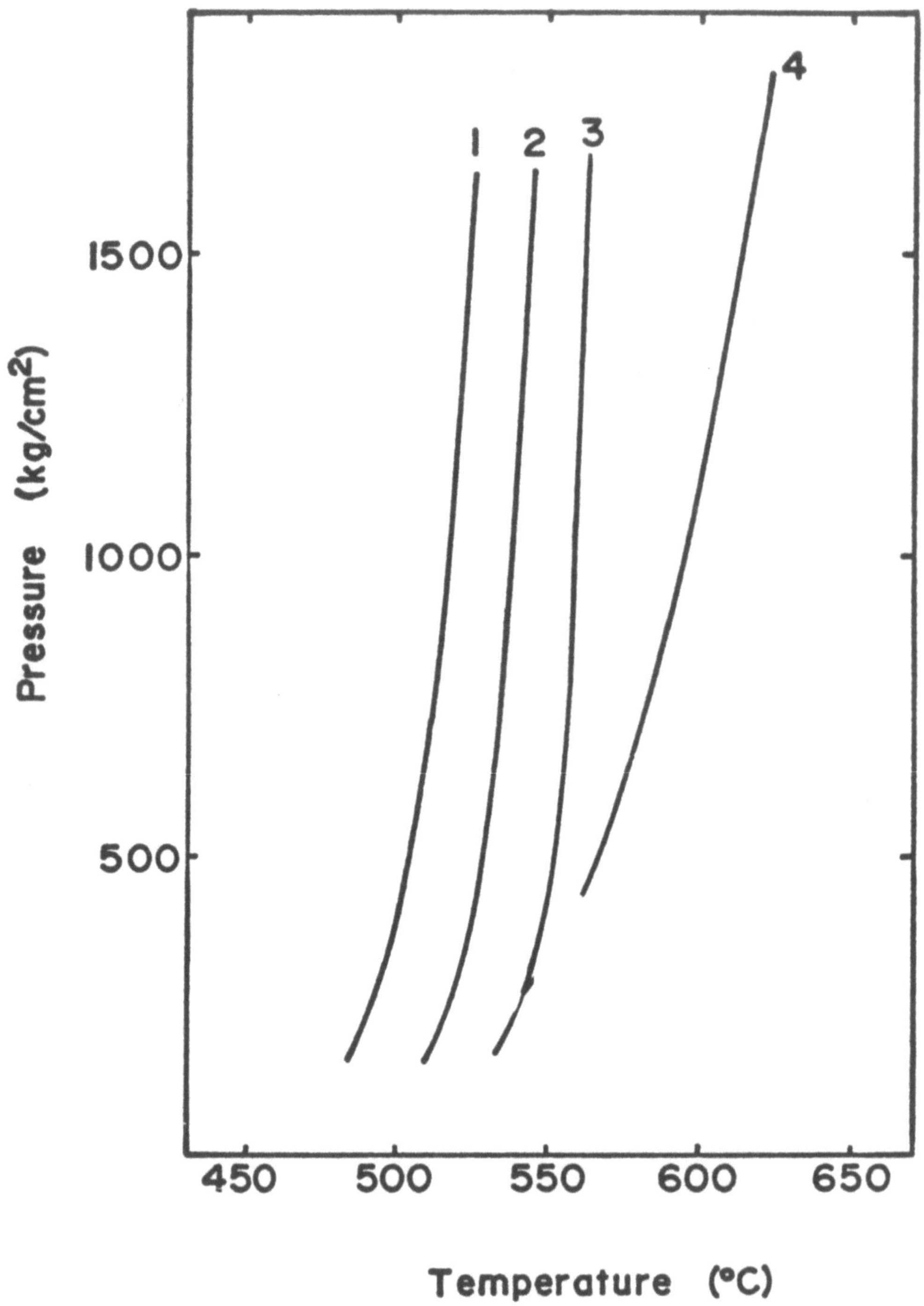

Fig. 4 P-T diagram of the system $MgO-Mg(OH)_2$ in 5 mol% LiCl soln.(1), 5 mol% KCl soln. (2), 5 mol% NaCl soln.(3) and pure water (4).

Reprinted from *Materials Research Bulletin*, **16** (1981) 327–333.

Phase Equilibria in the System Nd_2O_3–P_2O_5–H_2O at 500°C under 100 MPa and Synthesis of NdP_5O_{14} Crystals

Masahiro Yoshimura, Kazutaka Fujii and Shigeyuki Sōmiya
Laboratory for Hydrothermal Syntheses,
Research Laboratory of Engineering Materials, and
Department of Materials Science and Engineering,
Tokyo Institute of Technology,
4259 Nagatsuta, Midori, Yokohama 227, Japan.

(Received January 19, 1981; Communicated by R. DeVries)

ABSTRACT

Phase equilibria in the system Nd_2O_3-P_2O_5-H_2O was studied under a hydrothermal condition at 500°C under 100 MPa by a quenching technique using encapsuled samples. Equilibria among Nd phosphates; NdP_5O_{14}, NdP_3O_9 and $NdPO_4$, and liquid phase could be attained within 24 h. The stability fields for liquid (L), NdP_5O_{14}+L, NdP_5O_{14}+NdP_3O_9+L, NdP_3O_9+L, NdP_3O_9+$NdPO_4$+L, $NdPO_4$+L and $NdPO_4$+$Nd(OH)_3$+H_2O have been determined. The composition of the liquid phase with which NdP_5O_{14} and NdP_3O_9 or NdP_3O_9 and $NdPO_4$ could coexist was 3% Nd_2O_3-80% P_2O_5-17% H_2O or 4% Nd_2O_3-73% P_2O_5-23% H_2O respectively, at 500°C under 100 MPa. A phase diagram was constructed under this condition. Crystal growth of NdP_5O_{14} were tried under the equilibrium condition. Diamond-shaped crystals 0.3 mm in size were grown at 500°C for 4 days under 100 MPa. Crystals tended to show aggregation and surface roughness at lower growth temperatures. The crystals with polyhedral shape and maximum 0.5 mm in size were grown at 650°C.

Introduction

Since Danielmeyer and Weber(1) found an effective laser action in NdP_5O_{14} crystal, this material has been studied extensively for laser action(2-5), crystal structure(6-8), ferroelectric behavior(9,10) and crystal growth(11-16). NdP_5O_{14} crystals have been generally grown by dissolving Nd_2O_3 in hot orthophosphoric acid and subsequently heating this solution to several hundred degree Celsius in a covered crucible under atmospheric pressure. The species and pressures of vapors changed during heating in this method; that is, the vapors would be water-rich in earlier stages but phosphor-rich in later stages. The vaporization of the solvents has led to the precipitation and then growing of NdP_5O_{14} crystals. While large NdP_5O_{14} crystals up to 30 mm in cross section have been grown by such methods(15), the knowledge of the phase diagram in the system Nd_2O_3-P_2O_5-H_2O is important to understand the growing mechanism and to improve the quality of the crystal. Recently, Plättner et al.(16) studied the phase relations in this system in semi-covered crucibles and suggested a phase diagram by projecting the composition of solid and liquid phases during heating onto the plane Nd_2O_3-P_2O_5-H_2O. Similar studies have been done in the system of La_2O_3(17) or Eu_2O_3(18). True phase equilibria , however, have never been established in such works because of the continual vaporization from open crucibles during heating. In order to keep the composition unchanged, we have studied the system Nd_2O_3-P_2O_5-H_2O under a hydrothermal condition using sealed samples.

In present paper, we report the phase diagram of this system at 500°C under 100 MPa and some preliminary results of crystal synthesis under equilibrium conditions.

Experimental

The starting materials were Nd_2O_3†, P_2O_5††, 85% H_3PO_4 solution†† and pure water re-distilled by a quartz distillator after ion exchanging. Nd_2O_3 powder was calcined at 800°C for 1 day before weighing. Aqueous solutions of 10.0, 20.0 or 40.0% H_3PO_4 were prepared for reaction agents. To confirm the phase equilibria in the system, NdP_5O_{14} was also used as a starting material. It was synthesized by the reaction between Nd_2O_3 and $NH_4H_2PO_4$†† at 500°C for 3 days after pre-heating at 300°C for 17 h in an open crusible.

Weighed amounts of these materials were sealed hermetically into a gold capsule, 2.7 mm in inner diameter, 0.15 mm in thickness and 30 mm in length. The sealed capsule was put into a test-tube type pressure vessel of Stellite-25 and subjected to the hydrothermal reaction under desired pressure and temperature at the isothermal zone of an electric furnace. The temperature was measured with a portable DC Potentiometer* connected to two platinel thermocouples attached to the outside wall of the pressure vessel at two positions corresponding to the top and the bottom of the capsule. The pressure media was water, and the pressure was measured with a calibrated Heise bourdon gauge. The thermocouples were calibrated against the solidification point of Zn (419.58°C).

---

† > 99.9%, Shin-etsu Chem. Co. Ltd., Tokyo, Japan.

†† guaranteed reagent grade, Wakō Pure Chem. Ind. Ltd., Tokyo, Japan.

* Type 2727, Yokogawa Elec. Works, Ltd., Tokyo Japan.

At the completion of a run, the pressure vessel was quenched into cold water, then the capsule was weighed to check the leakage during the run. The products were washed with water, then examined with an optical microscope**, a scanning electron microscope*** and by X-ray powder diffraction with a diffractometer**** equipped with a rotating Cu anode and a graphite monochrometer. For the crystal growth of NdP_5O_{14}, about 24 mg Nd_2O_3, about 35 mg P_2O_5 and about 187 mg 85% H_3PO_4 solution were sealed in a gold capsule with 4.6 mm in inner diameter, 0.2 mm in thickness and 45 mm in length. NdP_5O_{14} powders were sometimes used. The sealed capsule was subjected to a hydrothermal growth for 4 days by the same procedure as the phase equilibria studies.

Results and Discussion

1. Phase equilibria in the system Nd_2O_3-P_2O_5-H_2O at 500°C under 100 MPa

NdP_5O_{14} ultraphosphate [U], NdP_3O_9 metaphosphate [M] and $NdPO_4$ orthophosphate [O], and $Nd(OH)_3$ neodymium trihydrate [T] were identified by X-ray powder diffraction in the products at 500°C under 100 MPa. In addition, liquid phase [L] existed over the range from orthophosphoric acid solution to polyphosphoric acid. $NdPO_4 \cdot H_2O$, which is stable at temperatures around a room temperature(16), has never been observed in the present work.

Phase equilibria including neodymium phosphates were studied by two methods ; synthesis of the phosphates from Nd_2O_3, P_2O_5 and 85% H_3PO_4 solution, and decomposition of the phosphates by water, 85% H_3PO_4 or P_2O_5 added H_3PO_4 solutions. The identical products were obtained for the samples with identical compositions regardless from either synthesis or decomposition after the equilibration for 24 h at 500°C under 100 MPa. It is, thus, apparent that the phase equilibria could be attained within 24 h in this condition. The phase boundaries were determined on the basis of these results as shown in Figure 1, where open symbols indicate one solid-phase region and filled symbols two solid-phase region. The stability regions for NdP_5O_{14}+L, NdP_5O_{14}+NdP_3O_9+L, NdP_3O_9+L, NdP_3O_9+$NdPO_4$+L, $NdPO_4$+L and $NdPO_4$+H_2O+$Nd(OH)_3$, were clearly established in this figure. The phase diagram in the Nd_2O_3-rich regions has not been determined since scattered results have been observed in the products by heterogeneous reactions due to too small an amont of H_2O in the capsule.

2. Liquid phase

The phase boundary between liquid and liquid + solid(s) was determined by microscopic observation of the neodymium phosphate crystals in the products. When only the liquid phase existed in the sample with given composition at 500°C under 100 MPa, quenching would lead to the precipitation of small phosphate crystals with irregular shapes. While the solid phase would coexist with the

** Vanox, Olympus Optical Co. Ltd., Tokyo, Japan.
*** JSM T-200, JEOL Co. Ltd., Tokyo, Japan.
**** RU 200, Rigaku Denki, Co. Ltd., Tokyo, Japan.

liquid, the solid could be grown more or less to larger crystals in the liquid. Thus, larger crystals could be observed in addition to the small crystals in the products after quenching. Only small crystals 0.01 ~ 0.05 mm in size have been observed in the products with the composition indicated by open diamonds (◇) in Figure 2, while some larger crystals up to 0.1 ~ 0.3 mm in size in the products and indicated by filled diamonds (◆). $NdPO_4$ seemed to be difficult to grow since the largest crystals were only 0.05 ~ 0.06 mm in size even in the products in the region of $NdPO_4$+L. As these crystals, however, were still larger and well-shaped than those precipitated from the liquid, the phase boundary between O+L and L could be determined.

The observed liquid phase field seemed to be similar but

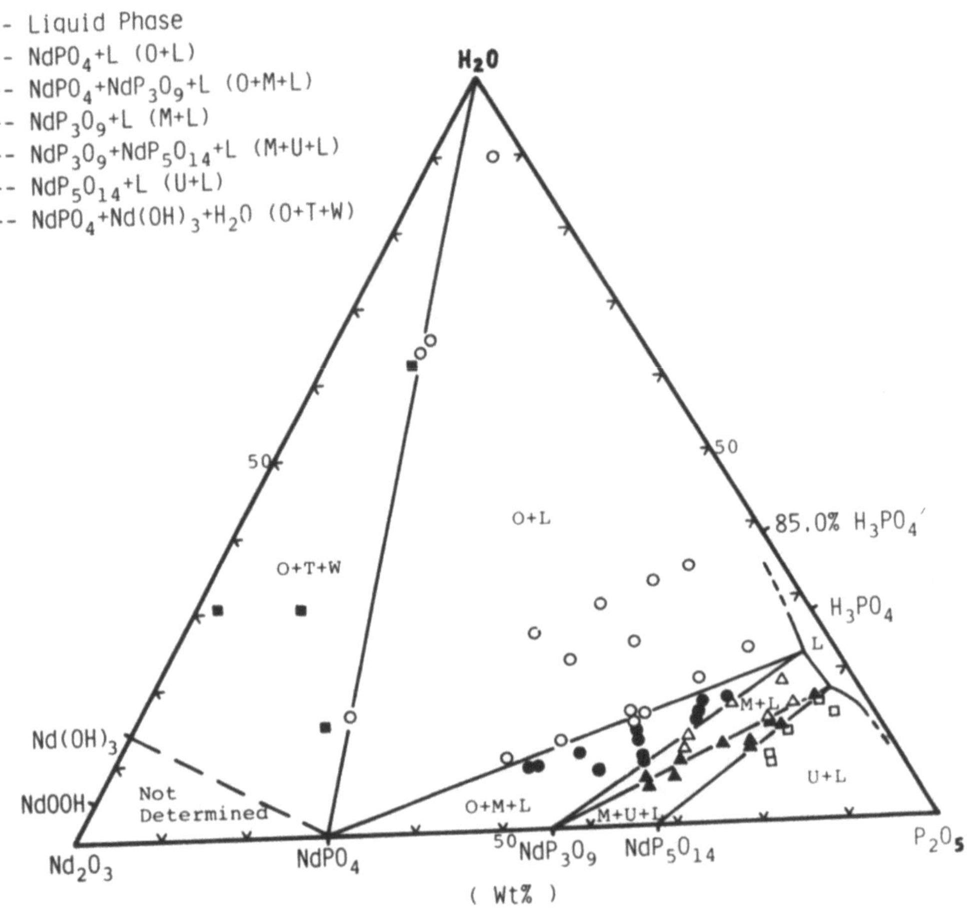

FIG. 1

Phase diagram in the system of Nd_2O_3-P_2O_5-H_2O at 500°C under 100 MPa narrower than that estimated by Plättner et al.(16) under atmospheric pressure(0.1 MPa) at temperatures from 110°C to 869°C, the azeotropic point of phosphoric acid. That is, the liquid composition with which NdP_5O_{14} and NdP_3O_9, or NdP_3O_9 and $NdPO_4$ can coexist is 4% Nd_2O_3-73% P_2O_5-23% H_2O, or 3% Nd_2O_3-80% P_2O_5-17% H_2O under 100 MPa at 500°C, whereas the liquid is 6% Nd_2O_3-70% P_2O_5 -24% H_2O at ≈ 120°C or 5% Nd_2O_3-77% P_2O_5-18% H_2O at ≈ 220°C, respectively under atmospheric pressure. These results may suggest lower solubility of neodymium phosphates in the liquid phase under high pressures where the main vapor species might be water rather than phosphoric species.

The studies of the influences of temperature and pressure on the phase equilibria in this system are in progress.

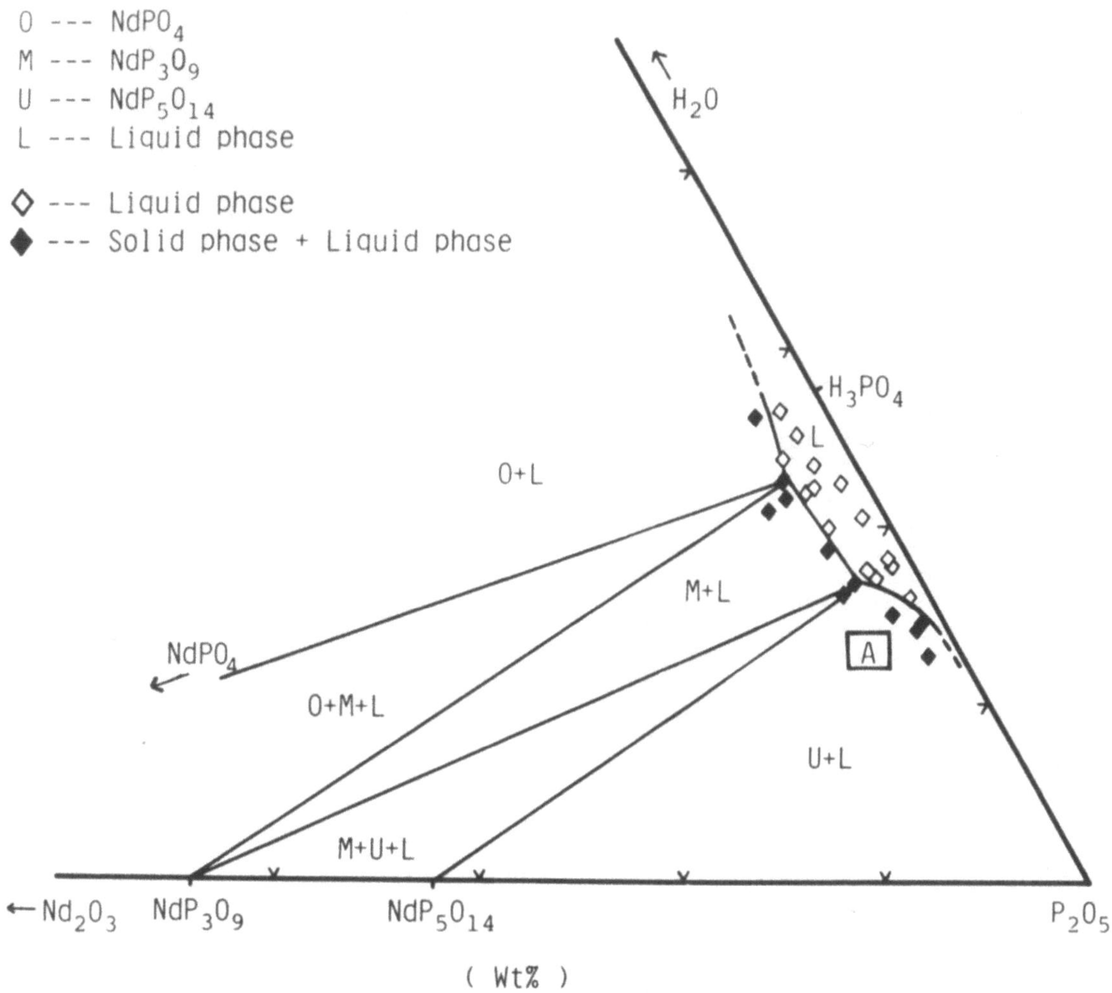

FIG.2

Phase diagram in the system of Nd_2O_3-P_2O_5-H_2O near the corner of P_2O_5 at 500°C under 100 MPa.

3. Synthesis of NdP_5O_{14} crystals under equilibrium conditions

As the phase diagram has been established, the growth of NdP_5O_{14} crystals has been carried out under equilibrium conditions at 500°C under 100 MPa. The starting composition was about 4.5% Nd_2O_3-82.5% P_2O_5-13% H_2O indicated as [A] in Figure 2. Many diamond-shaped crystals 0.1 ~ 0.2 mm (max. 0.3 mm) in size were synthesized after 4 days in a capsule with 4.6 mm ID, Figure 3 (a). They have smooth faces of (001), (100), (011) and (101) as seen in Figure 3 (b) and (c). This morphology was typical at temperatures of 350-500°C under 100 MPa. The lowering of growth temperature brought about aggregation and the formation of rough surfaces particularly in (100) surfaces.

Growth at 650°C leads to slightly larger crystals of about 0.3 mm (max 0.5 mm) with a polyhedral morphology by increased growth of (101) faces as seen in Figure 4. These crystals had generally smooth surfaces but sometimes contained inclusions as seen in this figure. No differences could be observed in grown crystals from the starting mixture of Nd_2O_3+P_2O_5+H_3PO_4 or NdP_5O_{14}+P_2O_5+H_3PO_4. This indicates that the above crystals were grown under equilibrium conditions.

The crystal growth under equilibrium conditions in a closed system appears to be favorable. Large crystals, however, can not be grown by spontaneous nucleation so that seeding and applying of a temperature gradient in a large container seem to be necessary to grow large crystals.

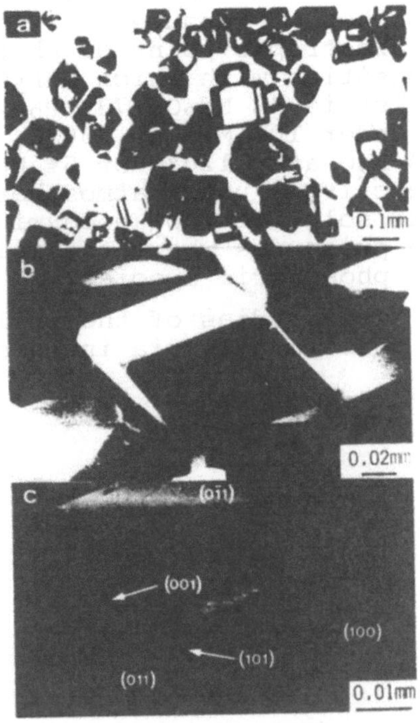

FIG. 3
(a) NdP_5O_{14} crystals grown at 500°C under 100 MPa.
(b) A typical crystal.
(c) Faces of the crystal.

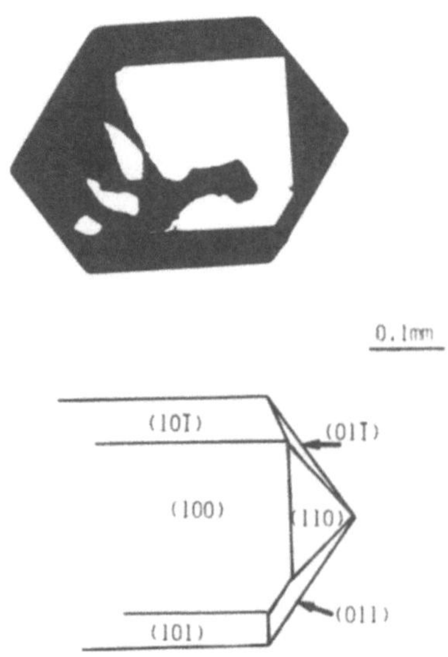

FIG. 4

A NdP_5O_{14} crystal grown at 650°C under 100 MPa.

References

1. H. G. Danielmeyer and H. P. Weber, IEEE J.Quantum Electron. QE-8, 805 (1972).
2. H. P. Weber, T. C. Damen, H. G. Danielmeyer, and B. C. Tofield, Appl. Phys. Lett. 22 534 (1973).
3. H. P. Weber, P. F. Liao, and B. C. Tofield, IEEE J. Quantum Electron. QE-10, 563 (1974).
4. G. Winzer, P. G. Möckel, R. Oberbacher, and L. Vité, Appl. Phys. 11, 121 (1976).
5. P. Möckel, Frequenz 32, 85 (1978).
6. K.-R. Albrand, R. Attig, J. Fenner, J. P. Jeser, and D. Mootz, Mat. Res. Bull. 9, 129 (1974).
7. H. Y-P. Hong, Acta Cryst. B 30, 468 (1974).
8. H. Schulz, K.-H. Thiemann, and J. Fenner, Mat. Res. Bull. 9, 1525 (1974).
9. J. P. Budin, A. Milatos-Rovfos, Nguyen Duc Chinh, and G. le Roux, J. Appl. Phys. 46, 2867 (1975).
10. H. P. Weber, B. C. Tofield, and P. F. Liao, Phys. Rev. B. 11, 1152 (1975).
11. B. C. Tofield, H. P. Weber, and T. C. Damen, Mat. Res. Bull. 9, 435 (1974).
12. D. C. Miller, L. K. Shick, and C. D. Brandle, J. Cryst. Growth 23, 313 (1974).
13. H. J. Danielmeyer, J. P. Jaser, E. Schoenherr and W. Stetter J. Cryst. Growth, 22, 298 (1974).
14. K. Shiroki and H. Kobayashi, Nippon Kagaku Kaishi, (J. Chem. Soc. Japan), 1979 (3), 342.
15. W. K. Zwicker, T. Kovats and S. R. Chinn, 5th Intern. Conf. on Crystal Growth, Boston, MA, 1977, Abstract No. 237.
16. R. D. Plättner, W. W. Krähler, W. K. Zwicker, T. Kovats, and S. R. Chinn, J. Cryst. Growth 49, 274 (1980).
17. N. N. Chudinova, L. P. Shklover and G. M. Balagina, Izv. Akad. Nauk SSSR, Neorg. Mater., 11 (4), 686 (1975) Eng. trans. 590 (1975).
18. N. N. Chudinova and G. M. Balagina, ibid., 15 (5), 828 (1979) Eng. trans., 650 (1979).

Reprinted from *Solid State Ionics*, **3/4** (1981) 477–481.

THE CONFIRMATION OF PHASE EQUILIBRIA IN THE SYSTEM ZrO_2–CeO_2 BELOW 1400°C

Masahiro YOSHIMURA, Eiji TANI and Shigeyuki SŌMIYA

Laboratory for Hydrothermal Syntheses, Research Laboratory of Engineering Materials, and Department of Materials Science and Engineering, Tokyo Institute of Technology, Nagatsuta, Yokohama 227, Japan

Critical review of the phase diagrams proposed previously suggested incomplete establishment of the phase equilibria in the system ZrO_2–CeO_2 below 1400°C. Application of hydrothermal techniques enabled us to attain the equilibria in this system. The tetragonal SS (solid solution) was no longer stable below 1000°C, where it decomposed into the stable monoclinic SS and cubic SS. The solubility limit of the tetragonal or cubic SS was 23 wt%(18 mol%) CeO_2 or 79 wt%(73 mol%) CeO_2, respectively at 1400°C. That of the monoclinic or cubic SS was 5 wt%(3.5 mol%) CeO_2 or 96 wt%(95 mol%) CeO_2, respectively at 800°C. Thermochemical calculation appears to make it possible to predict true equilibria in this system.

1. Introduction

The solid solutions in the system ZrO_2–CeO_2 are important refractory materials showing mixed conduction [1], which may be considered as electrode materials in MHD [2] and solid electrolyte [3] technology. The stability field and phase relationship of these solid solutions; cubic CeO_2 SS, tetragonal ZrO_2 SS and monoclinic ZrO_2 SS, have been studied by several investigators. Duwez and Odell [4] proposed a diagram as curve A in fig. 1, which was constructed from the phase analysis of 11 samples with every ten mole% composition, fired at 2000°C for 2 h, and annealed at 1375°C for 336 h or at 1100°C for 1000 h. The boundary between the tetragonal and the monoclinic SS was determined by measurement of thermal expansion. Curve B shows the results of Longo and Podda [5] who analyzed sâmples fired at 1700° or 1600°C for 8 h, at 1500°C for 150 h, at 1400°C for 1000 h, at 1300°C for 1200 h, or at 1200 or 1100°C for 1500 h, after prefiring at 1400°C for 250–1000 h. They determined the boundary between tetragonal and monoclinic SS by measuring electrical conductivity. They added the compound $Ce_2Zr_3O_{10}$ as a stable phase below 870°C [6] in this system. Later, curve C was proposed by Negas et al. [7]. They noticed partial reduction of Ce^{4+} to Ce^{3+} at elevated temperatures. The Ce^{3+} content reached 66% at the liquidus temperatures (>2300°C) of the system ZrO_2–CeO_{2-x} in air [8], but seemed to be low below 1600°C [5]. Thus, the ternary system

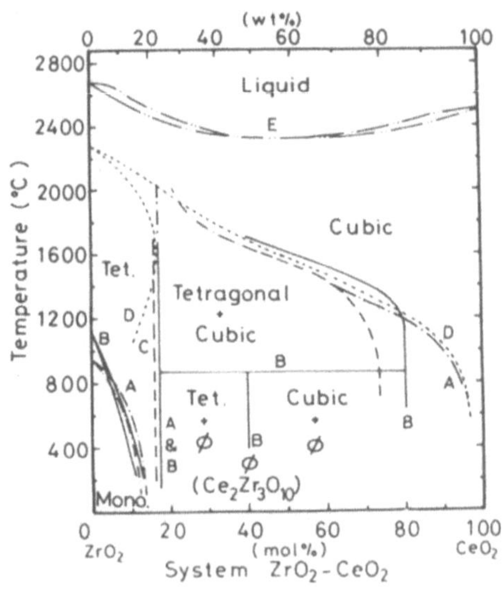

Fig. 1. Phase relations in the system ZrO_2–CeO_2 reported previously. Curve (A) is from ref. [4], (B) from refs. [5, 6], (C) from ref. [7], (D) is calculated in ref. [9], and (E) is from ref. [8].

ZrO_2–CeO_2–Ce_2O_3 can be regarded as a binary system ZrO_2–CeO_2 below 1600°C in air.

Among these three results reported, the stability field of tetragonal SS was in good agreement at around 25 wt% (19 mol%) CeO_2, and that of cubic SS was in agreement at temperatures between 1700 and 1400°C, but in disagreement at lower temperatures. In these experiments, samples were annealed for a long period after firing at higher temperatures as described above. However long annealing or heating cannot ensure the attainment of the phase equilibria, particularly at low temperatures. The phase assemblage in the samples fired below 1000°C was not necessarily identical with that annealed at the same temperature [9]. A thermochemical assessment [9] suggested that the stability field of tetragonal and cubic SS should become narrower than the experimental one at temperatures lower than 1400°C as illustrated by curve D in fig. 1, which was calculated on the basis of a regular solution model [10] using estimated thermochemical data.

One of the possible explanations for such discrepancies is the difficulty in attaining equilibria by solid-state reactions in this system at low temperatures. Thus high-temperature states might exist in metastable form at lower temperatures even after long annealing. It should be noted that cation diffusion is very slow in contrast to anion diffusion in fluorite-type structures [11].

In order to accelerate the reactions at low temperatures, this system has been studied under hydrothermal conditions. The present study deals with the effects of temperature, heating duration and starting materials on the phase relations in this system. A phase diagram is proposed on the basis of the experimental results and thermochemical calculations.

2. Experimental

Starting materials were (i) co-precipitated, (ii) mixed, or (iii) fired samples. The first was prepared by co-precipitation of hydroxides from an appropriate mixture of $ZnCl_4$* and cerium nitrate** solutions. They were washed and dried in air at 120°C for 7 days. For runs above 1000°C, they were calcined at 800°C for 1 h to convert into the oxide. The second was a mixture of ZrO_2† and CeO_2‡, and the last was prepared by firing the mixed oxide at 1600°C or 1400°C for 4 h in air.

These materials were sealed in a platinum capsule, 2.7 mm i.d. and 35 mm long, with water or aqueous solutions of 7–30 wt% LiCl, as a mineralizer. Other solutions were also used as mineralizers. The capsules were heated at the desired temperatures for extended durations; 200–700°C up to for 7 days, 800°C for 1 day, 1200°C for 12 h or 1400°C for 1 h, under a pressure of 100 MPa (1020 kg/cm^2) either in a test-tube-type pressure vessel below 800°C or in a high-gas-pressure apparatus [12] above 800°C. After completion of the heating run, the test-tube-type vessel was quenched into cold water, while the high-gas-pressure apparatus was cooled down to room temperature by ≈20°C/min. The products were decapsuled, washed, dried, then examined by X-ray powder diffraction and observations using a scanning electron microscope and a petrographic microscope.

3. Results and discussion

The first question to answer is whether the presence of high-pressure water influences the equilibrium or only increases the rate of reaction or crystallization in this system. Good agreement between the hydrothermal results in this work and previous results [4, 5, 7] by heating in air at high temperatures, as described later, indicates that the hydrothermal condition influences only the kinetics of the reaction in

* Hf-free $ZnCl_4$, Nikko Engineering Co. Ltd., Tokyo, Japan.

** $Ce(NO_3)_3{\cdot}6H_2O$, 99.9%, Nissan Kigenso Co. Ltd., Osaka, Japan.

† ZrO_2, 99.9%, Soekawa Rikagaku Co. Ltd., Tokyo, Japan.

‡ CeO_2, 99.9%, Shin-etsu Chemical Co. Ltd., Tokyo, Japan.

this system as well as in the systems ZrO_2–CaO and ZrO_2–Y_2O_3 [13].

After the hydrothermal treatment with 15 wt% LiCl solution at 1400°C under 100 MPa for 1 h, the 59 wt%(50 mol%) CeO_2 sample fired at 1600°C, which was single-phase cubic SS with lattice parameter a = 5.260 Å, was converted into a mixture of a tetragonal SS with a = 5.148 Å and c = 5.235 Å and a cubic SS with a = 5.340 Å as shown in fig. 2. The mixed oxide also yielded similar products; a tetragonal SS with a = 5.150 Å and c = 5.240 Å and a cubic SS with a = 5.318 Å. The co-precipitated 48 wt%(40 mol%) CeO_2 sample produced almost the same phases by the same hydrothermal treatment; a tetragonal SS with a = 5.146 Å and c = 5.236 Å and a cubic SS with a = 5.327 Å. As the products at 1400°C were identical in both directions, from high temperature (1600°C fired sample) or low temperature (mixed sample and co-precipitated sample), the phase equilibrium appeared to be attained by these hydrothermal treatments. Thus, the tetragonal SS with $\bar{a}$ = 5.148(2) Å and $\bar{c}$ = 5.237(3) Å and the cubic SS with $\bar{a}$ = 5.328(12) Å are the stable phases at 1400°C in the system ZrO_2–CeO_2. They correspond to the solid solution phases with 23 wt%(18 mol%) CeO_2 and 79 wt%(73 mol%) CeO_2, respectively, according to a linear relation between the composition and the lattice parameter [4,5].

Fig. 3 shows the lattice parameters of the solid solutions as a function of duration of treatment at 1200°C. Fig. 3 clearly indicates that phase equilibria appears to be established within 1 h and that the tetragonal and cubic SS were the stable phases in the equilibrium state at 1200°C. Small amounts of a monoclinic SS were sometimes observed in the products at 1200°C. This seems to be formed by the phase transformation from the tetragonal SS during cooling, because the tetragonal phase is not stable below 1000°C.

Treatment at 800°C for 24 h yielded the same products; a mixture of a monoclinic and a cubic SS, both from the co-precipitated and the fired samples as seen in fig. 4. No tetragonal SS could be detected. The monoclinic and cubic SS, therefore, are the stable phases at 800°C. The lattice parameter of the monoclinic phase were always identical, a = 5.156(7) Å, b = 5.220(4) Å, c = 5.323(7) Å and β = 99.2(1)°, while that of the cubic phase was scattered, a = 5.401 Å from

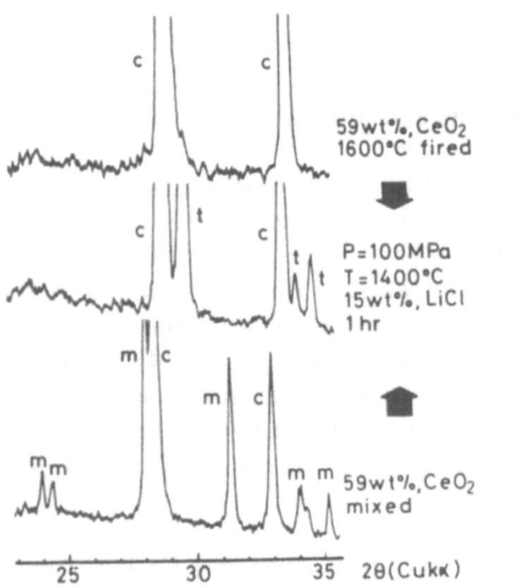

Fig. 2. Phase changes in the ZrO_2–CeO_2 samples after treatment at 1400°C for 1 h under 100 MPa with 15 wt% LiCl solution. c: cubic SS, t: tetragonal SS, m: monoclinic SS.

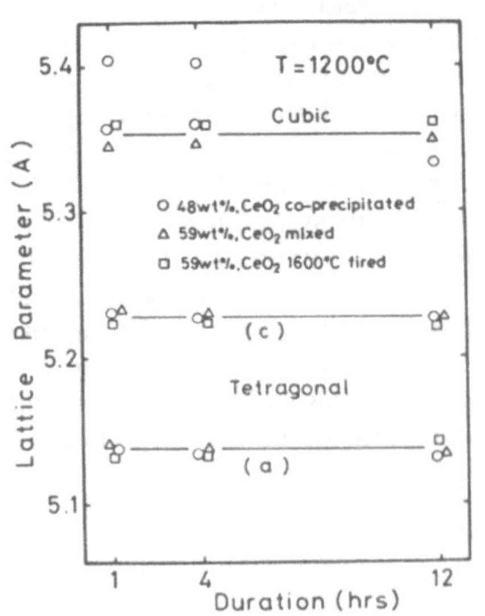

Fig. 3. Lattice parameters of the cubic and tetragonal phases produced by treatment at 1200°C from different starting materials.

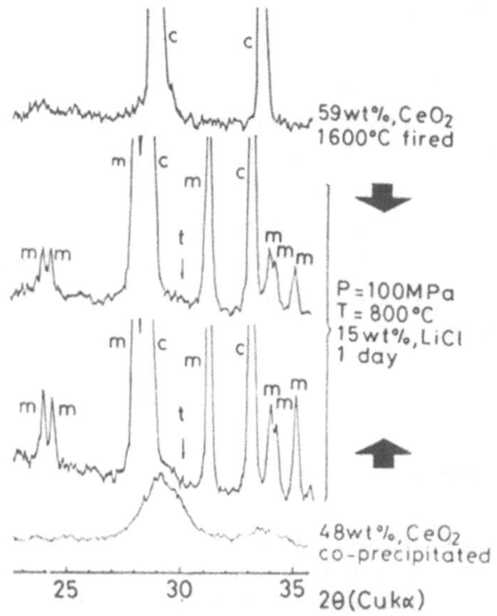

Fig. 4. Phase changes in the ZrO_2–CeO_2 samples after treatment at 800°C for 24 h under 100 MPa with 15 wt% LiCl solution.

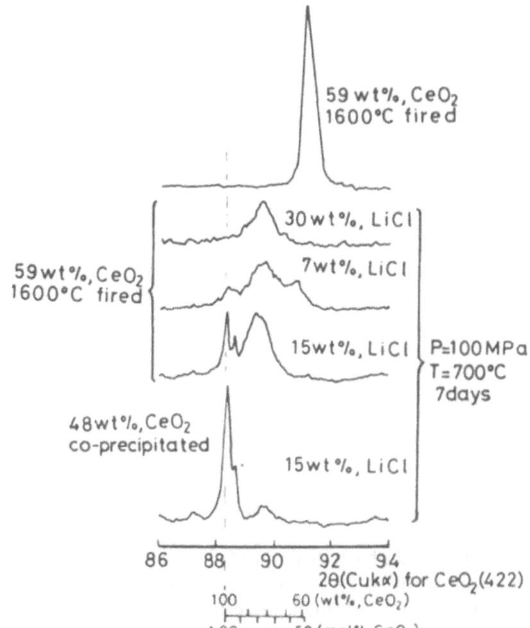

Fig. 5. Change of X-ray peak for (422) of cubic SS in the ZrO_2–CeO_2 samples after treatment at 700°C for 7 days under 100 MPa.

co-precipitated, 5.413 Å from mixed, 5.363 Å from fired at 1400°C or 5.373 Å from fired at 1600°C, respectively, indicating a different composition in CeO_2 content even after 24 h treatment at 800°C. These results indicate a slow kinetic process in these solid solutions even under hydrothermal conditions.

At lower temperatures, as the kinetics becomes much slower, the lattice parameters of the cubic SS were no longer identical among the products particularly in fired samples, which are less active due to heating at high temperatures. Fig. 5 showing the X-ray diffraction peaks for (422) lines in cubic CeO_2 SS after several treatments seems to indicate that the decomposition process was only half way to equilibrium even after treatment for 7 days at 700°C, and that such processes depend upon the concentration of mineralizer. The mixed samples were practically unchanged by the treatment at 700°C. The coprecipitated samples seemed to have the tendency to achieve equilibrium most easily. They, however, sometimes yielded a metastable tetragonal SS by hydrothermal treatment below 500°C. The confirmation of the equilibrium appears to be extremely difficult at such low temperatures, where the products depend upon thermal history of starting materials, and/or species and concentration of mineralizer [14].

In this regard, a thermochemical assessment has the potential of predicting true equilibria in such systems, though necessary thermochemical data are not always available. The phase boundaries among the tetragonal SS, tetragonal SS + cubic SS and cubic SS were calculated by a regular solution model using estimated thermochemical data:

$$\Delta G^{t\to c}_{ZrO_2} = 2573 - 1.0T , \quad \Delta G^{t\to c}_{CeO_2} = 0 - 1.0T ,$$
$$G^{c}_{ex} = 5150x(1-x) , \quad G^{t}_{ex} = 4850x(1-x) .$$

They are in good agreement with the experimental results determined from the average lattice parameter of each phase as shown in fig. 6. The detailed procedure of such calculations and extended experimental results will be given elsewhere [15]. Here, we would like to emphasize that the agreement between experimental and calculated values makes it possible

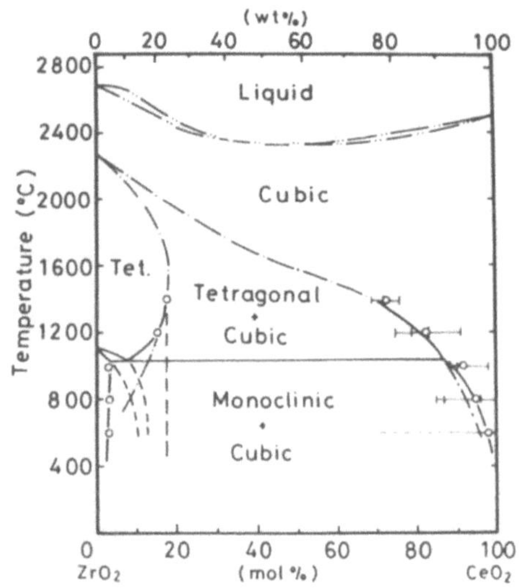

Fig. 6. Proposed phase diagram of the system ZrO_2–CeO_2 in the present study. —∘— Experimental results, —·—calculated boundary, --- metastable boundary, —··— after Rouanet [8]; they are the projection of liquidus observed and supposed solidus in air.

to predict a complete phase diagram by reasonable extrapolation. In this regard, the stability field of the tetragonal SS for the composition with 7–23 wt% (5–18 mol%) CeO_2 below 1000°C would be metastable, even though a reversible phase transformation with hysteresis could be observed during heating and cooling for the samples with less CeO_2 content [4, 5].

4. Summary

(1) Phase equilibria were attained in the system ZrO_2–CeO_2 by hydrothermal techniques at temperatures below 1400°C, particularly below 1200°C for the first time.

(2) The tetragonal SS was no longer stable below 1000°C and decomposed into the monoclinic SS and cubic SS which are stable in this temperature range.

(3) The solubility limit of the tetragonal or cubic SS was 23 wt% (18 mol%) CeO_2 or 79 wt% (73 mol%) CeO_2, respectively, at 1400°C but was smaller at lower temperatures.

(4) The solubility limit of the monoclinic or cubic SS was 5 wt% (3.5 mol%) CeO_2 or 96 wt% (95 mol%) CeO_2, respectively, at 800°C.

(5) Thermochemical calculation appears to make it possible to predict true equilibria in this system.

References

[1] S.F. Pal'guev, S.I. Alyamovskii and Z.S. Volchenkova, Russian J. Inorg. Chem. 4 (1959) 1158.

[2] B.R. Rossing and H.K. Bowen, in: Materials limiting problems in energy production, ed. C. Stein (Plenum Press, New York, 1976) pp. 335–356.

[3] K.W. Browall and R.E. Hanneman, G.E. Report, No. 75 CRD 012 (1975).

[4] P. Duwez and F. Odell, J. Am. Ceram. Soc. 33 (1950) 274.

[5] V. Longo and L. Podda, Ceramurgia 1 (1971) 83.

[6] V. Longo and D. Minichelli, J. Am. Ceram. Soc. 56 (1973) 600.

[7] T. Negas, R.S. Roth, C.L. McDaniel, H.S. Parker and C.D. Olson, 12th Rare Earth Research Conference, Vail, Colorado, July 1976.

[8] A. Rouanet, Compt. Rend. Acad. Sci. (Paris) C 266 (1968) 908.

[9] M. Yoshimura and H.K. Bowen, Bull. Am. Ceram. Soc. 56 (1977) 301.

[10] L. Kaufman and H. Bernstein, in: Phase diagrams, materials science and technology, Vol. 1, ed. A.M. Alper (Academic Press, New York, 1970) pp. 45–113.

[11] K. Ando, M. Akiyama and Y. Ōishi, J. Nucl. Mat., to be published.

[12] S. Sōmiya, S. Hirano, T. Fukuda and M. Sawada, Kōatsu Gasu 10 (1973) 368.

[13] V.S. Stubican and S.P. Ray, J. Am. Ceram. Soc. 60 (1977) 534.

[14] E. Tani, M. Yoshimura and S. Sōmiya, The 25th Jinko Kobutsu Toronkai (Symposium of Artificial Minerals), Nagoya, Oct. 16, 1980.

[15] E. Tani, M. Yoshimura and S. Sōmiya, to be published.

Reprinted from *Journal of the American Ceramic Society*, **66**(7) (1983) 506–510.

Revised Phase Diagram of the System ZrO_2-CeO_2 Below 1400°C

EIJI TANI, MASAHIRO YOSHIMURA,* and SHIGEYUKI SŌMIYA*

Research Laboratory of Engineering Materials, and Department of Materials Science and Engineering, Tokyo Institute of Technology, Nagatsuta, Yokohama, 227 Japan

The phase diagram of the system ZrO_2-CeO_2 was reinvestigated using hydrothermal techniques. Cubic, tetragonal, and monoclinic solid solutions are present in this system. The tetragonal solid solution decomposes to monoclinic and cubic solid solutions by a eutectoid reaction at 1050°±50°C. The solubility limits of the tetragonal and cubic solid solutions are about 18 and 70 mol% CeO_2, respectively, at 1400°C, and about 16 and 80 mol% CeO_2, respectively, at 1200°C. Solubility limits of the monoclinic and cubic solid solutions are about 1.5 and 88 mol% CeO_2 at 1000°C, and 1.5 and 98 mol% CeO_2 at 800°C, respectively. The compound $Ce_2Zr_3O_{10}$ is not found in this system.

I. Introduction

THE solid solutions in the system ZrO_2-CeO_2 are important for electrode materials[1] in MHD, fuel cells due to their mixed conduction,[2] or toughened ceramics.[3] The phase relations of this system have been studied by several investigators.[3–8] However, there are significant discrepancies in the diagrams, particularly below 1400°C, as shown in Fig. 1. Duwez and Odell[4] reported a diagram (Fig. 1, curve A) by long annealing of samples prefired at 2000°C. The tetragonal-monoclinic boundary was determined by thermal expansion measurements. Longo and Roitti[5,6] also studied this system with long annealing of the samples prefired at 1400°C (curve B). They drew the tetragonal-monoclinic boundary on the basis of electrical conductivity measurements[5] and added a compound $Ce_2Zr_3O_{10}$ as a stable phase below 870°C.[6] Later, curve C was proposed by Negas *et al.*[7] They pointed out that partial reduction of Ce^{4+} to Ce^{3+} can occur at high temperatures in air. The amount of Ce^{3+}, however, seemed to be small below 1600°C, so that the ternary system ZrO_2-CeO_2-Ce_2O_3 could be regarded as the pseudobinary system ZrO_2-CeO_2 below 1600°C in air. Recently, Lange[3] suggested the decomposition of the tetragonal solid solution to monoclinic and to cubic solid solutions at ≈270°C by a eutectoid reaction using fracture mechanics data.

Received June 28, 1982; revised copy received January 3, 1983; approved February 22, 1983.
*Member, the American Ceramic Society.

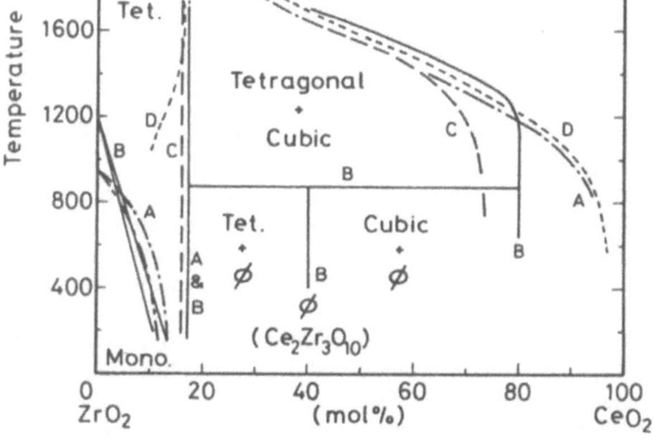

Fig. 1. Phase relations in the system ZrO_2-CeO_2 as reported previously. Curve *A*, Ref. 4; curve *B*, Refs. 5 and 6; curve *C*, Ref. 7; curve *D*, calculated in Ref. 10; and curve *E*, Ref. 9.

The phase boundaries (curves A, B, and C) in Fig. 1 agree at 1700° to 1400°C, but disagree at lower temperatures. These discrepancies indicate that even long annealing or heating could not always attain equilibrium at low temperatures, because cation diffusion is very much slower than oxygen diffusion.[8]

A thermochemical assessment suggested that the stability fields of tetragonal and cubic phases should become narrower than the experimental results below 1400°C, as illustrated by curve D in Fig. 1, which was calculated[10] on the basis of a regular solution model using estimated thermochemical data.

To establish the true phase equilibria, reaction rates must be accelerated at low temperatures. The authors studied this system under hydrothermal conditions[11] and found that mineralizers affected reaction rates without affecting the final products, and that LiCl was a preferable mineralizer for this system. Therefore, hydrothermal treatments influence only the kinetics of the reaction in the system ZrO_2-CeO_2 as well as in the systems ZrO_2-CaO[12] and ZrO_2-Y_2O_3.[13,14] A tentative phase diagram of this system under hydrothermal conditions has been reported,[15] using the sample containing 40 mol% CeO_2, which was in agreement with the calculated diagram.[10]

The present paper describes the phase equilibria in the system ZrO_2-CeO_2 attained by hydrothermal techniques.

II. Experimental Procedure

Starting materials were (i) "mixed," (ii) "fired," or (iii) "coprecipitated" samples. ZrO_2* and CeO_2† powders were mixed in an agate mortar with methanol for "mixed" samples. For the "fired" samples, "mixed" samples were pressed into pellets at 200 MPa and fired at 1400°C for 5 h. The fired products were crushed in a WC mortar and ground in an agate mortar with methanol. They were dried and pressed into pellets at 200 MPa and fired at 1600°C for 3 h. After the samples were quenched into cold water, they were again crushed in the WC mortar and reground in the agate mortar with methanol.

The "fired" sample containing 40 mol% CeO_2 was annealed at 810°C for 10 days to obtain the so-called $Ce_2Zr_3O_{10}$.[6]

The "coprecipitated" samples were hydrous zirconia-ceria solid solutions coprecipitated from a mixture of $Ce(NO_3)_3 \cdot 6H_2O$‡ and Hf-free $ZrCl_4$§ solutions with 3*N* NH_4OH.¶ The hydrous precipitates were filtered, redispersed in a 0.25*N* NH_4OH solution for washing, and then refiltered. After four washing and filtration treatments, they were dried in air at 120°C for seven days.

Appropriate amounts of the starting material and 15 wt% LiCl** solution were sealed in a Pt tube (2.7 mm ID, 0.15 mm thick, and 32 mm long), then treated under 100 MPa at 600° to 1400°C for 1 h to 3 days. The Pt tube was placed in a test-tube-type pressure vessel of a commercial alloy†† (6.5 mm ID) for the run below

*Soekawa Rikagaku Co., Ltd., Tokyo, Japan; 99.9%.
†Shin-etsu Chemical Co., Ltd., Tokyo, Japan; 99.9%.
‡Nissan Kigenso Co., Ltd., Osaka, Japan; 99.9%.
§Sn 680 ppm, Fe 620, Hf <70, Al <30; Nippon Mining Co., Ltd., Tokyo, Japan.
¶Guaranteed reagent, Wako Pure Chemical Industry Co., Ltd., Tokyo, Japan.
**Guaranteed reagent, Kanto Chemical Co., Inc., Tokyo, Japan.
††Stellite 25, Union Carbide Corp., New York, NY.

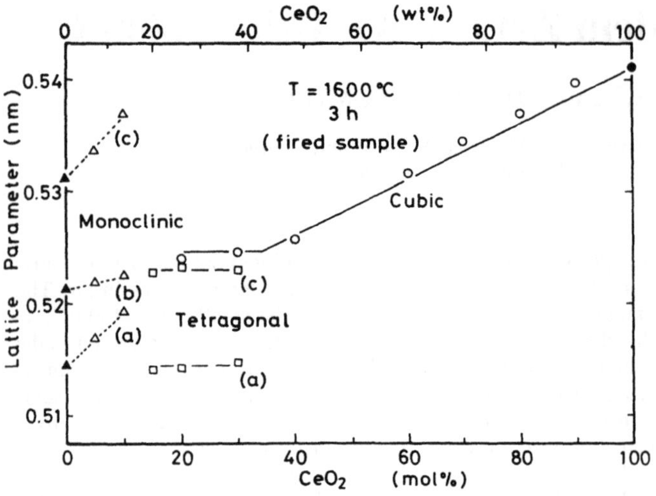

Fig. 2. Lattice parameter vs composition for fired samples; *a*, *b*, and *c* represent crystallographic axes.

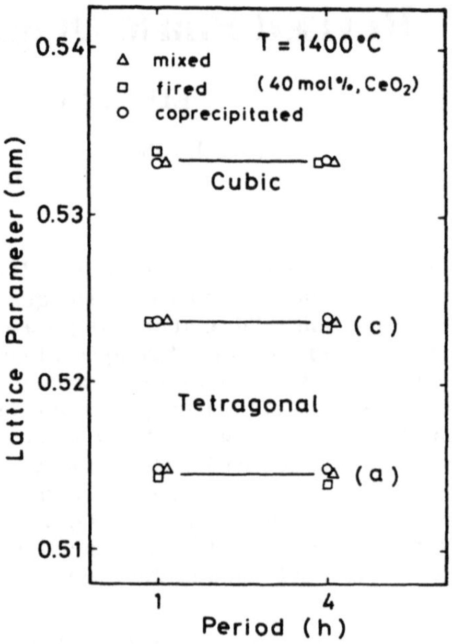

Fig. 3. Lattice parameter vs heat-treatment time for cubic and tetragonal phases at 1400°C under hydrothermal conditions.

800°C and a high gas pressure apparatus[16] for the run above 800°C. After hydrothermal treatments, the test-tube-type vessel was quenched in cold water, while the high gas pressure apparatus was cooled to room temperature at 20°C/min. The products were washed in distilled water and examined by an X-ray powder diffractometer.[‡‡] The lattice parameters were determined from the X-ray patterns obtained with slow scan rates of ¼ (2θ°/min) and a time constant of 5 s with Si[§§] as an internal standard.

III. Results and Discussion

Figure 2 shows phases and lattice parameters of the "fired" samples heated at 1600°C in air for 3 h. The phases present were monoclinic, tetragonal, and cubic solid solutions. Monoclinic phase was present in the specimens containing <10 mol% CeO_2. Since the ZrO_2-rich tetragonal phase cannot be quenched, the monoclinic phase in the specimens containing <10 mol% CeO_2 should be tetragonal at high temperatures. This lattice parameter-composition relation is in good agreement with the previous results.[4,5]

To confirm the phase equilibria, the "mixed," "fired," and "coprecipitated" samples of 40 mol% CeO_2 were treated hydrothermally under 100 MPa at 1400°, 1200°, and 1000°C for various periods. The phase equilibria could be ensured by coincidence of the products obtained from "fired," "mixed," and "coprecipitated" samples.

The phases produced from all samples were always tetragonal and cubic at 1400° (Fig. 3) and 1200°C, and their lattice parameters were identical, even after 1 h. These results clearly indicate that phase equilibria are completely established within 1 h, and the tetragonal and the cubic solid solutions are stable at 1400° and 1200°C under these conditions. At 1000°C, the products, however, consisted of monoclinic and cubic phases in all samples treated for 4 h. The "fired" sample, which had been a cubic solid solution, decomposed into the two phases: monoclinic and cubic. The "coprecipitated" samples, which seemed to be most reactive among these three samples, also yielded the monoclinic and cubic phases, with lattice parameters in good agreement with those of the phases produced from the "fired" samples. Therefore, the "fired" or "coprecipitated" samples were considered to attain equilibrium after 4 h at 1000°C.

Because phase equilibrium could be established under these hydrothermal conditions, samples with various CeO_2 contents were treated hydrothermally under 100 MPa at 1400°, 1200°, and 1000°C for 1, 4, or 12 h. The results of X-ray diffraction analyses and lattice parameter measurements of the products are summarized in Table I.

The products at 1400°C consisted of monoclinic, tetragonal, and cubic phases. The boundary of the two-phase tetragonal and cubic region at 1400°C should be slightly below 20 mol% and about 70 mol% CeO_2 from the lattice parameter results. Hydrothermal results at 1400°C are in good agreement with those obtained in air.[4,5,7] The presence of water in the hydrothermal conditions does not affect the equilibrium in the system ZrO_2-CeO_2.

The hydrothermal treatments at 1200°C yielded the same phases as produced at 1400°C; however, the two-phase region was wider at 1200° than at 1400°C. The two-phase tetragonal and cubic region is considered to extend from slightly above 15 to about 80 mol% CeO_2 at 1200°C.

At 1000°C, the tetragonal phase was no longer detected. Two phases, monoclinic and cubic, were observed at from 5 to 40 mol% CeO_2. The boundaries of these regions at 1000°C are considered to be about 1.5 and 88 mol% CeO_2.

Since no tetragonal phase was observed at 1000°C, a eutectoid reaction must exist between 1000° and 1200°C. The samples containing 15, 40, and 80 mol% CeO_2 were treated hydrothermally at 1100°C for 4 h. The results of X-ray analyses are given in Table I. The tetragonal phase was observed in the "mixed" and "fired" samples for 15 mol% CeO_2 or the "coprecipitated" and "fired" samples for 40 mol% CeO_2; therefore, the eutectoid temperature was considered to exist between 1000° and 1100°C.

Hydrothermal treatments of the "mixed," "fired," and "coprecipitated" samples always yielded monoclinic and cubic phases at 800°C, as described previously.[15] The lattice parameters of the products after hydrothermal treatments of the "coprecipitated" sample at 800°C were a=0.5149(1), b=0.5214(2), c=0.5319(1) nm, β=99.15(3)° for the monoclinic phase and a=0.54070(1) nm for the cubic phase, which correspond to about 1.5 and 98 mol% CeO_2, respectively. These values are almost the same as those (≈2 and 97 mol% CeO_2) estimated in the previous work[11] at 600°C.

The present study provided enough data to establish the ZrO_2-CeO_2 phase diagram below 1400°C. The phase diagram of Fig. 4 was drawn using the results obtained in this investigation and from thermodynamical calculation[10,15] based on a regular solution model.[18]

Both cubic and tetragonal regions are narrower than previously observed below 1400°C, which is reasonable from the thermo-

‡‡RU-200, Rigaku Denki Co., Ltd., Tokyo, Japan.
§§Kinzoku Kagaku Kenkyujo Co., Ltd., Tokyo, Japan.

Table I. Results of Hydrothermal Treatment in the System ZrO_2-CeO_2

Temperature (°C)	CeO_2 (mol%)	Treatment period (h)	Sample	Phases and lattice parameters (nm)
1400	5	4	Fired	M
				M; a=0.5169(2), b=0.5223(2), c=0.5336(2), β=99.13(5)°
	10	4	Mixed	M
				M; a=0.5188(5), b=0.5230(5), c=0.5364(6), β=98,9(2)°
		4	Fired	M(vs)+T(trace)
				M; a=0.5194(1), b=0.5228(2), c=0.5366(1), β=98.99(4)°
	15	1	Mixed	T
		1	Fired	T
		4	Mixed	T
				T; a=0.51406(2), c=0.52327(3)
		4	Fired	T
				T; a=0.51394(2), c=0.52295(3)
	20	4	Mixed	T(vs)+C(trace)
				C; a=0.53368(1)
				T; a=0.51470(3), c=0.52378(5)
		4	Fired	T(vs)+C(w)
				C; a=0.53381(5)
				T; a=0.51402(3), c=0.52307(4)
	40	1	Mixed	T(s)+C(s)
		1	Fired	T(s)+C(s)
		1	Coppt	T(s)+C(s)
		4	Mixed	T(s)+C(s)
				C; a=0.5332(1)
				T; a=0.51458(1), c=0.52366(1)
		4	Fired	T(s)+C(s)
				C; a=0.5332(3)
				T; a=0.51387(1), c=0.52316(1)
		4	Coppt	T(s)+C(s)
				C; a=0.5333(2)
				T; a=0.51490(2), c=0.52391(3)
1400	70	1	Mixed	C
		4	Mixed	C
				C; a=0.53359(2)
	80	4	Fired	C
				C; a=0.53583(1)
	90	4	Mixed	C
				C; a=0.53851(1)
		4	Fired	C
				C; a=0.53836(2)
1200	5	12	Mixed	M
		12	Fired	M
				M; a=0.5168(1), b=0.5222(1), c=0.5334(1), β=99.12(2)°
	10	12	Mixed	M(vs)+T(w)
				M; a=0.5194(2), b=0.5228(1), c=0.5365(1), β=99.00(4)°
		12	Fired	M(vs)+T(w)
		12	Coppt	M(vs)+T(w)
	15	1	Mixed	T(vs)+C(trace)
		1	Fired	T(vs)+M(trace)
		4	Mixed	T(vs)+M(trace)+C(trace)
		4	Fired	T(vs)+M(w)
		12	Mixed	T(vs)+M(trace)+C(trace)
				T; a=0.51387(4), c=0.52302(6)
		12	Fired	T(vs)+M(w)
	20	12	Mixed	T(s)+C(m)+M(w)
				C; a=0.53573(4)
				T; a=0.51367(5), c=0.5233(1)
		12	Fired	T(m)+C(s)+M(w)
	40	1	Mixed	C(s)+T(s)+M(trace)
		1	Coppt	C(s)+T(s)
		4	Mixed	C(s)+T(s)+M(trace)
			Fired	C(s)+T(s)+M(w)
			Coppt	C(s)+T(s)

Table I (continued)

1200	40	12	Mixed	C(s)+T(s)+M(trace) C; a=0.53594(3) T; a=0.5138(1), c=0.5229(1)
		12	Fired	C(s)+T(s)+M(w) C; a=0.53634(4) T; a=0.5141(3), c=0.5229(2)
		12	Coppt	C(s)+T(s)+M(trace)
	70	12	Mixed	C(vs)+T(w)+M(trace) C; a=0.53549(3) T; a=0.5137(2), c=0.5229(4)
		12	Fired	C(vs)+T(w)+M(trace) C; a=0.53568(2) T; a=0.5141(4), c=0.5233(8)
	80	1	Fired	C
		4	Fired	C
		12	Fired	C
	90	12	Mixed	C C; a=0.53848(2)
		12	Fired	C C; a=0.53855(5)
1100	15	4	Mixed	T(s)+M(m)+C(w)
		4	Fired	T(s)+M(w)+C(trace)
	40	4	Mixed	C(s)+M(s)
		4	Fired	C(s)+M(s)+T(trace)
		4	Coppt	C(s)+T(m)+M(w)
	80	4	Fired	C
1000	5	12	Coppt	M(vs)+C(trace) M; a=0.5147(4), b=0.5212(3), c=0.5316(4), β=99.2(1)°
	10	12	Fired	M(s)+T(trace)+C(m)
	40	4	Mixed	C(s)+M(s)
		4	Fired	C(s)+M(s)
		4	Coppt	C(s)+M(s)+T(trace)
		12	Fired	C(s)+M(s) C; a=0.53810(4) M; a=0.51545(9), b=0.5213(1), c=0.5318(1), β=99.19(3)°
1000	90	4	Mixed	C(vs)+M(trace)
		4	Fired	C
		4	Coppt	C
		12	Mixed	C
		12	Fired	C
		12	Coppt	C C; a=0.53844(1)
	95	12	Coppt	C C; a=0.5397(1)
	98	12	Coppt	C C; a=0.54051(1)

NOTE: M=monoclinic, T=tetragonal, C=cubic; vs=very strong, s=strong, m=medium, w=weak.

dynamical aspect.[10] Thus, the results obtained from hydrothermal treatments appeared to be the true equilibria. Phase equilibria could be attained by solid-state reactions above 1400°C in the system ZrO_2-CeO_2. The reaction rate, however, is very slow at low temperatures, and phase equilibria could hardly be attained with ordinary annealing in air.

The present study revealed that the tetragonal phase could not exist in equilibrium below 1000°C but that monoclinic and cubic phases are stable below 1000°C. Thus, the boundary between tetragonal and monoclinic phases for the samples containing 5 to 18 mol% CeO_2 (Fig. 1) should be metastable, even though a reversible phase transformation[4–6] has been observed with a thermal hysteresis during heating and cooling. The eutectoid reaction needs a large excess energy for long-range diffusion or mass transport. However, the tetragonal-monoclinic transformation, which is martensitic,[19] needs only a small excess energy without any mass transport. Therefore, the reversible monoclinic-tetragonal transformation could occur metastably at low temperatures, whereas the excess energy for the eutectoid reaction could not be obtained.

Roitti and Longo[6] and Longo and Minichelli[17] suggested that $Ce_2Zr_3O_{10}$ was a stable phase below 870°C in the system ZrO_2-CeO_2 on the basis of the annealing runs below 870°C for a long time. The X-ray pattern reported[17] contains some unindexed peaks, so that the presence of this compound appears to be questionable. The products of the "fired" sample containing 40 mol% CeO_2 annealed at 810°C in air for 10 days showed an X-ray pattern (Fig. 5) which somewhat resembles both the $Ce_2Zr_3O_{10}$ compound[17] and a cubic solid solution containing 40 mol% CeO_2. The hydrothermal treatment of this annealed sample at 800°C for 24 h yielded a mixture of a monoclinic phase (a=0.5155(1), b=0.5215(2), c=0.5322(1) nm, β=99.19(2)°) and a cubic phase (a=0.5380(1) nm). These results clearly indicate that $Ce_2Zr_3O_{10}$ does not exist in the system ZrO_2-CeO_2, or at least is not stable under hydrothermal conditions.

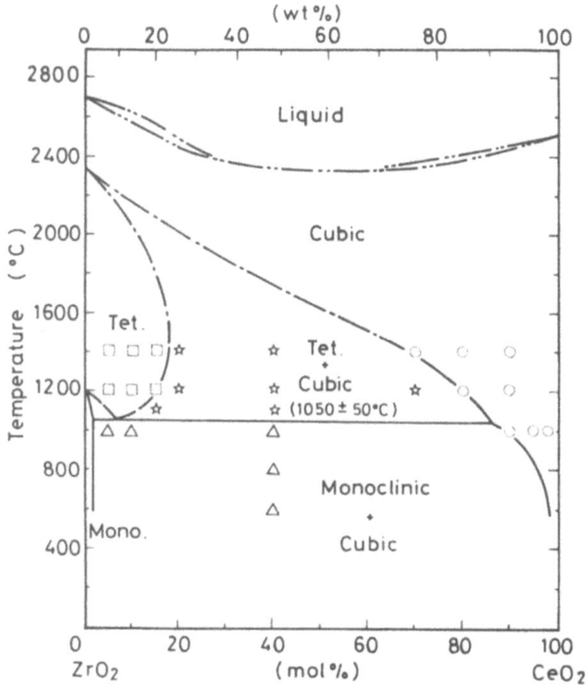

Fig. 4. Phase diagram of the system ZrO_2-CeO_2 according to present study; (———) experimental results, (– – –) Refs. 10 and 15, and (— - — -) Ref. 9.

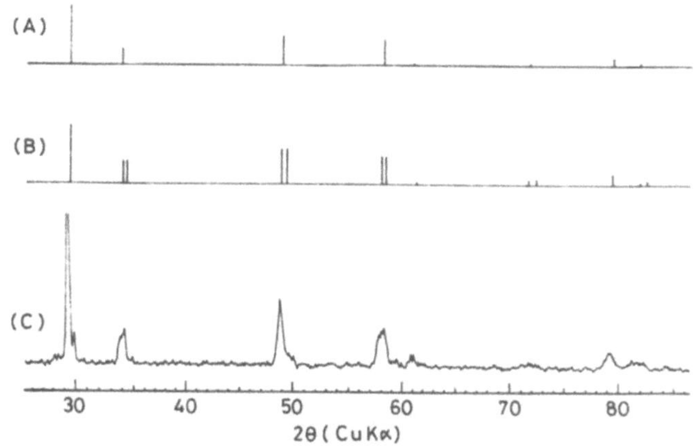

Fig. 5. X-ray diffraction patterns of (*A*) ZrO_2-CeO_2 cubic-type solid solution (40 mol% CeO_2), (*B*) $Ce_2Zr_3O_{10}$ (Ref. 17), and (*C*) "fired" sample (40 mol% CeO_2) annealed at 810°C for 10 days in air.

IV. Summary

(1) Phase equilibria have been attained in the system ZrO_2-CeO_2 by hydrothermal techniques below 1400°C.

(2) The solubility limit of the tetragonal or cubic solid solution was about 18 or 70 mol% CeO_2, respectively, at 1400°C, and decreased to about 16 or 80 mol% CeO_2, respectively, at 1200°C.

(3) Tetragonal solid solutions decomposed to monoclinic and cubic solid solutions by a eutectoid reaction between 1000° and 1100°C.

(4) The solubility limit of the monoclinic or cubic phase was about 1.5 or 88 mol% CeO_2, respectively, at 1000°C.

(5) The solubility limit of monoclinic solid solutions hardly changed from 1000° to 600°C; however, that of cubic solid solutions decreased to ≈97 mol% CeO_2 at 600°C.

References

[1]S. J. Schneider, T. Negas, and H. P. R. Frederikse, "MHD Materials," *Rev. Int. Hautes. Temp. Refract.*, **16**, 669–75 (1979).

[2]S. F. Pal'guev and Z. S. Volchenkova, "Electrical Conductivity and Transport Numbers in the CeO_2-ZrO_2 System," *Russ. J. Phys. Chem.*, **34**, 211–13 (1960).

[3]F. F. Lange, "Research of Microstructurally Developed Toughening Mechanisms in Ceramics"; p. 19, Rockwell International Tech. Rept. No. 12, ONR-N00014-77-C-0441, March 1981.

[4]P. Duwez and F. Odell, "Phase Relationships in the System Zirconia-Ceria," *J. Am. Ceram. Soc.*, **33** [9] 274–83 (1950).

[5]V. Longo and S. Roitti, "Solid State Phase Relations in the System CeO_2-ZrO_2," *Ceramurgia Int.*, **1** [1] 4–10 (1971).

[6]S. Roitti and V. Longo, "Investigation of Phase Equilibrium Diagrams Among Oxides by means of Electrical Conductivity Measurements. Application of the Method to the System CeO_2-ZrO_2," *Ceramurgia Int.*, **2** [2] 97–102 (1972).

[7]T. Negas, R. S. Roth, C. L. McDaniel, H. S. Parker, and C. D. Olson, "Influence of K_2O on the Cerium Oxide-ZrO_2 System"; pp. 605–14 in Proceedings of the 12th Rare Earth Research Conference, Vail, CO, July 1976.

[8]V. S. Stubican and J. R. Hellmann, "Phase Equilibria in Some Zirconia Systems"; pp. 25–36 in Advances in Ceramics, Vol. 3: Science and Technology of Zirconia. Edited by A. H. Heuer and L. W. Hobbs. The American Ceramic Society, Columbus, OH, 1981.

[9]A. Rouanet, "Zirconia-Cerium Oxide System at High Temperature," *C. R. Hebd. Seances Acad. Sci., Ser. C*, **266**, 908–11 (1968).

[10]M. Yoshimura and H. K. Bowen, "Phase Relations in the System ZrO_2-CeO_2-Iron Oxide in Air," *Am. Ceram. Soc. Bull.*, **56** [3] 301 (1977).

[11]E. Tani, M. Yoshimura, and S. Sōmiya, "Effect of Mineralizers on the Crystallization of the Solid Solutions in the System ZrO_2-CeO_2 under Hydrothermal Conditions," *Yogyo-Kyokai-Shi*, **90** [4] 195–201 (1982).

[12]V. S. Stubican and S. P. Ray, "Phase Equilibria and Ordering in the System ZrO_2-CaO," *J. Am. Ceram. Soc.*, **60** [11–12] 534–37 (1977).

[13]K. Nakamura, S. Hirano, and S. Sōmiya, "The System ZrO_2-Y_2O_3 Under Hydrothermal Condition at 1000 kg/cm^2," *Yogyo-Kyokai-Shi*, **83** [12] 570–74 (1975).

[14]V. S. Stubican, R. C. Hink, and S. P. Ray, "Phase Equilibria and Ordering in the System ZrO_2-Y_2O_3," *J. Am. Ceram. Soc.*, **61** [1–2] 17–21 (1978).

[15]M. Yoshimura, E. Tani, and S. Sōmiya, "The Confirmation of Phase Equilibria in the System ZrO_2-CeO_2 Below 1400°C," *Solid State Ionics*, **3/4**, 477–81 (1981).

[16]S. Sōmiya, S. Hirano, T. Fukuda, and Y. Sawada, "High Gas Pressure Apparatus up to 10 kb and 1500°C for Hydrothermal Syntheses and Hot Isostatic Pressing," *Atsuryoku Gijitsu (J. High Pressure Inst. Jpn.)*, **12**, 15–22 (1974).

[17]V. Longo and D. Minichelli, "X-Ray Characterization of $Ce_2Zr_3O_{10}$," *J. Am. Ceram. Soc.*, **56** [11] 600 (1973).

[18]L. Kaufman and H. Bernstein; pp. 45–112 in Computer Calculations of Refractory Metal Phase Diagrams; Phase Diagrams, Vol. 6–I. Edited by A. M. Alper. Academic Press, New York, 1970.

[19]E. C. Subbarao, H. S. Maiti, and K. K. Srivastava, "Martensitic Transformation in Zirconia," *Phys. Status Solidi A*, **21**, 9–40 (1974).

INDEX

MIX
Papier aus verantwortungsvollen Quellen
Paper from responsible sources
FSC® C105338

If you have any concerns about our products,
you can contact us on
ProductSafety@springernature.com

In case Publisher is established outside the EU,
the EU authorized representative is:
Springer Nature Customer Service Center GmbH
Europaplatz 3, 69115 Heidelberg, Germany

Printed by Libri Plureos GmbH
in Hamburg, Germany